DILLON, LAWRENCE S
THE GENETIC MECHANISM AND THE
000329619

QH430.D57

WITHDRAWN FROM STOCK
The University of Liverpool

The Genetic Mechanism and the Origin of Life

UNIVERSITY LIBRARY
LIVERPOOL

The Genetic Mechanism and the Origin of Life

LAWRENCE S. DILLON
Texas A & M University
College Station, Texas

PLENUM PRESS · NEW YORK AND LONDON

Library of Congress Cataloging in Publication Data

Dillon, Lawrence S
The genetic mechanism and the origin of life.

Bibliography: p.
Includes index.
1. Genetics. 2. Life–Origin. I. Title. [DNLM: 1. Genetics. 2. Biogenesis.
QH325 D579g]
QH430.D54 575.1 78-4478
ISBN 0-306-31090-2

© 1978 Plenum Press, New York
A Division of Plenum Publishing Corporation
227 West 17th Street, New York, N.Y. 10011

All rights reserved

No part of this book may be reproduced, stored in a retrieval system, or transmitted, in any form or by any means, electronic, mechanical, photocopying, microfilming, recording, or otherwise, without written permission from the Publisher

Printed in the United States of America

Preface

As shown in the text, there can be little doubt that the genetic mechanism is, for all practical purposes, equivalent to life itself. Consequently, it is unrealistic to seek knowledge of the origin of life and its subsequent evolution without simultaneously searching for an understanding of how this apparatus arose and evolved. Fortunately, the annual publication over the recent years of thousands of papers dealing with the genetic processes has brought the state of knowledge to a level where a synthesis of their major details in relation to life's history is feasible. Because of the voluminous body of literature, no single book can possibly treat all the ramifications of this fundamental subject; subdivision into multiple volumes is necessary. This volume, the first of a trilogy, explores the molecular aspects of the problem in connection with the precellular aspects up to the point of the origin of the cell. The second, currently in progress, is concerned with the subsequent evolution of the cell as revealed by the energy-related organelles and their genetic apparatuses and by ultrastructural details of other cellular parts. The third volume, as presently planned, deals with developmental, immunological, and other complexities at the organismic level and, in so doing, throws additional light on basic properties of the genetic processes themselves. Thus, the genetic apparatus provides the warp, and evolution the woof, of the intricate fabric that emerges.

Because this present volume is the first attempt at analyzing the genetic mechanism in context with the origin of life, it should be anticipated that several novel ideas might come forth that necessarily are at variance with prevailing dogma. At first one proposal may prove particularly disturbing, the disclosure that one of the two supposedly general principles of biology, the cell theory, is not truly universally applicable to living creatures. When given deeper consideration, however, it becomes apparent that the cell theory implies that life arose directly in the form of a cell without any precellular stages—an implication that appears most unlikely. Especially in light of the extreme com-

554949

plexity of the genetic mechanism in even the simplest of existing prokaryotes, the conclusion is inevitable that numerous precellular stages of life had to precede these.

The text's employment of viruses as models of those precellular stages may prove equally controversial for the moment. But on further contemplation, the realization must be reached, that, unsatisfactory in the capacity of protobionts as viruses may be, if they are rejected no actual evidence remains as to how life may have arisen. Without evidence, the problem of life's beginning consequently becomes reduced to one fit only for philosophical speculation and, accordingly, would no longer be suitable for investigation by scientists.

This employment of viruses as ancestral protobionts, based on the evidence presented by the genetic mechanism, necessarily invalidates the current practice of using synthetic colloidal particles in that role. But this should by no means be interpreted as implying that the explorations into life's advent by those methods have been meaningless. To the contrary, quite like the alchemists whose searches for the nonexistent lodestone laid the foundations for modern chemistry, explorations into the origins of life via synthetic coacervates, micelles, and proteinoid microspheres have contributed heavily toward an understanding of primeval life.

Whatever merit this study may possess must largely be attributed to those innumerable persons whose brilliant technological achievements and patient labors in the laboratory have revealed the countless details now known about the genetic mechanism. Although the sources of information employed in preparing the text are acknowledged in the usual fashion, it has been possible thus to cite only about 4000 papers, a small portion indeed of the total number of studies actually referred to during the course of this study. The vast majority of references not provided here, however, are to be found in those cited. To all these persons, named and unnamed, the author extends his deepest gratitude and admiration.

In addition, others have contributed in a more direct fashion by making available electron micrographs, light micrographs, and other matter for use herein, for which acknowledgment is made with the illustration as appropriate. In this connection, particular thanks are extended to Drs. Anna S. Tikhonenko, Academy of Sciences of the USSR, and Michel Wurtz, University of Basel, both of whom generously contributed a number of electron micrographs of viruses. Special acknowledgments are made, too, to Dr. Kenneth B. Marcu, of the Institute for Cancer Research, who very cooperatively furnished unpublished information on certain transfer RNAs, and Dr. H. R. Rappaport, Temple University, who supplied advance copies of his studies on ligand binding. Finally, my wife, as always, has collaborated in researching the literature, interpreting the data, and preparing the manuscript in all its stages.

LAWRENCE S. DILLON

Contents

If I can see farther than other men,
it's because I stand on the shoulders of giants.

ISAAC NEWTON

1

Origins of Life's Ingredients

The origins of life—being inextricably entwined with the origins of man—hold a fascination without equal in the legends and philosophies of humankind, and while the problem of life's beginnings has been the focus of attention for at least several millennia, only in the very recent past has enough precise information become available so as to permit discussion at the level of scientific investigation. The nature of the problem, however, places it beyond quick and immediate solution. Consequently, while much has already been written on the subject since the early 1950s, the coming years undoubtedly will witness the appearance in print of a still greater body of literature. This probability arises first through the realization that until reasonable solutions are found to those problems which its origins raise, life's own basic properties must in large measure remain incomprehensible (Keosian, 1968; Oparin, 1971). Among the facets which have been greatly illuminated by researches of the past three decades are those concerned with the nature of the biochemical substances of which living things are constructed. The results of these investigations into the origins of life's basic ingredients, first the molecules and later the polymers, provide the subject matter for the present chapter.

1.1. THE SETTING

In studies of the spontaneous origins of those fundamental biochemicals, a second aspect of the major topic becomes evident—as Wald (1974) has pointed out, any inquiry into the beginnings of life of necessity raises a number of broader questions. Not least among the more pressing of these is the nature of the earth immediately after its formation, for an understanding of the natural synthesis of the essential substances obviously requires a knowledge of the chemistry of the atmosphere and other surface features. To gain the necessary

Table 1.1
Time Scale of Life's Origins

Event	Approximate years ago (10^9 years)
Origin of the universe	19–20
Origin of our galaxy	15–16
Origin of the sun	5–6
Origin of the earth	4.6
Oldest rocks of earth	3.8
Origin of Life on earth	3.0
Oldest known fossils	2.86

factual background, the corresponding traits of other planets need investigation, and to comprehend those data, the origins of the solar system, the galaxies, and the universe in turn are requisite (Table 1.1). As solid data are available only from the earth and other planets, the present study is restricted to the members of the solar system.

1.1.1. Conditions on Other Planets

The spacecraft sent to explore other planets by the Russians and Americans have provided in a few short years more direct evidence about the members of the solar system than had previously been gained by earthbound investigators over several centuries (Rasool, 1972). As a consequence, it is becoming evident that new efforts need to be made toward reevaluating the nature of the atmosphere of the primitive earth. Moreover, the new information seems to have laid to final rest all hopes that life* would be found on our neighboring planets.

The Terrestrial Planets. Of the four terrestrial planets, Mercury, Venus, Earth, and Mars (Table 1.2), only the first remains scantily explored, but its atmosphere is believed to be thin and its surface temperatures high. The second, Venus, had long resisted telescopic investigations, for its surface features have been veiled from view by a permanent cloak of clouds. Beginning toward the close of 1967, several series of spacecraft were able to establish more clearly the nature of its atmosphere and surface conditions. Venera 4, which arrived at this planet on October 18, was followed in sequence by her sister craft, Venera 5 to Venera 10 (Keldish, 1977); together these have es-

* Here the term *life* is restricted to protein-containing forms, as these are the only actual living things known. Although somewhere in the universe silicon-bearing types or other variants may exist, the present usage appears reasonable and justifiable.

tablished that the Venusian surface temperature is 747° ± 20°K (474°C), and that the atmosphere has a pressure of 90 ± 15 atm. On 19 October of the same year, Mariner 5 began flying past Venus at a distance of 4100 km and explored the atmosphere by means of radio waves. By these means the atmospheric composition was determined as being 95% CO_2, 4% N, 0.4% O_2, and 0.1% H_2O. According to Rasool (1972), the quantity of water detected would, if fully condensed, be sufficient to provide a covering only 10 cm deep over the surface of the planet.

On Mars, the surface temperature and atmospheric pressure have now been shown to be as unfavorable for life as those on Venus, but in the opposite sense. During 1976 two Viking unmanned spacecraft assumed orbits around the planet, Viking 1 beginning on 20 June and the second on 7 August (Soffen, 1976). Preliminary results (Hess *et al.*, 1976b; Nier and McElroy, 1976; Owen and Biemann, 1976; Owen *et al.*, 1976) indicated an atmosphere less than 1.0% as dense as that on earth, at the surface having a mean pressure of 7.65 millibars and a temperature range of 188°K to 244°K during a summer day (Hess *et al.*, 1976a). As on Venus, this thin atmosphere consisted almost entirely of CO_2. Only small quantities of N_2, Ar, O, O_2, and CO were found in the earliest analyses, while Kr, Xe, and Ne were detected later. Much of the relatively scant H_2O that appears to be present on Mars is captured on the surface in the form of permanent polar ice caps; that at the north pole, at a summer temperature near 205°K, was estimated as ranging from 1 m to 1 km in thickness. This was not in the form of a continuous sheet but was extensively broken, at least in the peripheral region, to expose the underlying surface (Farmer *et al.*, 1976; Kieffer *et al.*, 1976). No evidence was found for the existence of a permanent CO_2 cap, although the development of a temporary one during the winter was not ruled out. In view of the severe cold, the relative lack of water, and the unfavorable composition of the atmosphere, the apparent absence of life thus far reported is scarcely surprising (Levin and Straat, 1976).

The Jovian Planets. In contrast to the four terrestrial planets, which consist largely of iron, nickel, and silicates, the outer or Jovian members of the solar system are made predominantly of hydrogen and helium. Hence their density is more like that of the sun, that of Jupiter being 1.33 g/cm^3, and Saturn only 0.71 g/cm^3; these values compare to the earth's 5.5 g/cm^3 (Table 1.2). As a result of the Pioneer 11 mission, which flew by the planet in December, 1974, much has been learned about Jupiter's atmospheric composition and temperature. Because of the clouds which conceal almost all of the surface, no firm data are presently available below heights of 40 km, however. In the zone beginning with that elevation and extending upward 100 km, the clouds were reported to consist of aqueous and particulate ammonia, ice, and ammonium hydrogen sulfide suspended in an atmosphere whose composition is mostly hydrogen and helium; these elements constituted 65% and 22% of the total, respectively. Above that zone hydrogen and helium also prevailed, apparently in

Table 1.2
The Planets and Their Atmospheres

Planet	Mean distance[a]	Diameter (km)	Density (g/cm^3)	Surface temperature (°K)	Atmospheric composition (%)[b]												Density (atm)
					H	He	CO_2	NH_3	O_2	N_2	H_2O	Ar	CH_4	C_2H_6	H_2S	Other	
Terrestrial																	
Mercury	57.9	4,840	5.2	700?													near 0
Venus	108.2	12,200	5.06	747			95		0.4	4	0.1						100
Earth	149.6	12,756	5.5	290			+		21	78	0.01	1					1
Mars	227.9	6,760	4.12	188–244			95		0.1–0.4	2–3	0.01–0.1	1–2				NO_2	0.008
Jovian																	
Jupiter	778.3	142,700	1.33	125	65	22		5?			1?		+	+	+		~3
Saturn	1,427	120,800	0.71	~100													
Uranus	2,869	47,600	1.15	~ 55													
Neptune	4,498	44,400	1.54														
Pluto	5,900		0.37–2.0	43	+	+							+			Ne?	

[a]Mean distance from sun in km × 10^6.
[b]Plus marks indicate the presence of minute or undetermined percentages.

the same ratios, but methane, methyl residues, acetylene, and ethane also abounded (Smoluchowski, 1975). The surface temperature at the equator was found to be approximately 160°K (Orton, 1975).

Although little is known of the conditions on Saturn, its atmosphere may be suspected to be largely of hydrogen and helium, plus ammonia and methane. Since water is present in its rings, it may also be presumed to occur to some extent in the atmosphere (Klein, 1972). There is no reason to assume that surface temperatures might be more favorable for life than those of Jupiter. Pluto, too, being the most remote, is the least known member of the solar system. Moreover, its small size, which recently was determined as being smaller than the moon (Cruikshank *et al.*, 1976), also contributes heavily to this relative absence of knowledge. However, the presence of solid methane has been detected on its surface where temperatures of about 43°K may prevail (Hart, 1974); its atmosphere could consist largely of neon.

As none of the other planets in our solar system has an atmospheric constitution and thermal conditions conducive to the formation of life (Margulis *et al.*, 1977), explorations into the beginnings must remain for the present to some extent in the realm of speculation. This new knowledge should nevertheless provide a firmer basis for deducing the probable composition of the earth's earliest atmosphere and its subsequent evolution. Present views are outlined below.

1.1.2. Primitive Terrestrial Conditions

To date, in experiments designed to elucidate possible origins of life's substances, it has been the consistent practice to assume that the atmosphere of the primitive earth had a composition similar to that of the Jovian planets (Urey, 1952, 1960). Thus it has been accepted as having consisted largely of hydrogen and helium, together with smaller amounts of ammonia, water vapor, and methane. Whether or not such an atmosphere prevailed on the earth immediately after its solidification is presently an unresolvable problem. If such an atmosphere did exist at first, however, the evidence from Mars and Venus strongly suggests that it must soon have become enriched with CO_2 and Ar by outgassing from the crust—argon is being released from the moon's crust even today, although there has been no volcanic activity for several billion years. Since helium and argon are inert at ordinary temperatures, they can well be omitted from experimental atmospheres, but if realism is a goal, the presence of reasonable percentages of CO_2 certainly should be requisite in future experiments.

The Primitive Atmosphere. As stated above, the Jovian type of atmosphere has provided the general basis for experimental syntheses under prebiotic conditions (Oparin, 1957, 1968). Although the constituents vary widely from laboratory to laboratory, the artificial models typically contain far larger proportions of H_2O, CH_4, and NH_3 than actually found on Jupiter by space probes (Miller and Urey, 1959; Urey, 1960). Hence they resemble the observed composition chiefly in containing H_2 in abundance, so that they are of a reducing rather than an oxidizing nature as the terrestrial atmosphere is today. According to some recent reports, O_2 was probably not completely absent after water had accumulated on the earth's surface to some extent but could have been maintained at a level of about 0.02% of the total composition. This supposition seems even more justifiable in view of the proportions now known to exist in the Venusian atmosphere. It is speculated that the oxygen resulted from the bombardment of the seas by certain wavelengths of ultraviolet light, which dissociate water into its elements. As it accumulated in the atmosphere, however, it would have been converted into O_3 by the same rays. As this ozone absorbs these rays, the physicochemical feedback mechanism would have resulted in an equilibrium between H_2O and O_2 at the atmospheric level cited above (Berkner and Marshall, 1965; Rutten, 1969, 1970).

Ammonia, too, is a necessary ingredient in the atmosphere, but probably never exceeded a partial pressure of 10^{-5} atm, at least after large seas were established. Being very soluble in water, it would have existed largely as NH_4^+ in the oceans if their alkalinity had been near pH 8, as it is at present (Miller and Urey, 1959). Under those conditions it probably never exceeded a concentration of 0.01 M in the seas, because of the absorption of NH_4^+ by clay minerals (Sillen, 1967; Bada and Miller, 1968); this concentration in solution would be in equilibrium with an atmospheric partial pressure of 10^{-5} atm of NH_3 (Miller, 1974). With the passing of geologic time, there was a gradual decrease in the amount of hydrogen in the atmosphere, and as hydrogen became scarcer, ammonia probably diminished to even smaller proportions, for it is unstable in the absence of hydrogen.

Methane is usually considered vital to the origin of organic compounds and, in the descriptions of experiments on prebiotic syntheses that follow, is frequently an important ingredient. However, some experimenters have found ethane more effective than methane. Carbon dioxide may have been no less significant in prebiotic syntheses, at least toward the end of the early stages, for reasons that become clearer in subsequent discussions. As suggested above, the atmospheres of the two planets whose orbits bracket that of the earth certainly strongly intimate the presence of ample concentrations of carbon dioxide. Although often omitted in such experiments, hydrogen sulfide must necessarily have formed a reasonable portion of the primitive atmosphere, for it too has an important role to play early in life's history.

Although the reducing atmosphere just discussed is the one most usually postulated to have existed, it is not universally accepted by any means. Some laboratories suggest that an oxidized atmosphere such as that of Mars could have prevailed on the primitive earth (Hubbard *et al.*, 1971, 1973; Tseng and Chang, 1975). Several experiments performed with such oxidized atmospheres as combinations of CO, CO_2, H_2O, and N_2 resulted in the synthesis of biochemicals when exposed to ultraviolet radiation. Consequently, free NH_3 is not an essential ingredient for the formation of these substances. An additional school of thought postulates the earth to have been molten during its period of accretion (Fanole, 1971; Armstrong and Hein, 1973). This point of view has received further support through certain results of the exploration of other planets by space probes, such as the granite found on the surface of Venus, the caldera-type craters discovered on Mars, and the basalt and anorthosite brought back from the moon (Shimizu, 1975). Had the earth been subjected to high temperatures during its formation, when it cooled its atmosphere would have been rich in H_2, CO, and HCN, while the seas would have contained formaldehyde, ammonia, and good concentrations of nitrate, sulfate, and phosphate ions.

The Primeval Seas. In contrast to the early atmosphere's complete dissimilarity from that of today, the primordial seas, although at first small and shallow (Rubey, 1951, 1955; Dillon, 1974), must have approached existing ones in regard to ionic content. To judge from the needs of primitive life, potassium, sodium, magnesium, phosphorus, calcium, iron, and chlorine probably comprised the major ions, along with NH_4^+ and a minute quantity of O^{2+}, as indicated above. A number of minor components, including manganese and zinc, were also present, but none of these appears to enter into widespread processes of fundamental importance.

In brief, then, the stage for the events preliminary to life's origin was set on a sterile earth devoid of large continents, studded with small, shallow seas, subjected to almost continual volcanism, and blanketed by a completely different type of atmosphere. But the earth itself was not then—nor has it been at any subsequent time—merely an inert stage upon which life underwent its origin and evolution. During every phase of life's development, the earth has been not just an active participant, but one of the main factors in the processes. Nowhere is this statement truer than in the earliest phases of life's history about to be described, for the earth served then as the ultimate source of all the interacting ingredients involved in the formative reactions, including much of the essential energy.

1.2. PREBIOTIC SYNTHESES (STAGE I)

It is in the earliest phase of life's origin, in which the fundamental biochemical molecules were synthesized from inorganic matter (Stage I; see Table 1.3), that experimentation has proven most valuable in elucidating feasible mechanisms. In broad terms, the experiments involve exogenous energy sources and various combinations of the gases (sometimes in solution) considered to have been prevalent in the primordial earth's atmosphere. Two major categories of experiments exist, based upon the nature of the product sought. In

Table 1.3
Stages in the Evolution of Life

Stage of evolution		Name	Characteristics
Cosmic	Organic[a]		
I	—	Atomic or elemental	Evolution of the elements; origin of sun, earth, etc.
II	I	Molecular; prebiotic synthesis	Origin of the basic biochemicals of living things
III	II	Prebiotic polymerization	Formation of peptides, simple proteins, nucleic acids, etc.
IV	III	Precellular or interacting systems	Precellular events
V	IV	Eucellular	Origin and evolution of the cell and advanced organisms

[a]Only this set of stages is referred to in the text.

one of these, the synthesis of amino acids is the principal objective; in the second, the formation of the nitrogenous bases of nucleosides. As a consequence of occasional overlap of results, the two types are not always sharply differentiated.

1.2.1. Syntheses of Amino Acids

The 20 amino acids found in proteins constitute the largest portion of the basic "alphabet" of biochemicals essential to life in general. According to Wald (1964), only nine other "letters" occur universally in living things, and of these, seven (five nitrogenous bases and two pentose sugars) enter into the formation of the two nucleic acids, DNA and RNA. Thus it is evident that living matter is characterized by a relatively small number of different molecular types. But through the capability of these few to polymerize into such long chains as proteins and nucleic acids, they are more than sufficient to provide the infinite variety found among living creatures. It was upon the amino acids that the first experiments of recent decades investigating possible pathways of life's origins were centered.

Electric Discharge Experiments. What is often considered the very first attempt to explore the origins of amino acids along modern lines was that of Miller (1953) at the University of Chicago. A closed system containing a mixture of four of the gases presumed present in the primitive atmosphere (ammonia, methane, hydrogen, and water vapor from boiling water) was subjected to both spark and silent electric discharges for a week. At the end of that time, a number of amino acids and related biochemicals were found to have been synthesized (Table 1.4). Among the former were included glycine, α-alanine, β-alanine, and aspartic and glutamic acids. These results were confirmed later by Pavlovskaya and Pasynskii (1959), although those workers replaced some of the hydrogen with carbon monoxide and employed a somewhat different type of apparatus.

Spark discharge, simulating lightning in the atmosphere, was widely used as an outside energy source by researchers in this field, who employed a great diversity of "primitive atmospheres." Abelson (1956, 1957) increased the complexity of the gas mixture by adding nitrogen, carbon monoxide, and carbon dioxide to Miller's four components, but a similar catalog of products resulted. More importantly, these studies demonstrated that ammonia could not be replaced by nitrogen and that carbon dioxide could not freely substitute for methane. In an attempt to synthesize the two sulfur-containing amino acids (cysteine and methionine), Heyns *et al.* (1957) added sulfur dioxide to the gas mixture but were unable to obtain the desired results. However, through the use of an entirely different set of reactants, Oró (1963a) succeeded in lengthening the list of amino acids produced by means of electric sparking. He employed various two- and three-carbon hydrocarbons in concentrated ammonium hydroxide and thus succeeded in adding valine, leucine, and isoleucine to the catalog of products synthesized under possible prebiotic conditions. Another amino acid, methionine, was added to this list through use of a mixture of methane, nitrogen, ammonia, water, and either hydrogen sulfide or methyl sulfhydryl (Ring *et al.*, 1972; Van Trump and Miller, 1972; Miller, 1974).

Another recent study replaced the electric sparking with high-frequency discharge (42.5 megacycles) at 1000 V and 200 mA. When mixtures of methane and ammonia were exposed to this energy source, a tetramer of hydrogen cyanide was produced in addition to a number of such common amino acids as glycine, alanine, and aspartic acid (Yuasa and Ishigami, 1975). The tetrameric hydrogen cyanide and related substances are of frequent use in prebiotic experiments and receive further mention later. A recent series of investigations has exposed various primitive-type atmospheres to silent electric discharge at low pressure (20 mm Hg) for several seconds. When mixtures of methane and ammonia or methane and nitrogen were thus treated, various amino compounds were formed, along with hydrogen cyanide, cyanogen, and cyanoacetylene (Toupance *et al.*, 1975). Only atmospheres rich in nitrogen gave rise to cyanogen and cyanoacetylene. In contrast, when methane and hydrogen sulfide were used, carbon disulfide was formed in yields of several percent if the H_2S concentration was high, that is, between 40 and 50%. Because thiols (CH_3SH and C_2H_5SH) resulted only at lower concentrations of H_2S, it was concluded that

H_2S had an inhibiting effect on the synthesis of hydrocarbons (Raulin and Toupance, 1975).

Experiments Using Ultraviolet Light. Among the most abundant energy sources on earth is solar ultraviolet light, which today even in the presence of an ozone layer provides nearly 150 times as many calories annually as lightning (Miller and Urey, 1959). If oxygen is presumed to have been absent from the atmosphere very early in earth's history, ozone also would have been wanting; consequently, the relative effectiveness of this type of radiation would then have been greatly enhanced. Accordingly, a number of investigators have utilized this type of energy in spite of the difficulties imposed by the narrow range of wavelengths provided by available artificial sources. Among the earlier attempts at using ultraviolet were those of Groth and his associates (Groth, 1957; Groth and Wyssenhoff, 1957, 1960). When they tried the 1848-Å wavelength provided by a mercury lamp on methane, ammonia, and water vapor, no amino acids were obtained; however, when the same mixture was exposed to the 1470-Å and 1295-Å wavelengths emitted by a xenon source, several were synthesized.

Because the usual gases employed do not readily absorb the longer ultraviolet rays, as demonstrated by the last experiments cited above, a small number of investigations have made use of water, which does absorb them. In one set of experiments, Ellenbogen (1958) bubbled methane through a suspension of ferrous sulfide in a solution of ammonium chloride exposed to a mercury lamp, which resulted in the synthesis of valine, phenylalanine, and methionine. Bahadur (1954; Bahadur *et al.*, 1958) permitted sunlight to act upon paraldehyde solutions, variously contained ferric chloride, ferric nitrate, or ammonia, and succeeded in producing serine, asparagine, and aspartic acid. Irradiating at a wavelength of 1848 Å produced glycine and alanine from formaldehyde, inorganic carbonates, and amonia; similar wavelengths have also been employed on hydrogen cyanide with comparable results (Noda and Ponnamperuma, 1971). Formaldehyde and ammonium salts in aqueous solution, exposed to the rays of a high-pressure mercury lamp, have produced serine, valine, and phenylalanine in addition to several simpler amino acids. Similarly Steinman *et al.* (1968) made use of a mercury lamp but irradiated ammonium thiocyanate with it. As a result, a sulfur-containing amino acid, methionine, was obtained, as well as a number of the more usual types.

A method for utilizing the longest ultraviolet rays ($\lambda > 2500$ Å) with gaseous mixtures employed the introduction of a suitable photon acceptor. Among those gases that have been used for this purpose is hydrogen sulfide, which absorbs a broad continuum of rays extending from wavelengths <2700 Å down to that of the vacuum ultraviolet. For one example, Sagan and Khare (1971) were able to synthesize a number of amino acids from various "primitive atmospheres" containing this substance; among those produced were alanine, glycine, cysteine, serine, and glutamic and aspartic acids. These workers pointed out further that, whereas the shock waves described below have been calculated to produce amino acids 10^4 times more efficiently than ultraviolet light in the presence of hydrogen sulfide, ultraviolet light of wavelengths >2600 Å was at least 10^3 times as abundant on the primordial earth. Thus these longer wavelengths could have been a highly effective agent in the presence of hydrogen sulfide for the prebiotic synthesis of amino acids. A somewhat related process has utilized high-energy hydrogen atoms generated by the photolytic breakdown of hydrogen sulfide to react with methane and produce amino acids (Hong *et al.*, 1974). Purportedly, this is particularly appropriate to such syntheses in interstellar space, as well as under primitive atmospheric conditions (Friedmann *et al.*, 1971).

Ionizing Radiation as the Energy Source. It is not likely, as Miller and Urey (1959) showed, that high-energy, or ionizing, radiation, such as X-rays and gamma rays, could have played a significant role in prebiotic syntheses, because the total level of energy now available on earth annually from those sources is exceedingly low. Nevertheless, a number of investigations into the origins of life's materials have employed this type of radiation. From one series of experiments of this sort, in which dilute aqueous solutions of ammonium acetate were irradiated with 2 MeV electrons, very small yields of glycine and aspartic acid were obtained (Hasselstrom *et al.*, 1957). Comparable minute quantities of glycine and alanine were synthesized by heavy doses of gamma rays acting on crystallized ammonium carbonate (Paschke *et al.*, 1957). Oró (1963b) used a 5-MeV beam on ^{14}C-labeled methane and obtained a number of amino acids and other molecules of biological importance. Previously, Palm and Calvin (1962) had exposed a mixture of methane, ammonia, hydrogen, and water vapor to a similar fast-electron beam and had identified glycine and aspartic acid among the products. In terms of yields, one of the more successful attempts employing ionizing radiation was that of Dose and Ponnamperuma (1967). These researchers exposed aqueous solutions of *N*-acetylglycine and ammonia to gamma rays and obtained a number of amino acids and derived compounds.

Experiments Employing Heat. A far more abundant energy source on earth is heat, derived from solar radiation, volcanism, and decay of radioactive elements. Because of this prevalence, a rather large number of experiments have employed thermality in one way or another. Among the earliest of these was a series by Fox *et al.* (1955, 1957a,b), who, by heating malic acid and urea to 150°C, obtained aspartic acid, α- and β-alanine, and several byproducts. Later trials, subjecting α-hydroxyglutaric acid or glucose together with such an ammonia source as urea to like temperatures yielded glutamic acid and glycine (Fox, 1960). Oró and Kamat (1961) demonstrated the formation of five amino acids, including glycine, alanine, aspartic acid, threonine, and serine, by simply heating formalin and hydroxylamine to 80–100°C for 40 to 60 hours. Lowe *et al.* (1963a) also used relatively low temperatures (90°C) on a

Table 1.4
Representative Results of Various Types of Amino Acid Syntheses, Expressed as Percentages

Product	Energy source					
	Electric spark[a]	Low temperature[b]	High temperatures			Shock waves[e]
			950°C[c]	1060°C[d]	980°C[d]	
Proteinogenic amino acids						
Glycine	13.0	40.0	60.3	1.0	59.0	71.0
α-Alanine	6.9	5.0	18.0	1.0	12.0	27.0
Aspartic acid	0.18	1.3	3.4	3.0	—	—
Glutamic acid	0.001	1.6	4.8	—	—	—
Valine	—	0.4	2.3	—	—	2.0
Proline	—	5.0	2.3	—	—	—
Serine	—	4.0	2.0	—	—	—
Leucine	—	0.5	2.4	—	—	0.2
Isoleucine	—	9.0	1.1	—	—	—
Phenylalanine	—	1.2	0.8	—	—	—
Threonine	—	—	0.9	—	—	—
Tyrosine	—	—	0.8	—	—	—
Other amino acids						
Sarcosine	1.02	—	—	—	—	—
β-alanine	3.04	—	?	90.0	28.0	—
N-Methylalanine	0.20	—	—	1.5	—	—
α-Amino-*N*-butyric acid	1.02	—	0.6	—	—	—
α-Amino-iso-butyric acid	0.02	—	—	—	—	—

solution of hydrogen cyanide and ammonia for just 18 hours and produced leucine, isoleucine, and glutamic acid, in addition to the five found by the Oró laboratories. More recently, Fox and Windsor (1970) synthesized no fewer than ten amino acids (Table 1.4) by heating mixtures of formalin, ammonia, and water. At even lower temperatures (75°C) over a period of 8 days, aminoacetonitrile in aqueous solution condensed to form glycine and perhaps other

Table 1.4. Cont'd.

Product	Energy source					
	Electric spark[a]	Low temperature[b]	High temperatures			Shock waves[e]
			950°C[c]	1060°C[d]	980°C[d]	
Alloisoleucine	—	—	0.3	—	—	—
β-Amino-*N*-butyric acid	—	—	—	1.0	1.0	—
Miscellaneous organic compounds						
Glycolic acid	11.0	—	—	—	—	—
Lactic acid	6.3	—	—	—	—	—
Succinic acid	0.81	—	—	1.5	—	—
Iminoacetic acid	1.12	—	—	—	—	—
Formic acid	47.0	—	—	—	—	—
Acetic acid	3.04	—	—	—	—	—
Propionic acid	2.87	—	—	—	—	—
Urea	0.41	—	—	—	—	—
Miscellaneous (or undetermined)	1.63	32.0	?	—	—	?

[a]Miller (1953).
[b]Fox and Windsor (1970).
[c]Harada and Fox (1964).
[d]Lawless and Boynton (1973).
[e]Bar-Nun et al. (1970a).

amino acids (Chadha *et al.*, 1971, 1975). Aminoacetonitrile itself has been a product of certain treatments of primitive atmospheres (Ponnamperuma and Woeller, 1967).

Under the impression that phenylalanine had not previously been produced experimentally under the hypothesized primitive earth conditions, Friedmann and Miller (1969) conducted a series of experiments designed especially to obtain that amino acid, employing a variety of energy sources. Among these were high temperatures, electric discharges, and ultraviolet light, all of which were separately applied to mixtures of simple hydrocarbons. Then by adding hydrogen sulfide, they successfully produced phenylalanine, along with small yields of tyrosine.

Very high temperatures have also proven successful. In the earliest high-temperature experiment, Harada and Fox (1964) bubbled methane through an aqueous ammonia solution, and the resulting gas mixture was then heated as it passed through silica gel, quartz sand, volcanic sand, or alumina in a Vycor glass reaction tube. Among the products at 950°C were 12 proteinogenic amino acids, of which glycine comprised 60% of the yield, plus several others (Table 1.4). However, a recent replica of this experiment was made, and more precise determinations of the products resulted from employment of gas chromatography either alone or in combination with mass spectrometry (Lawless and Boynton, 1973). At the highest temperatures (1060°C) only four proteinogenic amino acids were produced and all of these in very small amounts, 90% of the product being β-alanine (Table 1.4). At 980°C or 930°C only two proteinogenic amino acids, glycine and α-alanine could be found; at 980°C the former made up 59% of the yield, the latter 12%, and β-alanine furnished the bulk of the remainder. At 930°C only two amino acids could be detected in a ratio of 24 glycine to 1 of α-alanine. The great predominance of β-amino-acids resulting at high temperatures and the relatively few α-amino-acids that were found at any temperature casts some doubt upon the usefulness of very high temperatures in amino acid synthesis under primitive earth conditions. More moderate temperatures (240°C) converted glycine into a number of other amino acids when the experiment was conducted with alumina (Ivanov and Slavcheva, 1977).

Finally, thermal energy from an unexpected source was applied by Bar-Nun *et al.* (1970a) to produce amino acids. In a single-pulse shock tube, a mixture of methane, ethane, ammonia, and water vapor resulted in the formation of four simple amino acids (Table 1.4) upon exposure to shock waves. Yields calculated for some of these experiments were surprisingly high, in one set amounting to an efficiency of 36%. The production efficiency resulted from the high temperatures (1000–3500°K) momentarily attained and the rapid quench which followed. Thus the high-temperature thermodynamic barrier was bypassed. Consequently, it was proposed that thunder and shock waves, such as those produced by entry of meteorites and micrometeorites into the earth's atmosphere, may have served as the principal energy source for prebiological syntheses on the primordial earth (Hulett, 1970; Bar-Nun *et al.*, 1970b, 1971; Bar-Nun, 1975; Barak and Bar-Nun, 1975).

1.2.2. Syntheses of the Nitrogenous Bases

The nitrogenous bases that enter into the formation of nucleic acids are more widely diversified than the five usually mentioned might imply. Nevertheless, those five are far more frequent than any of the others, as together they form almost all DNA as well as one common type of RNA, and in addition they contribute heavily to the formation of the remaining RNAs. Hence, attention has centered upon the usual quintet, but two others (xanthine and hypoxanthine) also receive frequent mention here because of their abundance among the products synthesized. All of these bases, whether common or not, fall into two major categories, pyrimidines and purines. The usual pyrimidines, which have a single ring composed of two nitrogen and four carbon residues (Figure 1.1), are cytosine, thymine, and uracil, and a less frequent type, pseudouracil. The purine molecule consists of a ring like that of the pyrimidines plus a second consisting of one carbon and two nitrogen residues (Figure 1.1). To this category belong adenine and guanine, and two less frequent members, xanthine and hypoxanthine.

Syntheses of Purines. The overlap mentioned earlier that exists between the syntheses of amino acids and nitrogenous bases becomes apparent in what is probably the first successful attempt at producing a member of the latter category. When, as reported above, Oró and Kamat (1961) produced a number of amino acids by heating hydrogen cyanide and ammonia together, those products were present mostly in the form of a black insoluble polymer. Later, when this residue had been removed by filtration or centrifugation, it was determined that adenine was present in the remaining fluid (Oró and Kimball, 1962; Oró, 1965). The synthesis of this purine was soon confirmed by Lowe *et al.* (1963b). In addition, the latter study found that hypoxanthine also was formed from the ammonia, hydrogen cyanide, and the water employed.

Most of the remaining experiments on the synthesis of the nitrogenous bases have used either compounds related to cyanide in having a $-C{\equiv}N$ unit present or such intermediates as 4-aminoimidazole-5-carboxamide derived from the interaction of cyanide and ammonia in water (Fox and Dose, 1972). For example, Ferris and Orgel (1965) produced adenine when they added hydrogen cyanide to a complex substance related to those intermediates. However, hydrogen cyanide has itself proven an effective starting point for the formation of the nitrogenous bases. In turn, this substance has been produced abundantly by exposure of the usual "primitive atmospheric" mixtures to electric discharges (Miller, 1955) or to beta rays (Palm and Calvin, 1962), providing the proportions of hydrogen were kept small (Loew and Chang, 1975). Adenine and guanine have been reported as products of dilute solutions of hydrogen cyanide exposed to ultraviolet radiation (Ponnamperuma, 1965; Yang and Oró, 1971; Noda and Ponnamperuma, 1971). One system for producing adenine proved to be so efficient that it has been developed into an industrial process; this involves heating anhydrous hydrogen cyanide in liquid ammonia (Wakamatsu *et al.*, 1966).

Syntheses of Pyrimidines. As a whole, the pyrimidines are less stable than the purines and can be synthesized only under less violent conditions than those tolerated by the latter. Among the first pyrimidines to be produced was

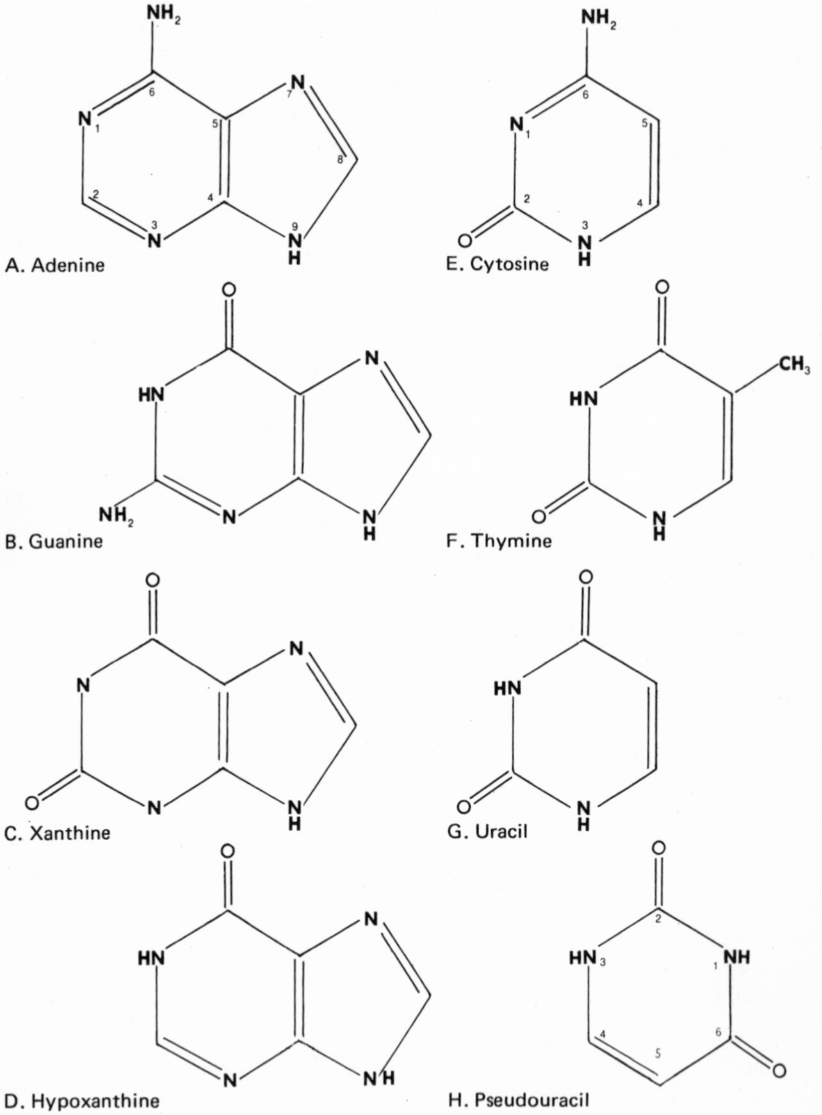

NH2 N 1 6 5 7 N 8 2 4 3 N 9 N H
A. Adenine

NH2 N 1 6 5 2 4 O 3 N H
E. Cytosine

O HN N NH2 N N H
B. Guanine

O CH3 HN O N H
F. Thymine

O N N O N N H
C. Xanthine

O HN O N H
G. Uracil

O HN N N NH
D. Hypoxanthine

O 2 HN 3 1 NH 4 6 5 O
H. Pseudouracil

Figure 1.1. The more important nitrogenous bases. Two related groups of bases are found in the nucleic acids of living things; in one group (A–D), the purines, the molecules are double rings, whereas in the second (E–H), the pyrimidines, they are single rings.

uracil, the formation of which resulted from heating a mixture of malic acid and urea to 100–140°C for 15–120 minutes in the presence of polyphosphoric acid (Fox and Harada, 1961). However, it is doubtful that polyphosphoric acid was widespread in occurrence on the primitive earth (Fox and Dose, 1972). A possibly more feasible method was proposed by Sanchez *et al.* (1966), employing a substance (cyanoacetylene) readily obtained by subjecting mixtures of methane and nitrogen to electric discharges. Cytosine resulted from the fusion of cyanoacetylene with urea, for example, under certain conditions, whereas asparagine and aspartic acid were formed under others. Cyanoacetylene can be utilized also to produce various purines (Ferris *et al.*, 1969, 1973a). More recently, cyanoacetaldehyde has been used with guanidine to form cytosine and uracil (Ferris *et al.*, 1974a). Thus far, however, thymine has been derived only by methylating uracil by means of formaldehyde and hydrazine (Stephen-Sherwood *et al.*, 1971). As a whole the results indicate that, although some substantial progress has been made, laboratory attempts at prebiotic synthesis of the nitrogenous bases still have not achieved the level of credulity attained in the laboratory with prebiotic syntheses of amino acids.

1.2.3. Syntheses of Carbon Sources

Quite frequently, experiments for the synthesis of either amino acids or nitrogenous bases form various hydrocarbons as by-products (Table 1.4), including formic, succinic, and acetic acids and other substances utilized as carbon or energy sources by numerous types of extant organisms. Consequently, less attention has been devoted to specific pathways for their formation than to those for the other ingredients of living things. Nevertheless, because of the obvious biological significance of sugars, a modest number of efforts at creating them have been made.

Experimentation with Formaldehyde. In most of the experiments designed for this purpose, formaldehyde has been employed as the raw material or has provided the focal point, for it has long been known that that substance can condense to form sugars (Boutlerow, 1861; Loew, 1886). Representative steps of the processes are shown in Figure 1.2. Some of the very first biochemical–evolutionary experiments, those of Groth and Suess (1938) and, somewhat later, those of Garrison *et al.* (1951), successfully produced formaldehyde from inorganic matter. Large yields of this substance have resulted

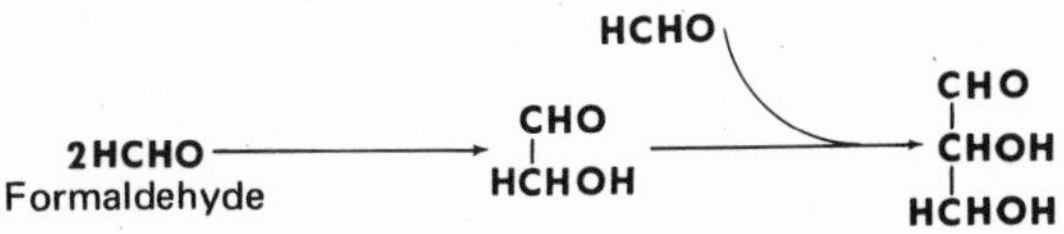

Figure 1.2. Synthesis of a simple sugar from formaldehyde. (Based on Euler and Euler, 1906.)

from gaseous mixtures that included methane and water vapor being exposed either to electric discharges (Miller, 1955; Calvin, 1956) or to gamma or beta rays (Palm and Calvin, 1962; Ponnamperuma and Flores, 1965). Furthermore, formaldehyde is still being formed today by photochemical activities in the atmosphere, for quantities up to 1 mg per milliliter of rain have been detected (Dhar and Ram, 1933; Calvin, 1967).

When aqueous solutions of this compound were allowed to condense in the presence of alkaline catalysts, a number of saccharides and related substances resulted, including fructose, sorbose, xylulose, and cellobiose (Euler and Euler, 1906; Schmitz, 1913). In more recent times, nearly thirty different monosaccharides were produced similarly by the alkali-catalyzed condensation of formaldehyde (Mariani and Torraca, 1953). Because of their greater stability, the majority of the monosaccharides formed were pentoses or hexoses, rather than such smaller molecules as trioses or such larger ones as heptoses. Aqueous solutions of formaldehyde (formalin) yielded the two preeminent pentoses, ribose and deoxyribose, when exposed to either ultraviolet or gamma rays (Ponnamperuma and Mariner, 1963; Ponnamperuma, 1965). To avoid the untenable proposition of the huge amounts of formaldehyde needed to provide a sufficient concentration in the oceans, Gabel and Ponnamperuma (1967) attempted to simulate a primitive hot spring containing the chemical. Accordingly they refluxed formalin at various concentrations, using a native clay (kaolinite) as a catalyst. At a concentration approximating 0.5 M, only 3-, 4-, and 5-carbon sugars (including ribose) were obtained; but at concentrations around 0.01 M, hexoses were synthesized in addition.

Experiments with Other Reagents. Not all experiments for the synthesis of sugars have started with formaldehyde. Sometimes other aldehydes, including glyceraldehyde and acetaldehyde, have been used instead, producing deoxyribose and deoxyxylose (Oró and Cox, 1962). Even gaseous mixtures, such as methane, ammonia, and water vapor, have been irradiated under an electron beam to produce ribose and deoxyribose (Ponnamperuma, 1965). Although further experimentation toward the syntheses of sugars is needed, and probably will be forthcoming, sufficient convincing data exist to suggest that saccharide formation could actually have occurred spontaneously in the environment usually postulated for the primordial earth.

1.2.4. The Formation of Phosphorylated Compounds

Phosphorylated compounds are so abundant in living things of all types that there can be little doubt that they arose early in life's history. For example, before a sugar molecule of any sort can be conducted through the cell membrane, or often before it can be metabolized, it must first be phosphorylated. Furthermore, many of the coenzymes involved in cell respiration, photosynthesis, and other vital processes are phosphorus-containing substances.

But, most importantly, the nucleic acids DNA and RNA are composed of phosphorylated compounds. In order for the nitrogenous bases to become coupled together, each pyrimidine or purine must first be united to a pentose sugar to form a nucleoside (Figure 1.3); then to each pentose, a phosphate residue must be attached to produce a nucleotide. The phosphate units can then serve in joining the nucleotides together to form the long complex chains that characterize DNA and RNA (Figure 1.3). These nucleic acids, especially the former, are the ultimate basis of all the vital processes of cells (Calvin and Calvin, 1964).

In the experimental search for prebiotic phosphorylating agents and the syntheses of phosphate esters under conjectured primitive earth conditions, two distinct approaches have been employed. One methodology utilizes heat in the absence of water to bring about reactions between inorganic phosphates of various types and the organic substances to be phosphorylated. The second involves use of water and reactive intermediates formed by the addition of inorganic phosphates to condensing agents. Before entering into a summary of the results of these two approaches, some facts concerning the element phosphorus and its compounds need to be recounted.

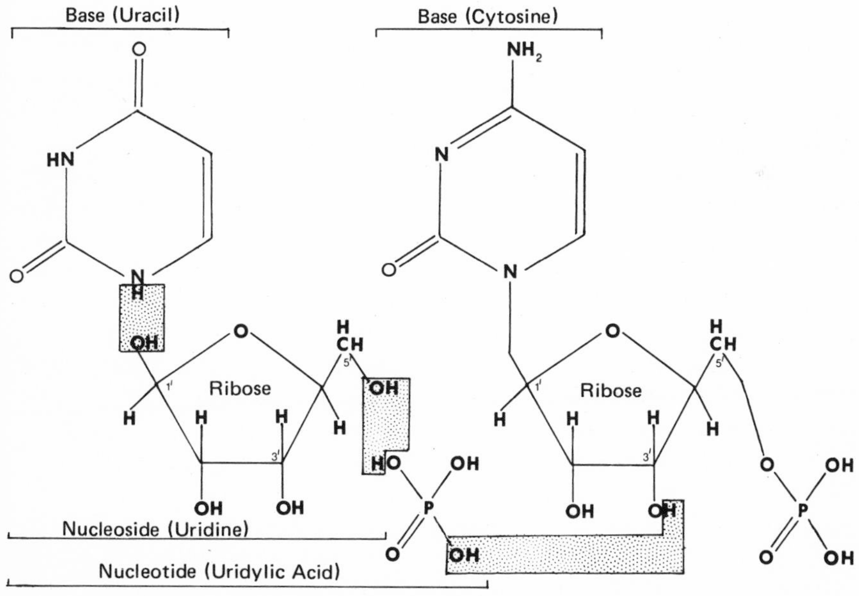

Figure 1.3. The formation of a nucleic acid. The combination of a nitrogenous base and a pentose (here ribose) is known as a nucleoside, and a singly phosphorylated nucleoside is known as a nucleotide. Polymers of nucleotides are referred to as nucleic acids. Note that all reactions involve dehydration.

Phosphorus and Its Compounds. First, this element has the capability of producing very large, complex molecules, the simpler of which fall into three major categories, orthophosphates, metaphosphates, and pyrophosphates (Figure 1.4). Depending upon conditions, these can be variously condensed into rings or chains of phosphates called polyphosphates, the tripolyphosphate ion being quite stable under ambient conditions (Van Wazer, 1958; Schwartz and Ponnamperuma, 1971). Second, the possible role of phosphorus in prebiotic syntheses poses two great difficulties, which derive from the element's relative rarity and the insolubility of minerals containing it. On the average, crustal rocks contain approximately 0.1% phosphorus; the phosphorus-containing mineral apatite is the principal single source, as it comprises about 0.6% of the total mineral mass of the earth's surface (Mason, 1966). In igneous rocks, this mineral is almost invariably in the form of fluoroapatite, but in sedimentary rocks, chloride, carbonate, or hydroxyl radicals may replace part of the fluoride (McKelvey, 1967).

Since the solubility-product constant of fluoroapatite is extremely low (one part in 10^{60}), phosphorus occurs in present-day oceans only in minute concentrations (Goldberg, 1954; Schwartz and Deuss, 1971). This dilution factor is of no great moment to living things today because of their complex transmembrane-transport systems; in attempts to reconstruct the origins of life's materials, however, the low concentration appears an almost insurmountable obstacle. The dilute quantities existing in the seas imply that the concentrations often employed in experiments are unrealistic (Hulett, 1969; Schwartz, 1972a). To avoid this objection, it has been proposed that suitable conditions may have existed locally, such as at hot springs, in temporary ponds (Schwartz, 1972b), or in evaporating tidal pools (Fuller *et al.*, 1972). Moreover, it should be understood that, whereas the earlier studies often failed to consider the geologic and solubility factors of phosphorus compounds, researchers more recently have usually attempted to make their experiments more relevant to primeval earth conditions (Schwartz, 1971; Schwartz *et al.*, 1975).

Experiments under Anhydrous Conditions. The approach in which anhydrous conditions were employed appears to a degree to antedate that employing aqueous media. Schramm and his co-workers (1962) used an organic solvent (dimethylformamide) in which low concentrations of adenine and ribose were dissolved. Then with diethyl cyclic polyphosphate (Figure 1.4) as a condensing agent, they were able to secure about a 20% yield of the nucleoside adenosine at temperatures between 55°C and 60°C. By using adenine and deoxyribose under similar conditions, a 40% yield of deoxyadenine was obtained. Later, Schramm (1965) determined that for the most successful syntheses of nucleosides, mixtures containing 70% tetraethyl-tetrametaphosphate and 30% tetraethyl-tetrapolyphosphate were necessary. Phosphorylation of nucleosides has been accomplished in organic solvents by treatment with inorganic thiophosphate (Stabaugh *et al.*, 1974).

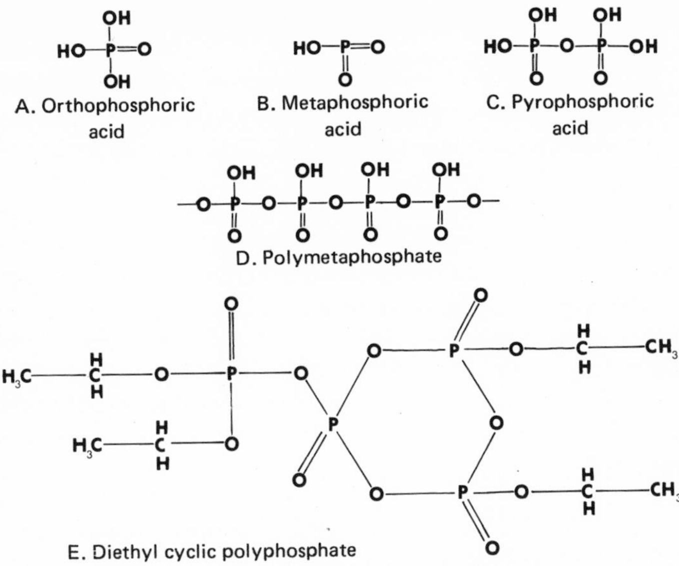

Figure 1.4. Some important compounds of phosphorus.

It has also been possible to produce polynucleotides on a purely inorganic basis by treating nucleosides with free polyphosphoric acid at 60°C (Schwartz *et al.*, 1965). Nucleotides have been synthesized successfully "dry" by heating nucleosides with inorganic phosphates, especially such acid salts as calcium dibasic phosphate [$Ca(H_2PO_4)_2$; Ponnamperuma and Mack, 1965; Skoda and Morávek, 1966; Beck *et al.*, 1967]. More recently, Lohrmann and Orgel (1971) proposed a dry phosphorylating system which appears to be feasible under such natural situations as deserts or warm beaches. They reported that at 66°C nucleosides became phosphorylated in a period of a few days when mixed with urea, ammonium chloride, and hydroxyapatite. In addition, phospholipids and simple membranes have been synthesized with the aid of silica and silicates under prebiotic conditions (Hargreaves *et al.*, 1977).

Syntheses in Aqueous Media. Experiments using aqueous solutions for phosphorylating are perhaps more frequently undertaken than those employing anhydrous conditions. Ponnamperuma and his colleagues (1963) irradiated dilute aqueous solutions of adenine, ribose, and ethylmetaphosphate with ultraviolet light at rather long wavelengths ($\lambda = 2400–2900$ Å) and produced the important energy carrier of the cell, adenosine triphosphate (ATP). In some cases such cyanide-related substances as cyanogen, cyanamide, and carbodiimide have been used as condensing agents. For example, Lohrmann and Orgel (1968, 1973) brought about the formation of uridine-5′-phosphate from uridine and inorganic phosphate by their use, though with small yields. Quite recently,

Schwartz (1972b; Schwartz *et al.*, 1973) synthesized nucleotides with cyanogen, water, and apatite, possibly as in an evaporating pond environment; somewhat similar studies employed the same approach but with orthophosphate and reducing sugars in place of the apatite (Halmann, 1975). Others have used inorganic trimetaphosphates to react with *cis*-glycols or nucleosides in alkaline solutions and have produced monophosphates or 2′- and 3′-phosphates of nucleosides (Schwartz, 1969; Saffhill, 1970). Since struvite rather than apatite is precipitated by phosphates from seawater containing more than 0.01 M ammonia, Handschuh and Orgel (1973a,b) suggested that the former mineral may have been abundant on the primitive earth. When it was heated with urea and nucleotides, good yields of nucleoside pyrophosphates were obtained (contrast McConnell, 1973). Hence, at the present time there exists a considerable body of evidence supporting a number of reasonable pathways to prebiotic phosphorylations that bypass the obstacles presented by the lack of abundance and insolubility of phosphorus.

1.3. PREBIOTIC POLYMERIZATION (STAGE II)

The prebiotic phosphorylations just described are transitional to Stage II in the origin of life in more than one sense. As with the Stage II processes, those processes dealt with the enhancement of molecular size and were frequently concerned with such condensing agents as cyanogen. Nonetheless, sharp distinctions do exist. In contrast to the end processes of Stage I, where the chemical union of no more than three separate compounds is involved, Stage II activities are concerned with the syntheses of gigantic molecules, referred to in biochemistry as macromolecules, in which tens or even hundreds of compounds of similar characteristics are joined together to form polymers. Attention now focuses on the possible pathways for the formation by spontaneous means of two major categories of substances, peptides and proteins on one hand, and nucleic acids on the other. As in the previous stage, overlap between the two categories can be noted on frequent occasions, for both share a common process in their production, the removal of water (Figures 1.3 and 1.5A). Sugars, too, are polymerized in this fashion (Figure 1.5C). While the end products that derive from the assembling of monomers to form these three classes of polymers are formidably complex, the processes involved are operationally so simple, relatively speaking, that sometimes they are referred to as "zipper chemistry."

1.3.1. Formation of Peptides and Proteins

The Basic Problem. Simple though the processes may be in a relative sense, a number of difficult problems center around them. First, the amino acids which comprise peptides and proteins either had been formed in the seas

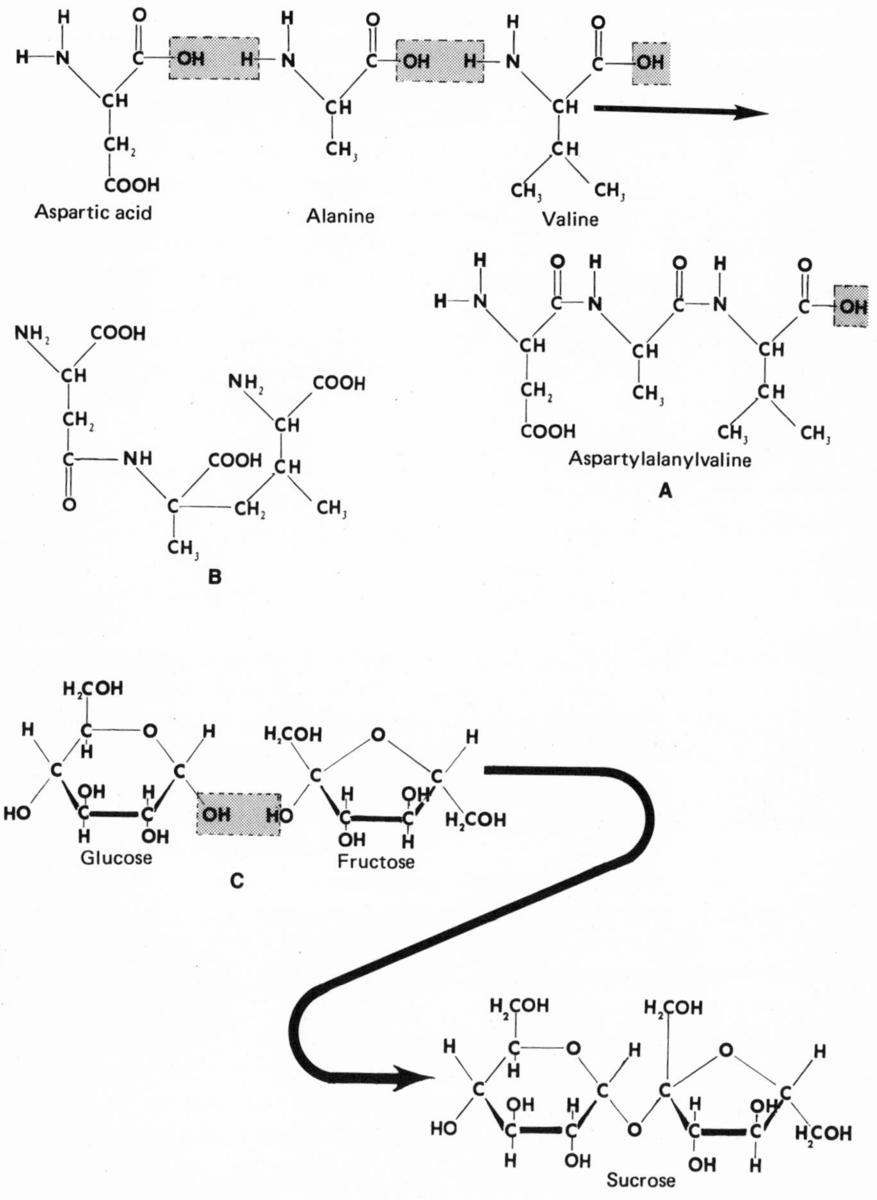

Figure 1.5. Monomers and polymers of biological importance. (A) Dehydration condensation (peptidization): in biologically formed polymers, amino acids are joined together between the α-amino radical of one and the α-carboxyl of a second. (B) Some nonbiological linkages: synthetic polymers often contain other linkages, such as the γ-to-α and carbon-to-carbon linkages illustrated. (C) The polymerization of sugars also involves the removal of water.

or had reached those bodies of water after formation in the atmosphere. Thus they most likely were in aqueous suspension together with the nitrogenous bases, saccharides, and various by-products of syntheses, transforming the early oceans into what is frequently known as the "primitive soup." Even though the earliest seas appear to have been generally small in comparison to existing ones (Rubey, 1951; Dillon, 1974), the soup must have been relatively dilute, a dilution that could only have increased as the oceans themselves gradually grew in dimensions and depth.

The difficulty which this dilution factor presents is best perceived through a condensed view of the energetics involved in peptide formation (Fox, 1974). The reaction has been expressed by scientists at one laboratory (Borsook and Huffman, 1944; Borsook, 1953) as:

$$\text{Amino acid X} + \text{amino acid Y} \rightarrow \text{dipeptide XY} + H_2O$$
$$\Delta G^\circ = 2\text{–}4 \text{ kcal}$$

In other words, to unite two amino acid residues by means of a peptide bond, between 2 and 4 kcal must be imparted to the reaction.

The difficulty is made even clearer if it is imagined that a polypeptide—one so small as to have a molecular weight of only 12,000—is to be synthesized spontaneously from natural suspensions bearing the necessary amino acids at 1 M concentrations. Then if these were permitted to interact, at dynamic equilibrium only 10^{-99} M concentration of the protein would be synthesized (Dixon and Webb, 1958; Fox and Dose, 1972). As the cited sources point out, the volume of the solution would have to be 10^{50} times as large as the entire earth in order to produce a single complete molecule of the polypeptide! In view of the oceanic volumes and the relatively small quantities of amino acids that might have been synthesized annually on the primitive earth, even the molar concentrations of amino acids visualized in the example are obviously far beyond the realm of possibility.

Constraints in Biological Systems. A second problematic consideration centers around the orderliness with which amino acids combine in biotic systems. In proteins of every extant organism, the amino acids are uniformly of the α-amine type, in which the carboxyl (–COOH) and amino (NH_2) radicals are attached to the first (α) carbon of the chain (Figures 1.5 and 1.6). Often when amino acids are synthesized under assumed primitive conditions, various other types are formed, such as β-alanine (Table 1.4) in which the amino radical is on the second (β) carbon of the chain. The amino acids of proteins also share another common trait, that of having a molecular structure of levorotatory configuration (Figure 1.6). Such a configuration may or may not actually rotate polarized light to the left; however, the molecular structure resembles that of certain substances which do possess that ability. A third constant feature of amino acids is that in forming proteins they are always united to one another by the processes (peptidization) illustrated in Figure 1.5A, in which only the car-

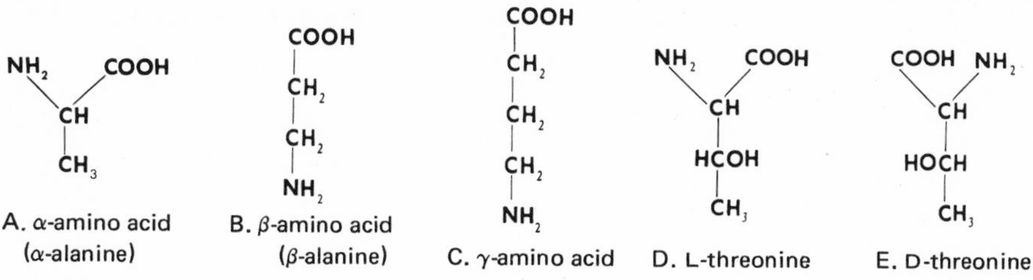

A. α-amino acid (α-alanine): NH_2, COOH, CH, CH_3

B. β-amino acid (β-alanine): COOH, CH_2, CH_2, NH_2

C. γ-amino acid: COOH, CH_2, CH_2, CH_2, NH_2

D. L-threonine: NH_2, COOH, CH, HCOH, CH_3

E. D-threonine: COOH, NH_2, CH, HOCH, CH_3

Figure 1.6. Natural and synthetic varieties of amino acids. All amino acids that enter into the formation of natural proteins are of the α-amino type (A) and have the levorotatory configuration (D). (D, E) Enantiomeric forms.

boxyl and amino radicals on the α-carbon of adjacent amino acids interact, despite the fact that other points of the molecules and other modes of attachment are chemically equally feasible.

1.3.2. Peptide Formation in Anhydrous Environments

A simple method of overcoming the thermodynamic barrier in any reaction is that of removing one of the products (in the present case, either the peptide or the water), so that equilibrium is no longer maintained. Accordingly, one of the first modes of attacking the present problem was that of removing the water by volatilization through heating mixtures of amino acids above the boiling point of water.

The Use of Heat in Experiments. Since few amino acids melt without breaking down at the temperatures required (150–180°C), the very early attempts at synthesizing peptides by heat under anhydrous conditions largely resulted in carbonized tar. Later, however, Fox and his co-workers (1956, 1957b) used two amino acids (aspartic and glutamic acids) that do melt when heated and discovered that these molten substances could serve as a solvent for other amino acids. In one case, these experimenters made a mixture of aspartic and glutamic acids and added an approximately equal quantity of a second mixture consisting of the other 18 amino acids and heated the combined mixtures to nearly 150°C. After the water had distilled out, the interacting mass was plunged into water, resulting in what was called a proteinoid (Fox and Harada, 1958, 1960).*

In still later experiments, they were able to carry out similar reactions at 100°C over a period of 150 hours in a polyphosphoric acid solution that was made by partial thermal evaporation of aqueous phosphoric acid (Fox and

*Proteinoids differ from proteins in that the bondings between constituents are not necessarily normal peptide bonds between α-amine and α-carboxyl residues. Such polymers are also known by the term polyamino acids.

Table 1.5
Composition of a Proteinoid (Amino Acids Arranged in Order of Abundance)[a]

Substance	Percent	Substance	Percent
Aspartic acid	51.9	Valine	1.33
Glutamic acid	13.3	Alanine	1.31
Phenylalanine	5.87	Proline	1.04
Tyrosine	3.87	Methionine	0.86
Leucine	3.44	Isoleucine	0.71
Glycine	2.93	Serine	0.63
Lysine	2.79	Threonine	0.55
Histidine	2.53	Ammonia	5.02
Arginine	1.83		
		Total	99.81

[a]Based on Fox and Yuyama (1963).

Yuyama, 1963). Whereas the original mixture consisted of aspartic acid 33%, glutamic acid 17%, and nearly 3% of each of 17 additional amino acids, the resulting proteinoid had the quite dissimilar composition shown in the analysis (Table 1.5). There it is at once apparent that the aspartic acid increased in proportion to 51% in the end product and that glutamic acid decreased to 13%. Similarly, in the end product the proportions of the other amino acids varied considerably from the original 3%. Since the percentage composition thus was not merely the random, statistical one expected, it was claimed that the proteinoid resulted from selection (Fox and Yuyama, 1963; Fox, 1965; Calvin, 1969). The selection in this case is chemical selection and should not be confused with the biological principle of natural selection, which results from differential survival rates. To avoid confusion, the principle involved in the creation of the unexpected ratio of amino acids in the proteinoid might better be recognized as nonrandomness of combination among amino acids. This principle, under any name, is of particular importance and enters into discussions on several occasions in later chapters. Dose and Rauchfuss (1972; Dose, 1974) have obtained comparable results with thermal polymers of amino acids.

Other Thermal Syntheses. Other laboratories have also succeeded in producing polyamino acids by thermal processes. Recently, polymeric materials were obtained in yields of up to 5.7% by weight when various mixtures of amino acids were heated to 175°C for periods up to 6 hours under a stream of nitrogen gas to retard decomposition (Saunders and Rohlfing, 1972). The study cited was of special interest in that the reaction mixtures were made in the same proportions as certain "prebiotic" syntheses reported in the literature or in those of lunar samples and meteorites. Here, too, nonrandomness in the combining of the amino acids was shown by certain of their results. For example, in what Saunders and Rohlfing refer to as the "Fox" and "Lunar A" sequences, three amino acids were represented in the starting mixtures respectively as follows: glycine 57% and 55%, alanine 9% and 14%, and aspartic acid, 10% and 13%. In the resulting polyamino acids, the same substances were represented in both "Fox" and "Lunar" experiments by 69%, 12%, and 12% to 13%, in the same sequence. The alanine was particularly notable in that it achieved approximately consistent levels in the respective end products by its proportions *increasing* in the "Fox" series and *decreasing* in the other. Further, more than simple bonding between amino acids was evidenced in the reactions thus carried out, for in several instances amino acids absent in the starting mixtures were found in the product. Similar nonrandomness of combination was found in the results of a later set of experiments designed to reduce the amount of change from the levorotatory form to the dextrostereo isomer (Rohlfing and Fouche, 1972). In these, heating was carried out under a nitrogen stream as before, but either glutamic acid or lycine was used with another amino acid in equal proportions or in a 4 : 7 ratio.

A recent thermal approach involved the use of imidazole as the catalyst to activate amino acids (or nucleotides) to form polymers (Lohrmann and Orgel, 1973). Basically, the processes more closely paralleled contemporary biological mechanisms than most others have in that stepwise syntheses were utilized. However, at present the applicability of the processes to prebiotic conditions is questioned, because of the requirements for quantities of urea and imidazole.

Experiments Using Condensing Agents. Somewhat less severe temperatures can be used when a substance called polymetaphosphate ethyl ester is employed as the dehydrating condensing agent. This reagent is prepared by a number of processes, and its mode of preparation plays an important part in determining the nature of its products. One method that has been followed for use in the synthesis of polypeptides is by refluxing phosphorus pentoxide in a mixture of 1 part chloroform and 2 parts ethyl ether for 12 hours (Schramm *et al.*, 1962). By means of the resulting polymetaphosphate ethyl ester and an organic solvent, arginine has been polymerized into polyarginine with a mean molecular weight between 4000 and 5000, and a polypeptide of mean molecular weight of 7400 has resulted from a mixture of the three amino acids tyrosine, alanine, and glutamic acid when treated with the reagent (Schramm and Wissman, 1958; Schramm *et al.*, 1961; Schramm, 1965). More recently a number of other polypeptides have been prepared by comparable techniques (Nooner and Oró, 1974). In a later section this condensing reagent also receives mention in relation to the formation of polynucleotides.

The Validity of the Anhydrous Approach. Although it has thus become evident that thermal condensations in the absence of water appear quite feasible, questions have been raised as to their validity under the natural environment of the primitive earth (Miller and Urey, 1959; Bernal, 1967; Steinman, 1971). The cell being largely aqueous and aquatic, these researches held that it appears logical for life itself to have arisen in the seas. Moreover, few cells can survive temperatures even approaching the boiling point of water, so that reactions requiring temperatures exceeding 80°C do not seem relevant. In defense of the thermal approach, however, Fox and Dose (1972) pointed out that, although the cell is indeed aqueous, organisms do not carry out peptide formation simply in dilute watery solutions, for the principal site of protein synthesis is the ribosome, a minute particle that is far from being largely aqueous. Furthermore, they suggested that organisms themselves' being aquatic does not necessarily signify that their precursor molecules could not have been formed in dry places. They also pointed out that the reason organisms cannot tolerate high temperatures is because of the water they contain, and further noted that many of the fundamental molecules of biology are, in formal chemistry, properly considered products of dehydrogenation and dehydration.

1.3.3. Peptide Formation under Other Conditions

Early Attempts. Peptide formation under aqueous conditions has been approached through many diverse techniques. Among the first successful syntheses were those of Bahadur and Ranganayaki (1958), in which sterile solutions of sucrose (2%) and glycine (0.1%) were exposed to sunlight. At the end of a month's exposure, short peptides such as glycylglycine, glycylalanine, and glycylnorleucine were found; although the yields were low, no peptides were detected in the controls maintained in the dark. As in certain of the thermal condensations, reactions other than simple peptide linkages evidently occurred, for the presence of at least two amino acids absent from the original mixture was noted (Dose and Zaki, 1971). Sucrose appeared to be an essential ingredient, for it was shown that amino acids either in a dry state or aqueous solution do not form peptides in the absence of sucrose when irradiated with ultraviolet or X-rays (Dose and Ettre, 1958; Dose and Risi, 1968). The latter study also reported that radiation broke down peptides in aqueous solution at nearly ten times the rate it formed them, a report that is difficult to understand in view of the net synthesis of peptides found by Bahadur and Ranganayaki. Furthermore, such radiolysis would apply to all peptides in a watery medium, regardless of whether they were formed dry, wet, or otherwise, once they had reached the surface layers of the seas.

Condensing Agents in Peptide Formation. A second major approach to peptide formation in the presence of water is through the use of certain of the very same condensing agents employed in the syntheses of the elementary biochemicals of life. Among the first to be so employed was hydrogen cyanide (Calvin, 1965), a substance produced in abundance by a number of prebiotic syntheses, including the earliest ones of Miller (1953, 1959). Even at quite low aqueous concentrations, such as may have prevailed in the "primitive soup," this substance has been demonstrated capable of forming polymers of amino acids at moderate temperatures (Matthews and Moser, 1967; Matthews, 1971; Ferris and Ryan, 1973; Ferris *et al.*, 1973a,b, 1974b). For instance, when mixed with ammonia (either anhydrous or in aqueous solution), it produced peptide-like solids if held at ambient or low temperatures for periods of between 4 days and 4 weeks in duration. Glycine was by far the predominant α-amino acid produced, but at least eleven others were detected in the polyamino acid. In fact, the sources cited suggest that such early experiments as Miller's may at first actually have produced peptides, which were then hydrolyzed, by the prolonged severe experimental conditions employed into the α-amino acids originally reported. This observation receives further support from Matthews's (1975) synthesis of heteropolypeptides directly from hydrogen cyanide and water.

One difficulty pointed out in the literature concerns the use of hydrogen cyanide as a condensing agent. This is the problem of hydrogen ion concentration: No condensation will result with hydrogen cyanide under the acid or neutral conditions now prevailing on earth, but only in such decidedly alkaline solutions as those indicated by a pH of between 8 and 9 (Ableson, 1966). Since it appears improbable that the early seas had a range much different from the pH 7.7–8.1 of the existing oceans (Sverdrup *et al.*, 1964), it is possible that suitable conditions could have prevailed in lakes underlain in part by basaltic rocks. Furthermore, another laboratory which attempted condensation of hydrogen cyanide in aqueous solution obtained only the amino acid citrulline, and peptide linkages could not be detected in the resulting proteinoid (Ferris *et al.*, 1973a).

Use of Cyanamide-Related Compounds. A number of cyanide derivatives have also been explored for use in peptide formation under prebiotic conditions, including hydrogen cyanide tetramer, cyanamide, dicyandiamide, and dicyanamide (Figure 1.7). Hydrogen cyanide tetramer has received limited attention thus far, but one dipeptide (glycylglycine) has been reported with yields approaching 4% (Chang *et al.*, 1969). Cyanamide has proven somewhat more successful, for when added to solutions of glycine and leucine and exposed to ultraviolet radiation, a number of dipeptides were secured (Ponnamperuma and Peterson, 1965). These included glycylglycine, leucylleucine, glycylleucine, and leucylglycine, the total yield being about 1%. In addition, the tripeptide diglycylglycine was obtained in a yield of 0.1%. Dicyandiamide was found by Steinman and his collaborators (1964) to produce the dipeptide alanylalanine when used in a solution of alanine in the dark, with a similar total yield. However, the largest yields have been obtained with dicyanamide under *acidic* conditions (Steinman *et al.*, 1965). This reagent is especially pertinent in prebiotic peptide formation because it induces nonrandom bonding of amino

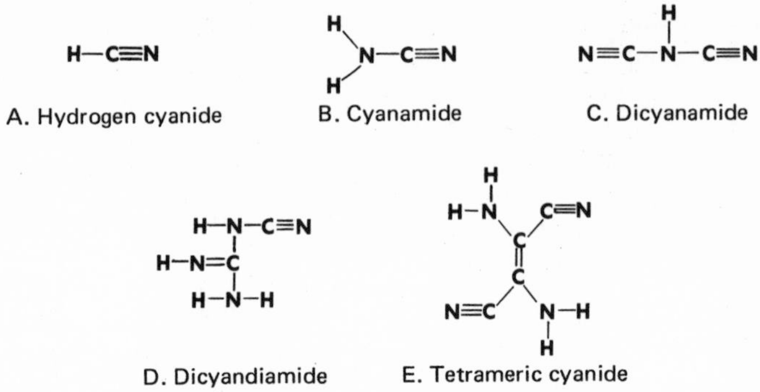

H—C≡N

A. Hydrogen cyanide

H_2N—C≡N

B. Cyanamide

N≡C—N(H)—C≡N

C. Dicyanamide

```
H—N—C≡N
    |
H—N=C
    |
  H—N—H
```

D. Dicyandiamide

```
    H
    |
  H—N     C≡N
     \   /
       C
       ||
       C
     /   \
  N≡C     N—H
          |
          H
```

E. Tetrameric cyanide

Figure 1.7. Cyanide-related condensing agents.

acids and thus synthesizes peptides having sequences similar to those of contemporary proteins (Steinman, 1971). In these experiments, it became clear that those amino acids which absorb ultraviolet rays were the most reactive in peptide condensations mediated by dicyanamide in the presence of light (Steinman and Cole, 1968). More importantly, it was discovered that when a protein was present (such as chemotrypsin), to a large extent the amino acids were sequenced in a manner corresponding to that protein (Steinman and Cole, 1967; Steinman, 1971).

Another use of preexisting polypeptides or simple proteins has been advanced (Okawa, 1954; Akabori, 1959) that to a degree is also based on hydrogen cyanide. This concept proposed that an intermediate compound, aminoacetonitrile (Figure 1.8A), first was synthesized from ammonia. formaldehyde, and hydrogen cyanide, and then became polymerized to polyglycine (Figure 1.8B). Later, according to Akabori, many of the hydrogen residues in this polypeptide were replaced in any number of ways by various suitable radicals to produce a great diversity of what he called ''fore-proteins'' (Figure 1.8C).

Other Aqueous-Medium Approaches. The foregoing accounts by no means exhaust the catalog of approaches that have been followed in peptide syntheses. Among additional ones is a quite novel methodology employed by Oró and Guidry (1960, 1961). In the earlier experiments, glycinamide was heated in an aqueous solution of ammonium hydroxide to 100°C for several hours, from which yields of polyglycine up to 25 mol. % were reported. Later, when glycine itself was heated in an aqueous solution of ammonium hydroxide at 140°C, polyglycine containing up to 18 residues resulted. Hence, the starting material reported above to have been conjectured by Akabori eventually to result in ''fore-proteins'' could have been synthesized by way of several alternate routes (Akabori and Yammamoto, 1972).

Still another approach combined atmospheric and moisture techniques and drew upon electric discharges, as in many of the early experiments. By subjecting a mixture of methane, ammonia, and water to electric discharges in the presence of amino acids, several investigations obtained small peptides (Noda and Ponnamperuma, 1971; Flores and Ponnamperuma, 1972). After being thus treated, the mixtures were stored at low temperatures for a month before being analyzed, in order to increase the yield of peptides. The peptide linkages were believed to result from the formation of either hydrogen cyanide tetramer or aminonitriles, but the actual processes were not determined. In a modification of this type of experimentation, radio-frequency electric discharge has been employed in connection with a mixture of methane, ammonia, and water vapor (Simionescu *et al.*, 1973). Under the cold plasma conditions thus used, a number of amino acids, sugars, and proteinoids were obtained; the latter were reported to display peptide linkages.

Combination Methodologies. The final major technique employed for peptide synthesis is really a dual one in that both dry and wet experiments were conducted. In all series, clay minerals, including kaolinite and montmorillonite, served as template for the polymerization of the amino acids (Degens *et al.*, 1970; Degens and Matheja, 1971). In the most successful attempts, solutions containing ten amino acids were added to kaolinite or montmorillonite and held for 7 days at 80°C. After adequate samples had been taken, the mineral suspensions were then evaporated and held under dry conditions at 140°C for 3 months. Kaolinite proved to be a much more effective polymerizer, even under aqueous conditions (Paecht-Horowitz, 1971, 1974). The condensation product from the solutions consisted of amino acids in proportions widely disparate

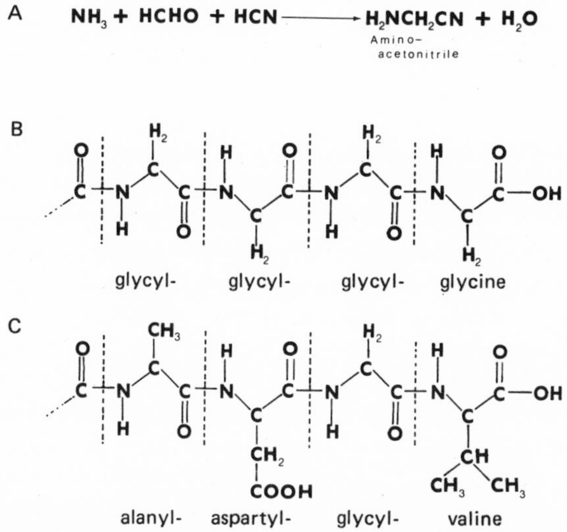

A $$NH_3 + HCHO + HCN \longrightarrow H_2NCH_2CN + H_2O$$

Amino-acetonitrile

B

glycyl- glycyl- glycyl- glycine

C

alanyl- aspartyl- glycyl- valine

Figure 1.8. A hypothetical method of producing polypeptides. (A) First step toward the formation of polyglycine. (B) Polyglycine. (C) Substituted polyglycine, a polypeptide.

from the original, aspartic acid, glutamic acid, cysteine, histidine, and arginine being far more abundant than the others. Proline proved to be nearly inactive in peptide formation by these means, less than 5% at the maximum entering into the reactions; in contrast, the majority of the amino acids polymerized at a level of about 60%. In the absence of the mineral, no polymerization occurred. The polyamino acids ranged between 500 and 5000 in molecular weight, the mode falling between 800 and 1400, equivalent to a maximum of 15 residues per peptide. Under the dry thermal conditions, much decomposition of the amino acids resulted, as well as the synthesis of peptides and other polymers. In the total polymeric structure from kaolinite, glutamic acid constituted the decidedly greatest proportion (46%), with aspartic acid and glycine ranking next (18% each), and the remaining amino acids providing small fractions each. With montmorillonite in place of kaolinite, the yield was much lower and the polypeptide quite different, glutamic acid, histidine, and serine comprising the greatest bulk, in a ratio of 10 : 2 : 1, respectively. The molecular weight of these polymers ranged between 1000 and 2000. Other minerals may be equally effective in peptide formation, for hydroxyapatite, combined with cyanate, has been employed successfully to produce a number of short polypeptides (Flores and Leckie, 1973).

1.3.4. Syntheses of Polynucleotides

In living things, the phosphate radical of nucleotides is uniformly attached to carbon 5 of the pentose residue (Figure 1.3), a position referred to in these substances as the 5′ (Ts'o, 1970). When nucleotides undergo polymerization in cells, the phosphate radical becomes linked to a second pentose at the latter's carbon 3 (that is, at the 3′ position); such 3′,5′-phosphate ester linkages consistently characterize natural nucleic acids (Calvin, 1961, 1969; Oró and Stephen-Sherwood 1974). Since those resulting from "prebiotic" experimentation often have other linkages, synthetic polymers of nucleotides are not usually referred to as nucleic acids, but only as polynucleotides.

Experiments Employing Polymetaphosphate Esters. The approaches and agents employed in syntheses of polynucleotides bear close similarity to those used in the formation of polyamino acids. As in those experiments, polymetaphosphate ethyl ester has been used successfully in polymerizing nucleotides (Schramm *et al.*, 1961, 1962; Schramm, 1965). Although either nucleosides or nucleotides have been used as reactants with this condensing agent, the latter gave somewhat superior results. When, as in one experiment, adenylic acid in polymetaphosphate ethyl ester was heated to 55°C for 18 hours, a 20% yield of polymer resulted. Subsequent sedimentation rates showed the mean molecular weight of this polymer [poly(A)] to approximate 21,000; the linkages between nucleotides were claimed to be predominantly of the 3′–5′ type. However, when deoxyribonucleotides were used in place of the ribonucleoside, pyrophosphate linkages prevailed. Poly(C), poly(G), poly(U), and poly(T) were also produced by these procedures; of these poly(U) formed the largest polymers, molecular weights as high as 50,000 being detected.

Although Schramm (1965) concluded that 3′-ribonucleosides in polymetaphosphate ethyl ester yielded polynucleotides comparable to those of organisms, a number of other investigators have failed to obtain similar results (Agarwal and Dahr, 1963; Gottikh and Slutsky, 1964; Jacob and Khorana, 1964; Kochetkov *et al.*, 1964). In contrast, their results indicated that polynucleotides produced by this technique largely contained unusual linkages, not to mention cross-linkages between chains, as well as branching. Further, one analysis of the polymers demonstrated that instead of equal numbers, between 5 and 15 times as many phosphoryl radicals as nucleotide residues were incorporated into the product (Hayes and Hansbury, 1964). In addition, pyrophosphate linkages and other bonds were shown to abound in such preparations (Fox and Dose, 1972).

As Fox and Dose (1972) have pointed out, however, the structural differences are of no real moment insofar as life's origins are concerned, providing the polymers demonstrate functional similarities. Such appears to be the case here. For example, a synthetic polyuridylic acid produced by the above processes incorporated phenylalanine in a cell-free system of *Escherichia coli* as efficiently as did the same material synthesized by means of phosphorylase. Through the observation of hyperchromicity, other polymers were demonstrated to resemble DNA in forming base pairs and in base stacking (Mahler and Cordes, 1967). Hypo- and hyperchromicity are terms given to changes in the extinction coefficient measured at the point of maximum absorption by nucleic acids in alkaline solution following incubation for 48 hours at 37°C. An increase in the coefficient (a hyperchromic shift) indicates a cleavage of hydrogen bonds (Michelson, 1959), whereas a decrease (a hypochromic shift) results from hydrogen bonding.

The Use of Various Polyphosphates. In a related methodology, Schramm (1971) and his co-workers found that phenylpolyphosphates showed an advantage over polymetaphosphate ethyl esters in that, unlike the latter, they neither ethylate nor cleave O=C bonds, but only phosphorylate. By means of this condensing agent, they were able to produce polyers of arabinonucleotides, which in many ways are intermediate between ribo- and deoxyribonucleotides. In one set of experiments uridylic acid was treated with polyphenyl phosphate, and the very stable compound polyarabinouridylic acid was obtained (Schramm and Ulmer-Schürnbrand, 1967). When treated with alkali or ribonuclease, the polymer proved resistant and thus resembled DNA; however, in acidic solution, the *N*-glucosidic linkage demonstrated resemblances to RNA, because of the inductive influence of the 2′-hydroxyl radical that is also characteristic of ribose. Chain lengths of around 7 units were produced, but among certain by-products (spongouridylates) polymers up to 20 units long were detected.

Another condensing agent for amino acids that appears useful in synthesizing polynucleotides is polyphosphoric acid or its salts. In one study, a mixture of 1 part of cytidylic acid and 2 of polyphosphoric acid was heated to 65°C for 2 hours under anhydrous conditions (Schwartz *et al.*, 1965). The resulting polymers showed alkaline hypochromicities of 16% or more, suggesting chain lengths of four to ten residues. Degradation of the product with various enzymes showed that a significant portion closely resembled biological di-, tri-, and tetracytidylic acid. However, chain branching and unusual linkages were also present in abundance.

Dry-State Approaches to the Problem. Another anhydrous thermal approach to this problem is even simpler than the foregoing ones in that it employs no condensing agent, but merely involves heating various mixtures to 160°C in the dry state (Morávek *et al.*, 1968). Uridylic acid, either alone or with uridine, or uridine and phosphate have been so treated, but so far the largest polymer produced has been only three nucleotide residues in length. The linkages, however, appeared to be predominantly of the 3′–5′ type, although some 2′–5′ linkages also were present. Deoxythymidine oligonucleotides up to four units in length have been produced from deoxythymidine 5′-triphosphate in the presence of cyanamide and 4-amino-5-imidazole carboxamide. In their synthesis warm aqueous solutions of the reagents were permitted to evaporate to induce the interactions (Stephen-Sherwood *et al.*, 1974).

Radiation in Nucleotide Polymerization. Since ionizing radiation has frequently been reported to depolymerize and otherwise destroy RNA (Scholes *et al.*, 1949; Scholes and Weiss, 1953a,b; Butler, 1959; Hems, 1960), it is rather surprising to find a report of its polymerizing action on mononucleotides. Contreras *et al.* (1962) reported that, on exposing ribonucleotides in aqueous solution to γ-irradiation, the resulting polymer showed alkaline hypochromicity. In their study, they used a cyclic process in which the polymer was moved continually on an ion exchange column and thus was protected against lysis by the radiation. Further, they found that a direct correlation existed between the size of the polymers formed and the dose of γ-rays employed and that phosphodiester bonds provided much of the linkage. The polymers thus produced displayed coding properties to a small extent.

Use of γ-rays as a condensing agent is open to many hazards, including the formation of unnatural bonds, particularly carbon-to-carbon linkages between pentose moieties (Barker *et al.*, 1962a,b). Recently it has been shown also that γ-rays induced the nitrogenous bases, whether free or contained in nucleosides and nucleotides, to bind to serum albumin and caused amino acids to unite to RNA (Yamamoto, 1972a,b, 1973a,b). Thus this type of radiation might have interfered with normal proceses rather than abetted them under primitive earth conditions.

Similar carbon-to-carbon bonds have been reported as having resulted from exposure of the nitrogenous bases to ultraviolet rays. Beukers and

Berends (1960) showed that, when thymine in frozen aqueous solutions was treated with ultraviolet light ($\lambda = 2537$ Å), dimers were formed by two adjacent carbons of one ring linking to the corresponding carbons of another in mirror-image fashion. In contrast, the same wavelength of light has been more recently applied to the synthesis of polynucleotides: Oró *et al.* (1969) successfully produced polymers of nucleotides in aqueous solutions, which resembled natural polynucleotides in possessing a significant percentage of 3′–5′ linkages.

Somewhat more convincing results with ultraviolet light were reported with the deoxyribonucleotides (McReynolds *et al.*, 1971). Neutral aqueous solutions of labeled deoxyriboadenosine monophosphate, or labeled triphosphates of any of the four deoxyribonucleosides, were irradiated at 2537 Å for 90 minutes, and the resulting products were separated by descending chromatography. The greatest chain lengths were of 10 to 13 units, the lengths increasing with the dosage. A predominance of the normal 3′–5′ linkages was indicated upon degradation of the polymer with various phosphatases.

The Employment of Aqueous Condensing Agents. Water-soluble condensing agents have received limited attention in the synthesis of polynucleotides. Sulston *et al.* (1968a,b; Orgel and Sulston, 1971) synthesized various short polynucleotides by means of a water-soluble carbodiimide, the processes of which were facilitated by using complementary polynucleotides as templates. For instance, polyadenylic acid polymerized more readily when polyuridylic acid was present, and the formation of polyguanylic acid was similarly accelerated by the presence of polycytidylic acid (Sulston *et al.*, 1969). One molar aqueous solution of ethylendiamene hydrochloride at 2°C and pH 8 yielded polymers up to 12 units in length in 5 days' time when poly(U) was used as a template (Usher and McHale, 1976). However, 95% of the bonds were 2′–5′ linkages. Cyanamide (Figure 1.7) has been used in place of carbodiimide as by Ibañez *et al.* (1971a,b), who employed it in the synthesis of deoxyribonucleotides. Low concentrations of labeled thymidine 5′-monophosphate were dissolved in aqueous cyanamide solutions at pH 7.3 and were permitted to react for 24 hours at 90°C. Polymers up to five links in length were reported, but the yields were measured in small fractional percentages. Moreover, the longest molecules were formed only in the presence of clays.

Such involvement of various types of clay minerals appears quite effective under certain conditions in polymerizing nucleotides held in aqueous solution (Otroshchenko and Vasilyera, 1977). Degens and his co-workers (1970, 1971) conducted a series of experiments with five varieties of clay minerals (kaolinite, illite, halloysite, montmorillonite, and vermiculite) and various combinations of reactants. All experiments involved temperatures of 60°C for 24 hours. In one series, low concentrations of four different purines and pyrimidines alone were used, but no polymerization resulted. Nor did any polymerization occur when the four bases together with various sugars were treated. When phosphate (in the form of potassium phosphate) was added to the base and sugar mixture, however, compounds of higher molecular weight assumed to be polynucleotides were observed. If amino acids were also added to the mixture, reaction products of even higher molecular weight were generated, but their nature was not further characterized. More recently, dinucleotides have been produced by first adsorbing nucleoside phosphoramidates on sodium and magnesium montmorillonite clays. Heating the dried mixture then resulted in small amounts of dinucleotides (Burton *et al.*, 1974).

Displacement Reactions. Still another approach to the formation of polynucleotides has been utilized, in which a displacement reaction on an activated carbon is involved. The method was originally developed by Elmore and Todd (1952) and later was further explored by Khorana (1964), but the first successful production of polynucleotides by this method was carried out by Nagyvary and Provenzale (1971). In the latter experiments, O^2,5′-cyclouridine-2′,3-cyclic phosphate was prepared from 5′-*O*-mesyluridine-2′, 3′-cyclic phosphate and *N,N*′-dicyclohexyl-4-morpholino-carboxamidine. These substances were reacted in dry dimethylformamide for two days at ambient temperature or overnight at 40°C.

In a second set of experiments (Nagyvary and Provenzale, 1971), a dimethylformamide solution of tri-*N*-butylammonium cyclouridylate was held at 80°C for 3 days; then the temperature was slowly increased to 100°C over a period of 3 days, at which level it was maintained for an additional 48 hours. Chain lengths ranging from 10 to 50 nucleotide units were reported. In what was claimed as the first solid-state polymerization of a nucleotide, cyclouridylates were exposed to amorphous lithium salts at an initial temperature of 100°C, which was gradually increased to 140°C over a period of 2 days. Chain lengths between 15 and 70 units were reported, with a major concentration near 17. No attempt was made to correlate the foregoing procedures with the conditions usually conjectured for the primitive earth.

1.4. SUMMARY

Although many difficulties are evident, which can be more appropriately summarized in a later discussion (Chapter 2, Section 2.3.1), the foregoing outline makes clear that much progress toward an understanding of the origins of life's materials has been achieved. Among the most clear-cut advances that have been made up to now are the following:

1. Methodologies for the prebiotic synthesis of all the proteinic amino acids have been developed, and in certain cases, several equally logical techniques have been proposed.
2. Less varied, but convincing, methods have been demonstrated for the production of the biologically important nitrogenous bases and pentose sugars.

3. Several procedures exist whereby the bases and pentose sugars can be phosphorylated to produce nucleotides, including mono-, di-, and triphosphorylated nucleosides.
4. Although difficulties still persist regarding the control of bonding, polypeptides and proteinoids that display a number of proteinlike properties can be produced under seeming prebiotic conditions.
5. Similar, but greater, difficulties exist in the control of linkages between nucleotides. Considerable progress has nevertheless been made toward production of polynucleotides which possess coding and certain other properties of biological nucleic acids, despite the irregularities in the linkage types.
6. In the synthesis of polyamino acids, nonrandomness in the amino acid sequences is a frequent characteristic.
7. In several instances it has been shown that the polymers (proteinoids and polynucleotides) are produced directly from the raw inorganic ingredients. In those cases the amino acids or bases are actually degradation products of the polymers, rather than intermediates. It is for this reason that the discussions of Stages I and II are here combined in a single chapter.
8. It is therefore apparent that so-called Stage II products actually may have *preceded* the formation of those of Stage I. Although the sequence usually proposed for these early events may appear logical on paper, they thus may not necessarily conform to the realities of prebiotic syntheses.

2

The Precellular, or Simple Interacting Systems, Level (Stage III)

Although in certain aspects of the prebiotic synthesis and polymerization of life's basic biochemical ingredients some problems still resist satisfactory solution, the number and diversity of procedures that have been successful promise that at least some portions of the theories proposed do approach reality. Nor should there be any real concern over which specific pathway was the one that had been followed exclusively on the primitive earth, for several, or even many, different processes may have been active during the billion or more years which the early stages seem to have occupied.

Beyond that primitive broth stage, however, a multitude of profound developments had to occur before the very first object recognizable as a cell could have come into being. Even the simplest organism is exceedingly complex in structure, for on one hand the electron microscope has established firmly that each contains numerous varied specialized parts (organelles). Moreover, biochemical researches have shown that a maze of enzymes, coenzymes, and other factors are intricately woven into interacting systems in every cell. Hence, it is to be expected that a number of intermediate stages must have existed between the polymeric level just described and the origin of the first cellular organism. What has been accomplished toward creating logical models of these intermediate forms is the focal point of this chapter.

2.1. SYNTHETIC MODELS OF PROTOBIONTS

In experimental attempts to model the intermediate levels of life's origins, four major avenues of investigation have been followed. Although each employs a distinctive set of procedures, the four agree that the earliest preliving things ("protobionts") consisted of one or more simple interacting systems of macromolecules. All also concur in assuming an aqueous environment to be an essential feature, at least to a certain extent. Hence, if anhydrous conditions had been extensively involved in the production of monomers and polymers, those products must have reached the seas before further progress was made. The models, too, have certain features in common, particularly those of being colloidal and in consisting of suspensions of macromolecules. However, each possesses individual properties and requires special modes of preparation, and each is supported at a high level of enthusiasm by a particular school of thought (Fox, 1976).

2.1.1. Coacervates

Among the first models of prebiotic particles to be synthesized was the coacervate droplet, originally developed by de Jong (1932, 1947). Upon mixing dilute solutions of gelatin and of gum arabic at moderate temperatures and under neutral conditions of acidity, it was found that the resulting solution frequently became turbid. On investigation, the turbidity was discovered to result from the clustering of the molecules of gelatin and gum into integrated particles. When the clusters had attained a size considerably greater than the original molecules themselves, usually between 2 and 670 nm, they separated from the solution in the form of droplets visible under the ordinary light microscope. The droplets have since been named *coacervates,* and the medium in which they are suspended has been termed the *equilibrium liquid*. After the coacervates had formed, the equilibrium liquid was consistently found to be nearly depleted of the organic polymers; thus these droplets represent an effective means of concentrating macromolecules from dilute aqueous suspensions.

Coacervates as Protobionts. For a time, studies on coacervates as prebiological agents (Oparin, 1957) of necessity employed such biologically synthesized substances as gum arabic and gelatin, but this procedure is no longer essential. Now it has been thoroughly established that coacervates can be developed from solutions containing two or more macromolecular components. Indeed, the products of prebiotic syntheses recently have been utilized directly to create coacervates. For example, products of hydrogen cyanide irradiated with ultraviolet light ($\lambda = 1849$ Å) have resulted in coacervate formation when combined with a gelatin sol (Noda *et al.*, 1975). The size of the droplets, as well as their ability to concentrate polymers, is known to be dependent upon the nature of the substances used, but even droplets formed of the same chemicals can vary extensively in size, the typical range being between 20 and 500 (Evreinova and Kuznetsova, 1959).

The ability to concentrate their contents relative to the equilibrium liquid likewise has a wide range; Evreinova and Kuznetsova (1959) demonstrated increases in the mean concentration from 1 to as many as 431 times that of the surrounding solution. Individual coacervate droplets in the same suspension also vary widely in concentration, the degree being inversely related to the diameter of the droplets and, therefore, also to their volume (Oparin, 1968, 1974; Evreinova *et al.*, 1972, 1974). Even within a given droplet, the concentration of material may not be uniform, for under the light microscope (Figure 2.1), opaque and clearer areas are often visible, and vacuoles are occasionally present (Evreinova and Kuznetsova, 1961, 1963). Particularly does this condition prevail when the components of the coacervate include nucleic acid and a histone (Evreinova, 1964). In the vacuoles, the original concentration of DNA and histone in one instance had been no more than 0.25%, but in the clumps within the droplets it reached 87% (Evreinova *et al.*, 1972). When lipoids are not employed, no membranous coat over the droplets is present, but the interface between the droplets and the equilibrium liquid is consistently sharply defined.

As a whole, coacervates are ephemeral; for instance, the droplets in one common coacervate system, consisting of serum albumin and histone, usually disappear within 30 minutes. Recently, however, more stable systems have been prepared; one consisting of polyphenoloxidase, histone, gum arabic, and pyrocatechol endured for over 3 years at pH 6.0 and ambient temperatures (Evreinova *et al.*, 1972, 1974).

To a small degree coacervates demonstrate possible beginnings of individualization from the environment in being able to absorb material selectively from the equilibrium liquid. When such dyes as neutral red or methylene blue are added to the medium containing gelatin and gum arabic coacervates, the droplets soon show concentrations of the dye many times greater than the original equilibrium liquid. Similar selectivity in absorption has also been demonstrated with various amino acids (Evreinova *et al.*, 1962), but the ability has varied with the amino acid added. The concentration of tyrosine, for example, increased more than 100-fold over the original, whereas that of tryptophan only doubled. In contrast, sugars and nucleotides were found not to be selectively absorbed, for the concentration remained the same in the droplets as in the surrounding liquid.

Open-System Experiments. The ability to concentrate material from the external environment is not performed in an actual lifelike fashion in coacervates, as Oparin (1968) pointed out, because equilibrium between the droplet and the medium is quickly reached. Hence, in the resulting static condition, the coacervates, in contrast to the open systems found in living cells, became closed systems in that they failed to continue to interact with the environment. In open systems, substances are continually withdrawn from the surroundings; then after undergoing modification within the system, some or all of the by-products are finally returned to the external environment, with equilibrium never being attained. To approach this type of activity in the coacerva-

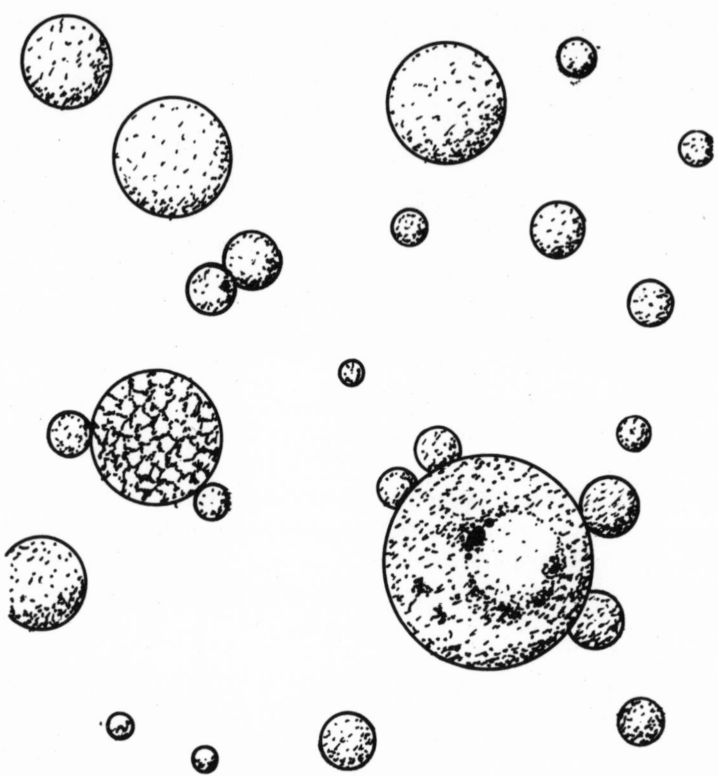

Figure 2.1. Coacervate droplets. The largest droplet shows the presence of a vacuole and inclusion bodies.

tes, Oparin and his colleagues (1962) added a phosphorylase (glycosyl transferase from potato), sometimes along with β-amylase, to the equilibrium liquids containing coacervates produced by gum arabic and histone and found that the enzymes readily became concentrated in the droplets. When glucose-1-phosphate was added to the medium, like other sugars it did not become particularly concentrated within the coacervates. That which did enter, however, was acted on by the phosphorylase present in the droplets and converted into starch (Figure 2.2), which could be detected by means of iodine. This reaction proceeded rapidly, so that the droplets nearly doubled in size within a half hour as the starch accumulated. Later, after β-amylase had also been added to the coacervates, the enzyme broke down the starch into maltose, which substance left the droplets to enter the equilibrium liquid (Figure 2.2). Thus there was a flow of glucose-1-phosphate from the medium into the droplets, and of maltose and inorganic phosphate from the droplets into the environment.

By analogous processes both the decomposition and synthesis of polynucleotides have been achieved (Oparin and Serebrovskaya, 1958; Oparin *et al.*, 1963). Simple oxidation–reduction reactions also have been induced in coacervates (Oparin *et al.*, 1964). In these latter experiments, a hydrogen-accepting dye, such as methyl red, was first introduced into the coacervates by way of the equilibrium medium. After this had become concentrated within the droplets, reduced nicotinamide-adenine-dinucleotide (NAD:H_2) was introduced into the droplets, where it transmitted its hydrogen to the methyl red. After being thus oxidized, the NAD as well as the reduced dye could reenter the medium. The energy released as heat in these reactions was mainly dispersed through the system. Similar models involving primitive photosynthesis have also been produced, through the employment of chlorophyll, methyl red, and ascorbic acid in the presence of light (Oparin, 1971; Serebrovskaya and Lozovaya, 1972).

Difficulties with Coacervates. In general, coacervates thus provide an interesting model suggesting the steps by which protobionts may have progressed or, at least, how the biological polymers may have become locally concentrated. Although under proper conditions these droplets are readily formed, as a rule they are quite unstable, and, as Kenyon and Steinman (1969) have pointed out, each type requires different specific conditions for its formation.

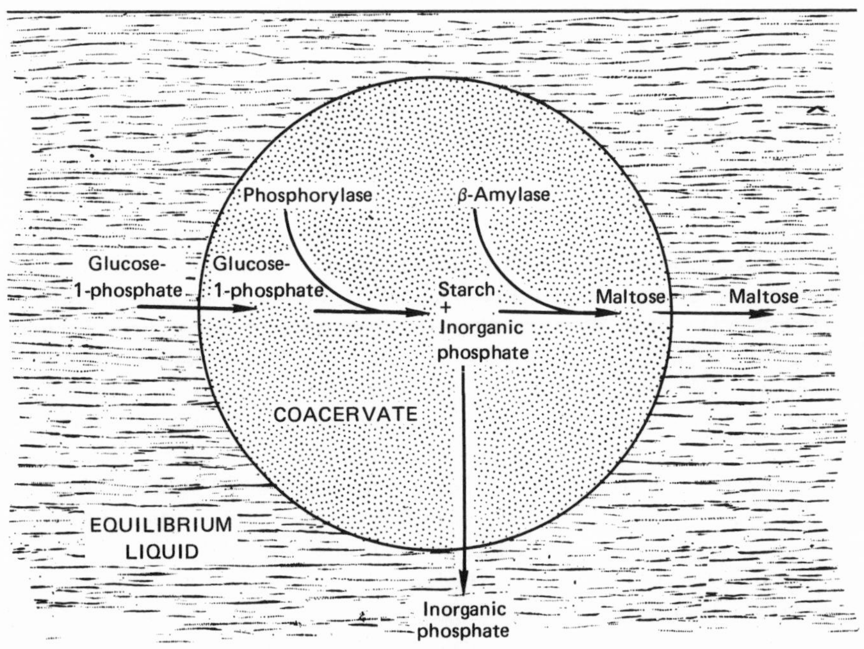

Figure 2.2. A coacervate droplet interacting with the environment.

Hence, at a given alkalinity or temperature, one kind might be favored over the others, so that the community of different types needed for life's elaboration could scarcely have existed. Further, considering these colloidal droplets as protobionts has been objected to on the grounds that typically they require for their formation substances from extant organisms, such as gum arabic and histone (Fox and Dose, 1972). Thus it would be difficult to conceive of them as forebears of life. In light of the above-cited more appropriate methods of synthesis recently undertaken, this objection may no longer be valid (Evreinova, 1964; Evreinova *et al.*, 1972; Noda *et al.*, 1975).

2.1.2. Proteinoid Microspheres

An entirely different type of colloidal droplet, the so-called proteinoid microsphere, holds several advantages (as well as a few disadvantages) over the foregoing type. Probably the most obvious advantage is in its being derived directly from materials exclusively prepared under presumed prebiotic conditions. The proteinoids (Fox and Harada, 1958, 1960; Vegotsky and Fox, 1959) are formed by heating amino acids in molten glutamic or aspartic acid, as discussed in Chapter 1, Section 1.3.2. When the heated mixtures have completely interacted, they are treated with boiling water and the clear supernatant solution is decanted. After the decanted liquid is allowed to cool without agitation, the presence of minute droplets may be detected by microscopic examination (Fox, 1965; Fox and Dose, 1972).

Other methods of preparation also may be employed. One modification involved heating a mixture of 18 amino acids on lava for 3 to 4 hours in an oven held at 170°C. The resulting amber-colored viscous liquid was collected by washing the rock with 1% bacteria-free sodium chloride solution to simulate tidal action of the early seas. This fluid was then found to be richly supplied with microspheres (Fox, 1964). Still other methods of preparing microspheres have resulted in products which approach coacervates in such properties as instability and uptake of dyes (Rohlfing, 1975). In still another approach, DL-lysine was heated at 170°C to form a solvent for such other amino acids as aspartic and glutamic acids (Fox and Suzuki, 1976). Peptide linkages as well as imido were detected in the resulting polymers.

Characteristics of Proteinoids. Even the first microspheres made had the size (0.5–7.0 μm), shape (Figure 2.3), and tendency to associate in clumps characteristic of coccoid bacteria (Fox and Yuyama, 1963). Within any given type, the size is far more uniform than that of coacervates. Depending on the amino acids employed in their formation, proteinoid microspheres may be either positive or negative to Gram stain, but usually Gram-positive microspheres result only from combinations of acidic and lysine-rich proteinoids. Perhaps the most remarkable feature of microspheres is the enzymelike property reported by Dose (1971, 1974).

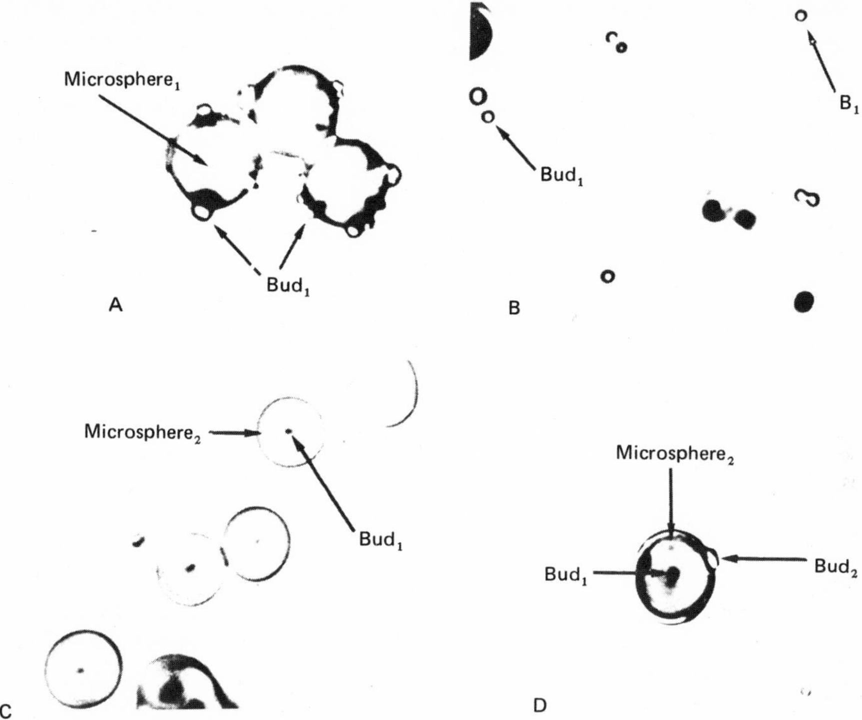

Figure 2.3. Proteinoid microspheres undergoing bud formation. (A) The original microspheres, 15 nm in diameter, produce buds which break free when the system is subjected to vibration or other mechanical shock (B). In turn, the released buds may grow into mature microspheres (C), which can produce a third generation by similar processes of bud formation, shock-release, and growth (D). (Courtesy of Professor Sidney W. Fox.)

Among the more outstanding of such catalytic activities is an ability to hydrolyze *p*-nitrophenyl acetate (Noguchi and Saito, 1962; Rohlfing and Fox, 1967; Usdin *et al.*, 1967), and to decarboxylate various substances, including glucuronic acid (Fox and Krampitz, 1964), pyruvic acid (Hardebeck *et al.*, 1968), and oxaloacetic acid (Rohlfing, 1967). More recently, light-enhanced decarboxylations by proteinoids have been reported (Wood and Hardebeck, 1972).

The list of enzymatic capabilities is rather long. Among the additional ones described was the amination of α-ketoglutaric acid and the deamination of glutamic acid, both of which reactions required the presence of copper ions (Krampitz *et al.*, 1967, 1968). Engagement in oxidation–reduction reactions has also been reported (Dose and Zaki, 1971), but the extent of the interactions was rather limited. Finally, when a polymer composed entirely of adenylic acid was introduced into a suspension of proteinoids, along with ATP and phenylalanine, peptides consisting of that amino acid were synthesized (Fox *et al.*, 1974; Harada and Fox, 1975).

As a rule, microspheres do not show a capacity for concentrating materials, nor do they possess the ability to grow to the same extent as coacervates; they merely shrink or expand when transferred to hyper- or hypotonic solutions, respectively. An outstanding trait, however, is that of engaging in a type of simple binary fission. Only acidic proteinoid microspheres demonstrated this ability (Fox, 1968) until recently, when calcium-containing varieties also were induced to cleave by use of pure water or heat. Fission was believed to derive from the effects of surface tension (Fox and Dose, 1972).

What have been referred to as *buds* appear spontaneously on microspheres when the latter are let stand in the original medium for a time (Fox *et al.*, 1967; Fox, 1975; Hsu, 1972). These buds (Figure 2.3), which have an appearance similar to those of yeast, are released when the suspension is heated mildly, subjected to sparking with a Tesla coil, or jolted mechanically. After release, the buds grow by accretion until the dimensions of the original microspheres have been attained. These second-generation particles have been shown in turn to be able to produce buds, capable of leading to a third generation.

Failings of Proteinoid Microspheres. Although proteinoid microspheres thus tend to display certain similarities to living cells, even to the point of possessing a surface membrane of sorts, they have features that are not desirable in prebiotic models. Their greatest flaw is their extreme sensitivity to changes in pH. Most must be maintained under highly acidic conditions, for upon an increase in pH of several units, they rapidly return to solution. Furthermore, they require unique environments, an objection that has been raised against coacervates. Since each type can be created only under highly specific conditions, the simultaneous existence of the diversity needed collectively to imitate the activities of even the simplest cell is difficult to visualize. Nevertheless, as an experimental model, the microsphere has proven invaluable (Dose, 1974; Hsu and Fox, 1976).

2.1.3. Micelles

Definition of Micelles. A third type of colloidal droplet which has been advocated as a possible model for certain prebiotic phenomena is the micelle. This class of droplets has particular value in simulating the origin of the unit membranes of cells. Such membranes are among the most distinctive features of cellular life—actually to the extent that it is difficult to conceive of living matter which could lack them. In the usual concept (Robertson, 1959, 1964), the unit membrane is a bilamellar structure, in which each lamella consists of a monomolecular layer of lipoproteins. When the two lamellae lie together as in the unit membrane, the nonpolar ends of the parallel row of long-chain fatty acids bond together and form the hydrophobic central region seen in cross-section as an electron-transparent band. The protein moieties thus lie toward each surface and provide hydrophilic coats over the membrane; with the usual preparatory techniques these become electron-opaque and provide the bilamellar appearance.

Characteristics of Micelles. Since neither the coacervates nor microspheres characteristically include both proteinaceous and lipoidal constituents, they fail to provide information regarding the possible origins of a cell structure of fundamental importance, the membrane. Thus the micelle serves to fill a major gap in the experimental evidence. One method by which models of the unit membrane have been created has been by spreading a monomolecular layer of oleic acid, which is insoluble in water, upon an aqueous solution of egg albumin. The film was then made to collapse by moving wax barriers toward one another; similar surface collapse could result in nature from wind, current, or wave action (Goldacre, 1958). When thus agitated, the surface film infolded to form air pockets; two or more of these frequently united, resulting in formation of spherical or cylindrical bodies coated with a bilamellar membrane (Figure 2.4). These micelles were reported to have diameters between 1 and 10 μm.

Micelles thus provide an additional means of creating droplets with a distinct composition isolated from the surrounding watery medium and thus of individualizing particles from the rest of the environment to a limited degree.

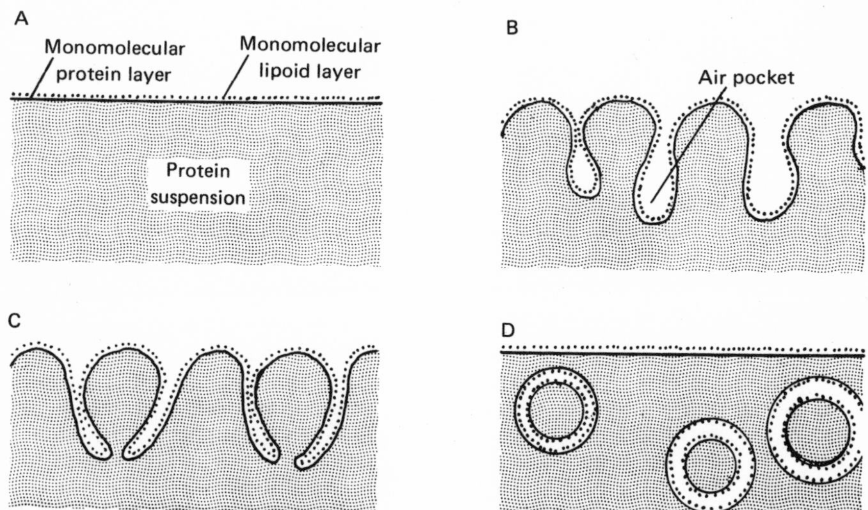

Figure 2.4. The formation of micelles. When a protein suspension covered by a lipoidal film (A) is subjected to vibration, the surface becomes agitated (B). The microscopic air pockets that result from the agitation may be deepened (C) in such a way that minute droplets are produced which are enclosed by double membranes (D) similar to those of cells. (Based on Goldacre, 1958.)

Very few studies, if any, seem to have been conducted on micelles to determine whether they, too, can carry out lifelike chemical reactions.

2.1.4. Sulfobes

Were a chronological sequence followed, the sulfobes would certainly have to head the discussion in any synopsis of the interacting-systems, or protobiont, stage of life's origins, for this relatively poorly explored type of particle was brought to light by A. L. Herrera in 1924. Although his studies are often overlooked in discussions on the subject, there can be no doubt that his researches on sulfobes were the very first to explore the biochemical origins of life.

Herrera began his line of investigation on the assumption that formaldehyde is an essential intermediate in photosynthesis by green plants, as was generally accepted at that time. Hence, his experiments involved reactions between formalin and various other reagents, those in which it reacted with ammonium thiocyanate being particularly successful. After ammonium thiocyanate had been dissolved in the formalin, the resulting solution was spread on an inert surface in very thin layers and permitted to react for several hours. Examination under the light microscope then revealed an abundance of microscopic particles; these were said to carry out activities analogous to those of living organisms, direct and mitotic divisions being among the activities reported (Herrera, 1942). Chemical analysis revealed that these sulfobes contained at least two varieties of amino acids, a minute quantity of starch, a proteinoid substance, and pigments.

The strongest points of the sulfobe undoubtedly are its proteinoid character and the morphological resemblances of the particles to bacteria, amoebas, diatoms, chromosomes, and even centrioles. Since Herrera himself stressed these superficial similarities to living things, sulfobes have been criticized as being unrealistic models, for they are unable to reproduce and lack an organized metabolism (Oparin, 1957). However, this point receives attention again later in this chapter.

2.2. AUTOCATALYSIS

Rather than search for biological properties in such colloidal droplets as are described above, several biochemists have approached the problem of developing interacting systems by way of autocatalysis. Autocatalysis corresponds in chemistry approximately to one of the critical properties of living systems, self-reproduction. Two major categories of autocatalytic processes have received attention, one proposing possible methods for the development of complex molecules (King, 1977), the other centering on the selectivity of biological systems for stereospecific compounds.

2.2.1. Autocatalysis of Proteins

In theorizing on the origins of complex molecules, biochemists have followed several widely divergent avenues. One school of thought proposes that the nucleic acids arose first before life had its origins, whereas another believes polypeptides to be the original basic substance. Both of these assumptions are based on the self-reproducing properties of the respective macromolecules.

It has long been known that DNA possesses at least a limited ability to reproduce itself by autocatalysis (Lederberg, 1960a,b; R. F. Fox, 1972). However, it is becoming increasingly apparent, as occasionally intimated in Chapter 1, that amino acids possess inherent sequence-directing abilities to the extent that orderly peptide chains can be synthesized even in the absence of nucleic acids of any sort (Kenyon and Steinman, 1969). The question arises then whether peptides or even simple proteins might not similarly possess some powers of promoting their own replication, that is, whether they may be capable of autocatalysis.

Nonrandomness in Amino Acid Association. In the first place, the interactions among amino acids of thermally produced proteinoids (Section 1.3.2.) show strong evidence of nonrandomness of association. The proportions of constituents in the product, it will be recalled, were reported as being far different from those of the original mixture (Table 1.4). Thus it became evident that the nature of each type of amino acid is unique, possibly as determined in part by its side chain, and their individual properties accordingly appear to introduce nonrandom constraints into the processes of peptide formation. The specific constraints often vary with the conditions of the reactions. For instance, when polymerization was carried out at 170°C, aspartic acid was more abundant in the proteinoid than glutamic acid, but at 160°C, glutamic acid was nearly eight times as abundant as aspartic acid (Krampitz, 1959; Fox and Harada, 1960, 1961; Fox *et al.*, 1963). Duration of the processes, too, exerted an influence, for the relative proportions of aspartic acid increased with length of time of preparation.

Aqueous condensations of amino acids promoted by such reagents as hydrogen cyanide, cyanamide, and dicyanamide also have given evidence of nonrandomness of combination. Steinman (1967a,b) explored the effect of various side chains of selected amino acids in the formation of polypeptides condensed through action of dicyanamide. Glycine in some sets and L-leucine in others were used as the initial amino acid, with which glycine, L-alanine, and L-leucine were permitted to interact and thereby lengthen the polymeric chain. With itself as the initial amino acid, glycine reacted twice as efficiently as alanine and three times as efficiently as leucine. With leucine as the chain initiator, glycine was three times more reactive than either alanine or leucine. Results were comparable, regardless of whether the first amino acid was in solution or was bound to resin. Thus, as before, the amino acid available at the active end of a growing peptide chain was shown to exercise a degree of constraint upon the amino acid that became attached to it.

These experiments involved only dipeptides, or at most tripeptides, as in a few of Steinman's. Undoubtedly, additional constraints could be expected to be imposed upon any amino acid that might be added to a longer polymer (Kenyon and Steinman, 1969). With sufficient increase in length, development of the accompanying secondary and tertiary structure also might be expected to influence additions to the chain by bringing certain internal residues into proximity with the growing end, as suggested by Pattee (1965).

Nonrandomness in Natural Proteins. The observed lack of randomness in synthetic proteinoids raises a question as to the relationships between adjacent amino acids in natural proteins. Steinman and his associates addressed themselves to the problem by comparing the frequencies of a dozen different dipeptides included in their synthetic material with natural proteins whose sequences had been determined (Steinman, 1967a,b; Steinman and Cole, 1967; Kenyon and Steinman, 1969). In the natural proteins the frequencies of the dipeptides, which involved various combinations of glycine, alanine, valine, leucine, isoleucine, and phenylalanine, were found even more disparate from those expected on a random basis than in the artificial. The nonrandomness was suggested to be correlated to increasing side-chain size and, hence, molecular weight.

To test the hypothesized interactions between amino acids further, an analysis of the frequency of occurrence of all 20 residues following an preceding each of six amino acids in 38 randomly selected proteins is here reported. The proteins are those whose sequences are conveniently summarized in readily available sources and include 11 cytochromes *c* (Dillon, 1978) from bacteria, yeast, mold, an insect, various vertebrates, and wheat, and 27 other types given by Jukes (1966). The latter group consists of human myoglobin and α-, β-, and γ-hemoglobin, fibrinopeptides A and B from six species of mammals, bovine chymotrypsinogen A and B, bovine trypsinogen, porcine elastase, tobacco mosaic virus protein (*vulgare* strain and mutants), corticotropin, α-melanocyte-stimulating hormone, and β-melanocyte-stimulating hormone from four different mammals. Included in the results of this study are the number of reactions involved with the six amino acids individually, the molecular weight of each amino acid, the total number of each found in the 38 proteins, and their respective percent frequency (Table 2.1). The frequency based on the number of triplets encoding each amino acid, which corresponds to the random-chance expectations, is provided for comparison.

Because glycine has been shown above to specify itself more frequently than it does any other amino acid, it appears advantageous first to examine its reactions with other residues in these natural proteins. The table shows a similar condition to exist here, in that a terminal glycyl residue already in the growing peptide chain specifies a second glycine about twice as frequently as expected according to its number of codons. The actual percent frequency is 13.3%, whereas the expected rate is only 6.5%. However, it appears to interact even more favorably with proline (20.0%). Its favorable reactions toward lysine (10.8%) and phenylalanine (15.6%) are still stronger in a relative sense, for these two amino acids are provided with only half the number of codons and thus only a 3.3% expected frequency on a random-chance basis. Comparable interactions can be noted in the frequency with which glycine is specified by each of the six amino acids studied. The most noticeable distinctions are that, whereas glycine specifies asparagine at a rate (2.8%) slightly lower than expected (3.3%), it is specified by the latter more frequently (8.8%) than expected (6.5%). Furthermore, it is specified by phenylalanine somewhat less often than on a random basis.

It is of interest to compare glycine with alanine, which has the same number of codons and is only slightly heavier. With the first three amino acids, its interactions are close to the expected. It is specified by glycine slightly less than at the normal level (5.0% against 6.5%), but specifies glycine at a much elevated level (9.0%). The first observation may be noted to disagree with the results based on the glycyl residue counts. Unlike glycine, it specifies itself at a level only slightly exceeding the expected. The remaining interacting amino acids, which are provided with only two codons each, specify alanine at the expected level of frequency. On the other hand, alanine strongly specifies asparagine as indicated by the observed frequency of 13.6% against an expected rate of 3.3%.

Since glycine interacts with itself at a high level of frequency, whereas alanine does not, a question arises as to how the other four interact with themselves. Lysine is seen to specify itself nearly six times more frequently than expected, at an actual rate exceeding 18.0% against 3.3% on a random-chance basis. In marked contrast, proline and asparagine appear slightly antagonistic toward themselves, and phenylalanine reacts strongly against association with itself.

Molecular Weight and Nonrandomness. Steinman is cited previously as invoking molecular size to explain the nonrandomness that he and his co-workers noted. The validity of this relationshp can be tested by comparisons of the molecular weights and frequencies of the respective amino acids in natural proteins, as summarized in Table 2.1. Number 1 rank is there assigned to the most frequent and a similar rank is accorded the lightest in molecular weight; successively higher numbers are given on the basis of decreased abundance on the one hand and increased weight on the other.

When the results are scanned, some correlation between the two parameters is immediately apparent, for glycine shows both the smallest molecular weight and the greatest frequency. But such close interrelationships are detectable only at three other points in the list: valine ranks fifth in both categories, aspartic acid tenth, and tryptophan last. In addition, approximate corre-

Table 2.1. Relations between

Associated amino acid	Number of codons	Expected % occurrence	Interacting amino acids: Actual frequency (% occurrence) following/preceding: Gly (75)	Ala (89)	Pro (115)	Asp (150)	Lys (146)	Phe (165)	Total reactions
Arginine	6	9.8	4.3/0.0	4.5/2.9	0.8/1.6	1.6/0.8	0.4/3.6	7.8/1.0	35/19
Leucine	6	9.8	7.0/4.5	6.1/7.5	7.2/5.6	12.8/2.4	3.6/8.6	8.8/14.8	80/79
Serine	6	9.8	5.3/5.9	7.2/9.0	5.0/6.4	5.6/2.4	3.3/3.3	10.7/6.8	63/60
Alanine	4	6.5	5.0/9.0	7.0/7.1	7.2/7.2	6.4/13.6	6.4/3.6	9.8/3.0	76/91
Glycine	4	6.5	13.3/9.4	10.1/9.4	20.0/16.0	2.4/8.8	10.8/9.0	15.6/5.9	139/126
Proline	4	6.5	6.6/11.0	3.0/3.8	5.6/5.6	11.2/8.8	2.5/6.4	4.9/1.0	60/77
Threonine	4	6.5	6.2/7.3	7.2/5.2	4.0/6.4	6.4/8.0	7.2/6.4	4.0/4.9	65/73
Valine	4	6.5	6.5/5.9	4.5/4.2	5.6/4.0	5.6/5.6	6.1/1.4	7.8/4.9	69/48
Isoleucine	3	4.9	6.5/1.0	5.8/4.7	3.2/9.6	4.8/4.0	5.4/4.7	3.0/16.2	61/60
Asparagine	2	3.3	4.0/9.3	9.8/5.7	9.6/11.2	5.6/6.4	6.1/6.8	0.0/4.0	70/68
Aspartic acid	2	3.3	7.6/1.7	4.7/9.0	4.8/7.2	3.2/1.6	3.6/2.8	3.0/4.0	57/56
Cysteine	2	3.3	1.3/3.4	0.9/6.2	0.0/0.0	0.8/1.6	3.3/1.1	4.9/0.0	21/28
Glutamic acid	2	3.3	6.3/5.9	6.0/2.8	4.0/0.8	10.4/10.4	3.3/6.5	3.0/8.9	65/60
Glutamine	2	3.3	3.3/3.4	5.2/5.7	1.6/0.8	2.4/0.8	1.8/4.3	3.0/4.9	35/41
Histidine	2	3.3	2.0/5.5	6.0/1.9	1.6/1.6	0.0/0.0	4.0/4.3	1.0/12.2	34/47
Lysine	2	3.3	7.6/9.0	3.3/8.5	11.2/4.8	15.2/16.8	18.7/18.9	10.7/3.0	127/126
Phenylalanine	2	3.3	4.0/3.1	0.8/4.2	0.8/4.8	3.2/0.0	1.4/4.3	1.0/1.0	24/37
Tyrosine	2	3.3	3.6/0.7	5.2/1.4	4.0/4.0	0.8/3.2	5.0/2.1	0.0/0.0	43/20
Methionine	1	1.7	0.0/1.4	0.8/0.5	0.8/1.6	1.6/0.8	4.7/1.4	1.0/2.0	19/14
Tryptophan	1	1.7	0.6/6.9	1.5/1.0	2.4/0.8	0.8/0.0	0.4/0.7	0.0/0.0	10/25
Total	61	99.9	306/298	222/222	124/125	124/122	276/276	101/102	1153/1145

[a]Rank by weight beginning with the highest and extending to the heaviest.

Amino Acids in Natural Proteins

In proteins			Rank		Codons	
Total number	Percent	Molecular weight	By weight[a]	By frequency	Numerical frequency	Adjusted rank
86	3.0	174	18	16a	15	20
209	7.1	131	8a	4	35	16a
181	6.2	105	3	5a	30	19
241	8.2	89	2	3	60	6
321	10.9	75	1	1	80	3
128	4.4	115	4	11	32	18
180	6.2	119	6	5b	45	11a
182	6.2	117	5	5c	46	10
118	4.0	131	8b	12	39	15
157	5.4	150	15	9	79	4
153	5.3	133	10	10	77	5
69	2.3	121	7	18	35	16b
165	5.6	147	13	8	82	2
86	3.0	146	11a	16b	43	13
97	3.4	155	16	14	49	8
276	9.4	146	11b	2	138	1
100	3.5	165	17	13	50	7
93	3.2	181	19	15	47	9
45	1.5	149	14	19	45	11b
42	1.4	204	20	20	42	14
2929	100.2%					

spondence can be noted at four other points: alanine, which ranks second in weight but third in abundance, serine, with ranks of 3 and 5 respectively, threonine, with 6 and 5, and histidine with 16 in weight against 14 in occurrence. Among the remaining 12 amino acids, however, no correlation between molecular size and frequency is evident.

Absolute frequency of occurrence as utilized in the foregoing analysis does not take into account the differences in the numbers of triplets that encode the various amino acids. On a statistical basis an amino acid (like arginine) which is encoded by six codons has a random-chance opportunity of being placed into a protein six times as great as does one (like tryptophan) which is provided with only a single triplet. The last two columns in the chart correct for this inequality by dividing the absolute frequency of each amino acid in the 38 proteins included here by their respective number of codons to provide a new relative frequency and a new abundance rank based on those figures. When these ranks are compared to those based on molecular weight, again no precise correspondence can be noted. Among the closest are glycine, which ranks third on this new basis and lightest in weight, glutamine with ranks of 13 and 11 in the same order, and arginine with 20 and 18, respectively. Hence, by and large neither codon number nor molecular weight correlates to amino acid frequency, so, at least in the proteins studied here, nonrandomness in association appears to be a marked feature of the components of natural proteins as well as those of synthetic ones.

CpG Deficiency in Vertebrate DNA. An observation related to the previous topic has been derived from analyses of vertebrate protein sequences including a total of more than 4000 sites. These indicated that amino acids having a high proportion of codons terminating in cytidine occur with significantly reduced frequency before amino acids whose codons begin with guanosine (Bullock and Elton, 1972). In turn this is reflected in the severe shortage in the vertebrate DNA of the nearest-neighbor base-doublet CpG (Josse *et al.*, 1961; Swartz *et al.*, 1962). Similar effects have not been shown by the amino acid sequences of related proteins from bacteria. Bullock and Elton suggested that the CpG deficiency of vertebrate DNA probably did not result from natural selection acting at the level of protein function or composition, but that it possibly reflected selection against that couplet at the level of translation (Subak-Sharpe *et al.*, 1966). Transfer RNAs and 5 S RNA sequences, it was pointed out, which are transcribed but not translated, do not show any deficiency in CpG frequency relative to comparable nonvertebrate sequences. However, translational differences between tRNA and rRNA on one hand and mRNA on the other fail to explain the observed deficiency of CpG in the DNA molecules of vertebrates.

Spontaneous Deamidation and Nonrandomness. Another indication that the sequence of association between amino acids is not random comes from a series of studies on the rate of deamidation of certain amino acids in peptides and proteins. Among the common amino acids which occur in natural proteins, only two (asparagine and glutamine) bear terminal amide groups ($CONH_2$). Even under mild chemical conditions, such as pH 7.4 in aqueous solutions at 37°C, these two amino acids have long been known to undergo hydrolytic nonenzymatic deamidation (Chibnall and Westall, 1932; Gilbert *et al.*, 1949; Schoffeniels, 1967). Similar deamidation has been reported to occur to residues of these amino acids in equine cytochrome *c*, human transferase, and a number of other proteins (Hartfenist, 1953; Flatmark, 1964, 1967; Flatmark and Sletten, 1968; Robinson *et al.*, 1973a). Recent studies on a large number of synthetic pentapeptides have shown that the half-life rate of deamidation varies widely with the sequence of amino acids (Robinson *et al.*, 1973b; Robinson, 1974). In these experiments, each pentapeptide began and ended in a glycine residue, and the deamidable residue (asparaginyl or glutaminyl) was located in the center; only the residues directly preceding and following the central site varied from one pentapeptide to another. Hence, the differencees in half-lives that were observed in this study stemmed from influences of adjacent bases, the range found for asparagine being 6 to 507 days and for glutamine, 96 to 3409 days.

Typical results for asparagine are perhaps best exemplified in those pentapeptides in which arginine preceded it, that is, was located in site 2. Thus the pentapeptide consisted of glycine–arginine–asparagine–X–glycine, with only the occupant of site 4 varying between tests. When alanine occupied this site following asparagine, the half-life was 18 days; when threonine was there it was 28 days, and when leucine was the occupant it became extended to 113 days. However, in another long series of pentapeptides in which alanine always occupied site 4 and the site 2 occupant varied, similar results were obtained. With arginine in that location, as pointed out above, the half-life span was 18 days, with histidine 45 days, glutamic acid 49 days, lycine 61 days, tyrosine 85 days, and leucine 217 days. Comparable results were notable in the series in which glutamine was the central degradable amino acid. Nuclear magnetic resonance studies have now been completed on comparable pentapeptides, in which restricted motion was consistently characteristic of the central residue (Keim *et al.*, 1973).

Autocatalysis of Tyrocidin. Although the nonrandomness documented above is suggestive of the involvement of autocatalysis in protein formation, clearer cut examples of this phenomenon can be cited. By means of cell-free preparations, several laboratories (Kleinkauf and Gevers, 1969; Lipmann, 1971, 1972; Lipmann *et al.*, 1971) in the absence of nucleic acids have been able to replicate tyricidin, a natural antibiotic consisting of ten amino acid residues. The amino acids (aa) were first activated with ATP as shown in the following incomplete equation:

$$\text{ATP} + \text{aa} \rightleftarrows \text{AMP} \cdot \text{aa}$$

After activation, the amino acids became fixed to enzyme-lined (E) sulfhydryl (SH) radicals:

$$\text{AMP} \cdot \text{aa} + \text{E} \cdot \text{SH} \rightleftarrows \text{AMP} + \text{E} \cdot \text{Saa}$$

and were thus carried into the reactions. In initiating the synthesis, the small enzyme A (mol. wt. = 100,000) bound and converted L-phenylalanine to the dextroenantiomorph. To this D-phenylalanine residue, the second, somewhat larger enzyme B (mol. wt. = 230,000) in turn attached proline and two more molecules of phenylalanine. The last of these amino acids was then coverted to the D-isomer (Kambe *et al.*, 1971) before the third enzyme commenced its activities. This enzyme C was by far the largest, having a molecular weight of 460,000; its function was that of attaching asparagine, glutamine, tyrosine, valine, and leucine in that sequence. Once the leucine residue had been added, peptide formation was terminated through that amino acid's being linked to the first member of the chain to form a cyclic molecule. If any amino acid was omitted from the substrate, polymerization proceeded as far as its site but there ceased (Roskoski *et al.*, 1970). Although the largest enzyme C could readily be charged at any time with 1 mol of each of the five amino acids it carried, no peptide formation occurred unless the intermediate enzyme B was present and then not until the initial tetrapeptide produced by enzymes A and B was provided. As Lipmann (1972) has pointed out, while the decapeptide was produced by three enzymes which had been made by way of the complex genetic apparatus, the synthesis of the antibiotic itself involved no such dependency on nucleic acids. Gramicidin S, a dimer whose peptide chains contain five amino acids each, has been shown also to be synthesized without direct aid from the genetic apparatus (Lipmann, 1971).

2.2.2. Self-Assembly of a Heme Enzyme

Among the common products resulting from exposure of water to ultraviolet light or ionizing radiation is hydrogen peroxide; hence, it is not surprising that among the nearly universal enzymes of organisms is one, known as catalase, that is capable of breaking down this substance. Catalase has a rather complex structure, for it consists of a protein and one or more heme units. Such hemes are composed of a ferrous ion linked to an intricate porphyrin molecule (Figure 2.5); beef liver catalase, for example, has a molecular weight of 250,000 and contains four heme radicals. Calvin (1956, 1969) has proposed an interesting model of how a complicated enzyme of this sort might have evolved under primitive earth conditions.

Although ferric ions in aqueous solutions can catalyze the decomposition of hydrogen peroxide, the efficiency of the catalysis is very low, having a coefficient of 10^{-5}. However, it is possible that the iron could have been incorporated into a porphyrin molecule (Figure 2.5) and thus produced a heme, which

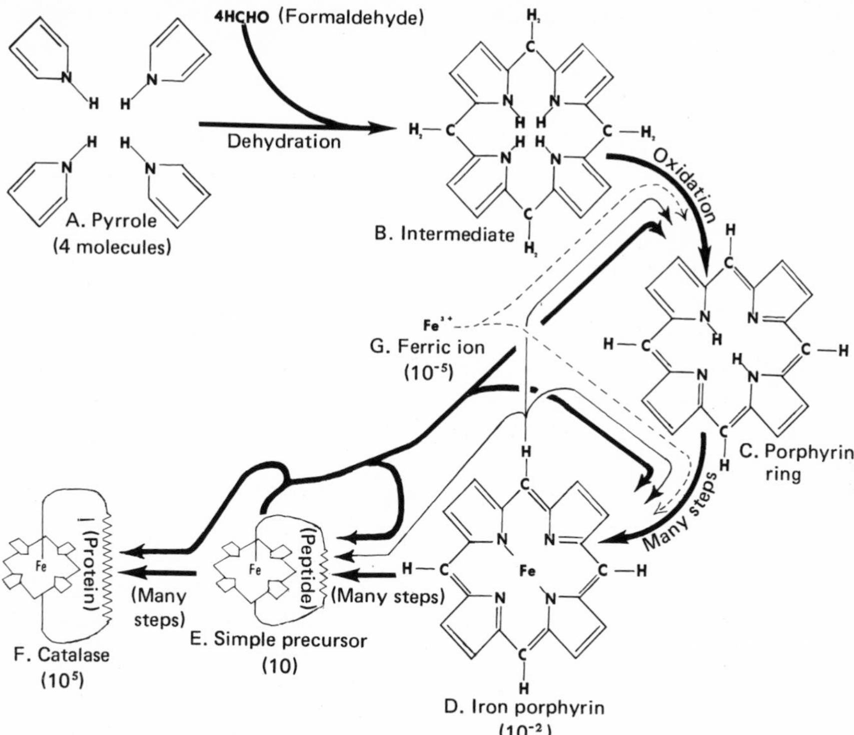

Figure 2.5. Autocatalytic development of a complex enzyme (catalase). Calvin (1956) conjectured that even complex enzymes might evolve in the seas through autocatalytic processes. Possibly such events could have commenced with ingredients probably present in the primitive seas (A), resulting in an intermediate compound (B); this might have undergone oxidation to a porphyrin (C). Since the only catalyst then available would have been ferric ions, the reaction would have proceeded slowly. By the iron's interacting with the porphyrin, a more efficient catalyst (D) would have been produced, which then would have accelerated the processes. Later, as polypeptides became added to the iron porphyrin, still greater efficiency and speed would have resulted with each improvement, until a product similar to catalase was finally brought into being.

acts upon hydrogen peroxide with an efficiency level 1000 times higher than that of ferric ions, as it has a coefficient of 10^{-2} (Calvin, 1956, 1969, 1975). The concepts involved in these processes are outlined below.

Prebiotic Synthesis of Porphyrin and Heme. Despite the porphyrin ring's being a very complex molecule, a method has been described whereby it might have been synthesized from primitive atmosphere components (Hodgson and Ponnamperuma, 1968). In the experiments concerned, electric discharges in a mixture of methane, ammonia, and water vapor produced the compound, possibly through the union of four pyrrole rings and four formaldehyde molecules. The latter might have provided carbon bridges between the pyrroles through dehydration and thus resulted in the formation of a porphyrin-type skeleton. By means of oxidation, possibly accelerated by ferric ions acting on hydrogen peroxide, six hydrogen residues could then have been removed from the preporphyrin molecule to produce the porphyrin ring system (Figure 2.5).

Autocatalytic Synthesis of Catalase. In the Calvin proposal, after a ferric ion had ultimately become incorporated into such a porphyrin ring and thus formed a heme molecule, eventually a simple, short polypeptide chain became attached to it. This development still further enhanced the efficiency of the catalytic reactions. Over eons of time, the original polypeptide chain continued to grow in length by a complex unpostulated series of steps until a molecule similar to modern catalase ultimately came into existence. Many of the individual steps in these conjectured events resulted in further increase of efficiency, until the level of catalytic activity found in catalase (10^4) ultimately had been attained. Thus the total efficiency increment is a billion times that of the primitive ferric ion: a single molecule of catalase can break down approximately 2 million molecules of hydrogen peroxide per minute. Each increased level of efficiency attained would affect not only the latest reactions in the series, but each and every earlier reaction that involved oxidation. Hence, the entire process would have been greatly accelerated by the accumulated effects. Because each new catalyst could thus serve throughout the whole series of reactions, Calvin suggested the term reflexive catalysis rather than autocatalysis.

2.2.3. Stereospecific Autocatalysis

As has already been intimated, most of the basic molecules of living things are optically specific, that is, they have a molecular configuration similar to those substances capable of rotating the plane of polarized light specifically to the right or to the left, as the case may be. All the amino acids (except glycine, which is neutral) and the nitrogenous bases are of a levorotatory configuration, whereas most of the sugars are dextrorotatory (West and Todd, 1961). In nature, the generation of any optically pure material requires the presence of organisms; in the absence of living things, a racemic system (an equimolar mixture of levo- and dextrorotatory material) is virtually the constant rule. Artificially synthesized materials and natural minerals are consistently racemic.

Hypothetical Enzymatic Stereospecificity. Several model systems for the production of stereospecific compounds have been advanced. One of them, based on a series of suppositions, begins by postulating the existence of compound A that could be converted to compound B by the reaction r. The conversion was visualized to proceed slowly when uncatalyzed, but at the rapid rate (K) when catalyzed by B. Both A and B were considered to be optically active. Furthermore, the stereoisomers of A were viewed as being converted into one another at a rapid rate, whereas those of B were converted only slowly. Finally, it was proposed that the products resulting from reactions catalyzed by either levo-B or dextro-B could be expected to have the same optic activity as their respective catalysts. Since compound B was absent initially, the first molecule of that substance obviously resulted from A by uncatalyzed reactions; hence, there is an equal statistical chance for either dextro-B or levo-B to have been the very first molecule of this substance produced. But once that original molecule had been formed, it would have rapidly catalyzed the reaction, so that the whole available pool of A would have been converted to the same enantiomorph of B as that randomly produced initial molecule.

As yet no biological system of the sort just described has been discovered, the closest approach being provided by the experiments of Allen and Gillard (1967). In these, peptide complexes coupled to a cobalt-containing substance yielded either levo- or dextro- compounds as the octahedral crystals developed. As Rutten (1971) pointed out, however, among other things, cobalt and related ambivalent metals are too rare on the surface of the earth for these results to hold much biological significance.

Recently, experiments for the study of the origin of optical activity in biochemicals were conducted in which polyamino acid chains were made from racemized amino acids (Thiemann, 1974; Thiemann and Darge, 1974). Three amino acids were used, α-alanine, α-amino-butyric acid, and lysine. All were thoroughly racemized before polymerization to ensure ideally racemic substrates. After polymerization into short chains had been achieved by way of conversion of the amino acids to the *N*-carboxyanhydrides, the polymers were found to rotate light to the left. Therefore it was assumed that the L-amino acids had polymerized to a greater extent than had the D-enantiomorphs. However, not all substances of levorotatory configuration actually rotate light to the left (West and Todd, 1961), so the observation remains inconclusive.

Minerals as Stereospecific Templates. For a comparable role in guiding the formation of stereospecific isomers, Rutten (1971) suggested quartz might be a far more likely candidate than cobalt, as it has always been among the most common minerals of the earth's crust and shows optical activity (Klabunowskii, 1959). Hence, it is theoretically possible that selective absorption of levo- and dextrorotatory isomers of biological chemicals could occur on the corresponding enantiomorphs of quartz crystals. This has now been demonstrated with alanine (Bonner *et al.*, 1974). To have been effective in prebiotic syntheses, however, it must be shown that one enantiomorph of quartz has been far more abundant in nature than the other, and as yet no such evidence has been presented.

A somewhat similar mechanism has more recently been advanced (Degens *et al.*, 1970; Jackson, 1971), but clay minerals, such as kaolinite, replace the quartz as the chemical-selective agent. Those workers reported that in mixtures of levo- and dextrorotatory and racemeric aspartic acid in 0.01 M aqueous solutions the various ingredients became polymerized at 90°C over a period of 4 weeks to different extents in the presence of kaolinite. According to the report, about 25% of the L-aspartic acid polymerized, 3% of the D-isomer, and 14% of the D,L-aspartic acid. Thus it appeared the polypeptide formation occurred asymmetrically on the clay mineral, with L-amino acids being polymerized preferentially to the D-enantiomorphs. However, some laboratories (Bonner and Flores, 1973; Flores and Bonner, 1974; McCullough and Lemmon, 1974) were unable to obtain similar results with phenylalanine in place of aspartic acid. In a variation on this general theme, solid samples of D-, L-, and D,L-leucine were exposed to a Sr^{90} source for 16 months; at the end of that time L-leucine was found to have decomposed slightly less rapidly than the dextroenantiomorph (Bonner and Flores, 1975). Consequently, the levo-amino acid would have accumulated to a greater concentration in the primitive soup.

Still another device has been employed for studies directed to the same end. However, these demanded the presence of a template, such as poly-L-leucine (Kenyon and Steinman, 1969; Darge *et al.*, 1973), or poly-D-uridylic acid (Schneider-Bernloehr *et al.*, 1968) for the preferential polymerization of the amino acid or nucleotide. Since neither of these studies provided any explanation as to the origin of the requisite template, the proposed mechanisms appear inapplicable to primitive earth conditions.

2.3. THE PRESENT STATUS OF THE LIFE ORIGINS PROBLEM–A CRITICAL ASSESSMENT

So many aspects of the problem of life's origins on earth have received tentative solution during the past quarter century that prospects for further progress appear bright indeed. On closer inspection, however, those achievements are noted to be concentrated almost exclusively within the synthetic and polymeric stages, while above those two elementary levels few convincing proposals have been advanced. The several types of colloidal particles that have been developed, while interesting and informative, fail to be convincing even as models of protobionts, their transitory nature being especially disturbing. When this ephemerality is considered together with the extreme sensitivity they display to mechanical disturbances and changes in pH, along with the unnatural conditions and often complex organic substances required for their formation, doubts inexorably arise concerning the suitability of these colloids for a role as protobiont ancestors. Perhaps what is called for at this point is a reassessment of the current status of the complete problem, in the hope that the resulting exposure of present weaknesses and strengths may accelerate progress in investigating the higher stages of life's beginnings.

2.3.1. Problems Centering around the Molecules of Life

Since Stages I and II in the origins of life are so closely interrelated, there is little reason to discuss them separately.

1. In general, nearly all the fundamental molecules of life have been synthesized, a number of them by a variety of methods, including the amino acids, sugars, and many of the nitrogenous bases. However, one category of such molecules which has rarely been synthesized includes the lipids (Allen and Ponnamperuma, 1967). Although this class of substances is not of particular importance as nutrients, its members play a very significant role in cells, especially in the formation of membranes, without which cells could not exist.

2. In the experimental production of these basic molecules, the conditions employed vary over broad limits, so that no general pattern of environmental requirements can be visualized, aside from a reducing type of atmosphere and sterile conditions. Extremes of heat and cold, or of wet and dry, do not present so great a problem as do other factors, for such extremes exist simultaneously on earth today. More difficult to comprehend are the requirements for strongly basic or acidic conditions, because the waters on earth today largely range between pH 5.5 to 8.5, and are most frequently mildly alkaline.

3. Some techniques appear unrealistic from a biological standpoint in their requirement of large quantities of specialized condensing agents, such as dicyanamide and complex phosphorus compounds, substances which do not occur today either in living organisms or on the earth's surface in comparable form and quantities. Had they been present in the primitive soup in the requisite concentrations, one would suspect that they would have become incorporated into protobionts and thus still persist at least in some extant organisms. Hence, if the evidence from existing living things is to be believed, such complex condensing agents either never existed in significant quantities or played a minor role in the formation of the primitive biochemicals.

4. The dilution factor, even with the oceans growing from relatively small bodies as sometimes visualized (Rubey, 1951, 1955; Dillon, 1974), imposes especially great difficulties upon the origins of the biological materials. This is a problem particularly with such dehydration reactions as peptide linkages and the bonding of nucleosides. Even avoiding the thermodynamic restrictions by carrying out the processes under anhydrous conditions, as has been done, merely changes the problem. The difficulty then is, first, providing large concentrations of pure aspartic or glutamic acid or other amino acid, and, second, melting that substance at a reasonable temperature, and, finally, adding the other amino acids. In the laboratory the large quantities of one or two particular amino acids, a widely diversified supply of others, and the sequence of the reactions are logical and simple procedures, but under natural conditions, the presence of the chemicals and the occurrence of the required physical condi-

tions in the proper sequence of steps could not have been other than a rare series of accidents indeed.

5. Much has been made of the specific optical properties of biological materials (Bernal, 1967), and there can be no doubt that the particular enantiomorphs of amino acids, nitrogenous bases, and sugars are peculiar to living things. Especially important in the context of life's origins over long periods of geologic times is one characteristic of amino acids that has recently been pointed out (Bada, 1972; Bada and Schroeder, 1972; Bada *et al.*, 1974): Under standard ambient conditions, a given enantiomorph undergoes molecular change toward its mirror-image form at a rate that is approximately constant. Hence, it was proposed that the proportions of L- and D-aspartic acid, for example, could be used to date fossil remains, the "half-life" in the example cited being ca. 15,000 years. In the present context, this process of molecular change signifies that, even though some method may be found to produce a given type of optically active material in quantity, in the seas it must be incorporated into living systems relatively quickly or else its purity of type would soon be lost. A basic principle which needs to be invoked here is one that has been found applicable to all levels of biological structure from cellular to organismic, and that is that structural identity implies common descent. And the molecular level should be no different in this regard.

6. The precision of the linkage formation found during the polymerization of nucleotides and amino acids has not been approached in the test tube as yet. In substance, this is another way of stating that, thus far, no proteins and no nucleic acids have been synthesized in the laboratory under prebiotic conditions even in reasonable percentages.

7. Phosphorylated compounds in general have been relatively poorly explored from the standpoint of their formation *in vitro*. Among the very important members of this category are such compounds as the triphosphorylated nucleosides, including ATP, GTP, and the others vitally essential in cells for the transfer of energy and possibly for the *in vivo* synthesis of nucleic acids.

2.3.2. Problems at the Interacting-Systems Level

Although a number of uncertainties thus still persist at the molecular and polymeric levels of life's origins, some of the problems are undoubtedly more apparent than real. For example, it is not necessary to ascribe to the earliest living things all of the molecular properties found among extant forms. Thus, the difficulties with mirror-image isomers as well as the irregularities of the linkages in polynucleotides could well be put aside, for these specialized conditions could logically have developed later in life's history. Unfortunately, with the problems at the interacting-systems level, the difficulties are several orders of magnitude greater.

Indeed the basic problem with all the models thus far advanced looms so large that an enumeration of their other, more minor failures is made meaningless. This major flaw is perhaps most easily demonstrated first in the model of the autocatalytic creation of the enzyme catalase described above. That model suggests that, as a consequence of the favorable selective value of increased catalytic efficiency, ferric ions first became incorporated into porphyrin rings, which eventually developed through heme molecules into a complex proteinaceous enzyme. Even if it is granted that the entire process is feasible outside of a cell, there appears to be no means by which many identical—or even fairly similar—molecules can be produced. Perhaps one can concede that a single molecule, or a very few, *might* be formed in the postulated manner, but certainly it is most improbable that any sizable population of identical molecules could result from random chance processes.* That model also fails to include a mechanism whereby the enzymes can be passed to subsequent generations of the protobionts—nor does it contain even a suggestion as to how the oxidation–reduction processes or their products were of any value to the early forebears of life. Thus it becomes evident that, only if the breakdown of hydrogen peroxide had been essential to the metabolism of the protobiont and if the random-chance growth in complexity visualized for the developing catalase had been under the guidance of an evolving genetic mechanism, could that enzyme have increased in abundance and survived to have become incorporated into living creatures. And these observations plainly signify that the catalase would have had to develop *within* living organisms, not free in the seas. Thus the enzyme must actually have resulted from ordinary biological evolutionary processes, not by autocatalysis carried out in the primitive environment.

Similar difficulties with the coacervates and other particles are also evident (Oparin, 1974). Although the various colloidal droplets are not considered to be alive, the nature of the experiments conducted on them intimates that they are models of how living things came into being. Realistically, though, they can be considered to be similar to actual primitive life only from the most superficial point of view. It is true that they can be induced to carry out enzymatic metabolic processes, but only in a mechanical fashion, for none of the products or the resulting energy is of any value, directly or indirectly, to the colloidal particle. As in the case of catalase, a genetic mechanism of some sort is required, both for the acquisition of metabolic processes essential to primitive life and to

*The unlikelihood of identical molecules' resulting from random-chance processes receives emphasis from several recent sources. These include a number of investigations into the biosynthesis of porphyrins, which indicate the great complexity of the actual steps involved in these processes (Russell, 1974; Frydman *et al.*, 1975; Jackson and Games, 1975; Shemin, 1975). In addition, a large number of isomeric forms are theoretically possible, which have been described as including 16 etio- and coproporphyrins and 96 meso- and protoporphyrins (Fischer and Orth, 1972; Aronoff, 1975). This diversity must then be multiplied by the number of polypeptides that could be formed by strict chance processes in producing, for example, a simple protein only 100 amino acid residues in length (20^{100}) to show the extreme unlikelihood of the proposed sequence of events occurring twice in identical fashion.

create the enzymes needed to carry them out. In other words, before photosynthesis, the enzymatic breakdown of a sugar, the synthesis of peptides such as polyphenylalanine (Fox *et al.*, 1974), or any other type of cellular process could have become meaningful in a protobiotic system, metabolic pathways for the utilization of the products of the particular processes must first have been established on a genetical basis. Hence, not only sulfobes, but the other models described in this chapter as well, are specious in that they bear apparent behavioral similarities to living things but lack the real requisites of even the simplest protobionts.

Thus, although the coacervates, micelles, and proteinoid microspheres have provided much information about the behavior of colloidal droplets and were worthy of exploration during the early period of the search for possible forebears of living things, they no longer can be considered meaningful in the context of the origins of life on earth. Neither they nor any other artificial systems are adequate to provide concrete evidence to show clearly how life and the requisite genetic apparatus actually did arise. Since the origins of inheritance are thus inseparably a part of the beginning of life, the first step toward an attempt at correcting the existing situation appears to be that of reviewing the essentials of the genetic apparatus in modern organisms.

3

THE GENETIC MECHANISM: I

DNA, Nucleoids, and Chromatin

Now that it has been made clear that the earliest protobiont had to possess a genetic mechanism before further evolution could occur, clues to two major questions will be sought as the basis for inheritance is reviewed in this present and the following two chapters. The first of these pertains to the nature of the genetic apparatus that probably existed in the very first forms of life, and the second involves the sequence of events that led to its present state of complexity. Fortunately data are available that indicate a probable series of developments, but to appreciate their significance fully, each aspect of the genetical processes must first be scrutinized. The most logical point for departure is an investigation of deoxyribose nucleic acid (DNA), the central ingredient in prevailing doctrines. Thus, this molecule, its mode of replication, and its organization in living cells are the topics of the present discussion, after some preliminary points have been established.

3.1. INTRODUCTION

A Paradox in Life's Origins. The requirement for a means of inheritance in the first living thing makes evident the existence of a paradox, as Woese (1970a,b) has pointed out. This seemingly self-contradictory condition is that, while genetic mechanisms can exist only within living things, the organisms themselves could not have come into being without such a device. Indeed, Bernal (1967) has proposed that a molecular type of replicating system was of necessity the precursor to the organisms. In another attempt at solving the riddle, the hypothesis was presented that a crystal of some unknown substance served in replicatory functions until the existing nucleic acid system arose (Cairns-Smith, 1966, 1971), but what kind of substance would be encoded by such a crystal was not made clear. Furthermore, the concept not only fails to solve the problem of how the current biological mechanism later arose and displaced the crystal, it also lacks support from any known factual evidence.

The Essentiality of Evidence from Biological Sources. Actually, a similar absence of compelling supportive evidence derived from extant living sources is a more serious flaw in the prevailing theories and protobiont models than is their failing to satisfy the requirement for a genetic mechanism. In Stages I and II of life's origins, the diverse paths that have been proposed become acceptable because of the similarity of the products to amino acids, bases, and polymers in the organisms of today. Thus their general acceptability persists despite their shortcomings in the form of the multitudinous procedures that have been employed and the wide divergency of environmental conditions they collectively require. No comparable level of confidence, however, has been engendered by microspheres, coacervates, or other colloidal models, largely because no similar feature can be demonstrated in extant organisms.

In science, as opposed to pure philosophy, observable facts are requisite; hence, to maintain the search for life's origins in a scientific framework, data derived from biological sources are quite as essential as in any other problem involving living things. No new approaches are required, as Miller and Orgel (1974) have proposed, only new analyses of the living things about us. Such reevaluations in no way obviate the need for further biochemical investigations. To the contrary, more extensive biochemical knowledge is requisite, for any approach to the study of the origins of life necessarily must be at the molecular level, with particular stress placed on the most primitive forms of living organisms.

3.2. THE FOCAL INGREDIENTS

As intimated in the preceding chapter, the basic substances whose synthesis necessitates the existence of a genetic apparatus are the enzymes which catalyze all metabolic activities. Since all known enzymes are proteins, the hereditary processes primarily are designed to ensure that the particular species of protein needed is actually produced. In brief, this is accomplished by guiding the processes so as to make certain that the proper amino acids become arranged in the correct sequence.

Structural Properties of Amino Acids. The 20 amino acids widespread among the proteins of living things today are of relatively limited diversity in that all possess a number of features in common. As pointed out earlier (Chapter 1, Section 1.3.1), all have an ammonia residue attached to the α-carbon, which also bears a carboxyl radical; in addition, their molecules uniformly are of the levorotatory configuration, except for glycine, which is optically neutral. They thus differ from one another only in that portion of the molecule beyond the α-carbon; the various and often complex modifications that exist are illustrated more appropriately in a later chapter (Figures 6.5 to 6.9).

Stability of Amino Acids. As a whole, the amino acids are quite stable, even in aqueous solution. Vallentyne (1956) provided a detailed study of five amino acids which indicated that, at ambient temperatures, alanine could persist in solution more than a billion years, and the most labile one studied, serine, could survive under the same conditions for about a million years. Alanine and glycine have, as a matter of fact, been secured in substantial quantities from Precambrian deposits exceeding 2 billion years in age (Pflug *et al.*, 1969). Although under aerobic conditions the life expectancy of an amino acid is shorter than given above by a factor of around 20 (Abelson 1957), under the reducing conditions usually postulated for the primitive earth their decomposition would not have been an important consideration. However, their racemization rate would have been. As mentioned before (Bada *et al.*, 1970; Bada and Portsch, 1973), the half-life of L-isoleucine is only about 1 million years at ambient temperatures (Bada, 1972). When united in the form of proteins, the rate of decay of the polymer is of a comparable order of magnitude, being between 100,000 and 1 million years (Fox and Dose, 1972). Because of this rate of spontaneous degradation, little light can be expected to be shed pertaining to the precise chemical configuration of early macromolecules through the study of very ancient deposits.

Proteins. The focus of the genetic apparatus is on the polymers of amino acids, not the components individually, for it is as peptides and proteins that the amino acids carry out most of their functions. In these polymers, the specific sequential arrangement of given amino acids is what results in one protein's being capable of breaking down hydrogen peroxide into water and oxygen, for instance, and another's being able to join particular nucleotides into a unique type of nucleic acid. These and all other capabilities of individual types of proteins are believed to stem in part from their characteristic overall configuration, that is, their tertiary structure (Figure 3.1). How the molecular shape, in combination with the location of various active sites, performs the special and often extremely complex activities has not as yet been fully elucidated. However, several approaches have now been utilized to create theories of protein structure, all of which draw upon amino acid sequence studies. Among these is one that viewed the folding of proteins as a transfer of information between two messages, the primary amino acid sequence and the biologically active conformation of the protein (Robson, 1974; Robson and Pain, 1974a,b,c). Another approach stressed the difference in orientation and chemical properties of the side chains of the amino acids (Jones, 1975), and expressed the differences in numerical fashion suitable for computer analyses. As a result, hydrophobicity of the amino acid was found to influence strongly the orientation of the side chain and also the dihedral angle ϕ, but not ψ, of the backbone, Although progress toward understanding protein functioning thus remains minimal, the

Pro-1
Pro-13
Gly-10
A

Pro-13
Pro-1
B

Figure 3.1. Tertiary structure of a protein. (A, B) Contrasting models of alamethicin. (A) Labquip model. (B) Space-filling model. (C) Balsa wood model of hemoglobin molecule at 5.5-Å resolution. (A,B, courtesy of Drs. M. E. Johnson, T. T. Wu, and the *Journal of Theoretical Biology;* C, courtesy of Dr. M. F. Perute.)

processes by means of which each particular sequence of amino acids is specified have been clarified to a large degree, as described below.

3.3. THE KEY MACROMOLECULE

According to current concepts, the sequence of amino acids in all proteins is specified by the primary molecular structure of the deoxyribose nucleic acid (DNA) of the cell. More than 90% of the DNA of the cell is contained in the nucleus, much of the remainder being found in such cytoplasmic organelles as the chloroplast and mitochondrion, and to a smaller degree in the nucleolus (Vincent, 1952; Monty *et al.*, 1956). Here the discussion is concerned primarily with the nuclear DNA, as the activities of the remainder are believed to be similar and are principally of interest in studies of the respective organelles in which they occur.

3.3.1. The Composition of DNA

The DNA of the eukaryotic nucleus and prokaryotic nucleoid is considered to be contained in the chromosomes, where it exists in the form of a complex

Table 3.1
Molar Ratios of Bases in DNA (from Various Sources)

DNA source	Pyrimidines: Cytosine	Pyrimidines: Cytosine modified	Pyrimidines: Thymine	Purines: Adenine	Purines: Adenine modified	Purines: Guanine
Beef thymus	21.2	1.3[a]	27.8	28.2	0	21.5
Beef spleen	20.8	1.2[a]	27.3	27.9	0	22.7
Beef sperm	20.7	1.2[a]	27.2	28.7	0	22.2
Rat bone marrow	20.4	1.2[a]	28.4	28.6	0	21.5
Herring testes	21.5	2.4[a]	28.2	27.9	0	19.5
Herring sperm	20.7	1.7[a]	27.5	27.8	0	22.2
Urchin sperm *(Paracentrotus)*	17.3	1.1[a]	32.1	32.8	0	17.7
Wheat germ	16.8	4.5[a]	27.1	26.9	0	23.1
Yeast	17.4	?	32.6	31.7	0	18.3
Escherichia coli	25.2	0.3[a]	23.9	26.0	0.6[c]	24.9
Mycobacterium tuberculosis	35.4	0	14.6	15.1	0.7[c]	34.9
T2 bacteriophage	0.0	16.7[b]	32.6	32.5	0.2[c]	18.2
φX174 bacteriophage	18.2	0	32.3	24.3	?	24.5
Transducing *bacteriophage* of *B. subtilis*	14.7	—	(35.9)[d]	35.9	—	13.4

[a]5-Methylcytosine.
[b]5-Hydroxymethylcytosine.
[c]*N*6-methyladenine.
[d]Uracil.

helical molecule consisting of two chains of nucleotides. As the nature of the nucleotides and the 3′–5′ linkages have already received mention (Chapter 1, Section 1.3.4), attention can be confined here to the manner in which the two chains are attached to one another, along with allied problems. It is the relations of the two strands that enable DNA to be replicated precisely and thereby induce genetic stability in organisms.

Ratios of the Bases in DNA. Among the important structural details is the observation that, when the proportions of the four bases in DNA are analyzed, the molar percent of one given pyrimidine is consistently approximated by that of one certain purine (Table 3.1). Thus the molar percent of the purine adenine is closely matched by that of the pyrimidine thymine, and the proportion of the pyrimidine cytosine (including its modified form) is uniformly approximated by that of the purine guanine (Chargaff, 1950, 1951). The correspondences between these pairs of bases are greatest, however, among the eukaryotes, especially such advanced ones as metazoans and the seed plants. Among bacteria, the disparities in molar proportions between members of these pairs may be noted to be larger; indeed, the results of analyses of bacterial DNA often vary from laboratory to laboratory (Hall, 1971). In viruses, even more pronounced discrepancies exist, as in that of ϕX174, which is listed in Table 3.1; in this case, however, the DNA molecule is single-stranded, not duplex (Sinsheimer, 1959). Other constants in base proportions, including the sum of the purines being equal to that of the pyrimidines, and the sum of the amino bases (adenine and cytosine) equaling that of the keto bases (guanine and thymine), are derived from these same correlations (Chargaff, 1951).

Asymmetrical Distribution of Bases. Base percentages of whole DNA provide only a part of the picture of base distribution, for frequently there are regions in the DNA molecule that are rich in one base or another, rather than the random pattern originally expected. In the first place, the two strands of the double helix differ in percent composition of base types. Characteristically in bacteria, for instance, one strand contains a higher ratio of pyrimidines, whereas the other has a greater proportion of purines (Rudner *et al.*, 1968, 1969). Since purines are much heavier than pyrimidines, one strand is accordingly designated as the light (L), the other as the heavy (H). These differences have now been suggested to be correlated to functional properties of DNA, and receive further attention later (Chapter 5). Moreover, double segments about 7000 nucleotide pairs long have been found to differ extensively in G + C content (Yamagishi, 1970; Yamagishi and Takahashi, 1971). More recently the DNA of *Micrococcus lysodeikticus,* which contains a high level of G + C pairs (about 74%), were compared segmentally to that of *Clostridium perfrigens,* which has a low level of the same base pairs (about 30%). In segments of any designated length, the G + C pairs in *Micrococcis* ranged from 70% to 77%, while in the second bacterium the range was from 27% to 34% (Yamagishi, 1974). Such extremes of doublet frequencies have given rise to a concept of "genetic code limit organisms" (Woese, 1967; Woese and Bleyman, 1972), in which the range in redundancy of doublets is viewed as limited by genetic code requirements (Elton, 1973a,b; Russell *et al.*, 1973).

Among mammals, a similar asymmetry has been noted in base distribution in the DNA, in that regions rich in adenosine or guanosine have been found by hybridization with tritium-labeled poly(U) and poly(C) (Shenkin and Burdon, 1974). In hamster cells, segments up to 130 nucleosides long were found to consist of nearly 95% adenosine, and others, usually less than 40 units long, contained between 70% and 90% guanosine. These enriched regions were not confined to any one chromosome or specific parts of chromosomes but were widespread throughout the genome. Comparisons with other species suggested that these adenosine- and guanosine-rich regions were more frequent among mammals than among prokaryotes or viruses. The DNA of higher plants also showed a similar distribution of concentrations of various bases (Pivec *et al.*, 1974), and the presence of pyrimidine sequences 25 to 200 units long has been suggested to be characteristic of eukaryotic DNA in general to a total extent of nearly 1% of the genome (Birnboim and Straus, 1975).

Repeated and Unique Sequences. Another situation involving bases centers on their distribution in a different sense, namely by sets, rather than by individual varieties. When the two strands of DNA are separated (that is, the molecule is "denatured"), and then allowed to rejoin, the renaturing of various sequences has been shown to proceed at differing rates. The more rapidly reassociating segments are thought to consist of the same sequence repeated a number of times and the slowly renaturing parts are considered to be unique sequences (that is, present only once in the haploid genome), which may represent the actual genes. Although bacterial DNA had been stated earlier to consist entirely of unique sequences (Britten, 1968), this now is known to be erroneous, for a short, cytosine-rich repeating unit has been isolated from *E. coli* DNA (Lin, 1974). Proportions of the two types of sequences vary among eukaryotes. In *Urechis* oocytes, for example, unique DNA accounted for at least 75% of the genome (Davis, 1975). While DNA of the mouse is only 60% unique under standard conditions of temperature (60°C) (Britten and Kohne, 1968: Brown and Church, 1971; Gelderman *et al.*, 1971), at 50°C, however, this unique DNA displayed considerable heterogeneity, yielding three subequal fractions and a minor one that differed in thermal stability and other properties. These four have been identified as nonrepetitive, moderately repeated, highly repeated, and satellite-like highly repetitive (Cech and Hearst, 1976).

In addition to the slowly associating unique DNA, two fractions generally are recognized among the repeating units. A fast fraction is diversified, as it includes the satellite DNAs (to be discussed later), hairpins (relatively short, complementary segments), and long repeating sequences of an unknown nature. The intermediate fraction consists mainly of genes for transfer and ribosomal RNAs and histones, plus others (Walker, 1969; Kuprijanova and Timofieva, 1974). Among such vertebrates as the loach, 50% of the DNA belongs in these two fractions, with the greater portion in the intermediate. Since comparable fractions have been reported for human, *Drosophila,* sea urchin, and slime mold genomes (Saunders *et al.*, 1972; Davidson *et al.*, 1974; Gall, 1974; Vorob'ev and Kosjuk, 1974), these categories of DNA appear to be a universal eukaryote feature. In fruit flies, for instance, structural genes and moderately reiterated sequences have been found adjacent in isolated DNA fragments (Georgiev *et al.*, 1977). The repetitive DNA is known to be transcribed and conducted to the cytoplasm, at least in frogs (Dobner and Flickinger, 1976).

In the intermediate fraction, some of the structural particulars have been identified. In *E. coli* DNA of this type, some of the 60 or more genes for the transfer RNAs have been shown to be clustered, probably in groups of three or four (Fournier *et al.*, 1974). A comparable condition for the rRNAs has been identified in the DNA of the toads *Xenopus mulleri* and *X. laevis*. DNA of these two species contains highly repetitive sequences in which the gene that codes for a 5 S rRNA alternates with a space-filling (spacer) sequence; about 24,000 copies are present in *X. mulleri* and 9000 in *X. laevis* (Brown and Sugimoto, 1973). During the developmental stages of the latter species of toad, the number of genes specifying ribosomal RNAs in general is increased by a factor of 2500; upon reaching sexual maturity, these genes are lost in males, but not in females (Kalt and Gall, 1974). In *Ascaris* germline DNA, highly repeated sequences are retained that are eliminated in somatic cells (Moritz and Roth, 1976).

Modified Bases of DNA. As shown in Table 3.1, a small proportion of the bases in the DNA molecule are modified, characteristically by the addition of a methyl radical. Usually eukaryote DNAs have only one modified base, 5-methylcytosine. In metazoans, this may amount to as much as 11% of the deoxycytidines, as in herring testes (Laland *et al.*, 1952), or as little as 0.5%, as in the brine shrimp, *Artemia salina* (Antonov *et al.*, 1962). The flowering plants show a higher range of methylation. For example, in cabbage (*Brassica oleifera*) about 10% of the total deoxycytidine is thus modified (Vanyushin and Belozerskii, 1959), whereas in a sunflower (*Helianthus annuus*), more than 37% is (Ergle and Katterman, 1961). Horsetail rush (*Equisetum arvense*), clubmoss (*Lycopodium clavatum*), and bracken fern (*Pteridium* sp.) DNAs fall within the same broad range, with 13.0%, 18.8%, and 23.5%, respectively (Thomas and Sherratt, 1956; Vanyushin and Belozerskii, 1959). While all of the higher eukaryotes have only this one modified base, more thorough investigation of unicellular types may demonstrate greater variation to exist at lower levels of advancement. In the dinoflagellate *Gynodinium,* for example, 5-hydroxymethyluracil was found to the extent of 37% of the thymidylic acid (Rae, 1973). In contrast, mitochondrial DNA appears to lack modified bases (Dawid, 1974).

The bacteria and viruses also display a limited variability. In a number of forms, including *Aerobacter, Proteus, Pseudomonas,* and bacteriophage T2r, only one modified base, N^6-methyldeoxyadenosine, is present (Wyatt, 1951; Dunn and Smith, 1958; Wainfan *et al.*, 1965). Most other bacteria and a blue-green alga (*Plectonema boryanum*), as well as many viruses, have both this base and that which occurs in eukaryotes (5-methylcytidine) (Kaye *et al.*, 1967; Hall, 1971). A very few viruses, notably bacteriophages λ and T1, have only the latter modified nucleoside (Klein and Sauerbier, 1965; Gough and Lederberg, 1966). In all cases, the level of frequency is low, rarely above 2% of the related major base. Greater diversification may be found among the prokaryotes and viruses as more extensive investigations are carried out, for one unexpected variant has been uncovered recently in the DNA of several strains of a bacteriophage that attacks *Bacillus subtilis.* In these, besides 5-hydroxymethyl-deoxyuridine, which completely replaced thymidine, a hexose displaced the usual pentose in about 1% of the nucleosides (Rosenberg, 1965). Since one strand of the DNA proved heavier than the other, it was concluded that much of the hexose was concentrated in one strand, but part of of the differences might also need to be attributed to the asymmetrical distribution of purines and pyrimidines noted earlier. Another extreme instance is phage XP-12 of *Xanthomonas,* which has been shown to have 5-methylcytosine completely replacing the cytosine (Ehrlich *et al.,* 1975).

The modified bases do not appear to occur randomly in the DNA but seem to be placed at particular sites by specific enzymes. Two such methylases have been isolated from *E. coli* and two from rat tissues (Gold and Hurwitz, 1964a,b; Kalousek and Morris, 1968; Sheid *et al.*, 1968). Earlier studies of *in vivo* methylation of the DNA in *E. coli* indicated that methylation occurred as the DNA is replicated, even in the absence of methionine, the usual cellular source of the methyl radical (Billen, 1968; Lark, 1968a,b). There observations seemed to support the hypothesis that had been advanced previously, that DNA methylation provided the cell with a means of distinguishing its own DNA from that of an invader, such as a virus (Srinivasan and Borek, 1966). Another suggestion along similar lines is that the base N^6-methyladenine blocks the action of restriction enzymes on DNA, because its conformation appears to interfere with coupling with thymine (Sternglanz and Bugg, 1973). But a more general function may be that of protecting the DNA from nuclease that induces single-stranded breaks (Marinus and Morris, 1974).

In eukaryotes, at least, it now has been established that newly synthesized DNA is not methylated, a lag of 2 minutes existing between synthesis and methylation in mouse adrenal cells (Kappler, 1970; Adams, 1974). Methylation enzymes then transform nearly 7% of the cytidines to 5-methylcytidine, after which treatment 0.1% of these modified bases are converted by a DNA deaminase to minor varieties of thymine (Scarano *et al.*, 1965; Scarano, 1969; Kappler, 1970). Since the conversion to thymine does not take place in isolated nuclei, it is believed to be under cytoplasmic control (Adams and Hogarth, 1973; Geraci *et al.,* 1974). However, it should be noted that the latter enzymatic alterations actually induce mutations of the transition type in the DNA, namely from a G+C base pair to an A+T (Volpe and Eremenko, 1974; Holliday and Pugh, 1975).

3.3.2. The Structure of DNA

Tertiary Structure of DNA. The standard base relationships, along with X-ray diffraction studies, enabled Watson and Crick (1953a,b) to deduce that the DNA molecule is double-stranded. Further, they were able to show that it was in the form of a right-handed helix, the two polynucleotide chains being wound around a common axis; union of the two is by means of hydrogen bonds between opposing bases (Figure 3.2). In the Watson–Crick model, base pairs

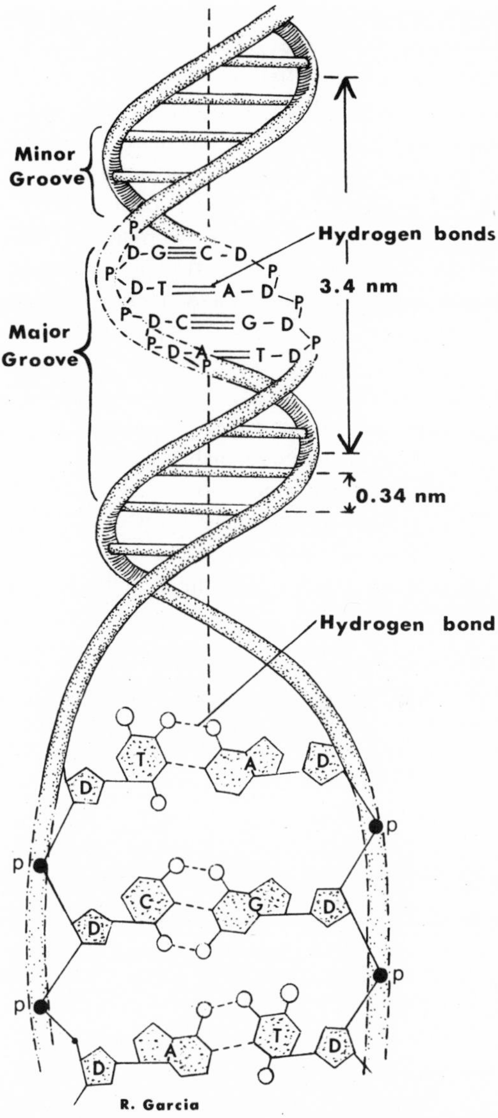

Figure 3.2. Structure of the DNA molecule. The double-stranded DNA molecule structure shown is that postulated by Watson and Crick. G, C, T, and A indicate the several common nitrogenous bases of nucleotides, D represents the deoxyribose pentose sugar moiety, and P the phosphate radical.

are spaced 3.4 Å apart in each strand; since there are ten base pairs per complete turn in the spiral, the periodicity is 34 Å, the diameter in the spiral being 20 Å. As shown in the figure, the two phosphate–pentose chains are not arranged on directly opposite sides of the helix, so that the two "grooves" between the opposing strands differ in width and are called *major* and *minor grooves* accordingly (Etkin, 1973; Rosenberg *et al.*, 1973).

Refinements in X-ray diffraction techniques later permitted the discovery to be made that the DNA molecules can assume three different forms (Langridge *et al.*, 1960a,b; Marvin *et al.*, 1961; Fuller *et al.*, 1965; Davies, 1967). As an examination of Table 3.2 shows, the Watson–Crick model most closely corresponds to the B structure, the form believed to be found in solutions of low ionic strength, as in the cell (Tunis-Schneider and Maestre, 1970; Fujita *et al.*, 1974). The A form is supposed to resemble that assumed by DNA–RNA hybrids and by double-stranded RNA in solution (Tunis and Hearst, 1968; Davidson, 1972; Day *et al.*, 1973). Thus, when DNA is being used as a template for making RNA, as seen later, the A configuration evidently must be adopted, possibly because of RNA's inability to assume the B form (Davidson, 1972). However, the correspondences between the DNA and RNA molecular configuration nowhere are complete, as shown in Table 3.2 (Sundaralingam, 1969). Various proteins, as well as organic solvents, have been suggested to change the configuration of DNA from the B to the A form (Brahms and Mom-

Table 3.2
Molecular Properties of Nucleic Acids

Nucleic acid	Length of period (Å)	Number of base pairs per period	Distance between bases in a strand (Å)	Diameter[a] of helix (Å)	Inclination of base to axis (°)
RNA-10	30	10	3.0	17.92	79
RNA-11	30	11	2.73	17.68	76
DNA-A	28.2	11	2.56	17.68	70
DNA-B	34	10	3.4	18.11	90
DNA-W-C[b]	34	10	3.4	20.0	90
DNA-C	31	9.3	3.32	18.11	90
Polyadenylic acid	30	8	3.8	11.90	80

[a]The "diameter" corresponds to twice the radius measured from the phosphorus residue to the helix axis.
[b]DNA-W-C provides the measurements given by Watson and Crick (1953a) for comparison.

maerts, 1964; Shih and Fasman, 1971). Apparently the C form is assumed only in concentrated salt solutions and in ethylene glycol (Tunis-Schneider and Maestre, 1970; Green and Mahler, 1971), but it may be the configuration assumed in chromatin and in certain viruses (Yang and Samejima, 1969; Fasman *et al.*, 1970).

Base Pairing. Aside from double-strandedness and configuration, the features proposed by Watson and Crick (1935a) as endowing DNA with its unique biological properties include the restraints that exist in the formation of base pairs. Because the pyrimidines thymine and cytosine always occur in approximately identical proportions as the purines adenine and guanine, respectively, it was suggested that during the formation of the double helix, pairing between bases was limited to combinations of adenine with thymine and to cytosine with guanine. Thus, in this concept, only four combinations could exist, A–T, T–A, C–G, and G–C. Hence, the two strands in the double helix are complementary, for the presence of a given base at a stated site in one chain always requires the complementary base at the corresponding point of the second.

By means of scale models, Watson and Crick showed that the base pairs could fit into the molecular dimensions indicated by X-ray diffraction studies only if they consisted of one pyrimidine and one purine. However, the more precise measurements of Sundaralingam (1969) intimate that this is not probable; other restraints possibly exist to produce the purine–pyrimidine combinations in DNA. Similar restraints do not exist in double-stranded RNA (Chapter 4).

However, the probable arrangement of the hydrogen bonds provided a promising explanation of the limitations in combinations, as shown in Figures 3.2 and 3.3. In Figure 3.2 it can also be noted that the two chains are of opposite polarity, the 5′-end of one being opposed by the 3′-end of the other, the strands thus having an antiparallel alignment. Within the double helix, the various base pairs lie parallel and may be stacked one above the other, much as a set of dinner plates. In these stacks, each base pair is offset to the left by 36° (in the B configuration) between adjacent members to form the right-handed helix and in the usual molecular configurations do not contact one another.

Other Models. From a stereochemical standpoint, it appears possible that DNA may assume a quadruple-helical structure under certain conditions (McGavin *et al.*, 1966). Such four-stranded molecules would have a period of 136 Å, with 20 double pairs of bases per period, and a pitch of 18° (McGavin, 1971a,b, 1973). Hence, the mass per unit length would be identical to that of the standard B-form DNA molecule. The four-stranded form would be distinguishable from the latter by X-ray diffraction studies should well-oriented specimens become available. Some evidence has been presented from certain proliferating sources for the natural occurrence of such four-stranded molecules (Cavalieri and Rosenberg, 1963). Moreover, synthetic three- and four-stranded

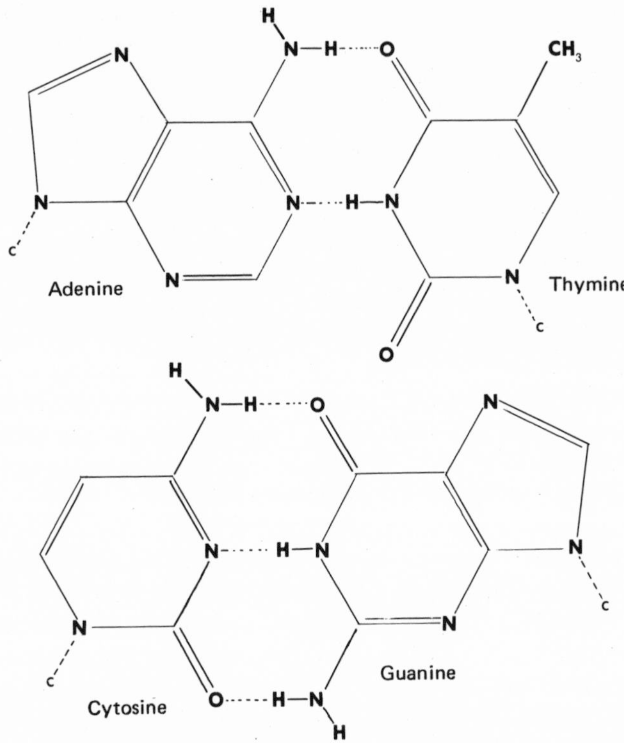

Figure 3.3. Hydrogen bond formation in the standard base pairs. Dotted lines indicate hydrogen bonds; the lowercase *c*s belong to the sugar moiety of the nucleotides.

polynucleic acids have been produced in the laboratory (Zimmerman *et al.*, 1975), and triple-stranded bovine DNA has been synthesized *in vitro* (Perlgut *et al.*, 1975).

Various proposals for hydrogen-bonded base pairs different from those of Watson–Crick have been advanced from time to time and have been the subject of a recent review (Voet and Rich, 1970). Refined X-ray diffraction techniques resulting in 3-Å resolution have permitted measurements of atomic positions within B-form DNA with a precision of a few tenths of an Ångstrom unit (Arnott and Hukins, 1972, 1973). These failed to disclose evidence for the existence of any other than the standard types of base pairs.

Although it was originally conceived (Watson and Crick, 1953a) that the hydrogen bonds between base-pair members were alone responsible for the stability of the helix, some questions have been raised as to the validity of the concept (Sinonağlu, 1968; Lesk, 1974). Instead, it was suggested that the stability is derived from base-stacking forces or apolar bonds; thus the role of the hydrogen bonding between base pairs is thought to be solely that of providing specificity in the recognition of complementary bases. This latter concept has been questioned, too, for recent experiments involving the complementation between adenine and thymine has cast some suspicion on the importance of hydrogen bond patterns in this role (Lezius and Scheit, 1967; Price *et al.*, 1967). After the keto group of the nucleotides had been replaced with sulfur residues, adenine still preferentially combined with thymine, although the number of bonds it could then form was reduced to one. These observations need further consideration, along with the fact that the DNA molecule has more space for freedom of base combination than does double-stranded RNA. That discussion is more pertinent, however, in connection with the transfer RNA molecule (Chapter 7).

Cyclic and Other Variant Forms. Among living things, the DNA molecule appears to occur in a small number of tertiary forms, but only the DNA of viruses, mitochondria, and chloroplasts is a sufficiently small molecule to be extracted intact to provide evidence of its actual form. In contrast, the DNA molecule from the nucleoid of most prokaryotes and that from chromosomes of eukaryotes are too large to be isolated without being degraded, so the actual configuration in most organisms remains undetermined.

Single-stranded DNA found in certain viruses and as a degradation product of double-stranded molecules may be either linear or cyclic (Figure 3.4B,C), and a similar statement is true for double-stranded DNA (Rastogi and Koch, 1974). Cyclic DNA, whether single- or double-stranded, occurs either as irregular circles or supercoiled. In the latter condition, the circular molecule is flattened so that opposite sides lie approximate to one another and the whole is then twisted about itself to form a helix (Figure 3.4A; Helinski and Clewell, 1971). Such superhelical coiling results from a difference between the total number of ordinary helical turns that would be present if the molecule were flattened and that which would have existed if a break in one strand had permitted unwinding. Thus the superhelix is viewed as resulting from stress (Pulleyblank and Morgan, 1975). However, certain polyamines, such as spermine, induce supercoiling, whereas others relax the molecule (Stevens, 1970), so that the "stress" may be the result of enzymatic action, not simple mechanics.

Plasmids, Episomes, and Polydisperse DNA. In all types of organisms, regardless of whether they are eukaryotes or prokaryotes, relatively short molecules of DNA are found that are double-stranded and covalently closed circles. Most frequently in eukaryotes these are confined to such organelles as the mitochondria and chloroplasts, but in bacteria they are free in the cytoplasm. Such extrachromosomal molecules are known as *plasmids,* if they replicate independently of the genome (Clowes, 1972; Meynell, 1973). During the past several years, an increasing number of bacterial properties have been reported whose behavior suggests that their genetic control is not by way of the chromosome, but derives from a plasmid. Among them are included colicin

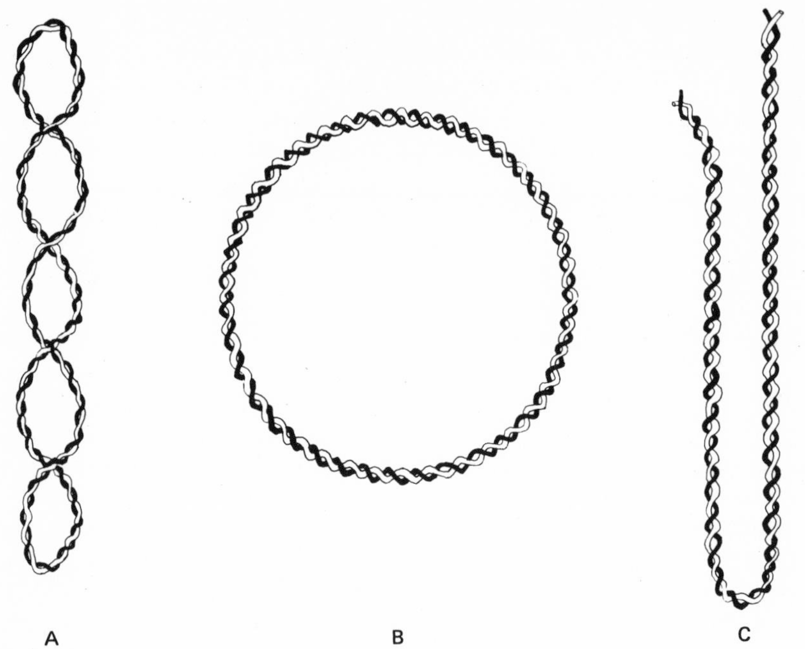

Figure 3.4. Variations in the molecular form of DNA. Smaller linear (C) molecules of DNA often assume a circular form (B), regardless of whether they are double- or single-stranded. Closed circles can become supercoiled (A), usually through enzymatic action.

production and multiple drug resistance in *E. coli,* both of which are well documented as deriving from plasmid activities (Meynell *et al.*, 1968; Novick, 1969). Another, whose molecular properties have recently been described, controls hemolysin in the same organism (Smith and Halls, 1967; Goebel *et al.*, 1974). In this case, actually two plasmids appeared to be involved, one (plasmid B) which had a sedimentation coefficient of 63 S, and a second (plasmid C) which had 55 S, corresponding to molecular weights of 41×10^6 and 32×10^6, respectively.

Although plasmids are usually separate from the nucleoid of bacteria, occasionally they may be joined to it and are then referred to as *episomes*—a term sometimes claimed to be unnecessary (Hayes, 1966). Plasmid replication is currently not clearly understood, with two diametrically opposed points of view being promulgated. One school of thought proposes that their replication is actually dependent on the genome, for two genomic products (*dna*A and *dna*C) have been demonstrated to be involved in plasmid replication (Goebel, 1974). On the opposite side, it has been shown that during replication, the new DNA is in the form of open-circular molecules produced discontinuously (Messing *et al.*, 1974).

In *Bacillus megaterium,* another type of double-stranded, covalently closed circular DNA of unknown function has been reported, which comprises up to 30% of the extractable DNA (Henneberry and Carlton, 1973). This has been named "polydisperse circular DNA" (Carlton and Smith, 1974) and may be present in other bacteria, including *E. coli* and *Salmonella typhosa* (Leavitt *et al.*, 1971).

Single-Stranded Regions in Eukaryotic DNA. At least in developmental stages of sea urchins, the DNA of the chromosomes does not appear to be undeviatingly double-stranded. Quite to the contrary, evidence has been presented that indicated the DNA to consist of double-stranded subunits joined together by means of single-stranded sectors (Case *et al.*, 1974; Case and Baker, 1975). When treated with an endonuclease (nuclease S_1 from *Aspergillus*) that acts only on single-stranded DNA, morula-stage DNA was degraded into molecules that sedimented predominantly at 23 S. These were thought to represent the opposing but similar ends of subunits estimated to sediment at 31 S, two 23 S particles being held together in life by a single-stranded region that was removed by the S_1 nuclease. The presence of other single-stranded gaps too short for attack by the endonuclease was also detected. On the average the double-stranded regions were stated to be about 9700 base pairs in length.

In the DNA of the blastula, the subunit appeared to be smaller, sedimenting at 27 S and consisting of a continuous molecule of duplex DNA about 3300 base pairs long. The gastrula, on the other hand, showed the presence of much longer double-strands of DNA, with only scattered regions being single-stranded. Single-strandedness has not been established in other organisms by these techniques, but the presence of comparable subunit structure in DNA has. In *Physarum polycephalum* and in three species of mammals (mouse, Chinese hamster, and HeLa cells), short strands of DNA of similar length were secured from gently lysed cells (Lett *et al.*, 1970; Brewer, 1972). Because Lett and his co-workers had found similar organization in the DNA of bacteria and plants, they thought that it probably prevailed in all extant organisms.

3.4. REPLICATION OF DNA

3.4.1. The Basic Features

The Cell Cycle. In all known eukaryotes, the DNA molecule is replicated at a definite time in the generation span of the cell. After cell division has occurred, each daughter cell commences to grow by means of protein synthesis, absorption of water, and so on. This first part of the cell cycle (Mitchison, 1971; Balls and Billett, 1973; Mazia, 1974), known as the G_1 phase, is measured from the termination of mitosis until the next stage begins. The second, or S, phase, is characterized by the replication of DNA; it is followed by a second gap (G_2), which lasts until the commencement of mitosis, or M phase, which terminates the cycle (Howard and Pelc, 1953; Stein and Baserga, 1972). Occasionally, the M phase is referred to as the D, that is, divisional stage (Cameron and Nachtwey, 1967). The relative proportions of the several phases vary from species to species; indeed, sometimes one or the other of the growth phases may be shortened to the point of extinction. For instance, in *Amoeba proteus,* the G_1 stage is completely absent (Prescott, 1973), and in yeast, DNA synthesis is greatest during meiosis, that is, the S and M phases run concurrently (Croes, 1966). Other variations are found among endomitotic cells, which do not divide and thus produce polyploidy, and in those which under-replicate and others containing genes that undergo amplification (Nagl, 1974).

In the earlier concepts, protein and RNA syntheses were believed to be totally confined to the two G phases, and DNA formation to the S phase, for it was held as self-evident that the DNA molecule could not be employed for transcription while it was being replicated. More recently, however, these ideas have been found greatly oversimplified. RNA synthesis has been demonstrated to increase gradually from G_1 through S into G_2, where it attains a peak level that continues into mitosis; during mitosis it drops to near the nil level at metaphase to commence an increase once more at telophase (Donnelly and Siskin, 1967; Nešković, 1968; Monesi, 1969).

Protein synthesis follows a similar pattern, increasing from G_1 through S to G_2, but it continues then through mitosis, although at a lower rate during metaphase and anaphase (Scharff and Robbins, 1966). The first studies reported that DNA synthesis ceased in the absence of protein synthesis in various eukaryotes, including *Physarum polycephalum* (Cummins and Rusch, 1966), *Tetrahymena pryiformis* (Hardin *et al.*, 1967), and HeLa cells (Seki and Mueller, 1975). Now it has been made clear that synthesis of protein in prophase is essential also, for in onions the cells failed to enter metaphase in its absence (Garcia-Herdugo *et al.*, 1974). Similar requirements, possibly for one or more specific proteins, have been demonstrated in yeast (Roth, 1973). In regenerating livers of rats, synthesis of specific enzymes related to DNA production, such as thymidine kinase and ribonucleoside diphosphate reductase, occurred at specific intervals of the cell cycle (Hwang *et al.*, 1974). Further details concerning macromolecular syntheses during the cycle are more pertinent to later discussions; those summarized here are sufficient to make the overall pattern clear.

Basic Requirements for Replication. As mentioned before, the unique structure of the DNA molecule is vital to its capacity for being replicated in a highly precise fashion and thereby providing genetic uniformity from one generation to another. Of special importance to this property are two molecular features, the double-helical structure and the complementarity of the base pairs. The ingredients usually deemed essential to the replicatory processes include the four common deoxyribonucleotide triphosphates, various enzymes, Mg^{2+}, and the DNA molecule to be replicated, but the actual value of some of these during the processes *in vivo* has been questioned, as noted below.

With these basic materials and the DNA structure usually postulated, the major steps in replicating the molecule appear few and self-evident. First the hydrogen bonds between members of the base pairs in the original DNA molecule must be broken, so that the two strands may be untwisted and separated from one another, and the separated strands may receive complementary triphosphorylated nucleosides one after another. After a pyrophosphate radical is removed from each nucleotide, the new additions become polymerized by enzymatic action to form the complementary chain. Finally, hydrogen bonds form between the members of each base pair, restoring the original double-helical configuration. As a consequence of these processes, two identical molecules are formed, each a precise replica of the original. Replication is thus semiconservative, because the original molecule is not preserved intact, as it would be were the processes fully conservative. As a result of the semiconservative processes, each new DNA molecule contains one-half of the original molecule; consequently, when cell division has been completed, the DNA complement of each daughter cell likewise consists of one-half old, and one-half new, DNA strands.

Unwinding the Strands. Virtually no information is available as to the mechanism involved in breaking the hydrogen bonds between base pairs, aside from the bonds' being weak and therefore readily broken. A second problem long recognized in the literature is concerned with the need for untwisting to permit the two original strands to separate (Levinthal and Crane, 1956; Longuet-Higgins and Zimm, 1960; Freese and Freese, 1963). In order to separate the two strands completely as required, the body of the molecule must make one turn about its axis for every ten base pairs present. Thus the 1-mm-long DNA molecule of *E. coli,* estimated to contain 3 million base pairs (Davidson, 1972), would need to revolve 300,000 times during the course of replication. At the usual cell cycle period of 20 minutes for this organism, the molecule would turn at the rate of 15,000 revolutions per minute, far faster than most electric motors.

Since the molecule in bacteria is perceived to be circular, the mechanism of unwinding is still further complicated (Crothers, 1969). One solution to the problem was proposed by Cairns (1963a,b), who assumed that duplication always began at one particular point of the helix and advanced in one direction toward the opposing end, where a swivel mechanism was postulated to exist. Because the swivel must serve both strands as a single unit, not separately, its nature is difficult to imagine.

More recently a protein ω has been isolated from bacteriophage T4 and from *E. coli,* that was demonstrated to convert superhelical DNA molecules to a much less twisted, covalently closed form. Hence the enzyme was suggested to serve as a dehelicase (Wang, 1971, 1973). In another virus, SV40 of primates, the nucleoprotein complex itself has been found to relax superhelical DNA (Sen and Levine, 1974). Proteins of a similar nature have now been isolated from eukaryotic sources, including *Drosophila,* embryonic mouse, and *Ustilago maydis* (Champoux and Dulbecco, 1972; Baase and Wang, 1974; Banks and Spanos, 1975). The mouse embryo and *Drosophila* proteins differed from ω of *E. coli* in being able to relax both positive and negative superhelical DNAs and in requiring a monovalent ion rather than Mg^{2+} or other divalent ion in order to function efficiently. The *Ustilago* protein ω, which had a molecular weight of 20,000, was reported to have a number of features suggestive of a structural, rather than an enzymatic, role. For example, of the approximately 3.0×10^5 molecules of protein ω reported to be present, each bound to not more than 7–10 single-stranded nucleotides.

Although electron micrographs and certain autoradiographic studies (Cairns, 1963a) appear to support the concept of swiveling in the bacterial genome, several problems remain. For one, the DNA molecule in those organisms has to be folded many hundreds of times to be contained within the restricted confines of the cell, a characteristic that would interfere drastically with free unwinding. Among eukaryotes, there is a difficulty associated with the unwinding of the DNA molecule which seems to have received only occasional attention in the literature. During the processes of meiosis, homologous chromosomes must align themselves during synapsis on a point-to-point basis; to accomplish this it seems requisite that the α-helix of the molecules become completely straightened, for two helices can contact one another only periodically (Moore, 1974), that is, at every tenth base in the B configuration.

The First Proposed System. *In vitro* synthesis of DNA, as pointed out above, utilizes the triphosphorylated nucleosides as substrates and is able to be catalyzed by several different polymerases. In the earlier model of DNA replication, the formation of the nucleoside triphosphates, rather than the polymerization of the DNA, was usually considered the level at which controls are exercised by the cell on these processes (Davidson, 1963; Blakley and Vitols, 1968). Usually the steps envisioned to be essential included (1) formation of ribonucleoside monophosophates from precursors; (2) modification of the phosphorylated ribonucleosides into deoxyribonucleoside monophosphates; (3) addition of phosphate to these deoxyribonucleoside monophosphates to form the corresponding triphosophates triphosphates means of kinases; and (4) polymerization of the triphosphates to DNA by means of enzymes involving the removal of pyrophosphate. As shown below, the actual *in vivo* processes are proving to be quite different, but this familiar *in vitro* system provides a firm historical basis for initiating discussion. Among the departures from this model that now have been documented is the presence of a number of enzymes, including a multiplicity of DNA polymerases, as shown in the bacterial system below.

3.4.2. The Bacterial System

Bacterial Polymerase I. In bacteria, three DNA polymerases have been found, each of which at first was considered capable of replicating DNA *in vitro* without the others. The first one demonstrated (A. Kornberg, 1957, 1966), referred to as DNA polymerase I or the Kornberg enzyme, is by far the best known and probably the most abundant in the organisms. This appeared to consist of a single polypeptide chain of a molecular weight of 109,000, containing only one sulfhydryl and one disulfide group and two tightly bound molecules of zinc (Setlow, 1974). The sulfhydryl group did not seem to be part of the active site (A. Kornberg, 1969), nor was there any evidence of a prosthetic group. According to A. Kornberg, at least five major sites are present in the active center, four of which are indicated in Figure 3.5. Among those proposed are sites for (1) receiving the template DNA, which is oriented with the polarity shown, (2) holding the growing chain, or so-called primer, which is oppositely oriented, (3) recognizing the 3′-hydroxyl group of the primer's terminal nucleotide, (4) holding the triphosphorylated nucleoside to be added, and (5) conducting 5′→3′ exonuclease activity.

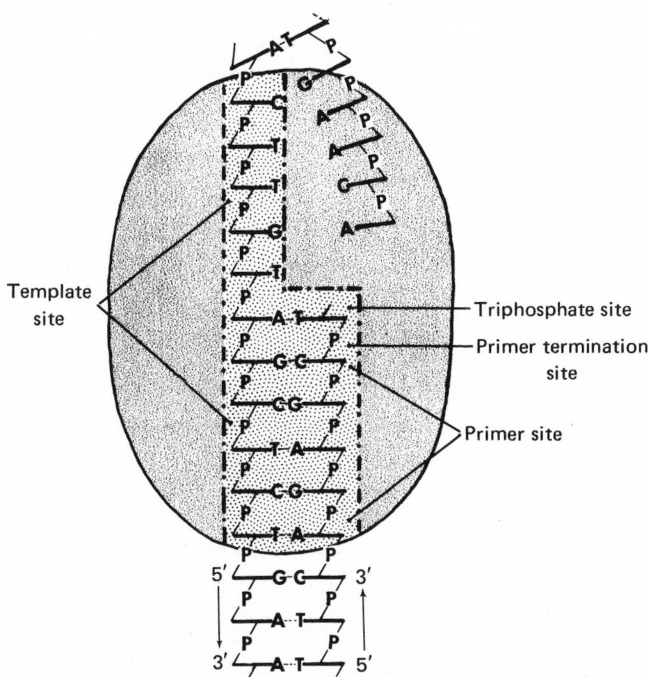

Figure 3.5. Active sites in the reactive region of DNA polymerase I (after Kornberg, 1969).

Two types of exonuclease* activity have been reported for polymerase I. Site 3 has been suggested to serve in $3' \rightarrow 5'$ breakdown of DNA, in addition to its synthetic activities (A. Kornberg, 1969). The enzyme bound double-stranded DNA molecules not only at breaks and ends but was active along the entire length of single-stranded DNA. This is tantamount to stating that the enzyme binds only single-stranded DNA (Setlow, 1974). By enzymatic treatment, the polymerase I molecule has been split into two fragments; the larger of these retained the polymerase and $3' \rightarrow 5'$ exonuclease activities, while the smaller portion showed only $5' \rightarrow 3'$ exonuclease properties. The enzyme also has been shown to have a role in the breakdown of trinucleotides (Deutscher and Kornberg, 1969a), but *in vivo* the enzyme seemed to be required primarily for the rapid conversion of short, newly synthesized strands into moderate-sized lengths (Olivera and Bonhoeffer, 1974; Tait and Smith, 1974).

The supposed $5' \rightarrow 3'$ hydrolytic activities reported by other laboratories (Klett *et al.*, 1968; Deutscher and Kornberg, 1969b) were absent in highly purified preparations from *B. subtilis* but not in those prepared from *E. coli* (Okazaki and Kornberg, 1964; Deutscher and Kornberg, 1969a). As this enzyme replicated only from internal breaks ("nicks") in one strand, A. Kornberg originally proposed that, in this system, polymerization of DNA involved an endonuclease to break one strand, polymerase I to replicate the main chain and the dangling branch of the broken strand, and a ligase to fuse the resulting new side segment to the main primer molecule.

In vitro studies have reported that *E. coli* and bacteriophage T4 polymerase I possessed "proofreading" properties in that mismatched bases at 3'-termini were excised by the exonuclease activity of these enzymes (Englund, 1971; Brutlag and Kornberg, 1972; Hershfield and Nossal, 1972). Only after the errors had been corrected were additional nucleotides added to the DNA molecule. Artificially added ribonucleotides, however, were not removed, so long as the pairing was in the standard Watson–Crick fashion (Kössel and Roychoudhury, 1974). A corrective function for polymerase I *in vivo* has also been demonstrated, for ultraviolet radiation dimers of pyrimidines were excised in wild-type *E. coli* but not in mutants deficient in this enzyme (Glickman, 1974).

According to a current investigation (Hendler *et al.*, 1975), none of the functions described above can be attributed solely to the polymerase. This polymerase, and perhaps the other two discussed below, apparently occurred in the form of a complex which actually carried out the synthesis of DNA. The complex had a molecular weight of 390,000 and consisted of polymerase I, deoxyribonucleoside triphosphates, and DNA deoxynucleotidyltransferase.

* *Exonucleases* are enzymes which degrade a particular nucleic acid stepwise from one end of the molecule, in contrast to endonucleases which attack the molecule at points within the chain.

Hence, the actual *in vivo* functions of all the polymerases need investigating anew, using natural complexes rather than purified enzymes.

The Other Bacterial Polymerases. DNA polymerase II, after its discovery in a polymerase I–deficient mutant of *E. coli,* was found to be attached to the cell membrane (DeLucia and Cairns, 1969; Smith *et al.*, 1970), a point of considerable importance. The purified enzyme had a molecular weight between 60,000 and 90,000 (Knippers, 1970) or 130,000 (Moses, 1974). *In vitro* it synthesized DNA in the $5' \rightarrow 3'$ direction in the presence of all four deoxyribonucleoside triphosphates, Mg^{2+}, and NH_4^{2+}, and DNA as template. Like polymerase I, this enzyme carried out synthesis at only a fraction of the rate found in living organisms, and thus far no nuclease activity has been reported for it. Also as in polymerase I, a template treated with an endonuclease, or such naturally interrupted DNA molecules as salmon sperm DNA, was required for synthesis (Moses, 1974). In addition, polymerase II specifically interacted with *E. coli* dehelicase *in vitro* (Ganesan *et al.*, 1973).

Still more recently, another type of DNA polymerase was found in a polymerase I–deficient *E. coli* mutant (T. Kornberg and Gefter, 1971). This DNA polymerase III, as it is called, is more sensitive to heat than polymerase II, and, unlike the latter, is quite readily inhibited by salts (Gefter *et al.*, 1971; Nüsslein *et al.*, 1971). The enzyme from *E. coli* has been reported to have a molecular weight of 180,000 and to consist of two components, one of molecular weight 140,000, the other of 40,000 (Livingston *et al.*, 1975). Further, it was shown to have two exonuclease activities, so that it can attack either terminus of single-stranded DNA (Livingston and Richardson, 1975). Insofar as they are now known, the other biological properties of this enzyme are enumerated later.

All three polymerases appear to be widespread among bacteria, for corresponding enzymes have been described from *Bacillus subtilis* and *Acinetobacter calcoacetinus* (T. Kornberg and Gefter, 1974; Ness and Kleppe, 1974; Low *et al.*, 1976). The DNA polymerase of bacteriophage T4 is quite distinct in that it requires single-stranded DNA, being inactive even on nicked double-stranded DNA (Lehman, 1974a).

Chain Initiation in Prokaryotes and Viruses. The current approach to the total problem of DNA replication is that of "dividing and conquering," that is, viewing particular aspects of replication individually, much as protein synthesis is (Chapter 4). Obviously, the first step in replication involves recognition by the polymerase of the site on the DNA molecule where the processes may be begun. Some light has now been thrown on the nature of *initiation,* but identification of the actual sites remains for the base sequencing of DNA to be accomplished.

A soluble enzyme has been extracted from uninfected *E. coli,* which catalyzed the conversion of the single-stranded DNA from bacteriophages M13 and ϕX174 to the double-stranded replicative form. The systems in the two bacteriophages differed somewhat from one another, but in both instances DNA synthesis was initiated by priming with RNA (Schekman *et al.*, 1973). Similar requirements have been shown to exist in *E. coli* (Helinstetter, 1974), but concurrent protein synthesis was an additional requirement in this species. The unexpected demand for RNA was demonstrated by DNA synthesis's becoming blocked when RNA synthesis was inhibited, by the need for *ribo*nucleoside triphosphates, and by the presence of phosphodiester bonds linking RNA to the replicative-form DNA molecule. Part of the RNA used as primer remained complexed with the DNA, probably by means of hydrogen bonds (Champoux and McConaughy, 1975). In addition to RNA, the formation of the double-stranded DNA of M13 required an RNA polymerase, a DNA polymerase, a dehelicase, and several unknown factors. Synthesis of the ϕX174 double-stranded DNA called for spermidine over and above those factors already listed, but this substance, like other polyamines, is now known to serve in unwinding superhelical molecules (Stevens, 1970).

Chain Elongation in Prokaryotes and Viruses. A thorough reanalysis of the roles of the three bacterial DNA polymerases *in vivo* has added extensively to an understanding of chain elongation. In *E. coli* cells, synthesis of DNA was reported to occur in three stages (Tait and Smith, 1974), and similar discontinuous DNA synthesis has been noted in *Bacillus subtilis* (Wang and Sternglanz, 1974; Massie *et al.,* 1975). The first pulse of DNA replication resulted in short strands (called Okazaki fragments) that sedimented at 20 S, the second round of strand elongation produced chains that sedimented between 70 S and 120 S, and the third round completed the long strands. Polymerase III proved to be the essential catalyst for the second stage of chain elongation, since strands longer than 20 S were produced only very slowly in its absence. It also was active in stage 3 of synthesis, but not essential. Similarly, polymerase I was shown to be active in stages 2 and 3, and greatly increased the rate of elongation; however, it was not essential for either process. Polymerase II served only in the absence of the other two enzymes and was not requisite at any stage. Some evidence indicated that the shortest strands resulted from replication in the 3′ to 5′ direction, and the longest ones from polymerization in the opposite sense (Louarn and Bird, 1974).

In addition to their roles in elongation of strands initiated by other enzymes, repair activities for the three polymerases have been demonstrated by a number of other investigations (Grossman *et al.*, 1973; Klenow and Overgaard-Hansen, 1973; Richardson *et al.,* 1973). The fact that the three polymerases act only on broken DNA, for example, even *in vitro,* is also suggestive of their involvement in repair synthesis (Werner, 1971b). However, other ligases are also present, one in *E. coli* having a molecular weight of 74,000 (Lehman, 1974b). This ligase catalyzed the covalent attachment of polydeoxyribonucleotides at the 5′-termini and of polyribonucleotides at the 3′-termini (Nath and Hurwitz, 1974).

Several viral enzymes active in DNA synthesis have recently been extracted and purified to homogeneity. These were produced by the double-stranded DNA bacteriophage T4, which infects *E. coli,* and showed direct involvement in DNA replication (Aposhian and Kornberg, 1962; Goulian *et al.*, 1968; Alberts and Frey, 1970; Delius *et al.*, 1972; Barry *et al.*, 1973). The product of T4 gene 43 has been shown to be a DNA polymerase (DeWaard *et al.*, 1965; Warner and Barnes, 1966a,b), while that coded by gene 32 has proven to be a DNA-unwinding enzyme (Alberts *et al.*, 1968). These two could combine in the form of a complex which increased the *in vitro* activity of a polymerase system by tenfold (Huberman *et al.*, 1971). Another complex, formed by the proteins specified by genes 44 and 62, together with an enzyme coded by gene 45, were active in DNA synthesis when purified, but their specific roles were not determined (Barry *et al.*, 1973).

Terminal cross-linking of DNA to produce hairpin loops has been described in a DNA polymerase system of bacteriophage T7-infected *E. coli*. To be active, three enzymes were required, including T7 DNA ligase (the product of gene 1.3), T7 DNA polymerase (coded by gene 5), and T7 exonuclease (specified by gene 6). In addition Mg^{2+} ions, the four deoxytriphosphates, and ATP (Sadowski *et al.*, 1974) also had to be present. This system was active on viral template DNA which possessed termini but not on that with nonunique (repetitive) ends.

Minicircle DNA Replication. In *E. coli* a relatively small cyclic molecule of DNA exists in addition to the very large one of similar form. This *minicircle DNA,* with a molecular weight of 1.5×10^6, contains 15 copies of plasmid DNA (Messing *et al.*, 1974), but otherwise its biological functions remain unknown. It exists in the organisms in two forms: form I is a supercoiled, double-stranded, cyclic molecule, whereas form II is an open cyclic molecule containing a discontinuity (nick) in one of the strands induced by action of an endonuclease associated with the plasmid. During replication, form II resulted from agregation of newly synthesized minicircular DNA and ultimately gave rise to form I molecules.

Similar minicircle DNA has been detected and isolated in such eukaryotes as *Leishmania tarentolae* (Salser *et al.*, 1973). To determine its base sequence to some extent, rGTP was substituted for dGTP during replication, so that treatment with alkali would cleave the product at guanine residues (Salser *et al.*, 1972). The data obtained appeared to rule out the existence of more than one type of minicircle DNA in this organism.

Present Status of Bacterial Replication. Explorations into the replication of the bacterial DNA molecule have been hampered by the fragile nature of the template and the interplay among a number of enzymes during the processes. More recently, sufficient advancement has been made, largely through methods permitting folded genomes to be obtained intact, that a summary of the current status of the problem can be presented. At least in *E. coli* and related forms, the following are the salient features of the bacterial system as understood now:

1. The folded, superhelical, intact DNA becomes attached to the cell membrane (T. Kornberg *et al.*, 1974).
2. Single-stranded regions must be present for initiation of polymerization (Roychoudhury, 1973; Dingman, 1974; Dingman *et al.*, 1974; Hamilton and Wu, 1974).
3. Divalent ions, such as Mn^{2+}, but especially Mg^{2+}, are required for maximum efficiency.
4. A number of enzymes are requisite for DNA synthesis, including RNA polymerase, DNA polymerase III (polymerase I substitutes to an extent), dehelicase (including spermidine), and two additional proteins, referred to as factors I and II or as copolymerase III (Hurwitz and Wickner, 1974; Schekman *et al.*, 1974).
5. For substrate, all four triphosphorylated deoxyribonucleosides, as well as the four similar ribonucleosides, are required (Hurwitz and Wickner, 1974).
6. Replication is bidirectional (Masters and Broda, 1971; Dingman, 1974; Sternglanz *et al.*, 1976).
7. Replicates are in the form of short chains, that is, the replicatory processes are discontinuous (Okazaki *et al.*, 1968a,b; Yudelevich *et al.*, 1968); the polymerase responsible for their formation remains unidentified.
8. Upon completion, these short, nascent replication fragments are joined together in two stages under the influence of polymerase III (Tait and Smith, 1974), but polymerase I may substitute under certain conditions. DNA ligase also seems to be required (T. Kornberg *et al.*, 1974).
9. A factor is present that blocks the synthesis of covalently linked, complementary sequences, such as those terminal cross-linked sites characteristic of several viruses (Barzilai and Thomas, 1970; Ihler and Kawai, 1971; Coulter *et al.*, 1974; Sadowski *et al.*, 1974).

3.4.3. The Eukaryotic System

Major Features. In the eukaryotes as a whole, replication of DNA occurs, in semiconservative fashion, along series of replication units arranged in tandem in the chromosomes (Huberman and Riggs, 1968; Newlon *et al.*, 1974). Because initiation occurs at the center of each such replication unit, known as a replicon, chain growth takes place bidirectionally toward the termini (Callan, 1973). After replication has attained the ends of a replicon, the newly synthesized chain fuses with those formed on the two adjacent units, so that at the close of synthesis, long continuous DNA molecules are the end products. According to a number of investigators, protein synthesis must occur concurrently in order to maintain the normal rate of DNA replication (Littlefield and Jacobs, 1965; Hand and Tamm, 1973; Weintraub and Holtzer, 1972; Gautschi and Kern, 1973). Others have found conflicting evidence, however (Young *et al.*, 1969; Fujiwara, 1972). The picture that appears to be emerging is that protein synthesis strongly promotes initiation, almost to the point of essentiality, whereas it is less effective on the other stages of DNA replication (Mueller *et al.*, 1962; Cummins and Rusch, 1966; Kim *et al.*, 1968). This has been demonstrated especially clearly in yeast (Hereford and Hartwell, 1973).

Eukaryotic DNA Polymerases. A number of DNA polymerases have been isolated from a wide diversity of eukaryotic sources; however, results are inconsistent thus far, for differing numbers of enzymes have been reported by various workers. In part the confusion results from the presence in the cytoplasm of polymerase activities that are associated with mitochondria and ribosomes (Kalf and Ch'ih, 1968; Baril *et al.,* 1973). At present, the consensus regarding polymerase activities appears to be as follows:

At least one, and possibly several, DNA polymerases are found in the cytoplasm, apart from those of the mitochondrion and other organelles. Often the major species is referred to as DNA polymerase C or C1, but the terminology has not as yet become stabilized. In the rat, polymerase C has been reported to have a molecular weight of around 100,000 and to display a low level of exonuclease, as well as strong polymerizing, activity (Poulson *et al.*, 1974; Tsuruo and Ukita, 1974). Further, it has been shown to be associated with the plasma membrane and to play an active role in the processes of chromosome replication, with its specific activity reaching its maximum during the S phase in rapidly growing tissues and in synchronized cells (Baril and Laszlo, 1971; Baril *et al.*, 1971; Chang and Bollum, 1972c; Chang *et al.*, 1973). Recently this polymerase has been resolved into three fractions, C1 and C2 having similar molecular weights (Hachman and Lezius, 1975). The third one, C3, was believed to represent a fragment of one of the others, while C1 was nearly twice as active in polymerizing on double-stranded poly(A-T) as C2. Three fractions have also been described from calf thymus (Yoshida *et al.*, 1974).

The two polymerases from the nuclei of rat tissue differed widely from one another in molecular weight; P-1 (or N-1) had a weight similar to that of the large cytoplasmic enzyme (C1), that is, close to 100,000, whereas P-2 (or N-2) had a molecular weight of 39,000 (Poulson *et al.*, 1974). Neither of these proteins demonstrated any exonuclease activity (Tsuruo and Ukita, 1974). Although only one polymerase was found in the yeast nucleus, the smaller species being wanting (E. Wintersberger, 1974; U. Wintersberger, 1974), the rat system may prove to be characteristic of the higher metazoans in general. It certainly seems to be universal at least among vertebrates, for similar enzymes have been reported for human cells (Srivastava and Minowada, 1972; Srivastava, 1974), calf thymus (Momparler *et al.*, 1973; Bekkaring-Kuylaars and Campagnari, 1974), and *Xenopus laevis* (Grippo *et al.*, 1975). Both polymerases of the nucleus, as well as those of the cytoplasm, may exist in the cell in the form of complexes, rather than free (Shiosaka *et al.*, 1974). Both also varied in abundance and activity during the cell cycle in avian erythroid cells (Wang and Popenoe, 1977). Invertebrate systems may prove somewhat different, for the single polymerase reported from sea urchin had a molecular weight of 150,000 (Fausler and Loeb, 1974).

On the basis of antigen–antibody reactions, all these polymerases appeared to be closely allied structurally among the various groups of eukaryotes. This commonness of morphology was found in mitochondrial DNA synthetases, too, whereas little affinity was shown with polymerases from prokaryote sources (Chang and Bollum, 1972a,b). Similarities of structure of mitochondrial and nuclear polymerases have also been indicated in the ciliate *Tetrahymena pyriformis,* for electron microscope autoradiography of cells labeled with [^{3}H]thymidine showed that the mitochondrion and peripheral nucleoli of the macronucleus were labeled more quickly than the chromatin granules of the macronucleus (Engberg *et al.*, 1974).

Chain Initiation in Eukaryotes. Evidence is accumulating which shows that DNA replication is at least initiated in association with the nuclear membrane (Comings and Kakefuda, 1968; Williams and Ocky, 1970; Mizuno *et al.*, 1971; O'Brien *et al.*, 1972) recalling replication in bacteria in which the cell membrane plays a similar role. In viral-infected animal cells, the viral DNA is similarly associated with the nuclear membrane; this has been demonstrated in adenovirus 2, 5, and 12, and KB and HeLa cells (Pearson and Hanawalt, 1971; van der Vleiet and Sussenbach, 1972; Wallace and Kates, 1972; Petterson, 1973; Yamashitsa and Shimojo, 1973). Actually the association of the DNA with the nuclear membrane is of a complicated nature, as shown by several recent investigations. In the first place, association was shown to be dependent on a protein M, which is produced in the mid-G_1 phase of mammalian cells, about 2 to 4 hours prior to S phase (Yamada and Hanaoka, 1973). Furthermore, in HeLa cells, the data suggested that the complexes of DNA and nuclear membrane (plus protein ?) formed during the first 10 minutes of the S phase remained throughout the cell cycle. The numerous species produced later in the S phase were of a more transitory nature (Cabradilla and Toliver, 1975).

The presence of various substances aside from the polymerases is necessary for initiation. In *Saccharomyces* four proteins have been shown to act in the following sequence in inducing initiation: the α-factor, then the product of gene *cdc 28,* followed by that of *cdc 4,* and finally that of *cdc 7* (Hereford and Hartwell, 1974). As in bacteria, RNA synthesis is also proving a requirement for DNA replication. Covalently linked RNA–DNA complexes have been described for a number of eukaryotes, including the slime mold, *Physarum* (Waqar and Huberman, 1973), polyoma-infected mouse fibroblast nuclei (Magnusson *et al.*, 1973), Ehrlich ascites cells (Sato *et al.*, 1972), HeLa cells (Neubort and Bases, 1974), and Burkitt cells (Srivastava and Bardos, 1973). Moreover, comparable complexes have been produced *in vitro* by DNA polymerases from chick embryo, human leukemia, and KB cells (Keller, 1972; Barker *et al.*, 1973). In initiating DNA replication, the RNA seemed to play a role as primer for the new strands (Scheckman *et al.*, 1972; Westergaard *et al.*, 1973). The evidence for the presence of such complexes in mammalian cells is as yet not fully convincing (Pearson *et al.*, 1976).

Discontinuous Replication in Eukaryotes. Discontinuous replication appears to be a universal feature of eukaryotic, as well as prokaryotic, DNA replication. For example, in five species of mammals investigated for initial events of DNA replication, short segments (Okazaki fragments) were found in each case (Hand and Tamm, 1974; Gautschi and Clarkson, 1975); however, the fragments differed in length from species to species. In the ox and man, the shortest mean lengths were observed, with values of 17 and 23 nm, respectively, while the rhesus and mouse had the longest segments, 42 and 45 nm, in that order. Some clustering of the active sites were observed in the mouse. These short fragments, which are viewed as intermediate products as in the bacteria (Burgoyne, 1972; Taylor, 1974), have been reported also from HeLa cells (Friedman, 1974), and from calf, human, and rat DNAs (Philippsen *et al.*, 1974), as well as from mitochondrial DNA (Wolstenholme *et al.*, 1973; Koike and Wolstenholme, 1974).

The replicating chromosomal DNA in *Drosophila melanogaster* cleavage nuclei has been seen under the electron microscope as a serial array of close-set replicated regions, created by pairs of diverging replication forks (Kriegstein and Hogness, 1974). In structure, these were stated to resemble the bidirectional replicating forks of bacteriophage DNAs, except that the single-stranded gaps had a length of less than 200 nucleotide residues and thus corresponded to the shorter Okazaki fragments of this organism. These short segments then were found to be joined into higher-molecular-weight fragments (Kowalski and Cheevers, 1976).

Not only metazoans, but the higher plants and fungi also, have been shown to replicate DNA in discontinuous fashion. Short-chain intermediates during DNA polymerization were demonstrated in *Physarum polycephalum* (Brewer, 1972) and *Vicia* embryos. In the latter, the rapidly labeled short strands had sedimentation coefficients of 10 S and 14 S which increased to 19 S and 22 S with time (Nagle, 1975; Sakamaki *et al.*, 1975).

Cytoplasmic DNA Replication in Eukaryotes. Among the unexpected results of explorations into DNA replication in eukaryotes was the discovery of DNA polymerase in the cytoplasm of several mammalian tissues, for it had been generally held that DNA replication and repair occurred solely in the nucleus, to which the chromatin is confined. The earliest finds of DNA polymerase in the cytoplasm were explained as deriving from contamination during cell fragmentation (Keir *et al.*, 1962; Keir, 1965) or as newly synthesized enzymes, not yet transported to the nucleus (Baril and Laszlo, 1971). But now DNA polymerase has been isolated from various mammalian tissues and from cultured cells (Lindsay *et al.*, 1970; Baril *et al.*, 1971; Wallace *et al.*, 1971; Weisbach *et al.*, 1971; Chang and Bollum, 1972a,b; Sedgwick *et al.*, 1972; Byrnes *et al.*, 1974a), including such anucleate cells as rabbit reticulocytes (Byrnes *et al.*, 1974b). The latter study also showed virtual identity of the cytoplasmic enzyme with that extracted from the nuclei of rabbit bone marrow erythroblasts, the precursor cells of reticulocytes. Similar close relationships were demonstrated in the DNA polymerase of the nucleus and cytoplasm of rat intestinal mucosa cells (Poulson *et al.*, 1974).

The occurrence of polymerization of DNA in the cytoplasm is not confined to the metazoans but has also been found in *Tetrahymena pyriformis* and in germinating wheat embryos (Andersen, 1974; Buchowicz, 1974). Through use of labeling techniques, it has been shown that in wheat seeds germinated for 6 hours, specific radioactivity of DNA in the cytoplasm was 20 times greater than that in the nucleus. As germination progressed, the radioactivity of the cytoplasmic DNA decreased and that of the nucleus increased. Thus it was proposed that the cytoplasm actively secreted a precursor DNA which accumulated in the nucleus. Chang and Bollum (1972b,c) had also observed a similar rapid activation of cytoplasmic DNA synthesis.

Cytoplasmic (Informosomal?) DNA. A little-known but highly controversial aspect of the eukaryotic genetic mechanism is the occurrence of DNA in the cytoplasm proper, aside from that contained in the organelles. Often this material had been asserted to be an artifact produced by degradation of nuclei during the preparatory processes, but other workers have presented evidence that the DNA is actually from the cytoplasmic fraction and have proposed that it functions in providing information. Thus it has been called "informosomal DNA" and "communication DNA" (cDNA) (Koch and Götz, 1972; Koch, 1973). In this role it is conjectured to represent a type produced by the multiple replication of certain genes (gene amplification) and transported thence to the cytoplasm, where it might subsequently be employed in protein synthesis (Koch and von Pfeil, 1972). In chick cells, the DNA has a molecular weight of 3×10^5 and in human cells approximately 3×10^6. A possibility exists that there may be a degree of relationship between this type of DNA and the cytoplasmic replication of DNA discussed above.

If the DNA recently described from the cytoplasm of yeast is representative of those from other sources, such a relationship must be remote indeed. In this organism, a type referred to as omicron DNA has been shown to consist of covalently closed circular molecules, which are confined strictly to the cytoplasm (Clark-Walker and Miklos, 1974). This *o* DNA, also reported as minicircle DNA from *Leishmania* (Salser *et al.*, 1973), was concluded not to represent a gene amplification product and therefore could not be of an informational function.

Satellite DNA. In many, or perhaps even most, eukaryotes, the DNA contains serially repeated sequences of nucleosides which vary both in length and complexity between species. Since some such sequences differ from the remaining DNA in G + C content, they appear as "satellites" when the DNA is banded on a cesium chloride density gradient or by other methods (Kit, 1961). At least a portion of these satellite molecules are located in the centromeric heterochromatin to be discussed below, as has been demonstrated by *in situ* RNA–DNA hybridization studies on mouse and *Drosophila* (Jones, 1970; Botchan *et al.*, 1971; Gall and Atherton, 1974).

Some progress has been made toward understanding the structural nature of satellite DNA, particularly in several rodents and in *Drosophila virilis.* In the guinea pig, satellite α has been shown to contain a large amount (30%) of a repeating hexanucleotide (Southern, 1970); moreover, the bulk of the DNA in this satellite seemed related to that basic unit by simple mutation and rearrangements. Similar conditions seemed to prevail in the two other rodent species studied, the house mouse and kangaroo rat, and in calf thymus (Walker, 1971; Fry *et al.*, 1973; Votavová and Šponar, 1974, 1975a,b). This supposition received emphasis by the report of a repeating heptamer which comprised 93% of a satellite DNA in *Drosophila virilis* (Gall and Atherton, 1974). Data from bovine kidney cell satellite I DNA also supported the conclusion but on a much larger scale, for it has been demonstrated to consist of repetitive units 1400 base pairs in length (Botchan, 1974; Mowbray *et al.*, 1975). Many satellite DNAs from various crustaceans also consist to a large degree of repeated units (Sueoka, 1961; Skinner *et al.*, 1970). More recently, when satellite DNA I and II from hermit crab were further analyzed, an unusual condition was noted (Skinner and Beattie, 1973; Skinner *et al.*, 1974). In satellite I, one strand contained no cytidine, the other no guanosine, the repeated units in the first case being $(\text{T–A–G–G})_n$, in the second $(\text{A–T–C–C})_n$. The second satellite also had disparate amounts of the two nucleosides, but to a lesser degree. In plants much variation in the quantity of satellite present has been described; among nine species of *Brassica,* for example, it ranged from nil in *B. oleracea* to 37% of total DNA in *B. nigra* (Berdize, 1975). *Cucumis melo* satellite DNA was shown to contain at least two components, one of which resembled that of the mouse in complexity, while the second was 6000 times as complex (Sinclair *et al.*, 1975). A satellite from the monocotyledonous genus *Cymbidium* was quite unusual for a plant source, in that it contained a major (A + T)-rich fraction (Capesius *et al.*, 1975).

As to the function of the satellite DNA, very little has been established with certainty. It presently appears most probable that satellites influence folding of the chromosome or are concerned with similar "housekeeping" activities (Walker, 1971; Votavová and Šponar, 1974). This idea is supported by the demonstration of satellite segments being localized in or adjoining the centromere of chromosomes (Jones, 1970; MacGregor and Kezer, 1971).

3.4.4. Common Features of DNA Replication

A number of features of DNA replication are either shared by both prokaryotes and eukaryotes or are too poorly explored as yet to warrant separate discussions. In view of the similarities being discovered in the major features of these processes, perhaps most of the remainder will eventually be found similar in both categories of organisms.

Precursors of DNA. During the replication of DNA in living *E. coli*, a different type of precursor appears to be utilized than during DNA repair, the former using labeled thymine, the latter thymidine (Werner, 1971a,b). Others have analyzed Werner's data mathematically and concluded that thymidine diphosphate was more likely than the triphosphate to be the direct raw material for DNA (Rubinow and Yen, 1972), but Friedland (1973), using human lymphoblast cells reached the opposite conclusion. Working with the slime mold *Physarum,* Bersier and Braun (1974) have supported the latter deduction, because they found the pools of all four deoxyribonucleoside triphosphates higher before and at the beginning of the S phase than in the middle of the G_2 phase. The pools of dATP and dCTP diminished by a factor of 5 during the S and early G_2 phases. Thus the earlier assumption that the deoxyribonucleoside triphosphates were involved as DNA precursors is substantiated. Two major patterns appear to be followed. Mammalian and fungal cells undergo a decrease in the level of macromolecular syntheses, respiration, and the ATP pool during cell division, whereas bacteria, *Schizosaccharomyces,* sea urchin eggs, and *Tetrahymena* maintain all these parameters fairly constant (Chin and Bernstein, 1968; Huzyk and Clark, 1971; Mitchinson, 1971; Nexø, 1975).

The extremely small size of the precursor pools has also been employed to provide clues to the nature of the replicating processes. Each deoxyribonucleoside triphosphate is present in *E. coli* only to the extent of 120,000 molecules per cell (Neuhard, 1966). In DNA synthesis, the recognition time of the nucleotides by the polymerase has been calculated as two orders of magnitude longer than in RNA synthesis (Maniloff, 1969). Hence it was suggested, in order to allow time for the sparse population of nucleotides to become associated with the DNA template, that the strands were separated for some distance ahead of the site of polymerization, by perhaps as much as 800 nucleotides (Werner, 1971b). Thus the polymerase would merely connect the already aligned nucleotides, and would not need to recognize, select, and place the correct complementary base at each step of the growing duplex helix.

Replication without Template. The DNA polymerase I of *E. coli* displays other interesting capabilities *in vitro*. If the enzyme is incubated in the presence of only deoxyriboadenosine triphosphate (dATP), deoxyribothymidine triphosphate (dTTP), and Mg^{2+}, in the absence of template DNA, a polymer of adenylic and thymidylic acids is formed (Schachman *et al.*, 1960). Polymerization takes place only after a lag period of several hours, but once initiated it proceeds rapidly until up to 80% of the two triphosphates has been incorporated. In the presence of the same two triphosphates and enzyme, the polymer thus formed can serve as template for the synthesis of an identical copy. The polymer has been shown to consist of alternating residues of adenosine and thymidine, ApTpApTpA . . . and is accordingly known as poly d(A-T).

When the experiment was replicated, but with cytidine and guanidine triphosphates instead, a polymer was produced after a similar lag period (A. Kornberg, 1961). However, instead of the polymer's being comprised of alternating bases, one entire strand of the duplex molecule consisted of guanidylic acid, the other of cytidylic acid residues. The two chains often differed in length, for the amount of cytosine incorporated was not consistently equal to the total guanine, nor were dGTP and dCTP necessarily found to the same extent in the product. Bromouracil and other base analogs have also been polymerized in comparable fashion (A. Kornberg, 1961), but the biological significance of these products is not evident.

Reverse Transcription. Another capability of polymerase I went virtually unnoticed for nearly a decade. Cavalieri (1963; Cavalieri and Carroll, 1970) had shown that the enzyme from *E. coli* could use an RNA template, such as poly r(A-U), to synthesize a strand of DNA poly d(T-A), but the applicability of the observation to living systems was not apparent immediately. Later, however, great excitement was aroused in scientific circles when two laboratories independently discovered that certain viruses contained enzymes which synthesized DNA from an RNA template (Baltimore, 1970; Temin and Mizutani, 1970; Viola, 1973; Wu and Gallo, 1975). The enzyme does not use ribonucleoside triphosphate as substrate, and activity is lost upon incubation with RNase. Apparently, the enzyme uses the single-stranded RNA of the viruses as a template and deoxyribonucleoside triphosphates as substrate for the synthesis of a single strand of DNA (Tavitian *et al.*, 1974). Thus the original product of this activity is a hybrid double strand, with a strand of DNA coupled to one of RNA, for hybridization experiments have shown that the synthesized DNA is complementary to the RNA. A DNA-dependent DNA polymerase is also present which later acts on the single-stranded DNA after its release from the RNA and converts it into a duplex molecule; this double-helical DNA can then be replicated repeatedly by ordinary DNA replicating systems. Apparently the enzyme requires poly(A) strands to be present on the RNA template (Getz *et al.*, 1974); histone mRNA, which lacks a poly(A) strand, is not normally transcribed by reverse transcriptase but can be when such strands are artificially added (Thrall *et al.*, 1974).

Initially only RNA viruses were believed to contain this RNA-dependent DNA polymerase, or *reverse transcriptase,* as it is sometimes called. The original sources were Rauscher mouse leukemia virus and Rous sarcoma virus, but a number of others in the same category were soon added to the list (Spiegelman *et al.*, 1970a,b; Stone *et al.*, 1971; Bishop *et al.*, 1973; Gallo *et al.*, 1973). In avian myeloblastosis virus, a protein has been isolated which stimulates the RNA-primed DNA synthesis; as far as could be determined, its action was specific for the species (Leis *et al.*, 1973). The reverse transcriptase of the virus has been shown to be a zinc metalloenzyme (Auld *et al.*, 1974) and loses its activity upon removal of the zinc (Poiesz *et al.*, 1974).

Then enzymes of similar capabilities were reported in mammary carcinoma and in human tumor cells from which RNA viruses were absent (Scolnick *et al.*, 1971). Now comparable enzymatic activity has been detected in normal mouse cells, and in both normal fetal and adult human and monkey cells, and appears to be widespread and of broad capabilities (Spiegelman *et al.*, 1971; Mayer *et al.*, 1974). More recently several RNA-dependent DNA polymerases have been isolated from such varied sources as normal rat liver, hepatomas, and cultured mouse fibroblasts; these enzymes responded better to synthetic double-stranded RNAs than to natural single-stranded ones (Ward *et al.*, 1972; Fry and Weissbach, 1973) and failed to respond to DNA templates. More importantly, another recently discovered RNA-dependent DNA synthetase may shed light on the immune response. Immune ribonucleic acid preparations were extracted from spleens of rabbits which had been immunized against infection with *Salmonella enteritidis*. Serial passive transfers with this extract successfully immunized the treated animals against the same organism, suggesting that the immune RNA might be able to be replicated in the recipient cells. A reverse transcriptase activity was proposed to explain the results (Kurashige and Mitsuhashi, 1973).

Polymerase I as a Reverse Transcriptase. Although, as cited above, *E. coli* polymerase was the first enzyme demonstrated capable of reverse transcription, it was not further investigated until the viral systems had been explored. During the past several years, however, progress has been rapid. Purified enzyme from *E. coli,* which is extremely stable, has been employed to transcribe DNA from such various natural templates as ribosomal RNA (28 S) from *Drosophila* and tobacco mosaic virus RNA (Loeb *et al.*, 1973; Gulati *et al.*, 1974). The four deoxyribonucleoside triphosphates were required as substrate, and the presence of Mg^{2+} ions was essential; activity peaked at a concentration of 6 mM of enzyme.

As in the virus polymerase, the bacterial protein also has been reported to require enzyme-bound zinc (Travaglini and Loeb, 1974). These same workers reported that, in *E. coli,* the RNA-dependent DNA transcriptive properties are intrinsic to the enzyme. Most probably the polymerase has a double-stranded RNA endonuclease activity which produced 3′-OH termini, thus enabling it to create additional initiation sites on RNA strands containing double-stranded regions.

Repair Synthesis. When DNA is damaged by radiation or by chemicals, an enzyme system is present, according to several studies (Strauss, 1968; Setlow, 1970; Hamilton *et al.*, 1974), which literally removes the damaged area and repairs the affected strand by replacing nucleotides to restore its original continuity. In several types of bacteria, specific enzymes capable of exercising the excision function on ultraviolet-radiation-damaged DNA have been isolated, including a UV-endonuclease and a UV-exonuclease from *Micrococcus luteus* and a UV-endonuclease from bacteriophage T4 (Friedberg and King, 1969; Kaplan *et al.*, 1969). In this same organism and also in *E. coli,* polymerase I has been reported to carry out this function when the DNA strand was broken by DNase I (Kelly *et al.*, 1969; Hamilton *et al.*, 1974). Although the same enzyme carried out repair in *B. subtilis,* a protein factor was detected which enhanced the priming activity of γ-ray-damaged DNA (Noguti and Kada, 1975).

When mutants deficient in various polymerases were studied, results suggested that *in vivo* either DNA polymerase I or III was requisite for complete repair of ultraviolet-damaged DNA and that polymerase II, in the absence of the others, mediated only a limited extent of repair (Tait *et al.*, 1974). However, polymerase III appeared to be the most active enzyme of the three in this function (Tomilin and Svetlova, 1974). One type of damage induced by γ-irradiation was degraded by endonuclease II in *E. coli* (Kirtikar *et al.*, 1975).

Similar repair systems have been demonstrated in a number of eukaryotes, including *Tetrahymena,* mouse, man, and seed plants (Brunk and Hanawalt, 1967; Lange, 1974; Ledoux *et al.*, 1974; Lieberman and Poirier, 1974a,b). In the mouse and human genomes, as in HeLa cells damaged by ultraviolet light, repair synthesis has been reported to be uniformly distributed among repetitive and unique sequences (Meltz and Painter, 1973). At present, nothing definite seems to be known regarding the enzymes specifically responsible for repair synthesis in these organisms. The ligases of repair and normal synthesis of DNA have received some attention, however (Söderhill and Lindall, 1976).

Photoreactivation. A second major repair pathway for irradiation-damaged DNA is present in most, if not all, organisms, which requires the presence of light, whereas the above system proceeds in the dark (Cook and McGrath, 1967). Ultraviolet light ($\lambda = 2200$–3000 Å) is postulated to induce dimers of the cyclobutane type to form between pyrimidines located in adjoining sites in a DNA strand (Michalke and Bremer, 1969; Lehman and Stevens, 1975). As a rule, the chief net effects are a decrease in total RNA synthesis, due to the production of shorter RNA chains and a deceleration of polymerization rate. This type of damage is believed to be repaired to a greater or lesser extent by means of a photoreactivating enzyme in the presence of visible light ($\lambda = 3000$–5000 Å) (Piessens and Eker, 1975).

The photoreactivating enzyme, generally considered to consist of a protein bearing a chromophore that absorbs visible light, seemed to repair ultraviolet-damaged areas in DNA by restoring the dimers to monomers (Eker and Fichtinger-Schepman, 1975). While the action of the system in reversing ultraviolet inactivation in *E. coli* by means of visible light was reported several decades ago (Kelner, 1949a,b), the responsible enzyme has been isolated and highly purified only recently (Saito and Werbin, 1970). This first purified preparation was made from the enzyme of the blue-green alga *Anacystis nidulans,* but now it has been described also from yeast, *E. coli* (Minato and Werbin, 1971; Sutherland *et al.*, 1973), and mammalian sources (Sutherland *et al.*, 1974). In human cells, its chief concentration was found in phagocytotic monocytes and polymorphonuclear leukocytes. In *B. subtilis* some evidence indicated that DNA polymerase I may be the active enzyme in this system (Noguti and Kada, 1975).

3.5. CHROMATIN AND THE CHROMOSOME

For the major part, attention in this section is devoted to the chromatin and chromosome of eukaryotes, for the so-called chromosome of bacteria has little in common with its eukaryotic counterpart. However, a closing discussion deals with the prokaryotic structure, which is more appropriately known as the nucleoid.

3.5.1. Chromatin Structure

In eukaryotes, the DNA does not occur as free molecules but only in the form of a complex, called either the chromatin, or, rarely, the nucleohistones. The term *chromatin* is broad and indefinite and refers to the mesh of threadlike material within the nucleus, which stains with many basic cytological stains (Brown and Bertke, 1974). During mitosis and meiosis, this substance condenses to form the rodlike bodies referred to as chromosomes (Bonner *et al.*, 1968). It has been found to consist of four groups of chemical constituents, DNA, RNA, histones, and nonhistone proteins (Monahan and Hall, 1974a,b). Much of the following discussion unfortunately must be based on the prevailing concept that the chromosome consists of one double-stranded DNA molecule, arranged as a single fiber stretching from one end of the chromosome to the other. No correlation has as yet been drawn between these simplistic ideas and the actual complexity shown later to exist. Since much attention is now being devoted to chromatin structure, perhaps the interrelationships will soon be clarified.

Euchromatin. Under the usual techniques of staining for light microscopy, chromatin has been found to form two major fractions, the light-staining euchromatin and the dark-staining heterochromatin. These two fractions alternate characteristically in many chromosomes to form a series of bands. As a general rule, the euchromatin is less compact and is considered to contain most of the unique DNA, that is, the actual genes; it replicates early in the S phase of the cell cycle, its histones having been acetylated during the G_1 phase by way of acetyl-CoA (Brown and Bertke, 1974). No differences in histone content seem to exist between the two types of chromatin, but the nonhistone proteins showed marked distinctions (Rodriguez and Becker, 1976).

Heterochromatin. In contrast, the heterochromatin is less active in the genetic processes, is more compact, and replicates late in the S stage (Harbers and Vogt, 1966). The heterochromatin that contains much of the satellite DNA makes up a major fraction of the total chromatin in all eukaryotes (DuPraw, 1968); in *Drosophila,* heterochromatin makes up the entire Y chromosome, the proximal third of the X chromosome, and the regions adjacent to the centromeres of the others (Pimpinelli *et al.*, 1975).

In the past some confusion has existed regarding the function of the heterochromatin, in part as a consequence of there being two types. One of these, known as facultative heterochromatin, is actually euchromatin that has become condensed under certain conditions; thus it contains the usual genes. On the other hand, constitutive heterochromatin is permanently condensed and may consist of redundant sequences of genes responsible for nonspecific activities (Darlington, 1942). However, evidence to the contrary exists, particularly in plants. By way of illustration, certain liverworts have been shown to be unable to survive in the absence of heterochromatin (Heitz, 1942; Lobeer, 1961), and in *Oenothera* the gene action of several complexes has been shown to be localized in heterochromatic regions (Japha, 1939). In strong contrast are the several recent proposals cited later that rRNA genes are located in the constitutive heterochromatin. Actually, the distinctions cannot be considered more than a functional state, for in female mammals one X chromosome is active genetically, and thus is essentially euchromatic, whereas the second becomes heterochromatic because it remains inert as the so-called Barr body (DuPraw, 1968), although the two X chromosomes are identical in composition and genetic properties.

3.5.2. The Constituents of Chromatin

As a whole, chromatin usually consists of approximately equal parts of protein and DNA, but the ratio is subject to variation, especially in metazoan reproductive cells. In bull spermatozoa, to cite one extreme example, the ratio has been shown to be 7 parts protein to 100 parts DNA (Marushige and Marushige, 1974). The proteins fall into two major categories, acid-soluble histones and acid-insoluble nonhistones (Stein and Stein, 1976). In addition there is a small amount of RNA.

The Histones. Histones comprise by far the larger fraction of chromatin proteins but appear to be poor in species number. All are relatively short molecules and are basic in reaction, for upon hydrolysis they yield a high percentage of the basic amino acids lysine and arginine. In this characteristic they grade into protamines, which are similarly rich in arginine, but these latter proteins are confined almost exclusively to the chromatin of sperm. One protamine, salmine, from salmon spermatozoa, was shown upon analysis to consist of arginine to the extent of 88% (West and Todd, 1961). Histones of sperm, such as that of the bull, may be abundantly supplied with cysteine as well as arginine; the former amino acid was reported to play a role in providing disulfide bonds between histone molecules. Thus it was proposed that the DNA of the chromatin was tightly packaged within the network of histone molecules provided by these disulfide linkages (Marushige and Marushige, 1974).

Chief Varieties of Histones. Among eukaryotes in general, five main varieties of histones are known to exist. In mammals at least, H1 (F1 or I) is the largest, having a molecular weight of 21,000; it is very rich in lysine (R. Kornberg, 1974; Monahan and Hall, 1974a,b). Contrastingly, H4 (F2A1 or IV) is the smallest (molecular weight 11,300) and is very rich in arginine. Another very arginine-rich type is H3 (F3 or III). Two other types, H2A (F2A2 or IIB1) and H2B (F2B or IIB2), are moderately lysine-rich; in addition, at least one tissue-specific histone is known, H5 (F2C or V), which appears to be confined to vertebrate erythrocyte chromatin (Brasch *et al.*, 1974; Tobin and Seligy, 1975). During development of the testes of rats, new varieties of histones H1 and H2B appear until maturity (Shires *et al.*, 1975). While most types are widely distributed among eukaryotes, the characteristics change from species to species. For example, the form corresponding to H1 in *Physarum polycephalum* has a molecular weight of 24,500 in contrast to the 21,000 reported from mammalian sources (Hockusch and Walker, 1974); most fungal types correspond closely to those from calf thymus, however (Felden *et al.*, 1976). Among a wide variety of higher plants, histones H4, H3, and H1 were identical or closely similar to those of the cow, but the remainder behaved differently electrophoretically (Spiker, 1975). Since H3 has now been reported in yeast (Brandt and von Holt, 1976), only H1 appears absent; H2 is not subdivided into subtypes, however (Franco *et al.*, 1974). All appear to be synthesized in this organism during the S phase of the cell cycle (Moll and Wintersberger, 1976).

In many cases, specific roles in chromatin behavior have been sought for the various histone species. Results of a circular dichroism study on calf thymus histones suggested that the two arginine-rich types, H3 and H4 are especially active in maintaining DNA conformation in chromatin (Vandergrift *et al.*, 1974). In birds, H5, the erythrocyte-specific histone, has been claimed to be both active and inactive in DNA conformation by different laboratories (Boffa *et al.*, 1971; Billett and Barry, 1974; Brasch *et al.*, 1974). Since the newer model of chromatin structure has been developed, as described below, descriptions of separate functions for each type may not be realistic; moreover, even the complete removal of selected histones exerts virtually no visible effect in chromatin structure (Brasch and Setterfield, 1974).

H1 does, however, have a special property deserving of mention. In mammalian cells and in *Tetrahymena,* much of the phosphorylation mentioned earlier that is associated with DNA synthesis appears to be actually the result of this histone's becoming phosphorylated (Balhorn *et al.*, 1971, 1972; Bradbury *et al.*, 1974; Gorovski *et al.*, 1974). Phosphorylation of distinct regions occurs in correlation to the cell cycle (Hohmann *et al.*, 1976; Jackson *et al.*, 1976). The study of this process in the ciliate was of significance in showing that phosphorylation could not have a role in chromosome condensation. The H1 histone of macronuclei became extensively phosphorylated in correlation to the cell cycle, as in metazoans, even though the chromatin of that nucleus does not form typical chromosomes during mitosis (Flickinger, 1965; Nilsson, 1970). A nuclear magnetic resonance study of this histone from calf thymus showed that the first 122 residues formed a globular head while the remaining 44% was in the form of a randomly coiled tail (Chapman *et al.*, 1976). It was suggested that the globular portion had a specific binding site on the chromosomal subunit.

Among vertebrates, H3 is the only histone that contains cysteine, but in echinoderms H4 has this amino acid instead (Wouters-Tyrou *et al.*, 1976). Calf liver H3 has recently been found to vary, 20% of that histone having serine in place of the cysteine (Garrard, 1976).

Histones in Chromatin Structure. In solutions, and possibly in natural chromatin, certain of the common histones have been found to associate as pairs, H4 joining with H3, and H2A uniting to H2B (Ilyin *et al.*, 1971). On the basis of these data, combined with the results of X-ray diffraction studies, it had been proposed that chromatin was a supercoil consisting of the usual duplex helix of DNA and a coiled covering of histones (Pardon and Wilkins, 1972). Further, the histone coat was conjectured to be in the form of a spiral longer and wider than that of the DNA, having a periodicity of 120 Å and a diameter of 100 Å. Various patterns of cross-linking between histones also had been proposed (Smythies *et al.*, 1974a,b).

These models were later found not to fit all the facts, for some of the histones were determined as actually occurring as the tetramer $(H3)_2(H4)_2$, together with oligomers of H2A2 · H2B (D'Anna and Isenberg, 1974; R. Kornberg, 1974). The fifth histone (H1) was supposed either to be added to these components or to occur elsewhere in the chromatin. Such combinations of histones have been reported from a number of higher eukaryotes, including *Xenopus* embryo, *Drosophila,* and pea chromatin (Fambrough *et al.*, 1968; Byrd and Kasinsky, 1973). The tetrameric and oligomeric histones were further viewed as uniting together to form complexes which formed a unique repeating structure with DNA.

Thus histones are appearing to be an integral part of a nucleoprotein molecule, not merely a covering over that of DNA. More specifically, the histonal complex is postulated to alternate with DNA strands about 200 nucleotide pairs long. This combination of substances in a single molecule might explain the observation that histone and DNA are synthesized together during the S phase of the cell cycle (Prescott, 1966), whereas the nonhistonal chromosomal proteins are produced independently of DNA synthesis (Stein and Borun, 1972). Throughout the cell cycle histones are continuously phosphorylated, but the level of this activity increased during the S phase (Balhorn *et al.*, 1975).

Interrelationships of Histones. Although only a few amino acid sequences of histones have as yet been established, those of five types from calf thymus permit comparisons to be made, at least on a preliminary basis. These five are aligned in Table 3.3, largely as in the computerized system of Temussi (1975), together with H4 from sea urchin (Wouters-Tyrou *et al.*, 1976). Despite Te-

Table 3.3

Comparisons of Calf Thymus and Sea Urchin Histones[a,b]

	−7 · 1 · 10 · 20 · 30 · 40
H2B	PEPAKSAPAPKKGSKKAVIKAQKKDGKKRKRSRKESYSVYVYKVLKQVHPDTGISS
H4	SGRGKGGKGLGKGGAKRHRKVLRDNIQGITKPAIRRLARGGVKRRISG
H2A	SGRGKQG---GKARAKAKTRSSRAGLQFPVGRVHRLLRKGNYAERVGA
S.U.[c]	SGRGK-GKGLGKGGAK_aRHRKVLRDNIQGITKPAIRRLARRGGVKRISG
H3	-ARTKQTARKSTGGKAPRKQLATKAAPKSAPATGGVKKPHRYRPGTVA
H1	SEAPAETAAPAPAEKSPAKKKKAAKKPGAGAAKRKAAGPPVSELITKAV

	50 · 60 · 70 · 80 · 90 · 100
H2B	KAMGIMNSFVNDIFERIAGEASRLAHYNKRSTITSREIQTAVRLLLPGELAKHAVS
H4	LIYEETRGVLKVFLENVIRDAVTYTEHAKRKTVTAMDVVYALKRQGRTLYGFGG$_{OH}$
H2A	GAPVYLAAVLEYLTAEILELAGNAARDNKKTRIIPKHLQLAIRNDEELNKLLGKVT
S.U.	LIYEGTRGVLKVFLENVIRDAVTYCEHAKRKTVAMDVVYALKRQGRTLYGFGG$_{OH}$
H3	LREIRRYQKSTELLIRKLPFQRLVREIAQDFKTDLRFQSGAVMALQEACEAYLVGL
H1	AASKERNGLSLAALKKALAAGGYDVGKNNSRIKLGLKSLVSKGTLVETKGTGASGS

	110 · 120 · 130 · 140 · 150
H2B	EGTKAVTKYTSSK$_{OH}$
H2A	IAQGGVLPNIQAVLLPKKTESHHKAKGK$_{OH}$
H3	FEDTNLCAIHAKRVTIMPKDIQLARRIRGERA$_{OH}$
H1	FKKLDKKAVEAKKPAKKKAAAPKAKKVAAKKPAAAKKPKKVAAKKAVAA$_{OH}$

[a] In part based on Temussi (1975).

[b] See Table 4.2 for abbreviations of amino acids; K_a, acetyllysine.

[c] S. U., sea urchin.

mussi's claim that all the calf histones showed close kinship, the comparisons seem more reasonably to indicate two groups of relationships. One includes the several subtypes of H2, between which the regions of correspondences indicate about a 25% level of identity. Histone H4 from the sea urchin (abbreviated S.U. in the table) appears almost identical to calf H4, except that occasional additions or deletions throw corresponding regions to the left or right here and there. However Fazal and Cole (1977) demonstrate that H1 occurs as four differing fractions in wheat germ histones, one of which migrated very close to H2B, and that other species showed distinct differences from corresponding types from calf thymus.

The second group of relationships includes H1 and H3, but only at a 12% level of identity; since the random-chance level is 5%, sequences of these histones from other sources are requisite to establish confidence that the kinship is real. The same statement is equally applicable to the identities which exist between H2A and H1, as these approximate an 8% level. Sequences of yeast, protozoan, and algal histones should be most helpful in elucidating the topic, especially if the protozoan sources were to include amoeboids and euglenoids, the algae, and chlorophytes and either xanthophytes or chrysophytes.

Subunit Structure of Chromatin. The novel model of chromatin structure based on histone behavior described above correlates rather strikingly with the results of degradation studies on rat liver chromatin by means of certain nucleases, in which the enzymes cleaved most of the DNA into segments about 200 base pairs long (Hewish and Burgoyne, 1973; Noll, 1974). Similar subunits of equivalent length have been reported also from trout testis nuclei, but not in those of trout sperm, which contain nucleoprotamine (Honda *et al.*, 1974). Even the chromatin of yeast, which does not condense into chromosomes, has been demonstrated to consist of subunits, which averaged only 135 base pairs in length (Lohr and Van Holde, 1975). Still shorter strands (100 base pairs long) have been reported in calf thymus chromatin (Clark and Felsenfeld, 1974; Van Holde *et al.*, 1974). However, in sea urchin sperm chromatin, the length of the subunit varied with the nuclease employed (Spadafora and Geraci, 1975), so it was suggested that there is a minimal DNA length, less than 200 nucleotide units long, rather than a constant length, that is capable of binding histone oligomers.

Electron microscope investigations also appear to support a concept of subunital structure for chromatin. Suitably prepared nuclei show linear arrays of spherical particles about 70 Å in diameter, which are connected by fine strands nearly 15 Å wide (Olins and Olins, 1973; Li, 1977). These particles have been variously named ν bodies (Figure 3.6), nucleosomes, and PS-particles (Van Holde *et al.*, 1974). Most recent studies have found the monomeric ν bodies to have a molecular weight of 300,000, to exhibit a protein : DNA ratio of 1.22 : 1 by weight, and to include a fragment of DNA having a molecular weight of 140,000 in each such particle (Senior *et al.*, 1975). Apparently the ν bodies are heterogeneous, falling into at least four classes that differ in

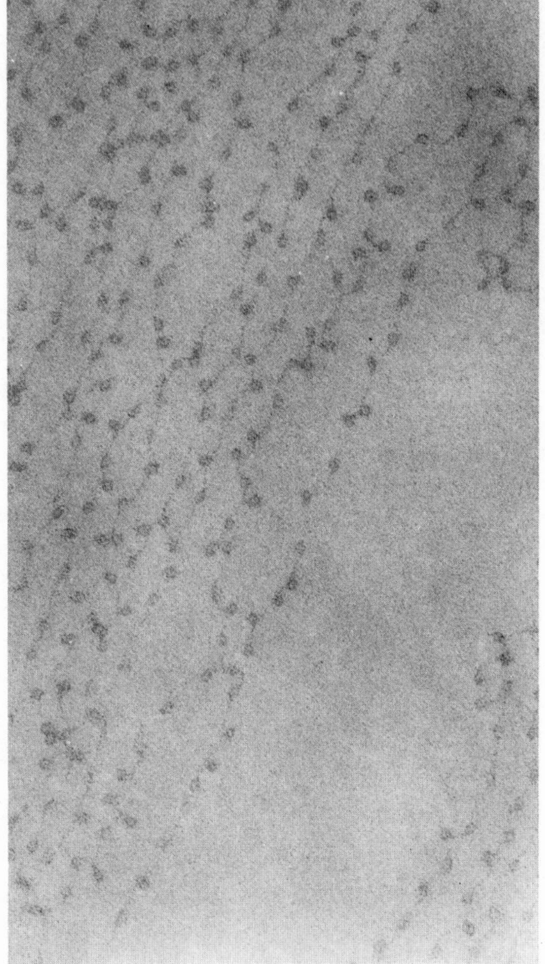

Figure 3.6. Nucleosomes or ν bodies. Chromatin fibers in eukaryotes consist of DNA interspersed with rounded protein bodies. 135,000 ×. (Courtesy of Drs. Ada and Donald Olins, Graduate School, Biomedical Science and Biology Division, Oak Ridge National Laboratories, Oak Ridge, Tennessee.)

length of the associated DNA segment and in the protein moiety (Pospelov *et al.*, 1977).

In a high-resolution electron microscopic study the ν bodies were shown to contain a central core and to be connected to the DNA strands at one side, not at opposite ends (Olins *et al.*, 1976). Apparently the bodies are permanent features of chromatin, for they have been demonstrated in metaphase nuclei and in mouse spermatids (Figure 3.7; Kurszenbaum and Tres, 1975). A recent in-

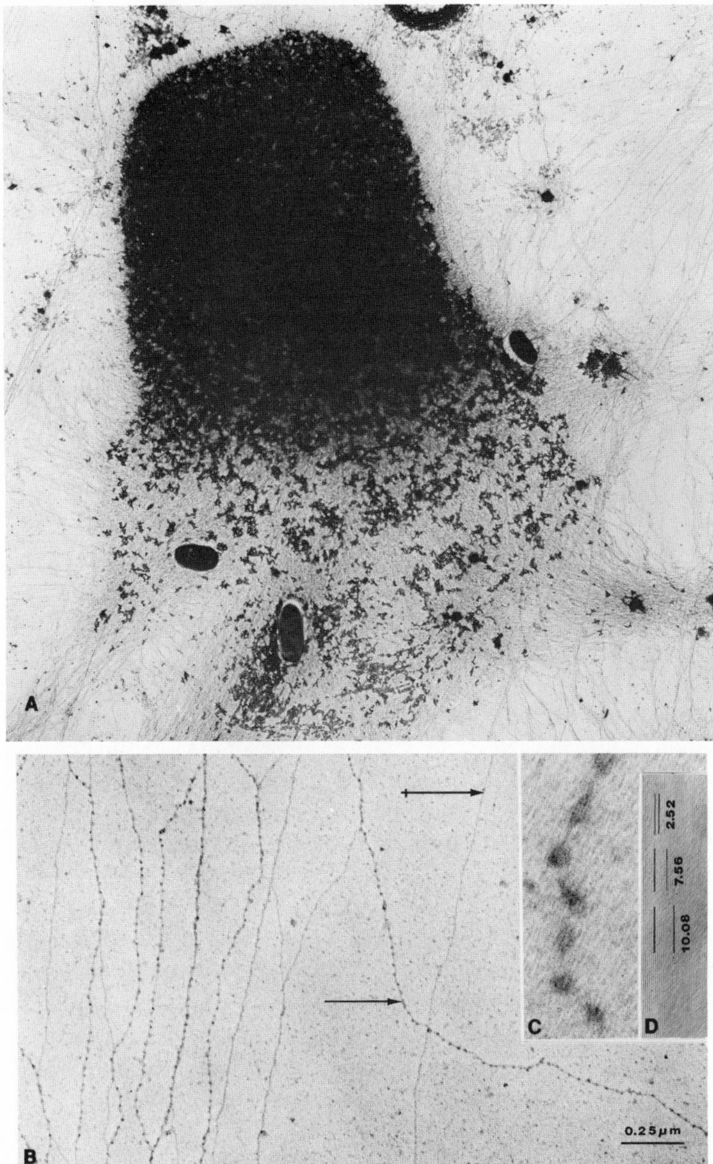

Figure 3.7. Chromatin structure in the spermatid of mouse testis. (A) While the anterior portion of the genome in mouse spermatids is so highly compacted as to be electron-opaque, the posterior portion remains dispersed. 6750 ×. (B) Although most of the fibers are seen at 58,000 × magnification to be heavily studded with υ bodies (arrow), a few are relatively smooth (crossed arrow). When magnified 400,000 × (C) and aligned with a copper phthalocyanin crystal lattice at the same magnification, the particles may be accurately measured (D). Lengths are in nanometers. (All parts of micrograph reproduced @ 65%.) (Courtesy of Dr. A. L. Kierszenbaum and the *Journal* of *Cell Biology*.)

terpretation of the organization of the bodies is also illustrated in Figure 3.8 (Bram, 1975; Bram *et al.*, 1975); the short length of DNA strand associated with the histone bead, referred to as a bridge, is about 50 base pairs long (Whitlock and Simpson, 1976). All eukaryotes thus far examined have similarly structured chromatin (McGhee and Engel, 1975; Jerzmanowski *et al.*, 1976). Recently it was possible to demonstrate that no clear-cut difference in transcriptional activity exists between the 200-unit DNA segments and the short bridges (Reeves and Jones, 1976).

Nonhistonal Chromatin Proteins. Rapidly accumulating evidence is presently indicating the acid-insoluble proteins of chromatin also to be of great importance in genetic functions. In particular they appear responsible for regulating gene expression and especially controlling transcription (Gilmour and Paul, 1970; Stein and Farber, 1972; Shea and Kleinsmith, 1973). A portion of the evidence supporting a regulatory roll for this class of proteins is derived from the significant variations that exist in the rates of their biosynthesis, turnover, and phosphorylations during defined periods of the cell cycle (Stein and Baserga, 1970; Platz *et al.*,1973). To the contrary, partial purification of the template-active fraction of chromatin has shown this 11% of the DNA to be more largely associated with acid-insoluble proteins (Gottesfeld *et al.*, 1974). Some of this nuclear protein remains after removal of the nucleoproteins and forms a matrix which maintains the spherical shape of the nucleus (Berezney and Coffey, 1975).

One method that has been employed to demonstrate their role in regulating transcription during the cell cycle is that of reconstituting chromatin with nonhistone chromatin proteins extracted during various specific stages of the cycle. In the continuously dividing HeLa cells studied, several classes of these proteins were found to be synthesized on stable mRNAs transcribed prior to mitosis, whereas others required the biosynthesis of mRNA during the G_1 phase (Stein and Matthews, 1973). Regulation has been assumed to be by binding of the proteins to DNA. Such binding appears to be optimal at 0.05 M concentrations of NaCl, increases with lowered pH, and is enhanced by the presence of Mg^{2+} ions. During the processes, approximately equal parts of the proteins bind reversibly and irreversibly (Sheehans and Olins, 1974). It has also been proposed that the protein may stimulate the activity of the nucleohistones, at least in part through the presence of phosphoproteins and protein kinases (Kamiyama *et al.*, 1972). Binding to the histones has now been reported to be decreased when the nonhistonal proteins were phosphorylated (Courtois *et al.*, 1975).

Figure 3.8. Interpretation of nucleosome structure. (A) Nucleosomes along a region of chromatin. 500,000 ×. (B) Rubber-tubing interpretation of nucleosome structure. According to high-resolution studies, nucleosomes have been indicated to be connected laterally to the DNA strands and to consist of a double coil, in which histones (B) are enclosed. (Courtesy of Dr. S. Bram; reproduced @ 90%.)

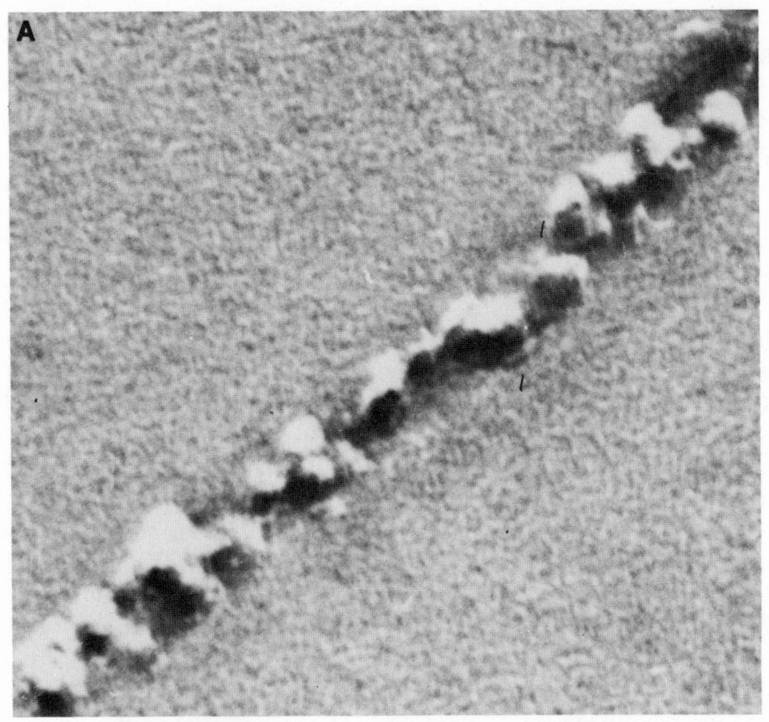

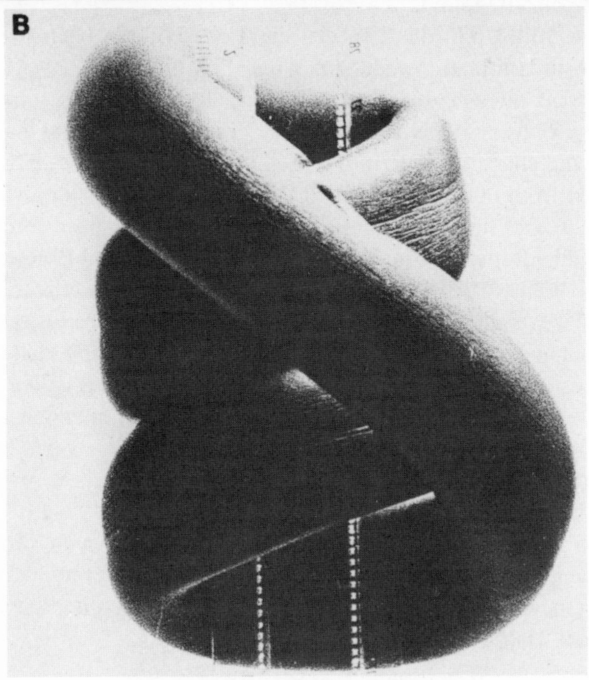

These proteins are highly diversified (Chiu and Hnilica, 1977). In rat liver, 13 to 14 types were at one time reported separable by gel electrophoresis, having molecular weights ranging from 13,000 to 150,000 (Warnecke *et al.*, 1973). This number was subsequently increased to between 20 and 25 different species in the same and other rat tissues (Gronow and Thackrah, 1973), and the range in molecular weight has been extended to between 10,000 and 200,000 (MacGillivray and Richwood, 1974; Kostraba *et al.*, 1975). Though the number of types present in a given tissue was thus rather constant, variation in relative proportions of the types was extensive from tissue to tissue.

Part of the variety reported stems from differing sources of the nuclear proteins. Cell proteins have been classified into three categories on the basis of their ability to migrate into nuclei: the N proteins were those that occurred predominently in the nucleus; the C class, those largely confined to the cytoplasm (hence, nearly nonnuclear); and the B group, those found in the nucleus and cytoplasm in nearly equal proportions (Bonner, 1975). Hence, part of the regulation of nuclear protein is exercised by the nuclear membrane through its selectivity in permitting entrance into and egress from the nucleus, but most of the nuclear proteins have been demonstrated actually to be synthesized in the nucleolus (Birnstiel and Hyde, 1963). Another method of fractionating the nuclear proteins revealed the presence of two groups differing in binding properties, one class being tightly, the other loosely, bound (Kostraba *et al.*, 1975). The latter group was shown to contain phosphoproteins as well as RNA, as described immediately below.

Chromosomal RNA. A special class of RNA has been suggested to be associated with chromatin, under the name "chromosomal RNA" (Holmes *et al.*, 1972). This nucleic acid does not appear to be mere degradation products of transfer or ribosomal RNA, as once suggested (Heyden and Zachau, 1971). Besides having characteristic eluting properties, chromosomal RNA (cRNA) is distinguished from both tRNA and rRNA in hybridizing with DNA to a much greater extent, and further, regardless of its source, has been found to contain 7% to 10% dihydropyrimidine (Huang, 1969; Mayfield and Bonner, 1972; Tolstoshev and Wells, 1974). This cRNA from mammals was reported to be of low molecular weight and was fractionated into 11 size classes, each of which probably contained a number of RNA species (Monahan and Hall, 1973); most of these types were found to be reversibly bound to proteins of chromatin. Similar properties were observed for cRNA from chick chromatin, but only six to eight size classes were separated (Marzluff *et al.*, 1975).

Recently, another fraction of RNA from the chromatin of mouse cells was reported to be associated with DNA (Monahan and Hall, 1974a,b); this was not covalently attached to the DNA nor was it associated with proteins. High-molecular-weight RNA also has been found in chromatin, but this was believed to represent newly synthesized RNA that had not yet undergone processing (Monahan and Hall, 1975).

3.5.3. Control Mechanisms

The Lac System of Prokaryotes. One enzymatic system involving those nonhistone proteins that interact with DNA which has received widespread attention is the lactose system of *E. coli*. Since the organism is a prokaryote, in a strict sense this system does not apply to the proteins of chromatin, especially in the familiar view that the DNA of bacteria is naked. Yet there can be little doubt that the proteins of actual chromatin may exercise similar functions in comparable fashion. Hence, in the absence of direct evidence from eukaryotic sources, the bacterial system can serve as a tentative model.

The lactose system, usually called the *lac* system, centers around the synthesis of the enzyme β-galactosidase which hydrolyzes lactose, and involves an operon, an operator, and a repressor. The *operon* is a cluster of genes, including that cistron which actually specifies β-galactosidase, two that code for related enzymes, and three others that affect the activity of the whole group of genes (Jacob and Monod, 1961; Harris, 1974). Among the last three is the *operator* cistron, which consists of a sequence of 21 nucleotide residues (Adler *et al.*, 1972; Kolata, 1974); mutations at the operator locus influence the expression of the entire cluster of genes. Another member of the trio is the *promoter* gene that appears also to regulate or facilitate transcription of the genes for the three enzymes. The expression of the entire set of genes (the *lac* operon) can be prevented by a protein, called the *lac repressor,* which exercises its action by binding to the operator.

In attempts at understanding the mode of attachment and repression by the protein, the sequence of the *lac* operator cistron has been determined (Gilbert and Maxam, 1973). As may be noted (Figure 3.9), the 21 nucleotides are symmetrically arranged, the components toward the 3′-end being complements in reverse sequence of those at the 5′-terminus. Because of this arrangement, it has been proposed that, under some circumstances, the DNA helix could unwind and form a hairpin loop in this region. The repressor thus would be enabled to distinguish and attach to this particular short strand, among the 6 million nucleotides in the *E. coli* DNA molecule, by means of the projecting loop (Gierer, 1966). Other equally credible models have also been proposed (Kolata, 1974). Now the nucleotide sequence of the transcript of the entire promoter–operator region has been determined also (Dickson *et al.*, 1974).

The system is still more complex than the essentials above suggest. Part of

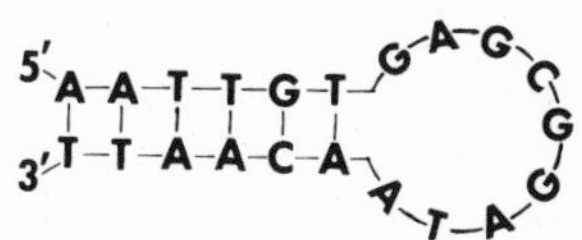

Figure 3.9. The nucleotide sequence of the lac operator gene. The 21 nucleotides of the DNA are symmetrically arranged, so that a loop might be formed. (Based on Gilbert and Maxam, 1973.)

the control mechanism of the lac system involves DNA sites where two other proteins are bound. One of the latter is the polymerase that transcribes the message from the enzyme genes, and the other is an enzyme that binds cyclic adenosine monophosphate, which must interact with the DNA before transcription of the lac genes can commence. Hence, three proteins are needed to control the transcription, operation, and repression of a single sequence of genes.

DNA Modification and Restriction Systems. Modification and restriction systems were discovered in bacteria several decades ago. These serve to degrade one type of DNA, usually of exogenous origin, in the presence of a second, typically that of the host, which is unaffected (Luria and Human, 1952; Luria, 1953). Basically, this system is believed to involve one or more modifying enzymes capable of methylating specific sites in DNA, and at least one restriction enzyme, really an endonuclease that cleaves DNA in a highly site-specific manner (Arber and Linn, 1969; Boyer, 1971, 1974; Meselson *et al.*, 1972). Thus those sites which are acted upon by the modifying enzyme are protected, while those which are not are rapidly degraded.

While as yet relatively poorly explored in eukaryotic cells, the existence of modification systems has been suggested, particularly on evidence derived from chloroplast inheritance in *Chlamydomonas,* a unicellular flagellate (Sager, 1972; Sager and Lane, 1972; Sager and Ramanis, 1974). The chloroplast genes of this organism show clear evidence of being transmitted from the female parent to all progeny, whereas those of the male are not transmitted, as they are destroyed in the zygote after fertilization. The molecular basis for this phenomenon has been proposed to be that of modification of the female DNA prior to fertilization, so that later in the fertilized egg the unmodified paternal DNA is degraded by restrictive enzymes (Sager and Kitchin, 1975). Similar mechanisms have been proposed to be responsible for such diversified traits as chromosome elimination, as in certain coccids and *Sciara* (Metz, 1938; Chandra and Brown, 1975), X-chromosome inactivation in the bandicoots and allied marsupials (Hayman and Martin, 1965; Walton, 1971), and mitochrondrial inheritance in yeast (Bolotin *et al.*, 1971; Dujon *et al.*, 1974). At least in these systems, it is implicit that the DNA is subservient to proteins and not vice versa.

3.5.4. The Structure of Chromosomes

Although chromosomes are generally viewed as chromatin condensed into rodlike forms, much more than mere compacting is involved in their formation, according to the complexity revealed by ultrastructural studies. In chromosomes, the distinctions between euchromatin and heterochromatin acquire significance, for their detection *in situ* is possible only in these structures. The location of the centromere, around which heterochromatin often centers, has itself been determined to be nonrandom (Imai, 1975).

Chromosomal Bands. The two major types of chromatin are made evident by a large number of staining techniques, no two of which give precisely identical results. Particularly effective is Giemsa staining, which is reported to demonstrate regions of chromosomes containing repetitive sequences of DNA (Pardue and Gall, 1970; Arrighi and Hsu, 1971); the resulting stained regions are known as C-bands. By means of such techniques, the satellite DNA of repetitive sequences, that is, pericentromeric heterochromatin, has been found universally present adjacent to the centromeres in animal chromosomes (Drets and Shaw, 1971; Comings and Mattoccia, 1972; Stack, 1975). For a time it appeared that comparable chromosomal banding was absent from plants (Vosa and Marchi, 1972), but now pericentromeric heterochromatin has been demonstrated in a number of flowering plants, including *Plantago ovata* and *Allium cepa* (Stack and Clarke, 1973a,b). Telomeres and nucleolar organizers stained especially heavily in these species. In mammalian cells, histones H1 and H2A, but not H2B and H3, have been found to be involved in chromosomal C-bands (Brown *et al.*, 1975).

Because the patterns of chromosomal staining vary with the technique employed, the existence of several kinds of vertebrate heterochromatin has been proposed (Bianchi and Ayres, 1971; Cooper and Hsu, 1972; Comings *et al.*, 1973). Comparable differences in banding have also been found in plants (Schweizer, 1973; Stack *et al.*, 1974). A more recently developed technique makes use of the fact that 5-bromodeoxyuridine quenches the fluorescence of a certain dye (#33258 Hoechst). In the absence of the bromylated uridine, the centromeric regions of mammalian chromosomes fluoresce brightly when thus stained. When the cells are grown for one full cell cycle in the bromodeoxyuridine, a remarkable difference in fluorescence between halves of the centromeres has been noted (Lin and Davidson, 1974). Because certain chromosomes can fuse into a chain in the mouse, it was suggested that the results indicated a polar continuity of the same DNA strand throughout the genome. In other words, the thymidine-rich strand of the satellite DNA is associated with the corresponding strand of ordinary DNA in each chromosome. Fluorescent staining techniques have also demonstrated the localization of 5-methylcytosine in heterochromatin (Miller *et al.*, 1974). Fluorescent stains that react with sequences rich in A–T base pairs produce Q-bands, which are complementary to C-bands (Lee and Collins, 1977). Those that have a greater affinity of G–C pairs result in R-bands (van de Sande *et al.*, 1977). Methylene-blue-stained regions are referred to as G-bands. A recent discovery made through use of trypsin staining techniques is of more than usual importance in that, in mankind, female autosomal chromosomes were found to be consistently longer than those of males (Kowalski *et al.*, 1976).

Heterochromatic regions have been reported to replicate to a lesser degree than euchromatic (Heitz, 1934; Berendes and Keyl, 1967; Hennig *et al.*, 1970). This has been found to be particularly true for redundant sequences, including ribosomal genes (Hennig, 1972; Spear and Gall, 1973). In the dipteran *Rhynchosciara americana,* the ribosomal genes were followed through the developmental cycle of the organism. Tissues which underwent only a low level of polyteny were found to have higher proportions of rRNA than those with a high level. Hence it was confirmed that ribosomal cistrons are not replicated to the same extent as other DNA regions in polytenic cells (Gambarini and Lara, 1974). Such a differential synthesis of DNA is further supported by changing patterns of replication reported for organisms as diversified as *Paramecium*, mammals, and *Lilium* (Stern and Hotta, 1973; Bernard *et al.*, 1974; Smith *et al.*, 1974).

The Structure of Chromosomes. Because of the size of the chromosomes relative to the DNA of which they are considered to consist, difficulties in correlating the differences have been experienced. Since the DNA double helix approximates 2 nm in breadth, and the chromosomes are between 100 and 200 nm, it is obvious that each must consist either of large numbers of DNA molecules (even more if separate than if they are combined with proteins) or a single DNA helix coiled in a complex fashion (Haggis *et al.*, 1964). Even in the lampbrush chromosomes of amphibian oocytes, the 20-nm-thick threads found in limited regions would have to contain 100 pure DNA molecules to attain that diameter. To explain these structures, Crick (1971) proposed a model consisting of two types of DNA, one fibrous, the other globular; these two varieties were then visualized as forming a continuous thread that was wound from one end of the chromosome to the other. The fibrous DNA corresponded in this model to euchromatin in consisting of the unique sequences and in being located between the dark bands of chromosomes. Hence, the globular variety resembled heterochromatin in forming those bands but was suggested to contain the histones and acidic proteins and thus to serve as a control mechanism. However, this model does not conform to what has been discovered recently about chromatin organization and chromosome ultrastructure, nor is it applicable to such genetically documented combinations of control and structural genes as the lac and similar systems.

Ultrastructure of Chromosomes. When viewed under the electron microscope, the chromosome is seen to consist of complexly woven fibers (Ris, 1959; Ris and Chandler, 1963). These fibers have been shown to be irregular in thickness, with diameters ranging between 20 and 30 nm, with a mean close to 23 nm (Figure 3.10). Originally thought to be artifacts, the fibers were confirmed as actual structural components through a demonstration of their pres-

Figure 3.10. Fine structure of mammalian chromosomes. (A) Portion of a banded region in a hamster metaphase chromosome, magnified 28,000 ×. The two bands are seen to interdigitate extensively. (B) At high magnification the fibers are visibly more densely packed than in interband regions; υ bodies are more readily discernible in less condensed areas. 68,000 ×. (Both A and B reproduced @ 70%.) (Courtesy of Dr. D. E. Comings and the *Journal of Cell Biology*.)

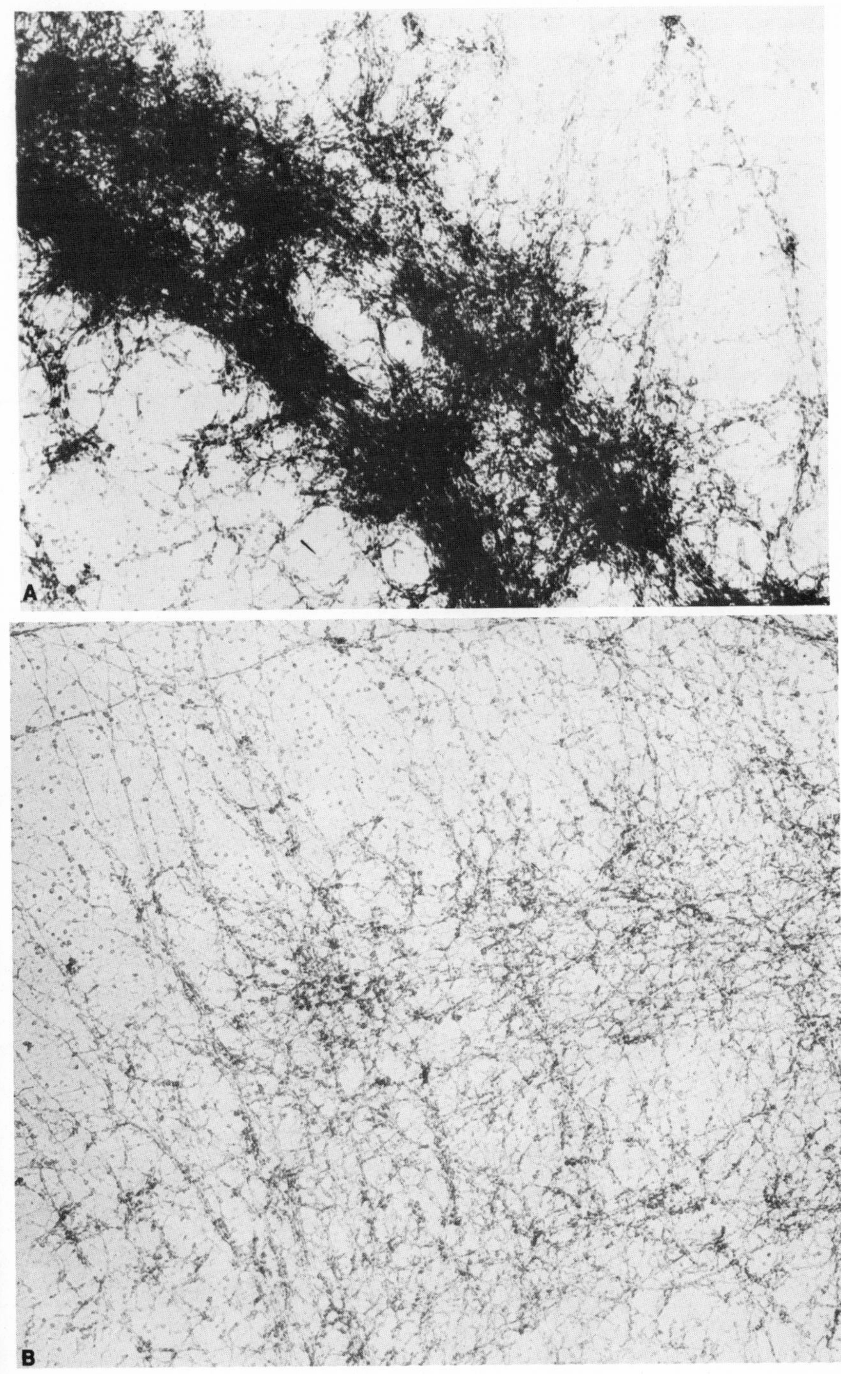
A
B

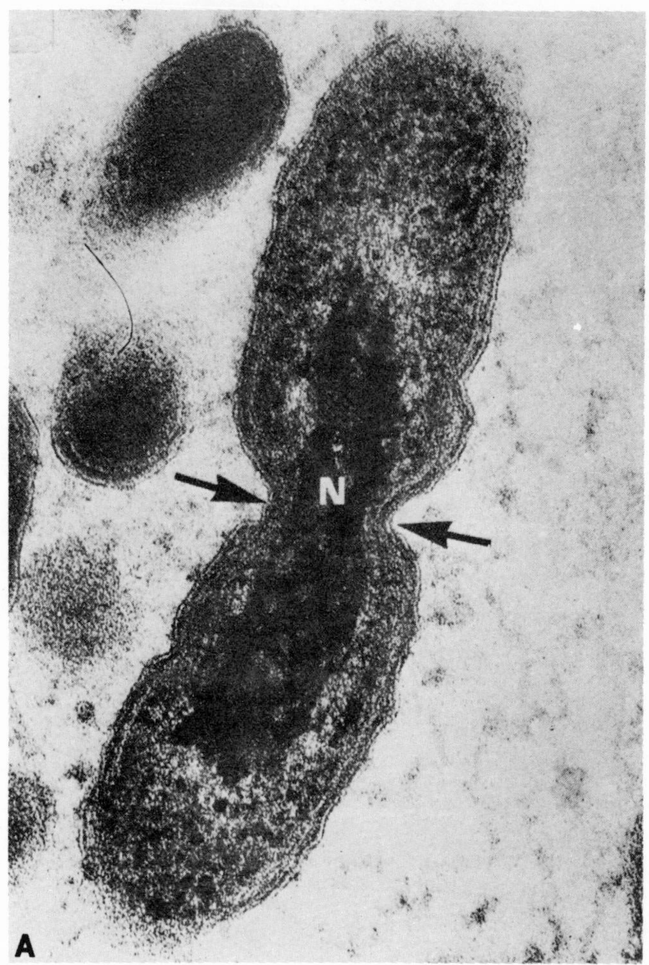

Figure 3.11. The nucleoid of bacteria. The DNA–protein complex of bacteria lacks a membrane covering and other features of the eukaryotic nucleus; consequently, it is most correctly referred to as a nucleoid. (A) Division of the nucleoid occurs transversely together with cell division when

ence *in vivo* in unsectioned nuclei and chromosomes of honeybee embryonic cells (DuPraw, 1965a,b). At first each individual thread was believed to consist of four fine fibrils coiled about themelves (Ris, 1955, 1959; Ris and Chandler, 1963), but currently the bulk of evidence favors the view that each fiber is a single component containing one duplex DNA helix (DuPraw, 1965a; Ris, 1966; Huberman, 1973).

On digesting these 20- to 30-nm so-called type B fibers with trypsin, it has

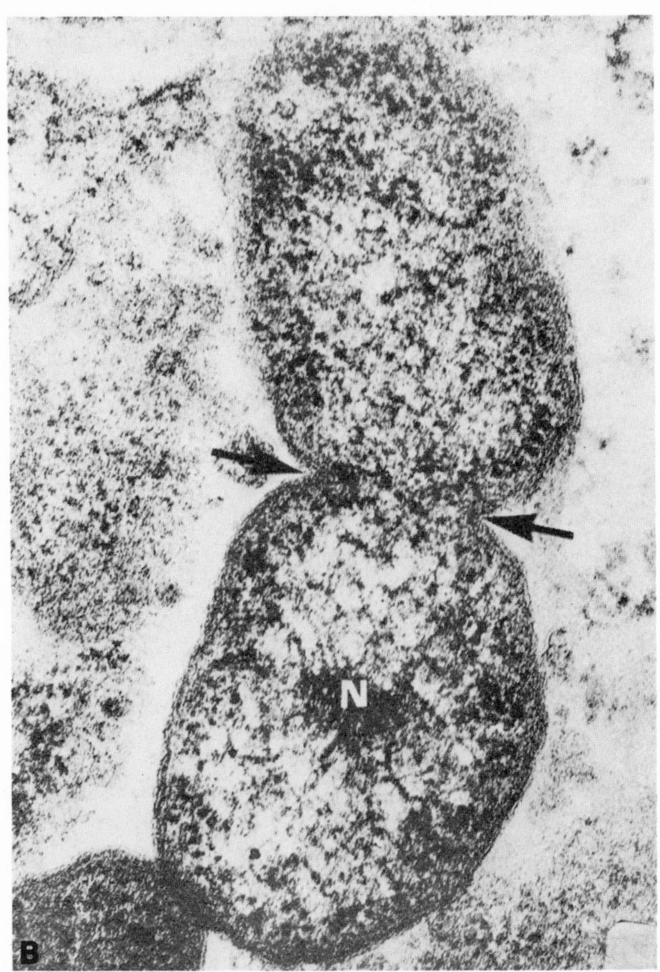

growth is at a slow rate. (B) When growth is rapid, the nucleoid divides well ahead of the cytoplasm; here it has completed division. (Courtesy of Dr. Gérard Devauchelle.)

been found that they were actually built up by the supercoiling of a fine fibril (Type A), 3.5–6.5 nm in diameter (DuPraw and Bahr, 1968). As the type A fibrils have a symmetrical mass distribution and a relatively high density, much of their content was deduced to be DNA. From these and data derived from electron-scattering measurements, it was calculated that for each nanometer of length of a type B fiber, there were 56 nm of DNA. This ratio would decrease at metaphase, for at that time, acid proteins and rRNA protein are added to the chromosomes (Salzman *et al.*, 1966). At this same stage of division, the type B fibril diameters increase and exceed 30 nm in human chromosomes (DuPraw, 1965a), due to these additions as well as contraction. Similar fibrous organization has been observed in sectioned chromosomes (Sotelo, 1969).

Based on these observations, DuPraw (1968) has offered an interpretation of the structure of a chromosome. In this model, he proposed that the type B fibrils wind back and forth across and around the chromosome, covering the latter's full length several times. Consequently the fibril may traverse the chromosome 200 or more times from end to end. If this model approaches reality, it explains how a DNA molecule 40,000 μm long can be contained in a chromosome only 4 μm in length—figures established many times over by genetic, cytologic, and biochemical means (Brown and Bertke, 1974).

The banding pattern that characterizes giant polytene chromosomes like those of *Drosophila* has also been investigated to a degree in unsectioned chromosomes under the electron microscope. As may be seen in Figure 3.10, the bands and interbands are composed of type B fibers (Rae, 1966; Comings and Okada, 1976). In the bands, the fibers are obviously packed much more closely than elsewhere, the interbands being loosely fibered and vesicular. The complexity of the winding and greatly enhanced number of type B fibers, in comparison with those of the human being, are self-evident features. These observations, as Brown and Bertke (1974) suggested, seem to require new correlations to be brought forward for existing biochemical and cytological data.

3.5.5. The Nucleoid of Prokaryotes

The nucleoid, the bacterial equivalent of the eukaryotic chromosome, is far simpler in composition than the complex fibroid type described above (Figure 3.11). Because of this evident simplicity, it has in the past been generally assumed to consist entirely of DNA. Currently some evidence has been advanced which intimates that this concept no longer may be tenable.

When the intact nucleoid of *E. coli* was isolated, proteins (in proportion relatively small in comparison to those of eukaryotes, but significant nevertheless), RNA, and, perhaps, lipids have been found associated with the DNA (Pettijohn *et al.*, 1973). The DNA to protein ratio was found to approximate 9 : 1 and the DNA to RNA ratio was 10 : 4. Labeled oleic acid occurred at a level of less than 100 : 1 relative to DNA, but the possibility that this represented contamination from membrane constituents could not be eliminated. As the electrophoretic bands of the protein fraction in part coincided with those of purified RNA polymerase and as 75% of the DNA-dependent RNA polymerase activity was associated with the nucleoid, it was assumed that the major protein constituents were simply polymerase. Nevertheless, other bands of considerable significance were also present. Most of the RNA was interpreted as newly formed RNA associated with polymerase activity. The existence of a DNA–*protein* complex in the nucleoid of *E. coli* has been confirmed by use of methylated albumin columns, studies that revealed a protein content of approximately 10% (Koslov and Svetlikova, 1974).

As isolated in the nucleoids, the DNA was complexly folded into a compact particle that sedimented at 1600 S (Stonington and Pettijohn, 1971), with a molecular weight exceeding 10^9. Comparisons of the folded and unfolded molecule showed that the unfolded DNA did not support as much RNA polymerase activity as did the folded. However, it was not establishable as to whether the higher efficiency could be attributed to the DNA conformation itself or to the possible presence of unknown proteins bound to the DNA. To maintain the

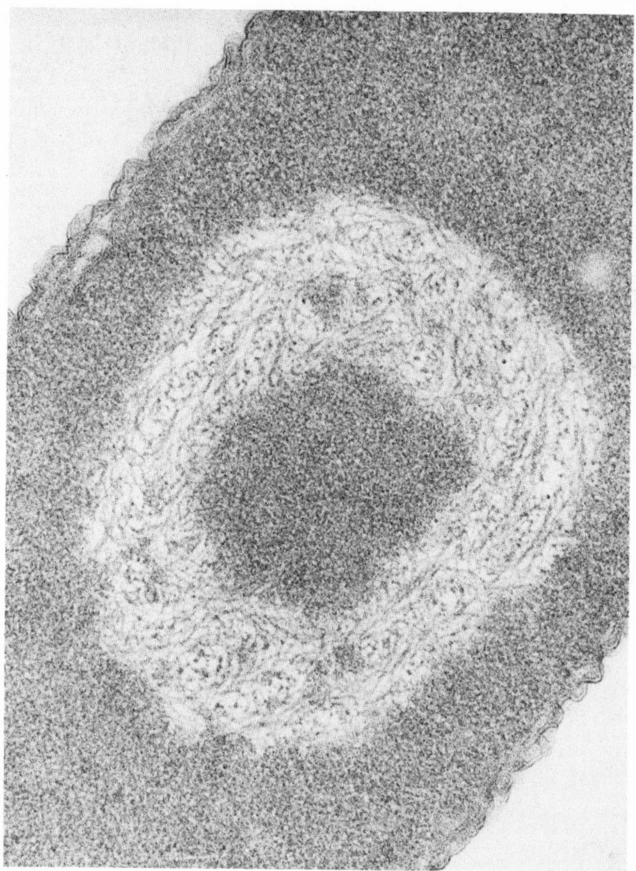

Figure 3.12. A condensed nucleoid of *E. coli*. The nucleoid becomes highly condensed before undergoing division. 150,000 ×. (Reproduced @ 48%.) (Courtesy of Drs. David Zusman and J. Haga.)

folding, the presence of an unidentified protein was essential, as was also an undetermined type of RNA. Similar complexes of DNA and proteins have been reported in the genome of the bacteriophage GA-1 from *Bacillus subtilis* (Arnberg and Arwent, 1976).

Other electron microscopic investigations of the *E. coli* nucleoid employed a heat-sensitive mutant that normally fails to divide above 40°C (Zusman *et al.*, 1973). Inhibitors of protein synthesis nevertheless induced cell division even at temperatures of 41°C. When protein synthesis was halted, the nucleoid underwent condensation in the fashion characteristic of normal division (Figure 3.12).

Thus it is evident that DNA of all organisms exists as complexes with proteins and that its replication requires a far more complicated set of events than heretofore visualized. The multitudinous strands that comprise the chromosomes and the segmental nature of those fibers require new concepts of how genes are arranged and how banding patterns result. Moreover, the processes of utilizing the informational content of the DNA need reexamination in light of the newer knowledge. The present status of such transcriptional activities is discussed in the two chapters which follow.

4

THE GENETIC MECHANISM: II

The Cell's Employment of DNA

Throughout the foregoing discussion, it became increasingly evident that the DNA molecule is completely dependent upon proteins for its every action. Replication was seen to proceed only in the presence of a series of enzymes, as did the synthesis of the nucleotides and their modification after incorporation. Even the structuring of the molecule into chromosomes or nucleoids required action from certain proteins. Hence DNA by itself is obviously inert, an attribute that cannot fail to preclude its being the first molecule of life—proteins certainly, and RNAs possibly, had to precede it in living systems. Although its absence of chemical activity bars DNA from further consideration in the search for the properties of the earliest protobiont, its presence should be expected in the genetic apparatus of the most advanced forms. Were the DNA molecule to participate actively in metabolism, it would of necessity undergo molecular changes. But through its inertness it remains stable, and through its stability it is capable of providing the same precise information to generation after generation of cells.

To continue the quest for the ancestral molecule, the products derived from the DNA informational content, the various types of RNA, are next examined. Perhaps one of these will prove sufficiently lifelike in its activities to intimate its having been the very first polymer of life.

4.1. THE TYPES OF RIBONUCLEIC ACID

In employing the genetic information contained in the DNA, the first step is that of copying, or *transcribing*, it into RNA by means of processes that appear to be carried out in complementary fashion (Forget *et al.*, 1975; Pieczenik *et al.*, 1975). Transcription results in several different types of RNA, which are named in accordance with their function or location in the cell. The structure and functions of messenger (mRNA), transfer (tRNA), and ribosomal (rRNA), are absolutely essential to further discussion. Heterogeneous nuclear RNA (hnRNA), however, does not require immediate attention.

4.1.1. Messenger RNA

The concept that a special type of RNA molecule carried a copy of the information contained in the DNA and served as a template for protein synthesis was first advanced by Brenner *et al.* (1961). In turn, this idea had been based on earlier work by Hershey *et al.* (1953), who had reported that RNA was synthesized rapidly in *E. coli* when infected with bacteriophage T2. Then Volkin and Astrachan (1956) labeled this RNA and concluded that in structural characteristics it resembled the DNA of the phage rather than the RNA of the bacteria. Thus the implication was obvious that it had been formed through use of the phage DNA as template.

Half-Life of mRNA. At the time of its discovery, mRNA was viewed as being of simple, single-strand structure and possessing a short half-life. It was further characterized as forming only a small fraction of the total RNA of a cell, as a rule between 2% and 3%, and as being of intermediate size—between 9 S and 19 S compared to the 4 S of tRNA and the 16 S to 45 S of rRNA. The short half-life has subsequently proven to be partially true for bacterial and viral mRNAs; for example, Levinthal and colleagues (1962) showed that the mRNA of *Bacillus subtilis* has a high turnover rate, corresponding to a half-life of only 2.5 minutes. The precise rate, however, varied with the individual species of mRNA and with certain other, largely unknown factors (Blundell *et al.*, 1972; Wice and Kennell, 1974). The presence of Ca^{2+} ions induced almost complete stabilization, half-lives of 20 to 45 minutes being indicated (Cremer and Schlessinger, 1974). While the sporulation mRNAs of *Bacillus subtilis* have been indicated to have short half-lives (Leighton, 1974), those of bacteriophage T7 have been demonstrated to be between 25 and 30 minutes. However, the mRNAs of this phage undergo rapid structural changes which quickly induce loss of function (Yamada *et al.*, 1974), so their effective half-lives are much shorter.

In contrast, the half-lives of eukaryotic mRNAs have been shown to vary widely, being quite stable in many cases. Among the diversity of examples available, the following few are representative of the range of stability: In sea urchin eggs, the mRNA has been reported to be synthesized prior to meiosis and to persist from fertilization through the blastula stage (Gross and Cousineau, 1973). Among such fungi as slime mold, the messenger for certain enzymes has also been shown to be extremely stable (Sussman, 1966), and, in the darkling beetle (*Tenebrio molitor*), mRNA for adult cuticular protein endured on ribosomes 4 or 5 days before it was translated into protein (Ilan *et al.*, 1966; Ilan, 1969). In contrast, HeLa Cell mRNA showed an average half-life of 3 to 4 hours (Penman *et al*, 1963). More recently, other workers have demonstrated that at least 40% of the mRNA present in the late G_2 phase of these cells persisted through mitosis and continued to function as the cell entered phase G_1 of the next interphase (Hodge *et al.*, 1969). In mouse kidney cells two types of mRNA were found; 57% of the newly synthesized messengers had a half-life of 6 hours, while the remaining 43% decayed at a half-life rate of 24 hours (Ouellette and Malt, 1976). Similar ranges of half-lives have been found in a number of other eukaryotic tissues (Bell and Rieder, 1965; Stiles *et al.*, 1976).

Lately some doubt has been cast on the validity of many of the long half-life estimates, as they were obtained through use of actinomycin D, which substance appears to have a prolonging effect on the durability of mRNAs. In a case involving a saturnid moth (*Hyalophora cecropia*), the half-life of a specific mRNA was determined with actinomycin D as exceeding 12 hours, whereas when cordycepin was used, it was found to be only 2 hours (Grayson and Berry, 1973). Furthermore, as in the chick lens messengers, two distinct classes of mRNA have been found in *Chironomus* salivary gland cell transcripts, which differed in half-life length as well as other characteristics (Edström and Tanguay, 1974). Changes in the length of the half-life of a number of mRNAs were demonstrated during the differentiation of myoblast cells (Buckingham *et al.*, 1974), and HeLa cells, mRNA with a half-life between 1 and 2 hours has been described in a current report (Puckett *et al.*, 1975).

As indicated by studies on bacteria, loss of the mRNA derives from enzymatic action in the cytoplasm, several enzyme systems being implicated. In the case of the lactose operon mRNA of *E. coli*, decay appeared to result from internal endonucleolytic cleavages (Blundell *et al.*, 1972); to the contrary, directional exonuclease activity seemed to be a predominant factor in the decay of the tryptophan operon mRNA (Morikawa and Imamoto, 1969; Wice and Kennell, 1974).

Structural Features. The original concepts of a simple, single-stranded structure for this type of RNA also are currently undergoing revision. The first indication that simplicity was not uniformly a characteristic of these molecules arose through base-sequence analyses of RNA viruses, the RNA molecules of which have now been established as serving largely in the capacity of messenger (Seigert *et al.*, 1973). In the earlier analyses, such as those of bacteriophages Qβ (Billeter *et al.*, 1969; Hindley and Staples, 1969; Goodman *et al.*, 1970) and R17 (Adams and Cory, 1970), only short segments of around 40 to 70 nucleotides in length were involved, and those were initiator fragments associated with ribosomes. Hence, the single hairpin loop indicated by the proposed secondary structure could be viewed as a functionally specialized region of the molecule. As the base sequences of segments exceeding 100 nucleotides in length were later determined, such as those in baeteriophage

MS2 (Contreras *et al.*, 1971; Fiers *et al.*, 1971), the existence of two or three loops at each end of the messenger was established, and similar findings were made in bacteriophage R17 (Sanger, 1971). As a result of these studies, especially the superb series undertaken at the University of Ghent, the nucleotide sequence of the three genes and inactive regions of the entire bacteriophage MS2 genome has been determined and a complex model (Figure 4.1A–C) for the secondary structure indicated (Min Jou *et al.,* 1972; Weissmann *et al.*, 1973; Fiers *et al.*, 1976). Direct physical evidence for such secondary structure has now been obtained also for a fragment of R17 mRNA (Gralla *et al.*, 1974).

This complexity of structure has been interpreted in natural selection terms (Ball, 1973; Riley, 1973), for the base pairing of loops obviously could place restraints on the randomness with which mutations might occur. These hairpin loops also have been suggested possibly to influence initiation at ribosome attachment sites (Ricard and Salser, 1974). In this connection, it should be noted that the RNA molecule is not only the messenger in these viruses but also is the sole bearer of genes. In other words, the genes of some of these viruses are translated into proteins without preliminary transcription into messenger nucleic acids. The codon–codon reactions involved in the hydrogen bonds of the loops have also been claimed to play a role in the nonrandomness of amino acid sequences (Polya and Phillips, 1976).

The mRNAs of prokaryotic and eukaryotic types alike do not lend themselves so readily to isolation, due to the complexity of the population of messengers existing at any given moment within a cell. However, it appears that at least some mRNAs of advanced organisms have double-stranded loops, for their presence has been determined in globin mRNA (Williamson *et al.*, 1971). Similar hairpin loops have been deduced to exist in the messengers for ovalbumin, human cytochrome *c*, and human hemoglobin α-chain, through analyses of the codons derived from the known amino acid sequence (White *et al.*, 1972), and by thermal denaturation technique (Van *et al.*, 1976). Furthermore, on a theoretical basis, messengers, like other RNAs, should be expected to form stable secondary structures (Gralla and DeLisi, 1974; Ricard and Salser, 1975).

Polyadenylic Acid [*Poly*(*A*)]. Another structural aspect of mRNA has come to light, the presence of long chains of polyadenylic acid at the 3′-end of the molecule. The earliest discoveries of polyadenylic acid [poly(A)] in RNAs were made in the nuclear contents of eukaryotic cells, but those results are discussed in Chapter 5 where they are more appropriate. In one early study using rat liver RNA, polypurine chains were found which showed a ratio of 5 adenosine to 1 guanosine residue (Lim *et al.*, 1970). More specifically determined RNA was then isolated from rabbit reticulocytes and established as being hemoglobin mRNA, with a sedimentation coefficient of 10 S. In purified preparations, it was demonstrated to possess a component 50–70 nucleotides long, with an adenylic acid content of over 70% (Lim and Canellakis, 1970; Burr and

Lingrel, 1971). Shortly afterward, it was shown that mRNA of HeLa cells similarly contained an adenylic-acid-rich region (Darnell *et al.*, 1971; Edmunds *et al.*, 1971). Recently, comparable poly(A)-rich segments have also been found in rat prostrate mRNA (Mainwaring *et al.*, 1974), and now have been identified in the messengers for many major proteins of yeast (Holland *et al.*, 1977).

Several unicellular types, including *Euglena gracilis* (Sagher *et al.*, 1974) and the common baker's yeast, *Saccharomyces cerevisiae* (McLaughlin *et al.*, 1973), have been reported to have a similar poly(A) sequence. In these two species the average poly(A) chain length was between 150 and 250 residues, but in the yeast, the results suggested that not all mRNAs carried this polyadenylate sequence. In contrast, in *Acanthamoeba* these segments were only 80 nucleotide residues long (Jantzen, 1974).

Among other multicellular eukaryotes which have been found to have polyadenylated messengers is the slime mold *Dictyostelium* (Firtel *et al.*, 1972; Jacobson *et al.*, 1974) and two flowering plants, *Zea mays* (Van de Walle, 1973) and *Vicia faba* (Sagher *et al.*, 1974), with 150–250 units in length. Mitochondria from HeLa cells were shown to contain such regions 50–100 units in length (Perlman *et al.*, 1973). In most cases studied, the poly(A) sequences from a single source were heterogeneous in length; this was true even for hemoglobin mRNA (Lim and Canellakis, 1970) and poliovirus RNA (Yogo and Wimmer, 1972), as well as for bulk poly(A)-containing RNAs extracted from eukaryotic cytoplasm (Gillespie *et al.*, 1973). Mouse globin mRNA, for example, contained two classes of pure poly(A) strands, one 50, the other 70, nucleotides long (Mansbridge *et al.*, 1974), but the size class distribution of this RNA species has been shown to be a function of time (Gorski *et al.*, 1975; Merkel *et al.*, 1976). Similar correlations had been reported for other mammalian cells (Jeffery and Brawerman, 1974) and sea urchin eggs (Mescher and Humphreys, 1974). One class of RNA tumor viruses, contrastingly, has been shown to contain long polyadenylic acid sequences which are subequal in length (Gillespie *et al.*, 1973). Because of their uniqueness, poly(A) sequences provide a convenient method of separating mRNAs and their precursors from other types in the cell (Hemminki, 1974); the only possible contaminants could be the genomes of RNA viruses, some of which have been shown to have similar termini, sometimes 190 units long (Rho and Green, 1974). By taking advantage of this method with HeLa cells, approximately 35,000 different poly(A)-containing RNA sequences were determined to be present, which fell into three distinct abundance classes (Bishop *et al.*, 1974). Thus a wide diversity of mRNA species appears to be present simultaneously in cells (Iatrou and Dixon, 1977).

The amount of polyadenylation varies also with development. In sea urchin eggs at cleavage, for instance, only 25% of the polysomal RNA bore poly(A) sequences (Fromson and Duchastel, 1975); this contrasts to 80% or

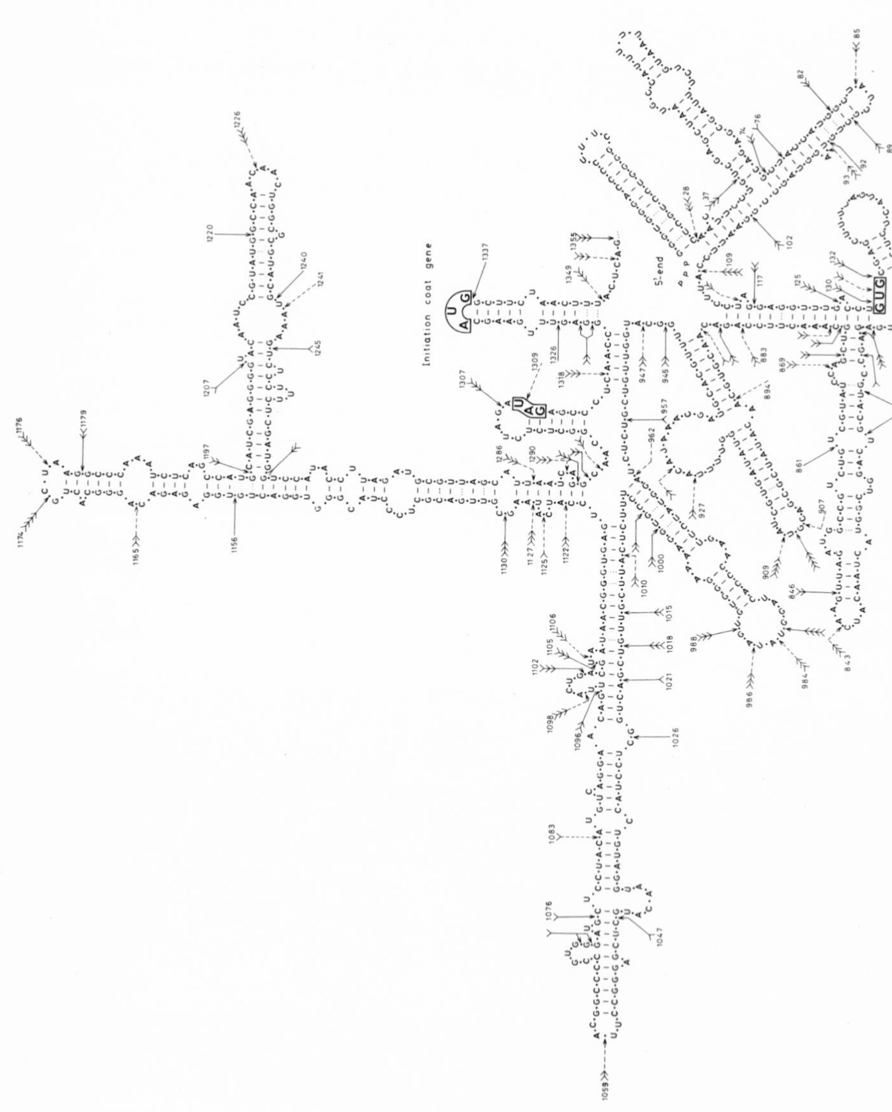

Initiation coat gene
5' end

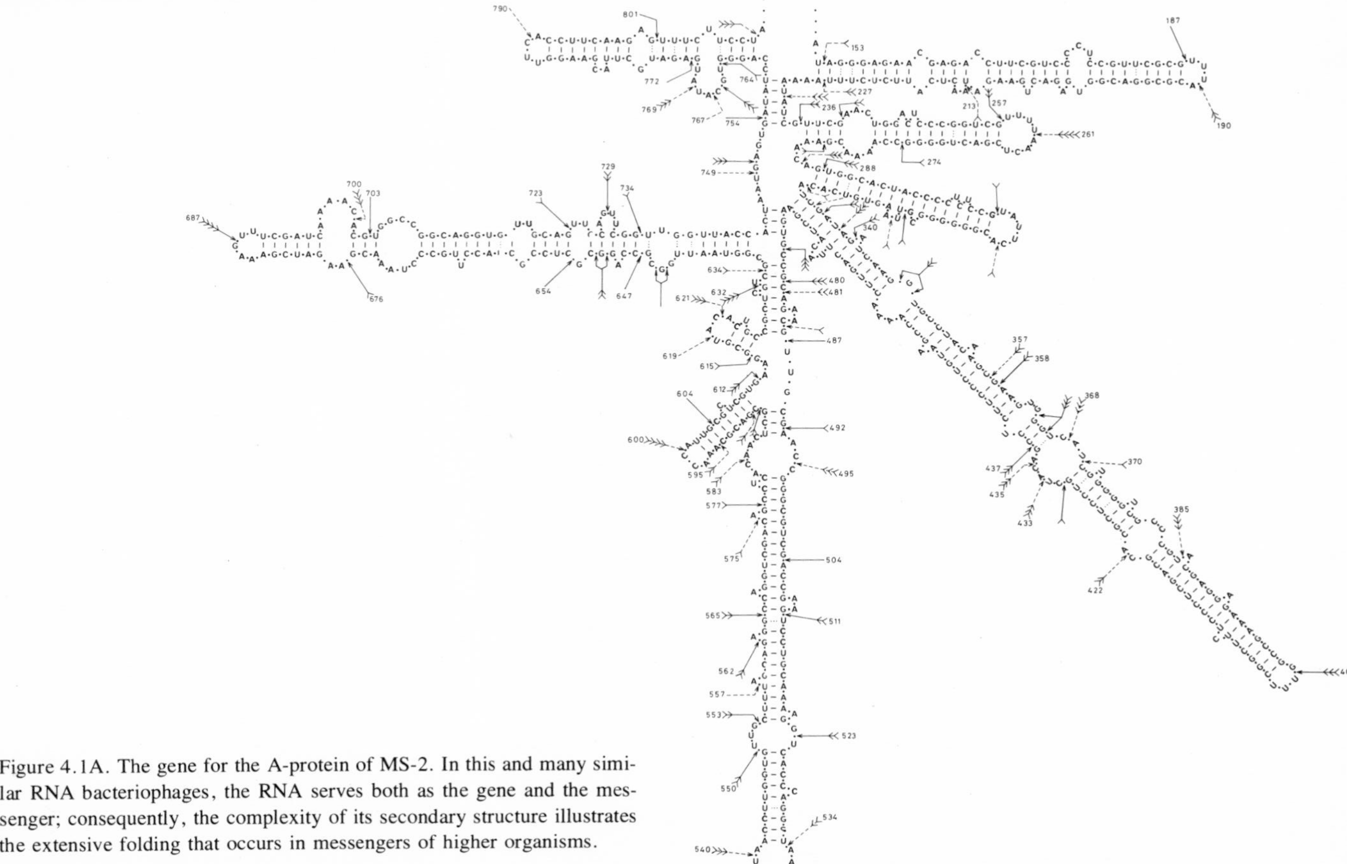

Figure 4.1A. The gene for the A-protein of MS-2. In this and many similar RNA bacteriophages, the RNA serves both as the gene and the messenger; consequently, the complexity of its secondary structure illustrates the extensive folding that occurs in messengers of higher organisms.

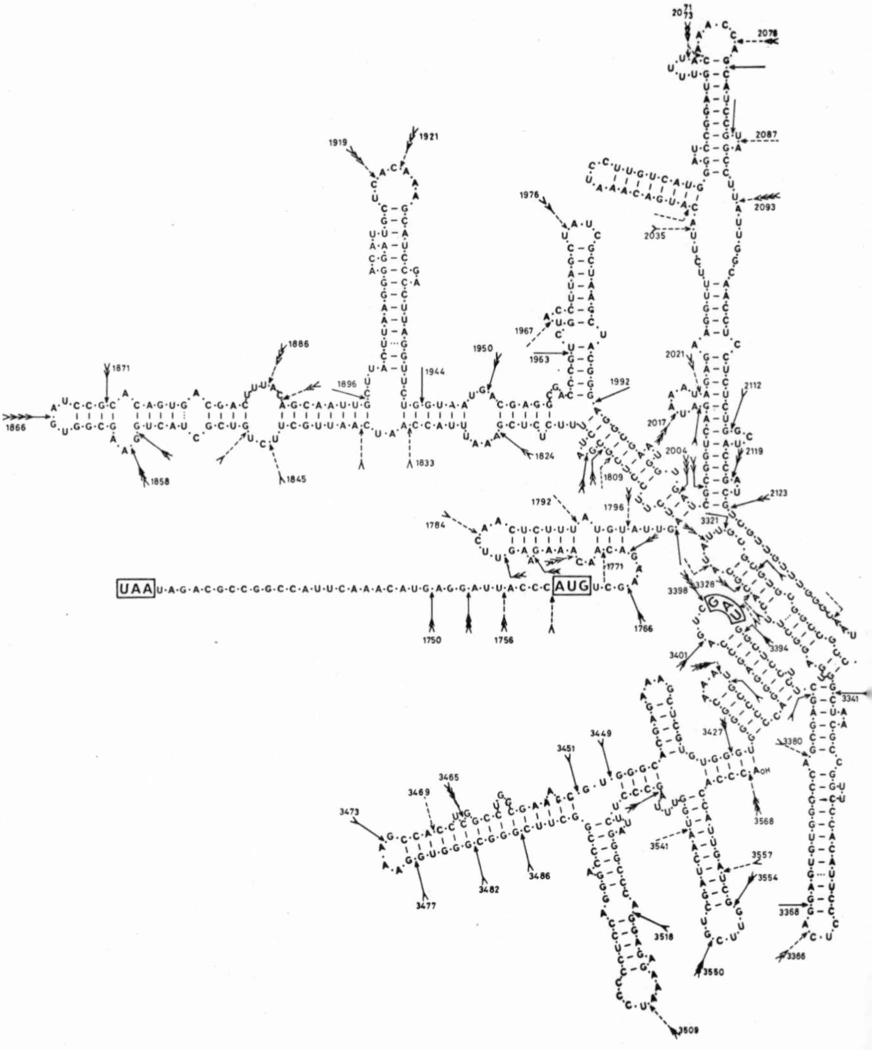

Figure 4.1B. The gene and messenger for the coat protein of MS-2.

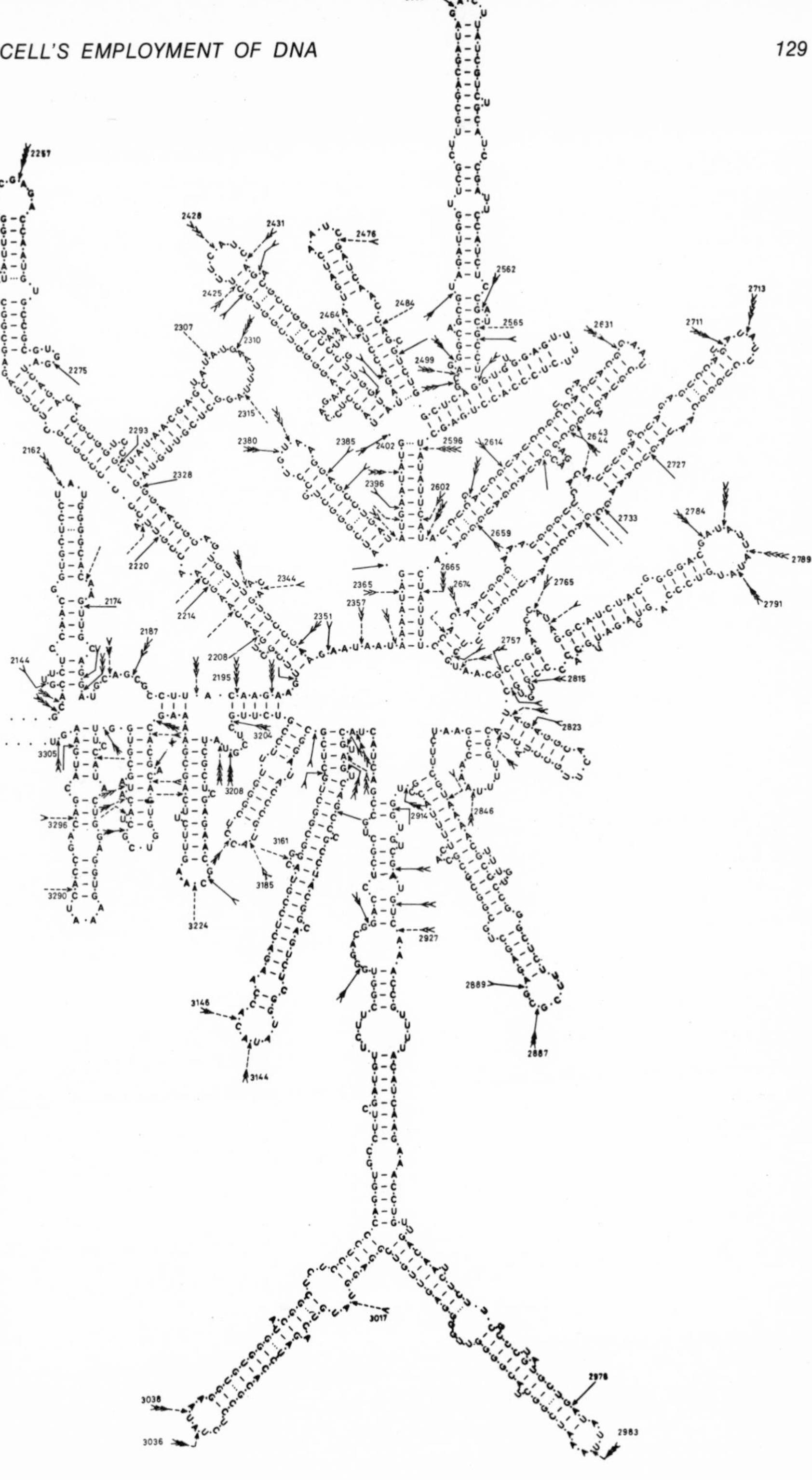

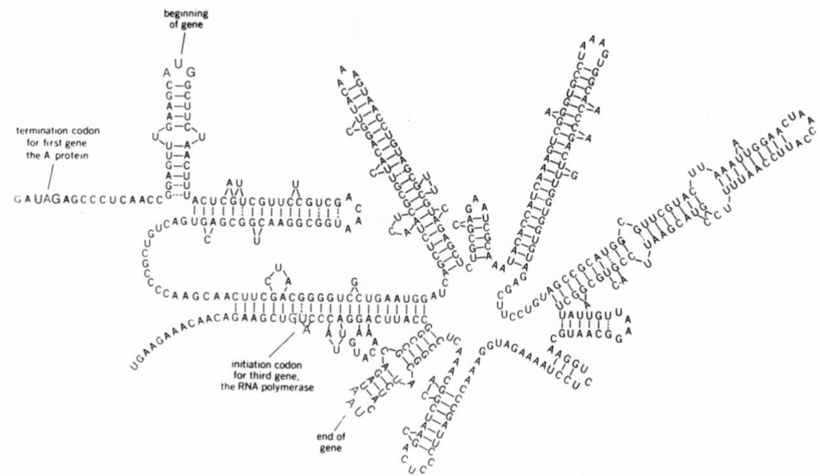

Figure 4.1C. The gene and messenger for the MS-2 RNA polymerase. This exemplary study makes clear the complexity of structuring present in long, linear nucleic acid chains. (A–C, courtesy of Dr. W. Fiers and co-workers and *Nature* **260:**500–507, 1976).

90% found among mammalian cells (Greenberg and Perry, 1972). This proportion then gradually increased with development, reaching 33% in the early blastula and 48% toward the end of that stage. Variation during the cell cycle likewise has been reported. In *Physarum*, 35% of the polysomal RNA was polyadenylated during the G_2 phase, whereas only 20% was in the S phase (Fouquet and Sauer, 1975).

The function of these poly(A)-rich regions still remains an open question. Early in their history, it was proposed that they were involved in the transport of mRNA from the nucleus to the cytoplasm (Philipson *et al.,* 1971), but the presence of similar sequences in mitochondrial mRNA (Perlman *et al.*, 1973) and the RNA of cytoplasmic viruses (Johnston and Bose, 1972; Yogo and Wimmer, 1972) precludes this from being likely. Furthermore, the increase in length of the poly(A) region within the cytoplasm (Slater *et al.*, 1973; Diez and Brawerman, 1974), an observation now substantiated by the discovery of a poly(A)-polymerase in the cytoplasm (Tsiapalis *et al.*, 1973), argues for an essential extranuclear role for these sequences. This role has been shown to be unrelated to translational properties, for mRNA functions as well when deprived of such regions as when they are present (Cann *et al.,* 1974; Fromson and Verma, 1976). Part of the problem may be correlated to the recent finding that not all poly(A)-bearing RNA enters into polysome formation (McLeod, 1975). A significant, but undermined, fraction of such RNA appeared to be an inactive (nontranslating) form of "mRNA"; thus it was not really messenger in function but may play some other role in the cell.

The Nonuniversality of Poly(A). The presence of polyadenylic-acid-rich sequences is thus widespread among eukaryotes, but they are not a universal feature. In addition to the above suggested absence in some yeast mRNAs, they are lacking in the histone mRNA fraction of HeLa cells (Adesnik and Darnell, 1972) and have been demonstrated to be absent in some nonhistone messengers of mammalian cells (Greenberg, 1976; Houdebine, 1976). Moreover, they have as yet not been detected in the bacterial viruses. Among other viruses, however, they have been demonstrated in the mRNAs. This feature has been reported present in the large DNA vaccinia virus (Kates, 1970), adenovirus (Philipson *et al.*, 1971), vesicular stomatitis virus (Ehrenfeld and Summers, 1972), Sanbis virus (Johnston and Bose, 1972), poliovirus (Yogo and Wimmer, 1972), Rous sarcoma and mouse Rauscher leukemia viruses (Lai and Duesberg, 1972), and reovirus (Stoltzfus *et al.*, 1973). An absence of the same material from prokaryotic mRNA has also been generally assumed (Perry *et al.*, 1972), but this has recently been shown to be erroneous (Nakazato *et al.*, 1975). In *E. coli* a few species of mRNA were found to have poly(A) segments 20 or more units in length, and a number of others had oligo(A) sequences of undetermined length. Segments of poly(A) of similar length have also been reported from *Caulobacter* (Ohta *et al.*, 1975).

Complete Structure of mRNA. From studies on mRNAs from insects (Ilan and Ilan, 1973a) and from developing embryos of the clawed toad, *Xenopus laevis* (Dina *et al.*, 1973, 1974), a fuller, but still confused picture of at least eukaryotic mRNA structure is beginning to emerge (Greenberg, 1975). It appears that each mRNA molecule contains three major regions, the poly(A) series at the 3′-end, between 50 and 200 nucleotides long, followed by the unique sequence of various lengths carrying the actual information transcribed from the DNA, and finally a repetitive fragment located at the 5′-end, about 50 to 60 nucleotides long. The latter region, too, is transcribed from DNA, apparently from repeated sequences. Indeed, these repetitive regions, rather than the unique sequences, have been proposed to carry the genetic information, at least for the keratin of feathers (Kemp, 1975). In HeLa cells about 6% of the total polyadenylated-mRNA preparation is transcribed from repetitive DNA sequences (Klein *et al.*, 1974). To the contrary, Campo and Bishop (1974) described two classes of mRNA in cultured rat cells, a minor fraction, that was of smaller size (< 18 S) and contained a higher proportion of repetitive sequence transcripts, and a major portion of larger size (> 18 S), with fewer redundant transcripts. Approximately 20% of the mRNA population appeared to have been transcribed from repetitive DNA.

Although more attention has been paid to the 3′- than to the 5′-termini of mRNAs, during the past few years several contributions to the latter have been published. Among the unexpected findings that have been reported is the presence of modified nucleosides and unusual phosphate-bonding patterns. Mouse L cells were the first found to have mRNAs with a low level of methylation, both base and ribose moieties being modified (Perry and Kelley, 1974; Wei *et al.*, 1976). Most prominent in the pattern of modification in mouse myeloma mRNA was N^6-methyladenosine, which occurred near each terminus. In addition, the 5′-termini had 7-methylguanosine and riboside-methylated nucleosides linked by three phosphate radicals joined by 5′ to 5′ bonds (Adams and Cory, 1975; Perry and Scherrer, 1975). Thus the last two or three nucleotide residues were reversed in direction relative to the rest of their respective molecules. Similar methylations and terminal structures were reported simultaneously by another laboratory in reticulocytes and in reovirus and vesicular stomatitis virus mRNAs (Murthukrishnan *et al.*, 1975); their study demonstrated also that these caps, as they are called, were essential for translation (Shimotohna *et al.*, 1977). According to a base-sequence analysis of the first 77 nucleotides from the 5′-end of galactose mRNA of *E. coli*, similar modifications were missing (Musso *et al.*, 1974), unless thay had been accidentally overlooked or removed during preparation. However, this messenger was reported to possess triphosphate termini (Toivonen and Nierlich, 1974). The light chain of an immunoglobin has been found to have the 3′-end blocked by a unique amino acid, pyroglutamic acid (Burnstein *et al.*, 1976).

The nature of the unique region appears to differ somewhat between prokaryotes and eukaryotes. Among the latter, at least in yeast, the mRNAs reportedly are monocistronic, carrying the information for only one polypeptide (Petersen and McLaughlin, 1973). Contrastingly, bacteria seem to have long messengers that are polycistronic, as well as short ones of monocistronic composition (Jacob and Monod, 1961; Ames and Hartman, 1963; Yanofsky and Ito, 1966).

Messenger RNA as Ribonucleoprotein. In addition to the above structural features, most species of polysomal mRNA seem to have two or more proteins bound to them and are thus ribonucleoproteins (Spirin, 1969; Georgiev and Samarina, 1971). This type of complex was noted in the reticulocytes of rabbits and ducks (Lebleu *et al.*, 1971; Morel *et al.*, 1971; Blobel, 1972), where the various polypeptide species were estimated to have molecular weights in the range of 5,000–130,000. At least one of the proteins was essential for the binding of message to the small ribosomal subunit (Lebleu *et al.*, 1971; Ilan and Ilan, 1973b; Dworkyn *et al.*, 1977). Eight distinct polypeptides with molecular weights between 44,000 and 100,000 were present in chick embryonic muscle mRNA (Bag and Sarkar, 1975). This nucleoprotein-complex nature of mRNA seems to be proving universal, at least among eukaryotes. Labeled mRNAs from chick brains, representing a large population of varied types, were determined to be consistently of this nature (Bryan and Hayashi, 1973), as were those from Ehrlich ascites tumor cells (Barrieux *et al.*, 1975). A similar complexed condition has been reported also from fish embryos (Spirin, 1969), sea urchin eggs (Monroy *et al.*, 1965), rat liver (Williamson, 1973), and higher plant embryos (Ajtkhozin and Akhanov, 1974). Further, specific proteins have been described as being associated with mRNA in the nucleus (Lukanidin *et al.*, 1972), in the cytoplasm (Spirin, 1969), in polysomes

(Ajtkhozhin and Akhanov, 1974), and bound to the poly(A) sequence (Kwan and Brawerman, 1972; Blobel, 1973; Rosbash and Ford, 1974), and in each class the proteins have been shown to be distinct (Liautard *et al.*, 1976). No function has been established for the proteins (Williamson, 1973), but the proteins have been shown to be site-specific within the mRNAs (Sundquist *et al.*, 1977).

By using ethidium bromide which becomes highly fluorescent when intercalated into double-stranded nucleic acids, it was found that 45% to 60% of duck globin ribonucleoprotein messengers were in the form of double helices (Favre *et al.*, 1975). Similar extensive secondary structure was reported also for rabbit globin mRNA–protein complex, and for that of calf lens and ewe mammary gland. Poly(A) regions comprised 4% to 9% of the messengers. While free mRNA showed as much secondary structure as the protein bound, the dichroic spectra of the two were significantly distinct. The poly(A) segments also were shown to bear proteins.

Purified Messengers. As a few messengers have been obtained in relatively pure form, some of their properties now are being disclosed. Mouse interferon mRNA was assayed biologically and found to consist of two forms, both of which were functional (Montagnier *et al.*, 1974). One retained by oligo (dT)-cellulose had a sedimentation coefficient > 18 S, was degraded readily by pancreatic RNase, and resisted RNase III; the other was not retained by oligo (dT)-cellulose, sedimented at values < 18 S, was resistant to pancreatic RNase but destroyed by RNase III, and appeared to be double-stranded.

Purified ovalbumin mRNA has been reported to exist in hen oviduct cytoplasm in a molecule far longer than required by the size of the protein molecule it codes (Haines *et al.*, 1974), contrary to reports suggesting that precursor ovalbumin mRNA did not exist (McKnight and Schimke, 1974). The purified messenger was shown to consist of between 1670 and 2640 nucleotides, whereas the 387 amino acids in ovalbumin should require the presence of only 1161. Hence, it must contain a large untranslated segment over and above the typical poly(A) sequence. This length exceeding 1600 nucleotides was confirmed by a second laboratory (Rosen *et al.*, 1975). From bacteria, purified mRNA for a specific lipoprotein has been obtained and found also to have a length (230– 250 nucleotide residues) exceeding the 180 nucleotides actually required (Hirashima *et al.*, 1974). Purified keratin mRNA from embryonic chicks indicated that between 25 and 35 different species existed in the embryonic chick feather and that 100 and 240 keratin cistrons existed in the chick genome (Kemp, 1975). Each of these genes consisted of both unique and repetitive sequences.

4.1.2. The Genetic Code

As stated above, the unique sequence of an mRNA molecule is a copy of one item of information that has been transcribed from the DNA molecule. This

Table 4.1
The Genetic Code as Transcribed into Messeger RNA

AAA	Lysine (Lys)	CAA	Glutamine (Gln)	GAA	Aspartic acid (Asp)	UAA	Stop (ochre)[b]
AAG	Lysine	CAG	Glutamine	GAG	Aspartic acid	UAG	Stop (amber)[b]
AAC	Asparagine (Asn)	CAC	Histidine (His)	GAC	Glutamic acid (Glu)	UAC	Tyrosine (Tyr)
AAU	Asparagine	CAU	Histidine	GAU	Glutamic acid	UAU	Tyrosine
ACA	Threonine (Thr)	CCA	Proline (Pro)	GCA	Alanine (Ala)	UCA	Serine (Ser)
ACG	Threonine	CCG	Proline	GCG	Alanine	UCG	Serine
ACC	Threonine	CCC	Proline	GCC	Alanine	UCC	Serine
ACU	Threonine	CCU	Proline	GCU	Alanine	UCU	Serine
AGA	Arginine (Arg)	CGA	Arginine (Arg)	GGA	Glycine (Gly)	UGA	Stop (opal)[b]
AGG	Arginine	CGG	Arginine	GGG	Glycine	UGG	Tryptophan (Trp)
AGC	Serine (Ser)	CGC	Arginine	GGC	Glycine	UGC	Cysteine (Cys)
AGU	Serine	CGU	Arginine	GGU	Glycine	UGU	Cysteine
AUA	Isoleucine (Ile)	CUA	Leucine (Leu)	GUA	Valine (Val)	UUA	Leucine (Leu)
AUG	Methionine (Met)[a]	CUG	Leucine	GUG	Valine	UUG	Leucine
AUC	Isoleucine	CUC	Leucine	GUC	Valine	UUC	Phenylalanine (Phe)
AUU	Isoleucine	CUU	Leucine	GUU	Valine	UUU	Phenylalanine

[a]The chain-initiating factor.
[b]The three "stops" serve as chain-terminating factors.

information is in the form of a series of three-letter code words, called codons, each of which codes for one given amino acid. Such triplet codons are required, because twenty different amino acids are commonly found in proteins, whereas only four principal types of bases are available in the nucleic acids. Thus a sequence of only two of these four possible bases would provide 4^2 combinations, not quite enough for the coding purposes (Crick *et al.*, 1961; Crick, 1966, 1967). Although the triplet sequence that is now well established as actually existing provides a plethora of code words (4^3), all 64 appear to be used by organisms. Hence, if the protein coded by a given mRNA contains 110 amino acids, as for example certain cytochromes *c* (Dillon, 1978), the message encoding it in the mRNA would consist of a base sequence 330 residues in length. This in turn would be a complementary copy of a sequence of identical length in a DNA molecule. Each such continuous sequence in a single molecule of DNA (or RNA in certain viruses) is a gene, currently often referred to as a *cistron*.

As shown in Table 4.1, 61 of these triplets code for amino acids, and the remaining three for stops, that is, any one of the latter indicates where in the long continuous molecule transcription is to be terminated (Garen, 1968). Since most of the 20 amino acids are each encoded by more than one triplet, the genetic code is said to be "degenerate." The term, however, is an unfortunate choice, for in biology it carries implications that the code was at one time more advanced or complex than it is now, but such is not necessarily the case. The term "redundant," applied by Wong (1975), appears far more appropriate. This redundancy is the first of a number of chemically unexpected phenomena which are encountered during this analysis of the genetic mechanism, processes which seem inexplicable on straightforward physiochemical terms, but which can be explained on an evolutionary basis, as Woese (1972) proposes.

As far as is known, the code is universal among extant living things; bacteria, yeast, seed plants, metazoans, and even viruses employ the same codons to specify any given amino acid. Various specific tests of this universality have been made from time to time. One study showed that amino acid coding in *Sarcina lutea* was identical to that of *Escherichia coli*, and that of the baker's yeast (*Saccharomyces cerevisiae*) was precisely the same as the preceding two (Groves and Kempner, 1967). Another investigation compared the codons recognized by amphibian and mammalian liver preparations to those of *E. coli* and likewise found no differences (Marshall, 1967).

This constancy has been even more convincingly demonstrated by the base-sequence analyses made of various bacteriophage cistrons which code for proteins for which the amino acid sequence had already been established. One such study was that of the bacteriophage MS2 coat protein gene cited earlier (Min Jou *et al.*, 1972), for which the amino acid sequence of the protein previously had been determined (Lin *et al.*, 1967; Weber and Koenigsberg, 1967; Vandekerckhove *et al.*, 1969). In fact, the latter sequence often provided clues for the deciphering of the nucleoside sequence. However, while close correspondence is shown in the illustration (Figure 4.1), some disharmonies were demonstrated to exist between the amino acids established at certain sites by means of the base sequence and those actually found by the amino acid analysis. In that same figure it is to be noted also that the supposed base sequence for the "read-through" protein contains only 52 bases, instead of a multiple of three such as 51 or 54 (Contreras *et al.*, 1973).

A topic of special pertinence in discussions of the origins of life is how the existing combinations of codons became assigned primitively to the respective amino acids. Although a number of approaches to the problem have been made, those studies are more appropriately reviewed in a separate chapter. Here attention will be confined to the recent analysis (Dillon, 1973) that revealed the nature of the nonrandomness in codon assignment which had been noted previously (Goldberg and Wittes, 1966). This lack of randomness is evident even from a cursory view of the codon catalog, especially when they are arranged, as in Table 4.2, according to the number of codons assigned to each amino acid.

In the tables and analysis of the pattern of codon assignment, stops, which frequently are also called "nonsense," are treated as though they represent one type of amino acid. Two features of the genetic code are especially noteworthy, the abundance of *sets of four* and *natural pairs* in the codon catalog. Natural pairs, lightly shaded in the tables, have the first two bases of each member codon identical, while the third position is occupied in turn by the common purines (for example, GA*A* and GA*G*), or, alternatively by the two usual pyrimidines (for example, UG*C* and UG*U*). Similarly in any given set of four the first two bases in all four codons are identical, while the third position is occupied in turn by one of the four common nucleotide residues, resulting in sets of combinations like GC*A*, GC*C*, GC*G*, and GC*U*. The presence of a few singles may also be noted, including AUA and UGG, neither of which forms a part of a pair or set of four coding one amino acid.

In the actual genetic code (Table 4.2), three amino acids are encoded by six codons apiece; each such collection of six may be seen to consist of a set of four plus a natural pair. Five amino acids are each provided with sets of four, and nine have natural pairs. Singles occur in the code assignments of only four amino acids, two of which (isoleucine and stops) are provided also with a natural pair each. Thus it is quite evident that the codons are not randomly assigned to amino acids, for 17 of the 21 are coded by such orderly combinations as pairs, sets of four, or combinations of pairs and sets.

In the analysis of this pattern of codon assignment (Dillon, 1973), random drawings were made of slips of paper bearing codons arranged in various combinations along with others bearing the name of an amino acid. In order to determine whether the existing pattern could have arisen by chance combinations of a single codon with an individual amino acid, slips bearing single codons

Table 4.2
The Genetic Code: Amino Acids Arranged According to Number of Codons[a]

Number of codons	Codons	Amino acid	Abreviation system 3-letter	1-letter[b]
6	AGA, AGG, CGA, CGC, CGG, CGU	Arginine	Arg	R
	CUA, CUC, CUG, CUU, UUA, UUG	Leucine	Leu	L
	AGC, AGU, UCA, UCC, UCG, UCU	Serine	Ser	S
4	GCA, GCC, GCG, GCU	Alanine	Ala	A
	GGA, GGC, GGG, GGU	Glycine	Gly	G
	CCA, CCC, CCG, CCU	Proline	Pro	P
	ACA, ACC, ACG, ACU	Threonine	Thr	T
	GUA, GUC, GUG, GUU	Valine	Val	V
3	AUA, AUC, AUU	Isoleucine	Ile	I
	UAA, UAG, UGA	Stops (nonsense)	—	—
2	AAC, AAU	Asparagine	Asn	N
	GAC, GAU	Aspartic acid	Asp	D
	UGC, UGU	Cysteine	Cys	C
	GAA, GAG	Glutamic acid	Glu	E
	CAA, CAG	Glutamine	Gln	Q
	CAC, CAU	Histidine	His	H
	AAA, AAG	Lysine	Lys	K
	UUC, UUU	Phenyl-alanine	Phe	F
	UAC, UAU	Tyrosine	Tyr	Y
1	AUG	Methionine	Met	M
	UGG	Tryptophan	Trp	W

[a]Natural pairs are lightly shaded, sets of four, heavily so.
[b]The system used by Dayhoff (1972) is followed.

Table 4.3
Results of Random Coding: Codons First Assigned by Sets of Four, Then Shared by Random Drawings

Number of codons	Codons	Amino acid
After first sharing processes		
4	AAA, AAC, AAG, AAU	Alanine
	GAA, GAC GAG, GAU	Asparagine
	UGA, UGC, UGG, UGU	Cysteine
	ACA, ACC, ACG, ACU	Glutamic acid
	AGA, AGC, AGG, AGU	Glutamine
	GGA, GGC, GGG, GGU	Isoleucine
	UUA, UUC, UUG, UUU	Lysine
	CGA, CGC, CGG, CGU	Methionine
	GUA, GUC, GUG, GUU	Phenylalanine
	UCA, UCC, UCG, UCU	Tryptophan
	CCA, CCC, CCG, CCU	Tyrosine
2	CAC, CAU	Aspartic acid
	CAA, CAG	Valine
2	CUC, CUU	Glycine
	CUA, CUG	Proline
2	UAC, UAU	Leucine
	UAA, UAG	Histidine
2	AUC, AUU	Serine
	AUA, AUG	Threonine
2	GCC, GCU	Stops

Number of codons	Codons	Amino acid
After final sharings		
6	UGA, UGC, UGG, UGU, AAA, AAG	Cysteine
	GUA, GUC, GUG, GUU, GGA, GGG	Phenylalanine
	CCA, CCC, CCG, CCU, CGA, CGG	Tyrosine
4	GAA, GAC, GAG, GAU	Asparagine
	ACA, ACC, ACG, ACU	Glutamic acid
	AGA, AGC, AGG, AGU	Glutamine
	UUA, UUC, UUG, UUU	Lysine
	UCA, UCC, UCG, UCU	Tryptophan
3	AUC, AUU, GCC	Serine
	CUG, CUU, CAA	Glycine
2	AAC, AAU	Alanine
	GCA, GCG	Arginine
	CAC, CAU	Aspartic acid
	UAA, UAG	Histidine
	GGC, GGU	Isoleucine
	UAC, UAU	Leucine
	CGC, CGU	Methionine
	CUA, CUG	Proline
	AUA, AUG	Threonine
1	GCU	Stop

were first drawn along with those carrying the name of an amino acid. After all the codons had thus been "assigned," no sets of four were found in the results, and only two natural pairs occurred. When the drawings then were made with slips bearing natural pairs of codons, a somewhat greater similarity between the experimental and actual code assignment pattern resulted, but still no sets of four existed and obviously no singles.

Slips bearing sets of four codons were then drawn. As a result, 16 amino acids became encoded with one such set apiece, while five others necessarily remained uncoded (Table 4.3). Thus, while the actual code catalog was approached more closely, further procedures obviously were required. In continuing the analysis, the names of two amino acids were drawn together, one of which had already been provided with codons while the other one had not; the codons of the former then were divided between the two in the form of natural pairs. After all amino acids had thus been provided with at least a natural pair each, further sharing of these pairs performed by a few similar random drawings resulted in the necessary singles and an exact duplicate of the existing pattern (Table 4.3).

Although the above steps involved randomness in the codon-assigning processes, in that mere slips of paper were drawn, the nonrandomness of the whole catalog of codons is made evident. From the results it is clear that in order to have produced the pattern characteristic of the existing code catalog, codon assignment must originally have been by providing 16 amino acids with a set of four each. Later, certain of those sets became halved, one amino acid retaining out of its original complement perhaps those members terminating in the pyrimidines, the second one becoming endowed with those ending in the purines. Still later, further partitioning resulted in subdividing two pairs so that four amino acids finally had a single codon, either alone or in company with a pair. A portion of this sequence of events had been proposed earlier by Jukes (1969), without his scrutinizing the processes in detail. The sharing of codons between amino acids is the second of the chemically inexplicable events found in the genetic mechanism and must be assumed to have been accomplished by evolutionary processes. These would include natural selective agents acting upon the various enzyme systems and nucleic acids involved in the genetic mechanism. It must be stressed that no other sequence of events seems to be available that can result in the pattern of the codon catalog, an observation that has a particular bearing in connection with any discussion of the origins of the coding mechanism, such as that in Chapter 6.

4.1.3. Transfer RNA

Although the genetic code ultimately provides a coding system for specifying a sequence of amino acids, it accomplishes this end in an indirect fashion. The triplets in mRNA actually interact with similar, but complementary, sets of nucleotides, known as *anticodons,* contained in transfer RNA (tRNA), of which there is a large diversity of types. Each type carries only one specific amino acid and each bears a distinctive sequence of three nucleotides in a particular portion of its molecule that serves as the anticodon. The sequence of amino acids for a given protein is thus indirectly determined by codons interacting with corresponding anticodons on the tRNAs, which molecules carry one specific amino acid each. Because of their relatively small size and great importance, tRNAs have received so much attention that a separate chapter is required to cover the subject adequately; consequently, detail here is limited to that required for understanding the nature of translation and transcription.

Characteristics. As a class, tRNAs are the smallest of the nucleic acids, with a molecular weight in the range of 23,000 to 28,000 and a sedimentation coefficient of 4 S (Brown, 1963). Each individual molecule consists of between 75 and 90 bases; serine tRNA ($tRNA^{Ser}$) from *E. coli* ranks among the longer ones with 89 bases, and $tRNA^{Ile}$ from the same source among the shorter, as it consists of only 76 nucleotide residues (Ishikura *et al.*, 1971; Yarus and Barrell, 1971). As the base sequence has now been determined for more than 90 different tRNAs (Tables 7.1–7.7), many features of the molecular primary structure are well established. Similarly it seems clear that the secondary structure of the molecule is in the form of a *cloverleaf,* several examples of which are shown in Figure 4.2. Four long arms are universal in the macromolecules, plus a fifth one of variable length. The stemlike one (arm I) is referred to as the *amino acid arm,* because of the firm evidence which indicates that it serves for attachment of the amino acid. To the left is a hairpin loop (arm II), known as the dihydrouracil arm, from the usual presence of one or more modified bases of that type. On the opposite side is arm V, the thymidine–pseudouridine (TΨC) arm, named from the nearly constant presence of those two nucleotide residues. The *anticodon arm* (arm III) is located at the lower end, and the last one (arm IV), is known as the extra arm (Cramer, 1969), which sometimes is so reduced as to be virtually absent.

The anticodon is located at the extreme tip of the anticodon arm and, from a chemical structural standpoint, is in reverse sequence to the codon (Figure 4.2). Thus the third base of the anticodon reacts with the first base of the codon, and the first base of the anticodon with the third codonal base. As the structure of this region and the remainder of the arm is particularly pertinent in the interactions with the mRNA, full discussion is held for the later chapter mentioned above.

Tertiary Structure. Although the tertiary, or three-dimensional, structure of tRNAs has not as yet been fully ascertained, it is consistently considered to involve extensive folding of the several arms. As a consequence, the TΨC arm appears to become partially concealed by or interconnected with certain others (Cramer, 1969; Levitt, 1969) so that the whole molecule approximates a bent

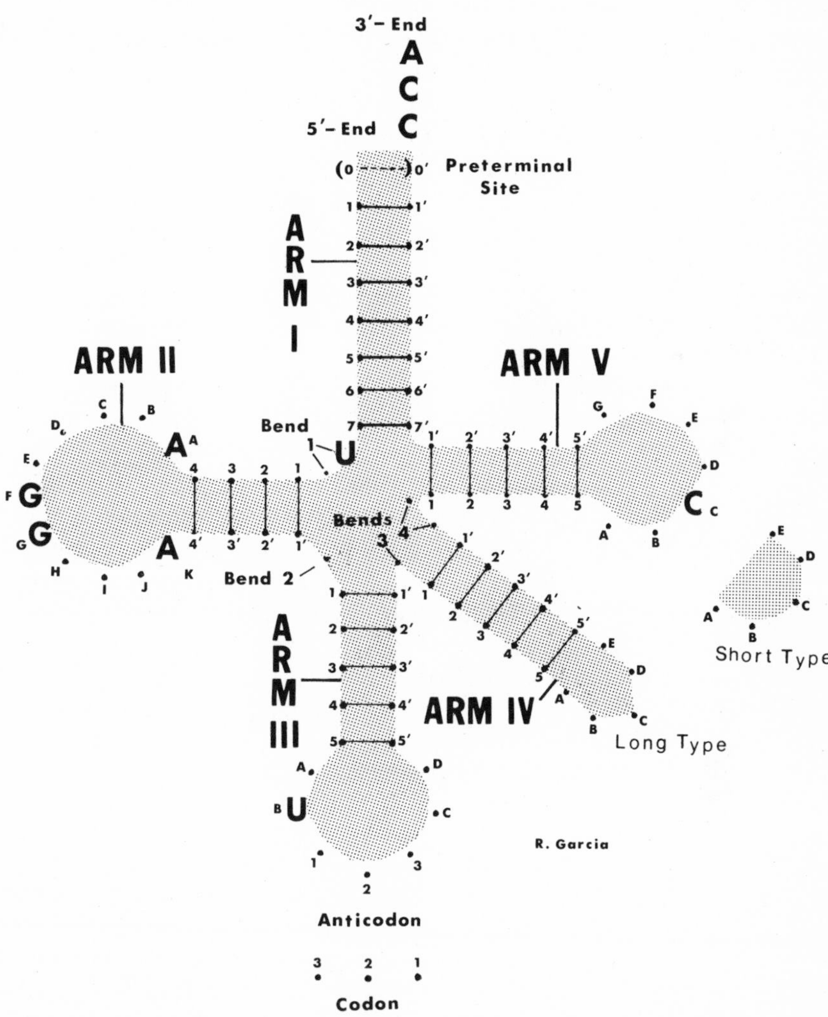

Figure 4.2. The secondary structure of transfer RNA. Transfer RNAs of prokaryotes, eukaryotes, and DNA viruses appear to be in the form of a "cloverleaf." Two major types are extant, one of which has arm IV long. The more usual condition is represented by the short type of arm IV, consisting of only four to six nucleotide residues. Bases universally present in homologous sites are shown in boldface.

elongate cylinder about 80 Å long (Figure 8.4), ovate in cross section, with dimensions of 25 × 35 Å (Kim and Rich, 1969). This configuration is maintained by means of tertiary base pairing between various components as described in Chapter 7 (Wong *et al.*, 1975). Because the anticodon loop is more than ten times as subject to degradation by endonuclease action as the others, it seems to be the most exposed portion of the molecule (Reid *et al.*, 1972). Abraham (1971) considered the –CCA end of the amino acid arm to be folded so as to approach the anticodon to some degree. Nuclear magnetic resonance spectroscopy studies showed that, at least in the phenylalanyl-tRNA analyzed, no detectable structural changes in configuration occurred when the amino acid became attached to the tRNA (Wong *et al.*, 1973). Extensive changes did occur, however, with extremes of cold and heat and at low salt concentrations; these involved loss of tertiary structure and even the cloverleaf pattern (Cole *et al.*, 1972; Cole and Crothers, 1972).

Activation Steps. Among the most important reactions involving this type of nucleic acid are the processes of activation, that is, charging the molecules with the amino acids. For the steps to occur, ATP (no other nucleoside triphosphate reacts), amino acids, Mg^{2+}, the tRNAs, and enzymes are necessary. The enzymes are correctly referred to as tRNA ligases (Chambers, 1971; Loftfield, 1972), although the older names, "aminoacyl-tRNA synthetases" and "amino acid–activating enzymes," still persist in the literature. Very little of a general nature can be stated for these ligases collectively, for they vary extensively in size and structure. A few are small, with a molecular weight of about 50,000, but others run to 260,000. Several, such as isoleucine and glutamine tRNA ligases from *E. coli,* consist of a single polypeptide chain; may others contain two identical subunits, and a very few are made of four identical subunits or two nonidentical pairs (Loftfield, 1972). One to four active sites may exist per molecule, but the active sites do not appear to interact with one another. Perhaps the sole generalization valid at present is that each ligase typically recognizes only one type of amino acid and only those tRNAs that are the right anticodons. But exceptions to this rule are widespread also, as pointed out below. Whether more than a single ligase is present in those cases where a number of isoaccepting tRNAs* exist in a cell has not been clearly determined (Hampel and Enger, 1973).

The generally accepted mechanism for charging a tRNA molecule is that (1) the enzyme reacts first with ATP, and then (2) with the amino acid, to form aminoacyl-adenylate. Next (3) it releases inorganic pyrophosphate, (4) binds the tRNA, (5) unites the aminoacyl-adenylate to the latter, and (6) finally releases the tRNA thus esterified. However, Loftfield (1972) presented evidence that, in the normal cell, the enzymes and tRNA existed as enzyme–tRNA

* Isoaccepting tRNAs are the various species of tRNA that carry the same given amino acid, such as $tRNA_1^{Leu}$, $tRNA_2^{Leu}$, $tRNA_3^{Leu}$, etc.

complexes and that the early reactions proceed at a sufficent pace only in the presence of the tRNA.

Several investigations completed recently serve to clarify the issue somewhat. For example, Maeliche *et al.* (1974) demonstrated clearly that ATP was essential for rat $tRNA^{Ser}$s to bind to yeast serine tRNA ligase. Whereas with yeast $tRNA^{Ser}$, a single binding site was found in the enzyme, as indicated by the 1 : 1 ratio that existed between the tRNA and ligase, in the absence of ATP and with rat $tRNA^{Ser}$, about five molecules per enzyme molecule were bound. However, when ATP was added, the normal 1 : 1 ratio was restored. Neither CTP nor UTP could substitute for ATP in these processes. Similarly, Lawrence and his co-workers (Lawrence, 1973; Lawrence *et al.*, 1973, 1974) have shown that ATP is essential in the aminoacylation of *E. coli* $tRNA^{Met}$ by the ligase and that adenosine inhibits the reactions. The inhibition probably involved the binding of adenosine to the site or sites usually occupied by ATP; analogs of adenosine produced a similar effect. Earlier Santi and his associates (Santi and Danenberg, 1971; Santi *et al.*, 1971) had reported comparable inhibition of phenylalanyl tRNA ligase activity by adenosine.

Not only do the ligases bind amino acids to tRNAs, they also remove them (Bonnet and Ebel, 1972, 1974; Bonnet *et al.*, 1972; Ebel *et al.*, 1973); hence, the interaction between ligase and tRNA may involve an equilibrium between aminoacylation and deacylation. In relation to this topic, it is pertinent to note that the deacylation reactions are totally nonspecific and that they occur principally at low levels of AMP and inorganic pyrophosphate (Bonnet *et al.*, 1972). Thus relative rates of recognition, acylation, and deacylation may all play important roles in the natural processes of supplying properly charged tRNAs for protein synthesis.

Activities of tRNAs. Besides their major activity in the genetic system, tRNAs participate in other reactions of lesser importance. In *Staphylococcus aureus,* $tRNA^{Lys}$ actively participated with an enzyme system in the synthesis of phosphatidylglycerol (Gould *et al.*, 1968; Nesbitt and Lennarz, 1968). In a second species of the same genus, *S. epidermidis,* two unusual species of $tRNA^{Gly}$ were found, and their base sequences determined (Roberts, 1972); this pair, both members of which lack the usual doublet –Tψ from loop IV, participates in the formation of peptidoglycans during the cell wall formation. Another activity of tRNAs involves protein synthesis, but without the aid of ribosomes and the remainder of the genetic mechanism. In 1965, Kaji and his associates found a soluble amino-acid-incorporating system that inserted leucine into certain proteins in *E coli*. Recently, as discussed more fully in Chapter 5, phenylalanine and histidine were reported to be joined to proteins in rabbit spermatozoa (Busby *et al.*, 1974), and an enzyme which catalyzed the transfer of leucine and phenylalanine from their tRNAs to the accepting proteins has been isolated (Leibowitz and Soffer, 1969). Thus tRNAs appear to be involved in many diverse activities in the cell and may eventually be demonstrated essential in nearly all cellular processes in which amino acids are utilized.

4.1.4. The Ribosome

As noted earlier, mRNA, after its formation on DNA, ultimately reaches the cytoplasm and becomes associated with the ribosomes, where actual protein synthesis occurs. Because these synthetic processes are very complex, as well as relatively scantily known, it is first necessary to examine the ribosome.

Structural Characteristics. Ribosomes originally were separated from cells in the form of membrane and particle combinations known as microsomes. Only in 1958 was the name ribosome applied to the particle that then became recognized as the active protein-synthesizing unit (Haguenau, 1958). Under the electron microscope, ribosomes appear as spheroidal electron-dense particles, about 20 nm in diameter. Frequently, series of ribosomes attached to what has been identified as mRNA strands have been isolated; these are called polyribosomes or, more typically, polysomes (Slayter *et al.*, 1968). Usually in eukaryotic cells, the ribosomes are attached to the rough endoplasmic reticulum, but in prokaryotes and in the mitochondria, chloroplasts, and nuclei of eukaryotes and sometimes in the cytoplasm of the latter as well, they are not attached to membranes. Although not associated with actual membranes, they have been reported to be bound to a nonmembranous cytoplasmic reticulum in *Bacillus megaterium* (Schlessinger *et al.*, 1965); unfortunately, this report appears to have escaped the attention of recent workers on these particles. Ribosomes also occur in the nucleus and nucleolus, but in these organelles have been studied to only a small extent (Sadowski and Howden, 1968). In *E. coli* the ribosomes have a molecular weight of around 2.7 million and a sedimentation value of 70 S; in eukaryotes their mass may be as great as 4 million daltons and the sedimentation factor up to 80 S. About 40% of the particle is protein, the remainder being RNA, although the proportions vary widely with the systematic position of the source organism.

Each ribosome is generally viewed as being a double body (Florendo, 1969), although it actually consists of three or four discrete units. The two principal subunits differ in size and specific functions (Garrett and Wittmann, 1973; Nanninga, 1973), in prokaryotes the sedimentation values of the larger and smaller subunits being 50 S and 30 S respectively and in eukaryotes, 60 S and 40 S in the same sequence.

Recently completed ultrastructural analyses of the eukaryotic particles have disclosed the morphology as being complex. First the large subunit of vertebrates was described as having a tail 50–60 nm in length (Meyer *et al.*, 1974), a feature that at first appeared to be absent in prokaryote and yeast ribosomes. A more complete study (Kiselev *et al.*, 1974) reported the large subunits of liver ribosomes to be rods 220 Å to 240 Å long by 70 Å to 95 Å wide. Near one end was a projection 60 Å long, followed by a depression filled with a substance of contrasting properties, behind which in turn was a second projection 30 Å high. When viewed from a different angle, the particles seemed to be subrectangular and nearly 150 Å wide. Small subunits were seen as elongated bodies 230 Å long, with a short head at one end; their width varied with the viewing angle from 75 Å to 110 Å. Comparable structural features have been noted in *Chromatium* ribosomes and in those of other prokaryotes (Wittmann, 1976); the ribosomal structure recently reported for *E. coli* is illustrated in Figure 4.3 (Lake, 1976; Cornick and Kretsinger, 1977).

When subunits from *E. coli* are extensively dialyzed against extremes of salt concentrations, a more open ribosomal structure is formed that sediments more slowly than untreated particles (Gesteland, 1966; Gormly *et al.*, 1971). This structural change has been termed "unfolding" and results in loss of the 5 S rRNA and proteins described later (Garrett *et al.*, 1974). Other, but less extreme, conformational changes were observed when 30 S subunits became associated with 50 S particles (Ginzberg and Zanier, 1975).

The association of the two subunits into monosomic ribosomes is dependent on the presence of Mg^{2+} ions, eukaryotic subunits in general requiring only low concentrations, and prokaryotic high. In this regard, *Chlamydomonas* subunits resembled prokaryotic ones, although the total monosomic size was like that in other eukaryotes (Sager and Hutchinson, 1967). Plant rRNAs, especially those of 23 S subunits, are quite sensitive to heating, cleaving into fragments of marked size classes (Grierson, 1974; Munsche and Wollgiehn, 1974). This thermal fragmentation has been shown to be correlated to the age of the rRNA. Precursors (to be described later) and newly synthesized mature

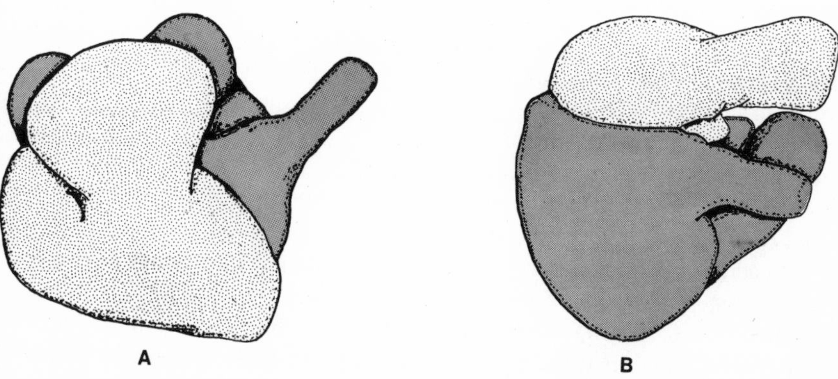

Figure 4.3. Ribosomal structure of *E. coli*. Recent electron microscope investigations show the ribosome to have a highly irregular configuration; here the small subunit is lightly stippled, the large one heavily so. The two parts may be oriented in two different ways, as indicated by the two views: (A) overlap orientation, (B) nonoverlap orientation. (Based on Lake, 1976.)

rRNA were resistant to heat, but breaks where cleavage had occurred were observed in 5-day-old molecules.

Ribosomal RNA. By means of various well-established techniques, the proteins can be stripped from each type of subunit to free the RNA molecule intact. In *E. coli*, the RNA of the larger subunit has been found to sediment at 23 S, whereas in mammals it does so in a broad range between 28 S and 34 S (Pace, 1973). The RNA of smaller subunits has a sedimentation coefficient of 16 S in *E. coli*, and between 18 S and 22 S in mammals. In addition, the larger subunit has a second RNA molecule associated with it, about 120 nucleotides long and with a sedimentation value close to 5 S; this has been reported from prokaryotic (Brownlee *et al.*, 1968; Morell and Marmur, 1968) and metazoan sources (Knight and Darnell, 1967; Forget and Weissman, 1969; Wen *et al.*, 1974), and receives special attention later. Metazoan cells have in addition a "7 S" (actually 5.8 S) RNA molecule associated with the 28 S, which is about 130 nucleotides long (Pene *et al.*, 1968).

Some insight into the actual configuration of the large bacterial subunit RNA has recently been gained. It appears that it has one side richly supplied with proteins, the other relatively deprived, so that opposing surfaces are quite different (Moore *et al.*, 1974). Although it has been shown that the small subunit RNA has all its proteins at the surface, not buried within the nucleic acid's tertiary structure, nothing has been established as to a possible inequality of distribution (Miller and Sypherd, 1973).

In rapidly dividing cells, such as bacteria and those of embryonic tissues, rRNA has a low turnover rate relative to the generation time of the cells. However, in mature metazoan cells having a greatly prolonged generation time, the existence of a definite half-life has been confirmed to a degree; in rat liver cells, for instance, one of approximately 5 days in length has been reported (Loeb *et al.*, 1965).

5 S rRNA. Because the 5 S molecule is the smallest rRNA, it is natural that it should be the most thoroughly investigated. In living cells, this species has been shown to bind to a specific site of the large subunit (Comb and Sarkar, 1967), the attachment being permanent, since no interchange between large subunits appears to occur (Kaempfer and Meselson, 1968). Only by unfolding the ribosome in *E. coli* can the 5 S rRNA be released (Morell and Marmur, 1968; Siddiqui and Hosokawa, 1968).

According to the researches utilizing chloramphenicol-induced protein-synthesis suppression, more than one type of 5 S rRNA may be present in *E. coli* (Forget and Jordan, 1969), but the structural differences thus far recognized are of a minor nature and seem to be confined to the 5′-terminal sequences. Whether any functional distinctions exist has not been ascertained. Biological differences have been found between prokaryotic and eukaryotic 5 S rRNAs, as might be expected from the numerous base distinctions between these types (Table 4.4). The 5 S rRNAs from *E. coli* have been demonstrated as recognizing ribosomal proteins from various prokaryotic sources, but those of eukaryotic origins failed to do so (Bellemare *et al.*, 1973). In contrast, the 5.8 S rRNA of eukaryotes has been reported to bind to two prokaryotic proteins (L18 and L25) and thus was believed to be functionally similar to the 5 S rRNA of prokaryotes (Erdman, 1976). The base sequences described below do not support such a view, however.

Considerable heterogeneity has been reported in the 5 S rRNA of *Xenopus laevis*. At least one sequence of the substance is restricted to the ovaries, whereas another occurs in the liver, kidney, spleen, and testes, as well as the ovaries (Wegnez *et al.*, 1972; Ford and Southern, 1973). Since extensive redundancy exists in the DNA cistrons which code for the 5 S rRNA in prokaryotes and in eukaryotes, including *Xenopus*, man, and *Drosophila*, heterogeneity might not be unexpected (Brown and Weber, 1968; Brown *et al.*, 1971; Hatlen and Attardi, 1971; Procunier and Tartof, 1976). Although the genes thus share the repetitive conditions, those that encode the substance vary widely in organization. Nevertheless, despite similar redundancy in the genes, no tissue-specific sequences could be detected in the 5 S rRNA from leaves, stems, and roots of *Vicia faba*, the broad bean (Payne *et al.*, 1973). It has also been established that, in animal cells, 5 S rRNA continues to be synthesized during mitosis, when the chromosomes are dividing, at a level of around 78% that of the interphase rate (Zylber and Penman, 1971).

Some of the steps involved in attachment of the 5 S rRNA to cores of ribosomes from *E. coli* have been demonstrated *in vitro* by Yu and Wittmann (1973). Three proteins of the large subunits, designated L5, L18, and L25 as discussed below, formed a complex with the 5 S molecule. This combination then could be united to a large subunit core lacking those three proteins, but in the uncomplexed form the 5 S rRNA did not react with that subunit core. Lately it has been shown that this molecule plays a vital role in the assembly of the large subunit (Dohme and Nierhaus, 1976).

Nucleotide Sequences of 5 S rRNAs. As the primary structure has now been determined for the 5 S rRNA from at least seven prokaryotes and ten eukaryotes, a comparative study provides much information (Table 4.4). This is especially true because the several prokaryotic and eukaryotic sources are highly diversified. Among the former are included the blue-green alga, *Anacystis* (Corry *et al.*, 1974a,b), and four eubacteria, *E. coli* (Brownlee *et al.*, 1967, 1968), *Pseudomonas fluorescens* (Dubuy and Weissman, 1971), *Bacillus megaterium* and *B. stearothermophilus* (Pribula *et al.*, 1974), *Proteus vulgaris* (Fischel and Ebel, 1975), and *Clostridium pasteurianum* (Pribula *et al.*, 1976). The eukaryotes are represented by three yeasts, *Torulopsis utilis* (Nishikawa and Takemura, 1974), *Saccharomyces cerevisiae* (Miyazaki, 1974), and *S. carlsbergensis* (Hindley and Page, 1972); the green alga *Chlorella* (Jordan and Galling, 1973; Jordan *et al.*, 1974), one insect, *Drosophila* (Benhamou and Jordan, 1976; Benhamou *et al.*, 1977); and five vertebrates, *Xenopus* (Brown-

Table 4.4

Nucleotide Sequences of 5 S rRNAs from Various Organisms, and 5.8 S rRNA from Yeast and Mammals[a]

	Loop 1A · Loop 2A · Loop 3A · Loop 3B · Loop 2B (positions 10, 20, 30, 40, 50, 60, 70, 80)
Anacystis	p--UCCUGGUGUC--UAUGG-C-GGUAUG-GAACCACUCUGACCCCAUCCCGA-ACUCAGUUGUGAAACAU-ACCUGCGGC
Clostridium	p--UCC-AGUGUC--UAUGA-C--UUAGAGUAAC-ACUCCUUCCC-AUUCCGA-ACACAGGCAGGUUAAGCUCUAAGUGCU
E. coli	pUG-CCUGGCGGC-CG-UAG-C-GCGGUG-GUCCCAC-CUGACCCCAUGCCGA-ACUCAGAAGUGAAACGCC-GU-AGCGC
Proteus	pUGU-CUGGCGGC-CA-UAG-C-CCAGUG-GUCCCAC-CUGAUCCCAUGCCGA-ACUCAGAAGUGAAACGU-UGU-AGCGC
Pseudomonas	pUGUUCU-UUGACG-AGUAG-UAGCAUUG-GAA-CAC-CUGAUCCCAUCCCGA-ACUCAGAGGUGAAACGA-UGC-AUCGC
B. mega	p---UCUGGUGGCGAUAGCG-A-AGA--G-GUCACAC-CCGUUCCCAUACCGA-ACACGGAAGUUAAGCUC-U-UUAGCGC
B. stear	p---CCUAGUGACAAUAGCG-AGAGA--G-GAAACAC-CCGUCUCCAUCCCGA-ACACGGAAGUUAAGCUC-UCCCAGCGC
Torula	ppp-G-GUU-GCGGCCAUAUCUAGCAGAAAG----CAC--CGUUCUCCGUCCGAUCAACUGUAGUUAAGC--UGCUAAGAGC
Sac. cer.	ppp-G-GUU-GCGGCCAUAUCUACCAGAAAG----CAC--CGUUCUCCGUCCGAUCAACUGψAGUUAAGC--UGGUAAGAGC
Sac. carl.	ppp-G-GUU-GCGGCCAUAUCUACCAGAAAG----CAC--CGUUCUCCGUCCGAUAACCUGUAGUUAAGC--UGGUAAGAGC
Chlorella	ppp-AUGCUACGUUC-AUA-CACCACGAAAG----CAC-CCGAUCCCAUCA-GA-ACUCGGAAGUUAAACG-UGGUU-GGGC
Drosophila	ppp-G-CCAACG-ACCA---UACCACG-CUG-AAUACAU-CGGUUCU-CGUCCGAU-CACCGAAAUUAAGCA-GCGUC-GGGC
Xenopus	p-G-CCUACG-GCCA---CACCACC-CUG-AAAGUGC-CCGAUCU-CGUCUGAU-CUCGGAAGCGAAGCA-GGGUC-GGGC
Chicken	ppp-G-CCUACG-GCCA---UCCCACCCCUGUAACG--C-CCGAUCU-CGUCUGAU-CUCGGAAGCUAAGCA-GGGUC-GGGC
Mammal	ppp-G-UCUACG-GCCA---UACCACC-CUG-AACGCGC-CCGAUCU-CGUCUGAU-CUCGGAAGCUAAGCA-GGGUC-GGGC
Yeast 5.8 S	p-AAACUUUUCAACGGAUCUCUUGGUUCUCGCAUCGAUGAAGAACGCAGCGAAAUGCGAUAACGUAAUGUGAAψUGCAGAA
Mammal 5.8 S	p-(C)GACUCUUAGCGGU$_m$GGAUCACUCGGCUCGUGCGUCGAUGAAGAACGCAGCGCUAGCψGCGAGAAUUAAUGψGGAAUUGC$_m$

	Loop 4A (90–100), Loop 4B (110), Loop 1B (140); positions 90 · 100 · 110 · 120 · 130 · 140 · 150 · 160
Anacystis	--AA--CGAUAGCUCCCGGGUAG-CCGGUCGCU-AAAAUA----GCUCGA---CGCC-AGGUC$_{OH}$
Clostridium	G-AU---GGUA-CUGCAGGGGAA-GCCCUGUGG-AGAAGUAGG---U------CGCU-GGG--U$_{OH}$
E. coli	C-GA--UGGUA-GUGUGGGGUCU-CCCCAUG-CGAGAGUAGGGAACUGCC--------AGGCAU$_{OH}$
Proteus	C-GA--UGAUG-GUGUGGGGUCU-CCCCAUGU-GAGAGUAGGGAACUGCC--------AGGCAU$_{OH}$
Pseudomonas	C-GA--UGGUA-GUGUGGGGUUU-CCCCAUGUCAAGAUCU-CGACCAUAG--------A-GCAU$_{OH}$
B. mega	C-AA--UGGUA-GUUGGGACUUU-GUCCCUGUGAGA-GUA-GGACGUUGC-------CAGGC$_{OH}$
B. stear	C-GA--UGGUA-GUUGGGGCCAGCGCCCCUGCAAGA-GUA-GGUUGUCGC-------UAGGC$_{OH}$
Torula	CUGAUCGAGUA-GUGUAG--U-GGGUG-----ACCAUAC-GCGAAACUCAGGUGCUGCAA-UCU$_{OH}$
Sac. cer.	CUGACCGAGUA-GUGUAG--U-GGGUG-----ACCAUAC-GCGAAACUCAGGUGCUGCAA-UCU$_{OH}$
Sac. carl.	CUGACCGAGUA-GUGUAG--U-GGGUG-----ACCAUAC-GCGAAACCUAGGUGCUGCAA-UCU$_{OH}$
Chlorella	UCGA-CUAGUA-CUG--GG-UUGGAGG--AUUACCUGAGUGGGAAC-CCC---GACGUAG-UGU$_{OH}$
Drosophila	GCGG-UUAGUA-CUU--AGAU---GGG--AG-ACC-GCUUGGGAACACCGCGUGUUGUUGGCCU$_{OH}$
Xenopus	CUGG-UUAGUA-CUU--GGAU---GGG--AG-ACC-GCCUGGGAAUACCAGGUGUCGUAGGCUU$_{OH}$
Chicken	CUGG-UUAGUA-CUU--GGAU---GGG--AG-ACCU-CCUGGGAAUACCGGGUGCUGUAGGCUUU$_{OH}$
Mammal	CUGG-UUAGUA-CUU--GGAU---GGG--AG-ACC-GCCUGGGAAUACCGGGUGCUGUAGGCUU$_{OH}$
Yeast 5.8 S	UUCCGUGAAUCAUCGAAUCUUUGAACGCACAUUGCGCCCUUGGUAUUCCAGGGGGCAUGCCUGUUUGAGCGUCAUUU$_{OH}$
Mammal 5.8 S	AGGACACAUUGAUCAUCGACACUUCGAACGCACUUGCGGCCCCGGGUUCCUCCCGGGGCUACGCCUGUCUGAGCGUCGCUU$_{OH}$

[a]Italics indicate double-stranded regions.

lee *et al.*, 1972), chicken (Pace *et al.*, 1974), and three mammals (kangaroo, mouse, and man). The latter are shown as a single sequence, as all are identical (Forget and Weissman, 1967; Williamson and Brownlee, 1969; Averner and Pace, 1972).

In preparing the table, the residues in the respective sequences were aligned by inspection to provide the greatest number of homologies, following the method of Sankoff *et al.* (1973). In these processes the several areas shaded in the table were of particular value, as they seemed to contain highly conserved short sequences. Without doubt, a number of small segments here or there might be moved to the right or left and thus increase the correspondences to a still higher level, but for the present that table appears to indicate a large measure of phylogenetic relationship throughout the living world.

This level of homology received further support through the appearance in print of a logical secondary structural arrangement for the 5 S rRNA of *E. coli* (Fox and Woese, 1975). The proposed structure (Figure 4.4) fitted all the data provided by earlier studies; some of these had indicated the existence of extensive base pairing (Cantor, 1967, 1968; Cramer and Erdman, 1968; Lewin, 1970) and others had described the recognition reactions with the 50 S ribosomal subunit (Bellemare *et al.*, 1972). The paired regions in *E. coli* molecules that have been suggested by Fox and Woese are shown in the table by italics and by numbering which is repeated in the illustration (Figure 4.4). Corresponding regions then are here shown in like fashion for all the remaining sequences.

Loop 1, involving the two terminal segments, is an especially consistent feature of 5 S rRNA in all the sequences, although its length and constituent residues are alike subject to considerable variation. Similar universality can be noted also for loops 2 and 3, the latter being especially uniform in relative position and in number of included nucleotides (Vigne *et al.*, 1973). Loop 4 also

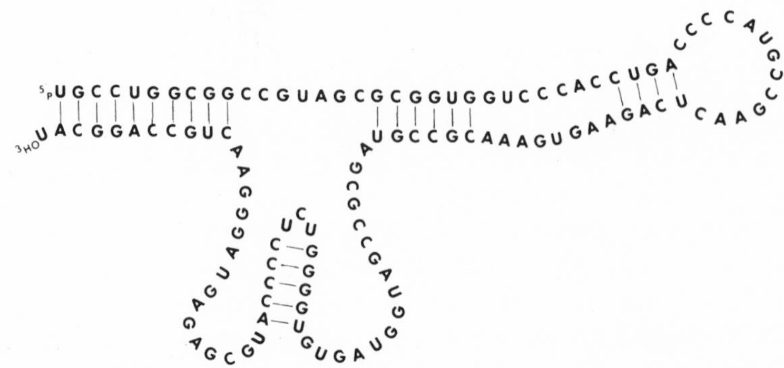

Figure 4.4. Secondary structure of 5 S rRNA. This diagrammatic interpretation of the secondary structure of the 5 S rRNA molecule of *E. coli* is based in large measure on Fox and Woese (1975).

is present throughout the living world, but the base pairing has become so reduced that Fox and Woese overlooked it. Here in the eukaryote loop 4 sequences, a number of GpG doublets exist which could with equal logic pair with the single CpC segment present, but the determination of the one that actually forms the base pairs must remain for the future.

The sequences should be noted to lack modified bases; the only exception that has been observed is the presence of pseudouracil in the molecule from *S. cerevisiae* near site 60 (Miyazaki, 1974). This absence of modified and unusual bases is in sharp contrast to the tRNAs and to the larger species of rRNA.

Closer examination of the table permits several observations to be made regarding phylogenetic relationships. The sequences from the three yeasts, while obviously indicative of close alliance, differ in four or five sites; thus on this basis they are more distantly related to one another than the marsupial molecule is to that of man or than either of these mammalian types is to the mouse, dolphin, or ox, for all the mammals are identical in this respect (Walker *et al.*, 1975). The bacteria, especially *Clostridium,* are shown by comparisons of these sequences also to be a highly divergent group, as are also the blue-green algae (Corry *et al.*, 1974b). This wide divergency shown to exist within these two groups of organisms had been indicated earlier by other cellular characteristics (Dillon, 1962, 1963). Further, the 5 S rRNAs of the yeasts are quite obviously more closely related to those of the other eukaryotes than to those of the bacteria (Jordan and Galling, 1973; Kimura and Ohta, 1973).

The great disparities between the bacterial sequences require further attention. For example, those of *E. coli* and *P. fluorescens* differ at 39 sites (Brownlee *et al.*, 1967, 1968; Dayhoff, 1972); thus the two molecules are nearly as distinct as those from *Chlorella* and *Homo,* which differ in 44 sites (Forget and Weissman, 1969). Even more interesting is the contrast between the several yeast sequences and that from *Chlorella,* which are as distinct with 62 differences as the yeast molecule is from that of human (64 differing sites). These observations are irreconcilable with relations postulated in two-kingdom and multiple-kingdom schemes of classification but are in complete harmony with the single-kingdom concept proposed elsewhere (Dillon, 1962, 1963).

The 5.8 S rRNA Molecule. By refinement of techniques, it has now been established that the rRNA species formerly known as the 7 S type actually has a sedimentation coefficient of 5.8 S (Rubin, 1973). This type is confined to the cytoplasmic ribosomes of eukaryotic cells, in which it is noncovalently bound to the RNA molecule of the large subunits, apparently by hydrogen bonding. Thus the large subunit RNA of eukaryotes is supplied with two additional ones of low molecular weight; prokaryote and chloroplast equivalents have only one appendaged type, the 5 S variety, and mitochondria appear to have none.

Thus far no secondary or tertiary structures have been suggested for 5.8 S rRNA, but the nucleotide sequences of this rRNA from yeast and mammals have recently been reported (Rubin, 1973; Nazar *et al.*, 1976). With the exception of a single pseudouracil residue (at site 73), this species resembles 5 S rRNA in lacking modified or unusual bases (Table 4.4); however, little correspondence in the nucleotide sequences of these two types from yeast can be noted. Further, three stable species of 5.8 S rRNA have been shown to occur in *Saccharomyces,* which differ only in the initial sequences (Rubin, 1974). Whereas the major species has simply AAAC–, the two minor forms have UAUUAAAAAC– and AUAUUAAAAAC–, respectively. Moreover, tissue-specific differences have been demonstrated in the methylation of 5.8 S rRNA among eukaryotes (Nazar *et al.*, 1975). Partial sequence analyses from several seed plants intimate that this species of rRNA is less conservative evolutionarily than the 5 S type appears to be (Woledge *et al.*, 1974; Cedergren and Sankoff, 1976), but the two eukaryotic types given in Table 4.4 show homologies only at the random-chance level of 25%.

The Larger rRNA Molecules in Bacteria. As a consequence of the extreme difficulties involved in analyses of large nucleic acid molecules, the two major subunit rRNAs remain relatively poorly known. Although still quite incomplete, base-sequence studies of bacterial components thus far has yielded significant results. First, the sequences which have been established demonstrate that both the 16 S and 23 S rRNAs are heterogeneous in a given cell, to a level of about 10% (Fellner and Sanger, 1968; Fellner *et al.*, 1970, 1972b; Muto, 1970). Such lack of complete homogeneity is not surprising in view of the number of genes that code for these rRNAs (Smith *et al.*, 1968; Kiss *et al.*, 1977) a matter that receives attention below.

On the basis of the high degree of competition that exists between 16 S and 23 S in hybridizing with *E. coli* DNA, it had earlier been proposed that the larger type might have arisen by a process of duplication of the cistrons encoding for the smaller molecule (Attardi *et al.*, 1965; Retel and Planta, 1968). Furthermore, several nucleotide sequences isolated from both types were similar in the pattern of methylation (Fellner and Sanger, 1968). However, now that about 95% of the base sequence of the 16 S rRNA has been established (Fellner *et al.*, 1972b; Noller, 1974; Noller and Herr, 1974; Shine and Dalgarno, 1974a; Wittmann, 1976), little correspondence can be noticed with that 10% of the 23 S molecule that has been sequenced (Ehresmann *et al.*, 1970, 1975; Fellner and Ebel, 1970; Fellner *et al.*, 1972a). Hence, the two are probably not related (Garrett and Wittmann, 1973). In contrast, a 12-nucleotide-long region of 23 S rRNA from *E. coli* has been shown to be complementary to a section of 5 S rRNA from the same organism (Herr and Noller, 1975). In view of the need for three proteins to bind the 5 S molecule to the large one, the significance of the complementarity is not clear.

Knowledge of secondary structure is similarly incomplete, but it has been demonstrated that both types are extensively arranged in multiple hairpin loops. In fact, the longest single-stranded regions appear to be between 7 and 8 nucleotide residues long (Garrett and Wittmann, 1973). By far the largest portion of the 35% that is single-stranded seems to be confined to the ends of the hairpin loops. Thus between 60% and 70% of each molecular type is in the form of double-stranded loops, having an average length of ten base pairs and the A-form molecular arrangement (Table 3.2; Cox *et al.*, 1968; Spencer *et al.*, 1969; Cox, 1970). A comparable degree of secondary structure exists in the 16 S rRNA when it is within the subunit as when it is free (Wittmann, 1976).

The Larger Subunit rRNAs of Eukaryotes. During the past several years, a degree of progress has been achieved in unraveling the sequences of the larger subunit rRNAs in eukaryotes. In a number of forms the largest rRNA (24 S to 28 S) has been shown to dissociate into several segments when treated with heat or dimethyl sulfoxide to disrupt the hydrogen bonds between the base pairs. This condition has been reported from various protozoans (Loening, 1968; Rawson and Stutz, 1968; Bostock *et al.*, 1971; Stevens and Pachler, 1972), and in a snail (Koser and Collier, 1971) and many insects (Shine and Dalgarno, 1973; Shine *et al.*, 1974), as well as in the chloroplast ribosomes of seed plants (Leaver and Ingle, 1971). The results indicated that the breaks in the molecule were present *in vivo* and were not artifacts of preparation. Further evidence has now been presented for insect 26 S rRNA by a study of the 3′-terminal sequences (Shine and Dalgarno, 1974b). Three sequences were found in equimolar amounts in each insect studied. For cultured *Antheraea eucalypti* cells, the sequences were $YGUGU_{OH}$,* $YCGU_{OH}$, and GCU_{OH}; for *Galleria mellonella* larvae, YGU_{OH}, YCU_{OH}, and GCU_{OH}; and for *Drosophila melanogaster*, $YCGA_{OH}$, $YGUA_{OH}$, and $GYUG_{OH}$. Hence, the termini appear subject to rapid evolutionary change.

To the contrary, the 3′-termini of the 16 S to 18 S rRNA of eukaryotes seem to be evolutionarily conservative, for the same sequence has been found in yeast as in *Drosophila* (Shine and Dalgarno, 1974b). Furthermore, less extensive data from the wax moth (*Galleria*) and monkey and hamster kidney cells were consistent with the same sequence. The octonucleotide section reported was $GAUCAUUA_{OH}$. This was noted to correspond (in complementary sequence from the –OH end) to the terminator codons UAA and UGA (but not to UAG, as implied in the article). Consequently, it was proposed that this rRNA plays a major role in termination of peptide synthesis. This suggestion fails to explain how the many mutant tRNAs described later (Chapter 7) could interact with these terminator codons, as documented by a number of investigators, if such a constant feature as this terminus were solely responsible for the action.

Modified Bases of rRNAs. A number of modified bases are known from these rRNA molecules, but only two are distinctive, each confined to a single type. One of these, dimethyladenine, occurred solely in the smaller

* Y is the abbreviation for any pyrimidine nucleotide.

ribosomal subunit RNA, while the other, 3-methyluracil, was confined entirely to that of the larger subunit (Klagsbrun, 1973). In wheat embryos, the large and small unit rRNAs alike had *O*-2′-methylpseudouridine present (Gray, 1974). Both the 40 S precursor and the mature 28 S rRNA of *Xenopus laevis* had a preterminal sequence containing three 2′-*O*-methylations (Slack and Loening, 1974); this four-nucleotide segment appeared to be $-G_mA_mA_mAp$. Recently a hypermodified nucleotide in mammalian 18 S rRNA has been identified as 1-methyl-3-γ-(α-amino-α-carboxypropyl) pseudouridine (Maden *et al.*, 1975).

Small Subunit Proteins of Bacteria. Knowledge of the protein complement of bacterial ribosomes has gained sufficient ground to indicate a highly complicated organization for these bodies (Maugh, 1975). In the smaller 30 S subunit of *E. coli,* 21 proteins, and in the larger, 34, are now known to occur, all those of the former being exposed on the surface (Miller and Sypherd, 1973). The 30 S subunit of *Bacillus stearothermophilus* was reported to be slightly larger in containing 23 proteins (Isono and Isono, 1975). The individual proteins are designated by simple numerals in one widely employed system, the sequence of the numerical assignments being based upon molecular size in descending order from the largest to smallest. Thus in *E. coli* the largest protein of the 30 S subunit, having a molecular weight of 65,000, is referred to as S1, and the smallest as S21, while the corresponding extremes of the larger subunit are L1 and L34 respectively. As a whole, it has been indicated that these proteins attach to the rRNA by their C termini (Daya-Grosjean *et al.*, 1974; Funatsu *et al.*, 1977). The individual proteins of the smaller subunit have been investigated more completely than have those of the larger (Garrett and Wittmann, 1973), and *in vitro* synthesis of a number of them has been reported (Kaltschmidt *et al.*, 1974). The syntheses were carried out in a cell-free system consisting of ribosomes, initiation factors, RNA polymerase, a cytoplasmic fraction containing unknown enzymes, and *E. coli* DNA.

In addition to these firmly attached proteins, a number of relatively large, mostly acidic types have been reported. At first these were suspected to be simply adhering proteins derived from the supernatant cytoplasm (Hardy *et al.*, 1969; van Duin *et al.*, 1972), but later it was proposed that they should be considered actual constituents of ribosomes (Subramanian, 1974). To settle the question, a study was made comparing the high-molecular-weight proteins from membrane-associated and free ribosomes. As only the latter were found to possess such proteins, it was concluded that their origin was in the supernatant cytoplasm as first believed (Brouwer and Planta, 1975).

Sequences of Proteins of the Small Subunit. The sequences of the ten proteins from the small ribosomal subunit of *E. coli* whose primary structure has now been determined (Table 4.5) represent nearly half of the 21 present. The longest established sequence is that of S4, whose 203 residues provide a molecular weight of 22,550 (Schiltz and Reinbolt, 1975). The shortest, S21, is 70 units long (Vandekerckhove *et al.*, 1975) and thus far exceeds

Table 4.5
:stablished Amino Acid Sequences of Proteins from the Small Subunit of E. coli Ribosomes

	10 20 30 40 50
4	$_{N}$ARYLGPKLKLSRREGTDLFLKSGVRAIDTKCKIEQAPGEHGARKPRLSDYGVQ
6	$_{N}$MRHYEIVFMVHPDESEEVPGMIERYTAAITGAEGKIHRLEDWGRRQLAYPINK
8	$_{N}$SMEDPIADMLTRIRNGQAANKAAVTMPSSKLKVAIANVLKEEGFIEDFKVEGD
9	NAENQYYGTGRRKSSAARVFIKPGNGKIVINQRSLEQYFGRETARMVVREPLEL
12	$_{N}$ATVNQLVRKPRARKVAKSNVPALEACPQKRGVCTRVYTTTPKKPNSALRKVCR
15	$_{N}$SLSTEATAKIVSEFGRDANDTGSTEVQVALLTAQINHLQGHFAEKKDHHSRRG
16	$_{N}$MVTIRLARHGAKKRPFYQVVVADSRNARNGRFIERVGFFNPIASEKEEGTRLD
18	$_{Ac}$ARYFARRKECRFTAQGVQEIDYKDIATLKNYITESGKIVPSRITGTRAKYQRQ
20	$_{N}$ANIKSAKKRAIQSEKARKHNASRRSMMRTFIKKVYAAIEAGDKAAAQKAFNEM
21	$_{N}$PVIKVRENEPFDVALRRFKRSCEKAGVLAEVRRREFYEKPTTERKRAKASAVK

	60 70 80 90 100
4	LREKEKVRRIYGVLERQFRNYYKEAARLKGNTGENLALLEGRLDNVVYRMGFGA
6	LHKAHYVLMNVEAPQEVIDELETTFRFNDAVIRSMVMRTKHAVTEASPMVKAKD
8	TKPELELTTLKFQGKVVAEISERVSRPGIRIYKLQDKPKRVMGDTRARKLQICV
9	VNMVEKLDLYITVKGGGISGQAGAIRHGITRALMEYDESLRSELRKAGFVTRDA
12	VRLTNGFEVTSYIGGEGHNLQEHSVILIRGGRVKXLPGVRYHTVRGALDCSGVK
15	LLRMVSQRRKLLNYLKRKDVARYTQLIERLGLRR$_{COOH}$
16	LDRIAHWVGQGATISDRVAALIKEVNKAA$_{COOH}$
18	LARAIKRARYLSLLPYTDRHQ$_{COOH}$
20	QPIVDRQAAKGLIHKNKAARHKANLTAQINKLA$_{COOH}$
21	RHAKKLARENARRTRLY$_{COOH}$

	110 120 130 140 150 160
4	FRAEARELVSHKAIMVNGRVVNIASYQVDPNSVVIREKAKKESRVKAALELAEQ
6	ERRERRDDFANETADDAEAGDSEEEEEE$_{COOH}$
8	AY$_{COOH}$
9	RQVERKKVGLRKARRPEFSKR$_{COOH}$
12	DRKQARSKYGVKRPKA$_{COOH}$

	170 180 190 200
4	REKPTWLEVDAGKMEGTFKRKPERSDLSADINEHLIVELYSK$_{COOH}$

Table 4.6
Amino Acid Sequences of Proteins from the Large Ribosomal Subunit of E. coli

Protein	Sequence[a]
	10 20 30 40 50 60 70 80
L10	ALNLQDKQAIVAEVSQVAKGALSAVVADSRGVTVDKMTELRKAGREAGVYMRVVRNTLLRRAVEGTPFECLKDAFGPVLIAYRS
L7, L12	S*ITKDQIIEAVAAMSVMDVVELISAMEEKFGVSAAAAVAVAAGPVEAAEEKTEFDVILKAAGANKVAVIKAVRGATGLGLKEAK
L16	M*LQPKRTKFRKMHKGRNRGLAQGTDVSFGSFGLKAVGRGRLTARQIEAARRAMTRAVKRQGKIWIRVFPDKPITEKPLAVR*MGK
L18	MDKKSARIRRATRARRKLQELGATRLVVHRTPRHIYAQVIAPNGSEVLVAASTVEKAIAEQLKYTGNKDAAAAVGKAVAERALE
L25	MFTINAEVRKEQGKGASRRLRAANKFPAIIYGGKEAPLAIELDHDKVMNMQAKAEFYSEVLTIVVDGKEIKVKAQDVQRHPYKP
L27	AHKKAGGSTRNGRDSEAKRLGVKRFGESVLAGSIIVRQRGTKFHAGANVGCGRDHTLFAKANGKVKFEVKGPKNRKFISIEAE$_{COOH}$
L29	MKAKELREKSVEELNTELLNLLREQFNLRMQAASGQLQQSHLLKLQLNTKQVRRDVARVKAGA$_{COOH}$
L30	AKTIKITQTRSAIGRLPKHKATLLGLGLRRIGHTVEREDTPAIRGMINAVSFMVKVEE$_{COOH}$
L32	AVQQNKPTRSLRGMRRSHDALTAVTSLSVDKTSGEKHLRHHITADGYYRGRKVIAK$_{COOH}$
L33	A*KGIREKIKLVSSAGTGHFYTTTKNKRTKPEKLELKKFDPVVRQHVYIKEAIK$_{COOH}$
L34	MLRTFQPSVLKRNRSHGFRARMATKNGRQVLARRRAKGRARLTVSK$_{COOH}$
	90 100 110 120 130 140 150 160
L10	MEHPGAAARLFKEFAKANAKFEVKAAAFEGELIPASQIDRLATLPTYEEAIARLMATMKEASAGKLVRTLAAVRDAKEAA$_{COOH}$
L7, L12	DLVESAPAALKEGVSKDDAEALKKALEEAGAEVEVK$_{COOH}$
L16	GKGNVEYWVALIQPGKVLYEMDGVPEELAREAFKLAAAKLPIKTTFVTKTKM$_{COOH}$
L18	KGIKDVSFDRSGFQYHGRVQALADAAREAGLQF$_{COOH}$
L25	KLQHIDFVRA$_{COOH}$

[a]A*, *N*-methylalanine; M*, *N*-monomethylmethionine; R*, unknown modification of arginine; S*, this serine is acetylated in L7 but not in L12; amino acid abbreviations are given in Table 4.2.

the shortest of the large subunit, L34, which is only 46 units in length. Unlike those of the large subunit described below, few methylated or otherwise modified amino acids have been detected, except for an acetylated alanine in site 1 of S18 (Yaguchi, 1975). Since this modification is absent in the homologous protein of *Bacillus stearothermophilus,* S19 (Yaguchi *et al.*, 1974), it does not appear to be essential for functioning. In addition, a residue of an undetermined nature occupies site 88 of S12 (Funatsu *et al.*, 1977).

Several of the proteins lack cysteine residues, among them being S6 (Hitz *et al.*, 1975), S9 (Chen and Wittmann-Liebold, 1975), S15 (Morinaga *et al.*, 1976), S16 (Vandekerckhove *et al.*, 1977), and S20 (Wittmann-Liebold *et al.*, 1976). In contrast, S12 contains four cysteine residues, whereas S4, S8 (Stadler and Wittmann-Liebold, 1976), S18, and S21 have one each. Those residues, however, are not situated in corresponding sites in the several proteins. Furthermore, an examination of the sequences discloses almost no identical sequences exceeding three or so amino acids in length. Although a large number of tripeptides occurs in more than one protein, the vast majority are confined to two. One such sequence, however, *K–A–G* (lysine, alanine, glycine) occurs in a number of the proteins, but not at similar sites in each case (Wittmann-Liebold and Dzionara, 1976a).

Proteins of the Large Ribosomal Subunit. The amino acid sequences of 12 proteins from the large ribosomal subunit of *E. coli* also have been established (Table 4.6). Of these, the longest one, L10, is 164 residues in length (Dovgas *et al.*, 1976), and the shortest (L34), only 46 (Chen and Ehrke, 1976). Four modified amino acids may be noted in the sequences, *N*-monomethylmethionine in site 1 of L16, and acetylserine and *N*-methylalanine in the same site of L7 and L33, respectively; in addition an unknown modification of arginine occupies site 81 of L16 (Brosius and Chen, 1976; Wittmann-Liebold and Pannenbecker, 1976). Unfortunately cysteine is absent from the majority of the sequences, for it frequently marks homologous points in proteins; it occurs only at site 70 of L10 and site 52 of L27 (Chen *et al.*, 1975).

Aligned on a numerical basis as the sequences are in the table, few homologous sites or regions can be noted, except between two or sometimes three of the twelve given, when L7 and L12 are omitted from consideration. At site 20, for instance, leucine (*L*) occurs in L16, L25 (Bitar and Wittmann-Liebold, 1975), and L27, and in the latter two this residue is preceded by arginine (*R*). But on further searching, one finds no additional correspondences in the regions preceding this pair of sites, nor in that which follows. Nor are any short sequences of three to five residues evident that are shared by even two members of the group. Triplets of alanine (*A*) are noted near sites 90 and 110 of L10, and at site 120 of L16, while a series of four is found near site 70 of L18 (Brosius *et al.*, 1975) and near site 34 of L7 and L12 (Terhorst *et al.*, 1973). However, in neighboring regions, further similarities between these sequences are notably absent; consequently, it must be concluded that no evidence now exists to suggest a common ancestry for these proteins. This observation corroborates conclusions drawn from immunochemical studies, which indicated a lack of extensive similarities in primary structure except between L7 and L12 and between S20 and L26. Of these, all except L26 has had its complete sequence established.

L7 and L12 are a special case, for they are identical except in the initial site, and this residue differs between the two only in being acetylated in L7 and not in L12 (Terhorst *et al.*, 1973). Consequently, the pair is considered to be two variants of the same protein, and as might be expected, they enter into the same reactions in the ribosome. Both have been reported to be involved in the GTP hydrolysis dependent on the two elongation factors, EF-G and EF-T. Others of the protein assemblage, at first on purely statistical considerations, were concluded to show relationships (Wittmann-Liebold and Dzionara, 1976a). Later comparisons with a series of proteins that were entirely unrelated, such as cytochrome *c,* chymotrypsinogen, papain, and phospholipase *A,* along with several artificially produced proteins, indicated that the ribosomal proteins as a whole showed a general lack of kinship (Brinacombe *et al.*, 1976; Wittmann-Liebold and Dzionara, 1976b). Before the problem can be completely resolved, the primary sequences of large and small subunit proteins from more primitive sources need to be established, particularly from such forms as *Clostridium, Beggiatoa,* and the blue-green algae.

The Prokaryotic Protein S1. The largest protein (S1) of the 30 S subunit is being demonstrated to be of particular importance in protein synthesis, and its complete amino acid sequence has been determined (Reinbolt and Schiltz, 1973). It probably plays a large role in binding the messenger to the small subunit (Fiser *et al.*, 1974). During these processes, the protein appears to become detached from the ribosome, for two forms of the 30 S subunit have been separated electrophoretically from *E. coli* preparations, which differed in its presence or absence (Dahlberg, 1974; Szer and Leffler, 1974). The type which lacked this protein could be converted to the other by the addition of S1 in the presence of mRNA; however, the affinity of the protein for the subunit was weak. Only that species containing S1 participated in the formation of 30 S initiation complex (Szer *et al.*, 1975). Furthermore, van Duin and van Knippenberg (1974) showed that poly(U)-dependent poly(phenylalanine) synthesis was totally dependent on the presence of this protein and that the resulting polysomes had one molecule of S1 per ribosome. S1 also appeared essential for the binding of formylmethionyl-$tRNA_F^{Met}$, even in the presence of natural messengers (van Knippenberg *et al.*, 1974). Comparable results were reported by Van Dieijen *et al.* (1975); however, this protein has been reported to be absent in *Bacillus stearothermophilus* (Isono and Isono, 1976).

Large Subunit Proteins of Bacteria. Partial digestion of the 50 S subunit from *E. coli* has permitted isolation of two long segments of the 23 S rRNA (Spierer *et al.*, 1975), and these in turn have made possible the localization of a number of proteins in this subunit. The 5′-terminal segment, containing 1200 nucleotides, proved to hold the specific binding sites for L4, L20, and L24, whereas the 3′-terminal fragment, 2000 nucleotides long, included those for L1, L2, L3, L6, L13, L16, and L23. Thus these two regions together contain the specific sites for 10 of the 34 proteins known to bind to the large subunit. At least eight of the large subunit proteins (L1, L4, L7–L10, L12, and L25) have been shown to be present in large pools within *E. coli* cells (Ulbrich and Nierhaus, 1975). Similar pools of several small subunit species also were found. At least L7 and L12 have been shown to participate actively in protein synthesis in *E. coli* (Glick, 1977).

Eukaryotic Ribosomal Proteins. Much less is known concerning the ribosomal proteins of eukaryotes. In ribosomes of human placental cells, the small ribosomal subunit (sedimentation coefficient 40 S) contained 30 proteins, with molecular weights ranging from 9600 to 37,600. The large subunit (60 S) had 36 proteins, with molecular weights between 8900 and 61,900 (Peeters *et al.*, 1973). It is interesting to note that whereas the number of proteins in the human large subunit thus differs from the 34 of the corresponding bacterial subunit (50 S) by less than 10%, the 30–32 in the small subunit represents nearly a 50% increase over the 21 of the small particle of the bacteria. Thus the smaller subunit has undergone greater evolutionary change in numbers of component enzymes than has the larger one. Similar results have been obtained from studies of rat liver ribosomes (Huynh-Van-Tan *et al.*, 1971; Welfle *et al.*, 1972; Terao and Ogata, 1975); but in rabbit reticulocyte ribosomes the large subunit had 39 proteins, the small one, 32 (Chatterjee *et al.*, 1973; Howard *et al.*, 1975). Pea seedling ribosomes were distinctive although in the same general range, for the 40 S subunit contained between 32 and 40 proteins, the 60 S between 44 and 55 (Gualerzi *et al.*, 1974).

The relatively slow pace of progress in investigations of the ribosomal proteins of eukaryotes, in part at least, stems from the heterogeneity that has been demonstrated in ribosomal structure and behavior (MacInnes, 1973). Differences in functional activities of ribosomes have been correlated to the age of the organ or animal (Johnson and Luttges, 1966); to the state of the ribosomes, that is, whether free or bound to membranes (Takagi *et al.*, 1970; Murthy, 1972); to the activity state of the tissue (Herrington and Hawtrey, 1971); and to the type of tissue (MacInnes, 1972). Moreover, a degree of species specificity appears to exist (Fujisawa and Eliceiri, 1975). In the last case, by means of hybridization studies, the 60 S subunit was identified as chiefly responsible for the variation, and polyacrylamide gel electrophoresis of liver and brain preparations demonstrated differing proportions of several proteins in both the 40 S and 60 S particles. At least two populations differing functionally and structurally *in vivo* as well as *in vitro* have been separated from mouse brain (MacInnes, 1973). Further, in HeLa cells, one protein has been found present only in, and another one solely absent from, membrane-bound 60 S ribosomal subunits in contrast to free ones, but 34 other separable proteins were identical in the two types (McConkey and Hauber, 1975). On the other hand, in the 40 S subunits, only 17–20 proteins were common to all, while at least 10 were heterogeneously distributed. Mitochondrial ribosomes from *Neurospora* differed from their cytoplasmic counterparts in almost all their proteins, despite the latter's being synthesized in the cytoplasm (Küntzel, 1969).

4.2. TRANSLATION AND PROTEIN SYNTHESIS

The complex ribosomes just described are the organelles most active in translating the encoded message contained in mRNA into the protein. With the assistance of tRNAs and nucleotides, they carry out this function in a series of steps, the nature of which is now barely being glimpsed. In the present view, the steps are considered to include (a) *binding* to the mRNA, (b) *initiating* the polypeptide chain, (c) *elongating* the chain, and (d) *terminating* and (e) *releasing* the protein molecule. Further, a cycle of activity is often ascribed to the ribosomes, and this, too, is an important aspect of the problem (Kurland, 1970; Hasselkorn and Rothman-Denes, 1973; Zalik and Jones, 1973; Bloemendal, 1977).

4.2.1. Binding and Initiation

During the onset of normal protein synthesis, two very different types of binding processes are found, one of which involves the fixing of an mRNA molecule to a given ribosome, a single event preceding the commencement of each complete round of translation. The second occurs between aminoacyl-tRNAs (tRNAs charged with their respective amino acids) and a ribosome, when the codons of the messenger interact with the anticodons of the tRNAs. This type of binding continues throughout translation, as long as the peptide chain is being elongated. In these processes, energy from cellular activities, including respiration and glycolysis, appears to be essential, because protein synthesis is retarded and mRNA is inactivated when energy-source levels are decreased (Westover and Jacobson, 1974).

General Aspects of Initiation. After its release into the cytoplasm, mRNA migrates to the ribosomes and becomes associated with the smaller subunit; this reactions's being confined to the smaller subunit has now been confirmed in eukaryotic as well as prokaryotic systems (Howard and Herbert, 1975). Protein S1 has been identified as being at the site at which the messenger attaches (Fiser *et al.*, 1974) and appears to be an active participant in this reaction, as discussed above (Szer *et al.*, 1975). In forming this association, it does not seem necessary for a subunit to thread itself onto the messenger by beginning at the 5′-end and moving along to an initiator site; attachment to the messenger apparently occurs directly at, or at least adjacent to, such a point (Bretscher, 1968, 1969; Modolell, 1974). Initiating sites are most usually marked by the sequence AUG, the sole codon for methionine (Lewin, 1970). The presence of this triplet where initiation begins has been confirmed by studies of three cistrons of the bacteriophage R17 genome (Steitz, 1969). By comparisons of the N-terminal amino acids of the respective proteins with the nucleotide sequences of the cistrons, it was shown that each of the latter began with the triplet AUG. Sequences from a number of other initiation sites also have shown the presence of AUG, including that of the *lac* operon of *E. coli* (Maizels, 1974).

A more recent analysis of the base sequence of MS2 RNA, however, shows that the cistron for its so-called A-protein has the initiator site marked by GUG, one of the four codons for valine (Volckaert and Fiers, 1973). Previously, two other laboratories had demonstrated with synthetic messages that methionyl-tRNA* responded to GUG in bacterial systems (Clark and Marcker, 1966; Ghosh *et al.*, 1967). Whereas in a yeast mRNA, GUG has been evidenced as not serving as an initiation site (Stewart *et al.*, 1971), rabbit reticulocyte globin mRNA in a cell-free system from *E. coli* functioned only after *N*-acetylvalyl-tRNA had been added (Laycock and Hunt, 1969). Whether this requirement for valyl-tRNA was dependent upon the presence of GUG at the initiating site is not clear.

In bacteria, only formylated methionyl-tRNA reacts with the initial AUG, and a special tRNA (identified in print by means of a subscript F) is present to carry the modified amino acid (Figure 4.5). What feature induces the formylated methionyl-$tRNA_F$ to react with the initiating site and not with any internal site is not known. It should be noted that certain other species, including $tRNA^{Ser}$ of *E. coli* and yeast, are also formylated (Kim, 1969). It is not understood why ordinary methionyl-$tRNA_M$ does not become attached at this site, since it responds to the same codon (McNeil and McLaughlin, 1974); hence, it has been proposed that some region of the initiator tRNA other than the anticodon is involved in recognition (Shine and Dalgarno, 1975). In this connection it should be noted that higher eukaryotic systems do not employ formylmethionyl-tRNA in this capacity, as discussed later. In both systems, after the initiator tRNA has first become bound to the smaller ribosomal subunit and then to the mRNA (Jay and Kaempfer, 1975), the larger subunit unites with the smaller and remains attached throughout the rest of the processes. The actual sequence of events is clarified later.

Ribosomal Binding Sites for tRNAs. In current models of ribosomal function, two binding sites for tRNAs are universally postulated, but the dispo-

* A standard convention should be noted here to avoid confusion. Methio*nine*-tRNA is a species of tRNA that is capable of carrying methionine, whereas methio*nyl*-tRNA is actually charged with (that is, carrying) that amino acid.

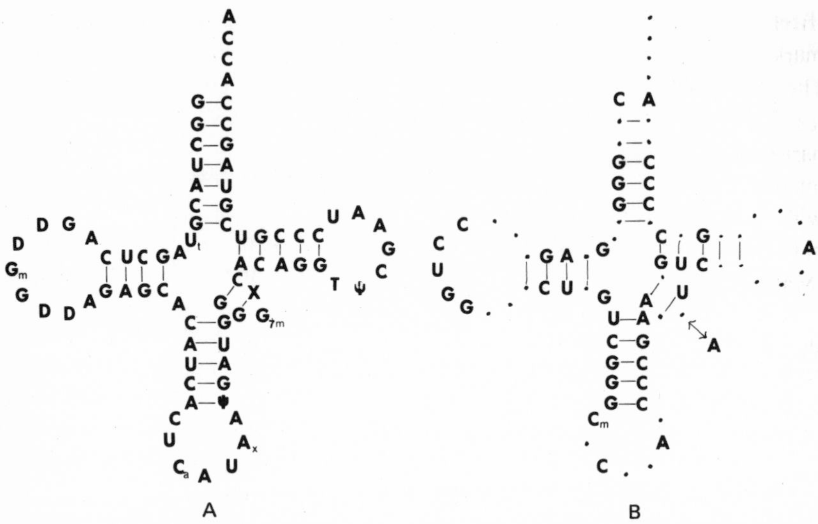

Figure 4.5. Methionine tRNAs from *E. coli*. (A) Internal tRNA$^{Met}_{M}$: the primary and secondary structure of the methionine tRNA that responds to the codon AUG within the body of the message but not at the initial site. (B) Initiator tRNA$^{Met}_{F1}$: the structure of the methionine tRNA that responds only to the AUG signal located at the beginning of the message. Sites corresponding to the noninitiating form are shown by solid circles. The A indicated by an arrow is the replacement found in tRNA$^{Met}_{F2}$.

sition of the pair varies widely from one proposal to another. One site is known as the A site, the point where the aminoacylated (charged) tRNAs enter; the binding of the tRNAs to this site has been demonstrated to be stimulated by polyamines (Igarashi *et al.*, 1974). The second one usually is referred to as the P site, where the amino acid of the tRNA has become linked to the growing peptide chain, that is, the site of peptidyl-tRNAs. Occasionally the P site is called the D, or donor, site (Busiello and DiGirolamo, 1973).

According to some accounts, both the A and P sites are present on the larger subunit (Williamson and Schweet, 1964; Gosh and Khorana, 1967; Bretscher, 1968; Davidson, 1972). In others, the P site is stated to be on the smaller subunit and the A site on the larger (Bretscher, 1966; Lewin, 1970). More recently, the P site has been convincingly demonstrated to be on the larger subunit, with the ribosomal proteins L27, L15, L2, L16, and L14 in its proximity (Czernilofsky *et al.*, 1974). It appears specifically to bind the –CCA 3′-terminus of the peptidyl-tRNA, which terminus is a characteristic feature of all known tRNAs. The A site binds at least part of the same terminus, but only of aminoacylated tRNAs; it has been confirmed as being on the smaller subunit and is known to have the proteins S2, S3, and S14 in its immediate vicinity (Garrett and Wittmann, 1973). S12 also has been indicated to be crucial in initiation reactions (Held *et al.*, 1974) and may be similarly located. Thus the aminoacylated tRNAs enter at the A site in the sequence specified by the messenger. After the amino acid of each newly entered tRNA has become united in turn to the preceding one by means of peptide bond formation, this growing peptide chain is transferred from the tRNA then at the P site by a postulated peptidyl-transferase (Ringer and Chládek, 1974). The new peptidyl-tRNA thus formed is transported to the P site, and the next round of activity begins (Busiello and DiGirolamo, 1973). The peptide transferal in *E. coli* is known also to require one of the proteins (L16) of the 50 S subunit (Moore *et al.*, 1975).

There are discrepancies between these observations and those on the cycling behavior of ribosomes. According to the latter concepts, mRNA is read at an initiator codon (AUG or GUG) by the 30 S subunit, and the initiator tRNA (formylmethionyl-tRNA in bacteria) is attached at the A site *before* the larger subunit attaches to the smaller (Ono *et al.*, 1968). Much of this behavior would be normal for all amino acids and thus would not provide any basis for the lack of response of initiator tRNAs to AUG codons occurring within the body of the messenger. It has also been indicated that the 5 S RNA of ribosomes functions in the binding of tRNAs through interaction with the TΨC arm (Forget and Weissmann, 1967; Dube, 1973; Erdman *et al.*, 1973; Richter *et al.*, 1973).

Other Factors Involved in Binding in Prokaryotes. To form the complex between a 30 S subunit, the mRNA, and the initiator tRNA, several other factors must be present. Among these is GTP, which is essential for all the binding reactions, including the uniting of the aminoacyl-tRNA to the second and subsequent codons of the messenger. In addition, three protein factors, variously called f_1, f_2, and f_3, or IF-1, IF-2, and IF-3, are present in 30 S subunits. The first two are active in the binding of formylmethionyl-tRNA to the 30 S subunit in response to the initiation codon and GTP (Anderson *et al.*, 1967), and the third appears necessary in the binding of the smaller subunit to natural mRNA (Brown and Doty, 1968; Wahba and Miller, 1974). However, the latter is stated by some investigators not to be involved in such binding with synthetic messengers like polyuridylate (Meier *et al.*, 1973), and by others is found to be essential (Schiff *et al.*, 1974). When purified, IF-1 proved to be comparatively small, with a molecular weight of 9000 (Hershey *et al.*, 1969), whereas IF-2 consisted of two subunits, IF-2a and IF-2b, with molecular weights of 91,000 and 82,000, respectively (Kolakofsky *et al.*, 1969; Fakunding *et al.*, 1974). Recently IF-3 from *Caulobacter crescentus* has been shown to have a molecular weight of about 25,000 (Leffler and Szer, 1974a,b).

That step, the binding of f-methionyl-$tRNA_F^{Met}$ to the small subunit of the ribosome, has been reported to be the first event in initiating translation (Jay and Kaempfer, 1975). This reaction occurred rapidly and in the absence of mRNA, but required IF-2, the activated $tRNA_F^{Met}$, and the 30 S subunit, which interacted to form a 34 S complex. Thus the formation of this complex preceded the binding of the mRNA; since it was an essential preliminary, it was proposed that initiator tRNA might play an active role in recognition of the initiation sites in mRNA. The second step, binding of the messenger, proceeded somewhat more slowly than did complex formation, and was dependent on the presence of IF-3. After these two steps had been completed, the 50 S subunit joined to form the complete initiator complex.

The above sequence of events is difficult to reconcile with data from other laboratories (Fakunding *et al.*, 1973; Kay *et al.*, 1973). In these reports the results suggested that IF-2 bound to intact 70 S ribosomes and that the binding of the initiator tRNA was dependent on IF-2 in the presence of proteins L7 and L12 from the 50 S subunits. Moreover, the resulting complex with these proteins also influenced the hydrolysis of GTP. More recently IF-2 has been shown to interact with the initiator tRNA without assistance from GTP, but only in the absence of Mg^{2+} (Sundari *et al.*, 1976). The situation is further complicated by the recent discovery that poly(A)-containing nuclear RNA has a different binding site than fractions of the same substance which lack such sequences (Grozdanovič and Hrader, 1975).

The specific roles of the three initiation factors are still being subjected to question. Of the trio, the most essential is undoubtedly IF-2. As shown above, this protein has been established as promoting the binding of formylated methionyl-tRNA to the small ribosomal subunits and none other, and as being requisite for binding the messenger (Lee-Huang and Ochoa, 1974a; Majumdar *et al.*, 1976). By itself, IF-1 is inactive and is not necessary at low temperatures; however, at ambient temperatures, it increases the amount of complex formation with IF-2. Hence, it appears to function in a stabilizing capacity (Mazumder, 1971). Currently it is proving to be a readily dissociable protein which increases the action of IF-3 and the exchange of ribosomal subparticles (Naaktgeboren *et al.*, 1977). In contrast to IF-1, IF-2 has been demonstrated to be extremely stable (Krauss and Leder, 1975).

Initiation Factor IF-3. IF-3 now has been determined as serving dual functions. First it plays an indispensible but unknown role in the formation of initiation complexes with natural and possibly with synthetic messengers. Then at the close of translation, it promotes the dissociation of the ribosomal subunits as described later (Sabol and Ochoa, 1974). Both activities are stated to result from its binding to the 30 S subunit. Because of changes in the pattern of its recognizing mRNA in *E. coli* when infected with bacteriophage T4, multiple forms of IF-3 have been suggested to exist (Steitz, 1973; Lee-Huang and Ochoa, 1974b), but some doubt as to the validity of this interpretation of the results has been cast through use of purified IF-3 (Schiff *et al.*, 1974). What appeared to be homogeneous but nonselective preparations of this initiation factor were resolved into two molecular species possessing different messenger-selecting capabilities (Lee-Huang and Ochoa, 1974b).

An unexpected property of IF-3 has been described—it released all charged tRNAs (except f-methionyl-$tRNA_F^{Met}$) from their complexes with 30 S ribosomal subunits and synthetic messengers (Gualerzi *et al.*, 1971). Although the biological significance of this activity remains obscure, a molecular basis for it has been reported. On addition of purified IF-3 to isolated 30 S ribosomal subunits, a change in conformation of those particles was noted (Paradies *et al.*, 1974). It was suggested that this property is perhaps also involved in its other functions, including the binding of mRNAs to the 30 S subunit. In contrast, another report proposed that these capabilities stemmed from IF-3 cross-linking to both the 30 S and 50 S particles (Hawley *et al.*, 1974).

Alternative Initiation Systems. Alternative systems have given somewhat different results. In one of these, natural mRNA was produced by phage T4 in infected *E. coli* cells, and soluble proteins of the bacterial ribosomes were extracted (Revel *et al.*, 1968a,b). Here again three factors, called A, B, and C, were found, but they can be only partly correlated with those given above. Factor A was shown to be essential for the binding of formylmethionyl-tRNA to the ribosome and is probably equivalent to IF-2, while factor B accelerated the formation of the first peptide bond between the initiator and the second aminoacyl tRNAs. Since factor C participated both in the binding of the 30 S subunit to the messenger and the attachment of the initiator tRNA to the smaller subunit, it was believed to consist of several proteins, one of which appeared to comprise part of factor IF-3 (Wahba *et al.*, 1968).

Binding of mRNA to the Ribosome. In general few details of the mechanism of mRNA binding to the 30 S or 40 S subunit have been ascertained, aside from the need for Mg^{2+}. As a rule no specific need for GTP appears to exist, as the requirement for a particular nucleoside triphosphate seems to vary with the mRNA in question. In a study of five T4-phage mRNAs *in vitro,* using a bacterial cell-free system, only one (that for deoxyribonucleosidemonophosphate kinase) could be initiated in the presence of a single nucleoside triphosphate, namely ATP (Natale and Buchanan, 1974). The other four needed at least two, uniformly including ATP among the requirements. Perhaps the need for ATP is correlated to the observations made on insect mRNAs (Ilan and Ilan, 1973a). The investigators found by labeling and treating a heterogeneous population of mRNAs with pancreatic ribonuclease that all the sequences began with A and were homologous at the 5′-terminus. They suggested also that the adenine might promote the binding of mRNA to ribosomes. A related suggestion is based on the partial sequence determined for the 5′-terminal of 16 S rRNA (van Knippenberg, 1975). Because of the resemblances of this 42-unit-long segment to portions of messengers from various bacteriophages, it was proposed that this end was involved in the recognition, binding, and relaxation of hairpin loops in mRNAs that contained AUG.

Another recent finding, pointed out earlier (Section 4.1.1), is that mRNAs have two or more different species of protein bound to them. Such messenger ribonucleoproteins have been found in a variety of cells (Spirin, 1969; Olsnes,

1970; Hellerman and Shafritz, 1975), including a heterogeneous population of mRNAs from chick embryo brains (Bryan and Hayashi, 1973). At least one of these proteins was believed to be essential to mRNA binding (Hellerman and Shafritz, 1975); some evidence has been presented which suggests that this necessary protein actually may be IF-3 (Ilan and Ilan, 1973b).

Binding and Initiation in Eukaryotes. As a whole, the requirements for binding of tRNA and mRNA to the ribosomes in eukaryotes differs little from those in prokaryotes. As there, the necessary ingredients include small ribosomal (40 S) subunits, mRNA, initiator tRNA, GTP, and the three initiating factors IF-1, IF-2, and IF-3 (Lucas-Lenard and Lipmann, 1971; Caskey *et al.*, 1972; Crystal and Anderson, 1972; Crystal *et al.*, 1974; Marcus *et al.*, 1974). In studies on mammalian cells, these factors are known also by the designations M1, M2, and M3 (Woodley *et al.*, 1974) and IF-M1, IF-M2, and IF-M3 (Cimadevilla and Hardesty, 1975). In addition, a substance called IF-EMG was identified in an ascites cell-free system (Strycharz *et al.*, 1974), and Levin *et al.* (1973) have isolated a protein from mouse fibroblast ribosomes that corresponds to *E. coli* IF-2 in forming a complex with rat liver initiator tRNA and GTP. The latter has proven not to be interchangeable with IF-2 from *E. coli,* however (Zasloff and Ochoa, 1971; McCroskey *et al.*, 1972). Potentiation of the messenger was demonstrated in mammalian systems, in which interaction of mRNA and the 18 S rRNA occurred (Kabat, 1975; Krystosek *et al.*, 1975).

Functionally the various enzymes appear to be similar to those of prokaryotes. IF-1 has been stated to mediate the GTP-dependent, template-independent binding of initiator tRNA to a 40 S ribosomal subunit in rabbit liver cells, and IF-2 to induce the transfer of an initiator tRNA to an initiation complex containing an 80 S ribosome, providing GTP, rabbit-globin messenger ribonucleoproteins, and a 60 S subunit were present (Cashion and Stanley, 1974). At the end of this complex formation, the $tRNA_F^{Met}$ supposedly was located at the P site of the 80 S ribosome. Normally IF-1 and IF-2 were found in the form of a complex, from which they could be separated by appropriate techniques; IF-1 from rabbit reticulocytes had a molecular weight of 150,000, and IF-2, 220,000. A factor which may correspond to IF-2 has been shown to be removed by salt washing of ribosomes (Smith and Henshaw, 1975). Much of its behavior seems to resemble that of prokaryote S1, especially in the creation of two subclasses of the small ribosomal subunit.

It has been demonstrated that in many animal cells the three ribosomal initiating factors, IF-1 to IF-3, can be replaced by soluble proteins from the cytoplasm; in the presence of these substances the system does not require GTP (Gasior and Moldave, 1972; Zasloff and Ochoa, 1973). Either spermine, spermidine, or putrescine can similarly serve as a substitute at low, but critical concentrations of Mg^{2+} (Teraoka and Tankak, 1973). Also it has been reported that, in pea cotyledons, extracted ribosomal initiating factors can form a complex with synthetic mRNA [poly(U)] and phenylalanyl-tRNA at 5 mM concentration of Mg^{2+}, in the absence of GTP and ribosomes (Wells and Beevers, 1974). A comparable observation had been made earlier for *E. coli* (Mazumder, 1972). Thus it would appear that the rRNA plays no role in these very early steps of protein synthesis. On the other hand, the presence of aminoacylated (and absence of uncharged) tRNA has been shown to be requisite in human cells (Vaughan and Hansen, 1973).

One principal difference between the prokaryotic and eukaryotic initial stages in the proteinogenic processes is the absence of formylated methionyl-tRNA among the latter organisms, although AUG similarly is the chief initiator codon. Recently the nature of the initiator tRNA of eukaryotes has been the subject of a number of investigations, the earliest of which dealt with yeast cytoplasmic initiator tRNA. The results showed that the sequence G–T–Ψ–C, nearly universally present in arm V of protein-synthesizing tRNAs, was absent in this instance, being replaced by G–A–U–C (Simsek *et al.*, 1973a,b). The presence of the latter sequence was later confirmed when the entire base composition of the tRNA was analyzed (Petrissant, 1973). On the basis of this evidence, it was conjectured that the eukaryotes had eliminated the need for formylating methionyl through the evolution of a more highly modified tRNA. A small number of other tRNAs, however, not involved in initiation have a similar sequence in this arm, so it can no longer be considered a specific recognition trait. Subsequently, it was reported that the cytoplasmic initiator tRNAs from a variety of sources, including wheat germ, rabbit liver, and sheep mammary gland, likewise lacked the same sequence (Petrissant, 1973; Simsek *et al.*, 1973b).

Specificity of Initiating Factors. Much evidence exists both in support of and against a concept that initiating factors possess a degree of specificity toward particular mRNAs. Originally the idea arose through investigation of bacteriophage-infected bacteria. Upon infection of *E. coli* with bacteriophage T4, synthesis of host protein ceases (Levinthal *et al.*, 1967; McCorquodale *et al.*, 1967), and, if the host had previously been infected with any RNA phage such as MS2, M12, or R17, further development of that virus became restricted when infection with T4 took place (Zinder, 1963). Furthermore, *in vitro* systems using ribosomes from T4-infected cells were able to translate T4 mRNA actively, but could not do so with RNA from R17, MS2, or f_2 (Hattman and Hofschneider, 1968; Hsu and Weiss, 1969; Schedl *et al.*, 1970). The lack of activity with f_2 mRNA and the retention of ability with T4 mRNA in T4-infected cells has been confirmed by Singer and Conway (1973), but contrasting effects resulted with higher concentrations of the messenger.

Among eukaryotes reports are even more contradictory. No need for homologous initiating factors has been demonstrated *in vitro* with a number of cell-free systems (Stavnezer and Huang, 1971; Berns *et al.*, 1972; Mathews *et al.*, 1972a,b; Metafora *et al.*, 1972; Pemberton *et al.*, 1972; Sampson *et al.*, 1972), as well as in the oocytes of the South African clawed toad (*Xenopus*)

(Gurdon *et al.*, 1971). However, in the translation of globin mRNA, reticulocyte ribosomal factors have shown preferential specificity for that messenger in the presence of Mengo virus RNA (LeBleu *et al.*, 1972). Initiating factors have been reported to react differentially to α- and β-globin mRNAs, and, in reticulocytes and ascites cells, factors have been isolated and purified that showed specificity for particular mRNAs (Nudel *et al.*, 1973; Wigle and Smith, 1973). In an ascites cell-free system using initiation factors from rabbit reticulocytes, viral mRNA could not compete with the messenger for either α- or β-globin (Hall and Arnstein, 1973). In contrast, it has been reported that no translational barrier existed between eukaryotes (Vaquero *et al.*, 1973), for rabbit globin mRNA was bound and translated by ribosomes from such distantly related forms as trout (liver) and kidney bean (root tips).

4.2.2. Elongation of the Peptide Chain

After peptide bonding has been initiated, the methionine residue appears to be removed, for only a small percentage of known amino acid sequences commence with methionine (Ishizuka *et al.,* 1974). Further additions to the forming peptide chain, however, do not depend on those factors that initiated the strand but draw upon several others as the ribosome moves along the mRNA (Leder, 1973). It appears possible, however, that some of the proteins involved with binding and initiation may also participate in elongation, so that the total number of factors in protein synthesis may eventually be reduced, although the present trend is in the opposite direction.

Elongation of the Chain in Bacteria. For the greater part, studies on peptide chain elongation in bacteria have utilized poly(U) as the messenger and phenylalanyl-tRNA as the amino acid source, in order to maintain polypeptide synthesis at the simplest possible level. Typically a complex including those factors is bound to the ribosomes before the search for elongation mechanisms is begun (Lippmann, 1969; Haenni and Lucas-Lenard, 1970). Some of the earlier inquiries into this topic demonstrated that the soluble fraction of *E.coli* contained two factors which induced polymerization of amino acids (Nathans and Lipmann, 1961; Allende *et al.*, 1964). These two, known as EF-T and EF-G, have been further purified and their probable role in poly(U) systems elucidated to a degree; a third factor, recently discovered, has been referred to as EF-P (Glick and Ganoza, 1976). EF-T has been found to consist of two fractions, the stable EF-Ts factor and the unstable EF-Tu (Lucas-Lenard and Lipmann, 1966; Leder, 1973). The functions of both EF-G and EF-T are dependent also on the ribosomal proteins L7 and L12 (Kischa *et al.*, 1971; Stöffler *et al.*, 1974; Koteliansky *et al.*, 1977; Visentin *et al.*, 1974). In addition, L10 is required for L7 and L12 to bind to the ribosome (Schrier *et al.*, 1973; Stöffler *et al.*, 1974) and L11 has been implicated in the EF-G–dependent breakdown of GTP (Schrier and Möller, 1975).

The role of EF-G was recently examined in a system in which guanosine 5′-triphosphate, 3′-diphosphate (pppGpp) was substituted for GTP (Hamel and Cashel, 1973). The substitute was effective in those initiation reactions catalyzed by IF-2, including the binding of formylmethionyl-tRNA on ribosomes; moreover, it also served efficiently in those reactions catalyzed by EF-T, including the binding of the second aminoacylated tRNA (AA-tRNA) and formation of dipeptidyl-tRNA, but it supported polypeptide synthesis very poorly.

EF-G has now been purified and crystalized and identified as the actual ribosome-dependent guanosine triphosphatase (Kaziro *et al.*, 1972). In the absence of mRNA and charged tRNA, it hydrolyzed GTP in the presence of ribosomes; neither the exchange of inorganic phosphate into GTP nor that of GDP into GTP was catalyzed by EF-G, however (Rohrbach *et al.*, 1974). Further, when this triphosphatase and GTP were added to poly(U)–ribosome complex having deacylated tRNA in the P site and a charged tRNA in the A site, the latter tRNA was moved to the P site from which the other was concomitantly released (Conway and Lipmann, 1964; Haenni and Lucas-Lenard, 1970; Inoue-Yokosawa *et al.*, 1974). The energy derived from breakdown of GTP is often considered to drive the protein-synthetic processes (Lucas-Lenard and Lippmann, 1971). An acidic protein (or perhaps two) is also involved with EF-G and EF-T in GTP hydrolysis (Terhorst *et al.*, 1973); one from yeast ribosomes has been found requisite for poly(U)-directed polyphenylalanine synthesis (Skogerson and Wakatama, 1976).

Thus EF-Tu also is associated with GTP hydrolysis, the two substances first uniting to form a complex. The resulting EF-Tu · GTP, together with ribosomal complexes having a charged tRNA in the A site, splits the GTP molecule (Yokosawa *et al.*, 1973; Kawakita *et al.*, 1974). Both sets of reactions involving GTP, that of EF-G as well as the EF-Tu-activated tRNA, appeared to occur at or near the A site (Otaka and Kaji, 1974). In addition, the Tu factor required the 50 S proteins L4, L7, and L12 for protein elongation, and the G factor needed L14 and L23 (Highland *et al.*, 1973, 1974). A partial amino acid sequence of Tu, 42 units in length, has now been determined (Wade *et al.*, 1975).

Further suggestion that EF-T is at least in part functionally identical to IF-1 or IF-2 comes from an investigation of recognition by EF-Tu of the GTΨC sequence in tRNA arm V (Richter *et al.*, 1973). It was found that the tetranucleotide GTΨC bound to 50 S ribosomal subunits and, in doing so, blocked further binding of aminoacylated tRNAs by EF-Tu, as well as inhibiting hydrolysis of GTP and ribosome-dependent synthesis of "magic spot I," ppGpp (Richter *et al.*, 1973; Glazier and Schlessinger, 1974). However, the actuality of the proposed identity of certain elongation and initiation factors was made doubtful by other studies which identified specific reactions in binding. These reactions appeared to be as follows (Leder, 1973; Miller and Weissbach, 1974):

$$\text{EF-Tu} \cdot \text{GDP} + \text{EF-Ts} \rightleftharpoons \text{EF-Tu} \cdot \text{EF-Ts} + \text{GDP}$$

$$\text{EF-Tu} \cdot \text{EF-Ts} + \text{GTP} \rightleftharpoons \text{EF-Tu} \cdot \text{GTP} + \text{EF-Ts}$$

$$\text{EF-Tu} \cdot \text{GTP} + \text{AA-tRNA} \rightleftharpoons \text{AA-tRNA} \cdot \text{EF-Tu} \cdot \text{GTP}$$

$$\text{AA-tRNA} \cdot \text{EF-Tu} \cdot \text{GTP} + \text{ribosome} \cdot \text{mRNA} \rightarrow$$

$$\text{AA-tRNA} \cdot \text{ribosome} \cdot \text{mRNA} + \text{EF-Tu} \cdot \text{GDP} + \text{P}_\text{i}$$

Polypeptide Chain Elongation in Eukaryotes. Similar ambiguities regarding the distinctiveness of the initiating and elongating factors exist among eukaryotes. One enzyme, called C, cited above as being considered as an eukaryotic initiator, is given by others as an elongation factor. Thus Seal and Marcus (1973), in exploring translation of satellite tobacco necrosis virus mRNA in a cell-free system from wheat embryo, referred to factors A and B as having functions in initiating protein synthesis, whereas C and D (stated to correspond to bacterial EF-G) were classed as elongation-oriented.

Actually there appears to be little close relationship between eukaryotic and prokaryotic elongation factors, if antigenic reactions are to be believed. Gordon *et al.* (1969) prepared antisera against electrophoretically homogeneous EF-T and EF-G and tested them against a number of factors from various bacterial and mammalian sources. Wide antigenic variability was found among the elongation enzymes from different bacteria, suggesting considerable evolutionary divergence to exist. What was referred to as EF-1 from Krebs II ascites cells was indicated to attach early to the ribosome (purportedly near the A site) early in translation and remained bound until termination (Grasmuk *et al.*, 1976). In contrast, EF-2 attached each time the messenger was moved and was released before another charged tRNA could be bound. Factors from mammalian sources, however, gave no detectable interactions with prokaryotic ones, indicating virtually complete absence of homology. There is also a body of evidence supporting the distinctiveness of elongation and initiation factors. For instance, cycloheximide retarded the elongation of peptide chains, whereas initiation rates remained unaffected (Pain and Clemens, 1973); as a result, the number of ribosomes per polysome was increased. Conversely, during normal mitosis, initiation was retarded, but the rate of elongation was unchanged (Perlman *et al.*, 1972).

In addition to these special factors, certain ribosomal proteins are also involved in elongation. Probably these will be found to include L40 and L41 which have been found in rat liver ribosomes to be homologous to *E. coli* L7 and L12 (Stöffler *et al.*, 1974).

Eukaryotic Elongation Factor 1. However, there is a strengthening trend toward recognition of two elongation factors among eukaryotes, EF-1 and EF-2. Along with GTP, EF-1 binds the charged tRNA specified by the messenger to the A site of the ribosome, the three substances forming a ternary complex (Jerez *et al.*, 1969). Multiple forms of EF-1 have been reported from various mammalian species, that have been indicated to represent different molecular weight polymers of a single repeated unit (Weissbach *et al.*, 1973; Drews *et al.*, 1974). In Krebs II mouse ascites cells, the dimer, tetramer, and hexamer forms showed similar affinities for GTP binding (Nolan *et al.*, 1974). Recently EF-1 from pig liver has been resolved into two complementary factors, EF-1α and EF-1β; moreover, EF-1α was itself found to consist of two differing molecular weight species (Iwasaki *et al.*, 1976). Comparable molecular weight polymers of EF-1 have been reported also from wheat embryos (Lanzani *et al.*, 1974); only two forms existed, however, the heavier having a molecular weight of 187,000, the lighter 51,000 (Bollini *et al.*, 1974). Poly(phenylalanine) synthesis was more active when EF-1 was added to a complex before EF-2 than when the additions were made in the opposite sequence (Grasmuk *et al.*, 1974).

Eukaryotic EF-2. The second elongation factor of eukaryotes, EF-2, has been purified from pig liver and found to have a molecular weight of around 100,000 and a sedimentation coefficient of 5.32 S (Mizumoto *et al.*, 1974). A short strand, 15 amino acid residues in length, has been sequenced and was stated to contain a residue X, a weakly basic substance that does not correspond to any of the common proteinogenic amino acids (Robinson *et al.*, 1974). GTP appeared to bind to EF-2 in a 1 : 1 ratio, but GDP served to modulate protein synthesis (Henriksen *et al.*, 1975a,b). Changes in the GTP : GDP ratios were suggested to induce conformation changes in the various associated factors, which led to alterations in reaction patterns throughout the course of proteinogenesis. Ribosomes have been shown to have a greater affinity for EF-2 than for EF-1; this interaction was further enhanced when aminoacyl-tRNA had become attached to the ribosomes (Nolan *et al.*, 1975).

Additional Elongation Factors. The danger of using a single model system for investigating any complex process, especially where synthetic reagents or *in vitro* systems are involved, is pointed out by other studies. When synthetic messengers other than poly(U) were used in these same *in vitro* systems, only small amounts of protein synthesis resulted (Nakamoto *et al.*, 1963; Nishizuka and Lipmann, 1966). For example, when EF-T and EF-G were sufficiently purified and employed with poly(A) as messenger, they were found incapable of carrying synthesis further after a single peptide bond had been formed in an *E. coli* system. Only when an additional factor called X was added could protein synthesis proceed beyond the dipeptide stage. This factor was shown to have a molecular weight of around 50,000 and to be antigenically different from both EF-T and EF-G (Ganoza and Fox, 1974).

4.2.3. Chain Termination

During the processes of protein synthesis, the active, growing end of the polypeptide chain is attached to a tRNA by the ester bond that exists between the most recently added amino acid and the adenine residue of the –CCA stem of the tRNA. When the last amino acid specified by the messenger has been added to the macromolecule, this link must be broken in order to release the finished protein. Breaking this bond and releasing the last tRNA have been shown to be active processes and that definite signals are required (Bretscher, 1965; Bretscher *et al.*, 1965).

Termination in Bacteria. In the study of protein synthesis termination in bacteria, *in vitro* systems derived from *E. coli* have been widely employed and, through their use, it soon was made apparent that three soluble proteins play as important roles in ending the processes as others did in the initiation and the elongation of the polypeptide chain (Scolnick *et al.*, 1968; Milman *et al.*, 1969; Traeger, 1970). Currently these three proteinaceous factors are known simply as RF-1, RF-2, and RF-3. The latter, originally called factor S, is heat labile and stimulates the actions of RF-1 and RF-2; it appears also to interact with GDP and GTP (Caskey, 1973). The other two, with molecular weights in the range of 40,000 to 50,000, effect release in response to specific codons in the messenger. RF-1 reacts to either UAA (*amber*) or UAG (*ochre*), whereas RF-2 is active with either UAA or UGA (*opal;* Caskey, 1973). Although *ochre* is often believed to be the only *in vivo* chain terminator signal (Person and Osborn, 1968; Lewin, 1970), the base sequence of the mRNA for the read-through protein from phage RS2 is terminated by UGA (Contreras *et al.*, 1973), as is that of the coat protein cistron in bacteriophage Qβ (Weiner and Weber, 1973). In prokaryotes and viruses *opal* was stated not to be as effective a termination signal as the other two, rarely attaining an efficiency of 98%. In mammal cells, the triplet UGA similarly may not be totally specific for termination, for serine-tRNA and a species of arginine-tRNA recognized that codon *in vitro* to a considerable extent and cysteine- and tryptophan-tRNAs did so to a lesser degree (Hatfield, 1972).

Under similar experimental conditions, the two ribosomal proteins (L7 and L12) that are active in chain initiation and elongation have been demonstrated to play an important role also in chain termination (Brot *et al.*, 1974). Because their activities throughout the processes involved GTP, it has been suggested that these two constituted a universal GTPase region of the ribosome (Highland *et al.*, 1973). Although GTP has not been directly implicated in chain termination, guanine nucleotides have been shown to be essential for RF binding to ribosomes.

Eukaryotic Termination. What little attention has been given mammalian termination processes has demonstrated them in substance to resemble those of bacteria, but only a single factor (RF) has been isolated, mainly from rabbit reticulocytes (Beaudet and Caskey, 1971; Tate *et al.*, 1973; Caskey *et al.*, 1974). As in prokaryotes, GTPase activity has proven to be necessary for RF functioning, and the same codons, UAG, UAA, and UGA, to serve as signals for terminating the chain.

4.2.4. The Ribosome Cycle

Since the initiation processes require only the smaller of the ribosomal subunits, it is evident that, after a ribosome has completed the translation of an mRNA, it must dissociate into its two subunits following termination and preceding the next initiation. Thus *in vivo,* the two ribosomal subunits do not couple on a permanent basis, but merely for a single round of protein synthesis (Nomura, 1973).

The Ribosome Cycle in Bacteria. By transferring bacteria having their ribosomes labeled with heavy isotopes into a new medium containing light isotopes for labeling, it was experimentally shown that the predicted cycling of ribosomes actually occurred *in vivo* (Kaempfer *et al.*, 1968). As the bacteria grew and multiplied, the original heavy ribosomes were gradually replaced by hybrid types, having one subunit heavy, the other light. In a cell-free system, it has been demonstrated that such subunit exchange is requisite for protein synthesis (Kaempfer, 1968). Because of the essential nature of dissociation, the existence of free 70 S ribosomes in live organisms has been questioned; it is thus believed that the complete ribosome may exist only on the messenger, never free in the cytoplasm. That this condition may be typical is suggested by evidence obtained through use of a rapid preparation of polysomes and ribosomes from fragile *E. coli* cells. The results indicated that virtually only polysomes and free subunits existed, and that less than 2% were in the form of free 70 S ribosomes (Mangiarotti and Schlessinger, 1966). Furthermore, in radioisotopic studies, it was observed that newly labeled RNA appeared in subunits and in polysomes, but not in intact 70 S ribosomes (Mangiarotti and Schlessinger, 1967).

In contrast, other workers have reported the presence of free 70 S ribosomes under various conditions. In cells deprived of amino acids or in those which had been treated with certain antibiotics (actinomycin or puromycin), translation of new messengers was halted, but that already in process was carried to completion. The ribosomes in such cells remained intact and did not dissociate into subunits (Kohler *et al.*, 1968). Later replicates of the experiment showed that the polysomal and freed ribosomes behaved differently, for the latter dissociated into their subunits at 1 mM concentrations of Mg^{2+}, whereas the former did so under much lower concentration levels of the ion (Ron *et al.*, 1968). Since behavior similar to the freed ribosomes was observed in the intact particles from controls, it was believed tht the former represented the normal condition upon complete translation of a mRNA.

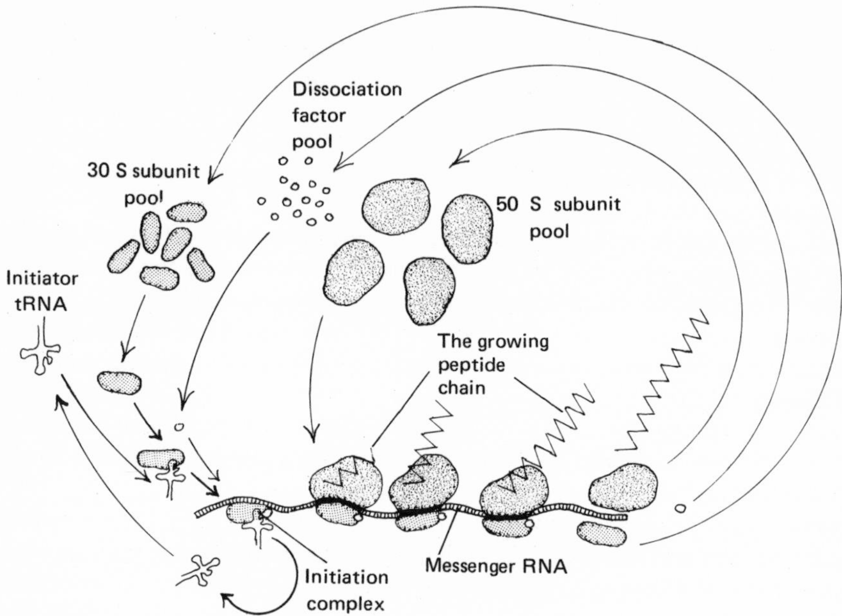

Figure 4.6. The ribosomal cycle in *E. coli*.

These and related observations led Kohler *et al*. (1968) to propose, as Eisenstadt and Brawerman (1967) had earlier, that a dissociation factor (DF) was needed for the 30 S and 50 S subunits to separate. Subsequently, Subramanian and his co-workers (1968) found that an extract from 30 S subunits contained a protein with the predicted properties of DF and that these properties were absent from 70 S ribosomes. When the DF extract was added to 70 S particles, the latter dissociated into the two usual subunits. The DF is now viewed as identical to one initiation factor (IF-3), as intimated in the discussion above (Sabol *et al*., 1970; Subramanian and Davis, 1970; Gottlieb *et al*., 1974). In bacteria the ribosomal cycle appears to approximate the events shown in Figure 4.6.

The Ribosome Cycle of Eukaryotes. As a whole, the ribosomal cycle of eukaryotes, insofar as it has been investigated, appears to differ little from that of the bacteria. First, it was established that the subunits disengaged and subsequently recombined as in the same cycle of prokaryotes (Kaempfer, 1969). Later, when a series of studies investigated the entire cycle, the subunits in mammalian cells were found to undergo rapid cycling as they carried out the major activities in the synthesis of a protein (Henshaw *et al*., 1973). As in the bacteria, monosomic ribosomes were relatively inert, and the free 40 S particles proved to bear a protein absent from the 60 S subunit (Hirsch *et al*., 1973).

Since this protein had to be released from the 40 S particle before recoupling could occur, it thus appeared functionally identical to RF.

A major distinction was that the proteins were attached to the 40 S subunit in two stages (Ayuso-Parilla *et al.*, 1973a,b). In the first stage, the subunit, referred to as the heavy form (40 S_H) acquired a density of 1.49 g/cm^3 upon association with a soluble protein of a molecular weight between 6 and 12×10^4. The light, or 40 S_L subunit was formed upon addition of a second protein of 7×10^5 in molecular weight, when its density became reduced to 1.40 g/cm^3. That is to say, the volume of the subunit was increased in two steps, while its density decreased. The ribosome also underwent changes in mass during the protein synthesis cycle; free cytoplasmic ribosomes have been demonstrated to sediment at 78 S, whereas those derived from polysomes sedimented at 85 S (Vournakis and Rich, 1972). The latter were greater in diameter than the former by 20 Å.

The relative state of the ribosomal cycle during the various phases of the eukaryote cell cycle has been investigated (Scharff and Robbins, 1966; Eremenko and Volpe, 1975). In HeLa cells, a certain equilibrium was noted among ribosomal subunits, whole ribosomes, and polysomes from G_1 to the end of the S phrase, but from the beginning of G_2 the proportion of polysomes was enhanced to the commencement of M. The total quantity of polysomes per cell was bimodal; a trough during S separated the pre- and postmitotic peaks, the former being much higher than that of the G_1. The incorporation of labeled leucine into nascent polypeptide chains on polysomes was also biphasic, but showed a higher peak in G_1 than in G_2. Young cells, moreover, were found to differ from old ones in having an abundance of small polysomes, rather than a lower number of large ones. In an earlier, less extensive study with hamster cells, polysomes were reported to be reassembled following mitosis, during which time they were absent. This G_1 reassembly occurred independently of *de novo* RNA synthesis and thus used preexisting mRNA (Steward *et al.*, 1968).

Like all other cellular components, ribosomes undergo a breakdown after a period of service. In rat liver, almost all the actual ribosomal proteins, of the large and small subunits alike, were degraded at the same rate as the rRNAs (Arias *et al.*, 1969; Tsurugi *et al.*, 1974). In other words, ribosomes were degraded as a whole. However, such associated proteins as initiation and elongation factors were found to differ in rates of breakdown.

5

THE GENETIC MECHANISM: III

Transcription, Processing, and an Analytical Synopsis

If this book were strictly faithful to the chronology of actual events, then transcription of the DNA molecule would have been among the first topics treated. Because of the structural and functional complexity of the various RNAs, however, much clarity is gained by delaying discussion of transcription of the DNA molecule and related events to this point. After these final aspects of the genetic mechanism have been viewed, an analysis then places the substance of the entire topic into context with the problem of life's origins.

5.1. TRANSCRIPTION OF THE DNA MOLECULE

Transcription of the DNA molecule brings into existence the several types of RNAs—actually their precursors—by processes that apparently are identical, regardless of the product. Hence, it is possible to view transcription as a single general procedure; only the posttranscriptional events need to be examined separately, for they are in part distinctive for each category of RNA.

5.1.1. The Polymerization of RNA

Two main systems for the polymerization of RNA exist in the cell, corresponding functionally to the two major types of polymerases that were seen to be active in replicating DNA. The most important and best known is that which provides the main topic of the section, the DNA-dependent RNA-polymerase system. Except in certain RNA viruses, it is this system alone that may be strictly stated to participate in *transcription*, for the second one, which centers around RNA-dependent RNA polymerization, actually must be considered to engage in *replication*.

DNA-Dependent RNA Polymerization in Bacteria. In bacteria, the synthesis of RNA directed by a DNA template is catalyzed by an RNA polymerase, also known as transcriptase (Geiduschek *et al.*, 1968). The enzyme used ribonucleoside 5′-triphosphates as substrates in the presence of Mn^{2+}, Mg^{2+}, and a DNA template, forming the usual 5′–3′ bonds and liberating pyrophosphate (Lewin, 1970). An excess of the polymerase was present in *E. coli*, some of it possibly in dormant form, for the study found a ratio of two to five enzyme molecules for each RNA molecule produced (Matsura, 1973). The complete enzyme (holoenzyme) has been purified from several species and was reported to be of comparable structure in each. In *E. coli* a molecular weight of 495,000 was determined (Matsura, 1973), in *Pseudomonas*, one of 485,000 (Zimmer and Millette, 1975), and in *Lactobacillus*, 425,000 (Stetter and Zillig, 1974). In the first two of these sources, the proteins were similar in other respects, such as subunit structure, whereas that of the third one was strongly disparate on several counts. In 6 M urea solutions, all the polymerases dissociated into four types of subunits. The α-polypeptide chain had a molecular weight of 40,000 (42,500 in *Lactobacillus*), the β, 155,000, and the β', 165,000, and the σ, 95,000. Contrastingly, the *Lactobacillus* β subunit was larger than the β', with molecular weights of 154,000 and 145,000, respectively (Stetter and Zillig, 1974), whereas the α subunit was much smaller than elsewhere, its molecular weight being only 52,500. As each of these polymerases have been demonstrated to contain 2α subunits, their composition may be represented as $\beta'\beta\alpha_2\sigma$ (Travers and Burgess, 1967). During protein synthesis in growing *E. coli*, the β and β' subunits were produced in a coordinate manner, possibly on a polycistronic mRNA (Matzura, 1973).

In *Lactobacillus* a second variety of polymerase was reported (Stetter and Zillig, 1974), which, while lacking the σ subunit, had a much heavier one named y in its stead. This polypeptide, which had a molecular weight of 84,000, differed further from σ in that it could not be separated from the β' subunit, as the latter can, but always remained as the complex $\beta'y$. Sigma and y never occurred together in the same molecules, yet enzymes containing y were stimulated to greater activity levels in the presence of the former. Consequently, it was proposed that the two subunits affected different steps in the transcription processes.

The configuration of the DNA plays an active role in transcription, for this activity has been found to proceed at a twofold higher rate on the supercoiled genome than on a relaxed molecule in *Pseudomonas* (Zimmer and Millette, 1975) and *E. coli* (Richardson, 1974, 1975). When the bacteriophage PM2 DNA employed in the *Pseudomonas* experiments was in the form of a supercoil, transcription proved to be a continuous process, with little chain termination being observed, whereas the relaxed circular DNA yielded discrete products. Hence, it was obvious that the one effect of configuration was a strong influence on chain termination. Recently the transcription complex (formed with template DNA, the growing RNA chain, and active RNA polymerase) has been isolated from lysozyme-treated *E. coli* cells (Naito and Ishihama, 1975). In addition to around 50 polymerase molecules per genome, the isolated complex contained other proteins that were not identified.

An Initiation Factor. The σ factor may be detached from the holoenzyme leaving the so-called core enzyme (Burgess *et al.*, 1969); although the latter catalyzed RNA synthesis *in vitro* on a heterologous DNA template, such as one derived from calf thymus, it was inactive on DNA from *E. coli* or phage T4. Since in the latter cases, addition of σ restored RNA synthesis, that substance was viewed as an initiation factor (Berg *et al.*, 1971; Burgess and Travers, 1971). Its presence enabled the resulting holoenzyme to read complete genes in T4 DNA in sequence, but in its absence, the core enzyme appeared to attach randomly, instead of at initiation sites. Consequently when σ was absent, varied lengths of products resulted from transcription, and the transcripts lacked the purine nucleoside triphosphate, ppGp . . . , which characteristically begins the RNA chains of *E. coli* and many phages (Davidson, 1972). After RNA synthesis has begun, the σ factor is released to join another polymerase molecule and initiate a new round of transcription.

Besides the *y* factor of *Lactobacillus* mentioned above, several other proteins have been described which supplement the activity of σ. A factor, *M*, stimulated transcription indiscriminately, whereas one called *psi* (ψ) appeared specific for *E. coli* DNA, as it was ineffective with templates from the other sources tested (Bautz, 1972). Another one from *E. coli*, distinguished as *X*, was somewhat heavier than *y* in having a molecular weight of 110,000 (Stetter and Zillig, 1974). This factor may prove identical with τ described earlier from the same organism (Burgess *et al.*, 1969).

Initiation Sites. The nature of the initiation sites on the DNA molecule is poorly understood at present. Since RNA chains usually begin with a purine residue, especially guanine, the proposal has been made that clusters of pyrimidine residues mark the initiation site on DNA strands (Szybalski *et al.*, 1966). The specific dinucleoside monophosphates, ApU, GpU, CpA, and UpA, were found more effective than nucleoside triphosphates in initiating transcription (Hoffman and Niyogi, 1973). Now that the nucleotide sequences preceding the initiation sites for the late and early transcripts of SV40 DNA have been determined (Dhar *et al.*, 1974; Zain *et al.*, 1974), it becomes obvious that this supposition does not receive support (Table 5.1). Nevertheless, when such analogs of ATP and GTP as formycin and 8-azaguanonine were substituted for the normal nucleotides, respectively, initiation of transcription was strongly inhibited (Roy-Burman, 1970; Darlix and Fromageot, 1974), although chain elongation

Table 5.1
Partial Transcripts of SV40 DNA[a]

	10 20
Early	GAAACAUAAAAUGAAUGCAAUUGUUGU
Late	UGUAACCAUUAUAAGCUGCAAUAAACA
	30 40 50
Early	UGUUAACUUGUUUAUUGCAGCUUAUAA
Late	AGUUAACAACAACAAUUGC
	60 70
Early	UGGUUACAAAUAAAGCAAUAGCA

[a]Based on Dhar et al. (1975) and Zain et al. (1975).

rate was unaffected. Initiation has been shown to be seriously inhibited also by amino acid starvation (Edlin and Broda, 1968; Ryan and Borek, 1971; Rogerson and Ezekiel, 1974).

Termination. Similarly, the means of indicating the sites where the RNA chain is to be terminated are not well understood. It has been established that a protein called the *rho* (ρ) factor, induced the RNA polymerase to recognize specific termination signals. This factor, which had a molecular weight of around 200,000, also appeared responsible for the release of the completed RNA molecule (Richardson, 1970; Roberts, 1970; Richardson *et al.*, 1975). The analogs of GTP and ATP mentioned above enhanced the efficiency of ρ in terminating transcription, so that chain length was drastically decreased. That is to say, the termination processes occurred prematurely, which is tantamount to stating that chain elongation was adversely influenced, although the rate remained normal (Darlix and Fromageot, 1974). The ρ factor has been claimed to dissociate the ternary transcription complex consisting of DNA · RNA-polymerase · RNA and also to modulate transcription (Darlix, 1974, 1975). In these processes RNase H also seemed to play a role, although it chiefly degraded the RNA moiety of RNA · DNA hybrids (Darlix, 1975). A DNA template, RNase III, and ρ factor sufficed *in vitro* for the production of transcripts similar to those of *in vivo* systems (Kapitza *et al.*, 1976). An inexplicable role associated with highly purified ρ from *E. coli* has been observed (Lowrey-Goldhammer and Richardson, 1974), in that the substance catalyzed an RNA-dependent hydrolysis of ribonucleoside triphosphates, including ATP, to nucleoside diphosphates and inorganic phosphate.

Control of Transcription in Prokaryotes. Very little has been established in prokaryotes concerning the control of transcription; however, a stringent control system has been described in *E. coli*, which regulated the synthesis of ribosomal components. Parts of the total mechanism appeared to determine the amount of mRNA for each of the various ribosomal proteins available for translation, which was presumed to operate at the level of initiation of transcription at the appropriate gene promoters (Dennis and Nomura, 1975).

DNA-Dependent RNA Polymerization in Eukaryotes. Although, as might be expected, studies on eukaryotic DNA-dependent RNA polymerization have been chiefly on mammalian subjects, other organisms also have now received some attention. While a number of polymerases have been purified (Jacob, 1973; Chambon, 1974; Hager *et al.*, 1977), few studies have as yet been conducted on their properties. At the moment, moreover, the existing state of knowledge is rather chaotic, as a result of the multiplicity of enzymes that exist and the unsettled state of the nomenclature for the polymerases. Another factor contributing to the confusion is the difficulty of isolating the enzymes; to free them from the deoxyribonucleoproteins (chromatin), the polymerases are usually solubilized by means of sonification (Roeder and Rutter, 1969, 1970). Furthermore, the relative proportions of the several types vary, along with specific qualities and levels of activities, from species to species and tissue to tissue and with the physiological stage of a given cell type (Schwartz *et al.*, 1974; Jaehning *et al.*, 1975; Sklar *et al.*, 1975).

The present trend with eukaryote enzymes appears to be toward recognition of three classes of DNA-dependent RNA polymerases, distinguished approximately as follows: Class I polymerases (often also called A, but the combination AI occurs in the literature) are insensitive to α-amantin, the chief species of which is found in the nucleolus, where it is presumed to transcribe mainly rRNA genes (Roeder and Rutter, 1970; Gissinger *et al.*, 1974; Weinmann and Roeder, 1974). Class II (or B) members are inhibited by low levels of α-amantin and occur in the nucleoplasm (Blatti *et al.*, 1970; Zylber and Penman, 1971; Reeder and Roeder, 1972). Here they are believed to synthesize hnRNA, much of which is converted to mRNA, as discussed later. Class III members, while less abundant than those in the preceding groups, may be fairly widespread among eukaryotes, having been reported from yeast, sea urchins, amphibians, and mammals including man (Lindell *et al.*, 1970; Roeder, 1974; Tsai and Saunders, 1974); they were not reported in *Drosophila*, however (Gross and Beer, 1975). Its members are inhibited only by higher concentrations of α-amantin; their activities are still largely speculative but seem to include the synthesis of tRNA precursors and 5 S rRNA (Marzluff *et al.*, 1974; Weinmann and Roeder, 1974). However, both class I and II enzymes from *Xenopus* transcribed rRNA and 5 S rRNA genes, although aberrantly in that they acted on the wrong strand (Honjo and Reeder, 1974); it was suggested that other, still undetermined factors may be needed for them to achieve accurate *in vitro* transcription. The enzyme C reported from rat liver and HeLa cells showed a similar sensitivity to high concentrations of α-amantin and appeared to correspond to a polymerase III (Seifart and Benecke, 1975).

Eukaryotic Polymerase I. Intact polymerases of class I typically show a molecular weight around 500,000 (Kedinger *et al.*, 1971); like those of the other categories, they have been found to be comprised of major and minor subunits. The major components of murine polymerases I had molecular weights of 117,000 and 195,000 (Sklar *et al.*, 1975), but the minor subunits were not determined; yeast polymerase A was found in two size class fractions, one of which lacked two polypeptides of molecular weights 48,000 and 37,000 (Huet *et al.*, 1976). With DNA from the virus SV40 as template and in superhelical coiled form, transcription by calf thymus polymerase I proceeded well in the presence of Mn^{2+}, and poorly with Mg^{2+}, and in the open-circular and double-stranded linear forms transcription was also poor (Chambon *et al.*, 1974). However, others have reported that Mg^{2+} and Mn^{2+} were equally efficient as stimulators of transcription with this enzyme (Furth and Austin, 1970; Keller and Goor, 1970; Monjardine and Crawford, 1970; Chambon *et al.*, 1973). In *in vitro* studies, spermine and spermidine were found to have a marked stimulating effect on the activity of polymerase I and II from kidney (Jänne *et al.*, 1975).

In rat liver three species of this class have been reported, A and B (or I and II) from the nucleolus and C (or III) from the nucleoplasm (Roeder and Rutter, 1970; Flint *et al.*, 1974).* Zinc appeared to be an important constituent in at least some of these proteases (Terhune and Sandstead, 1972). Yeast polymerase I has been shown to possess a molecular weight of $500{,}000 \pm 30{,}000$ and to consist of two large subunits (mol. wt. 190,000 and 135,000) and five smaller ones, with molecular weights ranging between 16,000 and 54,000 (Van Keulen *et al.*, 1975). The enzyme appeared to transcribe preferentially those genes for rRNAs in animals and plants alike (Gurley *et al.*, 1976; Holland *et al.*, 1977).

Polymerases of Class II. In class II, the polymerases in general have a molecular weight in the range of 570,000 (Kedinger *et al.*, 1974), and in some mammals appear to be confined to the nucleoplasm. Occasionally, as in rat liver, two members (A and B, or I and II) have been detected (Kedinger *et al.*, 1971; Flint *et al.*, 1974). Various numbers of subunits have been reported, the weights of which in mammals have not been as firmly established as have those for class I enzymes. In one case, one subunit was found to have a molecular weight of 140,000 but that of the second was either 170, 205, or 240×10^6 (Sklar *et al.*, 1975). A plant polymerase II, derived from cultures of parsley cells, was clearly shown to consist of seven subunits, with molecular weights of 16, 25, 26, 43, 140, 180, and 200×10^3, respectively (Link and Richter, 1975). That enzyme from wheat germ has been shown to contain seven tightly bound units of zinc per molecule (Petranyi *et al.*, 1977).

The activity of class II polymerases was reported to be regulated indepen-

* The confusion in the naming of these enzymes becomes evident here. The term polymerase II, for instance, may refer either to the α-amantin-sensitive class II of enzymes (or one of its species), or to a particular species of any of the classes A, B, or C, depending on the author.

dently of that of the others; autogenous regulation has been proposed as one possible mechanism for this phenomenon (Somers *et al.*, 1975), as has been done also for a corresponding bacterial enzyme (Goldberger, 1974). Because this α-amantin-sensitive class of RNA polymerases supposedly is active mainly in mRNA (or pre-mRNA) synthesis, a cell-free system has been developed for the transcription of globin mRNA from rabbit bone marrow chromatin (Steggles *et al.*, 1974). Quite unexpectedly, while both polymerase II from sheep liver and RNA polymerase (of unspecified type) from *E. coli* transcribed the foregoing *in vitro* system efficiently, neither proved active when rabbit liver chromatin was substituted for that from the bone marrow.

Polymerases of Class III. Very little is known concerning the members of class III, except that the major subunits are more nearly equal in size than are those of the others; for example, Sklar and his co-workers (1975) reported molecular weights of 138,000 and 155,000 for α and β subunits, respectively. This same team also found peptides with molecular weights of 29,000 and 19,000 associated with all classes of these RNA polymerases; another of 52,000 was found in all except class II, and a fourth, of 41,000, was confined to classes II and III.

Relatively few polymerases from other eukaryotes have received detailed attention. Among these few, two fungi, *Neurospora crassa* and *Mucor rouxii*, have been demonstrated to have a similarly complex set of polymerases present, with activities comparable to those of mammals (Timberlake and Turian, 1974; Young and Whiteley, 1975). In yeast, one investigation found that the cistrons for rRNAs were located in the light (L) strand of the satellite γ-DNA (Cramer *et al.*, 1972). Polymerase I of that organism was able to transcribe the L strand of γ-ribosomal DNA selectively *in vitro*, whereas polymerase II from the same source transcribed both the H and L strands with equal efficiency (Cramer *et al.*, 1974). In *Xenopus laevis*, genes for 5 S RNA were selectively and accurately transcribed by purified RNA polymerase III (Parker and Roeder, 1977).

Control of Transcription. Among the eukaryotes, investigations have been concerned more with mechanisms that control transcription than with the processes directly involved in that activity, at least in large measure because of the specializations in cell functioning that characterize multicellular organisms. Obviously since all the cells in the tissues of a mammal, for example, arise from a single fertilized egg, they must all have the same genetic composition, yet each tissue has different properties and often secretes distinctive products. In early attempts at solving the problem, evidence had been advanced that, in eukaryotic chromosomes, many genes are more or less permanently masked in a pattern that is specific for each organ. Many of the original data were derived from studies on the giant interphase chromosomes of gnats and other Diptera, which are readily studied under the light microscope. On these, the presence of enlargements (called puffs or Balbioni rings) had been observed, which were found to be specific for the different stages of the insect's development and for the various organs. More recently these puffs have been shown to be active sites of RNA synthesis, as in *Chironomus tentans* (Daneholt, 1973; Daneholt and Hosick, 1973) and in *Drosophila* (Tissières *et al.*, 1974). However, the RNA produced is of such large size (75 S) that it cannot qualify as messenger; moreover, its being conducted directly to the cytoplasm, as has been reported, is likewise atypical of mRNA behavior (Harris, 1974). Nevertheless, because the greatest portion of RNA synthesis appears to be restricted to the puffs, it is evident that the other genes in such chromosomes are repressed in some fashion. However, heat shock and ecdysterone treatment induced alteration in the band activity in *Drosophila* (Tissières *et al.*, 1974).

On the basis of similar studies on mammalian chromosomes, which lack puffs, proposals have been advanced that the histones mask the DNA of the chromosomes nonspecifically and completely. This masking by the histones was conceived to be counteracted by stereospecific interactions of nonhistone proteins with the DNA of the chromatin. Thus the exposure of specific sectors was conceived to permit the synthesis of organ-specific RNA (Paul *et al.*, 1970). The concept, however, does not appear reconcilable with the well-documented morphological features of whole chromosomes revealed by ultrastructural studies or with the molecular continuity which has been demonstrated to exist between DNA and the histones (Chapter 3). A control mechanism of a different nature has been described that operated during meiotic prophase in mouse testes cells to regulate transcription in the sex chromosomes and to provide differential rates of RNA synthesis in the various autosomal segments (Kierszenbaum and Tres, 1974).

Among the several chromosomal regulatory mechanisms which have been advocated recently is one demonstrated in the cytoplasm of two species of clawed toads (*Xenopus laevis* and *X. mulleri*). In these amphibians, the 40 S precursor molecule of rRNA was distinguishable by means of hybridization with the DNA; this reaction permitted it to be established that in interspecies hybrids the rRNA of *X. laevis* was synthesized preferentially over that of the second species (Honjo and Reeder, 1973). Repression of synthesis of the *X. mulleri* rRNA was nearly complete throughout early embryogenesis up to the swimming tadpole stage and thereafter proceeded only at a low level. However, in hybrid adults repression varied extensively, some examples synthesizing the rRNA in substantial quantities and others in no detectable amounts. From these data, it was concluded that *X. laevis* rDNA and maternal cytoplasm were similarly capable of repressing *X. mulleri* rDNA expression in hybrid embryos; the repression by the rDNA was deemed to be permanent, whereas that by the cytoplasm seemed to be transient and reversible.

Many studies on this subject do not lend themselves to broad application, as a result of their being based on specialized tissues. In cultures of human diploid fibroblast cells, for example, a change to fresh medium proved to stimulate the cells to proliferate. Transcription rates were then measured in an *in vitro* system for various combinations of chromatin, HeLa cell DNA, histones, and nonhistone chromosomal proteins from stimulated and unstimulated fibroblasts. From these results, it appeared that the increased template activity of stimulated cells was induced by undetermined changes that had occurred in the nonhistonal proteins (Stein *et al.*, 1972). Another study examined the mechanism by which estradiol affected quail oviduct protein synthesis and found that the estradiol receptor protein and RNA polymerase I were combined to form a complex (Müller *et al.*, 1974). After purification, the complex showed a sedimentation coefficient of 6.4 S and a molecular weight of around 130,000. Hence, in this organism the total complex was much lighter than polymerase I alone is in mammals. A comparable complex has been described in *E. coli* (Naito and Ishihama, 1975).

Other investigations frequently search for either single or whole classes of reactants that might regulate transciptive processes. One paper, for example, proposed that glycerol activated specific transcription promoters (Nakanishi *et al.*, 1974). Another instance was an investigation in *E. coli* of the transcription of genes whose products influenced the metabolism of various sugars. When glucose was present in the medium, only enzymes for the breakdown of that sugar were transcribed, but when glucose was absent, genes were activated for the metabolism of whatever sugar was present. In this case cyclic AMP was reported to regulate those cistrons (Pastan and Perlman, 1972). A third group of experiments was concerned with the control of actin and myosin synthesis in muscle cells; the results indicated that creatine selectively stimulated the formation of the heavy chain of myosin (Ingwall *et al.*, 1974). Still another demonstrated that the circadian rhythmic change in activity level of rat liver chromatin was at least partly under the control of glucocorticoids of the adrenal gland (Earp, 1974). Finally, certain fractions of the heterogeneous RNAs of the nucleus have been claimed to be responsible for regulating transcription (Kolata, 1974a,b).

The difficulty innate to all concepts of chemical regulation of transcription is that the level of the activating substance must in turn be under the control of a second activator, and that one under the influence of a third, and so on indefinitely. Perhaps more satisfactory as a general working principle is the proposal by Goldberger (1974) that each protein modulates the rate at which additional copies of itself are synthesized, as well as that of any other protein encoded within the same operon. While the idea is not without strong appeal, the mechanism by which each protein would act as its own regulator is difficult to perceive. The problem is that transcription produces various RNAs, and nothing else, from templates that consist solely of DNA. The transcripts, furthermore, first must be processed from large precursors before the functional messengers can be translated by an interacting set of proteins into the actual enzyme or structural protein that is assumed to be regulating its own quantities. How can such a product of another product of still other products regulate the template from which the original RNA was transcribed? On these points, little of a concrete nature has as yet been proposed; about all that is firmly established is that transcriptional activities are controlled (Dennis and Nomura, 1975).

Asymmetrical Transcription. Since it has long been known that double-stranded DNA is transcribed more actively than single-stranded (Hurwitz *et al.*, 1962), a question arises as to whether both strands of duplex DNA are transcribed simultaneously by RNA polymerase systems. Although *in vitro* systems typically indicated that both strands were transcribed (Chamberlin and Berg, 1962, 1964; Luria, 1965), under *in vivo* conditions the evidence was clear that only one strand actually served as a template for RNA synthesis (Hayashi *et al.*, 1963a; Marmur and Greenspan, 1963; Altman, 1971). Thus transcription in living things is asymmetrical.

Some of the supporting evidence for asymmetry comes from viruses. For example, in such DNA viruses as bacteriophages α and SP8, the two strands of the DNA are sufficiently different in pyrimidine/purine ratios as to be markedly distinct in weight. Although both the heavy (H) and the light (L) strands were transcribed under *in vitro* conditions, only the heavy one was in infected bacterial cells (Marmur and Greenspan, 1963; Altman, 1971). Contrary to the usual practice of considering the chain that was actively transcribed as the plus strand, in the bacteriophage ΦX174, which is a single-stranded DNA virus, that strand carried by the particles was referred to by this term. Since this plus strand was inactive in RNA synthesis, its complement, the minus, proved to be the carrier of genetic information in these viruses (Hayashi *et al.*, 1963a,b). Because DNA in organisms as a whole is known to serve only in this capacity, the typical minus strand thus appears to function only in replicating the plus, that is, in restoring the intact duplex molecule. Under *in vivo* conditions, the double-stranded DNA molecule is essential for the formation of functional RNA, for only the RNA polymerized by the double-stranded DNA can stimulate amino acid incorporation by a cell-free protein-synthesizing system (Wood and Berg, 1963; Jones and Truman, 1964; Roth, 1964). Why the presence of an inert strand is requisite for these processes remains undetermined, nor is it clear how the polymerase system in living cells distinguishes between the two types of strands, which are basically identical in composition.

The reason for the *in vitro* evidence's being inconsistent with the idea of asymmetrical transcription also remains obscure. For example, in a recent investigation of synthesis of RNA by polymerase I and II on monkey and mouse satellite DNAs (Maio and Kurnit, 1974), all the DNA fractions were copied *in vitro* by both polymerases from the two animals, with satellite and main band DNA fractions serving with equal efficiencies. Thus the *in vivo* absence of transcription of mouse satellite DNA was not found in this system. Moreover, mouse polymerase I transcribed the satellite DNA symmetrically, but polymerase II from the same source showed a fivefold preference for the H strand. When the strands were isolated as separate fractions and tested individually, both the L and H strands were found to be transcribed to equal extents by the two enzymes.

Overlapping Genes. Certainly the least expected result of current investigations into nucleic acid functioning was the recent confirmation that two or more genes sometimes overlap one another. One of the earlier reports, based on bacteriophage Qβ of *E. coli,* showed that when the coat protein gene in the phage RNA genome was translated, 97% of the product indicated that the gene approximated 400 nucleotides in length; however 3% of the products implied a length of 800 nucleotide residues. Consequently, the second, larger protein was at first viewed as a useless product resulting from the polymerase's accidentally missing the usual stop codon during translation (Kolata, 1977). Later, however, it was shown that this large protein actually was essential for the production of infectious virions, but only in small quantities; it now is referred to as the "read-through protein" because its formative processes do not obey the first terminal signal. Hence the gene for this enzyme completely overlaps that for the coat protein and then continues beyond.

A different sort of overlapping of genes has been reported more recently. In bacteriophage ΦX174, gene A nearly entirely overlaps gene B as above, but in this case transcription involves a reading frame shift (Weisbeek *et al.*, 1977), and the genome consists of DNA. Thus, transcription begins at one point for gene B and one or two nucleotides later (or earlier) for gene A. Similar gene overlap is suspected to exist in bacterial genomes but has not been clearly demonstrated as yet. Consequently, it no longer is valid with viruses, and perhaps not with prokaryotes, to predict the number of genes present in a genome solely on the basis of the DNA's molecular length.

RNA-Dependent RNA Polymerization. There is a small number of poorly known RNA-dependent RNA polymerases in prokaryotic and eukaryotic cells, which often are appropriately called RNA replicases. The one in bacteria has been considered identical to a DNA-dependent RNA polymerase, since in *Azobacter vinelandii* it has similar sedimentation characteristics (Krakow and Ochoa, 1963). Moreover, DNA-dependent polymerases of *E. coli* and other bacteria have been shown to respond to an RNA template (Davidson, 1972). In RNA viruses, this type of replicase is, of course, of prime importance in the replication of the genetic apparatus (Baltimore and Franklin, 1963; Cline *et al.*, 1963; Eason and Smellie, 1964).

Among eukaryotes, the system seems to be more distinct, for RNA-dependent activity was demonstrated in rat liver ribosomal preparations that did not respond to DNA templates, and which required Mg^{2+}, but not Mn^{2+}, ions (Wykes and Smellie, 1966; Wilkie and Smellie, 1968). An enzyme of this type from rabbit reticulocytes has recently been purified 2500-fold (Downey *et al.*,

1973). At least in mammals, some evidence has been presented that this system may play an important role in the immune response (Saito and Mitsuhashi, 1973).

Miscellaneous RNA-Associated Enzymes. A small number of diverse types of enzymes that act on RNA have been described which are of possible importance during the transcriptive processes. Among these were several ribonucleases, including one from *Bacillus subtilis*. The activity of this RNase was stimulated by Mn^{2+} and inhibited by Ca^{2+} ions and reached a maximum level about 2 hours after exponential growth terminated (Kerjan and Szulmajster, 1974). While it was able to degrade the product of RNA polymerase activity on phage DNA template, it failed to do so on the poly(AU) produced on poly d(AT). Another of these RNases, which occurred both extra- and intracellularly in *Streptomyces,* has been obtained in purified form. This one acted specifically on phosphodiester bonds of 3′-guanylic acid in RNA; it was reported to be a rather small, proline-rich polypeptide with a molecular weight of 11,900, which contained two cysteine residues and lacked tryptophan and methionine (Yoshida *et al.*, 1971).

A polynucleotide phosphorylase of *Micrococcus luteus* has been found to have a complex mode of action (Letendre and Singer, 1974). In its natural form, it induced the polymerization of ribonucleoside diphosphates to the corresponding polyribonucleotides (Klee, 1967, 1969; Moses and Singer, 1970). When preparations were highly purified, the presence of several protein species was detected, all of which could be converted to a single species, called form T, with a somewhat lower molecular weight than the original (form I) enzymes. In the absence of oligonucleotide primers, form T catalyzed polymerization only poorly, whereas form I was relatively efficient. The latter primer-independent form was converted to the form T by appropriate sulfhydryl reagents, and the form T primer-dependent type became independent when reduced with thiols. Thus a conformational change was deduced to be implicated, rather than molecule size differences. Polymerization of RNA appeared to involve additions to existing chains rather than production of totally new polymeric strands.

An RNA ligase isolated from bacteriophage T4-infected *E. coli* seems to have a counterpart in eukaryotes but only the former appears to have received attention (Cranston *et al.*, 1974). In a reaction that required ATP and Mg^{2+}, it catalyzed the conversion of labeled synthetic polyribonucleotides to an alkaline-phosphatase-resistant circular form, during which reaction ATP was converted to AMP and pyrophosphate. By use of purified preparations, an exchange between ATP and pyrophosphate (but not AMP) was demonstrated, with an enzyme–adenylate complex serving as an intermediate.

Polymerases active in the presence of Mg^{2+} on substrates consisting of single nucleotides have been isolated from various eukaryotic sources. One of these, from quail oviduct, acted on ATP to produce poly(A), doing so most effectively in the presence of a polynucleotide primer. The resulting poly(A) sequences were found to have a maximum length of 60 residues, which were covalently linked to the primer molecules (Müller *et al.*, 1975). Further, it displayed ordinary RNA-polymerizing activity also, but only at a low level of productivity. In contrast, a poly(U)-polymerase has been isolated from tobacco leaves that was highly specific for UTP and required Mn^{2+} and a polynucleotide primer, poly(A) being especially effective (Brishammar and Juntti, 1975). A molecular weight of 40,000 was estimated for the enzyme. Similar polymerases for poly(A), poly(G), and poly(U) have been reported from other sources (Eikhom and Spiegelman, 1967; Milanino and Chargaff, 1973; Winters and Edmonds, 1973). It should be especially noted that all these proteins synthesize single-stranded RNA without using a template, although at a low rate of efficiency.

5.1.2. Posttranscriptional Processing of rRNA

The posttranscriptional history of rRNA appears to be somewhat better established than that of other varieties, so it is expedient to describe its processing first. In contrast to many of the other steps in the genetical processes, eukaryotic mechanisms here have been better explored than the prokaryotic. In both types of cells, one major principle appears to be followed in the processing of an rRNA: After a large precursor molecule has been transcribed, it is split asymmetrically to form a large and a small rRNA molecule. These then undergo further modification to form the large and small ribosomal subunits, respectively.

Processing rRNA in Prokaryotes. In *E. coli,* as well as *B. subtilis,* the genes for the 16 S, 23 S, and 5 S rRNAs have been determined to lie in tandem, in that sequence, following a common promotor gene (Yankofsky and Spiegelman, 1962, 1963; Smith *et al.*, 1968; Pace, 1973; Hackett and Sauerbier, 1974). By means of hybridization experiments with *B. subtilis* RNA, it was shown that each genome of that species contained four cistrons for the 5 S rRNA and ten each for the 16 S and 23 S molecules. Because at least the two larger varieties are synthesized together (in all prokaryotes and in eukaryotes alike), the redundant genes for those species must alternate (Retèl and Planta, 1970; Weinberg and Penman, 1970; Kossman *et al.*, 1971; Pettijohn, 1972; Fujisaw *et al.*, 1973b; Pace, 1973; Zingales and Colli, 1977). In *E. coli* the number of gene sets for rRNA now appears firmly established at seven (Kiss *et al.*, 1977).

The original transcript of those sites was considerably larger than the combined mature molecules, as attested by its sedimentation coefficient of 30 S. This product has proven to contain one complete sequence of 16 S rRNA, one of 23 S rRNA, and about 1300 nucleotides in additional sequences (Hayes *et al.*, 1975). That of *Agrobacterium tumefaciens* was even larger (Grienenberger and Simon, 1975). Processing, which commenced while transcription still was in progress (Miller and Hamkalo, 1972; Dunn and Studier, 1973b; Nikolaev *et al.*, 1974), therefore involved removal of about 25% of the original transcript in prokaryotes and up to 50% in eukaryotes (Perry *et al.*, 1970b; Weinberg and Penman, 1970; Nikolaev *et al.*, 1974). In Gram-negative organisms, only a single short precursor of the 5 S rRNA occurred (Jordan *et al.*, 1971), but *Bacillus subtilis* and *B. megaterium* have proven to produce two precursors for the RNA species which were processed separately; one of these had 37, and the other 63, extra nucleotides (Pace *et al.*, 1973), located at both termini (Pace, 1973). The two precursors of *B. megaterium* have been demonstrated to be ribonucleoprotein particles sedimenting at 50 S and 30 S, respectively (Body and Brownstein, 1976). *B. licheniformis* had even larger, multiple precursors, with the extra sequences similarly located at both ends (Stoof *et al.*, 1974).

Processing studies to date, however, have concentrated on the largest precursor, that which ultimately results in the 16 S and 23 S rRNA. The first step detected was cleavage by RNase III (Nikolaev *et al.*, 1973), which produced two unequal portions, one sedimenting at 17.5 S, the other at 25 S (Hecht and Woese, 1968; Adesnik and Levinthal, 1969; Pace *et al.*, 1970); in addition there were fragments, one of which contained the sequence of 5 S rRNA (Hayes and Vasseur, 1976). Each of these cleavage products then was trimmed by processes catalyzed by RNase M (Meyhack *et al.*, 1974), thereby becoming reduced to the definitive 16 S and 23 S molecules (Lowry and Dahlberg, 1971; Sogin *et al.*, 1971). In these latter steps, fragments were removed from each end of the precursors (Dahlberg and Peacock, 1971; Sogin *et al.*, 1971). Correct maturation of the 5′-end of the 17 S precursor seemed to require a system in which protein synthesis could occur (Hayes and Vasseur, 1976).

Assembly of Prokaryotic Ribosomes. Although the evidence is still insufficient to provide a clear picture, some details have been established regarding the assembly of the rRNAs and the numerous proteins into the definitive ribosomal subunits. *In vivo* some of these steps seemed actually to take place while the unfinished subunits were attached to mRNA as polysomes (Mangiarotti *et al.*, 1974). Contrastingly, other investigations have isolated discrete precursor ribonucleoprotein particles from growing *E. coli* cells, as well as from *in vitro* systems. In the cultured bacterial cells, the growing intermediates of the 30 S subunit sedimented between 21 S and 27 S, and two precursor stages of the 50 S subunit sedimented between 30 S and 36 S and between 40 S and 43 S, respectively; however, the precursors in each case were in low concentration relative to mature particles, constituting only 2% of the whole (Lindahl, 1975). *In vitro* studies have found similar sized intermediates for both ribosomal particles, but most attention has been focused upon the 30 S subunit.

For the assembly of this smaller particle, the steps which seem to be suggested by the combined results of *in vivo* and *in vitro* researches are as follows: As the original 30 S precursor rRNA was still being synthesized, some ribosomal proteins bound to it (Mangiarotti *et al.*, 1968) and RNase III–induced cleavage occurred (Kossman *et al.*, 1971). Then additional proteins were gained in steplike fashion, leading to the several intermediate classes of ribonucleoprotein particles (Hayes and Vasseur, 1974; Nikolaev and Schlessinger, 1974). Methylation of the rRNAs occurred in two stages, the first of which took place early in processing (Sypherd, 1968), but before additional methyl residues could become united to the rRNA, four proteins (S4, S8, S15, and S17) of the six known to interact directly with 16 S rRNA had to be bound to these semimethylated precursors (Thammana and Held, 1974). As these in turn would not bind to the original cleavage product until the first stage of methylation had been completed (Helser *et al.*, 1972), conformational changes in the growing precursor molecule had been considered to be involved in these developments. During later maturation processes, several of the proteins of the large subunit also became methylated, L11 being particularly heavily modified (Chang and Chang, 1974, 1975). L1, L3, and L5 were intermediate in total methylations, with about one residue being added per two molecules of protein, while L7, L8, L9, L12, and L18 received only one residue per ten molecules. During these assembly processes, some of the ribosomal proteins became phosphorylated also, but only at a low level (Kurek *et al.*, 1972a,b).

The biosynthetic pathway of rRNA has recently been investigated in a blue-green alga (*Anacystis nidulans*) and was found to differ in a number of details from that of the bacteria (Seitz and Seitz, 1973). No large molecule containing precursors for both major rRNAs was detected, nor was any larger precursor found for the 23 S type alone. In contrast, a higher molecular pre-16-S rRNA was isolated, with a molecular weight of 660,000 compared to that of 540,000 for the mature product. This early form was unmethylated; preliminary explorations indicated that methylation did not occur until after trimming to final size had taken place.

Processing rRNA in Eukaryotes. Through the study of a variety of metazoan types, including rat liver, *Xenopus* mutants deficient in ribosomes, *Drosophila,* and, especially, HeLa cells, it has been well documented that the nucleolus plays the principal role in the transcription and processing of ribosomes among the eukaryotes (Jones, 1965; Liau *et al.*, 1968; Maden, 1968; Wimber and Steffensen, 1970). In *Drosophila melanogaster,* between 200 and 250 tandemly repeated copies of the cistrons that code for precursors of the 18 S and 28 S rRNAs have been reported to be contained in the nucleolus-organizer regions of the X and Y chromosomes (Ritossa *et al.*, 1966; Tartof and Perry, 1970; Tartof, 1971). Since meiotic recombination does not occur in normal XY males (Morgan, 1914), those two cistronic regions are genetically isolated, yet no apparent differences could be observed between the products of the X chromosomes and those of the Y (Maden and Tartof, 1974).

A similar location in the nucleolus-organizer region, but with a greater redundancy, has been demonstrated for these genes in *Xenopus laevis,* where between 400 and 500 copies of each existed in the haploid genome (Brown and Weber, 1968). The cistrons for the 5 S rRNA, on the other hand, were not found in this region; about 200 copies of these genes were reported to exist in *Drosophila* and between 20,000 and 27,000 in *Xenopus,* located on several chromosomes in the latter organism (Wimber and Steffensen, 1970; Pardue *et al.*, 1973). At least four different species of 5 S rRNAs were detected in this toad, differing in as many as seven sites in certain cases (Ford and Southern, 1973). Reiterated genes for the rRNAs also characterized *Saccharomyces cerevisiae* and *Euglena gracilis*; in the former organism about 140 copies per haploid genome were reported to be present, mostly on chromosome I (Kaback *et al.*, 1973), whereas 1000 copies per genome were found in the latter (Scott, 1973). In yeast the 5 S cistrons were linked to those of the larger rRNA species, but they outnumbered the latter by a ratio of 2 : 1 (Aarstad and Øyen, 1975).

Apparently in all eukaryotes, as in bacteria, rRNAs are transcribed as large precursor molecules, which in mammalian cells sediment at 45 S (Loening *et al.*, 1972; Hackett and Sauerbier, 1975); in rat liver nucleoli, they have been found to be even larger (Grummt, 1975). According to several laboratories, processing first involved the removal of a 24 S section from the 3′-end by enzyme action, followed by the cleavage of the remainder into a 32 S and a 20 S portion by similar processes (Willems *et al.*, 1969; Wellauer and Dawid, 1973a,b; Russell *et al.*, 1976). Finally, these intermediates were further reduced to the 28 S and 18 S molecules. An endonuclease, believed to be involved in the cleavage of ribosomal precursors, has now been isolated and partially purified (Winicov and Perry, 1974). In both the deleted and definitive portions of the precursor, a number of hairpin loops (Figure 5.1). not unlike the viral RNA illustrated in Figure 4.1, have been shown to be present. The foregoing three-step model may be slightly oversimplified, for another account of the processes reported five or six stages in the reduction of the 45 S precursor to the definitive rRNAs (Fujisawa *et al.*, 1973a,b). Accordingly, the proposed steps were as follows:

$$45\text{ S}\rightarrow 37\text{ S}\begin{array}{l}\nearrow 32\text{ S}\rightarrow 29.5\text{ S}\rightarrow 28\text{ S rRNA}\\ \searrow 21.5\text{ S}\rightarrow 18\text{ S rRNA}\end{array}$$

Another alternative route has been uncovered in mouse L cells, which differs from the foregoing in the size of the intermediate products (Winicov, 1976). Thus, much variation at the sites of trimming appears to exist among mammals. In *Drosophila,* at least the last steps in trimming the precursor molecule occur in the cytoplasm (Jordan *et al.*, 1976).

Among the metazoans as a whole, the nature of the original precursor molecule seems to show a degree of variation. In *Xenopus* cells, the original transcript sedimented at 40 S (Slack and Loening, 1974), while in *Chironomus* a 38 S particle was produced (Serfling *et al.*, 1974). An even smaller interme-

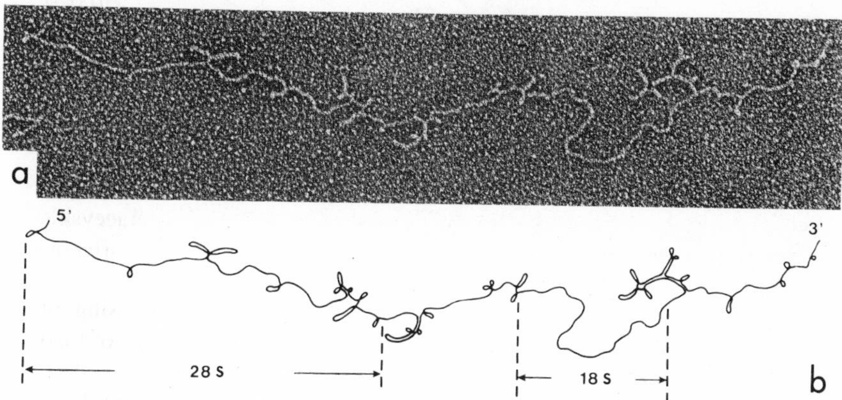

Figure 5.1. The 45 S rRNA precursor. The large precursorial molecule shown in the electron micrograph (a) is cleaved into precursors of the large and small rRNA subparticles and then trimmed. Although some of the secondary loops are removed (b), many of the hairpin loops are retained. (Reproduced through the courtesy of Peter Wellauer.)

diate was found in sea urchin embryos and oocytes, one which had a sedimentation coefficient of 32 S before being cleaved into 27 S and 21 S products (Giudice *et al.*, 1973). These investigations also ascertained that, in the presence of nuclear lysate of the embryos, even such a nonspecific endonuclease as bovine pancreatic ribonuclease catalyzed cleavage of the large pre-rRNA molecule specifically into the usual subunits. In the nuclear material, however, the enzyme did not split the precursor specifically. Very little appears to have been ascertained concerning the biosynthesis of the smaller rRNAs. However, it is now known that 5.8 S rRNA in yeast is derived from the larger precursor molecule (29 S) and that a 7 S molecule is an intermediate in the former's formation (Trapman *et al.*, 1975).

In plant cells, high-molecular-weight precursors to rRNA also have been identified (Leaver and Key, 1970; Rogers *et al.*, 1970; Grierson and Loening, 1974). For example, in the leaf cells of the mung bean (*Phaseolus aureus*), two large precursor rRNA molecules have been detected, that for cytoplasmic ribosomes showing a molecular weight of 290,000 and that for chloroplast ribosomes, 110,000 (Grierson, 1974). The cytoplasmic precursors were first cleaved to molecular weights of 100,000 and 145,000 and the chloroplast ones to 56,000; these products then were reduced further by an undetermined number of steps to result in the definitive rRNAs. Both the 26 S and 17 S cleavage products in yeast were already partially modified; among the modifications specifically noted was the presence of 32 pseudouridine residues in the larger and 14 in the smaller (Klootwijk and Planta, 1973a,b).

In vivo studies on rat liver suggested that feedback mechanisms may be active in the control of ribosomal RNA processing and maturation (Cooper and Gibson, 1969; Levitan and Webb, 1970; Rizzo and Webb, 1972). Possibly only 50% of the precursor rRNA left the nucleus in mature liver cells, but a greater percentage entered the cytoplasm when they were actively growing. *In vitro* experiments indicated that specific, but as yet unidentified, proteins from the cytoplasmic sol stimulated the release of the precursor from nuclei (Racevskis and Webb, 1974); in the presence of ATP these same enzymes also affected the degree of processing received by the early precursor.

Assembly of the Eukaryotic Ribosome. During processing of the precursors and assembling the various component parts, two stages of methylation of the rRNA take place, as in the bacteria. In a recent study, the early stage was observed to result in the methylation of 110 sites of the 45 S pre-rRNA molecule before the latter underwent cleavage. All of the residues appeared to have been added to sectors destined to become mature rRNAs, for the number remained unchanged as processing continued (Salim and Maden, 1973). The late activity resulted in the addition of six methyl radicals on the 18 S rRNA and one on the 28 S rRNA, so that the former ended with a total of 46 and the latter with more than 70, almost all of which were 2′-*O*-ribose substitutions, rather than base modifications (Maden and Salim, 1974).

Again as in bacteria, the precursor rRNA molecule received proteins immediately following transcription, so that the intact molecule extracted from the nucleolus carried a complement of around 60 proteins and sedimented at 60 S. Of these, 21 seemed to be retained on the mature large ribosomal subunit and 10 on the small (Prestayko *et al.*, 1974a,b). About 11 other proteins appeared to be added following cleavage of the molecule, as they were not detected in the precursor, and obviously a number of others were removed during the same period (Higashinakagawa and Muramatsu, 1974). Phosphorylation of some of the proteins also occurred during ribosome assembly, the level of which was found in rat liver, rabbit reticulocytes, and calf kidney cells to be higher than that of bacterial ribosomes (Kabat, 1970, 1971; Loeb and Blat, 1970; Walton *et al.*, 1971); similar levels of phosphorylation have been reported also in tumor cells (Olson *et al.*, 1974; Prestayko *et al.*, 1974a,b). At least two protein kinases that might phosphorylate ribosomal proteins have been detected in yeast cells, one associated with ribosomes, the other in the supernatant fraction (Grankowski *et al.*, 1974). Phosphorylation seemed to be confined to a narrow range of protein types, as only L7 and L12 received this kind of modification in reticulocytes (Issinger and Traut, 1974).

An antibiotic, camptothein, has been employed to demonstrate the close coupling that apparently exists between rRNA transcription and ribosomal processing (Kumar and Wu, 1973). At the low concentration (5 μM) used with the HeLa cells, this substance inhibited all transcription of high-molecular-weight RNA, including the 45 S precursor, but it had no detectable effect on protein synthesis for 3 hours. Although all previously synthesized 45 S pre-rRNA (that is, 60 S pre-rRNA · protein molecules) underwent normal processing in the presence of camptothein, one of the products of cleavage, 50 S subunits (28 S rRNA · protein) accumulated in the nucleoplasm. Upon removal of the inhibitor, transcription of 45 S pre-rRNA resumed, and the 50 S subunits were then rapidly transferred to the cytoplasm. The results were interpreted to imply that transcription of 45 S pre-rRNA must be proceeding actively for the later stages of ribosomal processing to continue.

5.1.3. The Processing of Messenger RNA

As in the case of the mistaken functional identity of DNA polymerases, *in vitro* studies on bacterial protein synthesis seem to have provided a false impression that mRNA requires no processing in bacteria (Darnell *et al.*, 1973). In light of the sequence determinations of several bacteriophage mRNAs and the fact that they obviously possess untranslated sectors, it is not unlikely that prokaryotes as a whole have comparable messenger structure. Perhaps when the matter receives the attention it needs, biogenesis of mRNA in prokaryotes will be found quite similar to that in eukaryotes, which has been studied more extensively.

The Processing of Eukaryotic mRNA. Like rRNA, mRNA is the product not of transcription alone, but of subsequent modification of large precursor molecules polymerized on the DNA template. Also as in rRNA the nucleolus appears to play a major part in the processing. This role was first suggested by the observation that there was a flow of mRNA from the chromatin toward the nucleolus (Rho and Bonner, 1961; Sirlin *et al.*, 1962). Later, mRNA was actually detected in the nucleolus (Brentani *et al.*, 1964, 1967; Brentani and Brentani, 1969), and material from the nucleolus was noted to be extruded into the cytoplasm (Telles and Coble, 1968).

Polyadenylated Regions. One of the most important features of precursor mRNA is in the presence of long sectors rich in adenosine, located at the 3′-end of the message region, like those described for definitive mRNA in Chapter 4 (Edmonds and Caramela, 1969; Weinberg, 1973). However, as in the mature mRNAs for histones, the corresponding precursors seemed to lack this feature (Darnell *et al.*, 1973). In length, the poly(A) sequences appeared to vary with the source site as well as species. For example, the nuclear dRNAs of Ehrlich ascites carcinoma cells contained polypurine segments only 8 or 9 units long, of which 6 or 7 were adenosine (Farashyan *et al.*, 1973); the pre-mRNAs had similar sequences between 150 and 250 nucleoside residues in length (Edmonds and Caramela, 1969).

The place, manner, and time sequence of adding these characteristic segments to the precursor are still controversial subjects. Originally the segments were considered to be synthesized on DNA templates and then appended to the early transcript by enzymatic action in the nucleus (Edmonds and Caramela,

1969; Philipson *et al.*, 1971; Darnell *et al.*, 1973; Hirsch and Penman, 1974). Later, after these precursor molecules had been transported from the nucleus or nucleolus to the cytoplasm, the polyadenylated sectors were thought to undergo reduction to half their maximum length (Kolata, 1974a). However, at least some evidence from sea urchin eggs has indicated a cytoplasmic rather than nuclear location for the processes (Slater and Slater, 1974). Fertilization of the eggs elicited a pronounced increase of poly(A), which response was not inhibited by either actinomycin D or ethidium bromide (Slater *et al.*, 1972). As labeled adenosine was found attached to unlabeled messengers, it was deduced that the abundant mRNA known to be present in sea urchin eggs had received the poly(A) in the cytoplasm (Slater and Slater, 1974).

Very few other features of the polyadenylation processes have come into full light as yet, but one study on sea urchin embryos reported results suggestive of a need for the concomitant occurrence of translation in order for them to proceed (Slater *et al.*, 1974). One particularly pressing problem centers on the nature of the recognition site by means of which the enzyme that adds these segments distinguishes nonhistonal precursor messengers from all the other types of RNAs present in the nucleus. Among the few investigations into this matter is one on rabbit globin mRNA (Proudfoot and Brownlee, 1974), which determined the sequence of the region preceding the poly(A). In this segment, which proved to be (5′)AUUGC–, the absence of a nonsense codon is an especially noteworthy feature.

In addition to this unique segment, precursor mRNAs appear to possess properties found among other types of intermediates of ribonucleic acids. Their molecules have been shown to contain a number of hairpin loops not unlike those of certain viral RNAs (Ryskov *et al.*, 1973, 1974). Moreover, mRNA now has been found to be methylated (Perry and Kelley, 1974), despite earlier reports to the contrary (Muramatsu and Fujisawa, 1968; Perry *et al.*, 1970a,b). Although the level of frequency was only one-sixth of that of rRNA, methyl residues were found both on the base and ribose moieties of mouse L cell and Novikoff hepatoma cells (Desrosiers *et al.*, 1974; Perry and Kelley, 1974). Similar modifications have been reported on SV40 mRNA in the nucleus as well as in the cytoplasm (Lavi and Shatkin, 1975). Apparently individual species of mRNAs are differently modified; in immature duck erythrocytes, the 9 S globin mRNA was relatively enriched in sequences containing two 2′-*O*-methyl nucleotides compared to other species synthesized by those cells (Perry and Scherrer, 1975).

Heterogeneous Nuclear RNA (hnRNA). Part of the difficulty of clarifying the picture of the processing of these molecules results from the pre-mRNA's being synthesized on the chromatin along with a number of other types of RNA. Together these various types are often classed as dRNA, from their compositional resemblance to DNA (Arion and Georgiev, 1967; Ryskov *et al.*, 1973). Another term frequently applied to the mixture is *heterogeneous* (or *heterodisperse*) *nuclear* RNA, abbreviated as hnRNA (Darnell *et al.*, 1973). These rapidly labeled species were first discovered in the nuclei of HeLa cells and duck erythrocytes (Scherrer *et al.*, 1963, 1966; Warner *et al.*, 1966), but subsequently have been reported from *Tetrahymena* (Prescott *et al.*, 1971), a slime mold (Firtel and Lodish, 1973), and diverse metazoans, including insects, echinoderms, and amphibians (Brown and Gurdon, 1966; Edström and Daneholt, 1967; Aronson and Wilt, 1969). As their presence has also been demonstrated recently in *Zea mays,* it is quite likely that they occur throughout the eukaryotes (Van de Walle and Deltour, 1974). In the ciliate and fungus, the molecules sedimented with coefficients <20 S; in the monocot the molecule averaged slightly heavier, while in metazoans the range was from 20 S to 80 S. Actual sedimentation rates were demonstrated to be highly sensitive to ionic concentration levels (Bramwell, 1974), and it was this sensitivity which led certain laboratories to consider hnRNA to be an artifact (Bramwell and Harris, 1967; Mayo and DeKloet, 1971).

A small number of other characteristics of this type of RNA have now been ascertained (Chan *et al.*, 1977). In sea urchins, at least 25%, and perhaps all, of the hnRNA was reported to contain repetitive sequences interspersed with nonrepetitive (Smith *et al.*, 1974), and in HeLa cells hybridization with DNA indicated a similar presence of repetitive segments (Melli *et al.*, 1975). Furthermore, poly(A) sectors, too, have been found in large fractions of the hnRNA (Torelli and Torelli, 1973; Weinberg, 1973).

The presence of such segments is not surprising, for at least a major portion of hnRNA is generally considered to represent precursor molecules to mRNA (Spohr *et al.*, 1974; Hames and Perry, 1977). This was especially clearly demonstrated by the addition of nuclear RNA from avian erythroblasts to a cell-free system (Knöchel and Tiedemann, 1975), when globins were obtained. However, the level of globin production stood at 20%, which contrasts to the 70% level found by addition of actual messenger (9 S mRNA) to the same system. These data are in harmony with the observation that the bulk of hnRNA was rapidly degraded in the nucleus, leaving only a small fraction to function as messenger in the cytoplasm (Attardi *et al.*, 1966; Soeiro *et al.*, 1968).

hnRNA as Ribonucleoprotein. In HeLa Cells, the hnRNA-containing molecules were shown actually to be complex ribonucleoproteins, the protein moieties of which ranged in molecular weights from 39,000 to over 180,000 (Pederson, 1974); some of these appeared to remain on the mature messengers as mentioned in the preceding chapter. Similar complexes have been found in a number of other sources, including rat liver (Ishikawa *et al.*, 1974; Schweiger and Schmidt, 1974) and the slime mold, *Dictyostelium discoideum* (Firtel and Pederson, 1975). Some evidence indicated that, during their formation, the hnRNAs became united to globulin protein particles, called informofers, to yield a polysome-like structure (Samarina *et al.*, 1968). Moreover, it has been proposed that the informofer particles may detach newly formed hnRNA from the chromatin and thus convert those molecules to the hnRNA-protein complexes (Lukanidin *et al.*, 1972). Although the actual formative processes still remain unknown, immunological studies have suggested that the foregoing proposal might hold some validity (Ishikawa *et al.*, 1974). On the other hand, it may prove simplistic, for the hnRNA has been found to be part of a larger complex. In ascites tumor and yeast cells, evidence was found that the hnRNA is in the form of a ribonucleoprotein network, consisting of 14% RNA, 0.4% DNA, 63% protein, and 22.6% lipid (Faiferman and Pogo, 1975). This network in turn was bound to the nuclear membrane.

One fraction of these ribonucleoproteins that has received particular attention sediments as 30 S particles. In these, the protein moieties have been indicated at least to include enzymes involved in the processing of the hnRNA (Niessing and Sekeris, 1972). Another characteristic of the proteins revealed by *in vitro* studies was an ability to bind additional RNA (Samarina *et al.*, 1966). More recently this RNA-binding capacity has been demonstrated in several proteins whose molecular weights ranged between 25,000 and 42,000 (Schweiger and Schmidt, 1974), some of which were phosphorylated.

Another RNA-protein fraction which has been investigated to some extent is that which bears a poly(A) sequence like that of mRNA (Lim and Canellakis, 1970; Edmonds *et al.*, 1971; Lee *et al.*, 1971). In these complexes, two distinct components have been reported, sedimenting at 15 S and 17 S, respectively (Quinlan *et al.*, 1974). The heavier of these contained six polypeptide species, with molecular weights between 17,000 and 30,000, whereas the lighter had four larger proteins, including one with a molecular weight of 80,000. A similar large protein (73,000) was found in *Dictyostelium* poly(A)-containing hnRNA (Firtel and Pederson, 1975).

The Processing of mRNA in Viruses. Although bacteria still remain unexplored in this regard, some knowledge has been gained concerning the processing of the messenger in RNA viruses. In two RNA types (mouse sarcoma and avian myeloblastosis viruses), the 70 S genomes were shown to contain about eight poly(A) sequences, which ranged in length to nearly 200 nucleotides (Green and Cartas, 1972). Another feature characteristic of eukaryotic mRNA occurs in viruses of the DNA group—the need for posttranscriptional processing. In spite of *in vivo* transcription of bacteriophage T7 DNAs' being continuous over the early region, five distinct, relatively short molecules were extracted from infected *E. coli* (Siegel and Summers, 1970; Studier, 1973), which have been demonstrated to be cleavage products of RNase III activity (Dunn and Studier, 1973a,b). Subsequently it was shown that this fragmentation of the original transcript was requisite for converting it into active mRNA (Hercules *et al.*, 1974). The same condition was shown to exist also in bacteriophage T3.

5.1.4. The Processing of Transfer RNA

The biosynthetic paths of tRNAs have been investigated rather more equally in viruses, bacteria, and eukaryotes than has been the case with the foregoing types of RNAs. Although most of the processes still are at an early stage of investigation, they obviously resemble those of the other transcripts, with a large precursor being reduced to the definitive size, followed by methylation or other enzymatic alterations or additives (Burdon, 1971; Littauer and Inouye, 1973).

Processing tRNAs in Viruses. A DNA virus that parasitizes *E. coli,* bacteriophage T4, has been the principal source of information regarding viral tRNA processing. During infection of the host, the genome directed the synthesis of at least eight species of tRNAs, the first product being high-molecular-weight RNA (Nierlich *et al.*, 1973). The genes for these precursors seemed to be clustered in a restricted region of the T4 genome (McClain *et al.*, 1972; Wilson *et al.*, 1972). The several species of large precursors have been shown usually to include the forerunners of two tRNA species each (Guthrie *et al.*, 1973; Guthrie, 1975). Because they began with the 5′,α-triphosphate, pppG, the precursors were presumed to be primary products of transcription, which were split *in vitro* by an *E. coli* endonuclease. In addition to carrying two types of tRNA, each precursor had a sequence of 41 nucleotides at the 5′-end and several at the other terminus, which apparently became removed during subsequent processing. This activity has been variously reported as requiring three, four, and five steps (McClain *et al.*, 1975; Schmidt, 1975; Seidman and McClain, 1975; Seidman *et al.*, 1975).

Lately the nucleotide sequences of the common precursor to $tRNA^{Ser}$ and $tRNA^{Pro}$ have been determined in the same bacteriophage (Barrell *et al.*, 1974). In addition to almost the entire base sequence of each of the two definitive tRNA species, including all except one of the minor and modified bases, the sequence pUUUAAUUUA was found at the 5′- and pAAU at the 3′-end. Also present was the short sequence pCU, located between the two tRNA molecules (Figure 5.2). In the illustration, the triplet –ACC always present at the 3′-end of tRNAs can be noted to be missing from both species; here, as elsewhere, this is added enzymatically following cleavage, as described below. In the $tRNA^{Ser}$, the 2′-*O*-methylguanine modification also appeared to be made after cleavage was completed. In this same virus four other tRNA species seemed to be formed by a similar series of steps, but the precursor of $tRNA^{Tyr}$ apparently lacked the dimeric structure and certain other features of those six (Guthrie *et al.*, 1973).

Nucleotidyltransferase. The addition of the –CCA terminus after cleavage has been completed seems to be carried out by similar processes in viruses, bacteria, and eukaryotes, in all of which an enzyme called tRNA nucleotidyl-

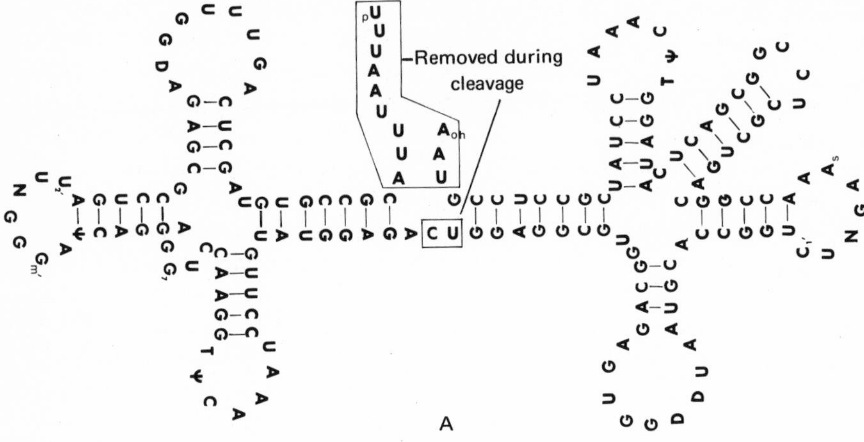

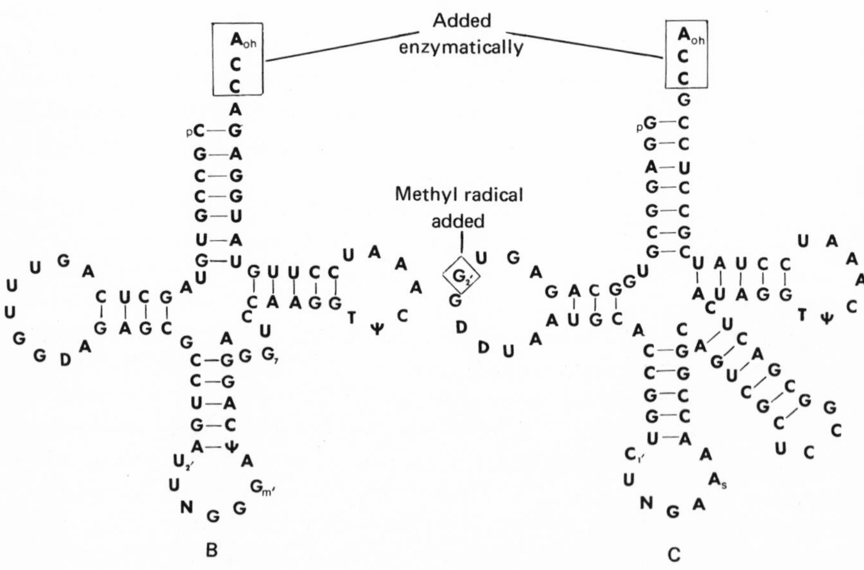

Figure 5.2. The precursor of two tRNAs of bacteriophage T4. (A) Precursor to tRNA^{Pro} + tRNA.^{Ser} (B) Mature tRNA.^{Pro} (C) Mature tRNA.^{Ser} In this virus, the double precursor undergoes modification immediately following transcription (A). Then after cleavage splits the molecule and trims off three regions (A), the terminal sequence ACC– is added to each part and one guanosine undergoes methylation (C). (After Barrell *et al.*, 1974.)

transferase uniformly serves as the catalyst (Deutscher, 1973; Deutscher *et al.*, 1974). In yeast, the enzyme used ATP, CTP, and tRNAs lacking the –CCA terminus as substrates, but not complete tRNA molecules (Rether *et al.*, 1974). In *E. coli,* the corresponding enzyme required Mg^{2+} ions for maximum efficiency, but Mn^{2+} ions served to a lesser extent (Carre and Chapeville, 1974; Carre *et al.*, 1974). With the necessary substrates present, the enzyme added the terminal –CCA to precursor tRNAs and to certain viral RNAs in the absence of template.

Precursors of tRNA in Bacteria. Although few details are available thus far, it has become evident during the past few years that tRNAs in bacteria are biosynthesized by processes corresponding to those of the viruses in being products derived by cleavage of large precursor molecules (Altman, 1971; Vickers and Midgley, 1971). The genes for these have been shown to be widely distributed throughout the cyclic DNA molecule of these organisms, at least in *E. coli* and *Salmonella typhimurium* (Sanderson, 1967; Gorini, 1970); in the former bacterium, some of them, however, were found to be clustered (Squires *et al.*, 1973), for the cistrons for $tRNA_2^{Gly}$, $tRNA_2^{Tyr}$, and $tRNA_3^{Thr}$ were found concentrated in one small region. Identical multiple copies of the gene for glycine tRNA were also detected, and in *E. coli,* two for $tRNA^{Tyr}$ have been closely linked and transcribed together (Ghysen and Celis, 1974). A third gene for this species of nucleic acid has been found to be separate (Landy *et al.*, 1974). Lately a population of precursor tRNAs have been isolated from *E. coli* X402, which fell into the two categories α and β (Dijk and Singhal, 1974). The members of the α group were about 200 nucleotide residues in length and appeared to be the direct primary products of transcription of the DNA. Cleavage converted these to the β type, of about 120 nucleotide residues in length, and the latter in turn were believed to be trimmed further to become mature tRNAs. Some smaller molecules also appeared to be formed by cleavage of the α group, but whether these were processed into tRNAs was not established, nor was the extent of methylation noted. The cleaving enzyme, known as endonuclease P, has been isolated from *E. coli* and purified (Altman and Smith, 1971; Robertson *et al.*, 1972).

Processing tRNAs in Eukaryotes. In eukaryotes, the overall picture of biosynthesis of tRNAs strongly resembles that of the bacteria and viruses. The genes that specify the tRNAs are highly redundant—in an early study a total of around 13,000 tRNA cistrons was estimated for rat liver cells (Quincey and Wilson, 1969). Some examples more recently described include a total of 7800 genes for about 43 basic tRNA sequences in *Xenopus laevis* (Clarkson *et al.*, 1973a) and an average of 700 copies of each tRNA cistron per genome in the rhesus monkey (Roufa and Axelrod, 1971). The redundant genes for a given species of tRNA appear to be clustered in large groups, with spacer DNA filling gaps between the cistrons (Clarkson *et al.*, 1973b).

In yeast, the precursors of tRNAs have received initial analysis (Blatt and

Feldman, 1973). Slightly larger and less extensively methylated RNA mixtures were isolated, which on further processing assumed the full characteristics of mature tRNAs in a series of about four steps.

The pattern of methylation has been the subject of an investigation in which comparisons were made between a number of diverse prokaryotic and eukaryotic organisms (Klagsbrun, 1973). First it demonstrated that the methylation of tRNA and of the several types of rRNA involved highly specific processes, for each major class contained bases peculiar to itself. Among the eukaryotes, the tRNA always exhibited a strikingly similar pattern of methylation, regardless of whether it had been derived from yeast, amoeba, chicken, mouse, monkey, or man. In each organism, N^2-dimethylguanine and 1-methylhypoxanthine were confined strictly to tRNAs, whereas dimethyladenine occurred solely in 18 S rRNA and 3-methyluracil in 28 S rRNA. Although the prokaryotic pattern of methylation of tRNAs differed from that of the eukaryotes to a high degree, it varied only slightly from one bacterial species to another.

5.2. AN ALTERNATIVE PROTEIN-SYNTHESIZING SYSTEM

Aside from the RNA-based system for the synthesis of proteins, at least one other has been found in both prokaryotes and eukaryotes. The so-called soluble system has already received passing mention in connection with the functions of tRNAs (p. 143), but not all such ribosome-independent syntheses involve those nucleic acids. Among prokaryotes, leucine was found to be incorporated directly into certain proteins of *E. coli* (Kaji *et al.*, 1965), and the enzyme which catalyzed the transfer of this amino acid and phenylalanine has been isolated in the same organism (Leibowitz and Soffer, 1969; Soffer, 1973). Methionine also has been found to be incorporated by a similar system (Horinishi *et al.*, 1975; Scarpulla *et al.*, 1976). The soluble system was first discovered in eukaryotes, when arginine was found to be incorporated into proteins in a ribosome-free fraction from brain tissue (Kaji *et al.*, 1963; Soffer and Mendelsohn, 1966; Gill, 1967; Soffer and Horinishi, 1969). As the reaction was sensitive to RNase, it was deduced to be dependent on tRNA. Since subsequent investigations have largely centered on eukaryotic systems, attention here is devoted principally to that taxon.

Results of Mammalian Tissue Investigations. Studies on various mammalian tissues have yielded significant results, brain, parathyroid, and sperm cells being especially productive. In addition to the arginine originally reported in brain tissue, tyrosine, too, has been found to be incorporated into proteins in the absence of ribosomes (Barra *et al.*, 1973b). Whereas both amino acids required ATP and Mg^{2+}, tyrosine incorporation, unlike that of arginine, did not need tRNAs, but K^+ had to be present. Arginine addition to proteins proceeded at a higher level in brain tissue but was also present in liver, thyroid, and kidney, whereas tyrosine incorporation occurred only marginally in the latter three tissues. In beef parathyroid, evidence was gathered that arginine was incorporated into either parathyroid hormone or a closely related protein (Kemper and Habener, 1974).

Phenylalanine was also found to be incorporated into proteins by a similar system. As this amino acid and tyrosine proved to be mutually inhibitory, it was believed that the two were involved in the same protein (Barra *et al.*, 1973a); both were established as being added to the –COOH terminus of the acceptor protein. The latter has now been shown to be closely similar in many respects to that of microtubules (Barra *et al.*, 1974) and to have a molecular weight of 54,000 when combined to tyrosine. The terminal amino acid to which the tyrosine became attached appeared to be glutamic acid (Arce *et al.*, 1975).

Amino Acid Incorporation in Sperm. Currently the results of studies on spermatozoa do not yield a unified picture. The earlier investigations found that ejaculated bull spermatozoa incorporated ^{14}C-labeled amino acids into material that was insoluble in trichloroacetic acid, in the nearly complete absence of RNA (Bhargava *et al.*, 1956; Abraham and Bhargava, 1963). On one hand, other researches showed that this phenomenon was localized in the sperm acrosome, but on the other, results of autoradiographic experiments indicated that it occurred in the cell debris, not in the spermatozoa (Martin and Brachet, 1959). More recently, rabbit spermatozoa were reported to incorporate all 20 of the amino acids at various rates; while about 75% of the label appeared in small unidentified molecules of less than 700 molecular weight, the remainder was in proteins (Busby *et al.*, 1974). Tryptophan and cysteine were incorporated far more extensively than the other amino acids, tyrosine and methionine ranking next, followed by arginine, phenylalanine, and proline, in that order.

5.3. AN ANNOTATED SYNOPSIS–SUMMARY AND ANALYSIS

Condensed though the foregoing account of the genetic mechanism may be, it nevertheless makes obvious that the inheritance of specific and individual characteristics is dependent on innumerable proteinaceous and nucleic acid factors that interplay in a most intricate fashion. In order to extract the main features from this maze of facts and conflicting details, a summary is essential to provide a basis for the conclusions that follow, and thereby to bring out factual evidence that may lead to a better understanding of the nature of early life.

The following summary will not repeat, like so many redundant genes, what has been presented in the preceding chapters but, rather, to polymerize the ingredients and processes into a single, more compact product. To provide this consolidated view, the sequence of topics followed in those discussions cannot be strictly adhered to, for categorization is a human device to simulate order, often where none actually can exist. Although that procedure is a valid one, its practice of necessity creates a checkerboard pattern of compartments, rather than exposing the original design.

5.3.1. The Chromosome and Nucleic Acids

Let is be clear that the genetic mechanism must be viewed as a virtual equivalent of life itself. Obviously, that apparatus is not solely concerned with the passing of individual and species characteristics from parent cells or organisms to their progeny, but, directly or indirectly, it is the basis from which every trait and activity derives, including shape, size, motion, excretion, reproduction, behavior, and all the rest. Its products, the proteins and ribonucleic acids, not only create and destroy all other molecules, including themselves, but also bind and release the energy essential for the life processes.

The DNA Molecule. In general, the DNA molecule is so structured that it effectively guides the creation of proteins and the several types of RNA and conserves the structural pattern of each product from generation to generation. The base sequence of its molecule provides a code which, relative to the complexity of the end products, is extremely simple, and, in theory, mechanically unerring. Through this stability of its primary structure, as well as from the inertness of the molecule, errors of replication are avoided and preservation of favorable traits achieved.

Although rarely mentioned in the literature, this inertness is undoubtedly the most salient characteristic of the DNA molecule, for nowhere during the cell cycle does DNA play an active role in any of life's processes. By itself, it makes no proteins nor carries out any other synthesis; rather it merely serves as a guide by means of which enzymes produce various RNAs. Even during its own reproduction, it plays a subordinate role, for proteins carry out the actual replicatory processes, while it lies passive. Certainly there is no indication *in vivo* that the DNA molecule is capable either of self-replication or of producing any other substance directly. A glance at the proteins involved in the genetic apparatus (Table 5.2) shows that this statement is sound, for at least 189 enzymes are required in prokaryotes and 279 in eukaryotes to replicate, transcribe, and translate the DNA code into its final products.

This observation is in direct opposition to statements frequently made in the literature and widespread in current dogma. However, the claim that DNA is self-replicatory is based on a peculiar concept that defines a self-duplicating molecule as including the system which carries out the actual replicating processes (Mills *et al.*, 1967; Spiegelman *et al.*, 1975). Thus in the first report cited, the "self-replicating nucleic acid molecule" was actually polymerized *in vitro* by a bacteriophage Qβ polymerase system. While such a concept may have some value in certain discussions, it is misleading when the fundamental properties of the macromolecules themselves are being considered. Similarly nonenzymatic syntheses of nucleic acids depend heavily on condensing agents and do not result in typical nucleic acids. For example, a nonenzymatic synthesis of poly(G) on a poly(C) template has been carried out in the presence of 0.2 M concentration of NaCl and a 0.1 M solution of 1-ethyl-3-(3-dimethylaminopropyl)-carbodiimide hydrochloride, but the resulting linkages were primarily of the 5′–5′ type (Sulston *et al.*, 1969). The fact that poly(U) has as yet not been produced even on poly(A) templates in the presence of similar condensing agents serves further to confirm the lack of reproductive powers of the nucleic acids. It is clear, then, that DNA is not a self-duplicating molecule; it is totally incapable of carrying out by itself any biological or chemical function.

As pointed out earlier (Chapter 4), this inactivity is one of the DNA molecule's greatest assets, for if DNA were active in metabolism it would of necessity undergo physicochemical changes as enzymes do in carrying out their functions, thereby increasing the chances that permanent changes might be induced that would decrease the precision of the genetic mechanism. At very least, the long, continuous DNA molecules would have to fracture into individual cistrons or sets of closely related genes, in order that the rates at which the various enzymes and RNAs were synthesized could be controlled.

Nucleoids and Chromosomes. When the state of knowledge was relatively scanty, the view that chromosomes were primarily composed of DNA coated with proteins was justified, but now that the dual composition of chromatin has been well established, adjustments to the concept are in order. To a lesser extent, the same statement holds true also for the bacterial nucleoid. Both these structures obviously consist of intimate combinations of DNA and proteins. While it is too early to perceive the precise roles played by the numerous histone and nonhistone proteins, it is time that new investigations be made into DNA replication and the mechanism that conserves the duality of the protein–nucleic-acid relationships. In eukaryotic cells, the role of the cytoplasm during replication of the chromosomes assumes greater stature, for the synthesis of DNA segments as well as the essential proteins seems to occur there. In other words, it is obvious that DNA replication is dependent on protein synthesis, not only for the creation of replicases, ligases, and other enzymes involved in the deoxyribonucleic acid duplication, but for the structural components that link the DNA segments into long chains.

The Ribonucleic Acids. In many ways, the ribonucleic acids are more amenable for occurring in the earliest protobionts than is DNA, for they show properties more suggestive of life. For instance, messenger RNA moves from the DNA to the ribosomes, ribosomes travel along the message and translate it, and transfer RNAs come in an amazing variety of forms, as one expects of living things. Yet on closer examination, too, these first favorable impressions gradually give way to misgivings. No type can be perceived that is capable of

Table 5.2
The Protein Constituents of the Genetic Mechanism

Nucleic acid	Activity	Proteins: Type	Proteins: Names	Proteins: Functions	Number in: Prokaryotes	Number in: Eukaryotes	Subtotals: Prokaryotes	Subtotals: Eukaryotes
DNA	Structure	Histones	H1, H2A, H2B, H3, H4	?	0?	4		
		Nonhistones		Regulate gene function and transcription	?	20+	Several	24+
	Replication	DNA Polymerase	I, II, III, Cl, etc.	Polymerize deoxyribonucleotides	3	5+		
		Modifiers	Methylases, etc.	Modifies bases	2+	4+		
		Dehelicase	ω	Unwinds the strands	2	2		
		Swivelase		Unwinds the strands	1	?		
		Reductase		Forms deoxyribose nucleotides	1+	1+		
		Ligase		Joins short segments	1+	1+		
		Nucleases	Exo- and endonuclease	Produce breaks and remove ends of strands	2+	2+	12+	15+
	Transcription	RNA polymerase	DNA-dependent IA, IIA, etc.	Polymerizes ribonucleotides	1+	8+		
		Initiation factor	σ, y	Initiates transcription	2+	2+		
		Termination factor	ρ	Terminates transcription	1	1		
		Nuclease	H, etc.	Aids in termination	1	1		
		Phosphorylase	Form-T, etc.	?	1+	1+		
		Ligase		Joins segments	1+	1+	7+	14+
mRNA	Structure	?	?	?	2+	2+	2+	2+
	Processing	Polymerase	?	Adds poly(A) segments	1+	2+		

		Nucleases	Endonucleases	Cleaves transcript	2+	2+		
			Exonucleases	Trims transcripts, etc.	2+	4+		
		Modifying enzymes	Methylase, etc.		2?	2+		
		Structural	?	Found on precursors	2+	8+	9+	18+
	Trans-lation	Initiating factors	IF-1 to IF-3	Aid in binding messengers to ribosome	3	3		
		Elongation factors	EF-G, EF-T, etc.	GTPases, etc.	4+	2+		
		Termination factors	RF-1 to RF-3	Terminate translation	3	1+	10+	6+
rRNA	Struc-ture	Small unit proteins	S1, etc.	Various, mostly unknown	21	30+		
		Large unit proteins	L1, etc.	Various, mostly unknown	34	36+	55+	66+
	Processing	Endonuclease	RNase III, etc.	Cleave precursors	2	4+		
		Exonuclease	RNase M, etc.	Cleave precursors	2	2+		
		Modifying	Methylases, phos-phorylase	Modify nucleosides	4	4+		
		Undetermined		Present on precursor only	8	10+	16+	20+
tRNA	Processing	Modifying	Methylases, thiolase, etc.	Modify nucleotides	33+	49+		
		Endonucleases	Endonuclease P, etc.	Cleave precursor	2+	2+		
		Exonuclease		Trim precursor	1+	1+		
		Ligases		Join segments	1+	1+		
		Transferases		Add -CCA terminus	1?	1?		
	Charging	Ligase	Aminoacyl-tRNA ligases	Bind amino acid to tRNA	40+	60+	78+	114+
Total proteins needed for DNA system to function							189+	279+

reproducing itself, for mRNAs, tRNAs, and rRNAs alike are transcribed into existence by proteins. In these processes, DNA renders vital service, but in its usual lifeless fashion, for it only provides the encoded message which enzymes read to create the various classes of RNA. Moreover, the original products of transcription are inactive also, for only after further treatment by other proteins do the molecules become ready for use.

And then in use, are the RNAs *active* participants in protein synthesis, or do they behave here, too, like DNA in being inert? Certainly mRNA does not translate itself to form polypeptides; rather it is translated by whole series of proteins—initiating, elongating, and releasing factors, large and small subunit proteins, and all the rest, known and unknown. Similarly rRNA has not been shown to play an active role in translation; only its proteins have been established as doing so. The tRNAs alone appear to be somewhat less passive, in that the anticodonal region of these nucleic acids plays a direct, essential role in the translational processes. However, they are charged by means of proteins, moved into position by others, and it may yet be established that even the codon–anticodon reactions are dependent upon still others.

In short, the RNAs are smaller and much more mobile molecules than DNA but are equally inert. Their chief value is in a mechanical capacity, whether as a carrier of a copied message, as in mRNAs, as a skeletal structure, as in the ribosome, on which proteins can integrate to perform their complex reactions, or as support for amino acids to simplify their transport and their peptidization by proteins. In another sense, their role is that of making DNA useful to the cell, for it is only through their being copied from DNA and carried into the cytoplasm that the DNA molecule can be employed in cellular activities. Hence, the RNAs had to exist prior to the advent of DNA, otherwise the latter could not have held survival or functional value to its possessors—only if the whole series of RNAs had already existed could DNA have been evolutionarily advantageous. Thus the genetic mechanism provides evidence that the RNAs, while older than DNA, could not have been of value at the very earliest stages of life. In any phylogeny of early life, the RNAs therefore have to be shown as arising at intermediate levels of advancement. The several types did not necessarily arise simultaneously, nor fully blown; each class might be expected to have had an evolutionary history of its own.

5.3.2. The Proteins

Throughout the discussion of the genetic mechanism, it becomes increasingly evident that the proteins are not only important constituents, but really the only active ingredients in its functioning. This is true at the levels of DNA replication and transcription, in attachment of the message to the ribosome, in the combining of amino acids to tRNA, in the joining of amino acyl-tRNAs to the message–ribosome complex, and in all the other events of translation. Thus the ultimate and only goal of the genetic mechanism is the production of proteins, an objective that is attained only by virtue of protein activities. Hence the evidence is compelling that the genetic processes are basically the means by which proteins produce more proteins.

Since proteins are essential to the replicatory processes of DNAs and RNAs alike, it follows that they had to be present at the advent of those nucleic acids, and therefore they had to antedate both types. Thus proteins must be considered the first ingredients of earliest life. It might be argued that proteins require the presence of DNA and RNA in order to reproduce themselves, and this observation is sound in general. However, it is not universally true, for certain polypeptides are today synthesized *in vivo* without direct intervention of nucleic acids. For instance, both tyrocidin and gramicidin S are known to be produced by enzymatic action alone, as pointed out in Chapter 2. True enough, the enzymes that synthesize these polypeptides are themselves made by way of the DNA–RNA–protein apparatus; nevertheless, here is evidence in living things that demonstrates the feasibility of protein-directed synthesis of proteins without nucleic acids. In contrast, no *in vivo* mechanism is known to exist for the protein-free, nucleic-acid-directed synthesis of another nucleic acid.

Moreover, the nonrandomness of association between amino acids that is clearly perceptible in proteins of today is suggestive of the former existence of a protein-synthesizing mechanism devoid of nucleic acids. For, if mutations in DNA occur randomly through chemical and physical agents, as is commonly accepted, then it is self-evident that any nonrandomness in protein structure cannot result directly from the nucleic acids, nor can it be correlated to amino acid structure, as shown earlier. Consequently, it remains a distinct possibility that some of this condition might be attributable to evolutionary causes. In other words, at least a major fraction of the nonrandomness may represent a relict condition derived from ancient ancestral molecules that replicated proteins by more direct processes than those extant in organisms today. The soluble system that incorporates various amino acids directly into proteins is suggestive of the same conclusion. If this system is actually a relict of a former primordial method of reproducing, it also provides some evidence which suggests an origin for the tRNAs prior to any other nucleic acid.

5.3.3. Conclusions

From the present genetic mechanism, the following major conclusions are derived:

1. The proteins are indisputably the class of biochemicals which arose first in life's history.
2. Tentatively it appears that the tRNAs were the earliest type of nucleic acid used by the evolving protobiont.

3. Later, if not concurrently with tRNA, mRNA was added, followed somewhat later by rRNA.
4. DNA was developed in life's history subsequent to mRNA's origin. Whether rRNA arose even later cannot be established, but it remains a possibility. This likelihood is derived from the lack of kinship that exists among ribosomal proteins, suggestive of their origins' having been in the cytoplasm, their becoming attached to a supportive structure of nucleic acid being a relatively recent event.

This point of view proposing proteins as the primordial substance is not original; only the method of establishing it is. Among writers who have previously advocated this early advent of protein are those who favor the coacervate as the earliest biont (Oparin, 1957, 1968, 1971; Smith *et al.*, 1967; Liebl and Lieblova, 1968; Akabori and Yamamoto, 1972). As pointed out in Chapter 2, the creation of coacervates requires several high-molecular-weight substances, among which polypeptides are considered distinct candidates. Another group that follows the present point of view is that which proposes the proteinoid microsphere as the ancestral stock (Fox *et al.*, 1959; Fox, 1960, 1971; Fox and Yuyama, 1963; Fox and Dose, 1972). Still others have advanced the idea purely on conceptual grounds (Mueller, 1955; Sagan, 1957; Needham, 1959; Hanson, 1966; Orgel, 1968; Calvin, 1969; Lacey and Pruitt, 1969; Lacey and Mullins, 1972), but no previous analysis of the genetic mechanism of extant living things has been directed toward this end.

With this preliminary evidence in hand, it becomes possible to seek corroborative data from particular aspects of the genetic processes. Such confirmatory information is sought first through close examination of the genetic code and its possible origins, in the next chapter, and then in the two following ones through detailed examination of a related subject, the tRNAs and their structural relations.

6

Micromolecular Evolution—The Origin of the Genetic Code

Aside from the conclusions just reached, the foregoing review of the genetic mechanism makes it evident that the system is totally dependent on a coding device, a code built on triplets of nucleotides. Thus an understanding of the origins of the entire apparatus appears to be contingent on a knowledge of the beginnings of the code catalog itself. Because of the importance of the problem, numerous attempts at solving it have been made along a diversity of avenues, which fall into four major categories: conceptual, mathematical, biochemical, and biological. Since assumptions made by studies in the first three of these groupings are often in conflict with the findings just summarized, the present status of the problem as outlined below may be viewed more objectively if those conclusions are held in abeyance momentarily.

6.1. CONCEPTUAL APPROACHES

Conceptual approaches involve the visualization of the possible nature of the earliest preliving things, often through use of experimentally derived models, although they are not requisite. Some investigators depict the original protobiont as consisting solely of nucleic acids; others see it as made up of proteins. Among those that have accepted proteins as the earliest molecules are the coacervate and microsphere schools of thought, whose approaches have been presented in a preceding chapter and need not be reconsidered in detail here. In substance, these and other adherents of this concept follow the sequence of ideas that (1) proteins are essential in the protobiont (microspheres or coacervates); (2) these protobionts increase in size and undergo "division" under mechanical stimulation or chemical change; and therefore (3) a genetical mechanism is unnecessary at the earliest stages of life's evolution. In some accounts, it has been proposed that, when the proteins replicated, those polypeptides present in the protobiont arranged the sequences of amino acids in nonrandom chains through interactions between the amino acids themselves. Also it has been considered that the ordering probably was not so precise as the residue-by-residue arrangements produced by the present genetic code (Fox *et al.*, 1959, 1971; Fox and Nakashima, 1967). Later, as a cellular construction was acquired, the polypeptide self-ordering mechanism was gradually replaced by a coded protein–nucleic-acid system (Jukes, 1966; Ishigami and Nagono, 1975). The evolutionary sequence has been viewed as: amino acids → proteinoids → polynucleotides → contemporary proteins (Fox *et al.*, 1971).

Perhaps the majority of conceptual studies have perceived the nucleic acids, rather than proteins, as the prime molecular ingredient of the earliest precursors of life, because of the essential role they play in protein synthesis today. In the earlier of these works, DNA was conceived as the first polymer of life, since it has been widely considered the self-replicating substance of the chromosome (Simpson, 1949; Huxley, 1963; Smith, 1966; Conrad, 1970; Krzanowska, 1970; Jukes and Gatlin, 1971; Schapp, 1971; Bishop *et al.*, 1972). More recently, however, the trend has been toward favoring RNA in this capacity, for a diversity of reasons. On one hand, it has been concluded that the RNA must have been combined with polypeptides to form such nucleoproteins as are found in ribosomes, because the ribosomal particles are essential for protein synthesis in extant organisms (Fox *et al.*, 1971; Mednikov, 1971). On the other hand, it has been conceived that tRNA was among the earliest of the nucleic acids, because of its involvement in transport of the amino acids (Jukes, 1969; Abraham, 1971; Dayhoff, 1971; Rich, 1974). It has also been proposed that mRNA appeared prior to any of the other types, because only this variety was then believed present in the simpler viruses (Salthe, 1972). In the latter point of view, amino acids were suggested to have linked directly to the mRNA by means of 2′-OH-aminoacyl esters (Lacey *et al.*, 1975). Such combinations of amino acids and nucleotide chains have been created in the laboratory, but so far no peptide linkages have been brought about between the attached amino acid residues. Since that time, as shown earlier (p. 199), the occurrence of tRNAs in these viruses has been demonstrated, thereby eliminating the factual framework on which this type of concept was built.

Another general proposal was that a precursor of some unusual sort preceded the development of the definitive nucleic acids. One idea was that the earliest genetical molecules consisted of random mixtures of RNA and DNA, the proportions of the two types varying with the relative abundance of ribose and deoxyribose in the environment and the functional properties possessed by the resulting mixture (Orgel, 1968). A second suggestion was that the first coding polynucleotides consisted largely of purines and pyrimidines arranged in alternating fashion. These sequences in turn were visualized as coding the formation of polypeptides in which hydrophobic and hydrophilic amino acids alternated (Orgel, 1972). Such polypeptide chains, but not the corresponding polynucleotide strands, have now been synthesized in the laboratory; these consisted of valine and lysine and formed stable β structures (Brack and Orgel, 1975). In aqueous solutions, the polypeptides formed bilayers having a hydrophobic interior and a hydrophilic exterior. Contrastingly, Woese (1973) has proposed that polypeptides in which alternate amino acid residues were basic had been essential in order to have polymerized the nucleotides into primitive nucleic acids.

Another conceptual framework has been constructed through speculations on possible interrelations between amino acids and short polynucleotides that might correspond to the codons, or between amino acids and a given mononucleotide (Melcher, 1970; Jett and Jamieson, 1971). It has been suggested that perhaps only two distinct bases occurred originally in nucleic acids, and that two more arose when four proved stereochemically possible in double-helical structures (Crick, 1968). The same source proposed that triplet sequences were necessary for structural and space considerations, for in extant organisms apparent limits exist as to the extreme ratios of $(G+C):(A+T)$ permissible in DNA (Woese and Bleyman, 1972). Restraints in the level of $G+C$ seem active below a minimum of 20 to 25% and above a maximum of 75%.

6.2. MATHEMATICAL CONCEPTS

To gain insight into the origin of the genetic code, biomathematicians have sought to explore its fundamental nature in generalized abstractions. Basically their approach proposes that, if the genesis of the code has been associated with physicomathematical laws rather than simply with random chance, it may be possible to find the basis for the code organization. Then the significance of the correspondence between triplets and amino acids may prove to conform with an entirely deterministic working hypothesis.

In such studies one important quantity is 64, the total number of triplet combinations of nucleotides contained in the code (Besson and Gavaudan, 1967; Schutzenberger *et al.*, 1969; Kaplan, 1971). Another numerical value is provided by the number of different amino acids that are commonly found in proteins, for though hundreds of varieties are either known or theoretically possible, only 20 are universal in occurrence. In some cases, this number has been extended to 23 in order to include the three nonsense codons (terminators or stops) which are not assigned to amino acids (Gavaudan, 1971a,b). These and other numerical analyses (e.g., Miklos, 1971; Gatlin, 1972; Ratner and Bachinskii, 1972a; Moore *et al.*, 1973; Papentin, 1973; Yockey, 1973; Hasegawa and Yano, 1975) provide models of physicochemical principles that might have been involved in the early development of the genetic code. One such study has even suggested a possible basis for the asymmetry that exists in the distribution of the number of codons assigned to the several amino acids (Ratner and Bachinskii, 1972b).

6.3. BIOCHEMICAL APPROACHES

The possibility that direct chemical relations exist between a given amino acid and its codons contained in nucleic acid has been explored by a large variety of experiments. In a number of such investigations, reactions between mononucleotides and polyamino acids have been examined comparatively, as well as those between monoamino acids and polynucleotides. One study using the former type of interaction showed that GMP reacted far more strongly with polyarginine than did any other mononucleotide (Woese, 1968). Another laboratory immobilized nine representative amino acids individually on a prepared chromatographic support. These were found to bind selectively to ribonucleoside 5′-phosphates, their binding behavior being dependent on the nature of the base and that of the specific amino acid (Saxinger and Ponnamperuma, 1971; Saxinger *et al.*, 1971). Comparable experiments, supplemented by molecular model investigations, predicted a correct codon (GGG) for glycine (Lacey and Pruitt, 1969).

In related studies, adenine and cytosine were coupled to a polystyrene column; then measurements were made of the relative efficiencies with which phenylalanine and glycine coupled to these bases (Harpold and Calvin, 1973; Calvin, 1975). Both amino acids reacted more efficiently to adenine than to the other base. A second laboratory reversed these procedures by attaching the amino acid to the polymer column and employing di- and trinucleotides (Saxinger and Ponnamperuma, 1974). Among the results noted were that both glycine and tryptophan bound more freely to the trinucleotide AAU than to any other polynucleotide tested. Proteinoids have similarly been subjected to solutions of polynucleotides and found to show some degree of selectivity (Fox, 1974). Lysine-rich proteinoids, devoid of arginine, formed microparticles (precipitates) with poly(C) and to a small extent with poly(U). In contrast, arginine-rich proteinoids reacted strongly with poly(G) and poly(I) and weakly with poly(A) and poly(U). Micelles also have been explored along comparable lines (Nagyvary and Fendler, 1974).

Photochemical reactions between amino acids and nucleic acids have likewise received attention (Lesk, 1970). The use of ultraviolet rays has induced cross-binding between DNA and protein in live cells (Smith, 1968), and cysteine has been added by similar means *in vitro* in polynucleotides and to nucleic acids (Smith and Meun, 1968). Comparable coupling was produced between various amino acids and [^{14}C]uracil by photochemical techniques (Smith, 1969); 11 of the amino acids tested were considered significantly reactive.

By means of proton magnetic resonance spectroscopy, the interaction between all 20 amino acids and neutral poly(A) and between a selected number of amino acids and poly(I) or poly(U) has been investigated (Raszka and Mandel, 1971, 1972a). When poly(A) was employed with either aromatic or aliphatic amino acids, a shift in the position of the C_2 and C_8 protons was noted; this reaction was deemed consistent with a destacking of the initially partially stacked polynucleotide chain due to intercalation of the side chain. Fluorescent studies on the interaction of aromatic amino acids with calf thymus DNA tended to confirm these results (Hélène, 1971). Models of single-stranded RNA seven units long and of a polypeptide chain of similar length have been constructed; when the latter consisted of the amino acids that corresponded sequentially to the codons, the two helices were so related in configuration that, in theory, they could have interacted (Carter and Kraut, 1974).

Although all the above experimental results doubtlessly hold significance in understanding the physical and chemical activities of living creatures, the question must be asked, as Raszka and Mandel (1972b) have, "Of what significance are they in the origin of the genetic code?"

To answer this question, they pointed out that, if the present genetic code represents preferred physical or chemical interaction between a given amino acid and a codon, studies on the four homopolynucleotides [poly(A), etc.] and the amino acid each such triplet encodes should provide clear-cut evidence to that effect. Thus phenylalanine should strongly interact either with its codon, poly(U), or its anticodon, poly(A), and, conversely, lysine should show decided reactions with either poly(A) or its anticodon, poly(U). To the contrary, the experimental evidence showed that both polynucleotides reacted more strongly with phenylalanine than with lysine, and that even the corresponding monomers did likewise. Further, whereas poly(G) and poly(C) should react preferentially with glycine and lysine, available data have shown them again to have favored the aromatic amino acids. In short, the results indicated that all codons would have encoded the aromatic amino acids, especially phenylalanine and tyrosine, and thus the entire genetic codon catalog would be totally meaningless. In part at least, the interactions noted were stacking phenomena, no or little hydrogen bonding having occurred (Arfmann *et al.*, 1974). Hence, it was clear that no direct physical or chemical relationships existed between the codons and the amino acids. This conclusion is strengthened by the evidence presented by Saxinger and Ponnamperuma (1974) which showed that direct reactions of different amino acids with oligonucleotides depend on the type of the amino acid and the relative length, not just the composition, of the oligonucleotide. Since the strictly conceptual and mathematical approaches also depend upon the existence of such physicochemical relations, they too lose real significance in the present context.

Furthermore, a fallacy exists in the theoretical base on which most of the biochemical experiments rest, which has not been noted to date. Most of these studies are directed toward ascertaining chemical affinities between the amino acids and polynucleotides which represent either the respective codons or anticodons for the former. It is obvious, however, from the review of the tRNA charging processes, that the specific tRNA ligase reacts first with the correct species of tRNA and then with the amino acid. Recognition of the tRNA by the ligase may or may not involve the anticodon—this point has not been finally established, but it is amply clear that there is no direct interaction between the tRNA and the amino acid, except at the 3′-terminus (−CCA), and even there, the attachment is brought about by the ligase. Were the interaction between amino acids and tRNAs a direct one, then no such enzyme intermediary would be required. Anticodons, aside from a possible role in ligase recognition, are known to react only with codons in the mRNA and then only while the latter is attached to the ribosome. Hence studies of amino-acid–nucleotide reactions appear to hold little chemical, as well as no biological, value insofar as the origin of life is concerned.

6.4. A BIOLOGICAL CONCEPT

The only alternative concept of the genetic code origin available to replace the physicochemical relationships theory, according to Raszka and Mandel (1972b), is that the code catalog is a frozen accident. This hypothesis assumes that all organisms had a common ancestry and that the code was universal as a consequence of its being established in the ancestral stock. Because its essential nature precludes any change as a result of the ensuing lethality, it has in a sense become "frozen." Although those writers did not state the logical conclusion of this alternative, it is implicit in their deductions that the genetic code has a biological basis for its structural organization, and no other.

In the only biological concept advanced to date regarding the origins of the genetic code (Dillon, 1973), an extant organism is employed as a model. The chemautotroph this study utilized as an example is one of the colorless sulfur bacteria and belongs in the genus *Thioploca*. It, along with certain other members of the Beggiotoales, has been shown to rank among the most primitive of present-day cellular life, both structurally and functionally (Breed *et al.*, 1957; Dillon, 1962, 1963; Maier and Murray, 1965). For growth and propagation, this organism needs light and only four substances in its milieu, namely, H_2O, H_2S, NH_3, and CO_2. No organic compounds of any sort, not even vitamins, need be present in the environment (Keil, 1912; Pringsheim, 1964). Because growing colonies of *Thioploca* and related genera deposit sulfur within their cells, it has been assumed that the light is employed to break down the H_2S, thus releasing H_2. This assumption receives support through the absence of sugar synthesis characteristic of other organisms having photosynthetic processes. The absence of sugar formation is similarly reflected in the complete lack of cell respiration and of even such respiratory macromolecules as the various cytochromes (Burton *et al.*, 1966). In these photosynthetic organisms, no chlorophyll or other photoaccepting pigment is present; however H_2S is well known as a photon acceptor (Chapter 1). Thus this compound may have served on the primitive earth, as in these colorless sulfur bacteria, as the earliest means of capturing energy from the sun.

6.4.1. Basic Steps

Since few details of metabolic processes other than the above have as yet been described for the model organism, those proposed were entirely speculative (Dillon, 1973), but with only the four inorganic substances to serve as the raw materials for the synthesis of amino acids and other fundamental biochemicals, alternative metabolic pathways are difficult to perceive. The major steps suggested as possible ones for the early biosynthesis of the amino acids were standard processes of organic chemistry involving reduction by H_2:

$$CO_2 \rightarrow {}^{-}COOH \rightarrow {}^{-}CHO \rightarrow {}^{-}HCOH \rightarrow {}^{-}CH_3$$

(Carboxyl) (Ketone or aldehyde) (Alcohol) (Methyl)

This basic series of reactions was speculated to have been used by the model organism in the synthesis of amino acids in combination with the NH_3 that is also necessary in its environment for growth. As in many familiar cellular processes, the NH_3 probably reacted with ketones to produce amines. It was therefore proposed that, in the initial reactions, two molecules of CO_2 became reduced by means of the H_2 liberated by the photosynthetic degradation of H_2S and thus became coupled together (Figure 6.1). This product next was further reduced by H_2 from the same source to produce glyoxylic acid. Amina-

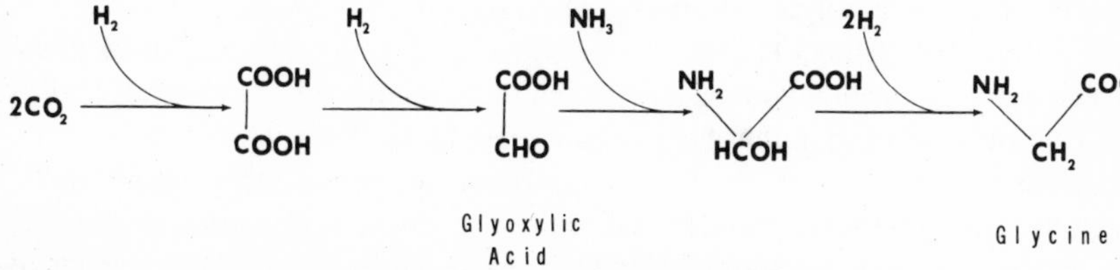

H_2 H_2 NH_3 $2H_2$

$2CO_2$ → COOH–COOH → COOH–CHO → NH_2–HCOH–COOH → NH_2–CH_2–COOH

Glyoxylic Acid Glycine

Figure 6.1. Hypothetical steps in the primitive synthesis of glycine. Several bacteria are known which fix CO_2 and produce glycine, perhaps by comparable steps.

tion of this ketone acid, followed by still further reduction by H_2, would then result in the amino acid glycine.

It is to be noted that all the inorganic materials employed in this hypothetical series of steps are among those frequently used in laboratory prebiological syntheses. In prebiotic syntheses, it might have been that CH_4 was used instead of CO_2 now employed by *Thioploca*, but the abundance of CO_2 now ascertained to be present in the atmospheres of neighboring planets suggests that this is unnecessary. Nevertheless, glyoxylic acid is actually involved in the simple cellular processes of *Thioploca* and may well also have been a common ingredient in the primeval seas. Furthermore, glycine consistently has ranked among the most abundant products of experimental prebiotic syntheses (Chapter 1) and thus appears readily synthesizable from inorganic sources.

What is of more importance relative to this subject is the now thoroughly documented biosynthesis by bacteria of acetate from free CO_2 (Ljungdahl and Wood, 1969; Sun *et al.*, 1969; Parker *et al.*, 1971, 1972; Schulman *et al.*, 1972, 1973). The first product of CO_2 fixation has been suggested to be formic acid in those species of *Clostridium, Butyribacterium*, and *Diplococcus* which have been most thoroughly investigated (Sun *et al.*, 1969; Schulman *et al.*, 1972); the fixing processes seem to have involved NADPH, but this point has not been established (Higa *et al.*, 1976). After acetic acid was produced from this first intermediate compound, the molecule was elongated to form butyric acid (Schulman *et al.*, 1972). Thus the molecular elongation steps outlined below may eventually receive support from these sources. But most notable of all the findings in the present context is that glycine was produced from CO_2 by *Clostridium cylindrosporum* and *C. acidic-urici* (Barker *et al.*, 1940; Barker and Beck, 1941; Sagers *et al.*, 1961; Ljungdahl and Wood, 1969). Thus the proposal that glycine was synthesized from CO_2 in protobionts is supported by solid evidence from actual processes in bacteria. It was this conclusion, that CO_2 could be fixed to produce gylcine and other amino acids, that was important in the hypothesis under review, not the tentative steps that were proposed.

6.4.2. Origins of the Group I Amino Acids

With only the four inorganic substances enumerated above present in its medium, the model organism must synthesize its amino acid requirements under severe constraints. Thus it was proposed that, in the absence of conflicting evidence, the steps hypothesized above or comparable ones that resulted in the synthesis of glycine might be accepted tentatively. Then with glycine as the basis, the same or similar processes could produce several other amino acids. Basically these syntheses merely require the successive lengthening of the carbon skeleton by stepwise additions of CO_2, followed by reduction of the added radical by means of H_2, as in the formation by bacteria of butyric from acetic acid mentioned above.

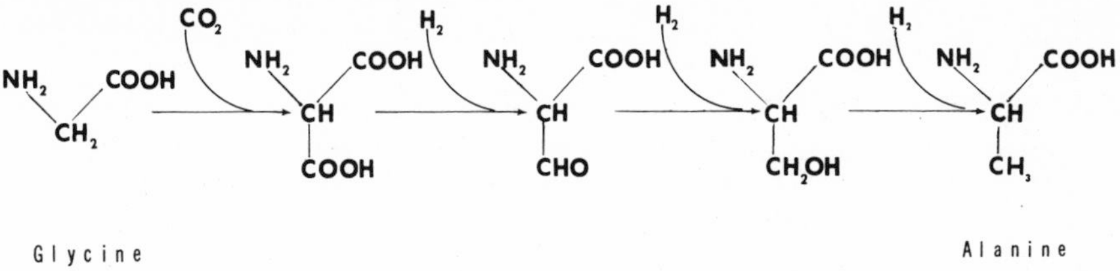

Glycine Alanine

Figure 6.2. Primitive synthesis of alanine. It has been proposed that alanine could be produced from glycine by elongating the carbon skeleton of the latter by additional CO_2 fixation.

It was proposed that the first elongation of the carbon skeleton resulted from the addition of a molecule of CO_2 to glycine. The resulting ^-COOH radical was then reduced in three steps through the carboxyl and hydroxyl to a methyl radical (Figure 6.2), thereby producing a second common amino acid, alanine. Since the basic steps in these reactions were identical to those used in glycine formation, it was suggested that, in simple ancestral organisms, perhaps much of the same enzyme system could have been employed for the biosynthesis of alanine.

A third amino acid was conjectured to be formed merely by the addition of another CO_2 to alanine, again as a carboxyl radical. This would have resulted in the creation of aspartic acid which possesses three carbons in the skeleton (Figure 6.3A). In the author's laboratory, continuous bubbling of CO_2 through a 0.1 aqueous solution of alanine at ambient and elevated temperatures for a week failed to produce any detectable aspartic acid (Dillon, unpublished). Hence, the addition can probably be made only through enzymatic mediation.

To carry the carbon-skeleton elongation processes still further, it was supposed that the carboxyl radical of the aspartic acid would first have been reduced to a methyl residue before the third CO_2 was added to the molecule. After the latter had then been added in the form of a carboxyl radical, it was presumed to have been reduced through the usual steps to a methyl radical, again possibly by the same enzymatic train. This last addition was deduced to have become attached to the β-carbon of the molecular skeleton, not the γ-carbon, so that the branched chain characteristic of valine was produced (Figure 6.3B). This branched configuration was believed to have interfered with further straight-line elongation of the carbon skeleton. Thus in this hypothesis a series of four amino acids could be biosynthesized by the model organism, as possibly in early protobionts, through a simple series of reactions perhaps largely catalyzed by a single enzyme system.

In view of the limitations of the nutrient intake of these colorless sulfur bacteria, the proposed sequence of major events is so straightforward as to carry a high degree of credibility. In addition to the thoroughly documented reports of bacterial synthesis of glycine by CO_2 fixation, support for a compara-

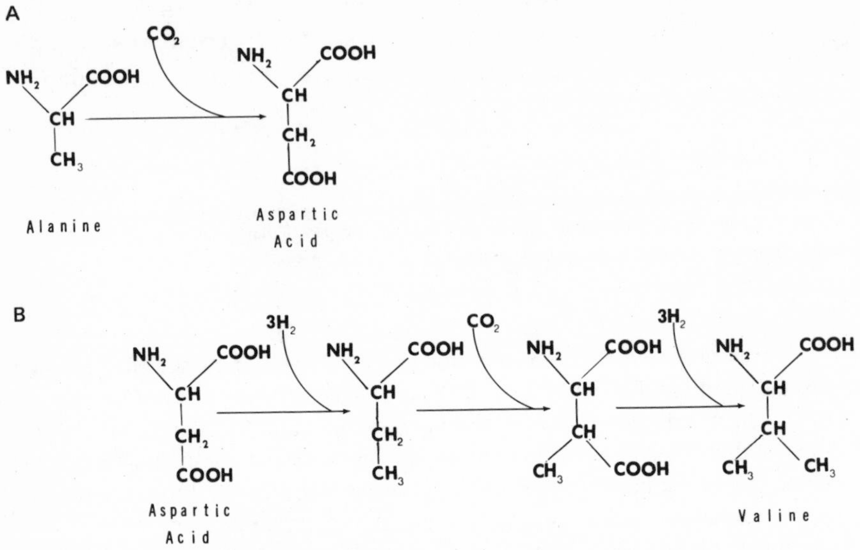

Figure 6.3. Primordial syntheses of other amino acids. (A) Aspartic acid might have been synthesized from alanine by the enzymatic addition of CO_2. (B) Hypothetical steps in the protobiontic production of valine from aspartic acid.

ble series of steps having existed in primitive life was advanced, derived from the genetic codon catalog itself. When the entire series of reactions itemized above was summarized in diagrammatic form (Figure 6.4) and the codons for each amino acid listed thereon, a high degree of correlation immediately could be noted. Each of the four amino acids, as was shown, was uniformly encoded by triplets in which the initial site was occupied by guanosine. As was pointed out in the original article, only two explanations are in evidence for this combination of facts: first, the obvious structural relationships that exist between the amino acids produced by this proposed chain of events, and second, their codons' containing the same base in the initial site. One alternative is that the two sets of correspondences resulted from mere random chance; in view of the well-established nonrandomness of codon assignments, however, this explanation seems highly improbable. The second possibility appears far more likely, namely, that these amino acids had evolved in primitive organisms together with the codons.

Before this alternative is given further consideration, however, another aspect of the possible metabolism of amino acids in early organisms needs to be examined. This facet centers around a second characteristic of the genetic code pattern pointed out in Chapter 4. It will be recalled that in order to approach that pattern, it was necessary first to use sets of four triplet nucleosides to encode 16 amino acids; then some of these original complements of codons had to be shared with those amino acids that were not thus provisioned, thereby producing the pairs, singles, and combinations that exist in the actual codon catalog.

Since the branched arrangement of the carbon skeleton of valine appeared to have circumvented additional direct biosynthetic pathways beyond that amino acid, it was proposed that further progress may have been made by using aspartic acid in a second biosynthetic path. After the carboxyl radical of aspartic acid had been reduced to a methyl residue as before, synthetic processes eventually seem to have developed that were capable of adding a CO_2 to the γ-carbon, thus producing glutamic acid (Figure 6.4). When this new amino acid thus arose, all the then-existing codon catalog had been assigned to the four original amino acids, glycine, alanine, aspartic acid, and valine. Hence, in order for the new amino acid to become provided with codons, sharing in those sets of four was necessary, and a pair, terminating in the purines, was derived for it from those of aspartic acid. Consequently, all the codons beginning with G occur in natural groups of four, except those with A in site 2, which are

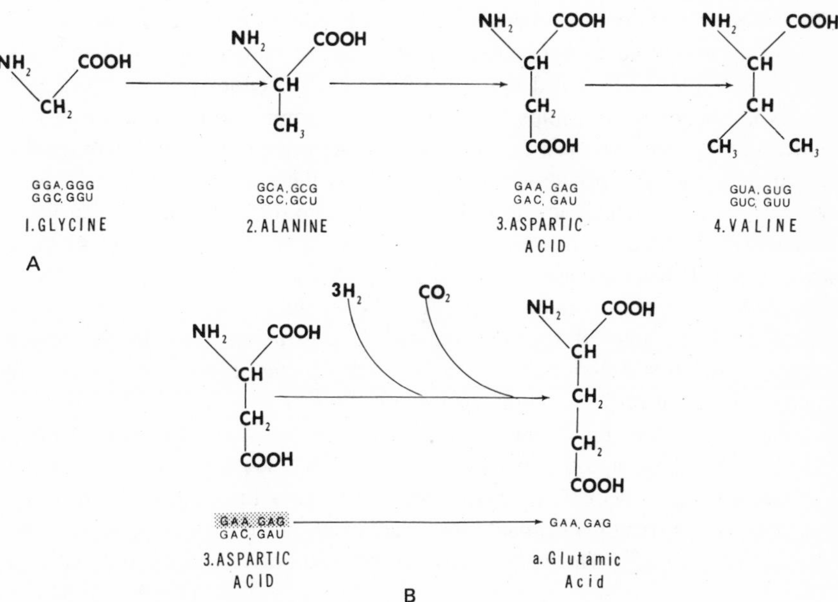

Figure 6.4. The five earliest amino acids. (A) The four amino acids conjectured to have been produced by earliest protobionts, together with their codon assignment. It should be noted that all the codons begin with G. (B) Glutamic acid appears to be a later derivative of aspartic acid and shares in the latter's original complement of codons.

equally divided between aspartic and glutamic acids. In support of this conjecture, it was noted that even in metazoan cells, close metabolic relationships existed between these two amino acids.*

Because of the extensive analysis of the genetic mechanism that has preceded the present discussion, it is here possible to make an observation that could not be made in the relatively short article under review (Dillon, 1973). There it could only be pointed out that the evolving amino acids and codons underwent their development in a close-knit interacting system, involving not just the enzymes that biosynthesized the amino acids, but the entire genetic mechanism. This mechanism, however, it now can be proposed, at that time probably consisted only of proteins and a relatively few transfer RNAs and perhaps their ligases (p. 142). Hence, it was comparatively simple for the then-existing genetic apparatus, including the coding system, to undergo evolutionary changes. In contrast, in the extremely complex mechanism of today, such changes in fundamental structure would be totally impossible because of the lethality that would likely accompany them.

6.4.3. Working Principles for Further Phylogenetic Tracings

Several working principles became evident as the phylogeny of this first group containing five amino acids was traced, which provided a basis for following the proposed evolutionary pathways of the remaining 15 amino acids. Those principles were stated to be: (1) The genetic codons provide clues as to the evolutionary relationships existing between amino acids, each initial base indicating a related set, just as guanosine in the first site is entirely confined to the members of group I amino acids. (2) Thus there are four groups of these substances, all of which appear to be derived from different members of group I. The latter group itself was derived from glycine, group II from glutamic acid, group III from aspartic acid, and group IV from alanine. (3) Valine probably failed to serve as a progenitor for the evolutionary derivation of other amino acids because of its branched-chain construction. (4) As the several groups arose in succession, the first four components in each were originally encoded by a set of four codons apiece, resulting in a total of 16 amino acids' being thus encoded. Functionally, sets of four are obviously the equivalents of doublets, but since a codon is merely a specialized sequence within a polynucleotide chain, even in short chains there could in actuality have been a third site present. Its specific occupant, however, would have been of no significance in a set of four. (5) Whenever a new amino acid was added evolutionarily to

* Metabolic relations between pairs of amino acids were employed recently as the basis for a concept on the origins of the genetic code (Wong, 1975, 1976). The chief conclusion parallels one of those of the study being reviewed, namely that the amino acids and genetic code evolved together. Others have reached similar conclusions on purely speculative grounds that the code evolved in early protobionts (Hartman, 1975; Jorré and Curnow, 1975; Lacey *et al.*, 1975).

any of the groups beyond the original four members, sharing of a set of four resulted in the production of two natural pairs of codons. (6) The evidence from aspartic and glutamic acids intimated that the new amino acids received from the donor that natural pair which terminated in the purines, while the donor retained the pair ending in the pyrimidines. This point, however, was stated to need further investigation. (7) On the basis of glycine's being encoded by the triplet GGG; it was supposed that the very first amino acid formed in each of the four groups may have been encoded by codons of identical bases, CCC, AAA, and UUU, but this was considered a minor point.

6.4.4. Evolution of the Remaining Amino Acids

The sequence of events that was proposed earlier was in the first place partly derived from the complementarity existing between G and C, for the members of the group that was treated as being the second to evolve are provided with codons beginning with cytosine. Secondary considerations of organization, however, also appeared to support the supposition. The sequence of origin of groups III and IV is less clear-cut, but in general the scant available data suggested that the group whose members were encoded by the codons beginning in A arose prior to the group whose component amino acids were provided with triplets beginning in U. The group arrangement of the amino acids and the concept of coevolution of amino acids and codons were viewed as being of greater moment in the present framework of discussion than the conjectures regarding the time sequence.

Evolution of Group II Amino Acids. Aside from the G–C complementarity relations in site 1, the probable early advent of group II amino acids receives support from its compositional similarity to group I. Here as in the latter, only five amino acids are included, and, as before, three of these are encoded by sets of four and two by natural pairs each. Thus the two groups are of obviously equal simplicity. In the phylogeny of group II, however, no simple skeletal-lengthening processes gave rise to new amino acids one after another in linear fashion, as in group I. Here one central parental substance was shown to have served as the starting place for four derivatives in a radiating manner (Figure 6.5).

On the basis of those premises outlined above, it was proposed that as group II members evolved, the β-carboxyl of glutamic acid became converted to a methyl residue by means of reduction processes similar to those employed throughout group I phylogeny. From the resulting α-amino-ketoglutaric acid, it was conjectured that four amino acids eventually were derived. Since proline's code complement includes CCC, and, especially, since relatively little molecular alteration is required to convert α-amino-butyric acid into proline, that amino acid was considered to have been the first synthesized from this source (Figure 6.5). The only modification necessary to produce the characteristic

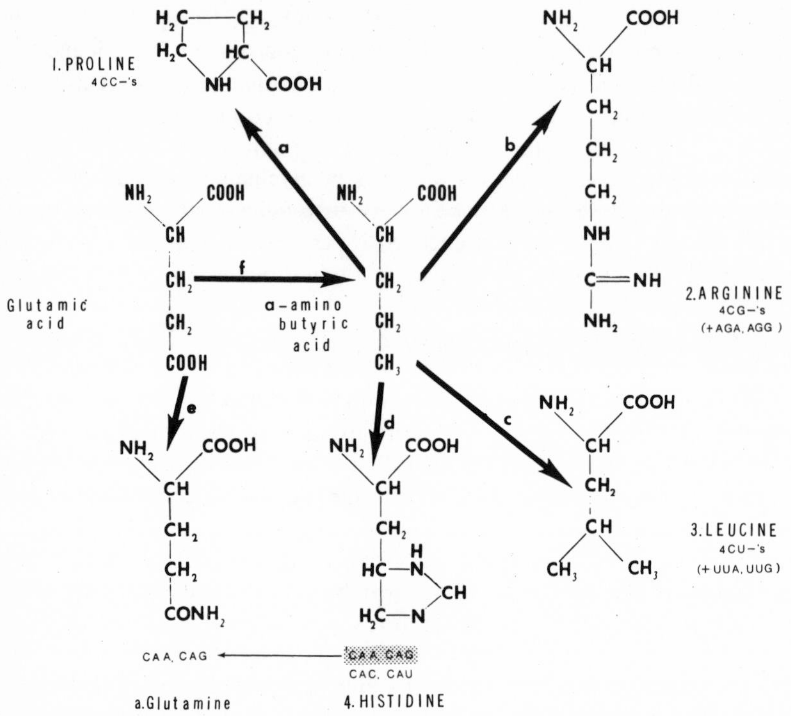

Figure 6.5. The phylogeny of group II amino acids. After the first four amino acids in this group arose by way of α-amino-butyric acid, they were each provided with a set of four codons with the initial base C. Later, when glutamine had been derived directly from glutamic acid, it was encoded by sharing in the codons of histidine.

cyclic configuration of that amino acid is the removal of one hydrogen residue each from the δ-carbon and the amino radical. In the formation of leucine, another probable early component in this group, the biosynthetic path could have been by steps resembling those of group I, particularly those of valine. If another CO_2 were to be added to α-amino-butyric acid on the γ-carbon and then reduced in the standard series of steps to a methyl radical, the branched-chain form of leucine would have resulted.

The importance of NH_3 in the formation of the remaining amino acids of group II was pointed out, for all three have nitrogenous additives. For example, one of the three, arginine, might readily have been biosynthesized from α-amino-butyric acid by addition of a guanido radical, a substance carrying three nitrogen-containing resudues (Figure 6.5). The possible origins of the second member of the trio, histidine, are much more obscure. Perhaps its biosynthesis involved combinations of the terminal region of the molecule with ammonia and methane to form the characteristic imidazone ring, but origin from arginine could have been a possible alternative. Steps in the latter metabolic pathway would have included partial deamination and shifts in the bonding pattern to induce the formation of the ring.

The four original components thus formed would have become provided, according to the basic principles, with sets of four codons each, as three of them still are in modern organisms. Glutamine, the fifth member of the group, was believed on the basis of several considerations to have arisen later. In contrast to the other four, it was likely not a derivative of α-amino-butyric acid, but more probably was developed directly from glutamic acid. This derivation would have required reduction of the γ-carboxyl of glutamic acid to a ketone, accompanied by amination, steps comparable to those occurring today in metazoan and other types of cells. Whatever its origin, it became provided with a pair of codons derived by sharing with those of the original complement of four possessed by histidine.

Whether the frequency of ammonia-bearing radicals in the later members of group II just discussed can validly be employed to deduce atmospheric changes is a moot question. If an increase in ammonia content had occurred during the time span involved in group II's evolution, then hydrogen would have had to be abundant, too, since ammonia is unstable in the absence of the latter element (p. 6). If any relationship can be justified between relative abundance in the environment and employment by organisms, then the proportions of ammonia may have been relatively high and continued at that level through the period during which the next group of amino acids evolved, for it too shows a high frequency of nitrogen-containing additions.

An Evolutionary History of Group III Amino Acids. Group III amino acids, all of which are encoded by triplets beginning with adenosine, were deduced to have been derived from aspartic acid by a diversity of routes. One of the suggested metabolic paths led directly from aspartic acid in a fashion similar to that of glutamine discussed above; that is, the β-carboxyl radical was first reduced to the aldehyde and then aminated to produce asparagine (Figure 6.6). As in the case of glutamine, it was pointed out, this series of steps is an active biosynthetic path in metazoan cells, thus substantiating the proposal to a degree.

In the second proposed pathway, the β-carboxyl of aspartic acid was considered to have been first reduced to a methyl residue by use of H_2 and the same series of steps hypothesized in group I phylogeny. The α-amino-propionic acid thus formed was stated then to have received a hydroxyl radical on the β-carbon, producing the amino acid threonine. Later, some of the threonine underwent demethylation to yield a third amino acid, serine.

The third conjectured biosynthetic pathway repeated one already established in group II, for, on the basis of the evidence that was advanced, α-amino-propionic acid seemed to have had a CO_2 attached to it, as in group I events, which subsequently became reduced to a methyl residue to form

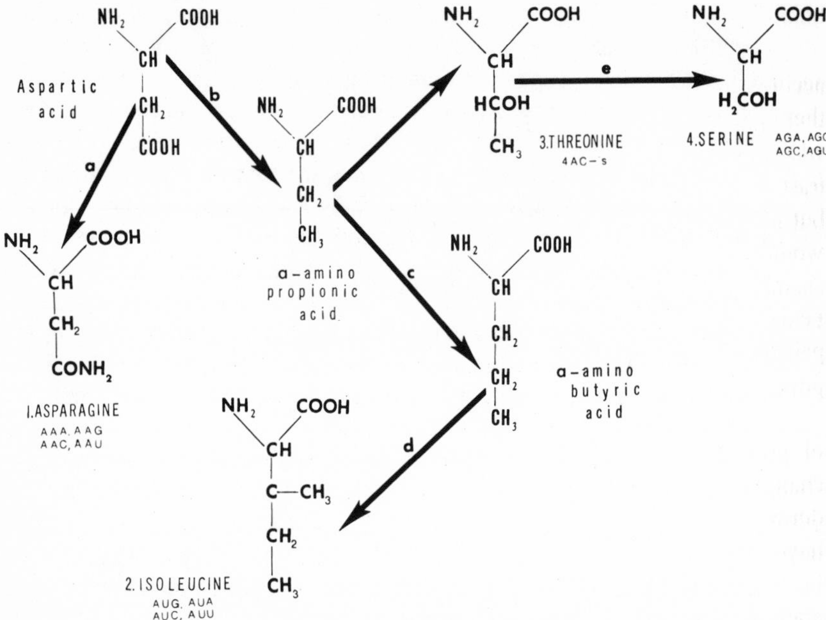

Figure 6.6. The hypothesized steps in the synthesis of the original group III components. In the phylogeny of the members several intermediate products are involved, with several pathways accordingly. Thus their origin required a larger number of enzyme systems than did those of groups I and II, and, accordingly, a more extensive genetic mechanism. All were encoded by sets of four codons, having the initial base A.

α-amino-butyric acid. This compound then served as the intermediate for further evolutionary development in group III as it had in group II, a point that receives additional attention below. In the present assemblage, this compound was speculated to have served only in one biosynthetic path. It seemed to have received a CO_2 on the β-carbon, which was reduced to a methyl radical by the usual route and thereby produced isoleucine.

With the creation of isoleucine, the original four components of group III had been biosynthesized, and each would then have been supplied with sets of four codons. Thus further additions, which were conceived to have been relatively extensive in the present group, became encoded by sharing in the sets of codons of the original members. Two of these later acquisitions were postulated to have been direct derivatives of α-amino-butyric acid. As in group II, one acquired a guanido radical on the δ-carbon of that substance to produce arginine, but the codons for this new amino acid appeared to have been acquired by sharing with serine (Figure 6.7). The source of the methylamine residue of lysine, which amino acid was suspected also to have been derived from α-

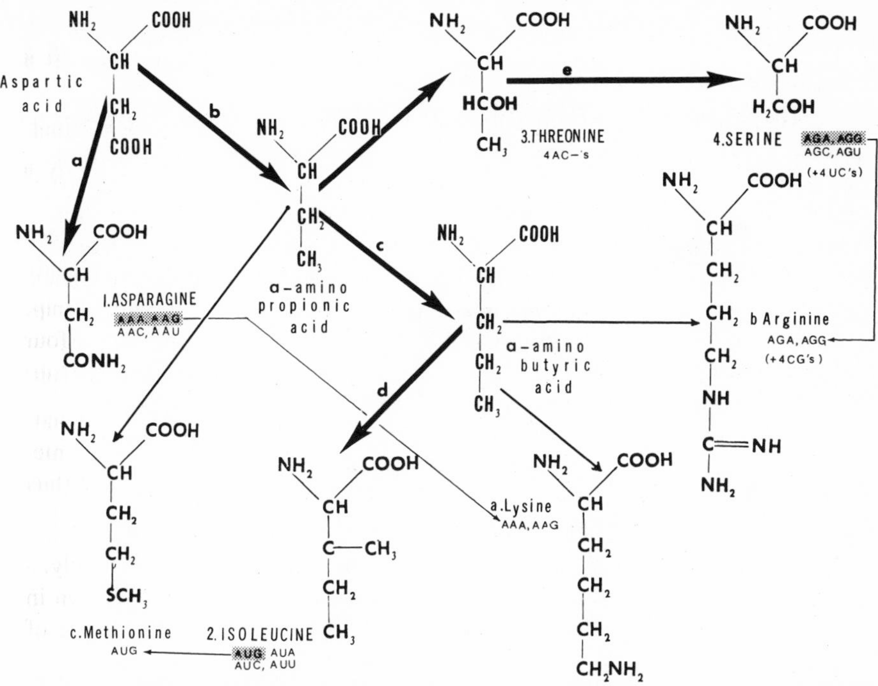

Figure 6.7. Group III's complete complement of amino acids. Far greater complexity of chemical relationships and of codon sharing is evident among the members of this group than among those of groups I and II. All have codons, however, commencing with the base A.

amino-butyric acid, was speculative, for no similar radical is found elsewhere among amino acids. The only evident route in the present context would have been by addition of a CO_2, reduction to a methyl radical, and amination. But this would have required complex enzymatic actions, for lysine in eukaryotic cells is among the amino acids least active in transamination processes (West and Todd, 1961). Therefore it was postulated that this substance was among the later components of the group. Consequently, it was supposed to have received a pair of codons from asparagine, rather than the latter's having been the recipient of codons from it.

The source of the terminal radical of the methionine molecule also had to remain speculative, although the remainder appeared to be directly from α-amino-propionic acid. Perhaps, it was suggested, a sulfhydryl radical became attached by enzymatic processes to the γ-carbon of the parent aspartic acid molecule, after which a methyl residue was substituted for the hydroxyl component of the sulfhydryl. At any rate, the late origin advocated for it appeared justifiable, because its manner of receiving a codon assignment by sharing (in this case with isoleucine) is unique. In place of the usual natural pair terminating in purines involved in most cases of sharing, it obviously received only a single codon, that one ending in G (Figure 6.7).

The origins proposed above for two amino acids, arginine and serine, need further consideration. A metabolic pathway for the first of these had already been suggested under the phylogeny of group II. As a member of that group, however, arginine was suggested to have become encoded by a set of four codons beginning with cytosine, the initial base of all codons in that subdivision. But in addition to that set, this amino acid also is encoded by the natural pair AGA and AGG. Thus it was proposed that it arose along two metabolic pathways, in one case using glutamic acid as its source so that it thus became a member of group II, and in the other aspartic acid, and thereby it became also a member of group III. In each case it was encoded appropriately.

This hypothesis receives some support from tRNA structure. As shown in detail in Chapter 7, the isoaccepting tRNAs for each comparable instance of double coding, as far as known, fall into two major groups on the basis of base-sequence studies. Undoubtedly the provision of separate sets of tRNAs also permits cellular control over the metabolic path of biosynthesis of these few amino acids. The second amino acid mentioned above, serine, represents another instance of this type, but had its primary origins in group IV. Its codon assignment, nevertheless, indicated that it, too, has had (and may still have in some organisms) a metabolic origin involving the demethylation of threonine.

Phylogeny of Group IV Amino Acids. The second proposed origin of serine was directly from alanine, the source compound of all the members of group IV. Throughout the present series, quite in contrast to group III, the processes are straightforward, serine for example being derived from alanine simply by substitution of a hydroxyl for a hydrogen residue (Figure 6.8). From this amino acid a second one seemingly was derived by the equally simple device of replacing the hydroxyl radical of serine with a sulfhydryl, thereby yielding cysteine.

On the basis of the coding assignments, phenylalanine was assumed to have been the first amino acid synthesized in this group, for UUU is among its codons. Its derivation from alanine was suggested to have involved the substitution of a phenol residue for a hydrogen in the β-methyl radical of alanine. From this substance, in turn, the fourth original member of this subdivision, tyrosine, appeared to have had origin by the equally simple process of substituting a hydroxyl radical for a hydrogen on the phenylalanine's aromatic residue. Hence two series of substitutions were deemed to have led to the creation of the four original members of the group, each of which would then have been provided with sets of four codons beginning in U.

The processes of sharing these early codon complements with later additions to the group were in several instances far from elementary or direct, and involved the introduction of a refinement into the codon catalog. The two

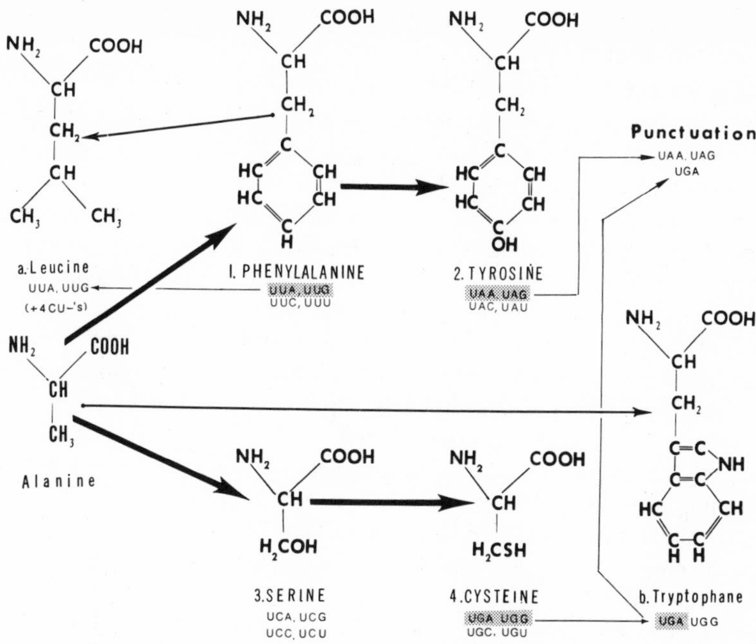

Figure 6.8. Chemical evolution among group IV amino acids. While the chemical relations between such members as phenylalanine and tyrosine on one hand and between serine and cysteine on the other are clear-cut, the coding interrelationships are exceedingly complex. All of the codons commence with the base U.

derived amino acids that arose later also were considered to have been formed by less direct procedures. Enzymatic fracturing of the phenol ring of phenylalamine, comparable to the splitting of the hexose ring during glycolysis, was claimed to represent a possible biosynthetic path for leucine. Its codons also appeared to have been derived from the same amino acid. Leucine thus came to resemble serine and arginine in being encoded by six codons, including as a member of group II a set of four with C as the initial letter, and in the present group the pair beginning with U donated by phenylalanine. The second later amino acid, tryptophan, appeared to have developed through the coupling of an indole radical to alanine, after a hydrogen had been removed from the latter's β-methyl residue. Its original codon assignment of a natural pair was speculated to have been received from cysteine (Figure 6.8).

Still later, the introduction of a refinement into the genetic coding system brought the code catalog to its present state of completion. This new feature, stops (that is, punctuation, often referred to as "nonsense"), was indicated to have come into existence along two pathways. The first of these involved the setting aside of a natural pair for terminating purposes which was derived from the set of four originally provided tyrosine. Finally, UGA, one of the two codons tryptophan had received from cysteine, also went for the same function.

Support for the proposed derivation of termination codons was presented, based on certain mutations in yeast and bacteria. As outlined in the paragraph above, the three stop triplets were postulated to have been derived from two predecessors, UGA from tryptophan, which had received it along with UGG from cysteine, and UAA and UAG from tyrosine, by the usual sharing processes (Figure 6.8). Several strains of *E. coli* have been reported that are marked by having suppressors of UGA in which that triplet in messages fails to serve as a terminator (Sambrook *et al.*, 1967; Zipser, 1967; Chan and Garen, 1969). In one such strain it has been demonstrated that suppression of termination by this codon was mediated by a tRNA which inserted tryptophan into the growing peptide chain wherever UGA appeared (Model *et al.*, 1969; Chan and Garen, 1970; Chan *et al.*, 1971; Hirsch, 1971). Thus it was suggested that the suppressor mutation had reversed the evolutionary processes to a degree by restoring to tryptophan one of its original pair of codons, as advocated above.

Comparable mutations are known in *E. coli* and *Saccharomyces cerevisiae* which suppress ochre (UAA) and amber (UGA) terminator codons. In certain of these mutants, tyrosine is inserted in growing peptide chains wherever those terminator codons occur (Goodman *et al.*, 1968; Smith *et al.*, 1968; Bruenn and Jacobson, 1972). Hence, again it was proposed that the mutations had similarly effected an evolutionary reversal by restoring codons to tyrosine which had originally encoded that amino acid. Further evidence favoring this concept was derived from unique structural similarities of yeast $\text{tRNA}^{\text{Amber}}$ to tRNA^{Tyr}; the same type of highly modified adenosine occurs in the identical location of site 39 of both tRNA molecules (Goodman *et al.*, 1968, 1970; Hashimoto *et al.*, 1969; Bruenn and Jaobson, 1972).

6.4.5. Micromolecular Evolution

The concept that the macromolecules of living things have evolved is today a widely accepted principle, substantiated by ample evidence in the form of a diversity of proteins. As this evolutionary picture of proteins has been alluded to on a number of occasions in earlier chapters, additional documentation of the principle is unnecessary. Further, if the proteins have evolved, then the nucleic acids encoding them also have come into existence in similar evolutionary fashion, for they of necessity must exhibit comparable homology. The base sequences and structure of tRNAs, discussed in Chapter 7, likewise substantiate macromolecular evolution, as do the base sequences of RNAs from such viruses as bacteriophages f2, MS2, and R17 (Steitz, 1969; Nichols, 1970; Jeppesen *et al.*, 1970; Min Jou *et al.*, 1972; Contreras *et al.*, 1973).

The evidence revealed by the analysis of the genetic code and model for the origins of amino acids thus merely extends this principle to the micromolecules from which many of the macromolecules are constructed. Indeed, in retrospect it is surprising that such a widely accepted and publicized principle as macromolecular evolution was not extended earlier in this fashion, for the common structural features of amino acid molecules are so pronounced that their categorization as homologous is self-evident. And every biologist knows that the presence of homology implies common descent. What the foregoing analysis does is simply to propose a series of routes intimating how these amino acids and codons may have come into existence. The proposed steps of origin may occasionally prove erroneous but the principal conclusion that amino acids and the genetic code coevolved seems irrefutable.

As pointed out elsewhere (Dillon, 1973), the homologies of molecular structure of certain other organic substances also strongly suggest that the various hexoses and comparable sugars likewise have come into existence by way of evolutionary processes. The bases of nucleosides, too, show pronounced structural similarities that imply common descent. Thus not only the macromolecules of living things, but all their basic micromolecules also have arisen and subsequently diversified by means of evolutionary processes.

At the moment this extension of the principle of evolution to the fundamental molecules of life only adds emphasis and complexity to the paradox mentioned previously. If a genetic apparatus is essential for the inheritance of features acquired by natural selective processes and such apparatuses are found only in living things, how could the very molecules out of which living things are constructed possibly have evolved into existence? Thus the substances on which life depends are suggested in this concept to have been developed within the living systems themselves! Nevertheless, if this conclusion is sound, as the homology of structure found among sugars, bases, and amino acids seems to substantiate, how can continued devotion to experimentation on coacervates, microspheres, or other laboratory products provide the solution to this paradox when only actual living things hold the answer?

6.4.6. Comparison of the Four Groups

Before providing additional evidence from tRNAs supporting the foregoing model of amino acid and codon evolution, some data substantiating the proposed sequence of events can be offered by comparison of the features of the four amino acid groups. These characteristics are summarized in Table 6.1.

In reviewing the groups, it should first be recalled that the only reasonable way of forming the existing pattern of codon assignments appears to be by providing 16 amino acids with sets of four (p. 139), followed by sharing of those original sets with other amino acids (Jukes, 1966, 1971; Dillon, 1973). Accordingly, four amino acids would have been encoded with codons bearing the initial letter G, four others with those beginning with C, and so on, which

Table 6.1
Characteristics of the Four Amino Acid Groups

		Members										
Group	Initial base of codons	Total number	With sets of four codons	With pairs of codons	With a single codon	In a second group	With a second $-NH_3$	With -OH	With -S	With aromatic residues	With hetero-cyclic residues	Nonsense codons
I	G	5	3	2	0	0	0	0	0	0	0	0
II	C	5	3	2	0	2	2	0	0	0	1	0
III	A	7	1	5	2	2	3	2	1	0	0	0
IV	U	7	1	5	2	2	1	2	1	2	1	3

in the present approach were considered to be the original components of four groups distinguished in that fashion. The simplest condition to be found in these groups accordingly would be indicated by the least extensive change from that original arrangement, and contrariwise, greatest evolutionary advancement would be represented by the most pronounced or numerous changes from the hypothetical ancestral condition.

On this basis, group I certainly appears to be the earliest, for it embraces one of the smallest assemblages of members (five) and has diverged only slightly from the original arrangement. Of its five components, three are encoded by sets of four, so that subdivision of an original complement has occurred only once, between aspartic and glutamic acids. Furthermore, none of its membership occurs in a second group, and the chemical composition is simple and repetitive throughout. No hydroxyl, sulfhydryl, or second amino radical occurs, nor is there any chemical complexity of any sort (Table 6.1). The carbon skeleton seems merely to have undergone successive steps of elongation (Figure 6.4).

Group II also qualifies as one of the simpler groups in including only five components, three of which are encoded by sets of four codons and the remaining two by pairs. Its members, however, indicate a greater degree of complexity for the group as a whole, because two of its components are also members of a second group. The molecular structure, too, is rather more diversified than that of group I, in that two members have a second amino radical and another contains a heterocyclic component (Table 6.1).

Above the level represented by the preceding amino acids, there can be little doubt that a higher degree of complexity exists. As Table 6.1 demonstrates, the two remaining groups show increasing degrees of advancement. Although they show many identical conditions, group IV is clearly more diversified. Thus the two groups are similar in the number of member amino acids (seven each) and in having only one member provided with a set of four codons. Each contains five that are coded by pairs and two by singles, and each shares two members with other groups. In molecular structure, both groups include amino acids having a second amino radical and others possessing hydroxyl or sulfur-containing residues. Additionally, group IV displays still greater advancement in having one heterocyclic and two aromatic residues and in including the nonsense or termination codons among its representatives. Therefore it appears that the sequence of origin proposed originally for the four amino acid groups (Dillon, 1973) is supported by their structural features, a conclusion that is evidenced also by the analysis of transfer RNA characteristics that follows in the next two chapters.

7

The Transfer Ribonucleic Acids

If not among all the universal components of organisms, then among the nucleic acids at least, the transfer RNAs are unique in having both a relatively constant molecular size and configuration. These consistencies are even more noteworthy because the class of substances (amino acids) they transport is totally lacking in both traits. Thus contrary to what might be expected from physicochemical considerations, no molecular relationships are in evidence between the carrier and transported compounds, a point that merits serious attention as the discussion proceeds. This chapter principally compares the characteristics of the base sequences of over 95 different species of tRNA. These results are then analyzed from an evolutionary standpoint in Chapter 8 in order to seek evidence relevant to the origin of life.

7.1. THE CHARACTERISTIC MOLECULAR FEATURES OF tRNAs

As mentioned above and described briefly in Chapter 4, the two consistent features of tRNAs are their small size, indicated by a sedimentation factor of 4 S, and a secondary configuration in the form of the familiar cloverleaf (Figure 4.2). Furthermore, the molecules, individually and collectively, contain a greater variety of modified bases than any other type of nucleic acid.

Functions of tRNAs. Another unusual feature of tRNAs that will be recalled is diversity of function (Griffin *et al.*, 1976). Although as vital to the genetic processes as are the other RNAs, tRNAs are active also in the formation of bacterial cell walls (p. 143) through peptidoglycan synthesis (Matsuhashi *et al.*, 1967; Petit *et al.*, 1968; Roberts *et al.*, 1968a,b; Stewart *et al.*, 1971; Roberts, 1972), and also participate in the soluble protein-synthesizing system

(p. 202); Kaji *et al.*, 1965a,b; Leibowitz and Soffer, 1969). Moreover, they are active in the regulation of amino-acid-biosynthetic operons (Schlesinger and Magasanik, 1964; Neidhardt, 1966; Roth and Ames, 1966; Silbert *et al.*, 1966), in the synthesis of aminoacyl-phosphatidylglycerol (Gould *et al.*, 1968; Nesbitt and Lennarz, 1968), and in the modification of lipopolysaccharides (Gentner and Berg, 1971). Still another unexpected function for these nucleic acids has been discovered recently—a $tRNA^{Trp}$ appears to be employed by the Rous sarcoma virus as the primer for reverse transcriptase activity (Maugh, 1974). In the *E. coli* soluble system of proteinogenesis, leucine apparently is carried by a special type of tRNA (Rao and Kaji, 1974), much as the glycine used in peptidoglycan synthesis has its particular species for transport (Roberts *et al.*, 1968a,b).

The Numbers of tRNAs. According to Crick (1966), a maximum of three different tRNAs should be required to provide completely for any species of amino acid that is encoded by four codons, and this number has proven to be the case with the isoaccepting species for glycine in *E.coli*. Of these, $tRNA_1^{Gly}$ responds only to the codon GGG, $tRNA_2^{Gly}$ to GGA and GGG, and $tRNA_3^{Gly}$ to GGC and GGU (see Table 7.11). Since in this example $tRNA_1^{Gly}$ is redundant, two different isoacceptors for a given amino acid could well suffice, and in those varieties encoded by but one or two codons, only a single species should be necessary. Thus while fewer than 40 tRNAs per organism should be more than sufficient, over 60 are known to exist in many forms of life (Söll, 1971; Yang and Novelli, 1971).

The excess number seems to stem from the large variety of isoaccepting species provided for certain amino acids (Jacobson, 1971). In *Drosophila*, serine, which is encoded by six codons, has been shown to be accepted by five major and at least two minor species of tRNAs (White *et al.*, 1975). When the coding properties of the major types were studied, both species 2 and 5 were found to respond to AGU and AGC, and both 4 and 7 responded only to UCG. In *E. coli*, leucine is carried by at least five isoaccepting species (Holmes *et al.*, 1975), and a similar multiplicity of isoaccepting forms specific for lysine has been found in chick embryos. In earlier studies of $tRNA^{Lys}$s, only two isoaccepting species had been found, as might be expected for this amino acid which is encoded only by AAA and AAG (Nishimura and Weinstein, 1969; Liu and Ortwerth, 1972; Woodward and Herbert, 1972; Pearson *et al.*, 1973). Refined techniques, particularly the use of benzoylated DEAE-cellulose (Gillam *et al.*, 1967) and reversed-phase chromatography (Pearson *et al.*, 1971), have now permitted the separation and purification of four isoaccepting species (Wittig *et al.*, 1973), but their codon recognition properties have not been completed as yet. Even the total tRNA extract from mitochondria, the DNA of which contains at most only 12 cistrons for tRNAs (Wu *et al.*, 1972), shows a comparable multiplicity of isoaccepting species for certain amino acids. This has been especially clearly demonstrated for methionyl-accepting types, which in mouse liver mitochondria include two major and two minor varieties (Wallace and Freeman, 1974).

Problems with numbers of isoaccepting species in the opposite sense also appear to exist, for rather than a plethora of forms, a deficiency may prevail in some organisms. In *Physarum,* for example, only a single tRNA was detected for several amino acids that are provided with four codons each, including glycine, valine, and alanine (Melera *et al.*, 1974). A total of only 44 species of tRNAs was found in this organism during the growth-phase mitotic cycle.

Methods Employed for Comparisons. To facilitate comparisons, the numerous nucleoside sequences now established for tRNAs are aligned in Tables 7.1–7.7 following the procedures of Dayhoff (1969) and Cedergren *et al.* (1972). Such obviously related features as the several bends, the GG and TψC sequences of loops II and V, respectively, and the anticodon are placed in corresponding vertical columns. To bring out homologies with the longer sequences to the maximum extent possible, gaps are inserted where needed in the shorter ones. Furthermore, the species are arranged by the groups proposed in the preceding chapter. As advocated by Cedergren *et al.*, isoaccepting tRNAs are identified on the basis of anticodon traits, rather than by their eluting characteristics; however, the system proposed in that paper for naming homologous types is abandoned in favor of the more familiar one of current usage. Thus major varieties are designated by subscript numerals, minor types by capital letters, and mutants by various subscripts, depending on the nature of the mutation.

Another device to make meaningful comparisons of such a large number of sequences is used here for the first time, that of devoting attention to one major region of the molecule at a time, beginning with the amino acid arm (Figure 4.2). In the analysis of each such region, the bends between arms are generally treated in turn together with the respective preceding portion. As these are examined, a search is made for sites of possible value in ligase recognition wherever appropriate, including pertinent suggestions made in the literature, but this topic receives fuller attention in the following chapter. It is interesting to note in this connection that, while extensive speculation exists regarding the nature of recognition of the various tRNA species by the ligase, no attention appears to have been given to ligase recognition of amino acids. Yet it is self-evident that comprehension of one of these interrelated phenomena without similar understanding of the other is virtually devoid of real significance. The occasional occurrence of mischarging amply testifies that the problem of amino acid recognition is a valid one for further investigation.

7.1.1. The Amino Acid Arm (Arm I)

In the chart of the amino acid arm sequences (Table 7.1), the triplet –CCA is omitted from the 3′-terminus, because it is universally present. This catholic-

Table 7.1
The Amino Acid Arm of Sequenced tRNAs[a]

	5′ Half 01234567	Bend 1 AB	3′ Half 7′6′5′4′3′2′1′	Preterm. base[b] 0′
GROUP I				
Glycine				
Salm$_1$	-p*GCGGGCG*	UA	*CGCCCGC*	U
Esch$_1$	-p*GCGGGCG*	U$_t$A	*CGCCCGC*	U
T2$_2$, T4$_2$, T6$_2$	-p*GCGGAUA*	UC	*UAUCGGC*	U
Esch$_2$	-p*GCGGGAG*	UC	*UGCCCGC*	U
Staph$_{2A,2B}$	-p*GCGGGAG*	U$_t$A	*CUUCCGC*	U
S. aur$_{2A,2B}$	-p*GCGGGAG*	U$_t$A	*CUUCCGC*	U
Esch$_{3,ins}$	-p*GCGGGAA*	UA	*UUCCCGC*	U
Sac$_3$	-p*GCGC$_m$AAG*	UG$_1$	*CUUGCGC*	A
Wheat$_{3A}$	-p*GCAC′$_2$CAG*	UG$_1$	*CUGGUGC*	A
Wheat$_{3B}$	-p*GCAC′$_2$CAG*	UG$_1$	*CUGGUGC*	A
Alanine				
Esch$_{1A}$	-p*GGGGGCA*	U$_t$A	*CGCUCCC*	A
Sac	-p*GGGCGUG*	UG$_1$	*CUCGUCC*	A
Tor	-p*GGGCGUG*	UG$_1$	*CUCGUCC*	A
Aspartic acid				
Esch$_1$	-p*GGAGCGG*	U$_t$A	*CCGUUCC*	G
Brew$_1$	-p*UCCGUGA*	ŪA	*UCGCGGA*	G
Glutamic acid				
Esch$_2$	-p*GUCCCCU*	UC	*AGGGGAC*	G
Sac$_3$	-p*UCCGAUA*	UA	*UAUCGGA*	G
Valine				
Salm$_1$	-p*GGGUGAU*	U$_t$A	*AUCACCC*	A
Esch$_1$	-p*GGGUGAU*	U$_t$A	*AUCACCC*	A
Esch$_{2A}$	-p*GCGUCCG*	U$_t$A	*CGGACGC*	A
Esch$_{2B}$	-p*GCGUCCA*	U$_t$A	*UGAACGC*	A
B. stear	-p*GAUUCCG*	UA	*CGGAAUC*	A
Sac	-p*GGUUUCG*	UG$_1$	*CGAAAUG*	A
Tor	-p*GGUUUCG*	UG$_1$	*CGAAAUC*	A
Brew	-p*GGUUUCG*	UG$_1$	*CGAAAUC*	A
Mus	-p*GUUUCCG*	UA	*CGGAAAC*	A

Table 7.1. Cont'd.

	5′ Half 01234567	Bend 1 AB	3′ Half 7′6′5′4′3′2′1′	Preterm. base[b] 0′
GROUP II				
Leucine				
$Esch_1$	-p*GCGAAGG*	UG	*CCCUCGC*	A
$Salm_{1,mu}$	-p*GCGAAGG*	UG	*CCCUCGC*	A
$Esch_2$	-p*GCCGAGG*	UG	*CCUCGGU*	A
Histidine				
$Salm_1$	p*GGUGGCUA*	U_tA	*UAGCCAC*	C_M
$Salm_{mu}$	p*GGUGGCUA*	U_tA	*UAGCCAC*	C_M
$Esch_1$	p*GGUGGCUA*	U_tA	*UAGCCAC*	C_M
Glutamine				
$T4_{1,psu}$	-p*UGGGAAU*	U_tA	*AUUCCCA*	G
$Esch_1$	-p*UGGGGUA*	U_tC	*UACCCCA*	G
$Esch_2$	-p*UGGGGUA*	U_tC	*UACCCCA*	G
Arginine				
$Esch_{1A,1B}$	-p*GCAUCCG*	U_tA	*CGGAUGC*	A
$Brew_2$	-pψ*UCCUCG*	UG_1	*CGGGGAA*	G
Proline				
T2, T4	-p*CUCCGUG*	UA	*UAUGGAG*	A
GROUP III				
Isoleucine				
$Esch_1$	-p*AGGCUUG*	UA	*CAGGCCU*	A
$Esch_2$	-p*GGGCUUG*	U_tA	*CAGGCCU*	A
Tor	-p*GGUCCCU*	UG	*AGGGACC*	A
Threonine				
Esch	-p*GCUGAUA*	UA	*UAUCAGC*	A
Serine				
$Esch_3$	-p*GGUGAGG*	U_tG[c]	*CCUCACC*	G
Rat_3	-p*GACGAGG*	UG	*CCUCGUC*	G
Arginine				
$Brew_3$	-p*GCGCUCG*	UG_1	*CGAGUGC*	G
$Brew_{3A}$	-p*GCGCUUG*	UG_1	*CGAGUGC*	U

Table 7.1. Cont'd.

	5′ Half 01234567	Bend 1 AB	3′ Half 7′6′5′4′3′2′1′	Preterm. base[b] 0′
Lysine				
Sac_2	-pψ*CCUUGU*	UA	*AψGAGGA*	G
$Sac_{\alpha s}$	-p*GCCUUGU*	UG_1	*ACAGGGC*	U
Methionine				
$Esch_M$	-p*GGCUACG*	U_tA	*CGUAGCC*	A
Ana_F	-p*CGCGGGG*	UA	*CCCCGCG*	A
$Esch_{F1,F3}$	-p*CGCGGGG*	U_tG	*CCCCGCA*	A
Bac_F	-p*CGCGGGG*	UG	*CCCCGCA*	A
Sac_F	-p*AGCCGCG*	UG_1	*CGCGGCU*	A
Mus_F	-p*AGCAGAG*	UG_1	*CUCUGCU*	A
Rab_F, $sheep_F$	-p*AGCAGAG*	UG_1	*CUCUGCU*	A
$Brew_3$, Sac_M	-p*GCUUCAG*	UA	*CUGGAGC*	A
Mus_4, Rab_4	-p*GCCUCGU*	UA	*ACGGGGC*	A
(Asparagine)				
GROUP IV				
Phenylalanine				
$Esch_1$	-p*GCCCGGA*	U_tA	*UCCGGGC*	A
B. stear	-p*GCUCGUG*	UA	*CGCGAGC*	A
Brew	-p*GCGGAUU*	UA	*AAUUCGC*	A
Sac	-p*GCGGAUU*	UA	*AAUUCGC*	A
Tor	-p*GCGGAUU*	UA	*AAUUCGC*	A
Wheat, pea	-p*GCGGGGA*	UA	*UCACCGC*	A
Rab, calf	-p*GCCGAAA*	UA	*UUUCGGC*	A
Tyrosine				
$Esch^+{}_{su}$	-p*GGUGGGG*	U_tU_t	*CCCCACC*	A
$Esch_1$	-p*GGGUGGG*	PU	*CCCACCC*	A
$Esch_2$	-p*GGUGGGG*	U_tU_t	*CCCCACC*	A
Sac_2	-p*CUCUCGG*	UA	*CCGGGAG*	A
Tor_2	-p*CUCUCGG*	UG_1	*CCGAGAG*	A

Table 7.1. Cont'd.

	5′ Half 01234567	Bend 1 AB	3′ Half 7′6′5′4′3′2′1′	Preterm. base[b] 0′
Serine				
T4	-p*GGAGGCG*	UG	*CGCCUCC*	G
$Esch_1$	-p*GGAAGUG*	U_tG	*CGCUUCC*	G
$Brew_{1A}$	-p*GGCAACU*	UG	*AGUUGUC*	G
$Brew_{1B}$	-p*GGCAACU*	UG	*AGUUGUC*	G
Rat_1	-p*GUAGUCG*	UG	*CGACUAC*	G
Rat_{2A}	-p*GUAGUCG*	UG	*CGACUAC*	G
Cysteine				
Sac	-p*GCUCGUA*	UG	*UGCGAGC*	U
Tryptophan				
Esch	-p*AGGGGCG*	U_tA	*CGCCCCU*	G
Rous	-p*GACCUCG*	UG_1	*CGGGGUC*	A
$Brew_{1A,1B}$	-p*GAAGCGG*	UG_1	*CCGUUUC*	A
Leucine				
T4	-p*GCGAGAA*	UG	*UUCUCGC*	A
Sac_3	-p*GGUUGUU*	UG	*AGCAACC*	A

[a]Italics indicate double-stranded regions. Bases with subscript numerals are methylated; subscript t or s, thiolations; m, methylation; a, acetylation; superscript 2, double modification M, unknown modification. For other conventions, see text. Ana, *Anacystis;* B. stear, *Bacillus stearothermophilus;* Bac, *Bacillus subtilis;* Brew, brewer's yeast; Esch, *Escherichia coli;* Mus, house mouse; Rab, rabbit; Rous, Rous sarcoma cells; S. aur, *Staphylococcus aureus;* Sac, baker's yeast; Salm, *Salmonella;* Staph, *Staphylococcus epidermidis;* T2, T4, T6, bacteriophage T2, etc; Tor, *Torulopsis utilis.*

[b]Followed by —CCA.

[c]Ish-Horowicz and Clark (1973) show an unmodified U in site A of band 1.

ity appears to derive from function (Tal *et al.*, 1972; Sprinzl and Cramer, 1973; Chinali *et al.*, 1974), for it is the end of this triplet that receives the amino acid and unites with the A site of the ribosome.

General Structure of Arm I. The amino acid arm is strikingly uniform in its pattern of organization, for it almost always consists of seven pairs of nucleosides, plus a solitary one before the –CCA terminus. Only three species are exceptional, all of which are $tRNA^{His}$s from *Salmonella* and *Escherichia*; these three have eight pairs of nucleosides and no unpaired site preceding the –CCA (Harada *et al.*, 1972; Singer and Smith, 1972; Singer *et al.*, 1972). Another universal feature is the unpaired region consisting of two nucleosides (bend 1), which follows the 5′-strand.

For ease of discussion, a new method of numbering sites is employed, the importance of which in avoiding confusion is especially clear in comparisons of this arm. Under the usual system, the base of site 1 of arm I may be said to pair with that in site 76 in one species, with that of site 83 in a second, and with that of site 79 in a third, depending upon the total number of nucleosides in the particular species of tRNA. In the present system, the two sites for each pair of nucleosides are assigned the same number, with that of the component nearer the 3′-end of the molecule distinguished by means of a prime. Thus the first pair of sites is always 1–1′, the second is 2–2′, and so on, regardless of the total length of the given molecule. In the unusual species mentioned above, the very first pair is designated as 0–0′ because it obviously results from pairing with the usually solitary preterminal base that occupies site 0′. In bends and those arms that contain loops, the unpaired members are assigned capital letters on a similar basis. In studies in which several regions are involved, this system can be utilized by employing combinations of the region and site number or letter. Thus site 3 of arm I would be I3, site B of bend 1 would be 1B, the postanticodon site would be indicated as IIIC, and so on. Consequently, the thymidine of the TψC arm always could be said to be in site VA, rather than in the diversity of site numbers now needed for identification in comparing several species.

The Preterminal Nucleoside. The preterminal nucleoside, which occupies site 0′ preceding the terminal triplet –CCA, is usually adenosine, as that nucleoside occurs in approximately 60% of the known sequences (Table 7.1). As guanosine ranks next in frequency, being found here in 20% of the species, the purines thus are decidedly the typical occupants. Uridine is nearly as frequent as guanosine, however, but cytidine is absent, except in modified form in the three $tRNA_1^{His}$s. This bias toward the purines, and especially that toward adenosine, would appear to mark this site as not universally involved in recognition by the ligase, for ideally a more nearly equal frequency for each of the four major bases would be expected. As shown later, the inequity results at least in part from the chemical requirements of the –CCA terminus for interaction with the amino acid.

Despite this obvious disparity, the occupant of the preterminal site has been proposed as one of a combination of points involved in recognition by the ligase. In addition to the present nucleoside, the proposed combination included the locus B of bend 1 and a certain sequence occupying the four pairs of sites located in the stem of arm II (i.e., II 1–1′ to II 4–4′). The evidence presented in support of the proposal was as follows: Yeast $tRNA^{Phe}$-ligase had been found capable of correctly charging the corresponding tRNAs of yeast, wheat, and *E. coli* and of mischarging *E. coli* $tRNA^{Val}$ with phenylalanine (Dudock *et al.*, 1969, 1970, 1971; Taglang *et al.*, 1970). Further experimentation showed that this same ligase could also mischarge a number of other species of tRNAs from *E. coli* with phenylalanine, including those for $valine_{2A}$, $valine_{2B}$, $methionine_{M}$, isoleucine, $alanine_1$, $alanine_2$, and lysine (Roe and Dudock, 1972). Comparisons of the base sequences of those tRNAs (except those of alanine) showed that each carried adenosine, both in site 0′ and in 1B. Moreover, each had the sequence –GCUC on the 5′-side of the stem of arm II coupled to –CGAG on the 3′-side. It was further stated that this yeast ligase had an absolute requirement for adenosine in I-0′, the preterminal locus. Hence in yeast cells, the ligase was suggested to require the cited combination of nucleosides for the recognition of yeast $tRNA^{Phe}$. However, since the yeast ligase was able to react with a wide variety of *E. coli* tRNAs containing this combination of nucleosides, that set of characteristics obviously cannot be of prime importance for the recognition of $tRNA^{Phe}$ by the corresponding ligase in the bacterial cell. Unfortunately, the cited report failed to indicate which of the tRNAs tested were not mischarged by the heterologous system. Additional reference to the role of this unpaired preterminal site in ligase recognition is made after the nature of the sequence of paired nucleosides is examined.

The Paired Sequence. In examining this sequence and other paired series that follow, it is sufficient to scrutinize only one side in detail because of the usual complementarity of base pairing. For the sake of consistency, the 5′-side is arbitrarily chosen to receive the bulk of attention. Throughout that side of the amino acid arm, a marked predominance of guanosine can be noted (Table 7.1), for it is the most frequent nucleoside in all except the sixth locus. Based on comparative counts, its percentage frequency in sites 1–7 of the sequences shown are, in order, 79%, 47%, 49%, 40%, 40%, 28%, and 60%. Obviously, only the frequency in the sixth site approaches the 25% level expected on the basis of random chance. Even that extra nucleoside mentioned above as being added at the beginning of this sequence in the bacterial $tRNA^{His}$s is guanosine (Table 7.1).

Cytidine is also relatively abundant on this side of the amino acid arm, greatly exceeding the expected ratio at sites 2, 3, and 6; however, it is always absent at site 7. Its relative frequencies are, in sequence, 6%, 43%, 30%, 21%, 13%, 30%, and 0%. Thus together guanosine and cytidine constitute 70% of the nucleosides on this side of the arm. In contrast to these two, whose frequencies tend to peak at the beginning of the arm and gradually diminish toward site 7, uridine is most abundant at the central site 4 and decreases toward both ends, as shown by its relative frequencies, 4%, 10%, 17%, 28%, 21%, 23%, and 17%. Adenosine is always absent in site 2 but is less rare toward site 7, occurring with the frequencies 9%, 0%, 4%, 11%, 19%, 19%, and 23%. In addition, pseudouridine has rarely been reported on this side of the arm, being found at locus 1 of yeast $tRNA_2^{Lys}$ and $tRNA_2^{Arg}$ (Madison *et al.*, 1972, 1974; Weissenbach *et al.*, 1975a,b), and only one modified nucleoside has been reported, methylated cytidine in eukaryotic $tRNA_3^{Gly}$ (Marcu *et al.*, 1973; Yoshida, 1973).

On the 3′-side of the arm, aside from the expected high frequency of cytidine, one salient feature is the scarcity of pseudouridine, which occurs solely at site 2′ in the yeast $tRNA_2^{Lys}$ noted above as having this same nucleoside at locus 1. Uridine is unexpectedly present in four tRNAs (Table 7.1) at site 2′, in each case being paired with guanosine.

Use in Ligase Recognition. The percentage frequencies of occurrences noted above strongly intimate that constraints against random substitution of nucleosides are actively operative in this arm. Omitting the few minor bases from consideration, the results tabulated show that only 8 of the 28 possible frequencies (4 bases × 7 sites) lie between 20% and 30%, that is, within 5% of the 25% frequency expected on a random basis. The constraints are so severe that at some sites certain nucleosides are totally absent and at several others they are virtually so. This situation is in complete contrast to what would be expected were a given region or the whole of the paired sequence consistently employed in recognition by the respective ligases. Were such a use widespread, then each base would be of nearly equal frequency at any given point of the active region in order to meet coding requirements, as they are in the codon catalog, for example.

The difficulties militating against firm proposals for a role in recognition by this arm are best brought out by an examination of the possibilities that exist. Since 13 species of $tRNA^{Gly}$ have been sequenced from a diversity of sources (Table 7.1), that group of isoaccepting forms appears ideal for illustrative purposes. At once it can be noted that, in the amino acid arm, the triplet GCG occurs nearly universally in these isoacceptors in the first three loci of the 5′-side; thus it is tempting to suggest that triplet as a recognition site, as has been done (Hill *et al.*, 1973; Stahl *et al.*, 1973).* Elsewhere in the table, however, the same triplet can be seen to occur in the same location in a number of other species, including *E. coli* $tRNA_{2B}^{Val}$ and $tRNA_A^{Leu}$, yeast and wheat $tRNA^{Phe}$, and bacteriophage T4 $tRNA^{Leu}$. Similar difficulty is experienced with

* Until the sequences of wheat $tRNA^{Gly}$s were reported, the triplet GCG always ocupied the first three sites in the then-known isoaccepting species.

the triplet GGG found in the two $tRNA^{Ala}$s from yeast (Table 7.1), for it is present in three others of diverse types. Even in those relatively rare varieties of tRNA that have A, rather than the usual G, in site 1, no evidence of its beginning a triplet signal can be detected. For instance, the triplet AGG occurs in two nonisoaccepting species from *E. coli*, $tRNA^{Ile}$ and $tRNA^{Trp}$. No matter from what sites in this arm a triplet may be chosen, its presence also in unrelated tRNAs can soon be noted.

Nevertheless, sparse evidence does exist that supports the belief that this arm plays some role in the recognition by the ligase. In *E. coli* a number of mutant $tRNA^{Tyr}$s are known, among which are included several amber suppressors that affect the amino-acid-charging specificity. A recent series of studies described results at the molecular level with single and double mutants that became mischarged with glutamine (Celis *et al.*, 1973; Smith and Celis, 1973; Ghysen and Celis, 1974; Inokuchi *et al.*, 1974).* The sites involved in this arm were sites 1 and 2 and/or their pairing sites, namely, 1′ and 2′, and, in one instance, the preterminal locus. The results are most provocative but scarcely conclusive. Mutant $tRNA^{Tyr}_{A1}$ in which the G in site 1 (bonded to C in site 1′) was mutated to A, inserted only glutamine; in this mutation, bonding was reported not to occur between the A of site 1 and the C of site 1′ (Figure 7.1).† As this change is accompanied by a substitution of C for the usual G in site 1 of the anticodon, effects from more than this arm are involved in the results. However, when the C normally present in site 1′ was replaced by a U, resulting in a G–U pair, glutamine was inserted instead of tyrosine, but only to a level of 20%. In a double mutant in which the entire first pair (G–C) was replaced by A–U, as in $tRNA^{Tyr}_{A_1U_{81}}$, glutamine was inserted to a level of 30%.

When the second pair in the sequence, also G–C, likewise underwent a similar double mutation to A–U (as in $tRNA^{Tyr}_{A_2U_{80}}$), no glutamine, but only tyrosine, was inserted. In a fifth mutant (G_{82}), in which the usual A of the preterminal site was replaced by a G, 100% glutamine was inserted. While these results thus intimate a role for this stem in recognition by the ligase, perhaps in combination with the anticodon (or perhaps just the base in the middle of the anticodon), it is difficult to perceive any absolute effect. The illustration in Figure 7.1 presents the normal tRNAs and the mutant amino acid stems. In each case a constant feature of the tRNAs which has not received notice previously also is indicated; this feature is absent from all other species of tRNAs which have been sequenced to date. Also related to the foregoing discussion is the ev-

* A somewhat parallel case has been reported for *E. coli* $tRNA^{Met}_{F}$, in which modification of guanosine was produced at a number of sites. In this arm only that at site 3′ influenced methionine-accepting activities; modification at site B of arm IV had a comparable result, however (Schulman, 1972).

† It should be noted that C–A bonds do occur, as described later when codon–anticodon interactions are described (p. 288).

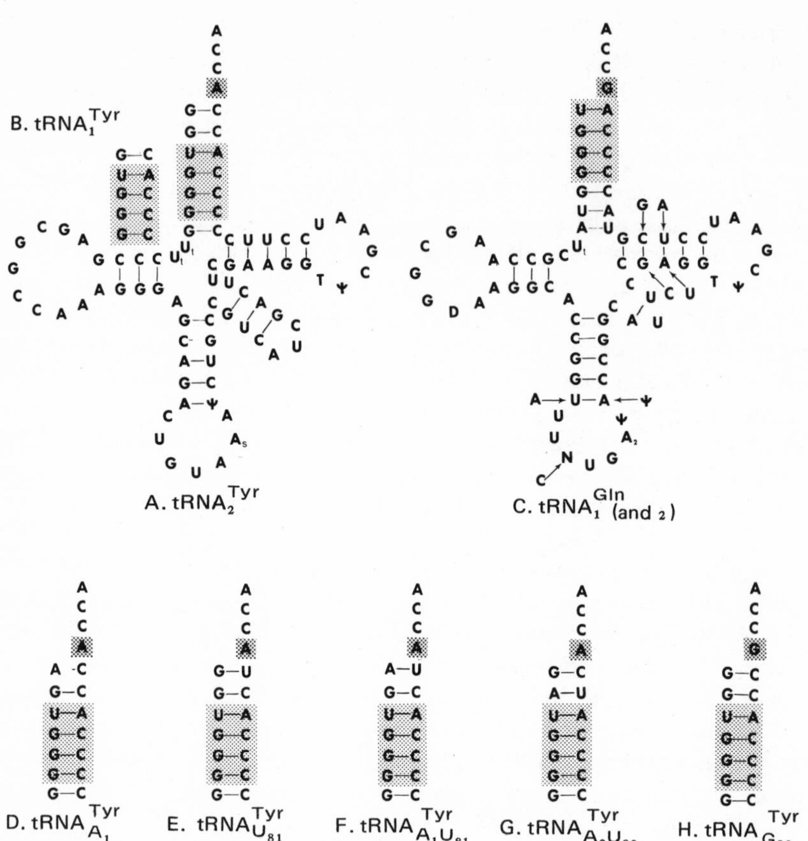

Figure 7.1. Normal and mutant tyrosine tRNAs. Some evidence exists which suggests that arm I of tRNAs may serve for recognition by the ligase. (A, B) Normal tRNA^{Tyr}s share a four-site sequence with normal tRNA^{Gln} (G) but the preterminal site nucleotides differ. (D–H) The five mutant tRNA^{Tyr}s differ in their recognition properties. D and H are charged only with glutamine, whereas G receives only tyrosine; 20% of the E type and 30% of F receive glutamine, the remainder being charged normally (with tyrosine).

idence which has been presented against the involvement of the anticodon in ligase recognition (Bhargava *et al.*, 1970), but this is a controversial point at present.

Certainly the foregoing mutant series indicates that the preterminal base plays an important role in recognition in some instances, and that the amino acid arm may do likewise in others. Whole regions, rather than single pairs, and combinations of regions with various other points, seem involved in the ligase-recognition reactions. But it appears clear that Chambers's (1971) conclusion, that structural analysis of isoaccepting tRNAs is of dubious value in identifying recognition sites, may no longer be wholly valid.

Bend 1. Between the amino acid stem and arm II is a short sector here referred to as bend 1 (Table 7.1). This feature consistently embraces two sites, the first of which (A) is always occupied by uridine. This nucleoside in prokaryotes is usually modified to 4-thiouridine and in one case (*E. coli* $tRNA_1^{Tyr}$) to the unidentified base P. Being ubiquitous, the uridine could scarcely be involved in recognition; its universality rather would suggest that the site was deep in the three-dimensional molecule, in the maintenance of which it perhaps plays a vital role. The modified nucleoside of prokaryotes (4-thiouridine) has been shown to interact with adenosine less strongly than the unmodified uridine does (Geller *et al.*, 1973); hence its survival value is rather obscure.

The second site of the bend is variously occupied. Cytidine occurs in this locus B only in three groups of isoaccepting species from prokaryotes, those accepting lysine, glutamic acid, and glutamine, and uridine appears to be restricted entirely to prokaryotic $tRNA^{Tyr}$s. Hence the two purines are the typical occupants of the site. Among prokaryotes adenosine is somewhat more frequent than guanosine, with the opposite prevailing in eukaryotes. The latter also have a modification here that does not occur in prokaryotes; this 1-methylguanosine is somewhat more characteristic than is the unmodified form.

7.1.2. The Dihydrouracil Arm (Arm II)

As a whole, arm II ranks among the most variable features of tRNAs. To begin with, although a stem is always present, it varies between three or four pairs of residues in length. The less constant portion, however, is the loop of unpaired nucleosides, which may contain anywhere between seven and twelve residues (Table 7.2); among the latter may be included as many as five, or as few as none, dihydrouridines.

The Paired Stem. When the paired sequence is examined, the presence of constraints becomes immediately evident, for guanosine nearly always is the occupant of site 1 of the 5′-side (Table 7.2). Four exceptional cases can be noted, one of which having uridine here is a $tRNA_{2B}^{Gly}$ from *Staphylococcus* that does not function in protein synthesis. In eukaryote tRNAs of groups III and IV, the nucleoside is often methylated to N^2-methylguanosine, but in prokaryotes it is always unmodified. The residue to which this guanosine is paired, that occupying site 1′, is typically its usual complement, cytidine, but uridine occurs here in four of the sequences. The remaining three of the four exceptions mentioned above have the reverse of the usual arrangement, with cytidine in site 1 and guanidine in site 1′ (Figure 7.1); this condition exists solely in the several sequenced *E. coli* $tRNA^{Tyr}$s, including those involved in the mischarging described above in the amino acid arm discussion. It is not evident, how-

Table 7.2

Arm II and Bend 2 of Sequenced tRNAs[a]

	1234	ABCDEFGHIJK	4'3'2'1'	Bend #2
GROUP I				
Glycine				
$Salm_1$	*GUUC*	AA--UG_2'G-D-A	*GAAC*	G
$Esch_1$	*GUUC*	AA--UG_2'G-D-A	*GAAC*	G
$T2_2$, $T4_2$, $T6_2$	*GUAU*	AA--UG_2'G-D-A	*UUAC*	C
$Esch_2$	*GUAU*	AA--UGGCU-A	*UUAC*	C
$Staph_{2A}$	*GUUC*	AA--UUU-D-A	*GAAC*	A
$Staph_{2B}$	*UUUC*	AA-CUUU-D-A	*GAAU*	A
S. aur_{2A}	*GUUC*	AA-CUUU-D-A	*GAAC*	A
S. aur_{2B}	*UUUC*	AA-C-UU-D-A	*GAAU*	A
$Esch_{3,\ ins}$	*GCUC*	AG-DDGG-D-A	*GAGC*	A
Sac_3	*GUUψ*	AG--DGG-D-A	*AAAU*	C
$Wheat_{3A,\ B}$	*GUCψ*	AG--DGG-U-A	*GAAU*	A
Alanine				
$Esch_{1A}$	*GCUC*	AG-CDGG-G-A	*GAGC*	G
Sac[b]	*GCGU*	AG-DGCG-D-A	*GCGC*	G_2^2
Tor	*GCGU*	AG-DDGG-D-A	*GCGC*	G_2^2
Aspartic acid				
Esch	*GUUC*	AG-DCGG-DDA	*GAAU*	A
$Brew_1$	*GUUψ*	AA--DGG-DCA	*GAAU*	G
Glutamic acid				
$Esch_2$	*GUCψ*	AG--AGGCCCA	*GGAC*	A
Sac_3	*GUGψ*	AA--CGGCD-A	*UCAC*	A
Valine				
$Salm_1$	*GCUC*	AG-CDGG-G-A	*GAGC*	A
$Esch_1$	*GCUC*	AG-CDGG-G-A	*GAGC*	A
$Esch_{2A}$	*GCUC*	AG-DDGGDD-A	*GAGC*	A
$Esch_{2B}$	*GCUC*	AG-DDGGDD-A	*GAGC*	A
B. $stear_2$	*GCUC*	AG-CDGG-G-A	*GAGC*	G
Sac	*GUCψ*	AG-DCGGDD-A	*UGGC*	A
Tor	*GUCψ*	AG-DDGGDC-A	*UGGC*	A
Brew	*GUCψ*	AG-DCGGDD-A	*UGGC*	A
Mus	*GUGψ*	AG--DGGDD-A	*UCAC*	G_2^2

Table 7.2. Cont'd.

	1234	ABCDEFGHIJK	4'3'2'1'	Bend #2
GROUP II				
Leucine				
Esch$_1$	*GCGG*	AA-DDG$_2'$GDAGA	*-CGC*	G
Salm$_{1,\ mu}$	*GCGG*	AA-DDG$_2'$GDAGA	*-CGC*	G
Esch$_2$	*GUGG*	AA-DDG$_2'$GDAGA	*-CAC*	G
Histidine				
Salm	*GCUC*	AG-DDGGD--A	*GAGC*	C
Salm$_{mu}$	*GCUC*	AG-DDGGD--A	*GAGC*	C
Esch	*GCUC*	AG-DDG$_2'$GD--A	*GAGC*	C
Glutamine				
T4$_{1,\ psu}$	*GCCA*	AG-DDGGD--A	*AGGC*	A
Esch$_1$	*GCCA*	AG-C-G$_2'$GD--A	*AGGC*	A
Esch$_2$	*GCCA*	AG-C-G$_2'$GD--A	*AGGC*	A
Arginine				
Esch$_{1A}$	*GCUC*	AG-CDGGD--A	*GAGU*	A
Esch$_{1B}$	*GCUC*	AG-CDGGU--A	*GAGU*	A
Brew$_2$	*G$_2$CCC*	AA--DGGDC-A	*CGGC*	G$_2^2$
Proline				
T2, T4	*GCUC*	AGUUUGGD--A	*GAGC*	G
GROUP III				
Isoleucine				
Esch$_1$	*GCUC*	AG-GDGGDD-A	*GAGC*	G
Esch$_2$	*GCUC*	AG-GDGGDD-A	*GAGC*	G
Tor	*G$_2$CCC*	AG-DDGGDD-A	*AGGC*	G$_2^2$
Threonine				
Esch	*GCUC*	AG-DDGGD--A	*GAGC*	G
Serine				
Esch$_3$	*GCCG*	AG-A-GGCDGA	*AGGC*	G
Rat$_3$	*GCCG*	AG-DUG$_2'$GDD-A	*AGGC*	G$_2^2$
Arginine				
Brew$_{3,3A}$	*G$_2$CGU*	AA-D-GG--CA	*ACGC*	G$_2^2$

Table 7.2. Cont'd.

	1234	ABCDEFGHIJK	4'3'2'1'	Bend #2
Lysine				
Sac_2	G_2CUC	AG-DCGGD--A	GAGC	G^2_2
$Sac_{\alpha s}$	G_2CGC	AA-DCGGD--A	GCGC	G^2_2
Methionine				
$Esch_M$	GCUC	AG-DDG$'_2$GDD-A	GAGC	A
Ana_F	GAGC	AGCCUGGD--A	GCUC	G
$Esch_{F1,\ F3}$	GAGC	AGCCUGGD--A	GCUC	G
Bac_F	GAGC	AGUUCGGD--A	GCUC	G
Sac_F	C_2CGC	AG-D-GG--AA	GCGC	G^2_2
Mus_F	G_2CGC	AG-C-GG--AA	GCGU	G^2_2
Rab_F, $sheep_F$	G_2CGC	AG-C-GG--AA	GCGU	G^2_2
$Brew_3$, Sac_M	G_2CUC	AG-DAGGD--A	GAGC	G^2_2
Mus_4, Rab_4	G_2CGC	AG-DAGGD--A	GCGC	G^2_2
(Asapargine)				
GROUP IV				
Phenylalanine				
$Esch_1$	GCUC	AG-DCGGD--A	GAGC	A
B. stear	GCUC	AG-UCGGD--A	GAGC	A
Brew	G_2CUC	AG-DDGGG--A	GAGC	G^2_2
Sac	G_2CUC	AG-DDGGG--A	GAGC	G^2_2
Tor	G_2CUC	AG-DDGGG--A	GAGC	G^2_2
Wheat, pea	G_2CUC	AG-DDGGG--A	GAGC	G^2_2
Rab, calf	G_2CUC	A_1G-DDGGG--A	GAGC	G^2_2
Tyrosine				
$Esch^+su$	CCCG	AG--CG$'_2$GCCAA	AGGG	A
$Esch_1$	CCCCG	AG--CG$'_2$GQCAA	AGGG	A
$Esch_2$	CCCG	AG--CG$'_2$GCCAA	AGGG	A
Sac_2	GCCA	AG-DDG$'_2$GDDDA	AGGC	G^2_2
Tor	G_2CCA	AG-DDG$'_2$GDDDA	AGGC	G^2_2

Table 7.2. Cont'd.

	1234	ABCDEFGHIJK	4'3'2'1'	Bend #2
Serine				
T4	GCAG	AG--UG$'_2$GDDUA	AUGC	A
$Esch_1$	GCCG	AG--CG$'_2$GDDGA	AGGC	A
$Brew_{1A}$	GCC$_a$G	AG-D-G$'_2$GDD-A	AGGC	G^2_2
$Brew_{1B}$	GCC$_a$G	AG-D-G$'_2$GDD-A	AGGC	G^2_2
Rat_1	GCC$_a$G	AG-D-G$'_2$GDD-A	AGGC	G^2_2
Rat_{2A}	GCC$_a$G	AG-D-G$'_2$GDD-A	AGGC	G^2_2
Cysteine				
Sac	GCGC	AG-D-GGD--A	GCGC	A
Tryptophan				
Esch	GUUC	AA-DDGGD--A	GAAC	A
Rous	G_2CGC	AA-C-G$'_2$GD--A	GCGC	G^2_2
$Brew_{1A,\ 1B}$	G_2CUC	AA-D-G$'_2$GD--A	GAGC	ψ
Leucine				
T4	GUCA	AA-DDG$'_2$GD-AA	AGGC	A
Sac_3	G_2CC$_a$G	AG-C-G$'_2$GDDCA	AGGC	G^2_2

[a]For an explanation of conventions, see Table 7.1, footnote *a*.
[b]The sequence given here is correct according to the digest fragments reported by Dolley *et al.* (1965b).

ever, how this unusual reversal could have any bearing on the misrecognition, as the glutamine tRNAs that are also involved there have the typical arrangement of nucleosides (Table 7.2; Figure 7.1).

The second pair, which occupies sites 2–2′ in the stem, also appears to be subjected to certain constraints, as either of the two common pyrimidines (Py) is the only occupant on the 5′-side, paired with the usual purine (Pu) complement in site 2′. However, there is one exception: an A–U pair occurs here in *E. coli* $tRNA^{Met}_{F}$ (Table 7.2), which type in prokaryotes serves as the initiator of translation. Thus this rare substitution may have functional implications. It can also be noted that, among group I tRNAs, the pair is usually U–A, but in groups II, III, and IV, C–G is by far the more frequent combination.

In the remainder of the paired sequences, effects of possible constraints are less evident. The third pair (3–3′) resembles the second in usually being

Py–Pu. Uridine is the pyrimidine in 40% of the tRNAs and cytidine in 30%; guanidine also is fairly frequent here, being the occupant in 25% of the tRNA sequences. Adenosine is found in only two cases, both of which are prokaryotic $tRNA_2^{Gly}$ (Table 7.2). A fourth pair (4–4′) may or may not be present; when it occurs, it has a similar pattern of Py–Pu pairing. Cytidine is by far the most prevalent nucleoside on the 5′-side, being the occupant in nearly 60% of the sequences, regardless of whether or not pairing exists. Pseudouracil likewise is a frequent occupant in this site but occurs only in group I tRNAs; thus it provides one piece of evidence that supports the group relationships presented in the preceding chapter. Guanosine and adenosine are present in a corresponding position only in the unpaired condition. In site 4′, cytidine is totally absent, whereas guanosine is the most prevalent nucleoside. Uridine occurs here only among group I tRNAs, a second piece of supporting data from this same region.

The Unpaired Region. As pointed out above, the unpaired region of arm II is one of the most variable features of the tRNA molecule; actually only one trait, the doublet GG, is totally consistent. The number of sites varies from a minimum of 7 to a maximum of 11, when 4 on each side of the loop are allotted to the stem region without regard to the presence of pairing in sites 4–4′, in order to maximize homologies as has been done here. If homologies are disregarded and all unpaired sites are included in the loop as a consequence, the maximum can be 12 sites. At site A (Figure 4.2), adenosine is nearly universally present, but several group IV tRNAs have guanosine instead. The basis for this treating of guanosine as part of the loop, rather than including it in the stem, is provided by *E. coli* $tRNA_1^{Tyr}$ (Table 7.2). But the latter could have resulted from the insertion of an extra base in the stem, as appears to be the case in another exceptional molecule, *Saccharomyces* $tRNA^{Ala}$, which has uridine in this site. At this time both points of view appear equally valid, but the first one is followed in Table 7.2.

Site B is universally occupied by a purine, guanosine being slightly more frequent than adenosine. Beyond this latter point in the loop there may be one, or sometimes two, 5,6-dihydrouridine residues before the doublet GG and one, two, or, rarely, three, following it. Frequently before the doublet, one dihydrouidine is replaced by another pyrimidine, usually cytidine, but occasionally adenosine is the replacement, as in *E. coli* $tRNA_2^{Glu}$ and $tRNA_3^{Ser}$ (Table 7.2). On the 3′-side of the doublet GG, the arrangement is highly diversified, random mutations seemingly having occurred. At site K, located just before the stem, there appears to be a universal need for adenosine, for that nucleoside invariably is the occupant. Here the value of considering the stem as being uniformly of four paired sites becomes evident, for otherwise the catholicity of adenosine at this locus would not be evident. Group I tRNAs never have another adenosine adjoining this one, group II and III members may have one adjacent to it, and group IV molecules may have three in sequence (Table 7.2).

Despite the variability in number of sites and in composition, the doublet GG consistently approximates a median location in the loop. The first member of the sequence frequently is 2′-*O*-methylguanosine, but since this modification is confined to the ribose moiety, the base may be correctly considered a universal feature in the strictest sense. Because the two nonproteinogenic species from *Staphylococcus* ($tRNA^{Gly}_{2A,B}$) have this doublet replaced by UU, it is tempting to suggest the importance of the GG pair in ligase recognition. However, it must be remembered that these two exceptional tRNAs also carry the amino acid glycine and therefore require recognition by a ligase, presumably the same one which aminoacylates the ordinary protein-synthesizing types. However, the GG doublet does play an important role in tertiary structure (Chapter 8).

Bend 2. Few generalizations may be drawn from a study of the composition of bend 2, despite its universality. It never consists of more than one nucleoside, with adenosine being present in about 40% of the cases. That nucleoside is especially characteristic of group I tRNAs and particularly those from prokaryotic sources. In group I and II tRNAs, cytidine is the occupant of this site to an extent of 10% of the total number. Guanosine is the most typical occupant, which in eukaryotes is characteristically modified to N_2,N_2-dimethylguanosine. This modification is not constant among eukaryotes, however; for example, in yeast $tRNA^{Asp}$ the guanosine is unmodified (Table 7.2). Uridine appears totally unacceptable here, and pseudouridine has been reported only in a $tRNA^{Trp}$ from brewer's yeast.

7.1.3. The Anticodon Arm (Arm III)

In marked contrast to that just described, arm III, which carries the anticodon, is unvarying in the pattern of its structure, for it always has a stem of five pairs and an unpaired loop almost undeviatingly of seven nucleosides. Only rarely do one or two additional nucleosides occur in the latter. Three (or in rare mutants four or more) central residues consistently form the anticodon (Figure 7.1), a feature which, because of its importance, is discussed in a separate section later.

The Paired Sequence. In the quintet of pairs that comprise the stem of arm III, various constraints may be operative that become increasingly severe toward the unpaired region. Even in the most proximal pair, they result in a complex situation that varies with the evolutionary level of the source organism. Among prokaryotes, the occupant of site 1 is usually cytidine, but guanosine is frequent in group IV tRNAs. Adenosine and uridine are known only in two $tRNA^{Gly}_1$s, and two $tRNA^{Gly}_2$s, respectively, and pseudouridine has never been reported. Its paired locus, site 1′, is consistently occupied by the Watson–Crick complement. Contrastingly, in tRNAs from eukaryotic sources, pseudouridine is by far the most prevalent nucleoside for site 1, usually paired to adenosine in site 1′, but yeast $tRNA^{Glu}_3$ has guanosine instead in the latter locus. Cytidine ranks second in abundance at site 1, but guanosine occurs there in yeast $tRNA^{Asp}$ and $tRNA^{Cys}$, with cytidine in site 1′ in the first instance and uridine in the second. Eukaryotic $tRNA^{Ser}$s often are distinct in having an A–U pair.

The second pair always consists of standard Watson–Crick combinations, except in a number of eukaryotic members of groups III and IV, in which pseudouridine replaces uridine on the 5′-side. Uridine and guanosine each occupy site 2 in 15% of the molecules, adenosine in ca. 20%, and cytidine in about 45% (Table 7.3). The situation changes sharply with site 3, cytidine with a rate of 12% being the least frequent nucleoside, except for pseudouridine which is absent. Adenosine is the most abundant, having a frequency near 44%, guanosine ranking next with 28%, and uridine being a poor third with 18%.

The fourth pair is nearly universally a combination of guanosine and cytosine, with G–C present in 72% of the molecules and C–G in 24%. The remaining 4% consists of U–A pairs, the A–U couplet being unknown here (Table 7.3). At site 5, adenosine and cytidine are the predominant nucleosides, being represented in 40% and 37% of the sequences, respectively. Guanosine is present in 15% and uridine in 6% of the molecules, whereas pseudouridine occurs only in mouse $tRNA_4^{Met}$. The latter, interestingly, is paired to another pseudouridine in site 5′. Despite the high frequency of adenosine in site 5, uridine has been reported only once in site 5′; in other words, the pair A–Ψ but not A–U is ordinarily acceptable in these two loci. The sole exception is in the *Salmonella* $tRNA^{His}$, from mutants of the class *hisT,* which charges normally but is ineffective in repression of the histidine operon (Singer *et al.*, 1972). The extreme unusualness of this combination here has not been previously noted. All other known pairings are standard Watson–Crick varieties.

Relative frequencies do not present the whole picture of this fifth pair. When it is viewed group by group, evidence is revealed that strongly supports the evolutionary relations proposed earlier. Among group I tRNAs, the fifth pair is nearly always C–G, one A–Ψ and two G–Cs being the only exceptions (Table 7.3). In contrast, among the members of group IV, A–Ψ is nearly universal, the U–A of *E. coli* $tRNA^{Ser}$ and $tRNA^{Trp}$ being the sole exceptions. Group II and III molecules are more variable, as might be expected of evolving populations; the A–Ψ combination is present in one-third or fewer of the member tRNAs. Since this pair occurs in such a wide diversity of major types, it can scarcely be supposed to serve a unique role in recognition. Hence, only the evolutionary derivation suggested here appears to offer a valid explanation for the pattern of occurrence noted.

The Unpaired Nucleosides. A similar conclusion results from comparisons of the respective unpaired doublets adjacent to the anticodon. Site B of the 5′-side is uniformly uridine and, thus providing an invariant feature of all tRNAs, does not need to enter further into this discussion. However, it should be borne in mind that in certain mutant tRNAs an additional nucleoside is inserted into this unpaired loop; this is usually considered as part of the anticodon (Riddle and Carbon, 1973; Atkins and Ryce, 1974). But in actual practice, codon-recognition activities may or may not be able to span more than the standard number of nucleosides. Site A among group I tRNAs is either uridine or cytidine in prokaryotic and eukaryotic types alike, with pseudouridine an occasional replacement in eukaryotic species. In group II, uridine and methylated uridine are the only reported occupants of this site. In group III, cytidine and 2′-methylcytidine alone occur, with one exception—uridine is present in *E. coli* $tRNA^{Thr}$. Cytidine and pseudouridine are common in group IV members but an unknown nucleoside (N) is found in *E. coli* $tRNA^{Ser}$. The most striking correlation to the group arrangement is that the nucleoside 3-methylcytidine is known only from this site and then solely from group IV members. It is to be noted that no purine occurs here.

On the opposite side of the anticodon in site C, the nucleoside present is often so highly modified and otherwise so important that special attention is given it in the next section. The occupant of site D is adenosine in 65% of the sequences recorded here, pseudouridine in 15%, cytidine in 11%, and uridine in 5.5%. Guanosine occurs only in those two $tRNA^{Gly}$s from *Staphylococcus* that do not participate in protein synthesis (Table 7.3). This exceptional characteristic, rather than the others which have been proposed in the literature, and which are discussed later, may serve to prevent recognition of these two tRNAs by ribosomes.

The Postanticodon Nucleoside. In site C, which follows the 3′-end of the anticodon, the nucleoside must serve a particularly vital function in protein synthesis, to judge from the extremely complex modifications that have evolved. Before the variations that exist here are examined, however, some common chemical knowledge must be brought to the forefront, because, although generally known, it is not always given proper consideration in discussions of tRNA bases. Any base (like any other chemical), regardless of whether it is a purine or pyrimidine, that is modified by methylation, thiolation, or by addition of any radical, no longer possesses the original properties. For example, adenine when methylated at the second site to become 2-methyladenine is not then simply adenine which happens to be methylated, but a new base. To make the point clearer, it is only necessary to recall that 5-methyluracil is known as thymine. Hence, if thymine is worthy of recognition as a distinct base, so is any other that is modified to a comparable or greater extent. This consideration does not necessarily apply in tRNAs to nucleosides in which the pentose moiety is modified, for the bases provide the more distinctive characteristics. It is most unfortunate that each variant does not have a distinctive name as thymine has, rather than only the descriptive chemical term. From this point of view it appears meaningless, for instance, to note the frequency of adenosine at a particular site, if its various modifications are included with it in the total count without further detailed notation.

At this postanticodon site, group considerations are of particular signifi-

Table 7.3
Anticodon Arm (Arm III) of Sequenced tRNAs[a]

	12345AB - CD5′4′3′2′1′
GROUP I	
Glycine	
$Salm_1$	*AGAGC*UU - AA*GCUCU*
$Esch_1$	*AGAGC*UU - AA*GCUCU*
$T2_2$, $T4_2$, $T6_2$	*UCAGA*CU - AA*ψCUGA*
$Esch_2$	*UCAGC*CU - AA*GCUGA*
$Staph_{2A}$	*CAUUC*CU - C*GGAAUG*
$Staph_{2B}$	*CGUUC*CU - C*GGAACG*
S. aur_{2A}	*CGUUC*CU - C*GGAACG*
S. aur_{2B}	*CAUUC*CU - U*GGAAUG*
$Esch_{3,ins}$	*CGACC*UU - AA*GGUCG*
Sac_3	*CAACG*ψU - A*ψCGUUG*
$Wheat_{3A,B}$	*GUACC*CU - AC_5*GGUAC*
Alanine	
$Esch_{1A}$	*CCUGC*UU - AC*GCAGG*
Sac	*CUCCC*UU - I_1ψ*GGGAG*
Tor	*ψUCGC*UU - I_1ψ*GCGAA*
Aspartic acid	
$Esch_1$	*CCUGC*CU - AC_2*GCAGG*
$Brew_1$	*GGCGC*ψU - GC_1*CGUGCC*
Glutamic acid	
$Esch_2$	*CCGCC*CU - AC_2*GGCGG*
Sac_3	*ψCACG*CU - AC*CGUGG*
Valine	
$Salm_1$	*CCUCC*CU - AA_2*GGAGG*
$Esch_1$	*CCUCC*CU - AA*GGAGG*
$Esch_{2A}$	*CCACC*UU - AU*GGUGG*
$Esch_{2B}$	*CCACC*UU - AU*GGUGG*
B. $stear_2$	*CCACC*UU - AG_6*GGUGG*
Sac	*ψCUGC*ψU - AC*GCAGA*
Tor	*ψCUGC*ψU - AC*GCAGA*
Brew	*ψCUGC*ψU - AC*GCAGA*
Mus_A	*ψUCGC*C_mU - AC_5*GCGAA*
Mus_B	*CUCGC*CU - AC_5*GCGAG*

Table 7.3. Cont'd.

	12345AB - CD5'4'3'2'1'
GROUP II	
Leucine	
$Esch_1$	CUAGCUU - G_mψGψUAG
$Salm_1$	CUAGCUU - G_mψGψUAG
$Salm_{mu}$	CUAGCUU - G_mUGUUAG
$Esch_2$	CUACCUU - G_mψGGUAG
Histidine	
Salm	CUGGAUU - A_2ψψCCAG
$Salm_{mu}$	CUGGAUU - A_2UUCCAG
Esch	CUGGAUU - A_2ψψCCAG
Glutamine	
$T4_1$, psu	UAGCACU - A_2CψGCUA
$Esch_1$	CCGGUUU_m - A_2ψACCGG
$Esch_2$	CCGGAU_mU - A_2ψψCCGG
Arginine	
$Esch_{1A,\ 1B}$	CUCGGC_tU - A_2ACCGAG
$Brew_2$	ψCUGGCU - AACCAGA
Proline	
T2, T4	CCUGAU'_2U - G_1AψCAGG
GROUP III	
Isoleucine	
$Esch_1$	CACCCCU - A_tAGGGUG
$Esch_2$	CACGACU - A_tAψCGUG
Tor	ψGGUGCU - A_tACGCCA
Threonine	
Esch	CACCCUU - AAGGGUG
Serine	
$Esch_3$	CUCCCC_tU - A_tAGGGAG
Rat_3	AψGGAC_3U - A_tAψCCAU
Arginine	
$Brew_{3,\ 3A}$	ψCUGACU - A_tAψCAGA

Table 7.3. Cont'd.

	12345AB - CD5'4'3'2'1'
Lysine	
Sac_2	$\psi\psi$CGGCU - A_tACCGAA
$Sac_{\alpha s}$	ψAUGACU - A_tAψCAUA
Methionine	
$Esch_M$	CAUCACU - A_tAψGAUG
Ana_F	UCGGGC'_2U - AACCCGA
$Esch_{F1,\ F3}$	UCGGGC'_2U - AACCCGA
Bac_F	UCGGGCU - AACCCGA
Sac_F	CAGGGCU - A_tACCCUG
Mus_F	CUGGGCC - A_tACCCAG
Rab_F, $sheep_F$	CUGGGCC - A_tACCCAG
$Brew_3$	ψCAGψCU - A_tAψCUGA
Mus_4, Rab_4	ψCAGψCU - A_tAψCUGA
(Asparagine)	
GROUP IV	
Phenylalanine	
Esch	GGGGAψU - A_sAψCCCC
B. stear	AAGGACU - A_sAψCCUU
Brew	CCAGAC_3U - YAψC_5UGG
Sac	CCAGAC_3U - YAψC_5UGG
Tor	CCAGAC_3U - YAψC_5UGG
Wheat, pea	ψCAGAC_3U - Y_WAψCUGA
Rab, calf	$\psi\psi$AGAC_3U - Y_MAψCUAA
Tyrosine	
$Esch^+su$	GCAGACU - A_sAψCUGC
$Esch_1$	GCAGACU - A_sAψCUGC
$Esch_2$	GCAGACU - A_sAψCUGC
Sac_2	CAAGACU - A_iAψCUUG
Tor_2	ψCAGACU - A_iAψCUCA

Table 7.3. Cont'd.

	12345AB - CD5'4'3'2'1'
Serine	
T4	$CCGGUC'_{2}U$ - $A_{s}AACCGG$
$Esch_1$	$CCGGUXU$ - $A_{s}AACCGG$
$Brew_{1A}$	$AAAGA\psi U$ - $A_{i}A\psi CUUU$
$Brew_{1B}$	$AAAGA\psi U$ - $AA\psi CUUU$
Rat_1	$A\psi GGAC_{3}U$ - $A_{i}A\psi_{m}CCAU$
Rat_2	$\psi\psi GGAC_{3}U$ - $A_{i}A\psi_{m}CCAA$
Cysteine	
Sac	$GCAGA\psi U$ - $A_{i}A\psi CUGU$
Tryptophan	
Esch	$CCGGUC_{3}U$ - $A_{s}AACCGG$
Rous	$\psi CUGAC_{3}U$ - $G_{1}A\psi_{m}CAGA$
$Brew_{1A,\ 1B}$	$\psi\psi CGAC_{3}U$ - $AA\psi CGAA$
Leucine	
T4	$CAGCACU$ - $A_{s}A\psi GCUG$
Sac_3	$CCUGA\psi U$ - $G_{1}C\psi CAGG$

[a]For an explanation of conventions, see Table 7.1, footnote *a*.

cance. Among group I tRNAs, adenosine is the characteristic nucleoside here, occurring among its members with a frequency of 66%. In this same group, cytidine, 2-methyladenosine, and methylated inosine each occur in about 4% of the molecules, whereas methylguanosine is found in a single sequence. The occurrence of cytidine and methylinosine in this group is especially noteworthy, for neither occurs at this locus in any of the others. The former nucleoside is confined to the two *Staphylococcus* $tRNA^{Gly}$s that are inactive in protein synthesis (Table 7.3); its presence here could, as suggested above for its neighboring nucleoside guanosine, be significant in the failure of ribosomal recognition. Among group II components, only methylguanosine, adenosine, and 2-methyladenosine have been reported in this postanticodon site, but the relatively scanty knowledge of this group could well be a contributing factor in this seeming constancy.

Hypermodified Bases of the Postanticodon Site. In contrast to the relative simplicity of the postanticodon nucleosides in groups I and II, those of groups III and IV are so strongly altered chemically that their bases have been termed "hypermodified." Among group III members (Table 7.3), a remarkable consistency in the nucleoside at this site has previously been reported by several laboratories (Ishikura *et al.*, 1969; Takemura *et al.*, 1969a,b; Powers and Peterkofsky, 1972). Accentuating the uniqueness of this trait, the base itself is hypermodified in a singular manner. The nucleoside containing this base (Figure 7.2A), abbreviated A_t in the table, has been determined to be *N*-[9-(β-D-ribofuranosyl)-purine-6yl-carbamoyl]threonine.* It has been established as being present in all members of this group which have been tested, not just those whose sequences are listed here. As might be expected from its complicated structure, its biosynthesis requires a train of enzymes (Körner and Söll, 1974). Only one exception has been noted, that of the initiator *E. coli* $tRNA_F^{Met}$. This single atypical case, nevertheless, is sufficient to indicate clearly that the hypermodified base of the others is not requisite for recognition of anticodons terminating in uridine, as has been suggested (Ishikura *et al.*, 1969). Although its consistent presence may provide greater effectiveness in codon–anticodon interactions (Parthasarathy *et al.*, 1974a,b) and thereby offer a selective advantage, it obviously cannot be an absolute chemical requirement.

A similar but more complex constancy exists at this site among the members of group IV (Table 7.3), along with an occasional exception. Here strong correlation to evolutionary advancement of the source organism also can be noted. Among prokaryotes, the nucleoside 2-methylthio-N^6-(Δ^2-isopentenyl)adenosine (Bartz *et al.*, 1970), abbreviated A_s, is unvariably present, while among eukaryotes it is simply N^6-(Δ^2-isopentenyl)adenosine (A_i) (Figure 7.3 B,C). The latter modified nucleoside apparently is the forerunner of A_s in prokaryotes, for it appeared in this site when *E. coli* tRNAs became undermethylated through methionine starvation (Isham and Stulberg, 1974). In certain seed plants, one of the methyl side chains may be hydroxylated to form ribosyl-*cis*-zeatin (Kamínek, 1974). At one time the presence of A_i in eukaryotes was believed to be as invariable as the prokaryotic counterpart (Armstrong *et al.*, 1969a,b; Nishimura *et al.*, 1969; Peterkofsky and Jesensky, 1969; Yamada *et al.*, 1971), but more recently several yeast species, including $tRNA_3^{Leu}$, have proven to contain 1-methylguanosine instead (Table 7.3). Serine tRNAs are of special interest in that those isoaccepting species recognizing A–– codons, being members of group III, have A_t in this site, while those recognizing U–– codons, like others of group IV, have either A_i or A_s following the anticodon.

Moreover, all eukaryotic $tRNA^{Phe}$s have been found to have this site occupied by Y, the base of which is so hypermodified as to no longer be either a purine or pyrimidine (Figure 7.3). It has been demonstrated that the base of Y

*A glycine analog, *N*-(purine-6yl-carbamoyl)glycine has been found in yeast tRNA, but it is not known whether this substance occupies a comparable site (Parthasarathy *et al.*, 1974a).

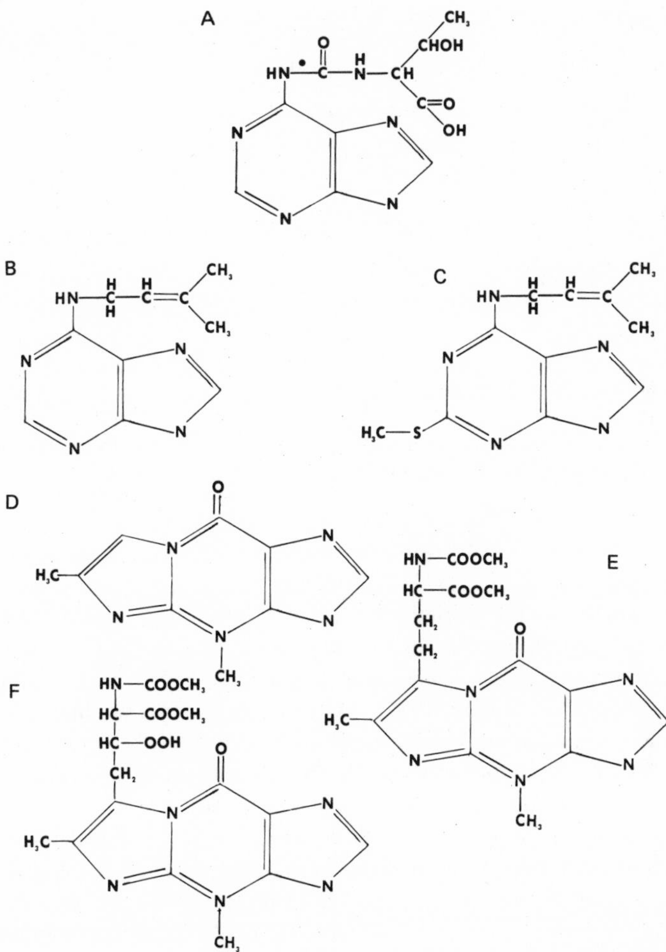

Figure 7.2. Hypermodified bases of the postanticodon site. (A) N-[9-(β-D-ribofuranosyl)-purine-6yl-carbamoyl] threonine (A_t), is characteristic of almost all known members of group III tRNAs (after Powers and Peterkofsky, 1972). (B) N^6-(Δ^2-isopentenyl) adenosine, A_i, is present in many eukaryotic, and (C) 2-methylthio-N^6-(Δ^2-isopentenyl) adenosine, A_s, in most prokaryotic, tRNAs of group IV. (D–F) In the $tRNA^{Phe}$s of many eukaryotes, various forms of Y occupy the postanticodon site: (D) Y^t of *Torulopsis* (Takemura *et al.*, 1973), (E) Y of yeast (Snoll, 1971), and (F) Y_m of mammals (Blobstein *et al.*, 1973).

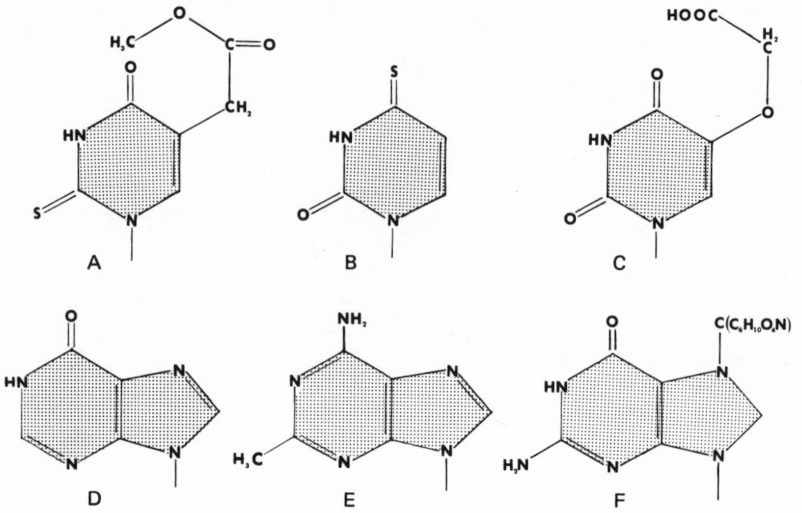

Figure 7.3. Uncommon bases frequently found in the anticodon site 1. (A, B) S type, (C) V type, (D) hypoxanthine (inosine), (E) N type, (F) Q type.

is produced in cells through fusion of an imidazole residue to guanine (Blobstein *et al.*, 1973; Li *et al.*, 1973; Thiebe and Poralla, 1973). As may be noted in the illustration, several varieties of Y are extant, one in *Saccharomyces,* a second in *Torulopsis,* another in the metazoans, and a fourth in the higher plants (RajBhandary *et al.*, 1968; Blobstein *et al.*, 1973, 1975; Takemura *et al.*, 1974; Münch and Thiebe, 1975). Moreover, this hypermodified nucleoside is not a universal feature of eukaryote $tRNA^{Phe}$, for it has been reported totally absent in crude preparations of tRNA from *Drosophila,* as well as in the $tRNA^{Phe}$ from the same source (White and Tener, 1973). As a consequence of these diversifications and absences, as in group III, no chemical relationships can be validly postulated to exist between the nucleoside of site C and that in either the first site of the codon or the last in the anticodon. Evolutionary relationships alone appear to provide a reasonable explanation; for even complete replacement of the base Y with proflavin has little effect on the activity of the $tRNA^{Phe}$ from yeast (Odom *et al.*, 1974).

7.1.4. The "Extra" Arm (Arm IV)

Whereas arm III containing the anticodon provides much convincing evidence of the naturalness of the group arrangement employed here, arm IV is burdened by so many confusing contrasts that its contribution could well be viewed as being of a negative nature. In the first place, as shown in Tables 7.4 and 7.5, two major types exist, one short, the other long. Second, closer exam-

Table 7.4
The "Extra" Arm (Arm IV) of Sequenced tRNAs, Short Type[a]

	ABCDE
GROUP I	
Glycine	
$Salm_1$	AU-AC
$Esch_1$	AU-AC
$T2_2$, $T4_2$, $T6_2$	UG-AU
$Esch_2$	UG-AU
$Staph_{2A}$	AG-GU
$Staph_{2B}$	AG-AU
S. $aur_{2A,2B}$	AG-GU
$Esch_{3,ins}$	GGG_{7}UC
Sac_3	GG--C
$Wheat_{3A,B}$	AG-AC_{5}
Alanine	
$Esch_{1A}$	AGG_{7}UC
Sac_A	AG-DC
Sac_B	AG-UC
Tor	AGGDC
Aspartic acid	
$Esch_1$	GGG_{7}UC
$Brew_1$	AG-AU
Glutamic acid	
$Esch_2$	UA-AC
Sac_3	AG-AC
Valine	
$Salm_1$	GGG_{7}UC
$Esch_1$	GGG_{7}UC
$Esch_{2A}$	GG$G_{7}$$U_{c}$C
$Esch_{2B}$	GG$G_{7}$$U_{c}$C
B. $stear_2$	AGG_{7}UC
Sac	ACG_{7}DC_{5}
Tor	AC-DC_{5}
Brew	ACG_{7}DC_{5}
Mus	AGG_{7}DC_{5}
GROUP II	
Histidine	
Salm	UUG_{7}UC
$Salm_{mu}$	UUG_{7}UC
Esch	UUG_{7}UC
Glutamine	
$T4_1$, psu	GAUGC
$Esch_1$	CAUUC
$Esch_2$	CAUUC
Arginine	
$Esch_{1A, 1B}$	CGG_{7}XC
$Brew_2$	AGADU
Proline	
T4	AGG_{7}UC
GROUP III	
Isoleucine	
$Esch_1$	AG$G_{7}$$N_{m}$C
$Esch_2$	AG$G_{7}$$N_{m}$C
Tor	AGADC_{5}
Threonine	
Esch	AGG_{7}UC
Arginine	
$Brew_{3,3A}$	AGADU
Lysine	
Sac_2	AUG_{7}DC_{5}
$Sac_{\alpha s}$	AGG_{7}UU
Methionine	
$Esch_M$	GGG_{7}XC
Ana_F	AGG_{7}UC
$Esch_{F1}$	AGG_{7}UC
$Esch_{F3}$	AGAUC
Bac_F	AGGUC
Sac_F	AUG_{7}DC_{5}
Mus_F	AGG_{7}DC_{5}
Rab_F, $sheep_F$	AGG_{7}DC_{5}
$Brew_3$, Sac_M	AGG_{7}DC_{5}
Mus_4, Rab_4	AGG_{7}DC_{5}
GROUP IV	
Phenylalanine	
Esch	GUG_{7}XC
B. stear	GUG_{7}XC
Brew	AGG_{7}UC
Sac	AGG_{7}UC
Tor	AGG_{7}UC
Wheat, pea	AGG_{7}DC
Rab, calf	AGG_{7}DC
Tyrosine	
Sac_2	AGADC_{5}
Tor	ACADC_{5}
Cysteine	
Sac	UGG_{7}DC
Tryptophan	
Esch	GUG_{7}UU
Rous	AGG_{7}CCU
$Brew_{1A, 1B}$	GGG_{7}DU

[a]For an explanation of conventions see Table 7.1, footnote *a*.

ination discloses that in each of these there is wide diversity in nucleoside content. Nonetheless, in spite of the difficulties it presents, this arm serves later to clarify an otherwise obscure discussion.

The Short Type. In the majority of the short types (Table 7.4), five nucleosides are present, but only four occur in many group I tRNAs and only three in yeast $tRNA_3^{Gly}$. Rous tumor $tRNA^{Trp}$, in contrast, has six, two of which are paired. Pairing is not usually indicated to take place between opposing members of the arm, but it is not unlikely that it occurs occasionally, as in the instance cited.

At several sites the existence of constraints may be detected. The occupant of site A typically is adenosine, which occurs in over 60% of the sequences. In group II, however, that nucleoside has been reported in this site in only 11% of those relatively few of its members which have been sequenced. Guanosine ranks next in importance, having a frequency of 20%, and uridine third, occurring in 15% of those known. As a rule cytidine is absent, but in prokaryotic members of group II, it is the typical occupant, with a frequency of nearly 50%. The site B nucleoside of all groups is most usually guanosine, for it is found in about 65% of the tRNAs; the next most frequent is uridine, with a 15% rate of occurrence. Cytidine and adenosine may be noted here in about 10% of the cases each.

The third site may be interpreted in several ways, but homologies are maximized in this loop by the procedures employed here. As may be seen in Table 7.4, it may be viewed as vacant in the major fraction of group I sequences, for only three or four nucleosides are contained in the arm among those tRNAs. When occupied, as is consistently the case in the remaining groups, 7-methylguanosine is the usual occupant, being here in nearly 60% of the known molecules, that is, in about 80% of this site when it is occupied. Unmodified guanosine, adenosine, and uridine are the only other nucleosides present at this locus, cytidine being consistently absent.

To a certain extent, site D nucleosides also are correlated to the proposed groups, particularly the first one. In this most primitive group the site is vacant in one instance (*Saccharomyces* $tRNA_3^{Gly}$), but more striking is the frequent (45%) occurrence of adenosine, a nucleoside which is absent in this location elsewhere. In the tRNAs as a whole, uridine or 5,6-dihydrouridine is the type present, each having a frequency of 33%. Guanosine has been reported once in each of the first two groups, but not at all in the others, whereas cytidine has been found once only, in group IV. A strange, complex nucleoside is rather frequent here in *E. coli* species, too, but so far has not been as common in group II members. This type, which had originally been indicated as X, has been identified as 3-(3-amino-3-carboxypropyl)uridine, and is abbreviated as U_c in Table 7.4 (Zachau, 1972; Nishimura *et al.*, 1974; Ohashi *et al.*, 1974).

Site F differs markedly from all the rest in having cytidine the most usual nucleoside, occupying 50% of the sites in unmodifed form and an additional

25% as 5-methylcytidine, which is restricted to eukaryotes. Uridine makes up the remaining fraction, being modified to 5,6-dihydrouridine only in group IV tRNAs, but no purine is known to occur here. In addition, one tRNA (Rous sarcoma virus $tRNA^{Trp}$) has a sixth site, similarly occupied by a pyrimidine; most interestingly, its base uracil is hydrogen-bonded to the cytosine located in site D.

The Long Type. Because of the extreme variability that is rampant in the long type of loop III, it is exceedingly difficult to find a valid basis for comparison. The list of the 17 whose sequences have been determined (Table 7.5) is arranged to provide maximum correlation, but many other alignments are possible. To clarify homologies further, however, numerous other tRNAs with a long arm IV need to have their nucleoside sequences determined.

The variation in the loop begins with the length, which ranges from a minimum of 9 to a maximum of 19 sites (Table 7.5). The number that comprises the paired stem is equally variable, with a range between three and seven pairs. The occurrence of this long loop IV is erratic, being absent among group I tRNAs and confined to one type each in groups II and III, whereas three major types in group IV possess it. Particularly noteworthy is the fact that those in groups II and III are of varieties also represented in group IV, that is, they carry those amino acids that are encoded by two different sets of codons as indicated by the initial nucleoside. Those that carry arginine, the third amino acid with six codons (two sets), however, have short arms. Thus no correlation exists between a large number of codons for an amino acid and the long extra arm in the tRNAs that carry it.

At first glance, the occurrence of long arms in isoaccepting species that are members of two different groups appears to regiment against the concept of the coevolution of amino acids and tRNAs. It would be expected, if that hypothesis were valid, that tRNAs for these multicoded species of amino acids would be of sharply contrasting structure, yet here they appear to share a highly distinctive trait. When closer comparisons of the long arms are made, however, it is found difficult to discern convincing affinities between isoaccepting members of different groups. For example, when the sequence of *E. coli* $tRNA^{Ser}_1$ from group IV is compared with $tRNA^{Ser}_3$ of group III from the same organism, a few similarities can be noted, especially the central unpaired sequences GAAA and CAAAA, respectively (Table 7.5). But in the paired regions no correspondences can be noted, for the one tRNA contains seven pairs, the other only five, and the base sequences are totally dissimilar. In fact, in the report giving the sequence of *E. coli* $tRNA^{Ser}_3$ (Ish-Horowicz and Clark, 1973), it was pointed out that few relationships existed between it and the previously described sequence of $tRNA^{Ser}_1$ (Ishikura *et al.*, 1971a,b). This lack of kinship between isoaccepting tRNAs appears explicable only on the basis of separate origins, or, at least, separate evolutionary histories subsequent to origin. This subject receives further attention later.

Table 7.5
The "Extra" Arm (Arm IV) and Bends 3 and 4 of the Sequenced tRNAs, Long Type[a]

	Bend #3	Arm IV 1234567	ABCDE	7'6'5'4'3'2'1'	Bend #4 ABC
GROUP I					
(None)					
GROUP II					
Leucine					
$Esch_1$	-	*-UGUCC-*	UUAC-	*GGACG--*	-U-
$Salm_1$, mu	-	*-UGUCC-*	UUAC-	*GGACG--*	-U-
$Esch_2$	U	*--GCCC-*	AAUA-	*GGGC---*	UU-
GROUP III					
Serine					
$Esch_3$	U	*AUGCGGU*	CAAAA	*GCUGCAU*	--C
Rat_3	U'_2	*--GψGC-*	$-UC_3U-$	*GCAC---*	$G-C_5$
GROUP IV					
Tyrosine					
$Esch^+_{su}$	C	*--GUC--*	-AUC-	*GAC----*	UUC
$Esch_1$	C	*-GGUC--*	-ACA-	*GACU---*	-UC
$Esch_2$	C	*--GUC--*	-ACA-	*GAC----*	UUC
Serine					
T4	C	*-AGUCGC*	UCCG-	*GCGACU-*	--C
$Esch_1$	C	*--GACCC*	GAAA-	*GGGUU--*	--C
$Brew_{1A}$	U'_2	*--GGGC-*	-UCU-	*GCCC---*	$G-C_5$
$Brew_{1B}$	U'_2	*--GGGC-*	-UUU-	*GCCC---*	$G-C_5$
Rat_1	U'_2	*--GGGG-*	$-UC_3U-$	*CCCC---*	$G-C_5$
Rat_2	U'_2	*--GGGG-*	$-UC_3U-$	*CCCC---*	$G-C_5$
Leucine					
T4	C	*--GGAA-*	UGAU-	*UUCC---*	UU-
Sac_3	-	*--UAUC-*	GUAA-	*GAUG---*	$--C_5$

[a]For an explanation of conventions, see Table 7.1, footnote *a*.

Bends 3 and 4. The bends adjacent to arm IV are of special interest, in that they may provide a model of a mechanism whereby loops in nucleic acids may be lengthened or diminished one nucleoside at a time, rather than by pairs. Thus the necessity for two simultaneous mutations to accomplish this end is obviated. To illustrate the processes, the two leucine tRNAs of group II are especially useful. In Table 7.5, it is clear that the stem regions of these two differ in length by one pair, and also that $\text{tRNA}_2^{\text{Leu}}$ has a bend 3 containing a single nucleoside and a bend 4 consisting of two sites, while $\text{tRNA}_1^{\text{Leu}}$ lacks a bend 3 and its bend 4 has only one nucleoside. Thus the conclusion is obvious that the situation in $\text{tRNA}_2^{\text{Leu}}$ could have arisen through the guanosine in site 2′ of $\text{tRNA}_1^{\text{Leu}}$ mutating to a uridine, thus resulting in a combination of uridines on opposing sides which do not pair. Hence, each of these now unbonded sites becomes part of a bend.

Evolution evidently could equally occur in the opposite direction. Were an unpaired nucleoside added at an appropriate end it would result in a bend, and if another were later acquired at the opposite end, it, too, would form a bend. Subsequent mutations to result in a pairable combination thus would bring about the formation of hydrogen bonds between them and the lengthening of the paired region accordingly.

Although the above model may have merit as a general concept, it appears simplistic in the present region when the other sequences listed in the table are also considered, especially *E. coli* $\text{tRNA}_3^{\text{Ser}}$. That species has the longest loop 3 of any that have been sequenced, so it could scarcely have experienced much shortening. Yet it has a single nucleoside in both bends, which, moreover, corresponds in type to those of other tRNAs. Since by and large all bases in the bends are pyrimidines, except for three guanosines in site A of bend 4, constraints of some sort are obviously active, and these would not permit random additions of bases from either the stem region or exogenous sources. The constancy of 5-methylcytidine in site C of bend 4 in all the eukaryotic tRNAs also is indicative of the presence of operational constraints.

7.1.5. The TΨC Arm (Arm V)

The last region of the tRNA molecule, arm V, has long been suspected to play a particularly vital role in tRNA functions, because a number of universal and near-universal features exist within it. Among these, the most outstanding is the extreme constancy of length, as all sequences have five sets of paired and seven unpaired nucleosides (Table 7.6). Certain features once believed to be essential to tRNA function, however, no longer can be considered absolutely requisite in light of recent base-sequence studies on normal proteinogenic types. But these are best considered in the discussions of the structural traits below.

The Paired Region. As pointed out above, the stem of arm V appears to be invariable in length, for all known sequences contain five pairs. At site 1 of

Table 7.6
TψC Arm (Arm V) of Sequenced tRNAs[a]

	12345	ABCDEFG	5'4'3'2'1'
GROUP I			
Glycine			
$Salm_1$	*GAGGG*	TψCGAUU	*CCCUU*
$Esch_1$	*GAGGG*	TψCGAUU	*CCCUU*
$T2_2$, $T4_2$, $T6_2$	*GUGAG*	TψCGAUU	*CUCAU*
$Esch_2$	*GCGGG*	TψCGAUU	*CCCGC*
$Staph_{2A}$	*AUAGG*	UGCAAGU	*CCUAU*
$Staph_{2B}$	*AUAGG*	UGCAAAU	*CCUAU*
S. aur_{2A}	*AUAGG*	UGCAAAU	*CCUAU*
S. aur_{2B}	*AUAGG*	UGUAAAU	*CCUAU*
$Esch_{3,ins}$	*GCGAG*	TψCGAGU	*CUCGU*
Sac_3	*CC_5CGG*	TψCGAUU	*CCGGG*
$Wheat_{3A}$	*C_5C_5GGG*	UψCGAU_1U	*CCCGG*
$Wheat_{3B}$	*C_5C_5GGG*	UGCGAU_1U	*CCCGG*
Alanine			
$Esch_{1A}$	*UGCGG*	TψCGAUC	*CCGCG*
Sac	*UCCGG*	TψCGAUU	*CCGGA*
Tor	*UCCGG*	TψCGAC_1U	*CCGGA*
Aspartic acid			
$Esch_1$	*GCGGG*	TψCGAGU	*CCCGψ*
$Brew_1$	*C_5GGGG*	TψCAAUU	*CCCCG*
Glutamic acid			
$Esch_3$	*AGGGG*	TψCGAAU	*CCCCU*
Sac_3	*C_5GGGG*	TψCGACU	*CCCCG*
Valine			
$Salm_1$	*GGCGG*	TψCGAUC	*CCGUC*
$Esch_1$	*GGCGG*	TψCGAUC	*CCGUC*
$Esch_{2A}$	*GGUGG*	TψCGAGU	*CCACU*
$Esch_{2B}$	*GUUGG*	TψCGAGU	*CCAAU*
B. $stear_2$	*GCUGG*	TψCGAGC	*CCAGU*
Sac	*C_5CCAG*	TψCGAU_1C	*CUGGG*
Tor	*CCCAG*	TψCGAU_1C	*CUGGG*
Brew	*CCCAG*	TψCGAU_1C	*CUGGG*
Mus	*C_5CCGG* (A)	UψCGAA_1A	*CCGGG*

Table 7.6. Cont'd.

	12345	ABCDEFG	5′4′3′2′1′
GROUP II			
Leucine			
Esch$_1$	*GGGGG*	TψCAAGU	*CCCCC*
Salm$_1$	*GGGGG*	TψCAAGU	*CCCCC*
Esch$_2$	*ACGGG*	TψCAAGU	*CCCGU*
Histidine			
Salm	*GUGGG*	TψCGAAU	*CCCAU*
Salm$_{mu}$	*GUGGG*	TψCGAAU	*CCCAU*
Esch	*GUGGG*	TψCGAAU	*CCCAU*
Glutamine			
T4$_{1,psu}$	*AAAGG*	TψCGAGU	*CCUUU*
Esch$_1$	*CCUGG*	TψCGAAU	*CCAGG*
Esch$_2$	*CGAGG*	TψCGAAU	*CCUCG*
Arginine			
Esch$_{1A,\ 1B}$	*GGAGG*	TψCGAAU	*CCUCC*
Brew$_2$	C_5*CAGG*	TψCAAG_1U	*CCUGG*
Proline			
T3, T4	*CAAGG*	TψCAAAU	*CCUUG*
GROUP III			
Isoleucine			
Esch$_1$	*GGUGG*	TψCAAGU	*CCAC*ψ
Esch$_2$	*GGUGG*	TψCAAGU	*CCAG*ψ
Tor	*AGCAG*	TψCGAU_1C	*CUGCU*
Threonine			
Esch	*GGCAG*	TψCGAAU	*CUGCC*
Serine			
Esch$_3$	*CGGGG*	TψCGAAU	*CCCCG*
Rat$_3$	*GUGGG*	TψCGAA_1U	*CCCAU*
Arginine			
Brew$_{3,\ 3A}$	*AUGGG*	TψCGAC_1C	*CCCAU*
Lysine			
Sac$_2$	*AGGGG*	TψCGAG_1C	*CCCCU*
Sac$_{\alpha s}$	*AGGGG*	TψCGAG_1C	*CCCCU*

Table 7.6. Cont'd.

	12345	ABCDEFG	5'4'3'2'1'
Methionine			
Esch$_M$	*ACAGG*	TψCGAAU	*CCCGU*
Ana$_F$	*AGAGG*	TψCAAAU	*CCUCU*
Esch$_{F1,\ F3}$	*GUCGG*	TψCAAAU	*CCGGC*
Bac$_F$	*GCAGG*	TψCAAAU	*CCUGC*
Sac$_F$	*CUCGG*	AUCG$\underset{1}{A}$AA	*CCG*$\underset{m}{A}\underset{m}{G}$
Mus$_F$	*GAUGG*	AUCG$\underset{1}{A}$AA	*CCAUC*
Rab$_F$, sheep$_F$	*GAUGG*	AUCG$\underset{1}{A}$AA	*CCAUC*
Brew$_3$, Sac$_M$	*GAGAG*	TψCG$\underset{1}{A}$AC	*CUCUC*
Mus$_4$, Rab$_4$	*GUGAG*	TψCG$\underset{1}{A}$UC	*CUCAC*
(Asparagine)			
GROUP IV			
Phenylalanine			
Esch	*CUUGG*	TψCGAUU	*CCGAG*
B. stear	*GGCGG*	TψCGAUU	*CCGUC*
Brew	$\underset{5}{C}$*UGUG*	TψCG$\underset{1}{A}$UC	*CACAG*
Sac	$\underset{5}{C}$*UGUG*	TψCG$\underset{1}{A}$UC	*CACAG*
Tor	$\underset{5}{C}$*UGUG*	TψCG$\underset{1}{A}$UC	*CACAG*
Wheat, pea	*GCGUG*	TψCG$\underset{1}{A}$UC	*CACGC*
Rab, calf	$\underset{5}{C}$*CUGG*	TψCG$\underset{1}{A}$UC	*CCGGG*
Tyrosine			
Esch$^{+}_{su}$	*GAAGG*	TψCGAAU	*CCUUC*
Esch$_1$	*GAAGG*	TψCGAAU	*CCUUC*
Esch$_2$	*GAAGG*	TψCGAAU	*CCUUC*
Sac$_2$	*GGGCG*	TψCG$\underset{1}{A}$CU	*CGCCC*
Tor$_2$	*GGGCG*	TψCG$\underset{1}{A}$AU	*CGCCC*
Serine			
T4	*AUAGG*	TψCAAAU	*CCUAU*
Esch$_1$	*CAGAG*	TψCGAAU	*CUCUG*
Brew$_{1A}$	*GCAGG*	TψCGAGU	*CCUGC*
Brew$_{1B}$	*GCAGG*	TψCAAAU	*CCUGC*
Rat$_1$	*GCAGG*	TψCG$\underset{1}{A}$AU	*CCUGC*
Rat$_2$	*GCAGG*	TψCG$\underset{1}{A}$AU	*CCUGX*

Table 7.6. Cont'd.

	12345	ABCDEFG	5'4'3'2'1'
Cysteine			
Sac	*CUUAG*	TψCGA$\underset{1}{\text{A}}$UC	*CUGAG*
Tryptophan			
Esch	*GGGAG*	TψCGAGU	*CUCUC*
Rous	*GCGUG*	ψψCGA$\underset{1}{\text{A}}$U	*CACGU*
$Brew_{1A}$	*GCAGG*	TψCAA$\underset{1}{\text{U}}$U	*CCUGψ*
$Brew_{1B}$	*GCAGG*	TψCAA$\underset{1}{\text{U}}$U	*CCUGU*
Leucine			
T4	*GUGGG*	TψCGAGU	*CCCAC*
Sac_3	*AAGAG*	TψCGAAU	*CUCUU*

[a]For an explanation of conventions, see Table 7.1, footnote *a*.

the 5′-end, guanosine is the nucleoside in over 50% and cytidine in about 30% of the species. Among eukaryotes the cytidine frequently, but by no means always, is modified to the 5-methylated condition. As uridine is found here only in the three $tRNA^{Ala}$s, adenosine is obviously the third most frequent occupant (16%).

Adenosine remains at about an equal frequency level at site 2, while uridine occurs there in nearly 20% of the molecules. The gain in frequency of the latter is largely at the expense of guanosine, which is reduced to a rate of 29%, while cytidine occurs in about 36% of the sequences at this position. At the third locus guanosine is nearly three times as frequent (about 50%) as any of the others, which occur at very nearly equal rates close to 17%.

In the last two loci, guanosine is by far the most abundant nucleoside, being the sole occupant of site 5* and having a frequency of 72% at site 4. At the latter point, the other usual purine, adenosine, occurs at a rate of 18%, so that the two pyrimidines are poorly represented there. On the opposing 3′-end of the loop, cytidine is universally present at site 5′, although uridine elsewhere frequently pairs with guanosine. For example, the proximal end of this stem shows a series of G–U pairs in the glycine tRNAs. As a whole, then, constraints on the bases appear to become increasingly severe with distance away from the proximal end of the loop, site 1 being exceptional to a degree.

Few unusual or modified bases have been reported in this stem, aside from the 5-methylcytidine mentioned above as occurring in site 1. The same modified base occurs in site 2 of the two $tRNA^{Gly}_{3}$s from wheat. Methyladenosine

* Adenosine has been reported here in a minor variety of $tRNA^{Val}$ from the mouse.

and methylguanosine occur together near the 3′-end of the stem in yeast $tRNA_F^{Met}$ and pseudouridine is found in the corresponding position of *E. coli* $tRNA_{1,2}^{Ile}$ and in $tRNA^{Trp}$ from brewer's yeast. Otherwise such bases are conspicuously absent.

The Unpaired Region. A number of relatively severe constraints occur also in the unpaired region, some of which were long believed absolute ones. The first trio of nucleosides, TΨC, was originally found consistently present in the nucleoside sequences as these became determined (Ofengand and Henes, 1969; Richter *et al.*, 1973), except in the initiator tRNAs of eukaryotes, where AUC occurred instead (Petrissant, 1973; Simsek *et al.*, 1973). These early exceptions were thus explained as modifications for their specialized functions. Later, when the molecular structure of the two nonproteinogenic $tRNA_{A,B}^{Gly}$s of *Staphylococcus* had been determined, the same explanation seemed to apply to sequence UGG that occurred at this location. Now the two wheat $tRNA_{3A,B}^{Gly}$s, shown to be active in protein synthesis (Marcu, 1973; Marcu *et al.*, 1973), have been reported to have UΨC and UGC in place of TΨC. A $tRNA^{Val}$ of mouse also has UΨC here (Piper and Clark, 1974a; Piper, 1975b), as does $tRNA^{Ile}$ from *Mycoplasma* sp. (Johnson *et al.*, 1970). The tRNA used in Rous-virus-infected cells in initiating transcription is unusual in having ΨΨC (Maugh, 1974). Although it had been reported that most species of tRNA in *Bacillus subtilis* lacked thymidine (Arnold and Kersten, 1973), this has since proven erroneous (Arnold *et al.*, 1975). Nevertheless, the only remaining universal feature in the trio is the cytidine at its third site, site C. According to one report (Randerath *et al.*, 1974), *E. coli* total tRNA extracts proved to contain 1 mole ribothymidine per mole of tRNA, whereas cytoplasmic tRNAs from eukaryotic sources had only half that amount. However, the known absence of this nucleoside from prokaryotic $tRNA^{Gly}$s which do not participate in proteinogenesis indicates possible species diversity so that generalizations appear unjustifiable. In contrast, mitochondrial tRNAs had only one-eighth as much thymidine as the prokaryote ones did but were more strongly methylated.

Adjacent to the constant presence of cytidine in site C is site D, which requires a purine. While adenosine is found in this site in about 20% of the molecules, guanosine is by far the more frequent purine. The adjoining site E also must be critical for tRNA functioning, for adenosine and 1-methyladenosine are its sole known occupants, the modified base being confined to eukaryotic species.

At site F in this region, some correlation to the group arrangement may be detected. In group I, adenosine and cytidine occur infrequently (10% and 4%, respectively), while uridine and guanosine are the usual nucleosides, as they have relative frequencies of 66% and 20%, in the same order. In group II, guanosine is the more typical nucleoside, with a 60% frequency, adenosine making up the remainder. In group III, however, adenosine becomes the prevailing nucleoside, being represented here in nearly 65% of the species. Concurrently, uridine is reduced in frequency to match that of cytidine, which remains rare as previously (at a rate of about 8%), whereas guanosine is found in approximately 20% of the types. In group IV, adenosine and uridine are of equal frequency and together account for about 80% of the site. Guanosine has a frequency of about 12%, whereas cytidine occurs only once.

Site C is almost completely reserved for pyrimidines, uridine being nearly four times as abundant as cytidine. However, adenosine does occur in four species of eukaryotic $tRNA_F^{Met}$ in group III. Since two other unique features are found in this region of these special-function tRNAs (AU in place of TΨ), this unpaired portion of loop IV may be of significance in recognition by the translational initiating complex, as well as in the maintenance of secondary and tertiary structure (Chapter 8).

7.2. CODON–ANTICODON INTERACTIONS

The interactions between the codons of the messenger and the anticodons of the tRNAs are so vital to the synthesis of error-free proteins that this region of the molecules requires close examination. Because of their essential nature, it comes as no surprise that peculiar constraints have evolved in anticodons, including the presence of many unusual bases (Hall, 1971) and the absence of one or more common ones. These rarer bases appear to offer advantages for introducing the subject.

7.2.1. Unusual Bases in the Anticodon

As shown in Table 7.7, the various unusual bases of the anticodon region of tRNAs are confined to the first site, except for the occurrence of pseudouridine at site 2 in yeast $tRNA^{Tyr}$s. Because of structural peculiarities, the biological basis of which is suggested later, tRNAs are built in a sequence opposite to that of mRNAs, that is, the 3′-end of the anticodon corresponds to the 5′-end of the codon. Hence, the base in site 1 of the anticodon pairs with that of site 3 of the codon when the message is being translated, and so on.

The Relative Abundance of the Several Watson–Crick Bases. When the first sites in the known anticodons are scanned (Table 7.7), it is readily seen that guanosine and cytidine occur frequently, sometimes in modified form in eukaryotic tRNAs. It also becomes immediately evident that adenosine in unmodified form has never been reported here and that uridine is confined to the several prokaryotic $tRNA_2^{Gly}$s. Furthermore, pseudouridine also is absent. To the author's knowledge, these constraints against three important bases at this site have not been noted previously in the literature.

One possible explanation for the rarity of uridine here that might seem valid is that, since this site is universally preceded on the 5′-side by a uridine,

Table 7.7
Known Anticodons of tRNAs[a]

GROUP I	
Glycine	
Salm$_1$	CCC
Salm$_{mu}$	CCCC
Esch$_1$	CCC
T2$_2$, T4$_2$, T6$_2$	U$_m$CC
Esch$_{2A}$	UCC
Esch$_{2A}$	$\overset{*}{U}$CC
Staph$_{2A,2B}$	UCC
S. aur$_{2A, 2B}$	UCC
Esch$_{ins}$	UCC
Esch$_3$	GCC
Sac$_3$	GCC
Wheat$_{3A}$	GCC
Wheat$_{3B}$	GCC
Alanine	
Bac	$\overset{*}{V}$GC
Esch$_{1A}$	VGC
Sac	IGC
Tor	IGC
Aspartic acid	
Esch$_1$	QUC
Brew$_1$	GUC

Glutamic acid	
Esch	SUC
Sac$_3$	SUC
Valine	
Salm$_1$	$\overset{*}{V}$AC
Bac$_1$	VAC
Esch$_1$	GAC
Esch$_{2A}$	GAC
Esch$_{2B}$	GAC
B. stear$_2$	GAC
Sac	IAC
Tor	IAC
Brew	IAC
Mus	IAC
GROUP II	
Leucine	
Esch$_1$	CAG
Salm$_1$	CAG
Esch$_2$	GAG
Histidine	
Salm	QUG
Salm$_{mu}$	QUG
Esch	QUG

Glutamine	
T4$_1$	XUG
T4$_{psu}$	XUA
Esch$_1$	XUG
Esch$_2$	CUG
Arginine	
Esch$_{1A, 1B}$	ICG
Brew$_2$	ICG
Proline	
T2, T4	XGG
GROUP III	
Isoleucine	
Esch$_1$	GAU
Esch$_2$	XAU
Tor	IAU
Threonine	
Bac	$\overset{*}{V}$GU
Esch	GGU
Serine	
Esch$_3$	GCU
Rat$_3$	GCU
Arginine	
Brew$_{3,3A}$	XCU

Lysine	
Sac$_2$	SU$_m$U
Sac$_{\alpha s}$	CUU
Methionine	
Esch$_M$	C$_a$AU
Ana$_F$	CAU
Esch$_{F1, F3}$	CAU
Bac$_F$	CAU
Sac$_F$	CAU
Mus$_F$	CAU
Rab$_F$, sheep$_F$	CAU
Brew$_3$	CAU
Mus$_4$, Rab$_4$	C$_2$AU
Asparagine	
Esch	QUU
GROUP IV	
Phenylalanine	
Esch	GAA
B. stear	G$_m$AA
Brew	G$_2$AA
Sac	G$_2$AA
Tor	G$_2$AA
Wheat, pea	G$_2$AA
Rab, calf	G$_2$AA

Tyrosine	
Esch$^+$su	CUA
Esch$_1$	QUA
Esch$_2$	GUA
Sac$_2$	GψA
Tor$_2$	GψA
Serine	
T4	XGA
Esch$_1$	VGA
Brew$_A$	IGA
Brew$_B$	IGA
Rat$_1$	IGA
Bombyx$_{2b}$	IGA
Rat$_2$	CGA
Bombyx$_{2a}$	GCA
Cysteine	
Sac	GCA
Tryptophan	
Esch	CCA
Rous	C$_m$CA
Brew$_{1A, 1B}$	C$_2$CA
Leucine	
T4	NAA
Sac$_3$	CAA

the presence of that same nucleoside in site 1 of the anticodon might induce misreading during translation. The resulting reading-frame shifts would have led to decreased accuracy of translation as a disastrous consequence. However, the occasional presence of uridine in site 1 of such normal proteinogenic nucleic acids as *E. coli* $tRNA_2^{Gly}$ (Table 7.7) lessens the probability of a requirement of the sort described. The absence of adenosine from site 1 might similarly appear to be correlated to the nearly universal presence of a purine at site C following the anticodon, where adenosine is the most frequent occupant. However, it is often there in highly modified form. Moreover, guanosine, which also occurs in the postanticodon locus C, is the most abundant nucleoside at site 1 of the anticodon, outnumbering cytidine by a ratio of 3 : 2 in the anticodons listed. Thus such proposals must be viewed as totally unsupportable by factual evidence.

Inosine. Because the earliest tRNAs to have their molecular primary structure determined contained inosine at the first site of the anticodon in place of uridine, it was proposed that the present nucleoside would totally replace the latter (Crick, 1966). Thus far, however, inosine has been reported predominantly from eukaryotic sources, the two varieties of $tRNA_1^{Arg}$ from *E. coli* being the only known exceptions. Nevertheless, the happenstance of its presence in the early sequences played a prominent role in the formulation of the "wobble hypothesis" (Crick, 1966), to be discussed below. Hypoxanthine (Figure 7.3D), the base of this nucleoside, bears sufficient similarity to uracil to pair freely with adenine, from which it seems to be derived after transcription of the tRNA (Dugré and Cedergren, 1974). This pairing reaction and those of other unusual bases found in this site are of such importance as to merit discussion in a separate section that follows.

Other Unusual Nucleosides. Occasional methylated guanosines and cytidines and one instance of acetylcytidine (in *E. coli* $tRNA_M^{Met}$) occur in site 1 of the anticodon (Ohashi *et al.*, 1972). Among these modified nucleosides is Q, which is a highly modified guanosine (Figure 7.3E). In addition, several other unusual types have been reported here that need separate consideration, including one referred to as "V" found in several prokaryotic species.

The base of V has been shown to be 5-oxyacetyluracil (Murao *et al.*, 1970; Nishimura, 1972; Morikawa *et al.*, 1974); the molecular configuration illustrated (Figure 7.3C) is advocated here, because it explains certain coding properties discussed below that would otherwise remain inexplicable. (No contradictory evidence has been presented thus far by X-ray diffraction studies.) It is known to occur in tRNAs for serine and alanine from *E. coli* and in $tRNA^{Val}$ from *E. coli* and *Salmonella*. A related but simpler nucleoside, 5-methoxyuridine, has been described from the anticodon of *B. subtilis* (Murao *et al.*, 1976), but little is known of its reactivity (Tables 7.7 and 7.8). An even more highly modified base than V, usually referred to as "S," has been determined to be 2-thiouracil-5-acetic acid methyl ester (Figure 7.3A, B) and related derivatives of 2-thiouracil (Baczynskyj *et al.*, 1968). Unlike V, the present base in various forms has been reported from both prokaryotic and eukaryotic sources (Nishimura, 1972), but notably in $tRNA_2^{Glu}$. It has been found also in yeast $tRNA_2^{Lys}$, probably as 5-methyl-2-thiouridine (Kimura-Harada *et al.*, 1971).

The last unusual nucleoside of moderate frequency in tRNAs is that known as "N," which in this location is simply 2-methyladenosine (Figure 7.3E). This nucleoside thus far has been reported only from prokaryotic tRNAs, including T4 and *E. coli* $tRNA_1^{Gln}$ and T4 $tRNA^{Leu}$. The N^+ has been employed in the anticodon of *E. coli* $tRNA_2^{Ile}$ for an undetermined nucleoside (Harada and Nishimura, 1974); in this position in Tables 7.7 and 7.10, but not at other sites, X is employed instead to indicate any unknown nucleoside in order to avoid confusion.

7.2.2. Constraints Active in the Anticodon

If the foregoing observations are reconsidered, it becomes apparent that constraints of some type must be active at the first site of the anticodon. According to general concepts, including mutation theory and the structure of the codon catalog, all common bases should occur at any given site in approximately equal frequencies, but it is quite evident that actual base determinations show strong disparities to exist. Guanosine is by far the most abundant nucleoside in this location, cytidine is a close second, and uridine a poor third. Adenosine occurs only in modified form and pseudouridine is totally absent, while in their place are the unusual nucleosides described above.

Moreover, some of these nucleosides have been shown to be restricted to certain combinations that include the occupant of the *second* site. For example, Q (a highly modified guanosine) was reported to occur only in prokaryotic tRNAs in which uridine was the nucleoside of the second site (Harada and Nishimura, 1972), as in QUC of *E. coli tRNA*^{Asp}, QUG of *Salmonella* $tRNA^{His}$, and QUU of *E. coli* $tRNA^{Tyr}$ (Table 7.7). This observation made evident to the author that the nucleosides occupying site 1 of the anticodon might be influenced by those of site 2 on a wide basis. Accordingly an analysis of each base thus far reported in site 1 has been made, along with those at site 2, the results of which are tabulated in Table 7.8. It should be noted that in the 98 codons listed, site 1 is occupied by cytidine 18 times and guanosine 18, nearly as might be expected on a random basis. Thus the population sampled here appears adequate to be suggestive, at least, but some exceptions to the generalizations made below undoubtedly will be found as more nucleoside sequences of tRNAs are reported along with their codon-recognition properties.

Constraints on the Common Bases. In the summary of results (Table 7.8), it can be noted that cytidine in site 1 occurs in combination with A, C, G, or U at site 2, but thus far has not been found with Ψ, and rarely occurs with G. Thus it would appear that, when either pseudouridine or guanosine is in site

2, it induces constraints on site 1 occupants, so that some other nucleoside seems to be specified where cytidine would be expected according to codon considerations (that is, where G terminates the coding triplet). Some of this correlation may stem from sampling error, for only nine anticodon sequences thus far contain guanosine in site 2, whereas close to 20 would be expected. Nevertheless, on the basis of present knowledge, the correlation given above appears valid. Guanosine seems to be less subjected to such constraints, for it has been reported to be present at site 1 whether A, C, G, Ψ, or U occupies site 2.

The other usual nucleosides are far from frequent in tRNAs at this first locus. Indeed, as pointed out above, adenosine is completely absent from site 1, and uridine is thus far known here only in several glycine tRNAs from prokaryotic and phage sources, in about 50% of which it is methylated.

Constraints on Unusual Nucleosides. One of the unusual nucleosides, V (5-oxyacetyluracil), occupies site 1 only when a common purine is the base at site 2. Another, inosine, is strongly favored by purines in site 2 but also occurs with cytidine there. Thus strong constraints are operative on these species, too. These are even more severe than those noted to be active on the common nucleosides, for inosine is confined largely to eukaryotic anticodons and V entirely to prokaryotic. To some extent the two are complementary, in that inosine occurs in eukaryotic valine and serine tRNAs and V in corresponding species from prokaryotic sources. The complementary relations are not universal, for both prokaryotic and eukaryotic $tRNA^{Arg}$s have I as the initial nucleoside of the anticodon.

Two of the remaining nucleoside types, S (2-thiouracil-5-acetic acid methyl ester and related compounds) and Q occur in both prokaryotes and eukaryotes. S has been reported from only three tRNAs so far, all of which have uridine in site 2. It is not unlikely that it will also be found in those with A at site 2, as is the case with Q (Nishimura, 1972). A similar situation is found with the N identified as 2-methyladenosine (Saneyoshi *et al.*, 1972), except that the latter apparently is confined to prokaryotes (Table 7.8).

Implications. Thus it is clear that, unexpectedly, interaction of some undetermined sort occurs between the bases of the nucleosides of sites 1 and 2. At the very least, the evidence is compelling that the nucleoside at site 2 strongly influences the nature of that which can occupy site 1, so that only certain combinations are of common occurrence. This implies that mutations in genes specifying tRNAs cannot take place completely randomly in the anticodon region, those that are under constraints being inviable if they occur. For instance, no transitions in the first site (neither from C to U nor from G to A) can survive, for A and U (except in glycine tRNAs) do not occur there. For successful transitions to take place, the proper modifying enzyme must already exist; by way of illustration, a C to U transition would be viable if the enzyme were present that could modify the uridine to V or S. Hence, if that enzyme

Table 7.8
Known Combinations of Nucleosides Occupying Sites 1 and 2 of the Anticodon of tRNAs

Site 1		Site 2		Number of combinations	
Nucleoside	Total occurrences	Nucleoside	Combination total	Prokaryote species	Eukaryote species
A, ψ	0	—	—	—	—
C	21	A	12	6	6
		C	4	4	0
		G	2	0	2
		U	3	2	1 (Yeast$_{\alpha s}$ Lys)
C_a	1	A	1	1 (*E. coli* Met_M)	0
C_m	1	A	1	0	1 (Rous Trp)
C'_{2m}	4	C	4	0	4
G	18	A	6	6	0
		C	7	2	5
		G	1	1 (*E. coli* Thr)	0
		U	2	1 (*E. coli* Tyr_2)	1 (Yeast)
		ψ	2	0	2
G_m	1	A	1	1 (*B. stear* Phe)	0
G_2	7	A	7	0	7
I	14	A	5	0	5
		C	3	2	1 (Yeast Arg_2)

N	1	A	1	1 (T4 Leu)	0
Q	6	U	6	6	0
S	2	U	2	1 (*E. coli* Glu)	1 (Yeast Glu)
S_m	1	U	1	0	1 (Yeast Lys_2)
U	6	C	6	6	0
U_m	4	C	4	4	0
V	4	A	2	2	0
		G	2	2	0
V^*	3	A	1	1	0
		G	2	2	0
X^a	7	A	1	1 (*E. coli* Ile_2)	0
		C	2	0	2
		G	2	2	0
		U	2	2	0
Totals		A=38 G=15			
	101	C=30 U=16	101	56	45
		ψ=2			

[a]As X represents any unknown nucleoside, the tallies shown will remain of no significance, until such time as determinations of the bases' nature have been made.

were absent, series of mutations would first have to take place elsewhere in the genome to bring that protein into existence. And until the C to U or G to A transition did occur, such an enzyme would obviously be useless to the organism. Since the necessary series of events would appear to be most improbable from an evolutionary standpoint, it is implicit in the observed data that certain of the modifying enzymes should be considered to have come into existence early in life's history.

The effects of the constraints probably are more clearly perceptible when focus is upon the site 2 nucleoside, rather than on that of site 1. As shown in Table 7.9, when adenosine occupies site 2 as it does in 31 of the anticodons, constraints are lax, for six of the eight types known from site 1 are acceptable. Constraints then operate only against uridine and S. When uridine is the occupant of site 2 as it is in 14 of the anticodons, it exercises slightly stronger constraint than adenosine, for only five of the eight appear permissible. In this case, inosine, uridine, and V do not seem acceptable. Cytidine (reported to be present in 25 anticodons) and guanosine (in 9) are still more restrictive when occupying site 2. The former is known only to permit cytidine, guanosine, inosine, and uridine in site 1, and the latter on limited information seems to constrain all except cytidine, guanosine, inosine, and V. Since pseudouridine has been reported from only two anticodons in site 2, it is not possible now to suggest whether or not it exercises constraints on the first-site occupant. Undoubtedly some, but certainly not all, of the other apparent constraints given above will be found nonexistent as more sequences are determined. For instance, it does not seem to be mere random chance that 10 anticodons of the 13 with inosine in site 1 have a purine in site 2.

Table 7.9
Limitations Imposed by Base of Second Site of Anticodons upon Bases Acceptable for First Site

Base of second site	Known acceptable bases of first site	Codons recognized
A	C, G, I, N, Q, V	No restrictions
C	C, G, I, U	-GG, -GC strongly -GA, -GU weakly
G	C, G, I, V	-CA, -CG, -CU in prokaryotes; all in eukaryotes
Ψ	G	-CC strongly, -CU weakly
U	C, G, N, Q, S	No restrictions

Constraints on Codons in Actual Messengers. Effects parallel to the constraints in the anticodons appear to be found in the triplets of mRNA, according to an analysis of the codons contained in the RNA of bacteriophage MS2 (Fiers *et al.*, 1976). In most cases, that study made it apparent that the frequency of occurrence of the several codons for a given amino acid departed widely from the random frequencies expected. Since over 1000 codons were involved in the total number actually transcribed, on a theoretical basis each triplet for a given amino acid should have occurred near the mean rate of that set. For instance, all the members of the set of four that codes for alanine (Table 7.10) approximated the mean rate of 23, and similarly those of the four that encode valine came reasonably close to the mean frequency of 19. In contrast, the GGU codon for glycine occurred 37 times in the mRNA compared to 16 times each for GGG and GGC and 12 for GGA—a difference of 300% between the most and least frequent members of this set of four. Scanning the table reveals that such deviation from the expected is the rule, not the exception, for only the codons for the two examples cited above and for cysteine approximated a random distribution.

An interesting statistic was pointed out by Fiers and his colleagues (1976). In all cases where codons terminating in –U and in –G, respectively, occur in the complement of a given amino acid, the former consistently greatly exceeded the latter in frequency of occurrence. While it is tempting to speculate that this commonness of uridine in site 3 of the codon might be correlated to its rarity in the interacting site 1 in the anticodon, that deduction completely fails to explain the total absence of adenosine in the latter locus. Thus the constraint against G and the marked preference for U must be concluded to be features of the coding part of the genetic apparatus, not of the translating portion. In this connection, it is especially pertinent to note that, in the entire message, these two nucleotides are the only ones that approach the mean rate of total occurrence in the third site of the codons—265 Gs and 276 Us may be counted there, against the mean value of 268. Contrastingly, C occurs in that terminal location 322 times, and A, only 208; this discrepancy thus provides another display of the nonrandomness of the codons in informational nucleic acids (Table 7.10).

Still another aspect of this phenomenon concerns those three amino acids that are provisioned with six codons each, that is, each is coded by one set of four, plus a pair having a different initial nucleotide residue. In leucine, the mean frequency (14) of the members of the pair approaches that of those in the set of four (16), so that only a slight preference for the latter group is expressed (Table 7.10). In arginine and serine, however, the preference for the components of the set of four codons is strongly marked. Those for serine have a mean frequency rate of 18 contrasting to one of 12 for the pair members, while in arginine the set of four codons average a rate of 16 each compared to a mere 7 for those of the pair. This observation correlates well with the suggestion ad-

Table 7.10
Codon Frequencies in the 1071-Codon-Long MS2 Genome

Codon	Frequency	
Glycine		
GGA	12	
GGC	16	
GGG	16	
GGU	37	
	81	
Proline		
CCA	9	
CCC	10	
CCG	13	
CCU	17	
	49	
Stops		
UAA	1	
UAG	2	
UGA	0	
	3	
Leucine		
CUA	15	65 (16)
CUC	26	
CUG	9	
CUU	15	
UUA	17	28 (14)
UUG	11	
	93	
Alanine		
GCA	21	
GCC	21	
GCG	23	
GCU	26	
	91	
Histidine		
CAC	9	
CAU	6	
	15	
Glutamine		
CAA	17	
CAG	22	
	39	
Lysine		
AAA	19	
AAG	26	
	45	
Asparagine		
AAC	28	
AAU	17	
	45	
Arginine		
CGA	10	62 (16)
CGC	20	
CGG	11	
CGU	21	
AGA	7	13 (7)
AGG	6	
	75	
Aspartic acid		
GAC	22	
GAU	28	
	50	
Glutamic acid		
GAA	16	
GAG	28	
	44	
Phenylalanine		
UUC	29	
UUU	19	
	48	
Tyrosine		
UAC	32	
UAU	9	
	41	
Isoleucine		
AUA	19	
AUC	25	
AUU	12	
	56	
Methionine		
AUG	18 (+2)	
Serine		
UCA	16	73 (18)
UCC	20	
UCG	22	
UCU	15	
AGC	16	24 (12)
AGU	8	
	97	
Valine		
GUA	16	
GUC	21	
GUG	18 (+1)	
GUU	21	
	76 (77)	
Cysteine		
UGC	6	
UGU	6	
	12	
Tryptophan		
UGG	23	
Threonine		
ACA	13	
ACC	21	
ACG	14	
ACU	19	
	67	

Terminal base total frequencies
A, 208; C, 322; G, 265; U, 276.

vanced in the preceding chapter that the pair of codons represents a secondary metabolic and biosynthetic path for each of the amino acids that belong to two groups.

7.2.3. Codon–Anticodon Interactions

Since the nucleosides in the codonal sites 1 and 2, and their anticodonal counterparts in sites 2 and 3 as well, almost universally have standard Watson–Crick bases, there is little need to doubt that their interactions involve the typical pattern of hydrogen bonding. However, the wide variety of types that occupy site 1 of the anticodon raises questions as to the possible survival value of their presence. One conceivable reason for the existence there of modified bases, as well as for the absence of certain standard ones, might derive from the configuration of the two molecules involved, since both the messenger and the transfer RNA molecules are helices. Obviously when two similar helices of like sign come into contact, the chains do not broadly coincide but necessarily merely cross one another at a single point in each given period. Thus if, as shown in Table 3.2, ten nucleosides comprise one period, then the three of either a codon or anticodon would span 108° of arc in the helical molecule. This implies that when the site 1 occupant of a codon in a messenger has reacted with that in site 3 of the anticodon, the hydrogen bonds would be viewed as being vertical to each nucleoside base; simultaneously, the bases of the opposite ends, in sites 3 and 1, respectively, would be inclined to each other at a 72° angle (Figure 7.4). Hence, it is evident, unless extensive unrolling of both molecules occurs, that considerable stress must be present during codon–anticodon interactions between the bases occupying the latter sites (Alden and Arnott, 1973).

Possible Effects of Mechanical Stress. The possible effects of such mechanical stress are made especially clear when adenosine occupies site 1 of the anticodon. In the nitrogenous base of this nucleoside, the only available hydrogen-donor site, an amino side chain, is located at one extreme of the interacting region of the molecule and the sole hydrogen-acceptor site is medial (Figure 7.5A). Hence, when the two RNA molecules interact, the mechanical stress imposed by their conformation may well be conceived to induce wobble relations between the bases at the third codonal and first anticodonal sites (Crick, 1966). Accordingly, the hydrogen-acceptor point of an adenine in this site would become situated adjacent to a similar acceptor residue of the normal complementary base (uracil) in the codon. Since adenine lacks a third active site of any sort, no hydrogen bonding between the opposing molecules could thus occur (Figure 7.5A); hence, adenine in site 1 of the anticodon might not be able to interact with uridine in site 3 of a codon.

In contrast, hypoxanthine, the base of inosine which Crick (1966) conjectured might displace uridine entirely, has its lone hydrogen-donor site located

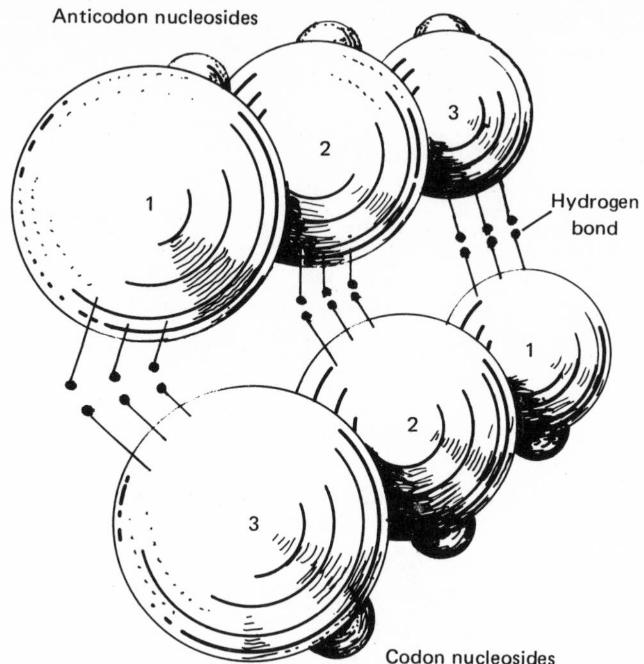

Figure 7.4. Mechanical stress in codon–anticodon interactions. Because both sets of nucleotides in the anticodon–codon interactions are parts of coiled chains, stress between anticodon site 1 and codon site 3 results when the other members are in apposition.

medially and a single acceptor site at one end of the interacting face of the molecule. Thus hypoxanthine and uracil cannot interact in their normal relations in RNA stands but could under stress-induced wobble (Figure 7.5B), as they appear actually to do. Hence, it seems reasonable that simple mechanical stress arising from the conformation of the two interacting helical molecules could be an important factor at the critical sites. That it is not the only such factor becomes evident below as specific codon–anticodon interactions are examined.

Guanosine and Q in Site 1 of the Anticodon. Throughout this section attention is given only to those interactions which have been thoroughly documented through (1) sequencing of the nucleosides in the anticodon of the tRNA and (2) establishment of the specific codons actually recognized by that anticodon through suitable experimental analyses. Some advantage is gained by beginning with an examination of the more common nucleosides, such as guanosine, and their roles in site 1 of the anticodon.

Unmodified guanosine has been found in the first site of the anticodon in a number of sequenced tRNAs, but its codon-recognition properties have been tested in only seven species, six of which are from *E. coli*, which the seventh is

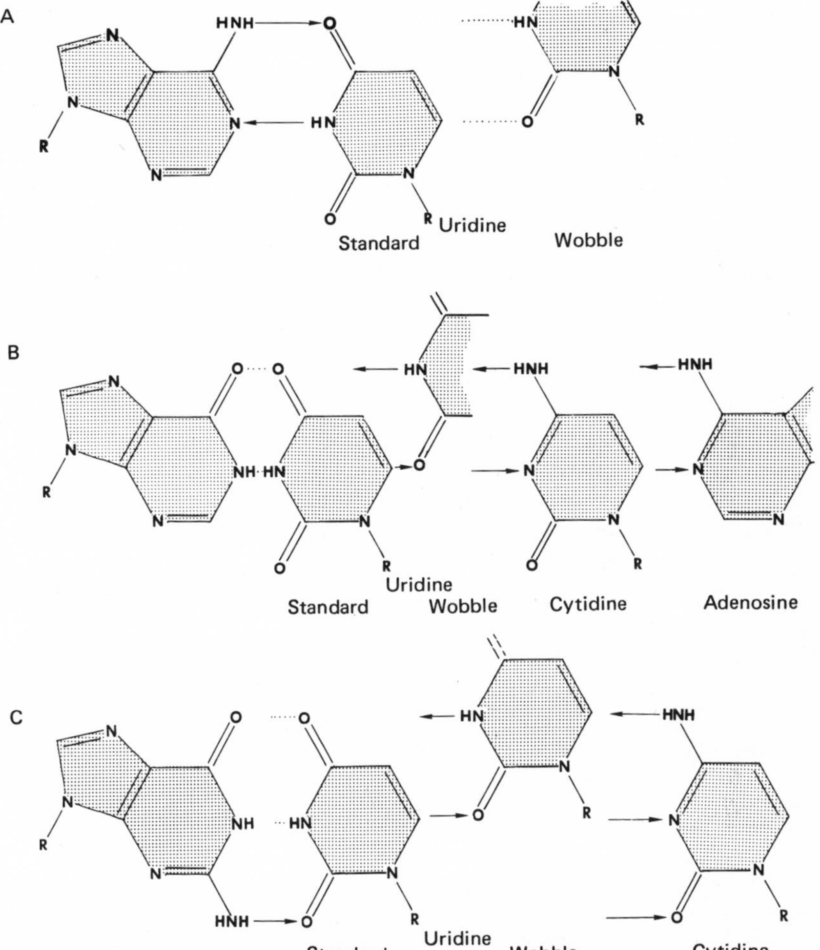

Figure 7.5. Codon–anticodon interactions. In each case, the nucleoside of site 1 of the anticodon is to the left, that of codon site 3 to the right. Successful hydrogen-bond formation is indicated by solid arrows, unsuccessful by broken lines. The arrows fly in the direction of hydrogen donation. (A) Thus adenosine in anticodon site 1 can form two hydrogen bonds with uridine in site 3 of the codon in standard relations, but can form none with it in wobble. (B) In contrast, inosine can form no hydrogen bonds with uridine in standard relations but can form two in wobble position. Cytidine and adenosine can bond twice with it in standard relations; the former can only form one bond with it in wobble position, and adenosine can form none. (C) In standard relations, guanosine can form only one bond with uridine but in wobble can form two; three hydrogen bonds can form between guanosine and cytidine in standard relations.

from the rat (Table 7.11). In all cases, recognition conforms to the patterns expected, base pairing being in standard relations with cytidine or in wobble with uridine (Figure 7.5C). With *E. coli* as the source organism the GCC anticodon of $tRNA_3^{Gly}$ recognizes GGC and GGU (Hill *et al.*, 1970, 1973; Carbon *et al.*, 1970; Squires and Carbon, 1971), the GAA of $tRNA^{Phe}$ reacts with UUC and UUU (Harada and Nishimura, 1972), the GGU of $tRNA^{Thr}$ with ACC and ACU (Clarke and Carbon, 1974), the GAU of $tRNA^{Ile}$ with AUC and AUU (Söll *et al.*, 1966; Caskey *et al.*, 1968; Yarus and Barrell, 1971), and the GAC of $tRNA_2^{Val}$ with GUC and GUU (Yaniv and Barrell, 1969, 1971). Similarly, the GCU anticodon of *E. coli* and rat $tRNA_3^{Ser}$s recognizes only AGC and AGU (Yamada and Ishikura, 1973; Rogg *et al.*, 1975). Although differences in reaction strength between G–C and G–U may have occurred in each case, they are reported only in the phenylalanine tRNA (Harada and Nishimura, 1972). UUC was recognized very strongly by GAA, whereas the recognition of UUU was weaker but still vigorous, a relative reaction indicated by the ratio 5 : 3 in the table. Hence, with the synthetic messengers that were employed, standard Watson–Crick reactions (Figure 7.5C) were indicated here to be stronger than wobble ones. Whether this would be equally true with long-chained natural messengers is unknown.

The modified guanosine known as Q has been studied in a whole series of *E. coli* tRNAs, in which it occupies site 1 of the anticodon (Harada and Nishimura, 1972). The species studied included $tRNA^{Asn}$, in which QUU interacted with AAU and AAC, $tRNA_1^{Asp}$ whose anticodon QUC recognized GAU and GAC, $tRNA_1^{His}$ with the anticodon QUG which reacted with CAU and CAC, and $tRNA_1^{Tyr}$ in which QUA recognized UAU and UAC (Doctor *et al.*, 1966). In all of the codons tested, those ending in U reacted five times more strongly with the Q than did those terminating in C. Thus it was pointed out that the modified nucleoside reacted in the opposite manner from the unmodified guanosine (Harada and Nishimura, 1972). Since the modification does not involve the hydrogen-bonding sites, bonding reactions would be expected to be identical to those of unmodified guanosine (Figure 7.5C). Thus, two conclusions from these observations can be drawn validly. First it is evident that Q is not merely guanosine that chances to be modified, but a substance with very different reactions. These differences would be more apt to be studied if the nucleoside were to receive a distinctive name; quonosine, for example, would have mnemonic value here and would also conform to current practice. Second, it is obvious that there is more to the interactions between nucleosides than mere hydrogen-bonding capabilities. This observation is strengthened by results of experiments in which two different species of tRNAs possessing complementary anticodons were united by means of anticodon–anticodon interactions (Eisinger and Gross, 1974). The association constant K was found to vary widely between modified and unmodified bases, even though the bonding patterns were similar.

Cytosine in Anticodon Site 1. In the ten tested tRNA species which have unmodified cytosine in site 1 of the anticodon, unlike the uniformity that

Table 7.11
Established Codon–Anticodon Interactions, Arranged Alphabetically by Amino Acid Specified

tRNA species	Group	Source	Anti-codon	Codons recognized	Relative reactivity
Asparagine	III	*E. coli*	QUU	AUU, AAC	5:1
Aspartic acid$_1$	I	*E. coli*	QUC	GAU, GAC	5:1
Glutamic acid$_2$	I	*E. coli*	SUC	GAA only	—
Glutamic acid$_3$	I	*S. cerevisiae*	SUC	GAA only	—
Glutamine$_1$	II	*E. coli*	XUG	CAA, CAG	(4:1)
Glutamine$_2$	II	*E. coli*	CUG	CAG, CAA	(4:1)
Glycine$_1$	I	*E. coli*	CCC	GGG only	—
Glycine$_1$	I	*Staphylococcus*	CCC	GGG only	—
Glycine$_2$	I	*E. coli*	UCC	GGA, GGG	None given
Glycine$_3$	I	*E. coli*	GCC	GGC, GGU	None given
Histidine$_1$	II	*E. coli*	QUC	CAU, CAC	5:1
Isoleucine	III	*E. coli*	GAU	AUU, AUC	—
Isoleucine	III	*E. coli*	N^+AU	AUA only	—
Leucine	IV	T4	NAA	UUA only	—
Leucine$_3$	IV	*S. cerevisiae*	CAA	UUG, UUA	UUA slight
Lysine$_1$	III	*S. cerevisiae*$_{\alpha 2880}$	CUU	AAG only	—
Lysine$_2$	III	*S. cerevisiae*	SUU	AAA, AAG	AAG weakly
Lysine	III	Rat	SUU	AAA only	—
Methionine$_M$	III	*E. coli*	C_aAU	AUG	Weakest
Methione$_M$	III	*S. cerevisiae*	CAU	AUG	Strongest 1:3
Methionine$_F$	III	*E. coli*	CAU	AUG	Intermediate
Phenylalanine	IV	*E. coli*	GAA	UUC, UUU	5:3
Serine$_1$	IV	*E. coli*	VGA	UCA, UCG, UCU	5:5:1
Serine$_3$	III	*E. coli*	GCU	AGC, AGU	None given
Serine$_{1,2}$	IV	*S. cerevisiae*	IGA	UCA, UCC, UCU	None given
Serine$_1$	IV	*Rattus*	IGA	UCA, UCC, UCU	None given
Serine$_{2a}$	IV	*Rattus*	CGA	UCG only	—
Serine$_3$	III	*Rattus*	GCU	AGC, AGU	None given
Threonine	III	*E. coli*	GGU	ACU, ACC	None given
Tryptophan	IV	*E. coli*	CCA	UGG, UGA	None given
Tyrosine$_1$	IV	*E. coli*	QUA	UAU, UAC	5:1
Tyrosine$_2$	IV	*E. coli*	CUA	UAG (amber)	—
Valine$_1$	I	*E. coli*	VAC	GUA, GUG, GUU	5:5:1
Valine$_2$	I	*E. coli*	GAC	GUU, GUC	None given

marked guanosine, a considerable range of variability is encountered. In a number of instances only codons terminating in guanosine are recognized. These include *Salmonella* and *E. coli* $tRNA_1^{Gly}$, both of which have the anticodon CCC and recognize only the codon GGG (Hill *et al.*, 1973), and rat $tRNA_2^{Ser}$ whose CGA interacted only with UCG (Rogg *et al.*, 1975). *E. coli* $tRNA_2^{Gln}$, with the anticodon CUG, has been stated to react only with CAG (Folk and Yaniv, 1972). In the latter example it is not clear whether or not this point had been definitely established by the actual experimental results, for examination of the data presented in that reference indicates a reaction had occurred with CAA which was one-fourth as strong as that with CAG. However, the CUU anticodon of *S. cerevisiae* αs 288C $tRNA^{Lys}$ and the CAU of *S. cerevisiae* $tRNA_M^{Met}$ and *E. coli* $tRNA_F^{Met}$ were restricted entirely to the codons AAG and AUG, respectively (Högenauer *et al.*, 1972; Smith *et al.*, 1973). Likewise rat $tRNA_2^{Ser}$, with the anticodon CGA recognized UCG only (Rogg and Staehelin, 1971a,b), and *E. coli* $tRNA_2^{Tyr}$, with CUA, reacted solely with UAG, the amber codon (Goodman *et al.*, 1968). In contrast, *E. coli* $tRNA^{Trp}$, with CCA as the anticodon, recognized both UGG and UGA (Hirsch, 1970, 1971) and *S. cerevisiae* $tRNA_3^{Leu}$ with CAA, correspondingly interacted with UUG and UUA (Lindahl *et al.*, 1967; Kowalski *et al.* 1971). The latter reference reported that the reaction of CAA with UUA was slight, but no differences were cited in the former case.

This capacity for interaction between C and A appears to be largely overlooked (including Crick, 1966). As indicated in Figure 7.6A, cytidine cannot bind with adenine in the standard relations but can form a single bond with it in wobble conformation (Table 7.11). This relatively weak interaction may account for the weakest of the reactions of CAA with UUA and UUG reported above, as well as for its possibly being overlooked in some studies on codon–anticodon reactions and related topics.

One modified cytidine at this site has been sufficiently described (Högenauer *et al.*, 1972). This single instance, in *E. coli* $tRNA_M^{Met}$, had the anticodon C^+AU and reacted only with AUG, as was the case also with *E. coli* $tRNA_F^{Met}$, with the anticodon CAU. However, the latter reacted about three times as strongly as did the former, so it was supposed that the modified nucleoside aided in the recognition of the initiator tRNA by the ribosomes. Unfortunately, the precise nature of this modification of the nucleoside has not been established.

Inosine in Site 1 of the Anticodon. Although inosine has been reported from the first anticodon site in a number of tRNAs, its specific interactions with codons have received notice in only one species from two sources. The species, $tRNA_1^{Ser}$, has been described as having the anticodon IGA, which recognized the codons UCA, UCC, and UCU in both yeast and the rat (Rogg and Staehelin, 1971a, b; Kruppa and Zachau, 1972). No differences in reaction strength with the several codons were reported.

The basis for this wide range of recognition capabilities on the part of

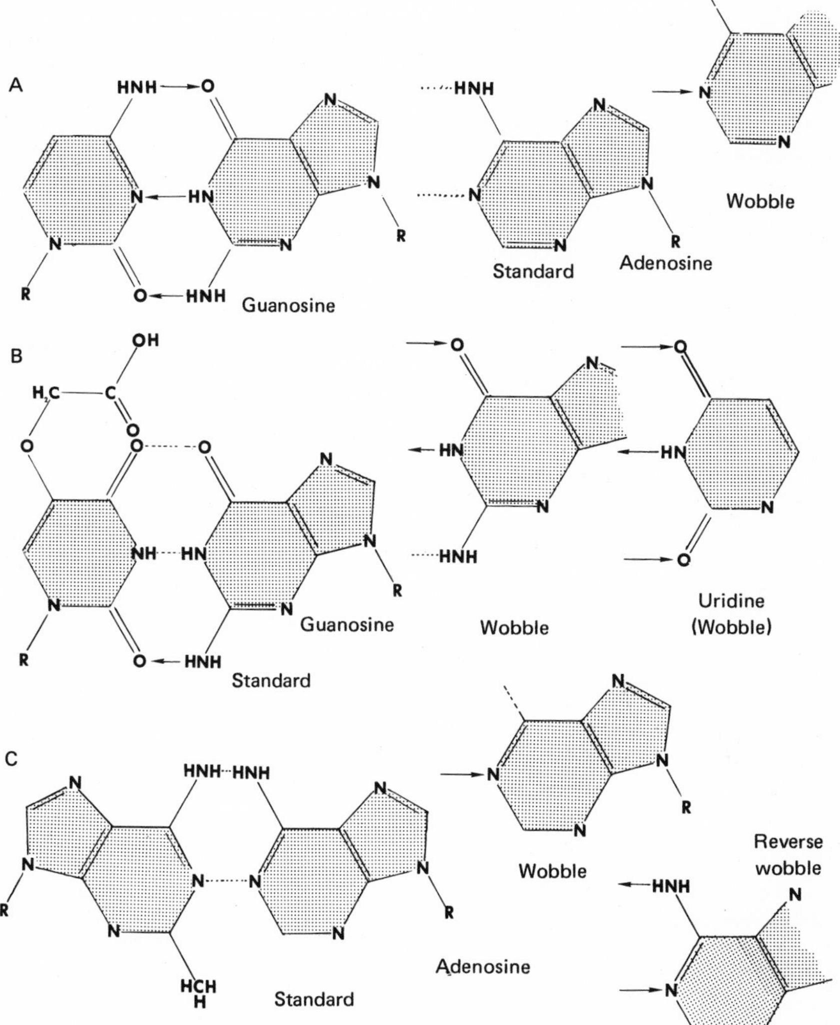

Figure 7.6. Bonding properties of occupants of anticodon site 1. (A) Cytidine can form three bonds with guanosine in standard and one in wobble relations. With adenosine no bonding can occur in standard relations, whereas one can do so in wobble. (B) To explain the observed interactions, the radical on C^5 of V is suggested to play a role in hydrogen bonding. (C) The bonding reactions between N and A are unclear at present; here it is proposed that reverse wobble might provide a reasonable explanation for the observed strong reactions with adenosine.

inosine was first suggested by Crick (1966). With adenine and cytosine, in standard relations the base hypoxanthine can form two hydrogen bonds. In contrast, it can produce none with uracil in the standard conformation but can form two with it in wobble position (Figure 7.5B). Accordingly Crick correctly predicted that inosine in the first site of the anticodon would recognize those codons that end in A, C, and U, as described above.

Uridine and V as Occupants of Site 1. Uridine is unusual in site 1 of the anticodon, for it is known only from $tRNA_2^{Gly}$s from various prokaryotic sources. Since in half of these, including those from bacteriophages T2, T4, and T6, it is modified in an undetermined fashion (Stahl *et al.*, 1973), the unmodified base is infrequent indeed. The codon-recognition properties of the anticodon UCC of this species of tRNA from *E. coli* has been experimentally tested (Carbon *et al.*, 1970; Squires and Carbon, 1971) and found to recognize those expected with wobble, namely GGA and GGG (Figure 7.5A,C). If different strengths of reaction were noted, none were reported.

The nucleoside V is little more abundant in this site than is the above, occurring thus far only in *E. coli* $tRNA_1^{Ser}$, $tRNA_1^{Ala}$, and $tRNA_1^{Val}$, as well as in *Salmonella* $tRNA_1^{Val}$. Fortunately the codon reactions have been studied in two species and proved to be identical. The VGA anticodon of $tRNA_1^{Ser}$ recognized UCA, UCG, and UCU (Ishikura *et al.*, 1971a,b), and the VAC of $tRNA_1^{Val}$ of *E. coli* reacted with GUA, GUG, and GUU (Kimura-Harada *et al.*, 1971). The latter study examined the relative strength of reactions in both species and found them similar in each case: V interacted equally strongly with A and G, but the reaction with U was only one-fifth as strong (Morikawa *et al.*, 1974). Thus while V, which is 5-oxyacetyluracil, reacts with another pyrimidine, it does so only weakly. This weakness of reaction, it should be noted, exists despite the formation of three hydrogen bonds in the wobble relation (Figure 7.6B); in contrast, the strong reaction with guanine also takes place in wobble and only two hydrogen bonds can form. These observations reinforce the statement made earlier that more than simple hydrogen bonding is involved in base-pair formation.

S in the Anticodon. A modification of uracil that is far more complex than V, known as S, occurs in the anticodon of a small number of known tRNAs. This is really a group of substances related to 2-thiouracil-5-acetic acid methyl ester and has been reported from both prokaryotic and eukaryotic sources; however, there are conflicting reports in regard to its presence in both groups as well as to other aspects. No disagreement exists that both yeast $tRNA_3^{Glu}$ and $tRNA_2^{Lys}$ have this nucleoside in site 1 of the anticodon (Yoshida *et al.*, 1971; Sen and Ghosh, 1973; Kobayashi *et al.* 1974), although that of the latter species has been indicated as being modified (Madison *et al.*, 1972). In *E. coli* $tRNA_2^{Glu}$, the presence of a complexly modified S was reported (Ohashi *et al.*, 1970; Singhal, 1971), which has been identified as 5-methylaminomethyl-2-thiouridine. The codon recognition properties of S-containing anticodons also are unclear at present. The SUC found in *E. coli* and yeast $tRNA_3^{Glu}$ purportedly related only to GAA (Sekiya *et al.*, 1969; Yoshida *et al.*,

1971; Kobayashi *et al.*, 1974), despite the differences in structure of the base. If this is actually the case, it is strictly in opposition to the wobble hypothesis which states that no tRNA will recognize solely those codons ending in A. In contrast, Woodward and Herbert (1972) demonstrated that $tRNA_2^{Lys}$ transferred the correct amino acid into all the lysine positions in hemoglobin when tested in a cell-free reticulocyte system. Thus its anticodon SUU recognized AAA and AAG equally well (Madison *et al.*, 1972, 1974; Madison and Boguslawski, 1974). Perhaps this difference in reaction pattern resulted from the undetermined modification of the S that was reported to occupy site 1 of this anticodon.

The Nucleoside N. The last nucleoside in this site of the anticodon whose codon-recognition properties have been studied is that one designated N, which is confined to prokaryotes. Unfortunately this same letter has been used in connection with various unknown nucleosides occurring elsewhere in tRNA molecules, such as in the extra arm of *E. coli* $tRNA_2^{Val}$ (Table 7.4; Yaniv and Barrell, 1971) as well as for an unidentified nucleoside in site 1 of the anticodon in *E. coli* $tRNA_1^{Gln}$, here replaced by an X to avoid confusion (Folk and Yaniv, 1972). Correctly used in the anticodon, N has been identified as 2-methyladenosine (Figure 7.6C). The reactions of this nucleoside with the terminal one in the codon are most unexpected, for it reacts either entirely or most strongly with codons terminating in adenosine. Only one species containing N is known from *E. coli*; that bacteria's $tRNA_1^{Ile}$, with the anticodon N^+AU, recognized only AUA (Harada and Nishimura, 1974), and one bacteriophage T4 species, $tRNA_1^{Ile}$, with the anticodon NAA, responded solely to UUA (Pinkerton *et al.*, 1972, 1973).

The possible basis for the specific interaction reported above between 2-methyladenosine and unmodified adenosine is of particular interest (Figure 7.7B). In standard positions, these two cannot form any hydrogen bonds and in wobble only one such bond can result. However, this same relationship could exist between two unmodified adenosines, and therefore it provides no biological importance to the methyl component. On the other hand, if it is envisioned that wobble could occur in the opposite direction, as indicated in the figure, then two hydrogen bonds could form. What remains unclear is why this nucleoside does not react with guanosine more strongly, for it is able to develop two hydrogen bonds in standard relations with this substance (Figure 7.7B). Perhaps the proximity of the methyl residue of the modified adenosine to the amine of the guanosine introduces constraints through space requirements, much as two keto radicals in apposition appear to do (Crick, 1966).

7.3. SUMMARY OF tRNA STRUCTURAL FEATURES

The structural features of tRNAs of possible functional significance may be summarized in two categories, those of actual, and those of virtual, universality (Table 7.12). In addition, a few others seem worthy of note, as they

Table 7.12
Characteristic Features of the tRNAs by Groups, Including Universal Ones[a]

Arm	Site	Group I	Group II	Group III	Group IV	Universal	Nearly universal
I	0′	U freq. (P)[a]	U abs.	U rare	U rare	U unmod., C abs.	—
	1	U rare; C, A, ψ abs.	U freq. (P); ψ(E) C, A abs.	U abs.; A freq.	U, ψ abs.	—	—
	2	A abs.	A abs.	A occas.; U abs.	A occas.	ψ abs.	—
	3	—	—	A abs.	—	ψ abs.	—
	4	A abs.; C_2 pres.[a]	—	—	—	ψ abs.	—
	6	G rare	G rare	G predom.	G predom.	—	—
	7	—	—	—	—	C abs.	—
(Bend 1)	B	C pres.; G abs.	C pres.; G rare	C abs.; G freq.	C abs.; G freq.; U, U_t pres.	—	—
II	1	U occas.; G_2 abs.	G_2 rare (E)	G_2 freq. (E)	C pres.; G_2 freq. (E)	A, ψ abs.	—
	2	—	—	A. occas.	—	—	Pyrim. pres.
	3	A pres.	—	—	C_a pres. (E)	—	—
	4	ψ freq.; G abs.; A abs.	A freq.; U abs.	U occas.	U abs.	—	—
	A	—	—	—	—	A (rarely A_1) pres.	—

	B	—	—	—	—	Purine pres.; rarely vacant	—
	C	—	—	Pyrim. occas.	—	—	Usually vacant
	D	Often vacant	—	G, A, U occas.	—	—	—
	E	Never vacant	Never vacant; A abs.	Often vacant	Often vacant; A abs.; U abs.	—	—
	F	—	—	—	—	G or G_2'	—
	G	—	—	—	—	G pres.	—
	H	Usually vacant	C abs.; U (rare)	Often vacant	G (freq.); Q (rare)	A abs.	—
	I	U; G (occas.), never vacant	A (rare); D abs.	C abs., usually vacant	Often vacant	—	—
	J	G abs.; A abs.	A abs.	—	D (rare)	—	—
	K	—	—	—	—	A pres.	—
(Bend 2)	-	A predom., C pres.	C pres.	C abs.	C, G abs.; ψ (occas.)	—	—
III	2	ψ abs.	ψ abs.	ψ pres.	ψ pres.; U abs.	—	—
	3	C_M pres.	—	—	—	ψ abs.	—
	4	—	—	—	—	A, ψ abs.	—
	5	C predom.; U abs.	—	U abs., ψ rare	A predom.; U occas.	—	—

Table 7.12. Cont'd.

Arm	Site	Group I	Group II	Group III	Group IV	Universal	Nearly universal
	5'	G predom.; A, U abs.	A, U occas.	A, U abs.	ψ_m rare; U abs.; ψ predom.	—	—
	A	ψ freq.	ψ abs.; U_m, C_t occas.	ψ abs.; C_2 pres. (P); C_3 rare.	ψ, C_3 freq.	—	—
	B	—	—	C occas.	—	—	U pres.
	C	A predom.; I_m, A_2 pres.	A_2 predom. (P)	A_t, A only	A_5(P); A_i usual (E); ψ pres.	—	—
	D	All bases + C_5 pres.	A, C, U, ψ, only	A only	A predom.; C rare	—	—
IV (short)	A	—	C pres. (P)	U abs.	—	ψ abs.	—
	B	A; C (occas.)	A freq.; C abs.	A, C abs.	A abs.; C (rare)	—	—
	C	Usually vacant; A abs.	U (occas.); G abs.	—	G abs.	—	—
	D	A freq.	—	—	C (rare)	—	—
	E	—	—	—	—	Pyrim. only	—
V	1	U occas.; C_5 freq.	C_5 abs.	C_5 abs.	C_5 rare	ψ abs.	—
	2	C_5 (rare)	—	—	—	ψ abs.	—
	3	C pres.	C abs.	C pres.	C abs.	ψ abs.	—

5	A rare	—	—	—	—	G usually pres.
1'	ψ abs.; A rare	ψ abs.	ψ occas.; G_m rare	ψ occas.	—	—
A	U pres.	T only	A occas.	ψ rare	—	T typical
B	G occas.	ψ only	U freq.	ψ only	—	ψ characteristic
C	—	—	—	—	C pres.	—
D	—	—	—	—	Purine pres.	—
E	—	—	—	—	A or A_1(E) pres.	—
F	U predom.	Purine only	A predom.	A, U predom.	ψ abs.	—
G	C, A pres.	C, A abs.	C, A pres.	C pres.; A abs.	G, ψ abs.	—

[a]P, Prokaryotic trait; E, Eukaryotic trait. Underline indicates unique feature of entire group.

characterize particular taxa or amino acid groups or have a known functional importance:

7.3.1. Characters of a Universal Nature

1. The "cloverleaf" pattern of secondary structure always possesses five arms, only two of which (II and IV) are variable in length.
2. The triplet –CAA is consistently present at the 3′-terminus of the molecule.
3. In the amino acid arm on the 5′-side, cytidine is absent from site 7 and adenosine from site 2. On the 3′-side, guanosine never occupies site 1′.
4. In the preterminal site 0′, preceding the –CCA terminus, cytidine is absent except when modified or when paired to a base added at site 0.
5. Uridine, in prokaryotes frequently modified to 4-thiouridine, is always the base in site A of bend 1.
6. The doublet GG is universally present in proteinogenic tRNAs in sites F and G located near the center of the unpaired region of arm II; it is absent only in the two $tRNA^{Gly}$s known to participate in cell wall formation.
7. Site K of the unpaired region of arm II is constantly occupied by adenosine.
8. Arms III and V always contain a stem of 5 pairs of nucleosides.
9. Site B in the unpaired region of arm III is invariably occupied by uridine.
10. In short types of arm IV, site C is never occupied by cytidine.
11. Arm V has a loop of seven unpaired nucleosides.
12. Guanosine is always the occupant of site 5 in the stem of arm V and cytosine uniformly occupies site 5′.
13. Cytosoine is universally present in site C of the unpaired sequence of arm V and a purine (usually guanosine) is uniformly adjacent to it in site D. Either adenosine or 1-methyladenosine consistently follows in site E.

7.3.2. Characters Nearly Universal in Occurrence

1. Seven pairs of nucleosides are usually present in the amino acid arm, the three prokaryotic $tRNA^{His}$s, with eight pairs, providing the only known exceptions.
2. A single unpaired base occupies site 0′, near the 3′-end of the molecule; histidine tRNAs from prokaryotes again are the sole exception.
3. In the stem of arm II, the first pair is nearly always G–C, but of the proteinogenic types, *E. coli* $tRNA^{TYR}$s are exceptional in having C–G.

4. The second pair of the same stem consistently is Py–Pu, but in the translation initiator of *E. coli* ($tRNA_F^{Met}$) the pair is A–U.
5. Site A of the unpaired region of arm II is nearly always occupied by adenosine; the exceptions appear to have arisen by insertion of an extra base in the stem.
6. Site A of the anticodon arm (III), is, with one possible exception, always occupied by a pyrimidine. The possible exceptional nucleoside, found in *E. coli* $tRNA^{Ser}$, is unidentified as yet.
7. Guanosine is absent from site D of the anticodon arm, except in tRNAs that do not function in protein synthesis.
8. In short types of arm IV, the usual occupant of site C is 7-methylguanosine.

7.3.3. Characters of More Restricted Significance

1. The translation initiator of prokaryotes has several noteworthy peculiarities. One of these is the presence of A–U as the second pair of the stem in arm II, in place of the otherwise consistent Py–Pu pair.
2. Group I tRNAs are unique in having: (A), Ψ frequently in the fourth site of the stem of arm II; (B), U frequently in site 4′ of arm II. Also they usually have C–G and never U–A in sites 5 and 5′ of arm III, the postanticodonal nucleoside (in site C) has either a standard base or at most is methylated. Only short types of arm IV occur, and these typically contain four (rarely three) nucleosides, whereas in the other groups they always contain five or even six.
3. In the relatively few group II tRNAs that have been sequenced, site D of the anticodon arm is occupied only by pseudouridine or uridine; the nuceloside in site C of the same arm is either methylguanosine or 2-methyladenosine.
4. Group III tRNAs always have adenosine in site D of the anticodon arm; site C of the same arm is always *N*-[9-(β-D-ribofuranosyl)-purine-6yl-carbamoyl] threonine.
5. Group IV tRNAs usually have A–Ψ in sites 5 and 5′ of arm III, U–A being the only other acceptable pair in those loci; site D in the same arm is nearly always occupied by adenosine, while site C is A_s in prokaryotes, and variously Y (or modifications thereof), 1-methylguanosine, or A_i in eukaryotes.

8

Reactive Sites and the Evolution of Transfer RNAs

Before the evolutionary relations of tRNAs can be examined, those specific sites in tRNA molecular structure that are of possible importance in a particular function need examination, in order to distinguish functional and evolutionary influences. Among the activities in which a given nucleotide may be especially significant are included the maintenance of a secondary or tertiary structure, recognition by the ligase, and attachment to the ribosome, as well as those codon–anticodon interactions that have already received attention (Chapter 7; Section 7.2).

8.1. REACTIVE SITES OF tRNAs

The universal and nearly universal sites of tRNAs have been especially subjected to suggestions as to their possible roles in recognition by the tRNA ligases. The amino acid arm in particular has received mention as being active in this function, largely because of certain uniformities of structure and its known participation in aminoacylation. For these same reasons, however, the arm has already been given adequate consideration and does not need to receive further attention here. Involvement in tertiary structure has been postulated recently for certain others of these constant features, and if these sites are examined now, it may be possible to eliminate them from serious consideration as active candidates for roles in recognition or comparable functions. In this way those that are actual participants in such activities may also be brought out more clearly.

8.1.1. Sites Active in Tertiary Structure

While little has been done on prokaryotic tRNAs, structural studies generally imply similar configurations for these nucleic acids regardless of their source. Since most investigations have centered on a single type, yeast $tRNA^{Phe}$, this generalization may not be found universally valid as studies become more diversified. Yet the numerous similarities of composition and structure that have been noted to exist among all tRNAs, regardless of whether they are from prokaryotic or eukaryotic sources, seems to suggest uniformity, at least to a large degree (Bolton and Kearns, 1975). This assumption has received further support through the discovery that bend 1 interacts with arm II in the same way in *E. coli* $tRNA^{Arg}$ and $tRNA^{Val}$ as in eukaryotic species (Wong *et al.*, 1973). Nevertheless, a few details in the interaction do differ from one tRNA species to another (Eisinger and Gross, 1975).

Although a variety of conformations was proposed for the tRNA molecule by earlier studies (Levitt, 1969; Abraham, 1971; Cramer, 1971), the consensus presently is that the native nucleic acids are in the form of an inverted L (Kim *et al.*, 1973, 1974a,b; Riesner *et al.*, 1973; Thomas *et al.*, 1973; Delaney *et al.*, 1974; Jones and Kearns, 1974; Suddath *et al.*, 1974). A modified diagram of this concept is shown in Figure 8.1. There also appears to be general agreement that this configuration remains unaltered by aminoacylation and in solutions of normal ionic strengths (Chatterjee and Kaji, 1970; Cole and Crothers, 1972; Cole *et al.*, 1972; Hashimoto *et al.*, 1972c; Wong *et al.*, 1973; Gamble and Schimmel, 1974; Yang and Söll, 1974), but this belief may not be completely sound (Caron *et al.*, 1976). Moreover, under acidic conditions at a pH lower than 5.5, a conformational change has been noted which remains stable when neutral conditions are resumed (Bina-Stein and Crothers, 1974).

Reactions between Arm II and Bend 1 Nucleosides. The foot of the inverted L (Figure 8.1) is formed largely by the amino acid stem of the tRNA molecule, while the upright consists of the remaining parts compressed by folding and spiraling. Most of arm II is suggested to be folded inwardly in such a fashion that the residues in sites 3–3′ react with the nucleoside in site B of bend 1. Simultaneously the occupants of arm II sites A and K, usually adenosine in the former locus and always so in the latter, interact with the nucleoside in site A of that bend, which is the universally present uridine in eukaryotes and the nearly constant 4-thiouridine in prokaryotes.

Interactions between Arms II and IV. As a result of the helical secondary structure typical of nucleic acids in general, the extra arm in the three-dimensional molecule (arm IV) lies adjacent to the basal region of the arm II stem (Figure 8.1). This arrangement enables interactions to occur between the typical purine occupying the extra arm's site 3 and the frequent G–C pair in arm II's sites 4 and 4′. This interaction does not appear to be universally possible, for, not uncommonly, unpaired nucleosides are in sites 4 and 4′, and on

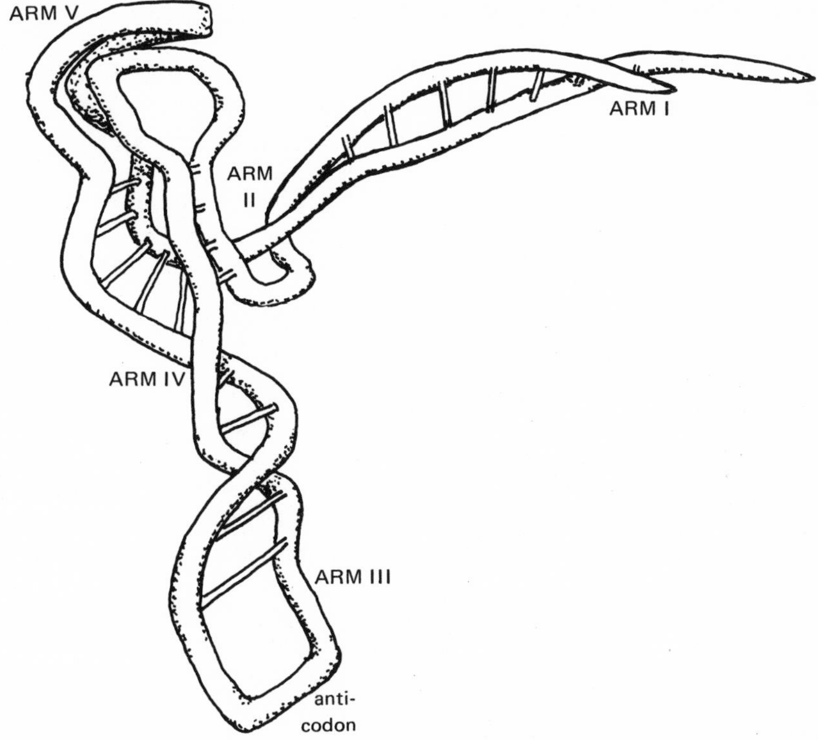

Figure 8.1. Three-dimensional structure of a tRNA. Most tertiary structural models of tRNAs suggest as here that arm I remains at a right angle to the rest of the molecule and that arm V is folded against arm II.

some occasions site 3 of arm IV appears to be unoccupied (see Tables 7.2 and 7.4), or may be occupied by a noncomplementary nucleoside. Thus the interaction may augment tertiary structural binding when a suitable base is present, but in view of its frequent absence must be looked on as nonessential.

In contrast, another set of interactants brought together by the helix must be deemed essential because of its universality. In site 5 of the extra arm, a pyrimidine is always present which has been shown to form hydrogen bonds with the equally catholic purine of site B in arm II. The complete constancy of this interaction would be dependent upon an as yet undemonstrated ability of 5-methylcytidine to form hydrogen bonds with adenosine as well as with guanosine, as in yeast $tRNA_2^{Tyr}$ and $tRNA_F^{Met}$, respectively. However, while not confirmed, the interaction of these nucleosides remains a distinct possibility, for hydrogen bonding between C and A is documented by known interactions of certain codons and anticodons (Chapter 7; Section 7.2.3).

Interactions between Bend 2 and Arm IV. Because of the same spiraling mechanism, the first nucleoside of arm IV is postulated to be brought into proximity to the sole site of bend 2 (Kim *et al.*, 1974a,b). In the species studied, yeast $tRNA^{Phe}$, this interaction was indicated as taking place between two purine nucleosides, adenosine and 2-dimethylguanosine. This conjectured interaction, if it proves actually to occur, must be considered of supplementary importance only, for in yeast $tRNA^{Val}_{A,B}$ the adenosine of arm IV would have to react with adenosine in bend 2 (Tables 7.2, 7.4). The proposed reactions receive some support by the frequency of purine in each of the sites, one striking example being provided by the two *E. coli* $tRNA^{Met}$. In the initiating species adenosine occurs in site 1 of arm IV and guanosine in bend 2; on the other hand, in the interior species, the two are transposed, with guanosine appropriately located in arm IV and adenosine in bend 2.

Unfortunately, few studies appear to have been conducted on any representative species possessing a long arm IV, aside from one investigation on those conformational changes in *E. coli* $tRNA^{Tyr}$s that occurred with melting (Yang and Crothers, 1972). When further researches have been conducted on this type, deeper insight may be gained into homologies between long and short arm IV nucleosides, without which their correspondences must remain a mystery.

Interactions between Arms II and V. In the Kim *et al.* (1974a,b) configuration, arm V is folded counterclockwise around the stem of the inverted L, and thus the tip approaches the unpaired region of arm II. Nuclear magnetic resonance studies of six tRNAs have suggested that these two arms are in a continuous helix (Schulman *et al.*, 1973). As a consequence, some of the universal and virtually constant features of the tRNA molecule, the dual guanosines of arm II and the TΨC of arm V, are brought into proximity. The guanosine that is a constant occupant of site G in arm II is indicated to bond to the cytidine in site C of arm V, whereas the first G (in site F) interacts with the pseudouridine usually found in site B of the latter arm. When pseudouridine is absent from this site in rare noninitiating proteinogenic species, such as wheat $tRNA^{Gly}_{3B}$ (Marcu *et al.*, 1973, 1974), it is replaced by a guanosine incapable of forming this bond. Hence, this interaction can be viewed as supplementing the hydrogen bonding provided by the guanosine of site G with the cytidine of site C; while dispensable, its general desirability is obvious from the near-catholicity of the pseudouridine in site B. Even in eukaryotic initiating $tRNA^{Phe}$s in which pseudouridine is replaced with uridine, a similar interaction with the guanosine occupant of arm II's site F can occur with the aid of wobble relations.

In addition the unpaired region of arm V has an internal structural interaction that may be of importance in recognition by the ribosomes. This involves the usual ribothymidine located in site A and the adenosine or 1-methyladenosine constantly present in site E. By thus pairing, the T–A or T–A_1 combination provides a fold that sets off the ΨCG triplet that lies between them in the form of a secondary loop; in this looplet, the G comes to lie adjacent to

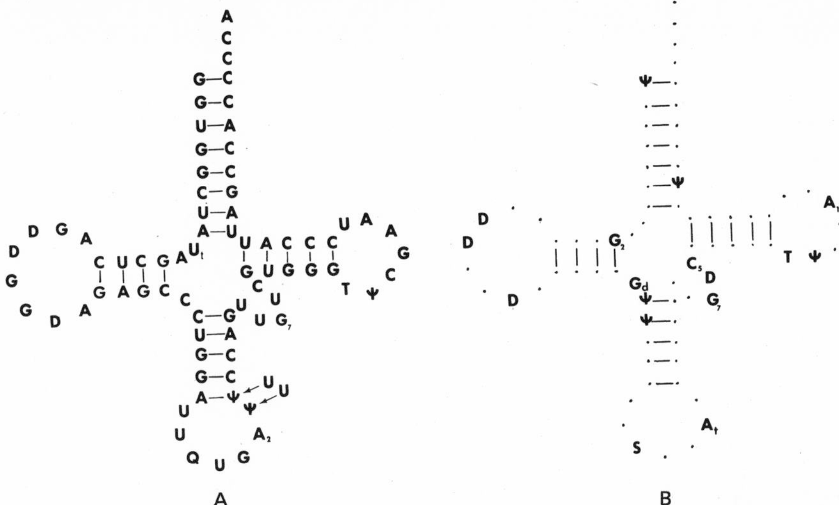

Figure 8.2. Modification of nucleotides in tRNAs. (A) Normal $tRNA^{His}$ from *S. typhimurium*: the sites indicated by $U \rightarrow \Psi$ fail to become modified in the mutant form *hisT1504*. (B) Modified nucleotides of yeast $tRNA_2^{Lys}$: only the nucleotides are shown that are modified posttranscriptionally. If each (or each group) of the 16 modifications is the result of the action of a separate enzyme, about 15 modifying enzymes may be necessary to mature the precursor of this tRNA.

the Ψ, so that wobble pairing possibly could occur between them. Thus four of the seven unpaired nucleosides of arm V are shown to be involved in either secondary or tertiary structure. The ribothymidine is occasionally replaced in normal tRNAs, including wheat $tRNA_{3A,B}^{Gly}$ and mouse $tRNA^{Val}$, but as the replacement is uridine (pseudouridine in Rous sarcoma $tRNA^{Trp}$), no bonding problems arise. In contrast, all determined eukaryotic initiating species ($tRNA_F^{Met}$) have an adenosine in place of the ribothymidine; hence no interaction with the nucleoside in site E is possible. The less extensive folding that ensues (Figure 8.2) may provide a valuable feature for recognition purposes by ribosomes in distinguishing initiating and internal types of $tRNA^{Met}$.

It is to be noted that much of the unpaired regions of both arms II and V and about half of the stem of the former are involved in structure, as are also all of bends 1 and 2 and 60% of short-type arm IV. On the other hand none of the nucleosides of either the amino acid arm or the arm III (anticodon arm) plays any role in tertiary configuration (Reid *et al.*, 1972).

8.1.2. Sites Possibly Active in Recognition by Ligases

It has long been established that translation of the codons contained in the messenger is totally independent of the species of amino acid that is attached to the tRNA (Chapeville *et al.*, 1962; von Ehrenstein *et al.*, 1963). Accuracy of translation depends entirely upon the specificity with which a tRNA ligase recognizes both the tRNA and the amino acid before it catalyzes the esterification of the former by the latter (Chambers, 1971). This observation, which is demonstrated further by misacylation studies reported later, reinforces conclusions reached in Chapter 6. There it was pointed out that biochemical investigations into the origin of the genetic code which proceed by direct interactions between polynucleotides and amino acids have no basis in reality.

The literature seeking to understand how the ligase carries out aminoacylation and recognizes the tRNA is quite extensive. Much of it is based either directly or indirectly on various genetic mutants involving tRNAs in bacteria and yeast, but another large segment derives from experiments designed specifically for the purpose. As a consequence, sites in every conceivable region of the tRNA molecule, with the exception of arm IV, have been proposed as being important in ligase recognition.

It should be clear that no single feature need be universally involved in these processes in each and every species of tRNA. However, it is self-evident that any such site or combination of sites must be unique in each case. For example, the triplet GCT in the 5′-side of the amino acid stem cannot be *the* site for recognition of prokaryotic glycine tRNAs by the ligase, because the identical sequence occurs also in *E. coli* $tRNA_2^{Val}$ and $tRNA_1^{Leu}$. Furthermore, some sort of evolutionary relationships in these vitally important sites should be perceptible between isoaccepting species from prokaryotic and eukaryotic sources. In view of the successes with heterologous systems in the aminoacylation of tRNAs and, especially, of the strong homologies readily discernible among the numerous tRNAs whose sequences have been determined, it is unlikely that sudden changes in recognition sites among isoaccepting species should be other than a rare occurrence.

The Amino Acid Arm and Aminoacylation

Adequate attention has already been given to the preterminal nucleoside and paired region of this arm and their possible roles in ligase recognition (Chambers, 1971; Celis *et al.*, 1973), but an essential step in these processes, the actual attachment of the amino acid to the –CCA terminus of the tRNA molecule, has not been reviewed. Since this short universal sequence has been the subject of much experimentation, it appears to be an ideal point at which to begin discussion of aminoacylation.

Attachment of the Amino Acid. Relatively few studies have been made of the possible biological basis for the particular sequence of nucleosides or of the underlying basis for the universal requirement of adenosine as the site of attachment. Probably the most important of those which have been completed is one in which either AMP or CMP was attached to Merrifield resin and the interactions which occurred with certain phenylalanine- or glycine-containing substances were studied (Harpold and Calvin, 1973). The results indicated that glycine was more reactive than phenylalanine and that both amino acids combined more readily with adenosine than with cytidine. It was not determined, however, at what site of the nucleoside the amino acid had become attached, nor were CMP and UMP tested. Formycin monophosphate has been found to substitute for ATP in this reaction, but its velocity of reaction is about 50 times slower than the latter's (Maelicke *et al.*, 1974). Although much slower, its fluorescent qualities offer an advantage for experimental studies.

The –CCA Terminus in Aminoacylation. A number of related researches have attacked the site of attachment problem by modifying the final adenosine of the –CCA segment. The exact position of the aminoacyl residue on the terminal nucleoside had long defied solution by ordinary chemical and biochemical means, because of the extremely rapid acyl migration on the vicinal *cis*-diol group of the adenosine (Wolfenden *et al.*, 1964). However, by using modified yeast $tRNA^{Phe}$s, a solution has recently been provided (Sprinzl and Cramer, 1973; Chinali *et al.*, 1974). One of the modified molecules had the terminal sequence –pCpCp2′-dA, whereas another had –pCpCp3′-dA, but a few additional modifications were also used. Several important conclusions were reached, one being that aminoacylation, at least in the test species, took place exclusively at the 2′-hydroxyl group of the native tRNA. However, it was further shown that during peptide formation, the aminoacyl residue migrated to the 3′-site. Under other experimental procedures, the 2′-hydroxyl attachment point for the amino acid has been similarly ascertained (Ofengand *et al.*, 1974).

In a complex series of modification experiments, a number of instructive details concerning this important triplet of nucleosides were brought to light (Tal *et al.*, 1972). Modification consisted of stepwise degradation of the 3′-terminal sequence. When reduced to –NCC,* AMP could be accepted but not CMP; –NC accepted both nucleosides in a 1 : 1 molar ratio, and –N did likewise but in a molar ratio of 2 CMP : 1 AMP. Thus constraints exist as to the proportions and numbers of nucleosides permissible in the sequence. Transfer RNAs ending in –NCA could not be aminoacylated, suggesting that the full sequence is essential to recognition of either the tRNA or the amino acid by the ligase.

Transfer RNA Ligases and the Charging Processes

As already pointed out in Chapter 4, the ligases which bind the amino acids to their respective tRNAs are highly varied as a group (Kisselev and Favorova, 1974). The variations commence with the number of subunits present, that is, the quaternary structure, four combinations having thus far been recognized: (1) The simplest type (α_1) is monomeric, representatives of

*N is any nucleotide, that is, either guanosine, adenosine, cytidine, or uridine.

which include yeast aspartyl-, isoleucyl-, and leucyl-tRNA ligases (Gangloff *et al.*, 1973; Hanke *et al.*, 1974); (2) the second type, dimeric (α_2), consisting of two copies of the same subunit; (3) another ($\alpha_1\beta_1$) is a combination of two different subunits; (4) while the last ($\alpha_2\beta_2$) contains two copies of each of two different polypeptides, as in phenylalanyl-tRNA ligase of yeast (Holler *et al.*, 1975).

Another widely variable feature is total size, which is not directly correlated to the quaternary structure. For example, two *E. coli* ligases of the α_1 type, those for proline and tryptophan, have molecular weights of 47,000 and 37,000, respectively, whereas two monomeric (α_1) varieties from the same organism, those for valine and isoleucine, have molecular weights of 100,000 and 118,000 in the same order (Chapeville and Rouget, 1972). Aside from molecular size and organization, a few other properties of tRNA ligases have been brought to light in recent years, but these are best presented species by species, after the general mode of charging has been reviewed.

The Processes of Charging tRNAs

The biosynthesis of aminoacyl-tRNA involves complex reactions between the ligase and three substrates, namely the amino acid, ATP, and the tRNA (Nazario and Evans, 1974). From the interactions, three substances are produced, the aminoacylated tRNA, AMP, and pyrophosphate. No single sequence of steps is followed universally, but two major patterns of interaction exist, with minor variations on one or the other in several cases. The more frequent sequence of events appears to be as follows, in which the first reaction results in activated amino acid (Allende and Allende, 1971):

$$\text{Amino acid (AA)} + \text{ATP} + \text{ligase} \rightleftharpoons \text{AA} \cdot \text{AMP} \cdot \text{ligase} + \text{PP}_i$$
$$\text{AA} \cdot \text{AMP} \cdot \text{ligase} + \text{RNA} \rightleftharpoons \text{AA} \cdot \text{tRNA} + \text{AMP} + \text{ligase}$$

The second major series of events is characterized by arginyl-tRNA of *E. coli* (Ravel *et al.*, 1965; Mehler and Mitra, 1967; Lapointe and Söll, 1972b), in which the steps outlined below are believed to be followed:

$$\text{tRNA} + \text{ligase} \rightleftharpoons \text{tRNA} \cdot \text{ligase}$$
$$\text{tNRA} \cdot \text{ligase} + \text{ATP} \rightleftharpoons \text{tRNA} \cdot \text{ligase} \cdot \text{ATP}$$
$$\text{tRNA} \cdot \text{ligase} \cdot \text{ATP} + \text{AA} \rightleftharpoons \text{AA} \cdot \text{tRNA} + \text{AMP} + \text{ligase} + \text{PP}_i$$

Whether Mg^{2+} ions are required throughout these steps is not yet clear, but a recent study has intimated that they are and that spermidine will not serve as a substitute (Thiebe, 1975).

Deviations from these two basic sets of interactions are detailed with the descriptions of several of the individual species which have been sufficiently investigated. Before doing so, it is pertinent to note that the ligases have been implicated in influencing the biosynthesis of their respective tRNA species as well as their amino acids (Steinberg, 1974).

Phenylalanyl-tRNA Ligase. One enzyme that has been relatively thoroughly explored is phenylalanyl-tRNA ligase from yeast, which is a tetramer of the $\alpha_2\beta_2$ variety (Fasiolo *et al.*, 1970; Schmidt *et al.*, 1971). When separated the subunits appear to possess no catalytic properties (Hennicke and Böck, 1975). The α and β subunits have molecular weights in the range of 74,000 and 63,000, respectively, corresponding to a molecular weight of 274,000 ± 8000 for the intact enzyme (Fasiolo and Ebel, 1974; Fasiolo *et al.*, 1974). Two bonding sites apiece were described for the $tRNA^{Phe}$ and for the aminoacyl, but those for the latter component were not of equal reactivity, one showing five times as much affinity as the other. The number of sites for ATP was not reported. Two thiol groups were demonstrated in the enzyme by another investigation; these proved to be highly reactive in the aminoacylation of the tRNA but not in the activation of phenylalanine (Murayama *et al.*, 1975a,b). In the processes of charging the tRNA, the substrates may bind to the enzyme in random sequence (Santi *et al.*, 1971).

One study recently completed on this ligase may be applicable to all species. In this research, $tRNA^{Phe}$ was modified by replacing the usual riboadenosine of the 3′-terminal site with either 3′- or 2′-deoxyadenosine (Hecht *et al.*, 1974). Both isomers could be aminoacylated by the ligase, and both could be bound to the A site as efficiently as the native tRNA, but only the 3′-aminoacyl-tRNA was able to accept the peptide, and neither could serve as a donor in the peptidyltransferase reaction.

Arginyl-tRNA Ligase. The ligase that binds arginine to $tRNA^{Arg}$ has been studied in *E. coli, Bacillus stearothermophilus,* and *Neurospora*. In each case, the enzyme appears to be monomeric (α_1). The *E. coli* and *Neurospora* ligases have been shown to have molecular weights of 72,000 and 70,000 respectively (Haines and Zamecnik, 1967; Hirshfield and Bloemers, 1969; Nazario and Evans, 1974). While thus similar in structure, the enzyme differed extensively in mode of action in the three organisms. With the enzyme from *E. coli,* the substrates interacted in random order, but all had to be bound before any product was released (Papas and Peterkofsky, 1972). In the other two types, highly ordered sequences were required, but the details of reactions differed greatly. That from *B. stearothermophilus* needed ATP and the tRNA before it could bind arginine (Parfait and Grosjean, 1972), whereas the *Neurospora* enzyme bound the $tRNA^{Arg}$ first, then the amino acid, and last the ATP (Nazario and Evans, 1974). In spite of the sequential requirements in the *Neurospora* system, however, the results indicated that arginyl-adenylate was an intermediate in the chain of reactions. This enzyme bound $tRNA^{Arg}$ or arginyl-tRNA from *E. coli* or *Saccharomyces* at only about one-tenth the rate it did in the homologous system (Evans and Nazario, 1974).

Tryptophanyl-tRNA Ligase. Although data from *E. coli* systems have suggested that only one aminoacyl-tRNA ligase existed for each amino acid (Novelli, 1967; Lengyel and Söll, 1969; Lapointe and Söll, 1972a), multiple molecular forms for some species have been described (Rymo *et al.*, 1972; Yem and Williams, 1973). Among the reports is a study on multiple forms of tryptophanyl-tRNA ligases from beef pancreas (Lemaire *et al.*, 1975). As a whole, this species of enzyme, regardless of source, has been found to consist uniformly of two identical subunits (α_2), which range in molecular weight from 35,000 in *B. stearothermophilus* to 54,000 from the ox and man (Gros *et al.*, 1972; Koch *et al.*, 1974; Penneys and Muench, 1974). The multimolecular species in beef pancreas seemed to be produced by partial cleavage of one subunit, resulting in total enzyme molecular weights of 82,000 and 85,000, in contrast to the native molecule's 108,000 (Lemaire *et al.*, 1975). Several short amino acid sequences centering around cysteine residues have indicated homologies of primary structure to exist in the enzymes from *E. coli* and man (Muench *et al.*, 1975).

Methionyl-tRNA Ligase. Methionyl-tRNA ligase from *E. coli* has received a fair amount of attention in the literature. Its quaternary structure proved unusual in consisting of four copies of a single unit having a molecular weight of 32,000 (Chapeville and Rouget, 1972). Occasionally the tetramer (α_4) was cleaved to two dimers, which proved physiologically active. Many studies on this enzyme, as well as on other types, have employed adenosine or 8-aminoadenosine and other analogs of adenosine to investigate the ATP-binding site (Lawrence, 1973; Lawrence *et al.*, 1974; Blanquet *et al.*, 1975), as the nucleoside competes with ATP for binding with the ligase. For this reaction the pentose moiety was essential, but the presence or absence of a phosphate radical had no effect. As methionine markedly enhanced the affinity of the nucleosides for the ligase, it was suggested that the sites for the amino acid and ATP were coupled, at least to some extent (Blanquet *et al.*, 1975).

This ligase is unique in its class because it catalyzes the aminoacylation of two very markedly different kinds of tRNAs, one of which serves solely in initiation, whereas the other reacts only with internal codons. In an *E. coli* system, no difference could be detected either in the rates of reaction or ion requirements in the interactions of the tRNA ligase and the two major varieties of tRNA (Lawrence *et al.*, 1973). Comparisons of the sequences of the two types from any single species of organism serve particularly well to elucidate the problems involved in discerning sites in tRNAs used in recognition by the ligases (Tables 7.1–7.7).

The Dihydrouracil Arm (Arm II)

Bend 1. Many experimental studies have implicated the dihydrouracil arm, along with the adjacent bend 1, as being active in recognition by the ligase. Since much of this arm remains exposed in the three-dimensional model, the proposal is not illogical; however, it is improbable that sites shown to be involved in tertiary structure could be available also for recognition. If the implication of these sites in structure is valid, as appears to be the case, and if the majority are universal as indicated by the existence of complementarity between reactants in known sequenced tRNAs, then bend 1 nucleosides would be unavailable for ligase recognition. This is also probable because of their internal location in the three-dimensional molecule.

The Paired Stem Sites. In addition, the paired nucleosides in sites 3 and 3′ of the stem and those of sites A, B, F, G, and K of the loop would have to be eliminated as possibilities for recognition points by the ligase, as would also the nucleoside of site 4′. This leaves only the paired occupants of sites 1–1′, 2–2′, and 4, and unpaired nucleosides of sites C, D, E, H, I, and J to participate in such processes. However, site C is rather rarely occupied and the sequence occuping H, I, and J is also usually missing in whole or in part, yeast $tRNA^{Tyr}$s being the only species in which all sites in the loop are occupied (Table 7.2). Thus the available operational framework for reactions with the ligase is severely restricted.

Comparisons of the more than 90 determined sequences provided in Table 7.2 reveal even severer limitations. As already shown (Chapter 7, Section 7.3.2.), in site 1 guanosine or 2-methylguanosine is present in the molecules of all known proteinogenic tRNAs, except prokaryotic $tRNA^{Tyr}$ in which cytidine is the occupant. The complement in site 1′ is almost always cytidine, with uridine replacing it in five sequences. The replacements fail to suggest recognition functions, however, for no clear pattern of occurrence is indicated. For example, yeast $tRNA^{Met}$ contains cytidine in site 1′, whereas the two corresponding species from mammals contain uridine. The only other exceptional nucleoside is guanosine in the three *E. coli* $tRNA^{Tyr}$s; thus the latter three have the usual bases in sites 1–1′, but their locations are the reverse of the standard ones.

Sites 2–2′ are similarly of doubtful value in providing a code for recognition. Only two combinations exist in general, U–A and C–G. The sole reported deviations from these pairs include one instance of C–A that occurs in yeast $tRNA_3^{Gly}$, three having U–G that represent one type of $tRNA^{Val}$ from three different species of yeast, and one having A–U in *E. coli* $tRNA_F^{Met}$.

The other sites in the stem present similar problems insofar as coding possibilities are concerned. Rather than examining the remaining two paired sites individually, the prevailing condition at these loci is perhaps made clearer by citing two contrasting situations found in *E. coli* tRNAs. On one hand, there is remarkably little variation from species to species. For instance, the 5′-strand of the stem consists of the sequence GCUC in nine species of *E. coli* tRNAs, namely those of glycine, alanine, valine, histidine, arginine, isoleucine, threonine, $methionine_M$, and phenylalanine (Table 7.2). On the other hand, within one family of isoaccepting species from *E. coli* are found three different stem sequences, all of which are generally supposed to be recognized by the same ligase. The three species and their 5′-strand stem sequences are $tRNA_1^{Gly}$ with GUUC, $tRNA_2^{Gly}$ with GUAU, and $tRNA_3^{Gly}$ with GCUC. Thus one ligase would have to recognize three different sequences in that number of isoaccepting $tRNA^{Gly}$s, one of which is identical to eight other nonisoaccepting types. Thus greater differences exist within a cluster of isoaccepting species than between vastly different ones. This is contrary to the rules derived by Bhargava (1971) from a theoretical analysis of ligase-recognition coding requirements.

The Case of the Serine tRNAs. In view of these data, proposals made on the basis of limited observation or experimentation lose conviction; nevertheless one study has possibilities. In this, comparisons of serine tRNAs were made from three sources, *E. coli,* yeast, and rat liver (Ishikura *et al.*, 1971a; Ish-Horowicz and Clark, 1973), among which the most striking similarities were found to occur in the dihydrouridine arm. The 5′-strand of the stem consisted of the triplet GCC, so it was proposed that this region possibly was active in ligase recognition, perhaps together with the preterminal base G and the first pair of nucleosides (G–C) in the amino acid arm, a characteristic of all three. Thus far this combination is unique to seven of the known serine-carrying tRNAs; however, $tRNA^{Ser}$ from bacteriophage T4 has GCA in the 5′-strand of the arm II stem. Moreoever, if the same sites are applied to the same end with other base sequences, some doubts are cast upon their widespread involvement in ligase reactions. For example, the combination GCU in the 5′-strand of the stem of arm II, G–C in sites 1–1′ of arm I, and A as the preterminal base is found in six *E. coli* species, $tRNA^{Ala}_{1A}$, $tRNA^{Val}$, $tRNA^{Arg}_{1}$, $tRNA^{Thr}$, $tRNA^{Met}_{M}$, and $tRNA^{Phe}$. Thus these sites, solely in the combination given by the cited article, are not widely applicable.

An Analysis of Phenylalanine tRNAs. Among other studies proposing an active role in interactions with the ligase is a series of investigations involving $tRNA^{Phe}$ and analogous species from a variety of sources. The majority in this series involves the ability of yeast phenylalanyl-tRNA ligase to aminoacylate a number of *E. coli* species, particularly $tRNA^{Val}_{1}$ from that source (Tagland *et al.*, 1970; Roe *et al.*, 1973; Williams *et al.*, 1974). Extensive homologies in primary structure have been reported to exist in the $tRNA^{Phe}$ from three eukaryotic sources (yeast, wheat germ, and rabbit liver) and one from *E. coli* (Keith *et al.*, 1973). The experimental studies showed that yeast phenylalanyl-tRNA ligase could aminoacylate wheat and *E. coli* $tRNA^{Phe}$ and seven other species from *E. coli,* including those for $valine_{1\,and\,2}$, $alanine_{1\,and\,2}$, lysine, isoleucine, and $methionine_{M}$. Comparisons of the sequences implicated the stem region of arm II and the preterminal nucleoside, which was adenosine in all instances. Two sites in the short extra arm were also found common to the entire set studied, but these are now indicated to serve in tertiary structural interactions. The stem region of arm II in all cases had GCUC or G_2CUC in the 5′-strand, with the standard complements in the 3′-strand. The extensive ability of the ligase to aminoacylate so many tRNA species in itself raises questions as to the appropriateness of the suggested regions actually as serving in recognition, otherwise all those that reacted would carry phenylalanine.

Recognition in Tyrosine tRNAs. A different approach to the problem also implicated the stem of this arm, along with much of arms III and IV. This employed *E. coli* tRNA$^{Tyr}_{1\,and\,2}$, which have long extra arms, and the appropriate tRNA ligase from the same source (Schoemaker and Schimmel, 1974). Mixtures of the purified tRNAs and ligase were permitted to form complexes, which then were exposed to ultraviolet irradiation. After the interacting sites had been fused by this treatment and the complexes had then been degraded with a specific nuclease, the 5′-side of the stem of the dihydrouracil arm, the anticodon, and the outer region of arm IV were found attached to the enzyme. Since no other regions of the tRNAs were thus attached, it was assumed that these parts, and only these, were involved in ligase-recognition processes.

The Remaining Arms

Limited Recognition by the Anticodon Arm Stem. A small number of other researches also have implicated parts of the anticodon arm in ligase recognition, particularly the anticodon or adjacent sites, but the evidence is often contradictory. Almost no data suggest that the paired stem region of the arm may be active in this role. The sole exception is provided by a study of the attachment areas of tRNA$^{Met}_{F}$ to its ligase from *E. coli* (Dube, 1973). After the exposed portions of the nucleic acid had been degraded by T_1 ribonuclease, the loop and 3′-strand of arm III, the greater part of arm V, and sites 3′ to 7′ of the amino acid arm were found still to be intact. Thus they represented the areas protected through attachment to the ligase. However, examination of the sequenced tRNAs (Table 7.3) shows problems similar to those found with arm II. Nonisoaccepting species sometimes possess identical sequences, such as *E. coli* tRNA$^{Ile}_{1}$ and tRNA^{Thr}, both of which have the sequence GGGUG in the 3′-strand of arm III. Also as in arm II, isoaccepting species from the same source organism frequently have different sequences, yet are recognized by the same ligase. For instance, *E. coli* tRNA$^{Gly}_{1,2,\,and\,3}$ have GCUCU, GCUGA, and GGUCG, in ascending order, and tRNA$^{Met}_{M\,and\,F}$ have ΨGAUG and CCCGA, respectively (Hashimoto *et al.*, 1972b).

Site D adjacent to this sequence and the adjoining site 5′ also have been indicated to be *inactive* in recognition by the ligase. A mutation of *Salmonella typhimurium* involved the pseudouridines occupying these sites in the wild type failing to become modified from uridines (Singer *et al.*, 1972). The mutant contained the same total quantity of tRNA^{His} as the wild type, and the ratio of charged to uncharged tRNA^{His} was identical *in vivo*. Hence, there was no change in ligase recognition. However, as the mutation induced an inability to repress the histidine operon, the two sites seemed to be important in other functions.

The Anticodon Region. The anticodon region's role in recognition was much disputed until around 1972 (Kisselev and Favorova, 1974), but now it is generally thought that it may play a role in some instances and not in others. For example, in the isoaccepting species of glycine tRNAs from prokaryotes, the anticodons have been suggested to be involved in this function (Carbon and Curry, 1968; Squires and Carbon, 1971; Hill *et al.*, 1973). A missense change (from G→U) in the third position of the anticodon of yeast valine tRNA induced an inability of the tRNA to become charged with valine (Chambers *et al.*, 1973). Similar results had been obtained earlier with a bisulfite-induced C→U change in the third position of *E. coli* $tRNA^{Glu}$ (Singhal, 1971), but because a number of other cytidines also were altered, the effects of the anticodon alteration were not distinguishable. Now further studies clearly showed that chemical conversion of C to U definitely affected the aminoacylation of the tRNA (Seno *et al.*, 1974; Singhal, 1974).

A genetic mutation in *E. coli,* known as *glyTsuA36(HA)* also induces a C→U alteration in site 3 of the anticodon of $tRNA_2^{Gly}$ (Roberts and Carbon, 1974), but gives contrasting results. In this case, the altered $tRNA^{Gly}$ was normally aminoacylated by the ligase, as it inserted glycine instead of arginine wherever the latter's codon AGA occurred in messengers. Still other evidence had been reported which contraindicated any involvement of the anticodon in ligase recognition. By splitting the molecule of $tRNA^{Tyr}$ from *Torulopsis* through the anticodon into 3′ and 5′ halves and then removing the terminal nucleotides by means of controlled nuclease action, the anticodons were removed. Then the halves were recombined and annealed (Hashimoto *et al.*, 1972a). These reconstituted tRNAs were able to be charged with tyrosine to about 25% of normal in spite of the absence of the anticodon.

Arms IV and V. The two remaining arms (IV and V) do not appear to have been reported to be active specifically in ligase recognition; thus the existing evidence shows no single site of the tRNA responsible for that purpose, either alone or in combination with other sites. Hence, the whole tRNA molecule, rather than its separate parts, perhaps together with tertiary or quaternary configurational characteristics, seem to be involved in recognition by the ligase. Kim (1975) proposed that the pseudoaxes of the interacting ligase and tRNA provide the basis of recognition, and Budzik *et al*. (1975) and Schoemaker *et al*. (1975) suggested that the entire region in which the two helical branches of the L-shaped tRNA molecule come together are active in this function. However, these explanations may yield to a more concrete one, after all the isoaccepting species for several amino acids from a single source, or better, from at least one prokaryotic and one eukaryotic source, have had their nucleotide sequences determined. Then single sites or short sequences variously located around the molecules may prove sufficient for recognition sites, that, in view of the current limited information, cannot be identified now.

8.1.3. Sites Utilized in Recognition by the Ribosome

At one time the then-existing evidence strongly supported the view that arm V and, more specifically the TΨC sequence of its loop, were undoubtedly sites employed in recognition by the ribosomes. For one thing, all known nucleotide sequences of tRNAs which entered into proteinogenic processes contained the triplet TΨC, whereas the two *Staphylococcus* $tRNA^{Gly}$s that do not react with ribosomes have UGC instead (Roberts, 1972, 1974; Roberts *et al.*, 1974). Experiments with ribosomes, tRNAs, and the polynucleotide sequence TΨCG demonstrated *in vitro* that the latter competed with tRNAs for ribosome binding (Ofengand and Henes, 1969; Grummt *et al.*, 1974). Further substance was added to the supposition when it was found that yeast $tRNA^{Met}_{F}$, which is employed only in initiation of translation, contained the sequence AUC (Dube *et al.*, 1968, 1969). This also proved to be the case later when the same isoaccepting species from what germ and mammalian sources had been sequenced (Petrissant, 1973; Simsek *et al.*, 1973a,b). The proposal was still further strengthened when mouse $tRNA^{Met}_{4}$, which enters into peptide elongation, not initiation, was shown to contain the standard sequence TΨC (Elder and Smith, 1973; Piper and Clark, 1974; Piper, 1975a).

However, several proteinogenic tRNAs from eukaryotic sources are now known that are deficient in ribothymidine, so that the above concept no longer can be considered valid. Two wheat $tRNA^{Gly}$s and three $tRNA^{Thr}$s have been shown to lack ribothymidine (Marcu *et al.*, 1973, 1974). Wheat $tRNA^{Gly}_{3B}$ proved to contain UGC, but the corresponding triplet in $tRNA^{Gly}_{3A}$ from the same organism was less aberrant, as it consisted of UΨC. Similar ribothymidine absence was also found in unidentified tRNAs from beef liver. In addition, mouse liver $tRNA^{Val}$ has been shown to have the sequence UΨC (Piper and Clark, 1974; Piper, 1975b). Thus the problem of the ribosome's distinguishing initiating from elongating tRNAs and nonproteinogenic from proteinogenic types remains unsolved. Nor is it clear how a ribosome reads the next codon in the message to receive a tRNA and then quickly selects and contacts the appropriate activated tRNA from the more than 60 different species of tRNAs each organism contains. The speed at which translation occurs raises doubts that a single enzymatic system situated at the elongation site could possibly carry out the processes of selecting the tRNAs without some still unrecognized assistance from the surrounding cytoplasm.

8.1.4. Sites Used by Modifying Enzymes

The problem of how modifying enzymes recognize the tRNAs that they employ as substrates has not been adequately addressed in the literature, so an in-depth analysis cannot be provided here. Nevertheless, what data have come to hand are highly instructive.

Modifiers of the Postanticodonal Nucleoside. As shown in Table 8.3, almost all group III tRNAs, which have anticodons with uridine in site 3, are followed by the hypermodified nucleoside *N*[9-(β-D-ribo-furanosyl)-purine-6yl-carbamoyl] threonine, abbreviated as A_t. Similarly most group IV tRNAs, those with adenosine in site 3, are followed by other hypermodified nucleosides, N^6-(Δ^2-isopentenyl)adenosine (A_i) in eukaryotes and 2-methylthio-N^6-(Δ^2-isopentenyl)adenosine (A_s). Like all nucleosides in tRNAs, those located here are standard bases in the early precursor molecule and are modified during maturation.

Some evidence suggests that the nucleoside in site 3 of the anticodon serves as the major, if not sole, character involved in recognition by the modifying enzyme. Perhaps the clearest example is a mutant variety of *E. coli* identified as *glyTsuA36*(*HA*) in which the $tRNA^{Gly2}_{A/G}$ undergoes a single mutation, a substitution of U for the usual C at site 3 of the anticodon (Roberts and Carbon, 1974). In the mutant form, the usual A of the postanticodonal site is modified to A_t or a derivative. Thus the U of the anticodon seems to serve as a recognition site by the modifying enzyme, which accordingly converts the adjacent A to A_t.

A comparable change, but involving a C → U transition at site 3 of the anticodon, has been reported for *E. coli* $tRNA^{Gly3}_{U/C}$ (Carbon and Fleck, 1974). In this case, the A in site C adjacent to the anticodon of the normal tRNA is converted to A_s, the typical nucleoside of group IV prokaryotic tRNAs. Thus again the evidence implicates the third site in the anticodon as sufficient in itself for recognition by the modifying enzyme.

Examination of the series of anticodon arm sequences (Table 7.3), however, suggests that more than that single site may be involved. Among group I tRNAs, all of which have G in site 3 of the anticodon, most known species from prokaryotes have A in the adjacent site C. Yet in a few species, such as $tRNA^{Glu}$ and $tRNA^{Val}_1$, the A is modified to 2-methyladenosine. In contrast, group II tRNAs, all of which have C in the third site of the anticodon, have the A when present methylated in a similar fashion. Finally, both *E. coli* $tRNA^{Met}$s which have been sequenced have the anticodon CAU, but $tRNA^{Met}_M$ has the postanticodonal nucleoside modified to A_t, whereas in $tRNA^{Met}_F$ it remains unmodified.

Pseudouridine Synthetases. The mutation *hisT1504* in *Staphylococcus typhimurium* cited above in another connection is of special importance in illuminating the nature of modifying enzymes to a degree. In the mutant form, the $tRNA^{His}$ is identical to the wild type, except that two uridine nucleosides in the anticodon arm fail to become transformed to pseudouridine (Singer *et al*., 1972). Thus the mutation really involves a loss of activity on the part of the modifying enzyme. If this is actually the case, one should suspect that other species, such as $tRNA^{Leu}_{1,2}$ and $tRNA^{Gln}_{1,2}$, should be equally defective, but these were not examined in the report cited. However, the pseudouridine of arm V is present as in the wild type. Thus it was proposed that separate modifiers of U → Ψ exist, one of which acts only on the uridine in site B of arm V, the other only on the uridines of sites D and 5′ in arm III (Figure 8.3).

An examination of Figure 8.3 makes it clear that each modifying enzyme

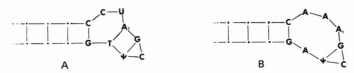

Figure 8.3. Possible intraloop bonding in arm V. If, as seems likely, hydrogen bonding occurs between members of the loop of arm V, the configuration of (A) those frequent types that contain a TΨC sequence would differ markedly from (B) the types that lack that sequence, as in the $tRNA^{His}_F$ of the mouse.

is highly specific in its locus of action, as well as mode. In the first place, only 7 of the total of 19 uridines present undergo modification, and those which do not become altered are as exposed in the three-dimensional molecule as those which do. For instance, four uridine residues are present in the loop of arm III but only that one situated in site D becomes modified. Obviously, the modifying enzyme of the uridine in site 1 of bend 1 also is highly specific, for that locus is the only one in any known tRNA in which 4-thiouridine (U_t) occurs. The dihydrouridines (D) also are equally the result of specific enzymatic action at specific points, as is also the ribothymidine (T) of arm V. Hence, there must be at least five enzymes which convert uridine nucleosides into their modified forms. The number involved in formation of the dihydrouridines is probably three instead of one, if the sequences in the three isoaccepting species of $tRNA^{Gly}$ from *E. coli* are to be believed. There it is to be noted that the unpaired nucleosides of arm II in $tRNA^{Gly}_2$ consist of AAUGGCUA, those of $tRNA^{Gly}_1$ are $AAUG_2GDA$, and those of $tRNA^{Gly}_3$, AGDDGGDA. Thus one contains two Us, but no D, another has one U and one D, and the third has three Ds but no U. Since $tRNA^{Gly}_1$ has two Us available, of which only one becomes modified, separate enzymes seem to be requisite. A few other species, such as *E. coli* $tRNA^{Arg}_{1A,B}$ suggest the same need (Table 7.2). Hence in the $tRNA^{His}$ cited above, the total number of uridine-modifying enzymes becomes seven, not just five, and each must have its own method of identifying the correct site in the right tRNAs.

The results of a recent investigation add considerable support for the foregoing deduction, even though they are derived from eukaryotic tissue. Glick and Leboy (1977) purified an enzyme 8000-fold from rat liver until under electrophoretic treatment it yielded a single band on polyacrylamide gel. This substance had a molecular weight of 95,000 and proved stable for at least a year. When $tRNA^{Glu}_2$ from *E. coli* was treated with this enzyme, the invariable adenine in the loop of arm V was converted to N^1-methyladenine, but no other adenine became modified. Since this enzyme, known as tRNA (adenine-1)-methyltransferase, thus was found to be highly specific in its activity, other modifying enzymes comparably restricted in substrate requirements are also most likely to be uncovered by future investigations.

The Number of Eukaryotic Modifying Enzymes. If this line of reasoning may be extended to the eukaryotes, whose modifying enzymes have not received study to date, the situation becomes much more complex. Especially is this the case in groups III and IV tRNAs, several of which have 16 or 17 modified nucleosides. The modified nucleosides of one such tRNA, that which carries lysine in *Saccharomyces,* are shown in Figure 8.1B.

8.2. EVOLUTIONARY RELATIONS OF tRNAs

Now that the structural features of the sequenced species of tRNAs have been examined for functional sites and found largely wanting of specific loci clearly indicated to serve in recognition, an evolutionary approach becomes feasible. The only functional sites uncovered to date appear to be those which, like the –CCA terminus, are universal in occurrence, or which, like the ribothymidine of arm V, are virtually so. Such universal or subuniversal features are involved either in attachment of the amino acid or in tertiary configuration stability. In addition, the anticodons obviously play a vital role in translation, but with them the list of sites known to be active in specific functions ends.

8.2.1. General Considerations

Dual Methods of Inducing Evolutionary Changes. One point the discussion of modifying enzymes clarifies is the dual mode whereby evolutionary changes have been produced in tRNAs. The first is the usual method involving mutations in DNA, the same processes employed in evolving proteins and other macromolecules. The second set of procedures, although not unique to tRNAs, is unusual, for its results indirectly from changes undergone by the modifying enzymes. Since it is these proteins that have undergone evolutionary alterations, most likely by the usual genetic-dependent route, their products, the mature tRNAs, thus have evolved by direct and indirect activities. This observation leads to the logical conclusion that other types of RNAs, ribosomal and messenger alike, also are products of dual methods of evolution working side by side, for they, too, undergo a degree of posttranscriptional modification by enzymatic action.

Alternative Pathways. Aside from the methodologies of inducing changes, two alternate paths appear to exist along which tRNAs may have evolved. The first is that, after the various ancestral species had been derived from the original type as proposed above, each isoaccepting species evolved independently within a functional framework first to prokaryotic, then to early eukaryotic, and eventually to advanced eukaryotic types (Figure 8.4A). Thus all glycine tRNAs, having a separate evolutionary history, would be expected to be more similar to each other than to alanine or other species, and each of

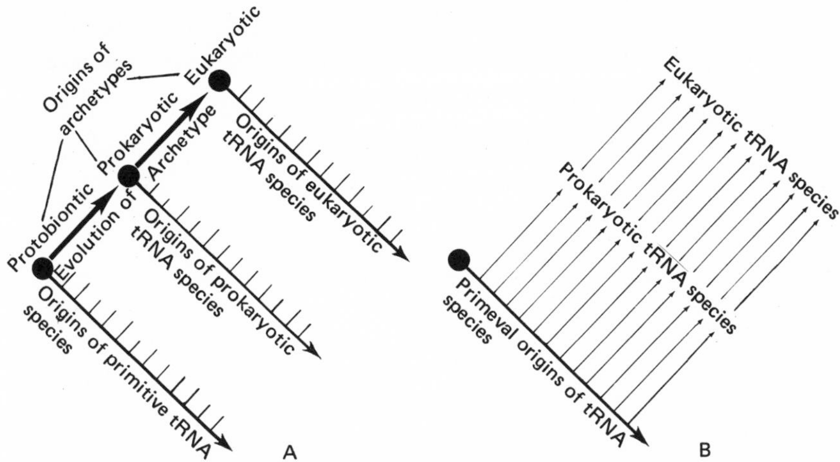

Figure 8.4. Two theoretically possible routes for the origin of transfer RNAs. (A) If only one archetypal tRNA arose, which alone evolved to higher levels while each descendant evolved independently into the various species found respectively in protobionts, prokaryotes, and eukaryotes, few intraspecies similarities would exist. (B) If the archetypal tRNA evolved early in the protobionts into the several species which later evolved as the organisms advanced, the species would resemble one another, as indicated by the actual tRNA sequences.

these in turn would resemble one another more closely within isoaccepting types than they would those accepting glycine, aspartic acid, or other amino acids.

The second, and only other evident, route would begin with an early (protobiontic) origin of a generalized tRNA, which would be passed to prokaryotic and eukaryotic descendants (Figure 8.4B). The development into the various isoaccepting species would occur independently at each major level of descent. Thus all prokaryotic tRNAs would share many features but would differ markedly from eukaryotic forms, which would have a number of features in common among themselves. In addition, eukaryotic tRNAs would share a number of traits with prokaryotic species, aside from universal or near-universal sites, which would reflect ancestral characteristics. These statements apply separately to enzymatically induced modifications, for the properties of the enzymes would show evolutionary influences likewise.

Other Proposals Concerning Phylogeny. Only a few attempts at tracing a phylogeny of the tRNAs have been made previously. Jukes and Holmquist (1972) compared the base sequences of 36 species of these nucleic acids, in which homologies were maximized by either eliminating certain regions or inserting blanks as necessary to bring all to a standard length of 74 nucleosides plus the –CCA terminus. By these processes, differences and correspondences between bases at a given locus could be noted, and a number of both constant and variable sites were detected. Among the conclusions drawn was the observation that the tRNAs for two different amino acids may have evolved from a common ancestor, first by gene duplication and subsequently in independent accumulation of mutational changes in nucleosides. Further, it was suggested that such evolutionary separation must include changes in the anticodon as well as the site employed for recognition by the specific aminoacyl-tRNA ligase. Because of the lethal consequences that probably would ensure from such changes' inducing mischarging and the like, it was proposed that a newly evolved tRNA might not actually be employed in transcription for a prolonged period until it was ready to carry the correct amino acid. No consideration was given to the fact that both the ligase and amino acid concerned would be functionless until the tRNA had acquired all the correct mutations. Nor was mention made of the fact that all known tRNAs are specified by redundant genes, which seem to vary little among themselves.

Another study (Gross, 1973) suggested that tRNAs have evolved by reducing the degree of variation in size that appeared to characterize prokaryote tRNAs. According to the concept, this reduction was achieved by more extensive modification of individual bases in the eukaryotes. Mullins *et al.* (1973a, b) advanced the idea that tRNAs arose from 5 S rRNA, because they found homology between three *E. coli* tRNAs (those for tyrosine, glutamine, and tryptophan), several eukaryote tRNAs, and *E. coli* 5 S rRNA in excess of chance expectation. This proposal was later questioned (Holmquist and Jukes, 1973), because of the methodology that had been employed in calculating chance expectation. These latter workers pointed out that accidental coincidences among bases, with only four variables, are far more frequent than between amino acids, with 20 variables. Hence, the application of methods for finding homologies between proteins could not validly be applied to nucleic acids. Mullins and his colleagues stated also that they found it impossible to construct a phylogeny of tRNAs because evolutionary divergences among these substances had proceeded to a point of equilibrium, so that no correlation between sequence similarities and phylogeny was to be expected.

However, two investigations did find traces of phylogenetic relationships. On the basis of nearly 30 tRNA sequences, one report (Cedergen *et al.*, 1972) detected ancestor–descendant relations in the valine family of tRNAs and suggested that among the isoaccepting species that carry leucine two different families of homologies exist. This two-family condition (or nonhomologous sets of isoaccepting tRNAs) was viewed as possibly arising through the degeneracy of the codons. The second report compared only 18 tRNAs (Dayhoff and McLaughlin, 1972); consequently, the phylogenetic tree was of necessity highly preliminary in nature. It estimated evolutionary distance in PAMs (Accepted Point Mutations) and suggested that after the various tRNA types had become differentiated in the protobiont, the eukaryotes and prokaryotes pursued separate paths of evolutionary divergence.

Procedures. Since the above reports appeared, the list of tRNA sequences has nearly tripled. Furthermore, the broad concept enlarged on in Chapter 6 now provides a model organization, so that widely divergent populations of tRNA types can be arranged into preliminary groups of possibly related species. From the complete tables provided in Chapter 7, a number of representative species have been selected to avoid nonproteinogenic types and to provide brevity by eliminating minor types (Tables 8.1–8.5). Those included in the tables are arranged by groups, in the proposed evolutionary sequence under each group, and prokaryotic and eukaryotic types are placed in separate columns. As in the preceding chapter, separate tables are provided for the sequences of each arm to facilitate comparisons as far as possible.

On the tables the darker shadings indicate identity (primary homology), either within an entire group or in several or all groups, while the lighter shading indicates partial, or secondary, homology, such as constancy of purines (Pu) or pyrimidines (Py), or of a given base variously modified. Thus invariables and transitions in the DNA molecules receive attention, but transversions do not. The latter, however, occur with notable frequency, despite the lack of a valid chemical basis to explain their origins, as tautomerism provides a basis for transitions; examination of the tables reveals G → C and C → G transversions to be especially prominent. First the sequences of the various regions are analyzed as to the evolution of the bases, including modifications; after each arm has received attention, an attempt is then made to suggest how the secondary evolutionary events (those derived from enzymatic modifications) may have occurred. In all instances *trends,* rather than hard-and-fast rules, are to be expected, at least until a greater diversity of organisms has been more deeply explored as to the nucleoside sequences of tRNAs.

8.2.2. Phylogenetic Changes in Arm I

As indicated by the scarcity of homologies (Table 8.1), the amino acid arm (Arm I) could with propriety be referred to as the variable arm, or the region devoid of evolutionary significance. Perhaps the lack of structural constancy also implies absence of functional importance, too. One might speculate that this absence of stability implies a relatively recent origin for this arm.

The Prokaryotic Representatives. Among the prokaryotic types, there appears to be a general trend from extreme variability in group I members toward greater uniformity in the later groups. This tendency may stem from absence of a sufficiency of sequenced species, however, rather than from real evolutionary influences. In group I, where ten types are included, sites 1–1′ are especially striking because of the constancy of the G–C pair occupying them. No comparable feature, except the very similar Pu–Py homologies of group IV, exists at these sites in the other groups.

In contrast, group III has a Pu–Py constant feature at sites 7–7′, a charac-

Table 8.1
Arm I of Sequenced tRNAs Arranged by Major Source Type[a]

	Prokaryotes					Eukaryotes			
	5′ Strand	Bend 1	3′ Strand	Preterm.		5′ Strand	Bend 1	3′ Strand	Preterm.
GROUP I					GROUP I				
Salm$_1^{Gly}$	-pGCGGGCG	UA	CGCCCGC	U	Sac$_3^{Gly}$	-pGCGC$'_2$AAG	UG	CUUGCGC	A
Esch$_1^{Gly}$	-pGCGGGCG	U_tA	CGCCCGC	U	Wheat$_3^{Gly}$	-pGCAC$'_2$CAG	UG_1	CUGGUGC	A
Esch$_2^{Gly}$	-pGCGGGAG	UC	UGCCCGC	U	Sac^{Ala}	-pGGGCGUG	UG_1	CUCGUCC	A
T4$_2^{Gly}$	-pGCGGAUA	UC	UAUCCGC	U	Tor^{Ala}	-pGGGCGUG	UG_1	CUCGUCC	A
Esch$_3^{Gly}$	-pGCGGGAA	UA	UUCCCGC	U	Brew^{Asp}	-pUCCGUGA	UA	UCGCGGA	G
Esch$_1^{Ala}$	-pGGGGGCA	U_tA	CGCUCCC	A	Sac$_2^{Glu}$	-pUCCGAUA	UA	UAUCGGA	G
Esch$_2$Glu	-pGUCCCCU	UC	AGGGGAC	G	Sac^{Val}	-pGGUUUCG	UG_1	CGAAAUC	A
Salm$_1^{Val}$	-pGGGUGAU	U_tA	AUCACCC	A	Brew^{Val}	-pGGUUUCG	UG_1	CGAAAUC	A
Esch$_1^{Val}$	-pGGGUGAU	U_tA	AUCACCC	A	Tor^{Val}	-pGGUUUCG	UG_1	CGAAAUC	A
Esch$_{2A}^{Val}$	-pGCGUCCG	U_tA	CGGACGC	A	Mus^{Val}	-pGUUUCCG	UA	CGGAAAC	A
GROUP II					GROUP II				
Esch$_1^{Leu}$	-pGCGAAGG	UG	CCCUCGC	A	Brew$_2^{Arg}$	-pψUCCUCG	UG_1	CGGGGAA	G
Esch$_2^{Leu}$	-pGCCGAGG	UG	CCUCGGU	A	GROUP III				
Esch^{His}	pGGUGGCUA	U_tA	UAGCCAC	C*	Tor^{Ile}	-pGGUCCCU	UG	AGGGACC	A
T4^{Gln}	-pUGGGAAU	U_tA	AUUCCCA	G	Brew$_3^{Arg}$	-pGCGC$\overset{(U)}{U}$CG	UG_1	CGAGUGC	G(U)

$Esch_1^{Arg}$	-p*GCAUCCG*	U_tA	*CGGAUGC*	A
GROUP III				
$Esch_1^{Ile}$	-p*AGGCUUG*	UA	*CAGGCCU*	A
$Esch_2^{Ile}$	-p*GGGCUUG*	U_tA	*CAGGCCC*	A
$Esch^{Thr}$	-p*GCUGAUA*	UA	*UAUCAGC*	A
$Esch_3^{Ser}$	-p*GGUGAGG*	UG	*CCUCACC*	G
$Esch_M^{Met}$	-p*GGCUACG*	U_tA	*CGUAGCC*	A
$Esch_F^{Met}$	-p*CGCGGGG*	U_tG	*CCCCGCA*	A
GROUP IV				
$Esch^{Phe}$	-p*GCCCGGA*	U_tA	*UCCGGGC*	A
$Esch_2^{Tyr}$	-p*GGUGGGG*	U_tU_t	*CCCCACC*	A
$Esch^{Ser}$	-p*GGAAGUG*	U_tG	*CGCUUCC*	G
$Esch^{Trp}$	-p*AGGGGCG*	U_tA	*CGCCCCU*	G
$T4^{Leu}$	-p*GCGAGAA*	UG	*UUCCCGC*	A

Sac_F^{Met}	-p*AGCCGCG*	UG_1	*CGCGGCU*	A
Mus_F^{Met}	-p*AGCAGAG*	UG_1	*CUCUGCU*	A
Mus_4^{Met}	-p*GCCUCGU*	UA	*ACGGGGC*	A
GROUP IV				
Sac^{Phe}	-p*GCGGGGA*	UA	*AAUUCGC*	A
$Wheat^{Phe}$	-p*GCGGGGA*	UA	*UCACCGC*	A
Rab^{Phe}	-p*GCCAAAA*	UA	*UUUCGGC*	A
Sac^{Tyr}	-p*CUCUCGG*	UA	*CCGGGAG*	A
Tor^{Tyr}	-p*CUCUCGG*	UG_1	*CCGAGAG*	A
$Brew_A^{Ser}$	-p*GGCAACU*	UG	*AGUUGUC*	G
Rat^{Ser}	-p*GUUGUCG*	UG	*CGACUAC*	G
Sac^{Cys}	-p*GCUCGUA*	UG	*UGCGAGC*	U
$Brew_1^{Trp}$	-p*GAAGCGG*	UG_1	*CCGUUUC*	A
Sac^{Leu}	-p*GGUUGUU*	UG	*AGCAACC*	A

[a]Italics indicate double-stranded regions. Bases with subscript numerals are methylated; subscript t or s, thiolations; m, methylation; a, acetylation; superscript 2, double modification; M, unknown modification. For other conventions, see text. The darker shading indicates primary homology, that is, the presence of the same base throughout. Secondary homology, indicated by lighter shading, involves the presence of only the same type (that is, pyrimidine or purine). Abbreviations for organisms: Ana, *Anacystis;* B. stear, *Bacillus stearothermophilus;* Bac, *Bacillus subtilis;* Brew, brewer's yeast; Esch, *Escherichia coli;* Mus, house mouse; Rab, rabbit; Rous, Rous sarcoma cells; S. aur, *Staphylococcus aureus;* Sac, baker's yeast; Salm, *Salmonella;* Staph, *Staphylococcus epidermidis;* T2, T4, T6: bacteriophage T2, etc.; Tor, *Torulopsis utilis.*

teristic that continues through group IV as well. In addition, the latter has a G–C consistent relationship at sites 5–5′, so that the five members shown are unified by one full and two secondary homologies in the paired region of the stem. The preterminal bases of both groups III and IV are uniformly purines; this trait would characterize all tRNAs beyond the most primitive level (represented by $tRNA^{Gly}$s) were it not for the very exceptional condition found in $tRNA^{His}$, in which a modified cytidine occurs in the preterminal site paired to guanosine located in an additional site, referred to as 0.

Bend 1 has its first site constantly occupied by a uridine nucleoside; in most tRNAs, however, this is modified into 4-thiouridine. No evolutionary trends are in evidence here or in site 2, which is occupied by the several major nucleosides except uridine. Only in group III is any homology to be noted, in that only purines occur in site 2.

The Eukaryotic Types. Even greater variability in this arm is at once noticeable among eukaryotic types, an instability that increases with evolutionary advancement of the tRNAs, in contrast to the condition prevalent among prokaryotes. In the paired stem, only sites 7–7′ of group I show consistent homology, although even that is merely the Pu–Py variety. Even the preterminal base is variable, except for a secondary homology among group I members. However, it is to be noted that in each of the two higher groups, a single exception upsets an otherwise consistent purine condition.

Unlike the prokaryotes, both sites in bend 1 show homologous relationships, the first one being occupied uniformly by uridine (always unmodified). The occupant of the second is either adenosine, guanosine, or 1-methylguanosine. These consistencies undoubtedly are associated with their involvement in tertiary structure. By analogy, one could deduce that in prokaryotes only site 1 is thus involved.

Comparisons of Types and Sources. As a result of this extensive variability, few if any relationships either within or between groups or sources can be detected. One possible set of relationships is exhibited by the prokaryotic tRNAs of group I. The stem of the $tRNA^{Ala}$ differs from that of $tRNA_1^{Gly}$ at only sites 2–2′ and 7, so the evolutionary proximity of these two types suggested earlier is again indicated.

Comparisons of sequences of this arm of isoaccepting species from heterologous sources are largely unprofitable. Only the yeast $tRNA_3^{Gly}$ is clearly suggestive of having had common relations with prokaryote $tRNA_3^{Gly}$. Elsewhere often short sequences indicate possible homology, but the presence of identical sequences in clearly unrelated types makes all comparisons meaningless here.

8.2.3. Phylogeny of Arm II Sequences

Just as obvious as the inconstancy of arm I sequences is the uniformity of those of arm II. Hence, this arm could as appropriately be named the "invariable'' as the ''dihydrouridine'' arm. Exemplifying this condition most clearly are the sequences in group I from both prokaryote and eukaryote sources. In these, only sites 3–3′, 4′, and J do not exhibit either complete or secondary homologies, except that in prokaryotes sites E and I also are variable. In short, 13 prokaryote and 15 eukaryote sites out of a total of 19 have identical or chemically closely related occupants. Doubtlessly this constancy is correlated to the active role played by this arm in tertiary structure. Whatever is the cause, however, the net result is similar to that induced by the variability of arm I—an absence of phylogenetic indicators. Nevertheless, some trends can be noted.

Phylogenetic Trends in Prokaryotic Species. In the paired stem (Table 8.2), complete homology throughout the entire prokaryotic series at sites 1–1′ is circumvented by *E. coli* $tRNA_1^{Tyr}$, in group IV. Similarly, only one exception spoils the secondary homologous relations at sites 2–2′, in this case $tRNA_F^{Met}$ of group III. At site 3, purines appear to diminish in frequency with evolutionary advancement, adenosine occurring only in group I members and guanosine only in groups II and III. In its paired site, 3′, the opposite condition prevails, with uridine confined to group I and cytidine to groups II and III. Site 4 shows group homologies only among the members of the first group, whereas the complement in site 4′ shows no uniformity in that group but exhibits secondary homology throughout the remaining prokaryotic types.

In the unpaired loop, the 5,6-dihydrouridines demonstrate several trends fairly well correlated to group relations (Table 8.2). Group I members tend to have relatively few, or even no, nucleosides of this sort, while group II components usually have two on the 5′-side of the doublet guanosines and one on the 3′-side. Group III shows a tendency toward having a condition opposite that of group II, and group IV frequently has cytidine in place of one or two of the 5,6-dihydrouridines.

Trends among Eukaryotic Species. Despite the widespread constant features, several marked trends among eukaryotic tRNAs (Table 8.2) tend to support the evolutionary concepts discussed earlier. In group I the consistent guanosine found throughout the eukaryotic types in site 1 is always unmodified and uridine is fairly common in site 1′. On the other hand, the guanosine in the remaining groups is almost always modified to 2-methylguanosine and cytidine is the nearly constant occupant of site 1′, mouse $tRNA_F^{Met}$ being the sole exception. Sites 2–2′ show similar relationships, with various Py–Pu combinations in group I tRNAs and strictly C–G pairs in the remainder. Although sites 3–3′ are without evident phylogenetic trends, sites 4–4′ provide interesting data, especially the former locus. Here in group I the nucleoside is always uridine or pseudouridine, but neither of these has been reported here in any of the other groups. Quite to the contrary, groups II and III constantly have cytidine in this site as far as currently known, while group IV tRNAs frequently have adenosine or guanosine. In site 4′, uridine also is frequent among the first group tRNAs but uniformly absent elsewhere.

Trends in the unpaired region are difficult to discern. Perhaps the most

Table 8.2
Arm II of Sequenced tRNAs Arranged by Major Source Type[a]

	Prokaryotes					Eukaryotes			
	1234	ABCDEFGHIJK	4'3'2'1'	Bend 2		1234	ABCDEFGHIJK	4'3'2'1'	Bend 2
GROUP I					GROUP I				
$Salm_1^{Gly}$	GUUC	AA--UGG_2D--A	GAAC	G	Sac_3^{Gly}	GCUψ	AG--DGG-D-A	AAAU	C
$Esch_1^{Gly}$	GUUC	AA--UGG_2D--A	GAAC	G	$Wheat_3^{Gly}$	GUCψ	AG--DGG-U-A	GAAU	A
$Esch_2^{Gly}$	GUAU	AA--UGGCU-A	UUAC	C	Sac^{Ala}	GCGU	AG-DCGG-D-A	GCGC	G_2^2
$T4_2^{Gly}$	GUAU	AA--UGG_2D--A	UUAC	C	Tor^{Ala}	GCGU	AG-DDGG-D-A	GCGC	G_2^2
$Esch_3^{Gly}$	GCUC	AG-DDGGD--A	GAGC	A	$Brew^{Asp}$	GUUψ	AA--DGG-DCA	GAAU	G
$Esch_1^{Ala}$	GCUC	AG-CDGG--GA	GAGC	G	Sac_2	GUGψ	AA--CGGCD-A	UCAC	A
$Esch_2^{Gln}$	GUCψ	AG--AGGCCCA	GGAC	A	Sac^{Val}	GUCψ	AG-DCGGDD-A	UGGC	A
$Salm^{Val}$	GCUC	AG-CDGG--GA	GAGC	A	$Brew^{Val}$	GUCψ	AG-DCGGDD-A	UGGC	A
$Esch_1^{Val}$	GCUC	AG-CDGG--GA	GAGC	A	Tor^{Val}	GUCψ	AG-DDGGDC-A	UGGC	A
$Esch_{2A}^{Val}$	GCUC	AG-DDGGDD-A	GAGC	A	Mus^{Val}	GUGψ	AG--DGGDD-A	UCAC	G_2^2
GROUP II					GROUP II				
$Esch_1^{Leu}$	GCGC	AA-DDG_2GDAGA	-CGC	G	$Brew_2^{Arg}$	GCCC	AA--DGGDC-A	CGGC	G_2^2
$Esch_2^{Leu}$	GUGG	AA-DDG_2GDAGA	-CAC	G					
$Salm^{His}$	GCUC	AG-DDGGD--A	GAGC	C					

tRNA				
Esch	GCUC	AG-DDGGD--A	GAGC	C
T4^{Gln}	GCCA	AG-DDGGD--A	AGGC	A
Esch$_1^{Gln}$	GCCA	AG-C-G$_2$GD--A	AGGC	A
Esch$_1^{Arg}$	GCUC	AG-CDGGD--A	GAGU	A
GROUP III				
Esch$_1^{Ile}$	GCUC	AG-GDGGDD-A	GAGC	G
Esch$_2^{Ile}$	GCUC	AG-GDGGDD-A	GAGC	G
Esch^{Thr}	GCUC	AG-DDGGD--A	GAGC	G
Esch$_3^{Ser}$	GCCG	AG-A-GGCDGA	AGGC	G
Esch$_M^{Met}$	GCUC	AG-DDG$_2$GDD-A	GAGC	A
Esch$_F^{Met}$	GAGC	AGCCUGGD--A	GCUC	G
GROUP IV				
Esch^{Phe}	GCUC	AG-DCGGD--A	GAGC	A
Esch$_1^{Tyr}$	CCCG	AG--CG$_2$GQCAA	AGGG	A
Esch^{Ser}	GCCG	AG--CG$_2$GDDGA	AGGC	A
Esch^{Trp}	GUUC	AA-DDG$_2$GD--A	GAAC	A
T4^{Leu}	GUCA	AA-DDG$_2$GD-AA	AGGC	A
GROUP III				
Tor^{Ile}	G$_2$CCC	AG-DDGGDD-A	AGGC	G_2^2
Brew$_3^{Arg}$	G$_2$CUC	AA--DGG-C-A	ACGC	G_2^2
Sac$_2^{Lys}$	G$_2$CUC	AG-DDGGD--A	GAGC	G_2^2
Sac$_F^{Met}$	G$_2$CGC	AG--DGG--AA	GCGC	G_2^2
Mus$_F^{Met}$	G$_2$CGC	AG--CGG--AA	GCGU	G_2^2
Mus$_4^{Met}$	G$_2$CGC	AG-DAGGD--A	GCGC	G_2^2
GROUP IV				
Sac^{Phe}	G$_2$CUC	AG-DDGGG--A	GAGC	G_2^2
Wheat^{Phe}	G$_2$CUC	AG-DDGGG--A	GAGC	G_2^2
Rab^{Phe}	G$_2$CUC	AG-DDGGG--A	GAGC	G_2^2
Sac^{Tyr}	G$_2$CCA	AG-DDG$_2$GDDDA	AGGC	G_2^2
Tor^{Tyr}	G$_2$CCA	AG-DDG$_2$GDDDA	AGGC	G_2^2
Brew$_A^{Ser}$	GCC$_a$G	AG--DG$_2$GDD-A	AGGC	G_2^2
Rat^{Ser}	GCC$_a$G	AG--DG$_2$GDD-A	AGGC	G_2^2
Sac^{Cyst}	GCGC	AG--DG$_2$GD--A	GCGC	A
Brew$_1^{Trp}$	G$_2$CUC	AA--DG$_2$GD--A	GAGC	ψ
Sac^{Leu}	GCCG	A---CG$_2$GDDCA	AGGC	G_2^2

[a]For explanation of conventions, see Table 8.1, footnote *a*.

valid one concerns the guanosine that occupies site F. Unlike prokaryotic tRNAs, in which 2′-methylguanosine occurs in scattered fashion throughout the entire series, this modification is strictly confined to those members of group IV phylogenetically later than $tRNA^{Phe}$. The latter species is characterized by having a unique series of three guanosine residues in sites F, G, and H.

Bend 2 Relationships. The nucleoside in the single site of bend 2 shows only rather loose correlation to groups. Among prokaryotic types, there appears to be a diminution in extent of variability with group advancement. In groups I and II all three nucleosides occur with equal frequency, uridine being absent here; but in group III guanosine usually is, and in group IV adenosine always is, the occupant of this locus. Similar relationships exist among eukaryotic species. Group I members show wide variation, but in the remainder dimethylguanosine is almost always the nucleoside. In group IV one adenosine ($tRNA^{Cys}$) and one pseudouridine ($tRNA^{Trp}$) are exceptional nucleosides. Nevertheless, the trend is otherwise highly consistent.

8.2.4. Phylogeny of Arm III Sequences

The known sequences of arm III, the anticodon arm, are so closely correlated with the evolutionary concept of genetic-code–amino-acid origins that they might be considered to be confirmatory evidence. They also offer the most clear-cut indications of the derivation of eukaryotic species from prokaryotic ancestors (Table 8.3). To obviate useless detail, the anticodons, whose structure largely derives from function, are omitted from consideration. For similar reasons, discussion in this section is by comparisons of paired and unpaired regions, rather than merely by the major source types.

The Paired Region. In the paired region, a number of group characteristics can be noted, mostly in the form of tendencies rather than sharply defined traits. The doublets occupying sites 1 and 2 in prokaryote group I members are AG, UC, CG, and CC, that of group II is mostly CA, but rarely CU or UA, in group III CA is the most frequent occupant, but CU and UC also occur, and in group IV GG, GC, CC, and CA are found, the frequent presence of guanosine in site 1 being especially noteworthy. The complements in sites 2′ and 1′ show comparable tendencies. Site 3 nucleosides also show trends among prokaryotes, A and U predominating in group I, A and G in group II, C in group III, and G (rarely A) in group IV. While sites 4 and 5 among prokaryotes do not display any clear correlation to group, these two demonstrate the sharpest tendencies among eukaryotes, the others being either too variable or uniform. Group I has either CG, CC, or GC in sites 4 and 5 respectively, group III has UG, GG, GA, or GΨ, and group IV seems to have GA consistently. The corresponding sites on the 3′-strand show comparable trends, with pseudouridine consistently the occupant of site 5′ in group IV. Thus in both prokaryotes and eukaryotes, the paired sequences strongly suggest that evolution of the tRNAs has been within four groups, each of which in turn had remote ancestry in a single progenitor type, probably $tRNA^{Gly}$.

Site A of the Unpaired Loop. The unpaired sequences of this arm likewise offer firm support for the group evolution concept in the prokaryotic and eukaryotic tRNAs alike (Table 8.3). Among the prokaryote species, uridine and cytidine occur with equal frequency in site A of group I, while in group II uridine is the more frequent occupant and the occasional cytidine sometimes is modified to 2-thiocytidine. In group III, 2′-methylcytidine is the usual nucleoside here, along with an occasional cytidine or uridine, and in group IV cytidine sometimes is modified to 3-methylcytidine, and uridine is replaced by pseudouridine and X.

This locus among the eukaryotes shows comparable group trends. Pseudouridine is somewhat more common among group I members than cytidine, and uridine occurs only in $tRNA^{Ala}$. In group III only cytidine seems to occupy this site, and in group IV, as among the prokaryote types, 3-methylcytidine is the characteristic occupant, along with an occasional cytidine or pseudouridine.

Site C of the Loop. As uridine is the universal occupant of site B, except in mammalian $tRNA^{Met}_{F}$, no phylogenetic clues are provided there, but site C occupants more than compensate. Among prokaryotic species in group I, unmodified adenosine is the prevalent nucleoside, but 2-methyladenosine appears in later members, such as $tRNA^{Glu}$ and $tRNA^{Val}$. The latter modification is also characteristic of group II members, along with an unidentified methylated guanosine in $tRNA^{Leu}$. Group III components are marked by the presence of *N*-(purine-6yl-carbamoyl)threonine (A_t), *E. coli* $tRNA^{Met}_{F}$ being the sole exception, and group IV members always have 2-methylthio-N^6-(Δ^2-isopentenyl)adenosine (A_s).

In large measure, the occupants of site C in the eukaryotes closely parallel those cited above. However, in group I the adenosine is always unmodified, but methylinosine and methylguanosine occasionally occur. Group III members have the identical complex base found in the prokaryotic species, *N*-(purine-6yl-carbamoyl)threonine (A_t), but most group IV tRNAs have N^6-(Δ^2-isopentenyl)adenosine (A_i). It should be noted that during the maturation of group IV tRNAs in prokaryotes, this substance is formed first at this site and later receives the methylthio modifications (Isham and Stulberg, 1974). Thus in this regard the prokaryotic types are more advanced than the eukaryotes. However, A_i is not universally present in eukaryotic group IV species. For instance, brewer's yeast $tRNA^{Trp}_{I}$ has an unmodified adenosine, and yeast $tRNA^{Leu}$ has 1-methylguanosine. Moreover, the $tRNA^{Phe}$s have the hypermodified nucleoside Y in several varieties. This base cannot be viewed as essential to the operational efficiency of those species that possess it, for in *Drosophila* it is lacking completely, yet its $tRNA^{Phe}$s must function normally (White and Tener, 1973). Moreover, chemical removal of the Y base had only a slight retarding effect on

Table 8.3
Arm III of Sequenced tRNAs Arranged by Major Source Type[a]

	Prokaryotes				Eukaryotes		
	12345	AB[b]CD	5′4′3′2′1′		12345	AB[b]CD	5′4′3′2′1′
GROUP I				GROUP I			
$Salm_1^{Gly}$	AGAGC	UU-AA	GCUCU	Sac_3^{Gly}	CAACG	ψU-Aψ	CGUUG
$Esch_1^{Gly}$	AGAGC	UU-AA	GCUCU	$Wheat_3^{Gly}$	GUACC	CU-AC_5	GGUAC
$Esch_2^{Gly}$	UCAGC	CU-AA	GCUGA	Sac^{Ala}	CUC_mCC	UU-I_mψ	GGGAG
$T4_2^{Gly}$	UCAGA	CU-AA	ψCUGA	Tor^{Ala}	ψUCGC	UU-I_mψ	GCGAA
$Esch_3^{Gly}$	CGACC	UU-AA	GGUCG	$Brew^{Asp}$	GGCGC	ψU-C_mG	GUCGG
$Esch_1^{Ala}$	CCUGC	UU-AC	GCAGG	Sac_2^{Glu}	ψCACG	CU-AC	CGUGG
$Esch_2^{Glu}$	CCGCC	CU-A_2C	GGCGG	Sac^{Val}	ψCUGC	ψU-AC	GCAGA
$Salm_1^{Val}$	CCUCC	CU-A_2A	GGAGG	$Brew^{Val}$	ψCUGC	ψU-AC	GCAGA
$Esch_1^{Val}$	CCUCC	CU-A_2A	GGAGG	Tor^{Val}	ψCUGC	ψU-AC	GCAGA
$Esch_{2A}^{Val}$	CCACC	UU-AU	GGUGG	Mus^{Val}	(C) ψUCGC	C_mU-AC_5	(G) GCGAA
GROUP II				GROUP II			
$Esch_1^{Leu}$	CUAGC	UU-G_mψ	GψUAG	$Brew_2^{Arg}$	ψCUGG	CU-AA	CCAGA
$Esch_2^{Leu}$	CUACC	UU-G_mψ	GGUAG	GROUP III			
$Esch^{His}$	CUGGA	UU-A_2ψ	ψCCAG	Tor^{Ile}	ψGGUG	CU-AA_t	CGCCA
				$Brew_3^{Arg}$	ψCUGA	CU-AA_t	ψCAGA

charging the tRNA, but recognition and function remained unchanged (Thiebe and Zachau, 1968; Odom *et al.*, 1974).

Site D of the Unpaired Region. Site D shows a curious series of consistencies and variables, as well as close parallelism between the prokaryotes and eukaryotes. In prokaryotes, group I members have unmodified adenosine (usually), cytidine (occasionally), or uridine (rarely), whereas the corresponding eukaryote species usually have cytidine, but 5-methylcytidine and

Table 8.3. Cont'd.

	Prokaryotes				Eukaryotes		
	12345	AB[b]CD	5′4′3′2′1′		12345	AB[b]CD	5′4′3′2′1′
$T4^{Gln}$	UAGCA	CU-A$_2$C	ψGCUA	$Sac_2{}^{Lys}$	ψψCGG	CU-A$_t$A	CCGAA
$Esch_1{}^{Gln}$	CCGGU	U$_m$U-A$_2$ψ	ACCGG	Rat^{Ser}	AψGGA	C$_3$U-A$_t$A	ψCCAU
$Esch_1{}^{Arg}$	CUCGG	C^2_tU-A$_2$A	CCGAG	$Sac_F{}^{Met}$	CAGGG	CU-A$_t$A	CCCUG
GROUP III				$Mus_F{}^{Met}$	CUGGG	CC-A$_t$A	CCCAG
$Esch_1{}^{Ile}$	CACCC	CU-A$_t$A	GGGUC	$Mus_4{}^{Met}$	ψCAGψ	CU-A$_t$A	ψCUGA
$Esch_2{}^{Ile}$	CACGA	CU-A$_t$A	ψCGUG	GROUP IV			
$Esch^{Thr}$	CACCC	UU-A$_t$A	GGGUG	Sac^{Phe}	CCAGA	C$_3$U-YA	ψCUGC
$Esch_3{}^{Ser}$	CUCCC	C$_2$U-A$_t$A	GGGAG	$Wheat^{Phe}$	ψCAGA	C$_3$U-Y$_W$A	ψCUGA
$Esch_M{}^{Met}$	CAUCA	CU-A$_t$A	ψGAUG	Rab^{Phe}	ψψAGA	C$_3$U-Y$_M$A	ψCUAA
$Esch_F{}^{Met}$	UCGGG	C$_2$U-AA	CCCGA	Sac^{Tyr}	CAAGA	CU-A$_i$A	ψCUUG
GROUP IV				Tor^{Tyr}	ψCAGA	CU-A$_i$A	ψCUGA
$Esch^{Phe}$	GGGGA	ψU-A$_s$A	ψCCCC	$Brew_A{}^{Ser}$	AAAGA	ψU-A$_i$A	ψCUUU
$Esch_2{}^{Tyr}$	GCAGA	CU-A$_s$A	ψCUGC	Rat^{Ser}	AAAGA	C$_3$U-A$_i$A	ψCCAU
$Esch^{Ser}$	CCGGU	NU-A$_s$A	ACCGG	Sac^{Cyst}	GCAGA	ψU-A$_i$A	ψCUGU
$Esch^{Trp}$	CCGGU	CU-A$_s$A	ACCGG	$Brew_1{}^{Trp}$	ψψCGA	CU-AA	ψCGAA
$T4^{Leu}$	GAGCA	CU-A$_s$A	ψGCUG	Sac^{Leu}	CCUGA	ψU-G$_1$C	ψCAGG

[a]For explanation of conventions, see Table 8.1, footnote *a*.
[b]Location of the anticodon.

pseudouridine also occur. Group II components mostly have pseudouridine in the prokaryotes, with cytidine and adenosine occasional substitutes. In groups III and IV of prokaryotic and eukaryotic types alike, adenosine is the only known occupant, except in yeast $tRNA^{Leu}$ in which cytidine is found.

Hence these loop components together make it implicit that tRNAs evolved in four groups from a common ancestor and that eukaryotic types are descendants from prokaryotic predecessors. It also appears clear that both branches have undergone modification since the eukaryotes diverged from the bacterial line.

8.2.5. Phylogeny of Arm IV Sequences

The variable arm (arm IV) is well named, for it shows few consistencies and scarcely any trends. Nevertheless, the contrasts between the short and long types afford an opportunity for speculation, supported by scant evidence, suggestive of how certain tRNAs may have become differentiated.

The Short Types. Among the short-type arm IV sequences, one trend is immediately evident in both prokaryotic and eukaryotic species, the existence of a reduced length in many group I tRNAs (Table 8.4). Whereas all other groups consistently have five sites, a number of group I components have only four. Indeed, *Saccharomyces cerevisiae* $tRNA_3^{Gly}$ has but three sites present; however, this may well represent an independent loss of an adenosine from site 4, on the basis of the isoaccepting species from wheat. Thus this arm appears to have been quite short primitively and gradually became lengthened until it stabilized with five sites. Consequently, eukaryotic species of tRNAs belonging to group I are clearly demonstrated to have been derived from corresponding prokaryotic types.

In site A purine nucleosides are predominant, but this tendency is far more marked in advanced groups than in groups I and II of the prokaryotes. In eukaryotic tRNAs, there is nearly absolute uniformity throughout, except for the uridine in this site in yeast $tRNA^{Cys}$. Hence, this site appears to have been highly variable primitively, becoming restricted to purines with advancement, until in the eukaryote species adenosine became almost the exclusive occupant. Similar, though less marked, trends can be detected for site B occupants.

Guanosine is the predominant nucleoside at site C, both in prokaryotes and eukaryotes. In the earliest types it remains unmodified, but is consistently modified to 7-methylguanosine with evolutionary advancement. Prokaryotic group II members appear frequently to replace the latter with uridine; in eukaryotes uridine never occurs, but adenosine is frequent in the three more advanced groups.

At site D marked contrasts between the prokaryotic and eukaryotic species exist, although both source types agree as to the major trends. For example, in both, adenosine occurs here among early types in group I and is gradually replaced by uridine and its modifications at higher levels. The strongest distinctions center around the types of modifications. In the prokaryotes, uridine becomes modified first to N,* which in group III is modified further by methylation. Another undetermined uridine derivative, X, appears in the three highest groups, but 5,6-dihydrouridine does not seem to be present here. To the con-

* In this site, N represents an unidentified modification of uridine.

trary, among eukaryotes the latter nucleoside is the only uridine modification acceptable at this site, for it is predominant here throughout all the groups. The only exceptions seem to be various early components of Group I and yeast $tRNA^{Phe}$ and $tRNA^{Trp}$ in group IV; the latter two nucleic acids pose some difficult problems as to the recognition requirements of the modifying enzymes.

On the basis of the above evidence, the most probable phylogeny appears to be the following sequence of events: Very early tRNAs, may have had adenosine in site D, which gradually became replaced with uridine in all later group I components and in the three higher groups. Before any modifying enzymes evolved to act on the nucleosides of this site, the eukaryotes diverged from the prokaryotes. Subsequently three modifying enzymes for this site evolved among the latter, namely, one for the formation of N, another for X, and a third to methylate N. Among the eukaryotes, only a single modifier for site D arose, that which modified uridine to 5,6-dihydrouridine. This enzyme could very likely have been a derivative of the corresponding enzyme which acts upon certain uridines in the loop of arm II. The dihydrouridine in site E of brewer's yeast $tRNA^{Trp}$ is here viewed as the result of action by this same enzyme, but why it acts on uridine in site E but not on that in site D in this tRNA is not clear.

Site E shows very little more of importance, for cytidine is by far its commonest occupant, with uridine an occasional substitute in both prokaryotes and eukaryotes. Among the latter, 5-methylcytidine is the characteristic nucleoside, with unmodified cytidine rather frequent here as well as uridine. At this site it is obvious that the modifying enzyme arose after the eukaryotes had separated from the prokaryotic stock, an observation that has some value in the discussion of the long type of arm IV below.

The Long Types. The fifteen known sequences of long-type arm IV are so highly diversified, even within groups, that few homologous sites are evident. The problem is most clearly presented by a comparison of the four sequenced $tRNA^{Leu}$s (Table 8.5). Even the two *E. coli* $tRNA^{Leu}$ of group II show fewer similarities than is typical of other arms; for example, only 7 of the 17 sites of this arm show the presence of identical nucleosides in these two species, whereas 12 of the 17 sites in arm V of *E. coli* $tRNA^{Gly}_{1\,and\,2}$ have identical occupants (Table 8.5) and 13 of the 17 are identical in arm V of these two $tRNA^{Leu}$s. Likewise almost no homologies exist between the latter pair and bacteriophage T4 $tRNA^{Leu}$ of group IV, or between the latter and the corresponding eukaryotic $tRNA^{Leu}$ also of group IV. Such wide diversity in the structure of this arm strongly intimates that this region of the tRNA may have in large measure arisen independently in the several species.

The *E. coli* $tRNA^{Tyr}$ is particularly clear in this regard, for the eukaryotic derivative is of the usual short-arm type. This observation accentuates the alternatives that are available: (1) The tRNAs have not evolved, but arose independently; (2) the long-arm type of tRNA is primitive, and gave rise to the short-

Table 8.4A

Arm IV of Sequenced tRNAs Arranged by Major Source Type: Short Type[a]

Prokaryote	ABCDE	Eukaryote	ABCDE
GROUP I			
Salm$_1^{Gly}$	AU-AC	Sac$_3^{Gly}$	-GG-C
Esch$_1^{Gly}$	AU-AC	Wheat$_3^{Gly}$	A-GAC$_5$ (U)
Esch$_2^{Gly}$	-UGAU	Sac^{Ala}	AGGDC
T4$_2^{Gly}$	-UGAU	Tor^{Ala}	AGGDC
Esch$_3^{Gly}$	GGG$_7$UC	Brew^{Asp}	A-GAU
Esch$_1^{Ala}$	AGG$_7$UC	Sac$_2^{Glu}$	A-GAC
Esch$_2^{Glu}$	UA-AC	Sac^{Val}	ACG$_7$DC$_5$
Salm$_1^{Val}$	GGG$_7$UC	Brew^{Val}	ACG$_7$DC$_5$
Esch$_1^{Val}$	GGG$_7$UC	Tor^{Val}	AC-DC$_5$
Esch$_2^{Val}$	GGG$_7$NC	Mus^{Val}	AGG$_7$DC$_5$
GROUP II			
T4^{Pro}	AGG$_7$UC		
Esch^{His}	UUG$_7$UC		
T4^{Gln}	GAUGC		
Esch^{Gln}	CAUUC		
Esch$_1^{Arg}$	CGG$_7$XC	Brew$_2^{Arg}$	AGADU

Prokaryote	ABCDE	Eukaryote	ABC
GROUP III			
Esch$_1^{Ile}$	AGG$_7$N$_m$C	Tor^{Ile}	AGA
Esch$_2^{Ile}$	AGG$_7$N$_m$C	Brew$_3^{Arg}$	AGA
Esch^{Thr}	AGG$_7$UC	Sac^{Lys}	AUG$_7$
Esch$_M^{Met}$	GGG$_7$XC	Sac$_F^{Met}$	AUG$_7$
Esch$_F^{Met}$	AGG$_7$UC	Mus$_F^{Met}$	AGG$_7$
		Mus$_4^{Met}$	AGG$_7$
GROUP IV			
Esch^{Phe}	GUG$_7$XC	Sac^{Phe}	AGG$_7$
		Wheat^{Phe}	AGG$_7$
		Rab^{Phe}	AGG$_7$
		Sac^{Tyr}	AGA
		Tor^{Tyr}	ACA
		Sac^{Cys}	UGG$_7$
Esch^{Trp}	GUG$_7$UU	Brew^{Trp}	GGG$_7$

[a]For explanation of conventions, see Table 8.1, footnote *a*.

Table 8.4B
Arm IV of Sequenced tRNAs Arranged by Major Source Type: Long Type[a]

	Bend 3	1234567ABCDE7'6'5'4'3'2'1'	Bend 4
Prokaryote			
GROUP II			
$Esch_1^{Leu}$	-	-UGUCC-UUAC-GGACG--	-U-
$Esch_2^{Leu}$	U	--GCCC-AAUA-GGGC---	UU-
GROUP III			
$Esch_3^{Ser}$	U	AUGCGGUCAAAAGCUGCAU	--C
GROUP IV			
$Esch_2^{Tyr}$	C	--GUC---ACA-GAC----	UUC
$Esch^{Ser}$	C	--GACCCGAAA-GGGGU--	-*C
$T4^{Leu}$	C	--GGAA-UGAU-UUCC---	UU-
Eukaryote			
GROUP IV			
$Brew_A^{Ser}$	U'_2	--GGGC--(U)(C)U-GCCC---	G-C_5
Rat_1^{Ser}	U'_2	--GGGG--UCU-CCCC---	G-C_5
Sac^{Leu}	-	--UAUC-GUAA-GAUG---	--C_5
GROUP III			
Rat_3^{Ser}	U'_2	--GψGC--UC$_3$U-GCAC---	G-C_5

Table 8.5
Arm V of Sequenced tRNAs Arranged by Major Source Type[a]

	Prokaryotes				Eukaryotes		
	12345	ABCDEFG	5'4'3'2'1'		12345	ABCDEFG	5'4'3'2
GROUP I				GROUP I			
$Salm_1^{Gly}$	GAGGG	TψCGAUU	CCCUU	Sac_3^{Gly}	C_5CCGG	TψCGAUU	CCGG
$Esch_1^{Gly}$	GAGGG	TψCGAUU	CCCUU	$Wheat_3^{Gly}$	C_5C_5GGG	UψCGA_1UU	CCCG
$Esch_2^{Gly}$	GCGGG	TψCGAUU	CCCGC	Sac^{Ala}	UCCGG	TψCGAUU	CCGG
$T4_2^{Gly}$	GUGAG	TψCGAUU	CUCAU	Tor^{Ala}	UCCGG	TψCGA_1CU	CCGG
$Esch_3^{Gly}$	GCGAG	TψCGACU	CUCGU	$Brew^{Asp}$	C_5GGGG	TψCAAUU	CCCC
$Esch_1^{Gly}$	UGCGG	TψCGAUC	CCGCG	Sac_2^{Glu}	C_5GGGG	TψCGACU	CCCC
$Esch_3^{Glu}$	AGGGG	TψCGAUU	CCCCU	Sac^{Val}	CCCAG	TψCGA_1UC	CUGG
$Salm^{Val}$	GGCGG	TψCGAUC	CCGUC	$Brew^{Val}$	CCCAG	TψCGA_1UC	CUGG
$Esch^{Val}$	GGCGG	TψCGAUC	CCGUC	Tor^{Val}	CCCAG	TψCGA_1UC	CUGG
$Esch_{2A}^{Val}$	GGUGG	TψCGAGU	CCACU	Mus^{Val}	C_5CCGG (A)	UψCGA_1AA	CCGG
GROUP II				GROUP II			
$Esch_1^{Leu}$	GGGGG	TψCAAGU	CCCCC	$Brew_2^{Arg}$	CCAGG	TψCAA_1GU	CCUG
$Esch_2^{Leu}$	ACGGG	TψCAAGU	CCCGU				
$Esch^{His}$	GUGGG	TψCGAAU	CCCAU				
$T4^{Gln}$	AAAGG	TψCGAGU	CCUUU				
$Esch_1^{Gln}$	CCUGG	TψCGAAU	CCAGG				
$Esch_1^{Arg}$	GGAGG	TψCGAAU	CCUCC				

Table 8.5. Cont'd.

	Prokaryotes				Eukaryotes		
	12345	ABCDEFG	5'4'3'2'1'		12345	ABCDEFG	5'4'3'2'1'
UP III				GROUP III			
h$_1^{Ile}$	GGUGG	TψCAAGU	CCACψ	Tor^{Ile}	AGCAG	TψCG$\underset{1}{A}$UC	CUGCU
h$_2^{Ile}$	GGUGG	TψCAAGU	CCACψ	Brew$_3^{Arg}$	AUGGG	TψCG$\underset{1}{A}$CC	CCCAU
h^{Thr}	GGCAG	TψCGAAU	CUGCC	Sac$_2^{Lys}$	AGGGG	TψCG$\underset{1}{A}$GC	CCCCU
h$_3^{Ser}$	CGGGG	TψCGAAU	CCCCG	Sac$_F^{Met}$	CUCGG	AUCG$\underset{1}{A}$AA	CCGA$\underset{mm}{G}$
h$_M^{Met}$	ACAGG	TψCGAAU	CCCGU	Mus$_F^{Met}$	GAUGG	AUCG$\underset{1}{A}$AA	CCAUC
h$_F^{Met}$	GUCGG	TψCAAAU	CCGGC	Mus$_4^{Met}$	GUGAG	TψCG$\underset{1}{A}$UC	CUCAC
UP IV				GROUP IV			
h^{Phe}	CUUGG	TψCGAUU	CCGAG	Sac^{Phe}	$\underset{5}{C}$UGUG	TψCG$\underset{1}{A}$UC	CACAG
h$_2^{Tyr}$	GAAGG	TψCGAAU	CCUUC	Wheat^{Phe}	GCGUG	TψCG$\underset{1}{A}$UC	CACGC
h^{Ser}	CAGAG	TψCGAAU	CUCUG	Rab^{Phe}	$\underset{5}{C}$CUGG	TψCG$\underset{1}{A}$UC	CCGGG
h^{Trp}	GGGAG	TψCGAGU	CUCUC	Sac^{Tyr}	GGGCG	TψCG$\underset{1}{A}$CU	CGCCC
eu	GUGGG	TψCGAGU	CCCAC	Tor^{Tyr}	GGGCG	TψCGA$\underset{1}{A}$U	CGCCC
				Brew$_A^{Ser}$	GCAGG	TψCGAGU	CCUGC
				Rat^{Ser}	GCAGG	TψCGAAU	CCUGC
				Sac^{Cys}	CUUAG	TψCG$\underset{1}{A}$UC	CUGAG
				Brew$_1^{Trp}$	GCAGG	TψCA$\underset{1}{A}$UU	CCUGψ
				Sac^{Leu}	AAGAG	TψCGAAU	CUCUU

r explanation of conventions, see Table 8.1, footnote *a*.

arm types along independent lines; and (3) the short-arm types are more primitive and gave rise to the long-arm types along independent lines. Obviously all the homologies of structure that exist between tRNAs make alternative 1 untenable, while the wide diversity of long-arm sequences combined with the high degree of homology within groups of short-arm types make alternative 2 unattractive. Therefore only alternative 3 can be considered probable, a suggestion that receives support from the phylogeny of the short arm IV sequences, which show that primitively four or even fewer sites existed and that this condition became stabilized with five sites at advanced levels of evolution.

Therefore it is reasonable to suggest that in the variable arm the sequence $AGADC_5$ of *Saccharomyces* $tRNA^{Tyr}$ may reflect the ancestral condition of the isoaccepting species to a degree. First it is apparent that the 5-methylcytidine of this short sequence corresponds to the same nucleoside in bend 4 of eukaryotic long sequences, and, by extrapolation, to the unmodified cytidine in bend 4 among prokaryotic types, including *E. coli* $tRNA^{Tyr}$. Examining the sequence of the latter species more fully, it is found that beginning with site F and extending through the paired sequence of the 3′-strand the sequence AGACUUC occurs. Thus by inserting the couplet CU and failing to modify the second uridine, the sequence $AGADC_5$ of yeast becomes AGACUUC of the *E. coli* tRNA. Comparable and complementary additions of the 5′-side would yield the complete sequence shown in Table 8.4. Admittedly the evidence is scant and other interpretations are equally valid; the present proposal is meant merely to suggest a possible mechanism for development of the long- from the short-type arm IV. However, it is here predicted that in larger viruses of the RNA type, rather than the T-even series which is much more advanced, will be found $tRNA^{Ser}$s, $tRNA^{Leu}$s, or $tRNA^{Arg}$s that have a short arm IV similar to the other tRNAs in comparable tRNA groups.

8.2.6. The Evolution of Arm V

In general, arm V sequences resemble those of arm II, in that many uniformities exist at numerous sites throughout the entire series because of their probable involvement in tertiary structure. Other sites, however, recall those of arm I in showing a wide diversification without clear-cut evolutionary ties. Nonetheless, a few generalizations are possible.

The Paired Region. Site 1 occupants (as well as those of site 1′) show interesting contrasts between the prokaryotes and eukaryotes. In the former, purine nucleosides predominate throughout all groups, while cytidine, absent in group I members, shows a tendency toward becoming more abundant with evolutionary advancement, particularly in group IV. Eukaryotes show just the opposite trend, for cytidine or 5-methylcytidine is nearly universal in group I and the purines become increasingly abundant with advancement (Table 8.5). Uridine occurs here only in group I, both in the prokaryotes and eukaryotes.

In the two major source types alike, sites 2–2′ and 3–3′ display increasing variability with advancing evolutionary development, group IV members being particularly diversified. All site 4 occupants are purine nucleosides, except in group IV eukaryotic tRNAs, in which pyrimidines are equally abundant. Sites 5–5′, being universally occupied by G–C pairs, fail to be of value in phylogeny.

The Unpaired Sequence. Because of their paucity of variability, none of the first five sites in the unpaired sequence provides any phylogenetic clues. However, site F occupant show a marked parallelism of the eukaryotic and prokaryotic species. In both types of organisms pyrimidines alone are found in group I members, except in certain $tRNA^{Val}$s. Groups II and III (and group IV also in prokaryotes) display purine nucleosides predominantly, while group IV members in eukaryotes are widely diversified.

The nucleosides in site G are not in accord in eukaryotes and prokaryotes. Among the latter, pyrimidines alone occupy this site, uridine seemingly being the sole occupant in groups II to IV. In the eukaryotes, the two common pyrimidines are found in all groups, except that uridine appears to be absent from group III members, while adenosine is a rare substitute (Table 8.5).

8.2.7. Evolution of the Modifying Enzymes

Although brief mention of phylogenetic alterations in modifying enzymes has been made where appropriate in preceding sections, a unified discussion discloses a clearer picture of the nature and extent of those changes. These are of two major sorts: (1) those that involve alterations only in the total number of modifying enzymes or of ties modified, and (2) those that involve changes in the kinds of modifications that exist.

Extent of Modification. It has long been noted that eukaryotic tRNAs have more modified nucleosides than do prokaryotic types. However, the total relative frequency between the two source types is seen to be not so disparate now that the primary structure of many tRNAs has been determined (Table 8.6). In that chart, counts are by site and modification; a later table summarizes the numbers of different major enzyme species.

Since the tables report only total counts, it is well to compare the state of relative knowledge of the several groups before the figures are examined. In group I, at least one tRNA carrying each of the five amino acid members has had its primary structure determined from both prokaryotic and eukaryotic sources. In contrast, group II tRNAs are the most poorly explored, for while all member amino acid species have been sequenced from prokaryotes, only a single type ($tRNA^{Arg}$) has had its nucleoside sequence established in eukaryotes. Thus the totals reported for this group cannot be considered truly representative, nor can those for group III, but to a lesser degree. Although at least a single tRNA sequence has been determined for all but one of the latter group's

Table 8.6
Phylogenetic Changes in Modifying Enzymes of tRNAs Counted by Site Enzymes

	Arm I	Bend 1	Arm II	Bend 2	Arm III	Arm IV	Arm V	Anti-codons	Total ancestral	Group total in prokaryotes	Group additions not in eukaryotes	New in eukaryotes	Group total in eukaryotes	Total known in group
GROUP I														
Prokaryote	0	1	6	0	2	2	2	4						
Also in eukaryotes[a] (early ancestral)	0	0	6	0	1	1	2	2	12	17	5	14	26	31
Eukaryote additions	1	1	0	1	6	2	3	0						
Total eukaryote	(1)	(1)	(6)	(1)	(7)	(3)	(5)	(2)						
GROUP II														
Prokaryote additions	1	0	6	0	5	1	0	2						
Total prokaryote	(1)	(1)	(4)	(0)	(7)	(2)	(2)	(3)						
Additions also in eukaryotes[a]	0	0	0	0	0	0	0	1	1	20	8	2	[12][b]	22
									(13)					
Eukaryote additions	1	0	1	0	0	0	0	0						
Total eukaryote	(1)	(1)	(3)	(1)	(1)	(1)	(3)	(1)						

Prokaryote additions	0	0	0	0	2	1	1	2						
Total prokaryote	(0)	(1)	(5)	(0)	(3)	(3)	(3)	(3)						
Additions also in eukaryotes[a]	0	0	0	0	2	0	1	0	3	18	3	6	25	34
Eukaryote additions	1	0	0	0	0	0	2	3	(16)					
Total eukaryote	(2)	(1)	(4)	(1)	(5)	(3)	(5)	(5)						
GROUP IV														
Prokaryote additions	0	1	1	0	5	0	0	0						
Total prokaryote	(0)	(2)	(6)	(0)	(6)	(2)	(2)	(5)						
Additions also in eukaryotes[a]	0	0	0	0	3	0	0	0						
Eukaryote additions	0	0	2	1	1	0	0	2	3	21	4	6	33	43
Total eukaryote	(0)	(1)	(8)	(2)	(8)	(3)	(5)	(4)	(19)					
Total shared (ancestral)	0	0	6	0	11	1	3	3						
Total prokaryote	1	2	7	0	13	4	3	8						
Total eukaryote	3	1	9	2	16	3	8	9						
Grand totals[c]	4	3	10	2	18	6	8	14	19	39	20	28	46	66

[a] At identical site, but not necessarily in the same group of tRNAs.
[b] Total based on only 1 sequence; hence, not significant.
[c] Group considerations are omitted to avoid including duplications in the totals.

components (asparagine), two of those reported from eukaryotes ($tRNA^{Arg}$ and $tRNA^{Lys}$) are unknown as yet from prokaryotes and similarly two from the latter ($tRNA^{Ser}$ and $tRNA^{Thr}$) have not been sequenced from eukaryote sources. Group IV is nearly as well known as group I, for all major types have been sequenced from both eukaryotic and prokaryotic sources, except $tRNA^{Cys}$ which has been determined only in the former. If punctuation types (nonsense) also have special tRNAs, which possibility has not yet been explored, these may eventually contribute to the totals.

In summary, groups I and IV members are well known, those of group III are reasonably well determined, but group II totals are not truly representative and frequently must be omitted from consideration.

In spite of the above weaknesses, several trends are clearly perceptible (Table 8.6). Among prokaryotes, there is a distinct tendency toward an increasing number of modified sites correlated to advancing evolutionary status of the groups. Group I contains only 17 modified sites, whereas group IV is known to have 21, a number that undoubtedly will be increased as further progress in established nucleoside sequences has been made. Of the 17 reported from group I prokaryotic tRNAs, 6 are not known from eukaryotes, and thus, on the basis of present knowledge, are to be viewed as having arisen after the eukaryotic stock diverged. Similar side-branch prokaryotic specializations can be perceived in each later group, as novel site modifications have been reported in all. Thus of the total of 39 site modifications described in prokaryotic tRNAs, 21 are unique; in other words, the prokaryotic types have undergone at least as much evolutionary change since the eukaryotes branched off as they had prior to that event.

A parallel series of changes is evident among the eukaryotic tRNAs. Whereas a total of 27 site modifications exist in group I members (Table 8.6), 33 have been reported in the less thoroughly explored group IV components. Of the 46 total site modifications found among eukaryotic tRNAs, 28 have as yet not been reported from prokaryotes and thus present acquisitions made subsequent to their divergence from the ancestral stock. Hence, it is obvious that the mitotic method of cell division has permitted a more rapid rate of change than possible in the nonmitotic prokaryotes, to the extent of about 33%.

The site modifications found in both prokaryotes and eukaryotes thus provide an insight into the late ancestral stock from which the latter diverged. In this conjectural group, about 25 site modifications appear to have existed just prior to the point where the eukaryotic stock diverged. If, as seems to be indicated by the available evidence, it is accepted that the group I tRNAs arose prior to the others, about 11 site modifications existed at the earliest level of prokaryote origins (Table 8.6), most of which (6) are in arm II.

Regional Comparisons. While the majority of the modifications at the earliest known levels are probably for structural purposes, as indicated by their

Nucleosides by Number of Sites[a]

	Arm III					Arm IV			Arm V				Anticodons		
	Stem		Loop						Stem		Loop				
	E	B	P	E	B	P	E	B	E	B	E	B	P	E	B
							1								
							2								
	2	1			2					1		1		1	
													1		
															1
				1											
															1
			1			1							1	1	
											1				
			1												
				1											
			1												
					1										
												1			
			1											1	
				1			1								
				1			1		2						
			1												
													1		
	1			1										1	
				1											
								1							
			1	1		1			1						
													1		
															1
	3	1	6	7	3	2	5	1	3	1	1	2	4	4	3
	4		16						4		3				
	20					8			7				11		

karyotic lines increased the number of site 1 modifications to eight, and the latter also gained one modification of site 2. Consequently in this respect, the evolutionary rates of the two groups of organisms have been almost equal.

Types of Modifications. The type of modification and its frequency are also of interest. Excluding those whose specific nature or structure remains to be elucidated, 27 varieties may be counted (Table 8.7), 11 of which occur in both major groups of organisms. Seven others are found solely in prokaryotes and nine varieties only in eukaryotes. By far the largest portion (16 or about 60%) of modified nucleosides exist at only a single site; six others are confined to two, and four are found in either three or four. Hence, modifications in general may be viewed as specializations of narrow scope.

Only two modifications occur at a large diversity of sites, and one of these, 5,6-dihydrouridine, is restricted to two of the five arms. Indeed five of its seven total occurrences are in the loop of arm II, while its other two are in arm IV of eukaryotes. The only modified nucleoside that really occurs widely in the molecules of tRNAs is pseudouridine, known from a total of 12 sites. Half of these, however, are confined to eukaryotes, so that its diversification among prokaryotes is restricted to those loci already modified in the ancestral stock.

Another interesting characteristic of modified nucleosides can be noted by comparing numbers of occurrences in loops and in stems. Because of the lack of consistency of base pairing in arm IV, modifications occurring there are omitted from consideration at this time. In stems, only a total of 15 sites have undergone modifications, whereas in loops and other unpaired regions such as the bends, 32 have been reported. These discrepancies in frequency may be accounted for in one of two ways. Either modifications are of more importance in unpaired regions, perhaps for structural purposes, or else they are more readily made there and have accordingly been acquired only for evolutionary reasons. Perhaps both these possibilities are valid, depending upon the specific region under consideration.

8.3. ORIGIN AND EVOLUTION OF tRNA

With what is now known about tRNA structure and functioning, an outline of possible and probable steps in its origin and subsequent evolution can be proposed. Although many of the early steps must remain obscure for the present, due to the absence of sequences from suitably primitive organisms, at least a skeleton of the events can be drawn up, the sequence of which perhaps is made clearer by Figure 8.2. Little reference is made in the following account to peripheral, although essential, events, such as the origins of ligases, mRNA, and so on, but these substances must be considered to have arisen and evolved concurrently with the tRNAs.

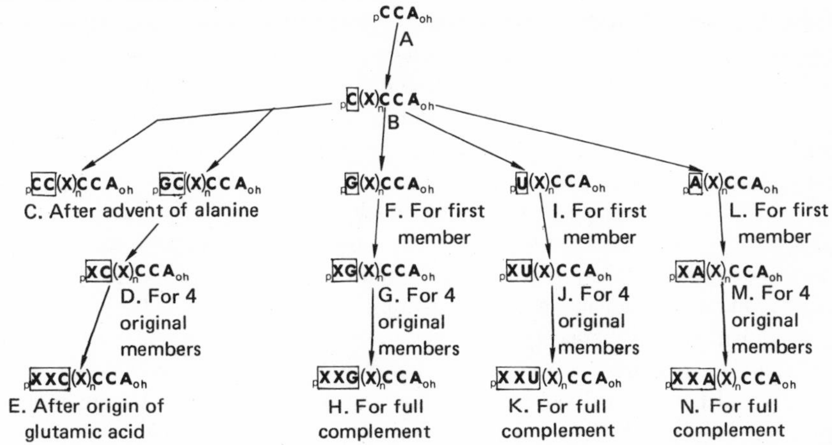

Figure 8.5. Proposed phylogeny of tRNAs. It is suggested that the ancestral tRNA consisted functionally of the –CCA terminus and transported all types of amino acids (A). Later, as the definitive L-amino avids were evolved, a coding end developed, first for glycine (B), then one by one for the remaining members of group I (C–E). Subsequently, in similar fashion codons evolved in turn for group II amino acids (F–H) and then for the other two group components (I–K and L–N).

Origins of tRNAs. In considering the possible origins of tRNAs, it is essential first to bring to mind the fundamental function in the cell economy which these substances provide that does not depend upon the presence of the rest of the genetic mechanism, and then to deduce the minimal structural features needed for carrying out this function. This basic action is one that is still of service in several systems of modern cells, that of simple transport of amino acids, which is utilized in the soluble system of protein synthesis, in peptidoglycan metabolism, and elsewhere (Chapter 4; Section 4.1.3). Then, following these simple origins, a biological system for the origin of the genetic code is incorporated; here that scheme advanced in Chapter 6 is followed.

The Earliest Stages. Since the 3′-terminus appears to be the universal point of attachment for amino acids, the –CCA found there must surely have been *the* essential nucleic acid sequence at the earliest stage of tRNA's existence. Perhaps the triplet bore an additional chain of nucleotide residues, but this brief sequence was sufficient in itself for the first definitive amino acid, glycine, to be bound to it by the ligase, and thereby to increase efficiency of the amino acid's transport and recognition. It is pertinent to note that even in metazoan and other advanced cells, an enzyme appears still to synthesize this terminal –CCA without template, and then adds it to the tRNA precursor during maturation.

This minimal tRNA molecule later increased in length, although it is not clear how such lengthening occurred. While the progenitory tRNA was thus becoming more elongate, the 5′-terminal apparently remained a cytidylic acid residue (Figure 8.2B), which provided a code for glycine, the only amino acid then carried by this means. Somewhat later, when the protobiont had successfully biosynthesized a second amino acid, alanine, doublets were consequently needed for coding. Two alternatives for the primitive formation of such doublets appear available. One is that the 5′-terminal region had originally consisted of at least two cytidines in tandem, although only the terminal one was then actually necessary; to code for alanine, the latter was mutated to G on the alanine-carrying species, yielding CC to code for glycine and CG for alanine. The second possibility is that the formation of this doublet could have involved the acquisition of a second C at this terminus to code for glycine, and of a G for alanine. In either case, it is evident that a simple messenger was also present and being employed in protein synthesis, otherwise the codons would have been devoid of survival value. Similar 5′-terminal mutations were necessary as the remaining original members of group I (aspartic acid and valine) arose (Figure 8.5D) and became carried by tRNAs. Then when glutamic acid was derived by modification of aspartic acid, another set of additions to the 5′-end of all the existing anticodons was needed, thus completing the first group of amino acids and bringing into existence a triplet system of coding.

Later Stages. At first, subsequent steps in the evolution of tRNAs were largely repetitions of those outlined above for group I components. The first amino acid of group II to be biosynthesized, proline, was eventually provided a codon by mutation of the original terminal C (that coded for glycine early in this phylogeny) to a G. Hence, it seems clear that the doublets in the foregoing series of events were created by additions to the chain and not by mere broadening of codon-recognizing sites toward the 3′end (Figure 8.5F). Thus, as new amino acids were derived from the α-amino-butyric acid, additions had to be made to the 5′-terminus of the primordial tRNA. Comparable steps thus must have occurred as each of the remaining two groups of amino acids arose and developed.

If this set of suppositions is valid, that series of steps makes clear one aspect of tRNA structure that previously had been obscure, in that it provides a logical basis for the anticodons' being constructed in an opposite sense to the codons. At least it explains why the third anticodonal site codes for the major signal of the codon, that of its site 1. The reason, of course, being, as outlined above, the 5′-terminus of the short primeval tRNA served as the first codon in each group, later members in turn requiring additions of nucleotides in the 5′ direction until a triplet code was attained. Thus it becomes evident simultaneously that the first messenger RNA probably consisted of a short sequence of Cs, when glycine was the only amino acid, and that alanine's advent required the insertion of Gs where needed. As these insertions into the messenger molecule were in the 3′-direction or at the 3′-terminus, codons ultimately developed to be the reverse of anticodons in structural direction.

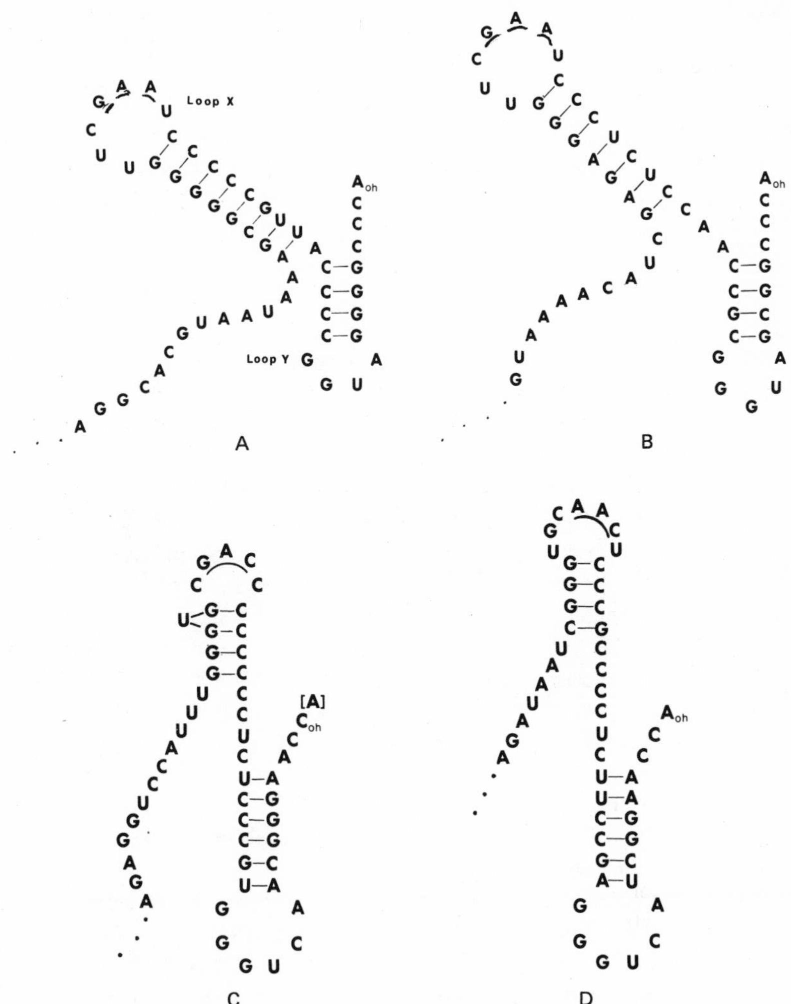

Figure 8.6. Primitive types of tRNAs from viruses. (A, B) The tRNA^{His} of two "strains" of tobacco mosaic virus lack anticodons for histidine and may prove to be mischarged by the usual heterologous systems employed. A possible anticodon in loop X (GAA) codes for phenylalanine (after Guilley *et al.*, 1975, and Lamy *et al.*, 1975). (C) Eggplant mosaic virus (Briand *et al.*, 1976) and (D) turnip yellow mosaic virus (Briand *et al.*, 1977) RNAs can be charged with valine; the probable anticodons are underlined in loop X.

Final Ancestral Events. After all amino acids had been encoded, further nucleotide additions were made to the tRNA molecules on both sides of the anticodon region. Hence it would not be surprising if at least some of the tRNAs reported to occur in simple RNA viruses (Chapter 9) proved to consist of far fewer nucleoside residues than those reported in bacteria and metazoans.

Indeed, the existence of such relatively brief tRNAs has been confirmed in two strains of tobacco mosaic virus (TMV), and in related viruses (Figure 8.6A,B). For several years, the 3′-end of the genome of these RNA types has been known to accept histidine in the presence of the histidyl-tRNA ligase (Öberg and Philipson, 1972). Now the determination of a 71-nucleoside-residue-long sequence at the 3′-end of each of two "strains" of TMV reveals an unusual secondary structure for the tRNA (Guilley *et al.*, 1975; Lamy *et al.*, 1975). In the illustration it may be noted that neither loop X nor Y contains an anticodon for histidine (GAG or UAG), so that the heterologous system used to activate this RNA may be mischarging it. Comparisons with two valine-carrying varieties from related plant viruses (Figure 8.6C,D) suggest that the anticodon is possibly GAA in loop X, which interacts with the codon UUC of phenylalanine. In any case, it is obvious that a primitive tRNA actually exists in these viruses, types that are far shorter than any other species known today. It is also to be noted that lengthening of the type hypothesized has occurred, as evidenced by the sequential series of changes indicated in loop Y. This in TMV begins with a sequence of only four nucleotides (Figure 8.6A), to which a third G is added in the second strain (Figure 8.6B). Then in the eggplant mosaic virus RNA, a C is inserted between the A and U (Figures 8.6C) (Briand *et al.*, 1976), a feature retained in the turnip yellow mosaic virus (TYMV) RNA (Figure 8.6D) (Briand *et al.*, 1977). The stem between the two loops also may be perceived to increase from 13 nucleotides in TMV to 16 in the TYMV, in which respect the other two are intermediates. In the figure it is also intimated why the numbering of sites had to begin at the 3′-end and proceed toward the 5′, quite opposite to today's procedures.

From the state represented by the above viral type, the lengthening processes continued until the amino acid stem had received seven pairing sites on the 5′-side, except in $tRNA^{His}$ which has received an eighth. Before the bacteria arose, many of the original nucleosides, all of which were of the common varieties as in the TMV tRNA, underwent modification, as new modifying enzymes arose. Most of these alterations occurred in arms II, IV, and V, and probably were involved primarily in tertiary structure. As shown in Tables 8.6 and 8.7, a total of 19 different enzymes existed prior to the advent of the bacteria. After eukaryotes diverged from these organisms, each branch gave rise to new modifying enzymes, but no major changes ever have been made to tRNA molecules since cellular life has existed.

9

The Genetic Mechanism of Viruses

The foregoing discussion of the phylogeny of tRNAs clearly indicates that the genetic mechanism had existed in living things prior to the advent of the prokaryotes. Thus the latter group per se throws very little light upon the origins and early evolution of the nucleic acids and other prominent features used in inheritance. To gain an understanding of those matters the search must be continued elsewhere: An exploration into the genetic processes in viruses may at least suggest models of the early events, even if not firm evidence of particular occurrence.

Since their discovery (Iwanowsky, 1892; Beijerinck, 1899), viruses have been viewed in several ways. At first they were considered primitive organisms, a point of view that gradually gave way to others as it became evident that they could exist and multiply only within living cells; the obvious conclusion of this observation was, of course, that they could not have existed prior to the advent of cellular life. As a result, on one hand the concept was advanced that the viruses are degenerate bacteria (Green, 1935; Burnet, 1945), and on the other that they are fragments of cells, particularly of metazoan cells (Bradley, 1971), or escaped genes (Luria and Darnell, 1967). At the present time, the latter concept appears to be the prevailing one.

Hence, as the genetic apparatus of viruses receives attention in this chapter in light of this point of view, close resemblances to metazoans or, at least, to other eukaryotes should be encountered frequently. It must be recognized, however, that the viruses, being well-adapted parasites, have often developed mechanisms for utilizing certain cellular processes of their host, but they do not merely use the intact host genetic system, as was formerly believed (Subak-Sharpe, 1968). In employing such cellular processes, the virus frequently modifies the host enzymes or nucleic acids to a greater or lesser extent: more rarely, they may even be utilized to a degree without modification. In a number of instances, however, macromolecular syntheses are carried out largely by virus-specified enzymes. In examining these viral functions, it is clearer to describe the DNA types (Fenner, 1976) whose genetic processes most closely resemble those discussed in preceding chapters, before scrutinizing the RNA varieties.

DNA VIRUSES

The DNA of viruses occurs in a diversity of molecular states. In a number of virus species, it exists as the standard Watson–Crick double helix, but some variation even on that theme is known, for the molecules may be linear, circular, or superhelical. In other types, the DNA may consist of a single strand, when typically it has a circular configuration. Still other variations in DNA are provided by the base composition, for often characteristically modified nucleosides are present or even completely replace certain standard ones.

In most DNA, as well as all RNA, viruses, the nucleic acid is located centrally and is referred to as the core, genome, or nucleocapsid. This is surrounded by a protein sheath, known as the coat, head, capsid, or, sometimes, envelope. As a rule, the coat or capsid is constructed of identical, more or less spherical molecules, variously called subunits or capsomers. Generally in virology the term virus is reserved for populations and species, whereas the individual constituents are referred to as virus particles or virions.

9.1. DOUBLE-STRANDED DNA VIRUSES

In double as in single-stranded viruses, the DNA molecule resembles that of bacteria and other prokaryotes in being essentially a naked nucleic acid polymer. No histones or other proteins appear to be incorporated into its structure as in eukaryotic chromosomes (Brown and Bertke, 1974). Hence, the name genophore, or even genome, is more appropriate than chromosome, in viruses and prokaryotes alike. Some characteristics of the better explored double-stranded DNA viruses are as follows.

9.1.1. The T-Even Bacteriophages

Among the biochemically best known viruses are the several T-even bacteriophages of *E.coli*, with T4 probably the most thoroughly investigated member of this group (Figure 9.1). Consequently, the following discussion refers largely to T4, even though that species may not be mentioned directly; only where T2 or T6 are known to differ do they receive individual mention.

The DNA in these forms is a typical double helix, except that the strands have "sticky ends." In each strand, the sequence of bases at one extremity is repeated at the other, a condition often referred to as "terminal redundancy" (Thomas and MacHattie, 1964; MacHattie *et al.*, 1967; Ritchie *et al.*, 1967; Hayes, 1968). Due to their complementary nature, the two redundant ends, which range from several hundred to a few thousands of nucleotides in length, unite to form a circle. The same condition exists in a number of such other bacteriophages as λ, T3, and T7. Viral DNAs are not known to assume the A configuration, but seem generally to remain in the B form. This has been confirmed by recent infrared linear dichroism studies, in which a transition to a predominantly C-like form was noted, but none to the A (Champeil and Brahams, 1974). Glucosylation was not a factor in the absence of the latter conformation.

Reconstitution of Virions. In the earlier days of modern virology, it was believed that the T4 virion, as others, consisted only of one or two coat proteins which surrounded the DNA. Hence, multiplication within the host, according to this concept, simply involved synthesis of the protein subunits and replication of the DNA; these parts upon completion assembled themselves spontaneously, as do many other proteins, such as collagen and bacterial flagella (Fraenkel-Conrat and Singer, 1957; Kellenberger, 1961; Abram and Koffler, 1964).

However, as knowledge has advanced, the reconstitution of virions has proven to be a far more complicated series of processes (Showe and Kellenberger, 1975). While the structure of the head protein subunit is specified by a single cistron, gene 23, many others also participate in assembly. For example, a certain mutation in gene 66 results in abnormally short heads, whereas one in gene 20 leads to extremely long tubular heads, a condition called polyheads. On the other hand, if a mutation occurs in gene 31, the head protein merely forms amorphous clumps, not a tubular structure of any sort. The product of gene 23 is itself not formed directly but involves a maturational step. First the immediate product of translation, having a molecular weight of 59,000, was found to be assembled into "τ-capsoids," smaller, less angular structures than the definitive capsid (Bijlenga *et al.*, 1974). In this location, it then underwent cleavage to a molecular weight of 47,500.

The self-association of the protein specified by gene 32 has recently received attention and has been reported as not being a simple indefinite association having a single combining constant (Carroll *et al.*, 1975). At least two association constants were necessary to explain most of the observed processes, but more were required to account for all of them. When the capsid is completed, the incorporation of DNA into the cavity depends on several proteins, including the product of gene 21, for mutations in that region of the DNA result in empty heads. Consequently, the assembly of these seemingly simple parts can be seen to be not just a mathematically predictable mechanical process, but one that involves an intricate series of interactions between numerous complex proteinaceous enzymes—some 40 different ones in T4. Sheath structure also has been shown to depend on other series of interacting proteins (Epstein *et al.*, 1963; Kellenberger, 1966).

In studies of the T-even bacteriophage DNA molecule, no cytosine has been found, for that nuceloside was entirely replaced by 5-hydroxymethylcytosine (Hershey *et al.*, 1953). The hydroxymethyl group typically was variously substituted with α- and β-glucose and the disaccharide, gentiobiose (Lehman and Pratt, 1960; Kuno and Lehman, 1962), substances which were essential for its normal functioning. The enzymes required for synthesis of the 5-hydroxymethyl-deoxyribocytidine and for its glucosylation after the new DNA has been synthesized were determined as being encoded by the phage DNA (Flaks and Cohen, 1959; Warner and Barnes, 1966). All the members of the T-even group had the hydroxymethylation and glucosylation occurring after the

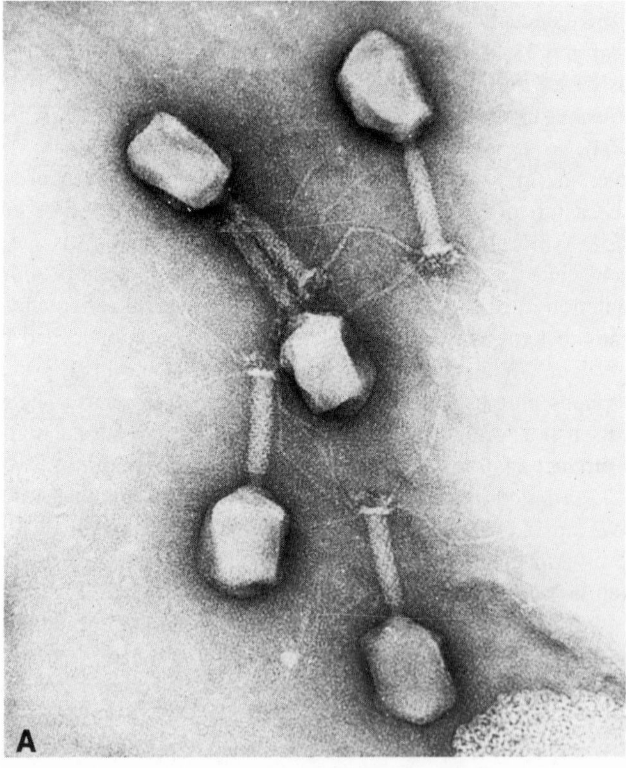

Figure 9.1. Examples of the T-even group of bacteriophages. The two forms illustrated differ in biochemical properties as well as in length of the so-called tail; all occur in *E. coli*. (A) Bacteriophage T2. Enlarged 150,000 × (courtesy of Dr. Anna S. Tikhonenko). (B) Bacteriophage T4. Enlarged 225,000 × (courtesy of Dr. Michel Wurtz).

DNA molecule had been synthesized, but the various types differed in amount and type of sugars bound to the hydroxyl radical (Lehman and Pratt, 1960; Cohen, 1967). In T4 all these residues were bound to glucose, 70% in the α and 30% in the β configuration. T2 and T6 agreed in having 25% of the hydroxyls free, but T2 had 70% bound to α-glucose, the rest to gentiobiose, whereas T6 had only 3% united to the first substance and 72% to the latter.

Postinfectional Metabolic Events. The first step in infection, attachment, is not a simple event, but involves at least two stages, adsorption and fixation. Not all the phages in a given population are capable of adsorption, for the T-even types have been shown by electronmicroscopy to exist in two forms, one having the tail fibers extended, the other with them contracted (Figure 9.2).

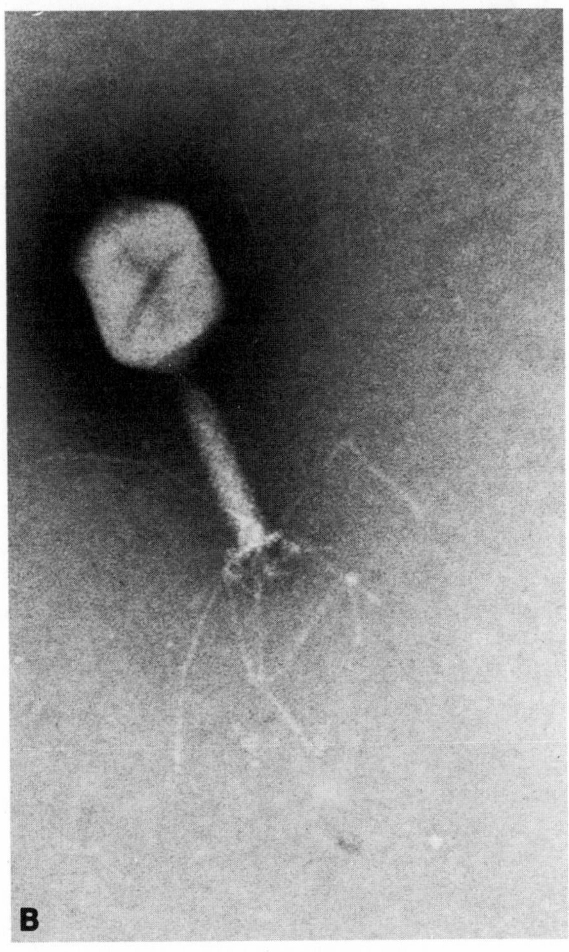

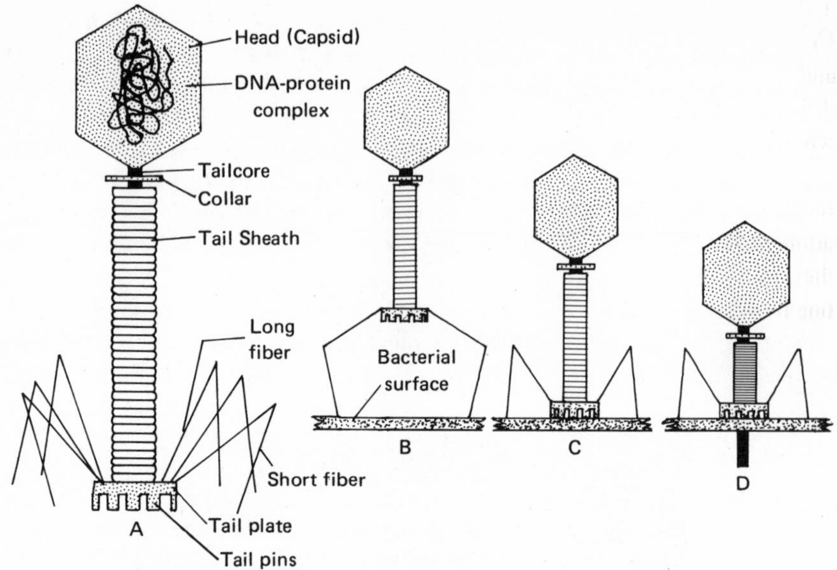

Figure 9.2. Bacteriophage T4 of *E. coli*. (A) Structure of the T4 virion. (B–D) Steps in infection: (B) The virion attaches to the surface of the bacterial wall by means of the tail fibers. After the tail pins have been brought into contact with the surface (C), the tail sheath contracts (D), forcing the core into the cytoplasm of the bacterium. Then the DNA–protein complex enters the host (based on Simon and Anderson, 1967).

Gamow (1969) has shown that at 8°C only 5% of the virions in a T4B preparation were active. Following attachment by means of the ends of the tail fibers (Simon and Anderson, 1967; Mathews, 1971; Broers *et al.*, 1975), the tail plate was brought to the bacterial wall surface, the sheath contracted as the core of the tail penetrated the wall, ejecting the DNA (Figure 9.2B–D). During contraction of the tail sheath, the ATP associated with that structure was released, either intact or as hydrolysis products (Wahl and Kozloff, 1962). Addition of ATP to contracted sheaths induced the latter to elongate (Sarkar *et al.*, 1964). Basically, contraction appeared to involve the interaction of a moderate number of proteins, particularly those of the long and short tail fibers and the seven or more of the baseplate (Yamamoto and Uchida, 1973, 1975).

The processes of transferring the DNA from the phage head into the bacterial cell remain unelucidated, but as a result of these activities the DNA molecule, 500 times as long as the head it occupied, unwound and passed through the core into the bacteria in less than 1 minute (Mathews, 1971). With it entered another substance also packed within the head, the so-called internal protein that has been shown to constitute between 3% and 7% of the total phage protein (Hershey, 1955; Minagawa, 1961). The T-even bacteriophages are highly virulent, for they possess some mechanism, still unknown, which rapidly and completely arrests all bacterial synthesis of DNA, RNA, and protein. Thus these viruses do not actually inhabit and replicate in living cells but only in freshly killed ones.

The DNA has been described as not penetrating to the interior of the host, but as remaining associated with the membrane, where transcription and replication take place (Rouviére *et al.*, 1969; Earhart, 1970; Miller and Kozinsky, 1970; Earhart *et al.*, 1973). It has long been known that transcription began immediately after infection and that two stages, early and late, were involved. Those products that were formed in the early stage were largely enzymes concerned with protein synthesis and DNA replication, and, consequently, were not incorporated to more than a small degree into the daughter virions, whereas the products of late transcription and the resulting proteins were (Hershey *et al.*, 1954).

Table 9.1 presents a list of approximately 25 enzymes thus far identified during the 10 to 15 minutes that constitute the early phase of T4 infection; unfortunately, neither the products of the late phase nor the processes involved in their production have been as thoroughly explored as the early ones. Approximately half of the early enzymes are concerned, directly or indirectly, with the biosynthesis of nucleotides, and the remainder with polydeoxyribonucleotide metabolism. Not all enzymes produced early in infection function immediately; two, deoxycytidylate hydroxymethylase and thymidylate synthetase, have been shown conclusively to become active later, the former being activated after about 5 to 8 minutes (Tomich and Greenberg, 1973; Tomich *et al.*, 1974). After activity has been gained, the enzyme functions increase exponentially for a period but then become linear. Concomitant protein synthesis is not required; however, DNA synthesis coincides in time with their activities and follows similar kinetics. Because of these correlations and the initial delay in activation, it was proposed that these enzymes are not functional catalysts until they have become part of an active complex. Recently, by use of chloramphenicol, it has been demonstrated that early synthesis occurs in two stages. That mRNA produced during the first 5 minutes following infection is known as immediate early, while that which is transcribed later is referred to as delayed early. Two sets of promoters are involved in the production of these two classes; recognition of the respective promoters is related to changes in the state of template DNA (Thermes *et al.*, 1976).

The polydeoxyribonucleotide ligase formed shortly after infection has been shown to differ extensively from the DNA ligase of the host, other prokaryotes, and the eukaryotes, as described in an earlier chapter (Chapter 3, Section 3.4.2). In addition to the usual activities (catalysis of phosphodiester bond formation between the 5′-phosphoryl and 3′-hydroxyl groups of short breaks or nicks in double-stranded DNA), the DNA ligase of T4 catalyzed the end-to-end coupling of double-helical DNA, even with fully base-paired termini (Cascino *et al.*, 1971). Furthermore, the enzyme acted also on DNA–RNA hybrid molecules (Kleppe *et al.*, 1970; Fareed *et al.*, 1971), and most recently has been demonstrated to join various short RNA strands end to end, at low temperatures (Sano and Feix, 1974).

Replication of T-Even DNA. As Mathews (1971) pointed out in a concise review of the topic, no completely satisfactory model of phage DNA replication has thus far been presented, not even for the much-studied T4. In this form replication begins 7 minutes after infection. Using tritium-labeled bromouracil techniques employed earlier for studies of bacterial DNA replication (Bonhoeffer and Gierer, 1963), Werner (1968a,b) determined that a number of growing DNA points appeared to be clustered on the replicating molecule, increasing from zero at 10 minutes to 60 at 30 minutes, after which it remained unchanged. From these and other data, he deduced that the originally linear double helix assumed a circular form, as shown in Figure 9.3A–G, by means of the redundant ends. Replication then was initiated at some specific site, and a growing point began to traverse the molecule, one strand being replicated, the other one gradually peeling off. After the first growing point had moved ca. 0.15 to 0.19 phage-equivalent lengths, a second growing point was formed at the initiation site. This process was repeated over and over again, resulting in a pinwheel pattern. A growing point did not necessarily cease activity after it had traversed the entire circular molecule but sometimes continued past the initiation site and thus produced concatemeric structures, that is, molecules linked end to end. Frankel's (1968) model modified that of Werner in suggesting that only one growing point developed on the parental circular strand at any given time and that others were initiated from the ends of the protruding strands (Figure 9.3H–K).

A third suggestion has become known as the ''rolling-circle model'' (Gilbert and Dressler, 1968); this is somewhat akin to the Werner model but contains several distinctive features. First the concept proposed that replication began while the DNA molecule was still linear; then, after one strand had been nicked, the 5′-end at the break became associated with a host membrane and replication began. Some of the short strands thus formed initially attached to one end of the remainder of the double helix and circularization of the molecule occurred (Figure 9.3L–P). Beyond that point no basic differences from the other proposals can be noted.

As Mathews (1971) has pointed out, all of the circular models of replication have certain weaknesses. Perhaps the most serious of these is that each one suggests that the DNA molecule from any given virion replicates independently of all others. Yet when bacteria were doubly infected with T4, once with phage labeled with bromouracil and then with phage labeled with ^{32}P, covalently bonded structures containing both labels were demonstrated to be present (Anraku and Tomizawa, 1965). Similar interactions between parental molecules have been found in heavily infected cells, in which cases electron micro-

Table 9.1
Some Proteins Induced by Bacteriophage T4

Period	Type	Proteins	Number known
Early	Reductase	Ribonucleoside diphosphate reductase	2+
	Reductase	Dihydrofolate reductase	1
	Kinase	Thymidine kinase	1
	Kinase	Deoxyribonucleotide kinase	1
	Kinase	Polynucleotide kinase	1
	Phosphatase	Deoxyuridine di- and triphosphatase	2
	Phosphatase	Deoxycytidine di- and triphosphatase	
	Phosphatase	5^2-polynucleotide phosphatase	1
	Nuclease	Deoxyribonuclease	5?
	Polymerase	DNA polymerase	1
	Synthetase	Thymidylate synthetase	1
	Ligase	Polynucleotide ligase	1
	Repair	DNA repair enzyme	2
	Deaminase	Deoxycytidylate deaminase	1
	Hydroxymethylase	Deoxycytidylate hydroxymethylase	1
	Transferase	Glucosyl transferase	2
	Methylase	DNA methylase	1
	Nucleotidase	3-deoxyribonucleotidase	1
		Total identified	25
Late		Lysozyme	1
		Dimer excision enzyme	1?
	Structural	Major sheath protein	1
	Structural	Minor sheath protein	1
	Structural	Major head protein	1
	Structural	Minor head protein	1
	Structural	Major tail fiber protein	1
	?	Internal polypeptide	?

graphs indicated that essentially all the replicating DNA present in a single host cell formed a complex (Huberman, 1968). Whatever the mechanism may ultimately prove to be, the processes of replicating the DNA are certainly far from simple, for the products of many genes are involved. Thus far known to be implicated in replication are the translational products of genes 32, 41, 42, 43, 44, 45, and 62 (Huang and Buchanan, 1974). The product of gene 43 is DNA polymerase and that of gene 32 is the DNA-dehelicase protein, but the precise roles of the others have not been established. Six of these proteins have been purified and used *in vitro* to catalyze the polymerization of DNA (Morris *et al.*, 1975).

Transcription of DNA. Some evidence has been advanced that not only early and late mRNA exist, but that four types are recognizable (Bolle *et al.*, 1968). Pre-early was viewed as being synthesized between 0 and 1 minute after infection, early from 1 to 5 minutes, quasi-late from 5 to 15 minutes, and late, thence through the latent period. Nor did the product of any given gene necessarily fall entirely within one of these categories. The gene for lysozyme, to cite one example, was transcribed within the first 3 minutes, then its transcription ceased until 7 minutes, after which time it continued throughout the latent period (Bautz *et al.*, 1966; Bautz and Reilly, 1966). Since the two strands of denatured T4 DNA bind poly(U) differentially, it has been found possible to separate the so-called heavy (H) and light (L) strands by density-gradient methods (Guha and Szybalski, 1968). It was then determined by hybridization and labeling techniques that all the mRNA made early in infection bound to the L strand and that, later, an ever-increasing proportion hybridized with the H strand. The former strand appeared to be transcribed in a counterclockwise direction, the latter in a clockwise (Stahl *et al.*, 1966); the nucleotide sequences of the species I RNA, which is 140 units long, have been determined both in T4 and T2 and have proven nearly identical (Paddock and Abelson, 1975a,b). The lysozyme gene was transcribed in a counterclockwise direction and therefore was an early gene, but its translational product did not become active until late in infection (Kasai *et al.*, 1968; Kasai and Bautz, 1969).

In order for transcription of the H strand to commence, DNA synthesis also had to be actively progressing (Riva and Geiduschek, 1969). Certain additional factors were required, particularly the as yet unidentified product of gene 55 (Snyder and Geiduschek, 1968). In the transcription of pre-early genes, the RNA polymerase of the host appeared to be employed without modification by the phage (Walter *et al.*, 1968). Later in the processes, some evidence has suggested that the host polymerase became modified, chiefly in the σ unit which may be involved in the local unwinding of the DNA (Chapter 4). After this unit had been altered, pre-early transcription ceased and early gene transcription began. All remaining genes, however, were transcribed by this same polymerase (Mathews, 1971), after DNA replication had begun.

Translation of mRNA. Somewhat less has been determined of the trans-

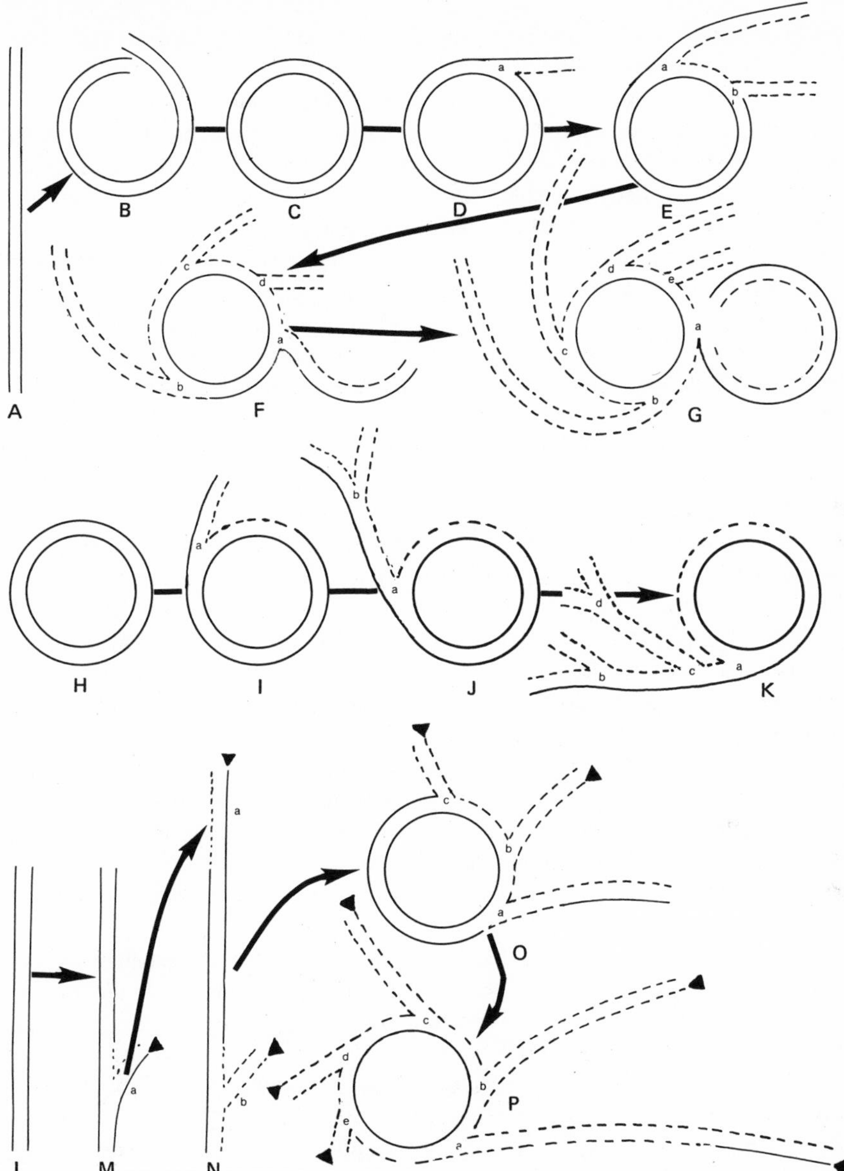

Figure 9.3. Various models that have been proposed for T4 DNA replication. (A–G) In the Werner model, the originally linear molecule (solid lines) assumes a circular configuration (A–C) before replication begins. New DNA chains (broken lines) are then formed around the circle, as the outer parental molecule peels off. (H–K) The Frankel model is similar, except that synthesis occurs as branches, rather than around the circle. (L–P) The Gilbert-Dressler, or rolling-circle, model is similar to the Werner but is indicated to initiate replication in connection with the host cell membrane (black triangles), to which 5′-termini bind.

lation mechanism of the T-even phages than of the transcriptional processes, but the overall picture appears similar. That is to say, the enzymes and nucleic acids seem to be of mixed derivations, some being of purely bacteriophage origin, others being from the host, and, in the latter case, either with or without modification by the parasite. Some investigations into translation of the messenger have also been aimed toward answering the question of how the host genetic processes are arrested following infection. Among these is a short series of studies into the possibility that the initiation factor (IF-3) of the host that binds mRNAs to ribosomes might become specific to messenger from phage. The early explorations into this aspect suggested that *E. coli* contained two species of this factor, IF-3α that promoted initiation-complex formation with host and T4 early mRNA, as well as with messenger from several other bacteriophages, and IF-3β that recognized only T4 late mRNA (Lee-Huang and Ochoa, 1971, 1973). However, later researches have consistently reported that no such change in specificity could be detected (Goldman and Lodish, 1972; Schiff *et al.*, 1974; Spremulli *et al.*, 1974).

Studies on polyribosome metabolism in T4-infected *E. coli* have thrown some light on both translation and transcription (Walsh and Cohen, 1974a,b). Ribosomes were found to separate in three fractions: (a) cytoplasmic, which constituted 55% of the total, were released when the cells were fractured, (b) DNA-bound, made up 30%, and (c) membrane-bound ribosomes approximated 15%, of the total. The DNA-bound ribosomes were released from the cellular membrane fraction on treatment with deoxyribonuclease, whereas the membrane-bound ones were not. The former appeared to be part of a complex engaged in both transcription and translation simultaneously; these more than likely moved from the cell membrane to the cytoplasm with mRNA. Of the newly transcribed mRNA, about 80% was found associated with ribosomes or their subunits, while the remainder apparently was not associated with any aspect of protein synthesis.

Further, it was demonstrated by other researches that host synthesis of ribosomes ceased immediately following T4 infection, although those ribosomes were used in T4 translation without modification (Brenner *et al.*, 1961). Since the rate of protein synthesis remained constant throughout infection (S. Cohen, 1948; P. Cohen and Ennis, 1965), ribosomal turnover rate was deduced to be rather slow. DNA replication and RNA synthesis, and a certain fraction of protein synthesis, proceeded concurrently in a complex associated with the host plasmalemma. Regulation of enzyme synthesis appeared to be by way of posttranscriptional events (Zetter and Cohen, 1974).

The tRNAs of T-Even Phages. The tRNAs, in contrast, are of mixed origins. The DNA of bacteriophage T4 has been shown to carry genes for eight different species of tRNA (Daniel *et al.*, 1970; McClain *et al.*, 1972). The first five types that were definitely identified were for species accepting glycine, proline, arginine, leucine, and isoleucine (Daniel *et al.*, 1970; Scherberg and

Weiss, 1972); later, others for serine and glutamine were found (Barrell *et al.*, 1974; Seidman *et al.*, 1974). In five instances, namely glycine, glutamine, proline, leucine, and serine, the nucleoside sequences have now been determined (Barrell *et al.*, 1973, 1974; Pinkerton *et al.*, 1972, 1973; Stahl *et al.*, 1973; Seidman *et al.*, 1974), which in general show close alliance to such prokaryotes as *E. coli*. For example, T4 $tRNA^{Gln}$ has the thiolated uridine at site 1 of bend 1 (Table 7.1), a trait found elsewhere only in prokaryotes. Another type from T4 shows a second hallmark of prokaryotic tRNAs, the presence of A_s in site C of the anticodon arm in $tRNA^{Leu}$ (Table 7.3); in the long extra arm of the same species the identical kinships are indicated by the absence of the C_5 which is consistently present in eukaryotes (Table 7.5). Furthermore, the T4 $tRNA^{Leu}$ has nucleoside N in the first site of the anticodon (Table 7.7), which has thus far been identified there only in prokaryotic species. Otherwise, comparisons by corresponding parts one by one show the relationships with *E. coli* to be rather remote (Tables 7.1–7.7). It is of interest to note, first, that the $tRNA^{Pro}$ of T2 and the $tRNA_2^{Gly}$ of both T2 and T6 are identical in nuceloside sequence to the corresponding species of T4, and, second, that a number of mutations involving tRNA structure have been identified in bacteriophage T4 (Wilson and Kells, 1972; Guthrie and McClain, 1973; McClain *et al.*, 1973; Comer *et al.*, 1974).*

Over and above the phage-specific types, many host enzymes and nucleic acids are used in T4 translation. Not only the host tRNAs but at least one of the tRNA ligases, that for valine, becomes modified by virus polypeptides. In T4, the host valyl-tRNA ligase was found to be modified by the τ factor, an early translational product (Marchin *et al.*, 1974; Müller and Marchin, 1975). Processing of precursors to tRNAs has been shown to be similar to that of bacteria, as pointed out in Chapter 8.

Assembling the Genetic Products. Although a number of details are still lacking, the basic features of the genetic processes are sufficiently known for bacteriophage T4 that flow diagrams of the metabolic processes involved in assembling the raw products have now appeared in print (Wood *et al.*, 1968; King and Wood, 1969; King, 1970; Poglazov, 1970). Early in infection, newly replicated DNA was seen by electron microscopy as very long fibrils (Figure 9.4). These then began to condense into packets, approximating the size and shape of the head (Mathews, 1971). However, they were not packed immediately into the head, but first appeared to be trimmed to a specific size and freed from the replication complex. Concurrently, the various parts of the tail were assembled in a specific sequence of events and spontaneously coupled to a DNA-packed head; when the tail fibers had then become added to the tail plate, assembling processes were brought to completion (Figure 9.4).

* Similar homologies among the three T-even bacteriophages have been shown to exist in other types of RNAs (Wilson *et al.*, 1972; Paddock and Abelson, 1973).

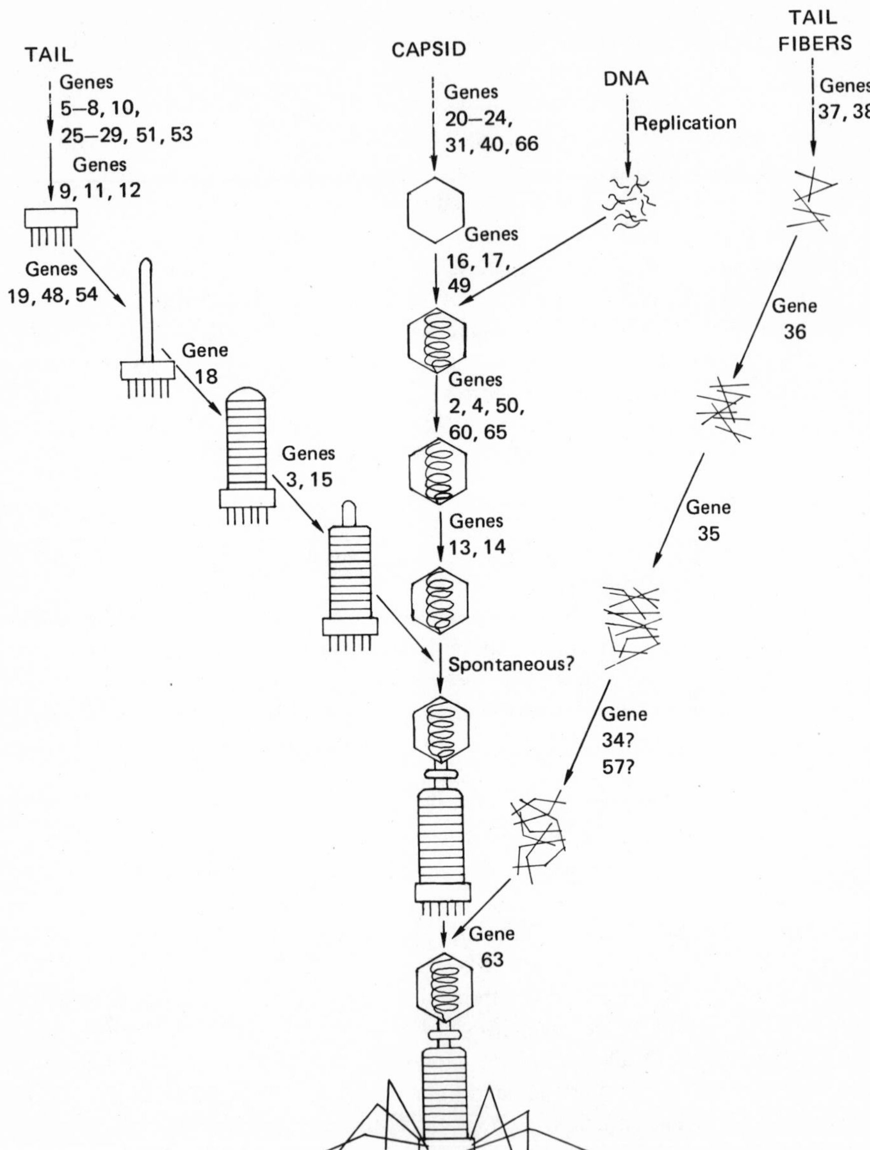

Figure 9.4. Morphogenetic pathway for bacteriophage T4. The complexity of these simple-appearing viruses is made evident by the number of genes whose products are involved in the numerous steps of replicating the virion (based on C. K. Mathews, 1971).

9.1.2. The T-Odd Bacteriophages

Since most features of the genetic processes that are widespread among viruses in general have been introduced in the above discussion, the remainder can be viewed in much less detail. The T-odd bacteriophages are simpler than the T-even group and are also far less homogeneous. Whereas the DNA molecule of the T-evens consistently ranged close to 130 million in weight, that of the largest member of the T-odd types (T5) has been shown to be around 80 million and that of the two smallest (T3 and T7) 25 million (Mathews, 1971; Studier, 1972). The T-odds also differ in having the tail noncontractile. One of the features shared by all members of the present group is that the DNA contains no modified bases of any sort. A second one is the lack of circularity in the DNA molecule, which arises through the absence of sticky ends, despite each strand's containing repetitive terminal sequences. None of the members has received as much attention as the T-even phages, T1 and T9 being especially poorly explored; consequently, most of the discussion here must be confined to the two better known of the T-odd bacteriophage species, T5 and T7.

Genetic Processes of T5. Of the five members of the T-odd group, bacteriophage T5 is the one which most closely resembles the T-even types, both in structure and in the replicatory processes. However, it differs in a number of traits. Perhaps the most outstanding distinction has been found during the infective processes, the injection of the DNA, which occurred in two distinct, interrupted stages. After attachment to the host, the phage injected a segment of DNA less than 10% as long as the whole molecule (Lanni and Lanni, 1966; Lanni, 1968); a period followed during which protein synthesis proceeded briefly, then the remainder of the DNA was injected. Three classes of proteins have been found: Class I included those formed between 1 and 6 minutes after infection; class II, those synthesized between 5 and 20 minutes; and class III, those produced from 14 minutes to lysis (McCorquodale and Buchanan, 1968; Szabo *et al.*, 1975). Transcription of the DNA employed a virus-modified form of the host RNA polymerase (Szabo and Moyer, 1975).

The double-stranded DNA molecule of T5 has proven to contain several single-strand interruptions located at specific points (Thomas and Rubenstein, 1964; Abelson and Thomas, 1966; Bujard, 1969), at least three, and more likely four, such nicks being present. According to Bujard, only one strand contained breaks, the other being continuous (Jacquemin-Sablon and Richardson, 1970; Knopf and Bujard, 1975). As might be expected, one break appeared to mark off the DNA that was injected during the first stage, a small segment which supposedly carried two genes. One of those cistrons induced degradation of host DNA, while the product of the second was essential for completion of DNA injection (Lanni, 1969; McCorquodale and Lanni, 1970). RNA has been found associated with DNA in the virions of T5 (Rosenkranz,

1973), but it has not been established whether this is injected into the host with the DNA.

Very little progress has been made toward understanding the processes of DNA replication of T5. Thus far it appears that no concatemeric structure is produced, accentuating the absence of circularity in the molecule (Smith and Skalka, 1966; Tanyashin, 1968). As in T4, different specific regions of the molecule have been reported to be transcribed at different times following infection (Lanni and Szybalski, 1969). Quite surprisingly, it was found that complementary regions of the two strands appeared to be transcribed, as more than 70% of the total DNA hybridized with the mRNA of the phage. What function the 20% that must represent "antimessengers" would perform was not evident, for it would seem most improbable that both a given locus and its complement in the second strand could code for useful proteins. Another dissimilarity from the genetic processes of the T-even bacteriophages was that T5 did not require DNA replication in order for protein synthesis to proceed normally (Hendrickson and McCorquodale, 1971), for in an *E. coli* infection with a T5 mutant that was unable to replicate DNA, all three classes of proteins were produced with times of onset and cessation of syntheses essentially as in wild-type infections. At least 14 different species of tRNAs appear to be induced by the T5 genome, $tRNA_{M}^{Met}$ and $tRNA_{F}^{Met}$ having now been specifically identified (Scherberg and Weiss, 1970; Chen *et al.*, 1975, 1976). A phage-specified DNA polymerase has now been purified (Fujimura and Roop, 1976).

Genetic Processes of T3 and T7. Since T3 and T7 are structurally similar in having the tail very short and in being small (Luftig and Haselkorn, 1968), it is well to discuss them together. In addition, their kinship is indicated by serological, genetical, and DNA cross-hybridization studies (Adams and Wade, 1954; Shildkraut *et al.*, 1962; Hausmann and Gomez, 1967), as well as by base-sequence homologies in their DNA (Davis and Hyman, 1971). Nevertheless, rather strong distinctions exist between them. As T7 has been far more thoroughly studied than T3, much of the discussion centers on that form, with notes on the latter being added when pertinent.

The DNA molecule of T7 resembles that of T5 in being linear and double-stranded and in having the genes arranged somewhat according to function. However, it is much smaller, as pointed out above, and does not seem to possess the series of nicks. Some 24 genes have been identified, which code for three classes of proteins (Table 9.2). Class I (pre-early) includes those synthesized between 4 and 8 minutes after infection, class II (early), those produced between 6 and 15 minutes, and class III (late), those synthesized between 6 or 8 minutes and lysis (Studier and Maizel, 1969; Studier, 1972). Some workers recognize only two classes, early and late, by combining classes II and III (Siegel and Sumners, 1970; Skare *et al.*, 1974). In the virions, the DNA surrounds a proteinaceous structure called the core, one end of which is attached to the capsid wall at the base of the tail (Serwer, 1976).

Table 9.2
Genes Found in Bacteriophage T7 and Their Functions[a]

Class	Gene	Function of protein coded	Approximate molecular weight
I. Pre-early	0.3	?	8,700
	0.5	?	40,000
	0.7	?	42,000
	1.0	RNA polymerase	110,000
	1.1	?	8,000
	1.3	Ligase	40,000
II. Early	1.7	?	17,000
	2.0	Reduces DNA synthesis	?
	3	Endonuclease	13,500
	3.5	Lysozyme	13,000
	4	Reduces DNA synthesis	67,000
	5	DNA polymerase	81,000
	6	Exonuclease	31,000
III. Late	7	?	14,700
	8	Head structure	62,000
	9	Head assembly	40,000
	10	Major head structure	38,000
	11	Tail structure	21,000
	12	Tail structure	86,000
	13	?	14,000
	14	Head structure	18,000
	15	Head structure	83,000
	16	Head structure	150,000
	17	Tail structure	76,000
	18	DNA maturation	?
	19	DNA maturation	73,000

[a]Based largely on Studier (1972, 1973).

Class I genes were transcribed by host RNA polymerase using only one strand (the *r* or *H*), but this did not normally function beyond gene 1.3 (Siegel and Summers, 1970; Hyman, 1971; Chakraborty *et al.*, 1974); after transcription was completed, the product then was processed by RNase III into functional messengers (Dunn and Studier, 1973a,b; Hercules *et al.*, 1974). During transcription, the σ factor governed the binding of the RNA polymerase to three specific sites within the early promoter region of the DNA molecule (Bordier and Dubochet, 1974), as in *E. coli.* Termination factor ρ played an undetermined role in termination and transcription (Darlix and Horiast, 1975). Some of the termination and initiation signals now have been determined. The first termination site appeared to be marked by a sequence that ended in guanosine and reinitiation occurred further along the molecule when a series of nucleosides beginning with cytosine was reached (Peters and Hayward, 1974a,b). That cistron was in turn terminated by guanosine and uridine. However, a later base-sequence study determined a portion of this initiator site to be UACGAUG (Pribnow, 1975), and revealed the existence of homologies with the genome of T7 in corresponding sections of those of A3, fd, SV40, λ, and P_L viruses (Table 9.3). A quite different sequence was obtained by Dausse *et al.* (1975).

The remainder of the genome was transcribed by the product of gene 1 (Chamberlin *et al.*, 1970; Summers and Siegel, 1970); the initiation site for this activity has been shown to be located within the pre-early region, just before gene 1.3 (Skare *et al.*, 1974). The T7 polymerase has now been purified and found to consist of three activity forms, one of which, with a molecular weight of 110,000 constituted about 80% of the total (Niles *et al.*, 1974). *In vitro* studies demonstrated that this form could transcribe all the unique mRNA species known to occur.

In the replication of the DNA molecule, host DNA was degraded and the resulting nucleotides were utilized (Labaw, 1951, 1953; Putnam *et al.*, 1952).

Table 9.3
Initiation Site Sequences in Various Viruses[a]

Viruses	Sequence
T7	UACGAUG
A3	UACGAUG
fd	UAUAAUA
SV40	UAUAAUG
Λ	GAUACUG
P_L	GAUACUG

[a]Based on Pribnow (1975).

Arising from the processes were products longer than the T7 DNA molecule itself (Kelly and Thomas, 1969), but the mode of formation of these intermediates was not clear. The mature DNA was believed to be formed from these precursors in a process which also produced the terminal redundancy that characterized the molecule. After infection, host DNA synthesis continued at the preinfection rate for about 5 minutes, along with synthesis of other macromolecules, but all then ceased before another period of equal length expired, when bacteriophage T7 DNA synthesis began. This process reached a maximum rate within another 5 or 10 minutes to cease just before or with lysis, that is, until about 30 minutes after infection (Hausmann and Gomez, 1967; Studier, 1969; Studier and Hausmann, 1969). In addition to the DNA polymerase coded by gene 5 (Grippo and Richardson, 1971), normal replication required the presence of the products of genes 2 and 4. Nothing seems to be known about the tRNAs used in the genetic processes.

In infections with phage T3, an enzyme was produced that markedly inhibited *E. coli* RNA polymerase activities but only slightly decreased T3-specific RNA polymerase, even at high concentrations (Mahadik *et al.*, 1972; Chakraborty *et al.*, 1974). To exercise this influence, it required the presence of the σ-factor (Mahadik *et al.*, 1974). This virus also was distinct in inducing the formation of an enzyme which cleaved *S*-adenosyl-methionine to thiomethyl-adenosine and homoserine; thus this enzyme destroyed the natural methyl-group donor used in methylation of DNA (Gold *et al.*, 1964; Gefter *et al.*, 1966; Hausmann and Gold, 1966). The enzyme did not appear to be essential, however, for mutants lacking the necessary gene replicated normally (Hausmann, 1967).

9.1.3. Bacteriophage λ and Other Temperate Forms

All bacteriophages of the T-even and T-odd groups are virulent, in that they induce the immediate generation of progeny and quickly lead to lysis of the host cell. Indeed, as pointed out earlier, the host cell is killed prior to lysis through the destruction of its ability to carry out synthesis of its own macromolecules. In bacteriophages of the so-called temperate types, including λ, two courses of progress are available upon infection: (1) In sensitive cells or under particular (unknown) conditions, phage replication may begin immediately and lead to lysis, as in virulent forms, or, (2) the phage genome may become associated with the host genetic apparatus and maintain itself intact, replicating in synchrony with the host DNA over long periods of time. Eventually, under certain conditions still not established, the phage genome, then known as the prophage, induces the formation of progeny and thus lysis. This delayed lysis is referred to as a lysogenic response (Mathews, 1971). Lysis may be readily induced in a lysogenic culture by X-ray or ultraviolet irradiation, thymine starvation, or other factors which interfere with normal nucleic acid metabolism.

Those temperate bacteriophages that have noncontractile tails similar to the T-odd phages fall into two major categories, the λ and the P2 families. In the first family are included bacteriophages λ, φ80, 21, 424, 434, and 82, whereas in the second only P2, 186, and 299 appear to be members. All the phages within a family are able to recombine with one another, and the cohesive ends of their DNA molecules can join; such reactions do not occur between members of different families, however (Baldwin *et al.*, 1966; Wang and Schwartz, 1967; Mandel and Berg, 1968; Murray and Murray, 1973).

Genetic Apparatus of Bacteriophage λ. Bacteriophage λ is a tailed form (Figure 9.5) of much simpler structure than the smallest of the T-odd phages and T4; its DNA has a molecular weight of 30.8×10^6 (Dyson and Van Holde, 1967; Davidson and Szybalski, 1971). This is a double-stranded molecule, with the 5′-end of each strand bearing a redundant terminus about 20 nucleosides in length (Kaiser and Wu, 1968). These termini, being complementary, can serve to unite the double helix into a circle, much as the replicative form described for T4 (Chapter 9, section 9.1.1). Recently, a 12-unit-long sequence of a portion of each end was established as being GGGCGGCGACCT and the set on the other strand being complementary (Wu and Taylor, 1971; Harvey *et al.,* 1973). A second peculiarity of λ DNA, as well as that of such other temperate phages as φ80, is that the base composition is not uniform along the whole molecule but contains segments rich in certain base pairs. One half (usually referred to as the left) was found to have a high percentage (55%) of guanine–cytosine pairs, while the other had a similar percent of adenine–thymine combinations (Hogness and Simmons, 1964). Later it was shown that about six segments of differing base frequencies could be separated, a central portion being especially rich in A–T pairs (Hogness *et al.,* 1967; Skalka *et al.,* 1968; Skalka, 1969). As in T7, there is a discrepancy in density between the two strands, so these, as there, are distinguished as H (or r) and L (or l) strands, respectively. It is now established that when packed in the virions the right-hand terminus of the DNA is connected to the proximal end of the tail (Saigo and Uchida, 1974), but the significance of this arrangement has not been uncovered.

Transcription in λ. Unlike T7, both strands of λ DNA are transcribed *in vivo*, but at different periods during infection. However, the periods and products have as yet not been sharply defined. It is fairly well established that a region *x* (in the H strand?) is essential in the regulation of early functions and that an adjacent segment, *y*, is evidently an initiation site for RNA polymerase (Eisen *et al.*, 1966; Pereira da Silva and Jacob, 1967; Roberts, 1969). The early genes, also apparently on the H strand (Figure 9.6), code for few well-documented proteins, an exonuclease and β-protein of unknown function being the only clearly established products (Korn and Weissbach, 1963; Radding *et al.*, 1967). There is evidence of others, however, including an endonuclease and a tRNA transferase (Schuster and Weissbach, 1968; Wainfan, 1968; Boyce

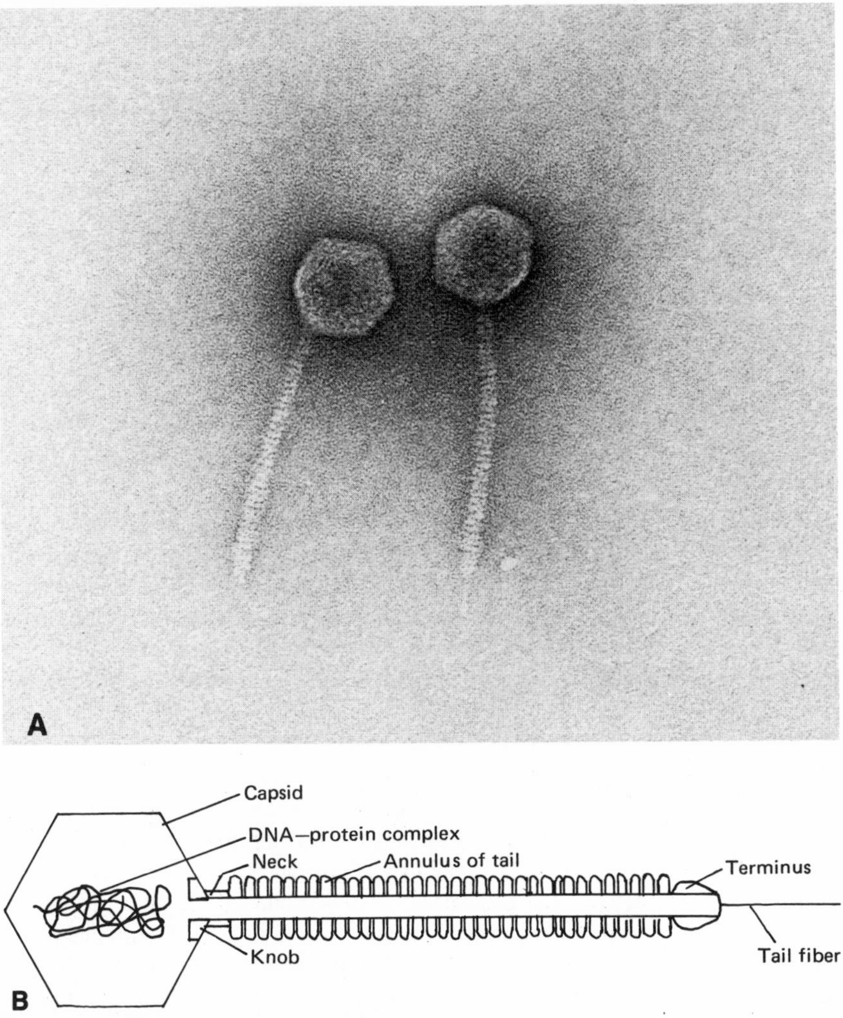

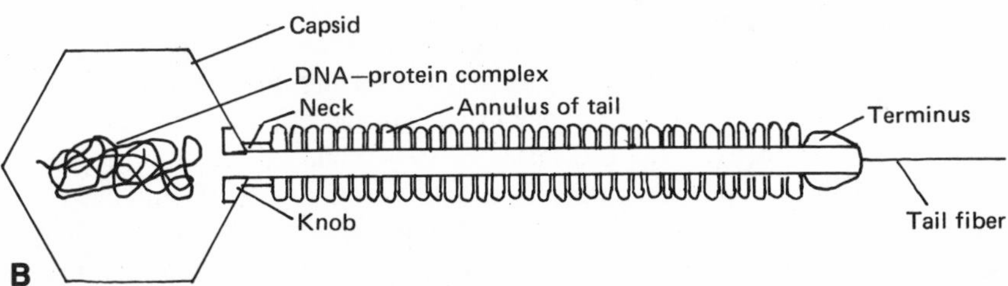

Figure 9.5. Bacteriophage λ. (A) This electron micrograph of bacteriophage λ is of a preparation negatively stained with uranyl acetate at pH 4.2. Magnified 225,000 × (courtesy of Dr. Michel Wurtz). (B) Structure of the virion.

et al., 1969). All members of the λ family seem to synthesize a similar exonuclease and at least ϕ80 induces a related β-protein (Pricer and Weissbach, 1967). The late proteins appear to be directed by genes on the L strand, where they are grouped in orderly fashion according to function. The head, tail, and tail fiber proteins are coded for by genes on the left half of the strand (Figure 9.6), in that sequence, and those that pertain to DNA replication and lysis oc-

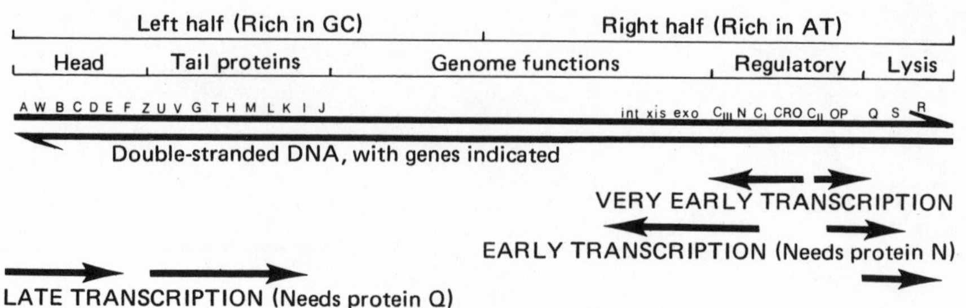

Figure 9.6. The genome of bacteriophage λ and its transcription.

cupy the right quarter (Mathews, 1971). Circularity and superhelicity of the λ DNA molecule have been shown to increase the amount of RNA synthesis, seemingly through increasing the number of initiation sites (Botchan *et al.*, 1973).

The product of early transcription (mRNA) has been reported to decay at a rate different from that of late (Gupta and Schlessinger, 1975). Whereas early mRNA had a half-life of 2.2 minutes at 42°C, the late messenger showed a complex decay curve; about 5% of the latter became degraded during the first 4 to 4.5 minutes, after which time the rate accelerated to a half-life rate of 1.5 minutes. No correlation was detected between the length of the mRNA and the half-life (Ray and Pearson, 1975). Transcription of early mRNA has been found to be bound to the operator sites of the early region of the DNA. One operator site and one promoter (initiator) region have now been sequenced; the former contained 79 and the latter 25 nucleotide residues (Pirotta, 1975; Walz and Pirotta, 1975). As yet control mechanisms for late protein synthesis have not been definitely determined. On one hand, posttranscriptional processes have been found to be active (Ray and Pearson, 1974), and on the other, repressor genes are claimed to provide a functional basis for regulation (Dottin and Pearson, 1973; Dottin *et al.*, 1975). Recently, functional inactivation of messengers for structural proteins have been noted (Ray and Pearson, 1975). At least some species of the RNA transcript have been found to bear poly(A) segments of variable lengths (Smith and Hedgpeth, 1975), which were stated not to be coded by the DNA template.

Replication of the DNA. During ordinary replication of λ DNA, both circular and concatemeric structures are produced (Smith and Skalka, 1966; Young and Sinsheimer, 1967, 1968; Kiger and Sinsheimer, 1969). The circular structures appeared only during the first half of the latent period; these then generated other circular DNA molecules by a rolling-circle method of replication (Takakashi, 1974). During the second half, a replicative intermediate of undertermined structure was formed, which ultimately resulted in the linear DNA of the virions (Kaiser, 1971). A single site for initiation of replication, located on the right, has been demonstrated, replication being from right to left (Makover, 1968; Tomizawa and Ogawa, 1968). Unfortunately, little is known as to the participation of host enzymes in λ DNA replication. Undoubtedly some host proteins are employed, but much dependence is placed also on phage products, including those coded by genes O and P (Kaiser, 1971).

When λ is incorporated into the DNA of the host as prophage, the replication of its DNA molecule is accomplished strictly by the host system along with the rest of the nucleoid, phage functions being repressed by the product of λ gene *c*I. Thus the prophage is replicated once each bacterial generation. Since the prophage is inserted into the host DNA in its circular configuration, its replication must involve some technical problems for the host DNA polymerase. While both insertion and excision from the bacterial DNA are under the control of phage products, no specific details are presently available pertaining to the actual mechanism (Gottesman and Weisberg, 1971).

Assembly of the Virion. The assembly of the numerous proteins and components of the virions is a complex process analogous to that of the T-even bacteriophages (Kellenberger and Edgar, 1971). To date greater progress has been achieved with the tail-assembling steps than with the capsid proper. The several stages are influenced by at least 11 genes, including G to M, T, U, V, and Z (Parkinson, 1968; Kühl and Katsura, 1975). The length of the tail is determined by the product of gene U, a protein with a molecular weight of 14,000 (Katsura and Kühl, 1974); in the absence of this enzyme a long-tailed (polytail) condition results, accompanied by a reduction in infectivity. If formation of the basal part of the tail is blocked, the major tail protein, which is encoded by gene V and has a molecular weight of 31,000, fails to assemble into a tail, even though the product of gene U is present. A third protein, the product of gene H, also has been reported to be involved in tail length (Katsura and Kühl, 1975a). Recently, a complete pathway for the morphogenesis of the tail has been proposed (Katsura and Kühl, 1975b).

Accounts of head assembly are still incomplete. The proteins that constitute the capsid are the products of genes B, C, E, X_1, X_2, and, possibly, A (Murialdo and Siminovitch, 1971, 1972; Hendrix and Casjens, 1974), but the genes and products concerned remain to be identified. After the empty capsid has been constructed, the DNA excised from concatemers is then inserted (Dawson *et al.*, 1975); excision is assumed to be under the influence of the product of gene A (Wang and Kaiser, 1973; Hohn *et al.*, 1974). The products of genes W and F, in that sequence, then must act upon the packed capsid before it can accept a tail (Casjens *et al.*, 1972). The DNA also undergoes maturational processing when being inserted into the heads (Syvanen, 1975).

Other Relatives of the λ Family. A large number of bacteriophages structurally similar to λ have been described, all of which have a polyhedral head and a long, flexible tail, which is not contractile but is densely covered with periodic striations. These characteristics place them in group IV of Tikhonenko (1970) and class B of Bradley (1967). One of the members, phage LV-1, attacks *Agrobacterium tumefaciens,* a Gram-negative rod which causes crown gall in many species of plants. Since it is a parasite of bacteria of quite different habits than *E. coli* and since its genetic mechanism has been studied to a small extent, it appears ideal for comparison to λ.

As in bacteriophage λ, the DNA of this temperate form proved to be double-stranded and of approximately the same molecular weight, 31×10^6 (Korant and Pootjes, 1970), probably coding for 30 to 40 proteins. Of these, four appeared to be major, and probably seven others were minor, structural proteins. Three classes of proteins have been described (Walls and Pootjes, 1975). Among the members of class I, early proteins which appeared as soon as 1 minute after infection, was one that reduced host protein synthesis; under its influence, depression of host functions gradually increased until 30 minutes, but never became complete. Fourteen others of this class appeared by 15 minutes. Class II included only five members, which were detectable after 30 minutes, while class III contained another five that appeared after 45 minutes. Among the latter was an endolysin that eventually induced lysis of the host cell.

9.1.4. Some Papova Viruses and Adenoviruses

The papova group includes a small number of viruses, whose molecular weight range of $3.6–5 \times 10^6$ is only one-tenth that of λ and its associates (Caro, 1965; Tai *et al.*, 1972). The polyoma virus of man and SV40 of monkeys represent the lighter end of the scale, along with the recently described virus BK of man (Howley *et al.*, 1975), whereas the heavier end is found among the papilloma (wart) viruses of mammals in general (Watson and Littlefield, 1960; Crawford, 1965). In all types, the DNA is a double-stranded, covalently closed circular molecule that is twisted into a superhelix (Tooze, 1973). No tail is present on the capsid, which uniformly is a skew icosahedral structure, having a triangulation number of 7 and consisting of 72 capsomers (Finch and Klug, 1965; Klug, 1965). With their diameters of around 45 nm, the polyoma and SV40 are among the smallest of the mammalian viruses. Proteins and nucleic acids, but no lipids, have been found in the virions (Takemoto *et al.*, 1971; Tooze, 1973).

Aspects of the Genetic Mechanism in SV40. Since the DNA of both SV40 and polyoma virus approximates 3.6×10^6, each contains about 6000 base pairs, sufficient to code for about ten proteins of a molecular weight of 20,000. It was found that a small smount of 5-methylcytosine was perhaps present, but no other unusual bases were noted in either type. Although these two viruses are related in general morphology, their DNAs displayed no homology in hybridization tests (Winocour, 1965; Benjamin, 1966). In the virions, the DNA has a chromatinlike structure, containing histones, and accordingly has been called a minichromosome. One of the histones has been identified as

H1; this and the others appear to be of viral origin (Varshavsky *et al.*, 1976). Replication appeared to be unidirectional and to involve a double-stranded circular RF (Robberson *et al.*, 1975).

During the infective processes, SV40 and polyoma virions are adsorbed onto the cell membrane, through which they penetrate, losing the capsid in the processes. In the case of SV40 at least, transcription of the DNA occurred in two stages, early and late (Sambrook *et al.*, 1973). Transcription of the early genes has been shown to be independent of DNA replication; in contrast, late mRNA synthesis was prevented when DNA polymerization was inhibited by such substances as cytosine arabinoside (Butel and Rapp, 1965; Sauer, 1971). However, once late transcription had begun, it continued even if DNA synthesis was subsequently inhibited (Brandner and Mueller, 1974). Certain mRNAs and their precursors showed the presence of four to six methyl groups per molecule (Aloni, 1975).

Early gene products included T-antigen and enzymes involved in DNA metabolism; these were transcribed from one strand of SV40 DNA, possibly by host RNA polymerase (Westphal and Kiehn, 1970). On the other hand, late gene products, which remain unidentified specifically, were transcribed from the second strand, tentatively suggested to be by means of modified host RNA polymerase II (Lindstrom and Dulbecco, 1972). In contrast, Aloni (1972) has presented results which suggested that transcription was symmetrical throughout and that one or both of the resulting strands was subsequently degraded. This regulation of gene expression in SV40 may involve extensive and highly regulated posttranscriptional processing of mRNAs. In cells transformed by this virus, transcription of the SV40 DNA was found correlated to the host cell cycle (Swetley and Watanabe, 1974).

Conclusions drawn from protein studies do not correlate well with the above results. Of the six DNA-associated polypeptides that have been partially characterized, numbers 1–3 were capsid proteins and contained tryptophan, all except 3 containing cysteine as well (Pett *et al.*, 1975); the largest, number 1, had a molecular weight of around 45,000 (Tooze, 1973). Numbers 4–6 were basic and lacked tryptophan, but only the first of this set contained cysteine. When the virus was grown in monkey kidney cells labeled prior to infection, these three proteins were labeled five times as heavily as number 1, indicating that they were partially derived from preexisting cell histones (Frearson and Crawford, 1972; Pett *et al.*, 1975). However, the conclusion of the latter citation that proteins 4, 5, and 6 were merely host histones H3, H2B, and H4, respectively, not viral products, does not seem warranted. Especially is this true in view of the results of the polyoma viral genetic apparatus described below. Translation made use of at least four viral-specific tRNAs, including those for asparagine, histidine, tyrosine, and aspartic acid (Katze, 1975).

Aspects of the Polyoma Viral Genetic Processes. The proteins of polyoma virus are similar to those of SV40, including the three histonelike ones. The latter have been shown also to be derived to a certain extent, but not fully, from preexisting host histones (Frearson and Crawford, 1972). Some progress has been made toward characterization of the messenger RNA, the synthesis of which appeared to be coupled to DNA replication (Glover, 1974). When labeling was carried out late in the infectious cycle, three size classes of polyoma-specific RNAs were found (Acheson *et al.*, 1971). These included giant species, with sedimentation coefficients between 30 S and 80 S, a second class which sedimented around 26 S (that is, about the size of a transcript from a single strand of polyoma DNA), and at least two others with coefficients between 16 S and 20 S. The first class was predominantly labeled by short (10- to 30-minute) exposures to [^{3}H]uridine, whereas the third class became apparent with exposures longer than 1 hour. More recently, the giant species have been demonstrated to be present only in the nucleus, whereas the two smaller classes were found mainly in the cytoplasm in association with polyribosomes (Buetti, 1974). A possible explanation of the size distribution may prove to be that the short strands were sent to the nucleus (Otto and Reichard, 1975), where the virions were assembled.

What is perhaps more indicative of the extent of autonomy of the viral genetic processes is a recent investigation of the isoaccepting species of $tRNA^{Lys}$ in normal and transformed cells (Juarez *et al.*, 1974). By use of a reverse-phase chromatographic system, five isoaccepting species had been detected in normal cells (Ortwerth and Liu, 1973; Ortwerth *et al.*, 1973); with the same procedures, seven isoaccepting species were found in transformed cells. Moreover, the major species of normal cells are determined as being relatively minor in transformed cells and vice versa. Thus it would appear that polyoma DNA must code for at least two species of $tRNA^{Lys}$ and must also strongly influence the transcription and perhaps the molecular structure of the others. It was clearly indicated, at least, that extensive investigations into the tRNAs of viral-infected cells would be highly profitable.

Comparisons with Adenoviruses. The adenoviruses, of which more than 50 different types have been isolated from mammalian and avian sources, are related to the papova viruses in several ways. One resemblance is their mutual lack of lipids and carbohydrates, and their undergoing DNA replication and virion assembly in the host cell nucleus provide others (Norrby, 1971; Tooze, 1973). However, the adenoviruses, which are the causative agents of many respiratory diseases, are much larger, having a total molecular weight of 175×10^6 and a DNA molecular weight of $20–25 \times 10^6$. The capsid (Figure 9.7) is constructed of 252 capsomers arranged on 20 equilateral triangular faces (Horne *et al.*, 1959; Norrby, 1966, 1969). Of this total 240 are bordered by six neighbors and are called hexons, while the remainder, having only five neighbors, are known as pentons. The 60 hexons that border the latter have slightly different properties than the other 180 (Smith *et al.*, 1965; Prage *et al.*, 1970). Each penton consists of a base anchored at one of the vertices of the icosahedron and a projecting fiber bearing a knob at its end (Valentine and Pereira,

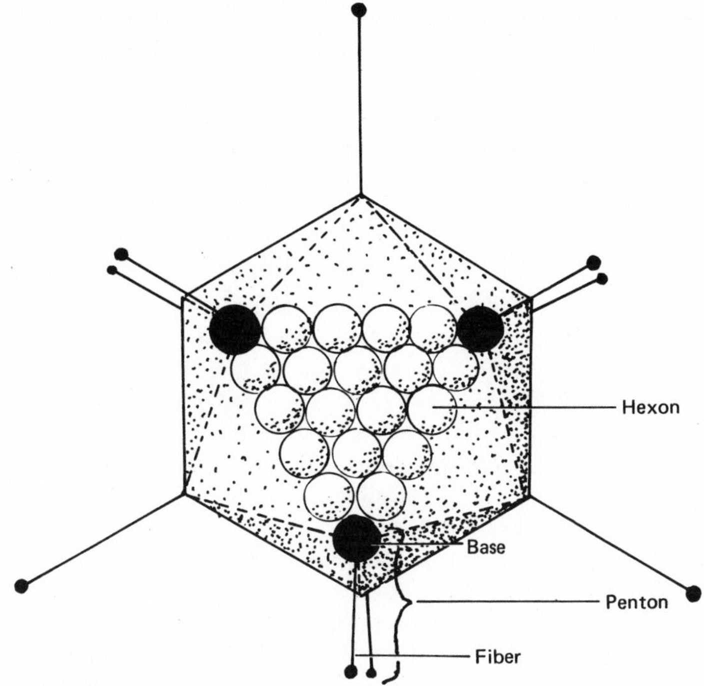

Figure 9.7. The adenovirus virion.

1965; Pettersson *et al.*, 1968). The fibers of serological subgroup I are around 10 nm long, those of subgroup II are of intermediate length, and those in subgroup III exceed 25 nm. These various capsomers may consist of a total of perhaps 13 structural polypeptides (Everitt *et al.*, 1973). Within the capsid is a dense core consisting of DNA and two polypeptides. The major of these core polypeptides has a molecular weight of 17,000 and contains ca. 21% arginine and 18% alanine, thus resembling certain histones (Prage and Pettersson, 1971); there is no evidence that this protein is derived from the host. The second of these is larger (mol.wt. =45,000) and is moderately rich in arginine (Laver, 1970).

The RNA transcribed prior to the onset of DNA replication differed in nucleotide sequences from those species transcribed late in the lyctic cycle. After transcription had been completed in the host cell nucleus, a segment of poly(A) ranging between 150 and 250 units long was added to all the viral transcripts (Philipson *et al.*, 1971). One of the late products, which sedimented at 5.5 S, was synthesized in abundance late in the infection period, but only a single copy of it appeared to be present per viral genome (Ohe, 1972). The entire base sequence of this mRNA molecule, 156 nucleotides long, has now been determined (Ohe and Weissman, 1970); its primary stucture indicated extensive base pairing that resulted in a complex secondary structure. After becoming attached to the ribosomes, this and other mRNAs have several proteins attached to them (Van der Marel *et al.*, 1975). The messengers have been shown not to be transcribed by a host-type RNA polymerase; that is, the transcriptase probably was entirely of virus origin (Price and Penman, 1972). Evidently the same polymerase transcribed the other light-molecular-weight RNAs also, including some with properties of tRNAs. The transcriptional processes, which were found correlated to the host cell cycle, were not sharply marked off into the early and late stages, but showed considerable overlap to occur (Taube *et al.*, 1974; Tal *et al.*, 1975). At 6 hours after infection of KB cells with adenovirus 2, before synthesis of viral DNA began, 5% of the poly(A)-containing RNA hybridized to viral DNA, but after 12 hours more than 80% hybridized with that genome. After 18 hours, four major size classes of cytoplasmic viral RNA containing poly(A) were detected, 27 S, 24 S, 19 S, and 12–15S. The first three of these proved to code strictly for late viral proteins, but those of the smallest class size were indicated to code for some early proteins as well as certain late ones (Philipson and Pettersson, 1973; Öberg *et al.*, 1975). Increased tRNA methylase activity has been noted in hamster tumors induced by adenovirus 12 (MacFarlane, 1969), but little else has been reported regarding the tRNAs of these viruses.

9.1.5. Various Groups of Complex DNA Viruses

Among the forms parasitizing the higher vertebrates, there are a number of major DNA viruses that have a lipid–glycoprotein envelope around the capsid. In addition, these types are larger than any of the foregoing and have other features indicative of a more complex nature. Since their macromolecular synthesizing processes are only partly known, brief summaries of several major categories will suffice for present purposes. Although each is quite imperfectly known, their distinctive features *in toto* cannot help but contribute greatly to the overall picture.

Herpesviruses. Pertinent information is available in the herpesviruses mainly for three human parasites, herpes simplex viruses (HSV)1 and 2 and Epstein-Barr virus (EBV), the latter being the causative agent of infectious mononucleosis. In each case the DNA is linear and double-stranded, lacks unusual bases, and is arranged as a toroid around a central plug of protein (Figure 9.8; Kieff *et al.*, 1971; Furlong *et al.*, 1972). The DNA of HSV-1 and 2 has been determined to have a molecular weight of 104×10^6, and that of EBV, one of 101×10^6 (Kieff *et al.*, 1971; Pritchett *et al.*, 1975). When placed in alkaline sucrose gradients, HSV-1 DNA broke into seven size classes of molecules, which were rather constant in molecular weights (Roizman and Spear, 1971; Frenkel and Roizman, 1972; contrast Fleckenstein *et al.*, 1975). Thus the

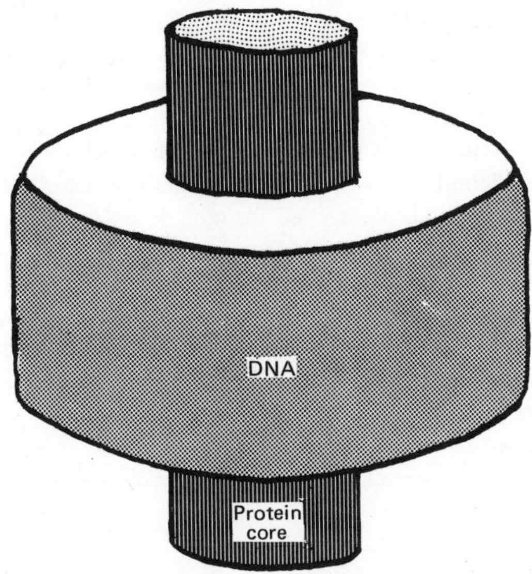

Figure 9.8. Herpesvirus virion DNA. The DNA is in the form of a toroid, through the central hole of which passes a rodlike protein core (based on Furlong *et al.*, 1972).

presence of specific nicks in both strands was indicated, but whether the breaks played a role in replication or transcription has not been established.

The outer capsid is in the form of an icosahedron and consists of 162 capsomers, most of which are hexamers, prisms approximately $19.5 \times 10 \times 10$ nm in dimensions (Wildy *et al.*, 1960). One protein in the capsid, identified as spermine, has been estimated to be present in an amount sufficient to neutralize 50% of the phosphate in the DNA (Gibson and Roizman, 1971). The outer capsid, which enclosed two others, was covered by an envelope consisting of lipids, spermidine, and 11 glycoproteins (Heine *et al.*, 1972; Savage *et al.*, 1972; Spear and Roizman, 1972). This envelope played an important role during penetration of the host cell, the processes of which remain obscure. After infection, the capsid and envelope were lost, and the DNA–protein complex moved into the host cell nucleus, where transcription and replication occurred.

Transcription was in two stages; early RNAs were transcribed beginning 2 hours after infection, and late ones 8 hours and thereafter. The mRNAs of HSV-1 have been shown to differ greatly in their functional lifetimes, early ones having much shorter half-lives than late (Ward and Stevens, 1975). In addition, various species of early and late mRNAs also differed in their average functional lives, but how these differences were achieved was not determined. Among the early enzymes were included thymidine kinase, DNA polymerase (Keir *et al.*, 1966; Buchan and Watson, 1969), and probably a DNA exonuclease (Keir, 1968); 24 species of structural proteins have been cataloged, most of which were produced by early processes (Spear and Roizman, 1972). After transcription, the mRNAs underwent cleavage, and poly(A) was added (Bachenheimer and Roizman, 1972). The RNA polymerase involved in transcription may be contained in the core DNA–protein complex.

Replication of the DNA has been only slightly investigated, but what has been learned seems to indicate a need for concurrent protein synthesis (Roizman and Roane, 1964). Some of the deoxyribonucleotides for the DNA might have been derived from breakdown of host DNA, but others were produced *de novo* (Levitt and Becker, 1967; Hay *et al.*, 1971). Assembly of the viral capsid and addition of the core appeared to be carried out largely in the nucleus, but some questions remained as to the site where the viral envelope was added. The envelope proteins became bound to cell membranes, where they were glycosylated (Spear and Roizman, 1970), but actual formation of the envelope might have occurred in the inner nuclear membrane or in nearly any membrane of the cell.

Vaccinia Virus. Vaccinia virus, one of the most complex metazoan viruses, contains a linear double-stranded DNA molecule with a molecular weight of 167×10^6 (Sarov and Becker, 1967; Bergoin and Dales, 1971). That of the related fowl pox is even larger, having a weight of 192×10^6 (Hyde *et al.*, 1967).

Replication was formerly thought to occur entirely within the cytoplasm of the host cell (McAuslan, 1971), but a significant fraction now has been demonstrated to occur within the host cell nucleus (LaColla and Weissbach, 1975). After entering the host cell, the envelope was shown to be degraded in a series of steps, releasing the so-called core, which appeared to consist of the inner capsid and its contents (Figure 9.9). This core contained DNA and proteins, one (or more) of which was an RNA polymerase which transcribed the viral DNA *in situ;* all of the transcripts made within the core were virus-specific early mRNA sequences (Kates and McAuslan, 1967a,b). These early activities

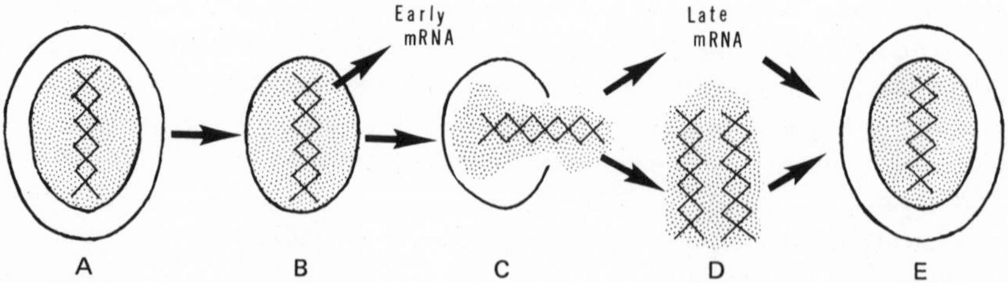

Figure 9.9. Stages in the replication of vaccinia virus: (A) Virion. (B) Core and early messenger synthesis. (C) Release of genome and enzymes. (D) Replication, transcription, and protein synthesis. (E) Assembly into daughter virions.

that proceed thus while still within the virion hold a special significance in a later discussion. After a period of time, the DNA molecule was released from the capsid and commenced to replicate, initiating the transcription of late sequences and retarding early mRNA synthesis (Oda and Joklik, 1967). The deoxyribonucleotides may have been derived in part from host chromatin, but the ribonucleotides appeared to be synthesized under virus control, for an ATPase has been found within the viral core (Gold and Dales, 1968).

The mRNAs have been demonstrated to undergo cleavage during processing (Katz and Moss, 1970) and bear poly(A) segments (Sheldon and Kates, 1974). The viral polymerase that produced these segments had a molecular weight of about 80,000 and consisted of two subunits (Moss *et al.*, 1975). Just the reverse of the processes in the herpes simplex virus, the early mRNAs were longer lived and much more strongly diversified in functional lifetimes than the late ones (Jungwirth and Joklik, 1965; McAuslan *et al.*, 1965; Kates and McAuslan, 1967c; Sebring and Salzman, 1967). The mRNAs also have a complex 5′-terminal segment, including methylated ribose moieties (Urushibara *et al.*, 1975), and the tRNAs in vaccinia-infected cells have been demonstrated to undergo more extensive methylation than uninfected ones (Klagsbrun, 1971).

9.2. SINGLE-STRANDED DNA VIRUSES

Two major classes of single-stranded DNA bacteriophages, but no plant or animal viruses, have been recognized, "spherical" (actually icosahedral) and filamentous. In the icosahedral group are found ϕX174, S13, ϕ1, and relatives and dϕl and kindred forms (Tikhonenko, 1970), whereas the filamentous type consists of bacteriophage fd, F1, M13, Zj/2, Ec9, AE2, and Pf. Although the genetic processes of the two groups are similar, the few distinctions that exist necessitate attention to those of each type separately.

The Icosahedral Group. The virions of this group are naked icosahedral capsids, composed of four species of structural proteins (Figure 9.10). Three of these enter into formation of a short prong or spike located at the 12 vertices, while the faces of the icosahedron appear to consist of only one type (Edgell *et al.*, 1969; Gelfand and Hayashi, 1969). In diameter, the particles average around 250 Å, to which each spike adds an additional 50 Å. These prongs are hollow and appear to play an important role in attachment to the host. Infection is semilethal to the host, bacterial DNA synthesis being inhibited within 12 to 14 minutes under the influence of a virus-induced protein (Lindquist and Sinsheimer, 1967a). Little or no effect by infection has been noted on host protein and RNA synthesis, except at high dosages of the virus (Stone, 1970).

The DNA molecule is single-stranded and circular, with a molecular weight of about 1.7×10^6 (Edgell *et al.*, 1969); thus it contains enough coding

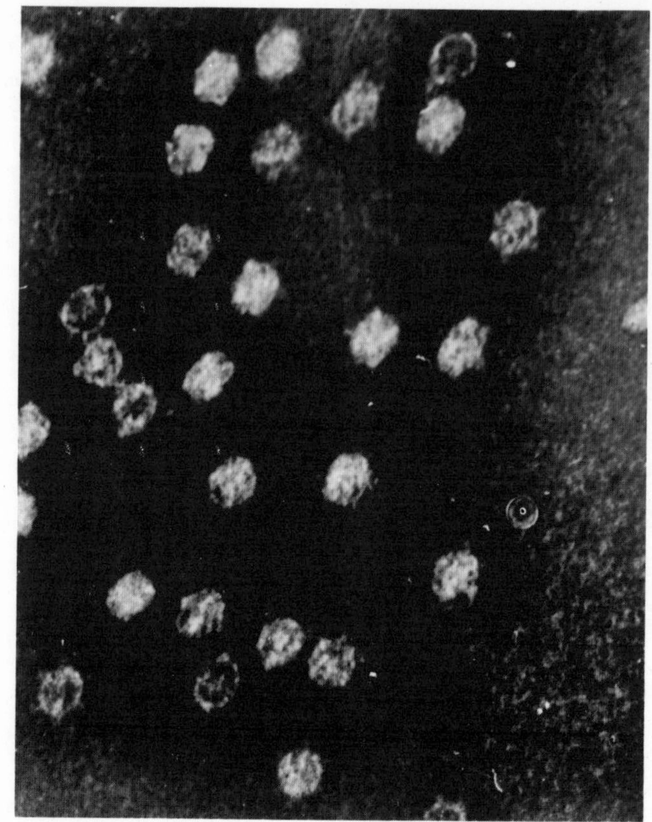

Figure 9.10. Two single-stranded viruses. (A) Bacteriophage ϕX174, and (B) bacteriophage 1ϕ7.

abilities to induce the formation of about eight or ten proteins. The bases are the usual ones, except that one unit of 5-methylcytosine is present per strand (Razin *et al.*, 1973). Replication of the DNA has proven far more complex than might be expected of the simple parental molecule, for it occurred in three stages (Sinsheimer, 1968; Sinsheimer *et al.*, 1968; Schröder *et al.*, 1973). In stage I immediately following infection, the single-stranded, circular plus strand of the bacteriophage was converted to a double-stranded circle by acquisition of a minus strand. Stage II involved the replication of this parental replicative form (RF) in a semiconservative fashion, resulting in the formation of 20–50 progeny RF molecules. These were of two types: RF1, which was supercoiled and had both strands covalently closed, and RF2, which was relaxed and had at least one single-strand break. In stage III, single-stranded progeny viral DNA was produced by asymmetrical semiconservative replication of the RF molecules.

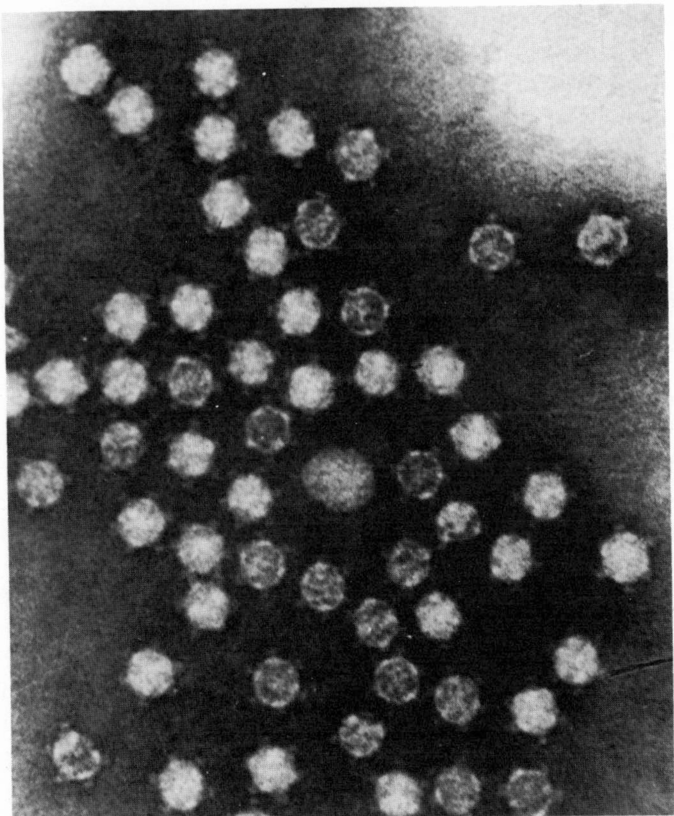

Both electron micrographs × 260,000 (through the courtesy of Dr. Anna S. Tikhonenko).

A number of authors have presented evidence that replication proceeds according to a rolling-circle model (Dressler and Wolfson, 1970; Baas and Jansz, 1972a,b; Francke and Roy, 1972). Other recent experiments have suggested the details of the processes to be as follows (Schröder and Kaerner, 1972): (1) The minus strand of the RF duplex remained a closed circle, while the plus strand was nicked and became continuously elongated on the template provided by the minus strand. (2) The original plus strand was peeled off from the rolling circle as a single-stranded tail, on which (3) a new minus strand was initiated and polymerized. (4) However, before the product of the third step had been completed, the replicating complex split into a circular portion and a linear tail. The former was a complete RF2 molecule, consisting of the old circular minus strand and the new linear plus strand, whereas the tail was comprised of the old linear plus strand and the new incomplete minus strand. (5) The daughter circular components were found to participate in replication continuously.

(6) The tails developed into RF1 molecules by first circularizing the plus strands; then when the minus strands were converted to the circular configuration, a gap was left, accounting for the RF2 type of molecule (Schröder *et al.*, 1973). More recently, the RF2 DNA has been shown to possess two types of gaps, one of which could be converted to RF1 molecules by action of ligase and polymerase I, whereas the other could not. Initiation of replication appeared to be at a locus at or near gene *A* (Johnson and Sinsheimer, 1974), where a gap was located at a specific site (Eisenberg and Denhardt, 1974). At the termination of the replicatory processes, when single-stranded viral DNA was produced, methylation of a linear DNA strand longer than one phage genome occurred at one specific site (Razin *et al.*, 1973). Although host polymerase probably played an important role in DNA synthesis, RF molecule formation involved at least one viral-specific protein (the product of gene VI), and single-strand DNA synthesis was dependent on products specified by genes V and VIII (Lindquist and Sinsheimer, 1967b; Sinsheimer *et al.*, 1967; Knippers *et al.*, 1969; Taketo, 1973; Wickner and Hurwitz, 1974).

It has been found that single-stranded, but not the native double-stranded, DNA of ϕX174 bound to ribosomes (Robertson, 1975). This action mimicked that of mRNA binding and had similar requirements, including GTP, formylmethionyl tRNA, IF-1, and IF-2. The DNA of the filamentous virus F1 described below behaved in similar fashion, but the regions that interacted with the ribosomes differed in nucleoside sequence. What the normal function of this interaction may be was not understood, but it has been employed to isolate fragments of DNA for sequence analysis. Through its use fragments of ϕX174 DNA containing the initiation sites of four genes have been isolated and the sequence established (Barrell *et al.*, 1975; Air *et al.*, 1976; van Mansfeld *et al.*, 1976), as has an additional binding site in a segment of the genome that apparently is not translated (Brown and Smith, 1977). More importantly, a segment of the DNA about 5400 nucleotides long has been sequenced; this sequence identified many of the genomic features active in transcription or translation (Sanger *et al.*, 1977). Undoubtedly the most important outcome of this accomplishment, however, was the discovery that two genes occupy the identical region of the DNA. During transcription, different reading frames are employed to make mRNA for two entirely distinct proteins. Another pair of genes (*A* and *B*) were found by genetic means also to occupy identical regions (Smith *et al.*, 1977). If similar conditions are found in the genomes of other viruses or in cellular organisms, the definition of a gene as a locus on a chromosome may need reevaluation.

Transcription of ϕX174 DNA has been investigated to only a small extent. The processes are totally asymmetric, for RNA prepared from pulse-labeled infected cells hybridized only with RF forms, not single-stranded parental DNA. Thus only the minus strand was transcribed (Hayashi *et al.*, 1963). The transcripts were of various lengths, falling into 18 size classes and ranging up to lengths greatly exceeding the viral DNA molecule (Hayashi and Hayashi, 1970; Clements and Sinsheimer, 1975). Hence, it was obvious that many of the classes were polycistronic and that the largest messengers were produced by continuation of synthesis on a circular DNA molecule past a termination site, if one existed; certainly none were indicated to be present at the ends of the individual genes. Early and late stages have not as yet been reported.

Filamentous Types. The filamentous representatives, such as M13, P1, and Xf, are very long and slender, with dimensions of 50–60 Å in diameter and 8000–10,000 Å in length. The DNA has been shown to be in the form of a single-stranded circle, having a molecular weight of 2×10^6 (Edgell *et al.*, 1969); the filament consists of only a single kind of protein, except at one end where there is a second type which has a molecular weight of 70,000 (Henry and Pratt, 1969). Infections with filamentous phages are not lethal, and host syntheses of macromolecules continue at near normal levels. The most significant change in the bacterial host is a lengthening in the division time by about 25% (Brown and Dowell, 1968a). Only male strains of the bacteria are subject to infection, as the bacterial sex-pili serve as entry points for the phage DNA and A protein (Marvin and Hohn, 1969; Marvin and Wachtel, 1975).

The basic features of DNA replication resemble those of ϕX174 to an extent, in that a circular double-stranded RF molecule is formed first by synthesis of a minus strand. This then is similarly replicated, and single-stranded synthesis of viral DNA is likewise the final stage (Marvin and Wachtel, 1975). However, only two genes code for enzymes necessary for DNA replication (Ray *et al.*, 1966; Brown and Dowell, 1968b; Mechler and Arber, 1969; Ray and Schekman, 1969). DNA replication has recently been further explored in *E. coli* mutants variously deficient in DNA polymerases I and III and infected with bacteriophage M13 (Staudenbauer, 1974). Host polymerase III was shown to be essential to the production of RF molecules and also to the formation of single-stranded viral DNA. Moreover, since in those mutants deficient in polymerase I, RF with gaps in one strand accumulated, polymerase I seemed to be needed to fill gaps in newly synthesized molecules.

Assembly of the cylindrical capsid has recently been described (Marvin and Wachtel, 1975). In these forms, the capsid was 60 Å in diameter and between 10,000 Å and 20,000 Å long, depending on the strain. A single major species with a molecular weight of 5000 comprises about 99% of the viral protein and is largely α-helical. During assembly, the capsomers overlap one another like shingles to form a hollow cylindrical coat, within the lumen of which the DNA lies.

One unusual observation has recently been made on coliphage P1. Although unable to infect the myxobacterium, this phage introduced genes from *E. coli* to *Myxococcus xanthus,* despite the latter's being classed in a taxonomic group far removed from the taxon to which the genus *Escherichia* belongs. The virus was adsorbed efficiently to the myxobacterium and injected its DNA into that organism (Kaiser and Dworkin, 1975).

RNA VIRUSES

Large numbers of RNA viruses attack plants as well as animals and bacteria, and are proving to be even more diversified than the DNA viruses. Like the latter, they fall into two major categories, single-stranded and double-stranded, but contrast in that the former group is more abundant than the latter.

9.3. SINGLE-STRANDED RNA VIRUSES

Since single-stranded RNA viruses are of such frequent occurrence, it comes as no surprise that representatives occur in both the prokaryotes and eukaryotes and in plant and metazoan cells alike. As there are a number of advantages in examining the RNA viral parasites of prokaryotes first, discussion of the more complex viruses of eukaryotes is deferred for a later section.

9.3.1. RNA Bacteriophages

The RNA bacteriophages are placed variously in three to five main groups (Watanabe *et al.*, 1967; Sakurai *et al.*, 1968; Miyake *et al.*, 1971; Weissmann *et al.*, 1973). Of these, only two have been sufficiently characterized to warrant consideration here, group I, represented by MS2, R17, and f2, and Group III, in which bacteriophage Qβ has been most thoroughly investigated. Even though placed in a separate group, the latter phage has many features in common with the three of group I. In all cases the RNA molecule is small, has a length in the range of 3500–4500 nucleotides, and is enclosed in an icosahedral capsid consisting of 180 coat protein subunits, plus at least one molecule of a second protein known as the A, or maturation, protein (Vasquez *et al.*, 1966; Steitz, 1968, 1970; Knolle and Hohn, 1975). All are confined to male strains of bacteria. During the infective processes, only the RNA enters the bacterium, entrance being gained through the base of the sex-pilus, not at the tip as in the filamentous types discussed above (Knolle and Kaudewitz, 1963; Valentine and Wedel, 1965). Synthetic complexes consisting only of protein A and the RNA have now been shown capable of infecting *E. coli* (Shiba and Miyake, 1975).

Molecular Aspects of Group I. The RNA molecule in group I RNA bacteriophages has a molecular weight of 1.1×10^6 and consists of about 3300 nucleotides. The coat protein, with a molecular weight of 14,000, contains no histidine; as already shown in Chapter 4, the complete amino acid sequences of this protein from phages f2, fr, MS2, and R17 have been determined (Weber, 1967; Weber and Konigsberg, 1967; Wittmann-Liebold and Wittmann, 1967; Min Jou *et al.*, 1972; Min Jon and Fiers, 1976), as has also the complete genome of MS2 (Fiers *et al.*, 1976). The maturation (A) protein in contrast contains histidine and has a molecular weight of 37,000. By means of reconstitution experiments, this protein has been shown to be essential to the spontaneous *in vitro* production of normal capsids, as well as for infectivity (Steitz, 1970). In what way it performs its role in the latter function remains to be elucidated. Recently a technique based on fluorescamine that made possible the analysis of nanomole quantities of protein fragments has permitted the establishment of the amino acid sequence of the maturation protein of MS2 (Vandekerckhove and Van Montagu, 1974; Fiers *et al.*, 1975). Since the nucleotide sequence of MS2 RNA has already received attention, along with its flowerlike secondary structure (Chapter 4, Section 4.1.1; Figure 4.1), no further discussion on this aspect appears necessary (Ball, 1973; Vandenberg *et al.*, 1975), except to note that portions of the RNA of R17 have similarly been sequenced (Adams, 1972; Rensing and Schoenmaker, 1973).

In addition to the structural and A proteins, the viral RNA molecule codes for one other known polypeptide, the β subunit of the RNA-dependent RNA polymerase (Fiers *et al.*, 1973). This subunit appears to combine with three host factors to form the active replicase (Haruna and Spiegelman, 1965; August *et al.*, 1968). Of these three, the γ and δ subunits have been demonstrated to be the dual parts of a host elongation factor (EF-Tu and EF-Ts, respectively), whereas α is factor *i*, a bacterial protein that inhibits polypeptide chain initiation at certain cistrons (Groner *et al.*, 1972a; Kamen *et al.*, 1972).

Replication of the viral RNA occurs in two stages (Weissmann and Ochoa, 1967; August *et al.*, 1968; Spiegelman *et al.*, 1968; Stavis and August, 1970). Using the flowerlike RNA of the virus as a template, the polymerase synthesizes a single minus strand in a reaction that also calls upon one, or perhaps two, host initiation factors, IF-1 and/or IF-2. In the second stage, only the polymerase is active; it uses the minus to replicate a new plus strand. Although the existence of double-stranded replicative forms (RF) was doubted for a time (Mills *et al.*, 1967; Feix *et al.*, 1968), they now are generally accepted in these phages (Knolle and Hohn, 1975). The plus strands, after being freed from the replicative complex, serve as messenger, so that transcription is unnecessary.

Genetic Processes in Group III RNA Viruses. As a whole, the typical representative of group III, bacteriophage Qβ, is rather similar to the members of group I, but differs strongly serologically and in being slightly larger (Strauss and Kaesburg, 1970). Its RNA includes around 4500 nucleotides and thus has a molecular weight of nearly 1.3×10^6 (Valentine *et al.*, 1964; Gussin, 1966; Horiuchi and Matsuhashi, 1970). This larger molecule codes for one protein not found in group I members, designated IIb or A_1, which arises by read-through from the coat gene into the following one that codes for the β

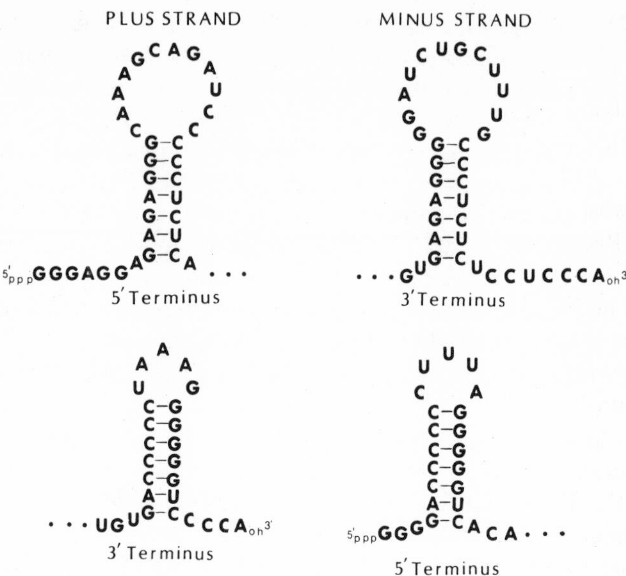

Figure 9.11. Termini of the RNA molecules of Qβ. The terminal A of each 3′-end is added by the replicase, although not coded for by the template (based on Goodman *et al.*, 1970).

subunit of the RNA polymerase (Horiuchi *et al.*, 1971; Moore *et al.*, 1971; Weiner and Weber, 1971). Hence there are really four genes present. This protein, which is found in various numbers in the virions, is essential for infectivity (Radloff and Kaesburg, 1973) and for the formation of viable Qβ virions (Hofstettler *et al.*, 1974).

Parts of the RNA molecule from Qβ, but not an entire cistron, have been sequenced, including both terminal regions of plus and minus strands (Goodman *et al.*, 1970). As may be seen in Figure 9.11, the 5′-terminus of the minus strand is complementary to that of the 3′-terminus of the plus strand; it should be noted that other interpretations of secondary structure may be equally plausible. Also it is evident that each 3′-terminal region ends in *–CCCA* and that the opposite end terminates merely in GGG– (Weith and Gilham, 1969).

These features are involved in a peculiarity that characterizes Qβ RNA replication. During these processes, the RNA polymerase, which resembles that of group I, begins synthesis at the 3′-end of the template, regardless of whether it is the plus or minus strand (August *et al.*, 1968; Spiegelman *et al.*, 1968). However, the terminal A of the –CCCA sequence is omitted, thus the GGG– at the 5′-terminus of the strand is synthesized first (Billeter *et al.*, 1969; Goodman *et al.*, 1970). When the strand is approaching completion, the terminal GGG– at the 5′-end of the template merely directs the incorporation of –CCC in the new molecule, but the enzyme itself adds a final A residue, thus producing the characteristic –CCCA sequence (Rensing and August, 1969; Weber and Weissmann, 1970). The replicase of Qβ, as well as those of MS2 and R17, shows a high specificity for viral plus and minus strands, failing to replicate RNAs from other viruses or most other sources which have been tested (Haruna and Spiegelman, 1965; Feix *et al.*, 1968; Haruna *et al.*, 1971). It is especially noteworthy that despite their having a similar –CCCA sequence at the 3′-terminus, the RNA molecules of MS2, R17, and f2 do not serve as templates for Qβ replicase (Weissmann *et al.*, 1973). However, the replicase did react with cytosine-containing synthetic polymers (Hori *et al.*, 1967), and, when both Mn^{2+} and Mg^{2+} were present, it synthesized complementary strands with a number of heterologous RNAs (Palmenburg and Kaesburg, 1974). In the latter experiments, recognition by the replicase depended upon the presence of –CCC at the 3′-terminus and an identical triplet at a definite internal (but unestablished) position; when these requirements were met, Qβ transcribed any oligo- or polynucleotide (Küppers and Sumper, 1975).

Since more copies of the coat protein were made than of the replicase, some control mechanism must reside in the RNA molecule. One control has been found in the coat protein (Skogerson *et al.*, 1971), which served to depress replicase synthesis. In an amber mutant in the coat protein, which induced the formation of fragments of that polypeptide, replicase accumulated throughout the infective stages, whereas normally that enzyme decreased in quantity (Palmenberg and Kaesberg, 1973).

9.3.2. Small RNA Viruses of Eukaryotes

Aside from the RNA bacteriophages, few small RNA viral types have been studied to the point where a satisfactory picture of the replicatory processes can be presented. Consequently, the genetic mechanisms in remaining types are of necessity often sketchy or fragmentary; nonetheless, even these incomplete accounts contain details pertinent to the major topic.

Picornaviruses and Arboviruses. One group of animal viruses, the picornaviruses, appears to be somewhat allied to the RNA bacteriophages. This group includes such forms as the poliovirus, mengovirus, and encephalomyelocarditis, Maus Elberfield, and foot-and-mouth disease viruses (Montagnier, 1968; Rueckert, 1971), all of which are lipid-free and of relatively small size. The rhinoviruses, too, seem to belong here, for an amino acid analysis of their proteins indicated no significant differences to exist (MacGregor and Mayor, 1971). However, the RNA molecule is much larger, that of rhinovirus 14 having a molecular weight of about 8.4×10^6 (Stott and Killington, 1972). Whether the turnip yellow virus and allies should also be placed close to the picornaviruses is still an unanswered question (Rueckert, 1971). The RNA molecule alone enters the host cell, by unknown processes, and is believed to serve as a plus strand during replication and protein synthesis.

Poliovirus, whose RNA has a molecular weight of 2.6×10^6 is undoubtedly the best known member (Granboulan and Girard, 1969; Tannock *et al.*, 1970), and encephalomyocarditis virus, with RNA of 2.7×10^6, ranks second (Porter *et al.*, 1974).

The RNA molecule is translated into four proteins that enter into the formation of the capsid and at least ten virus-specific noncapsid polypeptides. It has been shown that some of the structural proteins were cleavage products of larger precursor proteins and that these and the RNA were synthesized in the cytoplasm by virus-specific enzymes (Montagnier, 1968; Summers *et al.*, 1971). Thus the overall picture of poliovirus replication may prove similar, but more complex, than that of the RNA bacteriophages. Mengoviral, but not polioviral, RNA has been found associated with polysomal structures, but no double-stranded resistant structure could be found (Tobey, 1964). Thus it was suggested that, at least in the latter virus, double-stranded RNA may be absent or at least transient during replication; it is not clear, however, why *replicating* RNA should necessarily occur attached to ribosomes. The minus RNA strand has a poly(U) sequence 65 units in length, located at the 5′-terminal (Yogo *et al.*, 1974). Double-stranded RF forms were reported in encephalomyocarditis virus in an early paper (Montagnier and Sanders, 1963b), but their existence has since been questioned (Thach *et al.*, 1974). A long poly(C) tract has been found within the RNA, 85–90 cytidine residues being contained in a region 95–100 nucleosides long (Porter *et al.*, 1974). In what was reported as a double-stranded RF molecule, the plus strand contained long poly(A) sequences, whereas the minus strand correspondingly had poly(U) (Yogo and Wimmer, 1975; Goldstein *et al.*, 1976). The former sequences have been found to average close to 40 residues in length (Frisby *et al.*, 1976). The size of the synthesized RNA usually approximates that of the parental molecule, although some larger as well as shorter strands occur in most picornaviruses as in the bacteriophages (Montagnier and Sanders, 1963a; Sonnabend *et al.*, 1967).

The arbovirus group, in which Sindbis virus and other vertebrate arthropod-borne viruses are placed, contains slightly larger forms, for which a molecular weight of 3×10^6 has been determined (Montagnier, 1968; Casals, 1971). All members are enclosed in a lipoprotein envelope, the protein for which is newly made after infection (Pfefferkorn and Hunter, 1963).

Rhabdoviruses. A group of forms more complex than those of either of the two foregoing ones is the rhabdovirus type (Hummeler, 1971). Among its members is the vesicular stomatitis virus (VSV), which has a single-stranded RNA molecule of molecular weight approximately 4.2×10^6 (Huang and Wagner, 1966). The molecule is enclosed in a rodlike capsid that provides the basis for the group name; actually the capsid is more bullet-shaped than rodlike, for one end is hemispherical and the other planar (Cartwright *et al.*, 1970). Around the capsid is a proteinaceous membrane, formed by the M proteins; this in turn is densely covered by short, knobbed spikes made up of the lipoprotein G, which is active in infectivity (Sokol and Koprowski, 1975). Within the capsid is a cavity that opens on the planar end. In turn the cavity contains a tightly wound helix consisting of a complex containing RNA and the protein N (Breindl and Holland, 1975); two minor proteins, L and NS, are also universal among the rhabdoviruses, which are of widespread occurrence throughout the Metazoa and even the seedplants. Among them is one of importance in medicine, rabies (Crick and Brown, 1970), and several in agriculture, including lettuce necrotic yellows, plantain A and B, maize mosaic, and wheat striate mosaic viruses (Hull, 1970).

Of these, vesicular stomatitis virus appears to be the most thoroughly investigated, but, unfortunately, not all reports are in complete agreement as to the details of the protein-synthesizing processes (*cf.* Newman and Brown, 1970; Baltimore *et al.*, 1971). Nor do the descriptions of the proteins concur with analyses of the RNAs (Wagner *et al.*, 1969b; Mudd and Summers, 1970a; Wagner and Schnaitman, 1970). One point, however, appears clearly established—none of the proteins is of host origin (Kang and Prevec, 1969; Wagner *et al.*, 1969a); one of these, the transcriptase, is known to be present in various varions (Baltimore *et al.*, 1970; Bishop and Roy, 1971). In the Newman and Brown account, the RNAs from VSV-infected cells are widely heterogeneous in size. One major size class, sedimenting at 38 S, was presumed to correspond to the RNA of the infective virus component. In addition, four other major peaks were observed, with sedimentation coefficients of 28 S, 20 S, 12 S and <4 S, all of which were ribonuclease resistant. To explain the observations, it was proposed that the basic unit of the viral RNA is not the 38 S molecule usually found in the virions but a complex consisting either of four 18 S or of eight 11 S molecules.

In contrast, the Baltimore *et al.* (1971) study accepted the 38 S molecule as that of the virions and proposed that the messenger RNA was transcribed from this as the 28 S and 13 S components found in VSV-specific polyribosomes (Huang *et al.*, 1970). In turn, the 13 S size class was shown to consist of a number of different RNA species. These RNA molecules approximated the size needed to code for known VSV proteins (Mudd and Summers, 1970b); hence, cleavage in the formation of the viral peptides was demonstrated to be absent (Wagner *et al.*, 1970). These mRNAs were complementary in nucleotide sequence to the viral RNA (Huang *et al.*, 1970) and were claimed to be transcribed from the viral RNA by a RNA-dependent RNA polymerase carried in the virions. Thus the implication is clear that the RNA of this rhabdovirus is the minus, not the plus, strand as in the preceding RNA types. The transcripts have been shown to require methylation by virion-associated RNA nucleotidyltransferases before translation is possible (Both *et al.*, 1975). Regardless of the extent of the original infection, about 500 transcriptive intermediates can be present per host cell. These are responsible for the accumulation of over 10,000 viral complementary genome equivalents per cell (Flamand and Bishop, 1974).

In addition to VSV, five animal rhabdoviruses have been demonstrated to contain an RNA-dependent RNA polymerase within their virions, including Cocal, Chandipura, Piry, Kern Canyon, and pike fry viruses (Aaslestad *et al.*, 1971; Chang *et al.*, 1974; Roy *et al.*, 1975). Consequently, this feature may soon prove to characterize all the members. In the pike fry virus, the RNA molecule has a molecular weight of 3.8×10^6, and the polymerase serves similarly to that of VSV in synthesizing viral-complementary, heteropolymeric RNA. In addition, a protein kinase has been found in the virion, along with glycoprotein, a membrane protein, a nucleoprotein, an L protein, and a phosphoprotein (Roy *et al.*, 1975).

Bromoviruses. At present only a small number of types are classed as bromoviruses, all of which are plant pathogens. The three known members are brome mosaic, broad bean mottle, and cowpea chlorotic mottle viruses (Lane, 1974). All have a spherical capsid composed of 180 repeating subunits, each of a molecular weight around 20,000. The RNA molecule is single-stranded but consists in each case of three subunits; a fourth component detected by some techniques is often considered a fraction of one of the larger ones. The total molecular weights of the RNA in the three viral types have been reported to be as follows: brome mosaic virus, 1.05×10^6 (Hurst and Incardona, 1969; Sela and Antignus, 1971); cowpea chlorotic mottle virus, 1.1×10^6 (Bancroft *et al.*, 1968), and broad bean mottle virus, 1.15–1.25×10^6 (Paul, 1961; Yamazaki *et al.*, 1961). However, all the foregoing totals may be found oversimplified if brome mosaic virus is typical of the entire group. In this virus, three types of structurally similar virions are found, all of which were necessary for infectivity; the principal, if not sole, difference between them was in their RNA content. Type I contained an RNA molecule with a molecular weight of 1.09×10^6, type II, of molecular weight 0.99×10^6, and type III had two molecules, one weighing 0.75 and the other 0.28×10^6 (Lane and Kaesberg, 1971).

The viral RNA molecules appear to be minus strands, the transcripts of which serve directly as messengers. A feature not noted earlier among the RNA viruses is the ability of each of the four RNA components to become charged with tyrosine and thus serve as tRNAs (Hall *et al.*, 1972; Lane, 1974); however, only in brome mosaic virus has active participation in tRNA function been noted (Shih and Kaesberg, 1973; Shih *et al.*, 1974). The messengers code for an undetermined small number of proteins, including a virus-specific RNA polymerase in each type of virus (de Zoeten and Schlegel, 1967a,b) and a coat protein. All the capsid proteins, which range between 184 and 199 amino acid links in length, have been reported to be extremely rich in alanine; that of broad bean mottle virus was distinctive in lacking tryptophan (Gibbs and McIntyre, 1970). During replication of the RNA, both double- and partly double-stranded intermediates have been observed (Semal and Kummert, 1971).

9.3.3. Larger RNA Viruses

As will readily be observed, there is no sharp line of demarcation between the larger and smaller RNA viruses. This is especially true in regard to the RNA molecules, which in some cases will be noted to be no heavier in these larger types than those of small virus groups. Here, as above, representatives are found in plants as well as animals.

Paramyxoviruses. Among the paramyxoviruses are included such forms as mumps, measles, Newcastle disease, Sendai, and simian 5 viruses (Finch and Gibbs, 1970). All have irregularly globular or ovate capsids, enclosed in a spike-covered lipoprotein membrane reminiscent of the rhabdoviral structure. Also similar to the latter group's condition is the ribonucleoprotein molecule that fills the interior of the capsid in the form of a tightly coiled helical filament. The RNA molecule has a molecular weight of 6.5×10^6 (Nakajima and Obara, 1967; Compans and Choppin, 1968; Klenk and Choppin, 1969) and a sedimentation coefficient of 50 S.

Among the more thoroughly investigated members of the group, as far as macromolecular syntheses are concerned, is Newcastle disease virus. The single-stranded RNA of the virions has been demonstrated to be a minus strand; after infection, this molecule was replicated by means of a virus-specific, RNA-dependent RNA polymerase into a variety of plus strands which accumulated in the cytoplasm (Huppert *et al.*, 1970). However, these were produced by way of a double-stranded precursor, which occurred in small amounts in infected cells. The polymerase was present in the virions and has been reported to be closely related to that of the vesicular stomatitis virus, from which it differed only in being much larger (Huang and Wagner, 1966; Kingsbury, 1966; Huang *et al.*, 1971). Replication occurred in the cytoplasm, while assembly and release appeared to be membrane-associated functions. Messenger RNAs have proven to be methylated following transcription (Colonno and Stone, 1975).

Myxoviruses. Included in the myxovirus group are a number of influenza types and the fowl plague virus. In general, the structure and macromolecular syntheses are similar to those of the paramyxoviruses but with several distinct differences (Compans and Choppin, 1971). The capsid is enclosed in a lipoprotein envelope, with a dense covering of 120- to 150-Å-long spikes on the surface. In some electron micrographs the envelope has been observed to have a typical unit-membrane construction (Compans *et al.*, 1970). In part, the protein for some of the spikes appeared to be of host origin, for the antigenic properties varied in accordance with the host species in which they were cultured (Isacson and Koch, 1965; Rott *et al.*, 1970). Other spikes or portions of the membrane, however, showed the hemagglutinin and neuraminidase activities that characterize these viruses. Maturation apparently occurred exclusively in the plasma membrane, for no viral particles were found in the cisternae of the endoplasmic reticulum nor in the Golgi complex (Compans *et al.*, 1970).

The RNA and the proteins with which it is complexed do not form a single nucleocapsid but a multiple set, five different species being recognizable (Duesberg, 1968a,b; Pons and Hirst, 1968). While in the virion, the various components seemed to form a single tightly wound helix, similar to the nucleocapsid of paramyxoviruses and others; however, it was highly unstable and fragmented readily into its component parts (Almeida and Waterson, 1970). These segments ranged in sedimentation coefficients from 21 S to 8 S (Nayak and Baluda, 1967; Pons and Hirst, 1968). During assembly into virions, the actual types and numbers of segments entering into formation varied from particle to particle, resulting in high levels of genetic recombination and a very low rate of infectivity because of the frequent deficient genomes (Simpson and Hirst, 1968; Almeida and Waterson, 1970).

The ribonucleoprotein has been claimed to be formed within the nucleus of the host cell, from where it migrated to the cytoplasm and cell membrane for assembly (Breitenfield and Schäfer, 1957). Following infection, host RNA synthesis was *increased* and continued at a high level until virus release began after 4 hours, when a decline set in (Mahy, 1970). The early increase was accompanied by a similar rise in DNA-dependent RNA-polymerase activity (Borland and Mahy, 1968). In addition, an RNA-dependent RNA polymerase appeared after 2 hours and reached a maximum at 6 hours. A similar enzyme induced in the cytoplasm has been shown to be virus-specific. During replication, both complementary single-stranded RNA and a double-stranded replicative form could be detected in influenza-infected cells, in addition to parental-

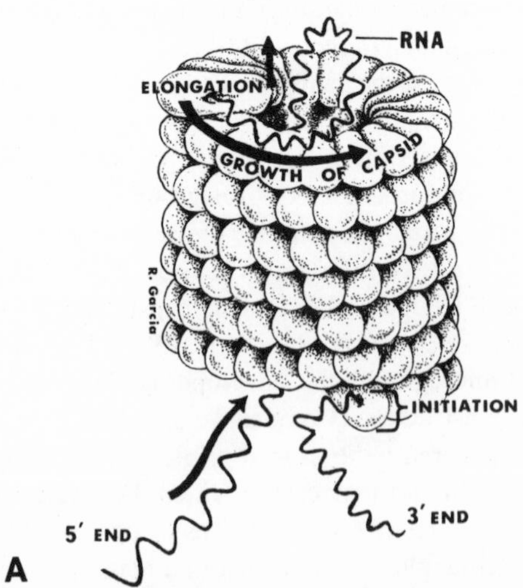

Figure 9.12. Tobacco mosaic virus. (A) Model of a TMV virion, with the RNA genome partly spiraled within the protein subunits and partly extending through the central lumen. (B) Growth of capsids involves elongation at one pole, while one end of the RNA molecule at the opposite pole gradually shortens. × 72,000 (courtesy of Dr. G. Lebeurier).

type single-stranded molecules, syntheses of all of which continued throughout infection (Duesberg and Robinson, 1967; Pons, 1967, 1972). Rather than being the minus strand as frequently believed, the RNA of the virions may prove to be the plus strand, because in an *in vitro* system, virion RNA has been translated into proteins similar in antigenic properties and molecular weights to those of the virions (Siegert and Hofschneider, 1973; Siegert *et al.*, 1973).

The Rigidoviruses. The two most thoroughly investigated members of the rigidoviruses, or rodviruses, are pathogens of tobacco, namely tobacco mosaic (TMV) and tobacco rattle viruses (TRV) (Hirth, 1971). Although this pair shows many similarities, particularly in structural features in general, they also exhibit a number of distinctions. In both types, the capsid is an elongate, slender rod, composed of a series of repeating protein subunits (Figure 9.12A) arranged as a tightly wound helix around a hollow core (Franklin *et al.*, 1957).

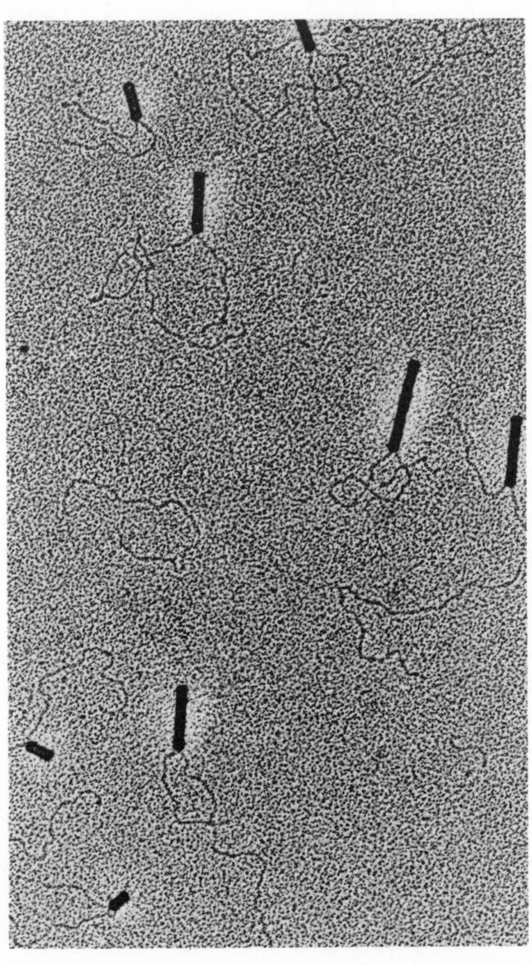

In TMV there are 2200, and in TRV 1830 subunits, each with molecular weights of 17,000 and 23,000–24,000, respectively (Finch, 1965; Offord, 1966). It has been clearly established that one end of the RNA molecule extends through the lumen of the rod, whereas the other is included between the subunits; in TRV nearly four nucleotides interact with each subunit. Since this RNA molecule also serves as the genome, it is clear that the interactions between it and each protein subunit cannot be specific in nature. In TRV the RNA of long particles contains 7100 nucleotides and has a molecular weight of 2.1×10^6, whereas in TMV, the molecular weight is 1.94×10^6 (Boedtker, 1959). While thus the RNA in each case is a regular helix due to its relations to the capsid coat protein, when in solution TMV RNA forms a random coil that is double-stranded, except for 30% which remains single-stranded.

Perhaps the most outstanding difference between these two viruses is that, in TRV, two classes of rods of distinct sizes are normally present (Hirth, 1971). The properties given above are for the long particles; in the short ones, the RNA molecule is only about one-half as long, having a molecular weight of 1.0×10^6. Each size class contains a different fraction of the genetic information. The long particles contain the information for RNA replication and are able to induce necrotic lesions if inoculated alone in host leaves. These, however, give rise to free viral RNA but no virions. When short particles, which lack infectivity by themselves, are added to long ones, then the lesions resulting from the inoculations contain virus particles of both size classes. Thus the short rods seem to carry that part of the genome involved with coat protein synthesis. Questions still remain unanswered as to how the small and large molecules of RNA are replicated.

Replication of the RNA of TMV involves a virus-specific polymerase, and RNA must be present for the assembly of virions, which is a polar process (Ohno *et al.*, 1971; Thouvenel *et al.*, 1971; Stussi *et al.*, 1972). While elongation of the capsid proceeds at one pole, electron micrographs indicate that both ends of the molecule project outward at the opposite pole (Figure 9.12B). Thus one of the projecting ends shortens as the capsid becomes longer (Lebeurier *et al.*, 1977). The 5′-end of the RNA molecule has been shown to bear a reversed terminus similar to those reported for certain other viruses, including the Rous sarcoma virus; however, in the present case, as in bromoviruses, the terminus was 7-mG(5′)ppp(5′)G (Keith and Fraenkel-Conrat, 1975). The initiation site for viral assembly has been shown to be located near this same end, and a fragment containing that site has been sequenced (Guilley *et al.*, 1974). Nearly 70% of the segment was in a long loop, preceded and followed by short single-stranded regions 15 nucleoside residues in length; about half of the latter and one-fourth of the loop regions were comprised of adenosine. A complex disk of proteins (Figure 9.13) has recently been found that is crucial in the recognition of the RNA and in the assembly of the rod (Champness *et al.*, 1976); this structure is formed prior to the development of the rod. The interconnections between subunits are especially noteworthy.

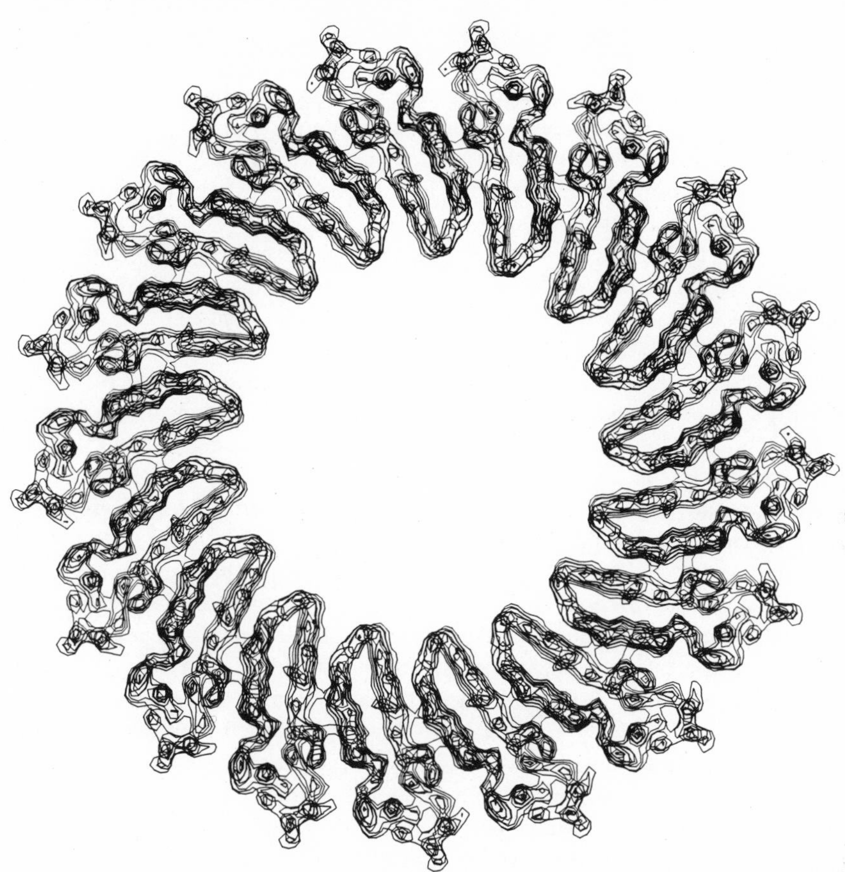

Figure 9.13. Section through a protein disk of TMV. Before the rodlike capsid of TMV is assembled, a double disk of protein is formed, each plate of which contains 12 subunits. The diagram illustrates the method of packing α-helices within and between subunits (courtesy of Dr. J. N. Champness and *Nature*).

The 3′-end of TMV RNA has been demonstrated to accept histidine when acted upon by an appropriate aminoacyl-tRNA ligase (Öberg and Philipson, 1972; Litvak *et al.*, 1973; Carriquiry and Litvak, 1974), and if cleaved by plant endonuclease the fragments accepted methionine and serine (Sela, 1972). A sequence from the 3′-end 71 nucleosides long has now been determined from TMV and one 74 residues long from GTAMV, a so-called strain of TMV (Guilley *et al.*, 1975; Lamy *et al.*, 1975). As pointed out in the preceding chapter, these represent ancestral types of tRNA (Chapter 8, Section 8.3; in Figure 8.6 it is especially noteworthy that modified bases are completely absent). A similar segment that accepts valine has been isolated from eggplant mosaic virus RNA and its sequence determined (Briand *et al.*, 1976).

9.3.4. Oncogenic RNA Viruses

RNA viruses capable of inducing either benign or malignant neoplasms or leukemias are of known widespread occurrence among birds and mammals, but probably also affect other vertebrates as well (Howatson, 1971). Despite their wide range of hosts and their medical importance, however, their genetic processes have been relatively poorly explored. One feature of replication which characterizes the known members of this group is the reverse transcriptase that synthesizes DNA from the RNA template carried by the viruses (Chapter 3, Section 3.4.4). In discussions of these viruses, they are frequently grouped according to their host's classification, a practice that also holds certain conveniences here. In addition, a rather small group, known as the arenaviruses, apparently needs to be placed close to the present types. Their kinship is reflected in the similar sedimentation (70 S) characteristics of the single-stranded RNA complex and in the common presence of ribosomes (Pedersen, 1971, 1973; Farber and Rawls, 1975). Among the more important members of the group are Pichinde and lymphocytic choriomeningitis viruses.

Avian Tumor Viruses. The various leukemias of chickens form a complex of neoplastic diseases involving blood cells, along with their precursors, and the reticular cells lining circulatory organs; these complexes are often associated with a diversity of sarcomas. The proper identification of particular avian leukemias is a difficult problem, because two or more types may be present simultaneously in a single bird, or a bird infected with one type may develop symptoms of a second. Probably the most thoroughly documented member of this group is the avian myeloblastosis virus, which induce the presence of large numbers of myeloblasts in the peripheral bloodstream (Gross, 1970). Another well-known virus closely allied to the foregoing is the Rous sarcoma virus, infections of which do not induce leukemia. In all cases, after penetration of the host cell has been achieved, by an unknown process, the virus is uncoated as its activities begin (Tooze, 1973). Progeny are produced only in host cells which proceed through mitosis, not just through an S phase (Humphries and Temin, 1972), for blocking nuclear division inhibits the production of progeny viruses (Temin, 1967). However, only one round of mitotic division is necessary to establish transformation (Rubin and Colby, 1968; Weiss, 1970).

The avian myeloblastosis virions (Figure 9.14A) consist of at least 17 different polypeptides, two of which are glycoproteins, probably associated with the envelope that surrounds the capsid (Bolognesi, 1974). Within the capsid is enclosed the RNA, various proteins, including the reverse transcriptase, and a small amount of DNA (Řiman and Beaudreau, 1970). Because the DNA has proven not to be related to the viral RNA, it is assumed to be of host cell origin (Levinson *et al.*, 1972). The RNA, which is largely single-stranded and complexed with protein, consists of a number of separate molecules, that fall into

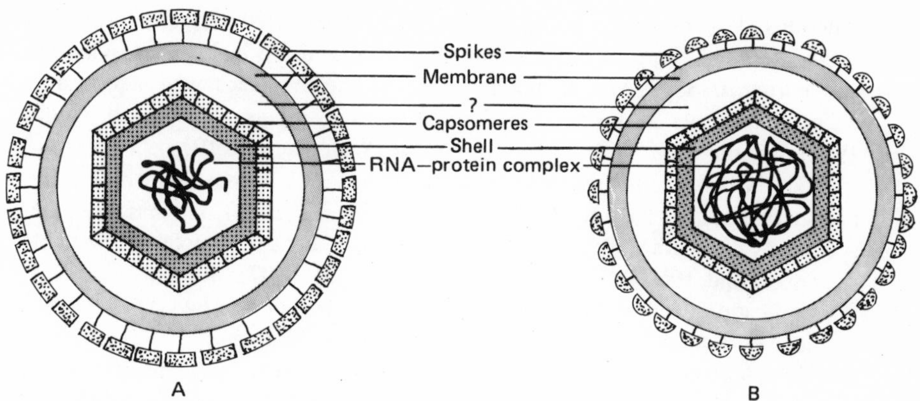

Figure 9.14. Two classes of RNA tumor viruses. (A) The avian type differs from (B) the mammalian primarily in the shape and size of the spikes and, especially, in the smaller size of the RNA–protein complex.

five size classes. The largest component, sedimenting between 60 S and 70 S, probably represents the genome (Bonar *et al.*, 1967); it in turn consists of three or four subunits (Duesberg, 1968a). In addition, the virions contain 27 S and 17 S (ribosomal), 7–9 S (of unknown function), and 4 S (transfer) RNAs. All types (including host derivatives) appear to contain long sequences (200 nucleotides in length) of polyadenylic acid (Green and Cartas, 1972; Lai and Duesberg, 1972; King and Wells, 1976), located at the 3′-terminal of the molecule (Phillips *et al.*, 1974; Drohan *et al.*, 1975; Wei and Moss, 1975).

The rRNA, which forms an average of two ribosomes per virion, has been suggested to be derived from the host (Bonar *et al.*, 1967; Bolognesi and Obara, 1970; Obara *et al.*, 1971; Říman *et al.*, 1972). In forming the ribosomes, the 27 S and 17 S rRNAs are processed within the virions from larger precursor molecules (Říman *et al.*, 1972), an observation strongly suggestive of viral rather than host origins for these two size classes of RNA. In this connection it is of interest to note that the tRNAs of the virions were similarly thought to be of host cell derivation (Bolognesi, 1974), but there is evidence to the contrary (Bonar *et al.*, 1967; Carnegie *et al.*, 1969; Trávníček, 1969; Elder and Smith, 1973). Moreover, purified virions contain aminoacyl-tRNA ligase activity and a specific tRNA methylase (Erikson and Erikson, 1972; Gantt *et al.*, 1972; Trávníček and Říman, 1973). Among the ligase activities which have been reported are those for arginine, tryptophan, cysteine, and lysine (Erikson and Erikson, 1972). It will also be recalled that the nucleoside sequence of a tryptophan tRNA employed by Rous sarcoma virus in initiating transcription has been determined (Tables 7.1–7.7); while this species of tRNA shows indications of being derived from host sources, it also is highly modified for virus functions (Waters *et al.*, 1975; Eiden *et al.*, 1976). Methionine tRNAs, both formylated and elongating types, have been found to be included in avian oncornavirus 70 S RNA. The initiator type of these species was at one time believed to serve as primer for the reverse transcriptase (Elder and Smith, 1974), but the $tRNA^{Trp}$ has proven to do so (Waters *et al.*, 1975).

The virus-specific reverse transcriptase of AMV (RNA-dependent DNA polymerase) is a zinc metalloenzyme (Auld *et al.*, 1974) and has been described as consisting of two polypeptides, α and β, having molecular weights of 65,000 and 105,000, respectively (Kacian *et al.*, 1971; Rho *et al.*, 1975). For replication of viral RNAs, in addition to the polymerase two other proteins are needed, RNase H, and a stimulatory protein (Mölling *et al.*, 1971), as well as an RNA template. Since the requirements for template lack specificity, the polymerase has been found useful for synthesizing DNA from a wide variety of RNA species (Spiegelman *et al.*, 1971; Gulati *et al.*, 1974; Seal and Loeb, 1976). The first product of polymerase activity was a hybrid RNA–DNA molecule (Verma *et al.*, 1971; Leis and Hurwitz, 1972); the hybrids then released single-stranded DNA molecules which were referred to as provirus (Temin, 1964). These were soon converted to a double-stranded helix, which ultimately must provide the template for RNA synthesis; however, little is known of the actual processes (Temin and Baltimore, 1972; Vigier, 1974). A similar set of events has been shown to occur also in Rous sarcoma virus (Leis *et al.*, 1975); in this form, transcription of viral DNA into RNA has been suggested to be mediated by host polymerase II (Rymo *et al.*, 1974). In addition, it has been reported that dAMP is covalently attached to the 3′-terminal adenosine of an RNA molecule in initial DNA-polymerase activity (Faras *et al.*, 1973; Keith *et al.*, 1974).

Attempts to elucidate the genetic mechanism of AMV have included investigations of interactions between the DNA polymerase and various synthetic polynucleotides, some of which have proven to be better templates for that enzyme than the endogenous virus genome (Spiegelman *et al.*, 1970). The polymerase has been shown to be very active in response to poly(A) and poly(C) as primed templates but inactive in the presence of poly(U) (Baltimore and Smoler, 1971a,b; Erickson *et al.*, 1973). Chemical modification of the sugar or base moieties were alike in drastically altering the template or inhibiting activities of a given polymer (Erickson and Grosch, 1974; Erickson, 1975). Transcription in AMV was shown to be accurate to the extent of about one error per 600 nucleotides synthesized (Battula and Loeb, 1975, 1976).

Mammalian Tumor Viruses. In structure and in a number of details, the mammalian tumor viruses of type C are similar to the avian viruses; some types appear of particular importance in possibly being responsible for leukemia in man (Wong-Staal *et al.*, 1976). The capsid is double-layered as there, and rounded projections cover the surface of the outer membrane; moreover, the

RNA–protein complex is segmented (Green, 1970). However, there are enough differences even in structure to indicate that, although doubtlessly derived from a common ancestral stock, the two groups have been separated for many eons. In the mammalian types, the surface projections are less prominent, lack spike-like bases, and are more loosely attached (Figure 9.14B). The capsid often has the capsomers more distinct (Nermut *et al.*, 1972), and within it the RNA–protein complex occupies a greater portion of the lumen. Moreover, this complex is not so compact, but, as in the viruses from birds, it is in the form of a whorl-like strand (Luftig and Kilham, 1971). Instead of the $tRNA^{Trp}$ of the bird viruses, proline tRNA appears to serve as primer for reverse transcription (Waters, 1975). The RNA portion of the complex approximates a total of around 10×10^6 (Hawatson, 1971; Tooze, 1973).

The macromolecules also display resemblances and dissimilarities. In the first place the polypeptides found in primate viral infections differ in showing fewer (three) major components plus several minor ones (Schäfer *et al.*, 1971, 1972; Gilden and Oroszlan, 1972; Moroni, 1972); however, those of murine and feline types had four major proteins as in those from avian sources (Roy-Burman *et al.*, 1974). In Friend and feline leukemia viruses, the molecular weights have been found to range from 10,000 to 15,000 for the three major polypeptides and 31,000 for one large, minor species (Bolognesi, 1974). The smallest of the major polypeptides (P1) is rich in arginine (Moroni, 1972) and appears to be related to P2 of avian types, for they both stain red with Coomassie blue, while the remainder of the four or five in both cases stain blue; the two also are similar in being associated with the RNA–protein complex. On the other hand, the second-smallest protein of mammals, P2, resembles the smallest in the birds (P1) in being associated with carbohydrate (in the present instance to an extent of 5%) and in being located on the surface of the virion (Witter *et al.*, 1973). These structural proteins have been found to be phosphoproteins, in the mammalian and avian species alike (Pal and Roy-Burman, 1975), being in part synthesized with the aid of a virion-associated protein kinase. In addition, two glycoproteins have been identified, one having a molecular weight between 60,000 and 80,000, and the other one between 40,000 and 50,000. These are believed to form the rounded knobs of the surface and to provide the basis for the hemagglutinin reaction (Duesberg *et al.*, 1970; Moroni, 1972; Schäfer *et al.*, 1972).

In the nucleic acids, a number of size classes have been recognized, paralleling those from avian hosts, but few as yet are sufficiently well known to compare in detail (Green, 1970). The 70 S RNA of Mason-Pfizer monkey virus, with a molecular weight of 8×10^6, now has been dissociated into three smaller class sizes, approximating 100,000, 35,000, and 25,000 (Schochetman and Schlom, 1975). As in the avian types, a 4 S size class has been recognized as including tRNAs (Jarrett *et al.*, 1971; Watson, 1971; Tronick *et al.*, 1972) and is discussed below. Similarly, ribosomes have been noticed within the virions (Sato *et al.*, 1971), which are sometimes claimed to be derived from host sources (Fidanian *et al.*, 1975).

Enzymes likewise are less thoroughly documented in mammalian viruses than in those of birds. As in those forms, a reverse transcriptase has been isolated, with a molecular weight of 70,000, which contrasts strongly with the much larger one from avian viruses (Ross *et al.*, 1972; Tronick *et al.*, 1972). Also as there, an RNase H has been recognized (Leis *et al.*, 1973; Wang and Duesberg, 1973). In addition, nucleases and the protein kinase mentioned above have been described (Hatanaka *et al.*, 1972). Several proteins of murine type C virus have recently been purified, including the envelope glycoproteins, a 30,000-molecular-weight major core protein, and a 15,000-molecular-weight internal protein (Strand and August, 1976). All were specified by the virus genome.

In rat ascites hepatoma cells, three different DNA-directed DNA polymerases have been claimed to be present (Tsuruo *et al.*, 1972a,b, 1975; Tsuruo and Ukita, 1974). Polymerase C from the cytoplasm had a molecular weight of around 142,000, whereas P-1 and P-2 from the nuclear-membrane-chromatin fraction had molecular weights of 117,000 and 44,000, respectively. Double-stranded DNAs were copied well, but synthetic polydeoxyribonucleotides were not. In contrast, single-stranded bacteriophage fd DNA served as a template for the larger polymerases when accompanied by RNA synthesis. Similar properties have also been described for DNA polymerase from feline leukemia and sarcoma viruses (Roy-Burman, 1971).

A more important feature of DNA synthesis in all oncogenic RNA viruses, regardless of source, has recently been pointed out (Sirover and Loeb, 1974). In that report it was demonstrated that these viruses showed an unusually high error rate during DNA synthesis, as signified also in an earlier study (Springgate *et al.*, 1973). This was determined on the basis of frequency with which noncomplementary base-paired nucleosides were incorporated by the DNA polymerase, a finding possibly relevant to its mode of action.

Little of a specific nature has been learned about the tRNAs of these viruses, except that the overall picture of these nucleic acids in virus-infected cells differs from that of normal tissue. For example, in a comparison of tRNAs from Morris hepatomas 7777 and 5123D with rat liver, the modified-base composition was found to be significantly distinct. The hepatoma 7777 tRNAs contained less pseudouridine, 5-methyluridine, and 3-methyl- and 5-methylcytidine, and more dihydrouridine and 1-methylguanosine than the liver (Randerath *et al.*, 1974). Furthermore, the Y base so characteristic of $tRNA^{Phe}$s of eukaryotes has proven to be absent (Grunberger *et al.*, 1975). New species of $tRNA^{His}$, $tRNA^{Tyr}$, and $tRNA^{Asn}$ had been reported earlier from Novikoff hepatoma (Baliga *et al.*, 1969); now these have been shown to differ functionally as well as in composition. Sharma and co-workers (1975) found that

Novikoff hepatoma tRNAs specifically influenced ovalbumin synthesis in a controlled system and that the major protein or proteins synthesized in their presence had higher electrophoretic mobility than ovalbumin. Thus it appears certain that the viral genomes strongly affected the structure of many tRNAs during protein synthesis.

The type B viruses of mammals diverge widely in protein constitution from those of type C. Five major polypeptides have been reported in appropriate infections ranging from 23,000 to 90,000 in molecular weight (Nowinski *et al.*, 1971; Sarkar *et al.*, 1971). The smallest of these (P1), as in the type C viruses, is associated with the RNA–protein complex, while one with a molecular weight of 52,000 is a glycoprotein of the outer membrane (Teramoto *et al.*, 1973).

9.4. DOUBLE-STRANDED RNA VIRUSES

The relatively few known members of a group of viruses distinguished by having double-stranded RNA as genomes nevertheless have a wide range of hosts. Reoviruses I, II, and III are common in man and other mammals, cytoplasmic polyhedrosus (CPV) and bluetongue viruses also infect mammals, and wound tumor, rice dwarf, and maize rough dwarf viruses attack various seedplants, to which they are transmitted by leafhoppers (Hull, 1970; Millward and Graham, 1971). More recently a number of species has been described from fungal hosts, including *Penicillium* (Szekely and Loviny, 1975), so the prospects are that this will eventually prove to be a large and important class.

Structural Features. Structurally the members appear to fall into two distinct subgroups. In reovirus and wound tumor virus, the structural proteins are arranged in the form of a double capsid, the nucleocapsid or inner component of which, often called the core, contains the double-stranded RNA (Millward and Graham, 1971). The outer capsid, which often exhibits a polyhedral form in electron micrographs, consists of around 92 capsomers (Hull, 1970). Among other members of the group, however, such as cytoplasmic polyhedrosis and bluetongue viruses, the outer capsid is lacking. In all types which have been studied, the RNA is not a single molecule but consists of a number of separate components that fall into three size classes.

Genetic Mechanism in Reovirus. In reovirus, the most thoroughly investigated member of this group, the total RNA present amounts to about 16×10^6 in molecular weight, while in wound tumor virus, it amounts to 15×10^6 (Bils and Hall, 1962; Dunnebacke and Kleinschmidt, 1967). In reovirus, the three size classes of RNA found in virions have molecular weights of 0.89, 1.55, and 2.66×10^6, respectively (Millward and Graham, 1970, 1971). It has been established that the virion molecule consists of a number of small units held together by noncovalent bonds and that all three size classes consist of three segments, except the largest which may contain four (Shatkin *et al.*, 1968).

During replication no DNA synthesis is required, so this group of viruses is not allied to the oncogenic RNA-containing type (Loh and Soergel, 1967). Two classes of progeny molecules are produced, double- and single-stranded. As the latter is found associated with polyribosomes, it is clear that it is mRNA (Prevec and Graham, 1966). Three size classes of these messengers occur, corresponding to the three size classes of parental RNA (Watanabe and Graham, 1967). As in vesicular stomatitis virus, the mRNAs must be methylated before they can be translated (Both *et al.*, 1975); this is accomplished by virion-associated RNA nucleotidyl-transferases (Furuichi *et al.*, 1975). The methylation is part of a blocked complex 5′-terminal structure (Miura *et al.*, 1974; Furuichi *et al.*, 1975, 1976). Production of both single- and double-stranded RNAs is dependent upon protein synthesis, but that requirement in the former is transient, whereas that of the latter is continuous. Thus replication of viral-type RNA is mediated by a viral-specified polymerase of short life span, and transcription is carried out by one with a long life. However, it is not known whether host polymerase serves initially or whether one or both polymerases are included in the virions.

According to analytical studies, the reovirus capside consists of seven different species of polypeptides, which fall into three size classes, corresponding to those of double-stranded RNA (Loh and Shatkin, 1968; Smith *et al.*, 1969). Thus only three of the ten total functions coded in the viral genome appear unaccounted for, providing each is monocistronic as appears to be the case with the structural proteins. In light of more recent developments, these conclusions appear somewhat simplistic, for, in addition to the two replicases, at least two other virus-induced enzymes have been discovered. Both reovirus and cytoplasmic polyhedrosis virus (CPV), as well as vaccinia virus, have been demonstrated to code for a phosphohydrolase that removes the 5′-terminal phosphate from viral mRNAs (Gold and Dales, 1968; Munyon *et al.*, 1968; Storer *et al.*, 1973). Initiation of transcription is coupled to the methylation of the mRNAs in CPV (Furuichi, 1974), but this feature is lacking in the fungal-virus RNAs, which contain a pyrimidine nucleotide at the 5′-end (Szekely and Loviny, 1975).

PROTEINACEOUS VIRUSES

During the last decade it has been discovered that a peculiar type of virus exists, of extremely small size (0.2×10^6 in total molecular weight), and with a diameter of only 14–15 nm (Cho, 1976). The most remarkable feature of the viral type is that it seems to lack nucleic acids of any sort. In short, its members are comprised entirely of protein (Alper *et al.*, 1966; Pattison and

Jones, 1967). Among these is the causative agent of scrapie disease, which can infect goats and rodents as well as sheep (Chandler, 1963; Zlotnik and Rennie, 1965). The absence of nucleic acids is indicated by the virus's failure to lose titre when treated with RNase and DNase, as well as with other appropriate reagents (Gibbons and Hunter, 1967). Furthermore, the scrapie virus is much more resistant to the action of ionizing and ultraviolet radiation than any other known form, facts also indicative of the absence of nucleic acids (Alper *et al.*, 1966, 1967; Gibbons and Hunter, 1967). Considerable speculation exists as to how this and related slow viruses replicate themselves, but little firm information is available to date (Gajudusek, 1967; Griffith, 1967). Most of the infectivity has been found to be associated with plasma membrane of host cells (Clarke and Millson, 1976).

SUMMARY AND CONCLUSIONS

Even though space limitations compelled the exclusion of a number of less important viral types from consideration in the foregoing summary, the amount of diversity described there becomes bewildering. Some groups contain DNA, others RNA, and still others no nucleic acid of any sort. Among these major types, certain groups have single-stranded genomes, while others have a duplex coding mechanism. In turn, a few RNA types can use their nucleic acid immediately for transcription, whereas others first must make a complementary replica. Sometimes replication of the RNA can proceed directly, but in many types DNA or some other replicating intermediate is required. The genomes may be continuous or fractured into a series of segments—the variations on the simple theme of replication seem nearly endless. Against this background of knowledge about these macromolecules and the complex paths by which they are synthesized, the concepts concerning viruses offer a strange contrast. Despite this extensive knowledge that has been gained by the most sophisticated technology available and through highly innovative insights, the viruses themselves are still viewed within a framework of dogma developed when almost nothing was known of their structure and functioning.

Undoubtedly the most widely accepted concept of these viruses is that they are escaped genes—small bits of host genome that have broken free, that now draw upon the cells in which they live to replicate their kind, often killing the host cells in the process (Luria and Darnell, 1967). A recently proposed variation on this theory is that the RNA tumor viruses might have arisen by RNA transcripts of particular segments of the host genome being incompletely processed (paraprocessing), so that long chains of poly(A) are left on the molecules (Gillespie and Gallo, 1975). As long as these viruses were known merely as ultrasmall particles, capable of passing through the finest filters, whose way of living and reproducing themselves were totally unexplored, such points of view would not be illogical, but in face of present knowledge they no longer can be considered acceptable.

All the viruses whose structure has been examined under the electron microscope, in the first place, show themselves too highly organized to be chance fragments. As has been seen, the capsids are often made of several distinct proteins with spines precisely oriented on a series of geometrically arranged plates and consistently comprised of a constant number of subunits. What lower metazoan, let alone a mammal, would have need in their genetic mechanism for genes that encode such proteins as these? Not only would each of the capsid proteins have had to be coded by part of the host genome, if viruses are escaped genes, but the polymerases, stimulating and assembling enzymes, methylases, and other virus-specific polypeptides would likewise, not to mention the occasional distinctive tRNA or rRNA that has been identified. In addition, bacterial and other prokaryotic genomes would need to be suspected of encoding for tails, plates, and other parts of the complex viral particles, if bacteriophages were once part of a normal genetic mechanism. Then, if one inquires as to what survival value such an elaborate set of genes would provide that it would have endured for countless millions or even billions of years within the prokaryotic host, no answer is forthcoming. Finally, when the escaped gene concept is viewed more closely, with its undertones of spontaneous production of active organic particles within cells, whether as escaping genes or cell fragments, a distinct parallelism to spontaneous generation concepts becomes apparent, ideas that prevailed prior to the time of Redi and Pasteur.

Fortunately, a small number of virologists already seem to accept a point of view that appears to harmonize with present knowledge to a larger degree. These are those who describe viruses as organisms, or who at least write about them in such terms (Fraser, 1967; Huang and Buchanan, 1974; Baltimore, 1976). True, the viruses cannot be viewed as a close-knit group of interrelated forms, nor in many respects do they behave like any other known taxon of living things. The explanation that appears to apply best to present knowledge of their morphology and metabolism is that they represent a long line of organisms of varied complexity. Being basically simple in comparison to bacteria and blue-green algae, they have become highly adapted for parasitic lives within their great diversity of hosts during the billions of years of their existence.

Take the DNA bacteriophages belonging to the T-even group, along with their relatives, as an example of an evolving complex (Figure 9.15). The series begins perhaps with phage S_d of *E. coli* and P22 of *Salmonella typhimurium,* in both of which the tails are scarcely as long as wide but nevertheless bear the end plate that is the hallmark of this sequence (Anderson, 1960; Tikhonenko, 1961; Yamamoto and Anderson, 1961). On a phage attacking *Chlorella pyrenoidosa,* a slightly longer tail is present (Tikhonenko and Zavarzina, 1966; Brown, 1972). Whether these early members of this lineage have a contractile sheath is not clear, but later ones share that characteristic as the tail increases in

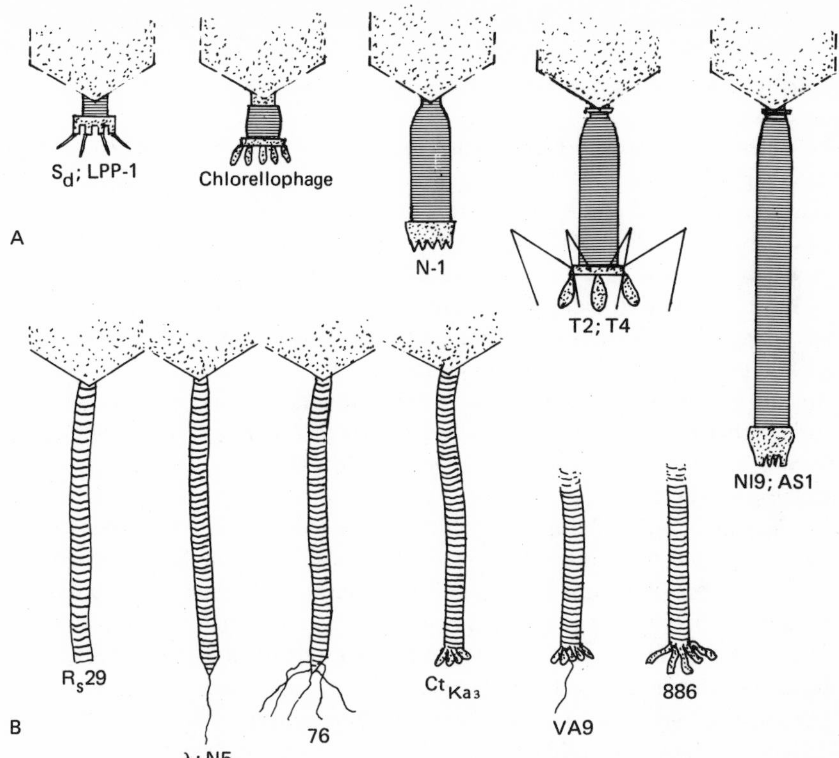

Figure 9.15. Sequential changes in virion structure. (A) A series showing successive stages in elongation of the tail. S_d, T2, and T4 are from *E. coli*. Chlorellophage is from *Chlorella*, and LPP-1, N1, and AS1 are from other algae, while N19 is from *Bacillus mycoides*. (B) The flexible-tailed group shows extensive variation in tailplate structure. R_s29 attacks *Corynebacterium diptheriae;* λ, *E. coli;* N5, *Bacillus mycoides;* and VA9, *B. anthracis;* 76, *Caulobacter vibrioides;* Ct_{Ka3}, *Caryophanon tenue;* and 886, *Aerobacter aerogenes*.

length. In T2 it is moderate in length; in T4, as in RZh of *Streptococcus,* it is markedly longer; in N17 of *B. mycoides,* it is quite long; and in No. 1 of the same host, it is even more elongate. All these forms share a number of structural traits, as well as size and nature of their genome and metabolic characteristics; for instance, the base sequence of the $tRNA_3^{Gly}$ of T2 is identical to that of T4 and T6 and its $tRNA^{Pro}$ is the same as that of T4, but the former, at least, differs from the same species in the host. Although the sequence of organisms cited as an example is admittedly simplistic, with focus largely being on only a single trait, it does clearly indicate an evolving condition within an obviously interrelated cluster of viruses, and *only organisms possess the capability for undergoing evolutionary changes of this type*. Similar evolutionary changes are also evident among those viruses that have noncontractile tails, in the serially increasing complexity of both RNA and DNA genomes, in total capsid size and structure, and in any suitable characteristic that may be examined (Figure 9.15B).

Hence, if viruses are organisms, as seems to be self-evident, and if they are simpler than any of the other known organisms, which is likewise obvious, this then implies that they arose prior to the cellular prokaryotes. And therein lies a problem, a difficulty which to this time has appeared insurmountable. For how can viruses have arisen and evolved before the advent of the cellular organisms that they need today in order to reproduce? But this seeming obstacle to the acceptance of viruses as being primordial organisms no longer needs to exist, for, if the cells themselves can be believed to have arisen from the organic matter suspended in primitive seas, the conditions certainly would have been sufficient to provide for these simpler forebears. To make this possibility clearer, the ensuing chapter suggests some of the steps that may have occurred in these processes.

Before leaving this argument for those evolutionary events, one consequence of accepting viruses as primitive organisms must be fully realized. Up to this point in biological progress, all recognized living things have proven to consist of one or more cells and have therefore been in harmony with the cell theory. When viruses are considered to be organisms, that is, living things, this formerly all-embracing concept no longer is universally applicable, and hence it must lose its status as one of the two unifying concepts of biology. This same conclusion, however, although never before drawn, was inherent in the concepts that viewed the ciliated protozoans as acellular (Hyman, 1940). Its loss is not a grave one, for while it was of immeasurable service in the earlier days of growth in biological knowledge, the cell theory no longer provides great insight into research problems. Even in this light, it still remains true for the cellular level of organic evolution, becoming meaningless only for the precellular stages.

10

The Origin of Early Life

The review of the genetic apparatus at both the precellular and cellular levels of life provides the information necessary to suggest what steps actually took place during the development of organisms beyond the primitive broth stage. The existence of that stage can scarcely be doubted, because of the biological requirements for it and the demonstration by biochemists of numerous paths for creating its organic ingredients. Although it may not be possible to establish firmly the exact species, or even types, of amino acids that had been present in the primeval soup, it is obvious that a sufficiency of those biochemicals had been created to permit their polymerization into early polyamino acids. The formation of these polyamines is also beyond question, for the presence of some type of proteinoid is an absolute requirement for life, as was revealed by the analysis of the gentic apparatus in Chapter V. In addition, nucleosides and nucleotides of various sorts probably had existed in the primeval seas, but they do not appear to have been essential during the earliest period of life's development.

10.1. A PRELIMINARY DEFINITION OF LIFE

The above statements immediately raise a question as to what else might be required for living things to have arisen, to answer which a minimum definition of the term life must first be sought. In attempting this, it becomes clear at once that there are different levels of life. For example, when a frog is decapitated, the animal as an entity is obviously dead, yet its heart continues to beat, sometimes for hours, before the cardiac muscle cells themselves die and become incapable of contraction. Hence "life" in the intact multicellular organism must be defined in one way, in the cells in quite another. However, only the definition of the term at the cellular level is of fundamental significance, for multicellular organismic levels cannot exist in the absence of cellular life. Even in cells, the word might logically be defined in various ways, depending on the complexity of the organism in question. Thus for present purposes the definition arrived at must be concerned with a *universal* requirement of living things; in other words, a statement of life at the minimal level, of absolutely catholic occurrence, must be sought.

As a basis for such a minimal definition, the ability of a cell to undergo division, or even to replicate its DNA, is inappropriate, for many long-lived cells of mammals and other metazoans cease to carry out those processes often prior to birth of the animal. Nor does the ability to break down carbohydrates by glycolysis or cell respiration apply universally, for certain bacteria, including *Beggiatoa* and *Clostridium*, do not have the necessary enzymes (Burton *et al.*, 1966). Because enzymes provide the basis for all cellular activities, including the replication and transcription of DNA, it appears essential that an appropriate definition should include them, yet the mere presence of intact and functional enzymes does not suffice, as these substances can often function after removal from the cell and in cells that have been fixed in gluteraldehyde or other suitable reagent. Since the enzymes and structural proteins in active living cells are undergoing continual cyclic destruction (Schoenheimer, 1942), a universal activity of organic things must be the ability to synthesize normal proteins in sufficient quantity at least to replace those that are catabolized by ordinary vital processes. Most cells and cellular organisms actually synthesize more than this requirement, so that growth results as this and the other ingredients of protoplasm accumulate.

Thus the minimal definition of life proposed here is as follows: *Life, basically, is the capability of synthesizing proteins in at least sufficient quantity to replace those that are catabolized by normal processes.* In this definition, the words ‘‘capability of synthesizing’’ are employed, rather than just ‘‘the synthesis of,’’ in order to embrace spores, seeds, cysts, and comparable bodies that are alive but inert for a time. Even when inactive, such bodies possess the capacity for producing proteins under favorable environmental conditions; cells, as well as seeds and spores, lack life (i.e., are dead) only when they no longer possess that essential ability under any natural set of conditions.

Although the foregoing definition may eventually prove unsatisfactory to some degree, any replacement or extension of it must surely include in its terms the ability to replace proteins. This observation thus reemphasizes the fundamental nature of proteins disclosed by the review of the genetic apparatus. That point receives further stress by the impossibility of finding a reasonable definition of life based solely on DNA or on the nucleic acids collectively, largely as a consequence of their relative inertness, pointed out in earlier chapters.

10.2. THE DISTINCTIVE CHARACTERISTICS OF VIRUSES

In utilizing this definition to investigate the problem of the origin of life, the viruses are here explored as models. To date, viruses have not received favorable detailed attention in this connection, largely because concepts regarding them have not kept pace with the increasing knowledge of their biochemistry and ultrastructure. For example, as recently as 1971, Lwoff and Tournier listed the following ten traits they considered to be distinctive of the viruses:

1. A single nucleic acid is present, that is, either RNA or DNA, but not both.
2, 3. They lack the ability to grow and to undergo division.
4. Reproduction is from the genetic material (nucleic acids) only.
5. The enzymes for energy metabolism are absent.
6, 7. Genetic information coding for enzymes of the energy cycle is wanting, as is that for the synthesis of ribosomal proteins.
8, 9. Ribosomes and genetic information for the synthesis of ribosomal RNA are lacking.
10. Genetic information for the synthesis of transfer RNA also is absent.

Many of these supposed distinctions have now been shown to be erroneous; in fact, a number were known to be false several years prior to the foregoing report. For example, ribosomes had been demonstrated present in avian myeloblastosis viruses in 1967 (Bonar *et al.*) and in Bittner virus in 1970 (Gay *et al.*). Thus so-called distinctions Nos. 7, 8, and 9 of the above list must be deleted. In 1972, McClain *et al.* reported that the genome of bacteriophage T4 coded for eight species of transfer RNAs, the sequences of four of which have now been determined, as cited elsewhere. As will be recalled, transfer RNAs have been found also in tobacco mosaic and a number of other viruses. Consequently, No. 10 must be removed from the foregoing list. Furthermore, the presence of DNA has been detected in RNA viruses and RNA is present in the DNA coliphage T5 (Rosenkranz, 1973), removing number 1 as a distinctive trait. Since the absence of enzymes of the Krebs cycle is also characteristic of such cellular forms as chlamydiae and colorless sulfur bacteria (Burton *et al.*, 1966; Weiss and Kiesow, 1966), Nos. 5 and 6 are not unique to the viruses, and therefore need to be removed. Only their lack of growth (No. 2) and inability to undergo cell division (No. 3) thus remain as truly characteristic features of viruses in general; reproduction solely from genetic material (No. 4) is valid in addition for certain types during their initial period of activity.

That duet or trio of traits is indeed distinctive and accordingly must be considered critically. Obviously the absence of growth and cell division stems from the lack of the cell that characterizes higher organisms; thus, in viruses as a whole, the capsid does not correspond to a cell. Rather the capsid is functionally closer to a spore envelope, for it serves to provide a means of dispersal for the genome and to protect the latter while it is in an unfavorable environment. Hence, virions represent the inert dissemination stage in the life cycle of viruses, thereby corresponding to the spores of bacteria, ferns, and fungi to that extent. Like those spores, the virions are alive as long as they retain the *capability* for duplicating their proteins when introduced into a favorable environment.

The real distinctive characteristic of viruses, then, lies in their method of reproduction, but not so much in its being from the "hereditary material alone" as stated in the foregoing list (No. 4) as in other ways. The actual basic distinctions are that, in these organisms, the genome undergoes transcription and replication in a naked condition, devoid of cellular coverings, with the proteins accordingly being transcribed in direct contact with a foreign environment. In context of life's origins, however, might these two traits not be those that should be expected in the primitive protobionts? The two alternatives in evidence for protobiontic structure appear to be that (1) the cell or its counterpart arose without having a genetical means of perpetuating it, and (2) a simple genetic mechanism arose in the seas, capable of reproducing its own kind, but was at first devoid of a cell-like covering. In short, either a cell or a genetic mechanism had to be the first to arise. Since a cell lacking a method of perpetuating itself obviously would quickly pass out of existence, the second option alone appears feasible. Thus during the most primitive stages of life, protein synthesis and genome replication had to occur directly in the seas until a cellular covering eventually could be acquired by evolutionary processes.

From this standpoint, then, present-day viruses appear to possess replicatory and synthetic processes quite appropriate for primordial organisms. Nevertheless, their status as such can be tested further, for if they are really primitive as supposed here, then they as a group should display at least a reasonably ordered array of changes from very simple to highly complex types. On the other hand, if they are actually escaped genes or other bits of cellular matter that have arisen independently throughout the organic world, a nearly complete lack of order should be evident. As a result it would be impossible to align the various types into a logical phylogenetic system. Although the attempt made below to arrange the viruses into a reasonable phylogeny is clearly incomplete and probably not free of errors, the numerous major groups do appear amenable to a methodical sequential arrangement, thereby suggesting that the primordial-organism alternative is more probably correct.

10.3. POSSIBLE STEPS IN THE ORIGINS OF EARLY LIFE

10.3.1. The Earliest Events

With these several critical points established reasonably firmly, a proposal can be made as to the nature of the earliest form of life, along with an actual living model to represent it. This suggestion, or any of a similar nature, needs to embrace four fundamentals that have now been demonstrated, namely (1) the essentiality of polyamino acids (proteinoids), (2) the immediate need for a genetic system, (3) a definition of life at the minimal level, and (4) a virus or viruses or other suitably primitive organism to provide corroborative evidence from a biological source. Perhaps the stress repeatedly placed upon these points and the summary of virus molecular biology have caused the author's conclusion to be anticipated: The first protobiont consisted of a primitive polyamino acid system which was capable of replicating itself. Whether the simple system consisted of a single polyamino acid chain or two complementary molecules is not clear at present, but primitively the replication of the single or dual components involved the superpositioning of individual amino acids upon the parent proteinoid chain and their subsequent polymerization. Actually the dual interacting system appears to hold more credibility than the self-replicating single molecule, because the creation of a second protein by an enzyme in living organisms has been documented (Chapter 2, Section 2.2.1.), whereas self-replication has not. However, this particular is not critical and may well be held for later investigation, for either system fulfills the essentials enumerated above. Each is entirely proteinaceous, self- or reciprocal-replication provides a genetic mechanism, and, as both thus provide the capability for producing proteins, either one would be living. Thus the paradox (Chapter 3, Section 3.1) first pointed out by Woese (1970) disappears, for life and the basic genetic mechanism are seen to be identical terms, each being the capacity to synthesize proteins.

A Model Organism. A living example of this early stage in life's history seems to be the scrapie disease agent, which has been reported to consist entirely of protein. Doubtlessly, self- or mutual-replication of protein molecules would be an inefficient process, so, if the scrapie virus proves to be entirely proteinaceous as claimed (p. 402), there is valid reason for the low rate of multiplication that has earned it and related forms the designation of "slow viruses" (Sigurdsson, 1954).

For those who wish to explore the nature of early life experimentally, the scrapie agent and other members of its group should prove promising subjects. Although there is no reason to suspect that these organisms would in today's host species use any but the orthodox α-L-amino acids, it might prove possible for them to employ other types *in vitro*. Two related problems that call for investigation are, first, the mechanism whereby their proteins are replicated, and, second, whether or not tRNAs are utilized. If tRNAs do participate in these processes, it needs to be ascertained whether they are actually requisite for protein synthesis in these simple viruses. The organisms need also to have the amino acid sequences of their proteins determined, for they may prove to be descendants of a stage that had not yet evolved the full complement of amino acids. This would be especially likely if certain major types of tRNAs are found not to be utilized during the biosynthesis of polypeptides.

The Early Sequence of Events. Based on the evidence presented by the nature of the amino acids, tRNAs, and simplest viruses, the sequences of events during the earliest portion of life's history apparently is as follows:

1. The synthesis of amino acids and their polymerization into polyamino acids was an essential prelude to the advent of life. It is entirely possible that the first polymers were comprised of various types of amino acids (that is, they included some unusual forms which are not involved in proteinogenesis today) and that patterns of bonding other than peptidization were utilized to some extent. This issue probably can never be resolved, because the amino acids most readily available today for viral use obviously are the standard type.
2. Over the eons of time, as the shallow oceans then prevalent (Rubey, 1955; Dillon, 1974) gradually became enriched with organic molecules and polymers, two polyamino acid chains accidentally came into contact which were mutually compatible in such a manner that each could replicate the other (or, alternatively, a single self-replicating peptidoid molecule arose). As the amino acids for the respective daughter molecules were necessarily derived directly from the seas, replicaton was probably slow indeed. Furthermore, it is not unlikely that inaccuracies of transcription were frequent, so that many of the products were inviable, that is, unable to replicate other polyamino acid chains.
3. This first accidental combination of two short interacting polyamino acid chains thus was the first living organism, according to the minimal definition for life provided above (Figure 10.1).
4. Even though replication was slow, so that perhaps as long as two hours was required for each complete replication, and even though as few as 50% of the replicates were viable, by the end of one year, 10^{200} replicates theoretically could have been produced. Obviously the supply of the necessary amino acids would quickly have become a limiting factor, for the original supply in the seas would gradually have been depleted.
5. Thus this first minute biont would have quickly filled the seas to the density permitted by the availability of newly formed and recycled amino acids. Like extant organisms of today, this first species would have been subject to mutations, genetic drift, and natural selection by environmental factors, and would have begun the processes of evolution.
6. Later, as evolution proceeded, interaction eventually occurred between one of the proteinaceous biont species and the carbon dioxide, ammonia, light, and other factors of the environment. Hence, as suggested in the analysis of amino-acid–codon interaction (Chapter 6), it became capable of producing the amino acid glycine. Through sub-

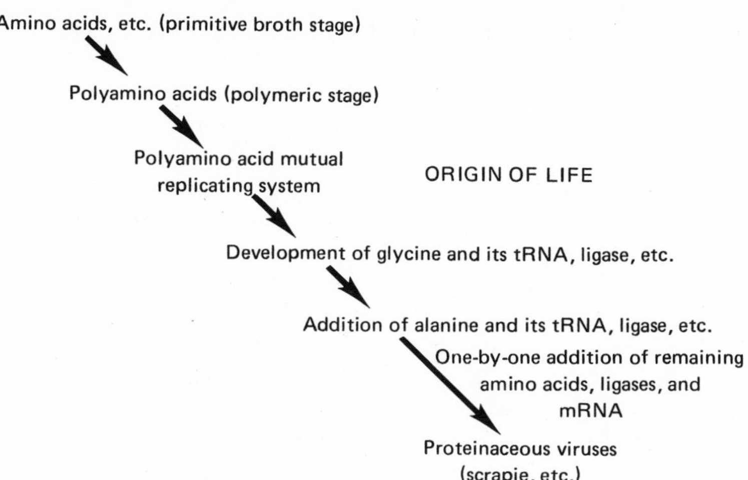

Figure 10.1. The proposed earliest series of stages in life's origin.

sequent interaction between the polynucleotides of the seas and a simple protein (the first tRNA ligase), this amino acid became coupled with the prototype of a transfer RNA. Unboubtedly this combination greatly accelerated the rate of incorporation of glycine and accordingly had high survival value.

7. Over geologic time, subsequent evolutionary progress led in many descendant species to further development of the amino acid complement of organisms and their transfer RNAs, as suggested in Chapters 6 and 8. With each addition to the amino acid catalog, greater accuracy of replication was acquired, and further decreasing dependency on the environment for amino acids occurred. However, in view of later developments,* it is not unlikely that the amino acids of the seas continued to be used also, as in extant viruses. But those available in the environment would have become increasingly like those of modern organisms, for, even at primitive levels, biosynthetic pathways of organic matter are far more efficient than the nonbiological random processes of the atmosphere. Thus α-amino acids, with levorotatory configuration, would ultimately have become the prevailing types in the environment.
8. As the transfer RNAs and amino acids increased in diversity, messenger RNA developed too. This consideration is an essential codevelopmental feature with the other two substances, for the elaboration of the anticodons built into transfer RNAs would be meaningless in the

* The evidence for this is suggested by the evolutionary development of the citric acid cycle, a subject that is beyond the scope of the present study (Dillon, unpublished).

absence of those polymers of codons which constitute the messenger RNAs.

9. Thus messenger RNA developed in organisms as the amino acids and other tRNAs became perfected, not necessarily on a precisely concurrent schedule as those products, but closely following them nevertheless. Each additional codon provided that much greater survival value in the form of increased accuracy of protein replication. Hence, fewer inaccuracies ensued, accompanied accordingly by decreased lethality in the replicates.
10. As each new species of amino acid was added to those already biosynthesized, it was necessary not only to evolve a new variety of tRNA, but also a novel tRNA ligase in order to recognize and bind those two other new substances. Therefore it is not unlikely that when a standard-type amino acid was first biosynthesized, its employment during replication of proteins for a while involved superpositioning it upon the parent protein or adding it to the terminal residue, as in the soluble system of modern cells. Then, after its tRNA and the necessary ligase had evolved to carry the amino acid into position, the existing early messenger could then have been readily extended or otherwise altered to include this later arrival through interaction of its nucleotides with those of the anticodon in the corresponding tRNA. Or perhaps the suggestion made by Yčas (1974) that earlier enzymes may have been less specific and catalyzed whole classes of reactions, rather than individual ones, is applicable here. Thus one ligase could have served for all the tRNAs.
11. The ligases, other enzymes, and possibly some or all of the tRNAs originally were contained *in the seas*, despite their having been synthesized by the primitive bionts, for no cells were as yet in existence. Thus the cells employed as hosts by modern viruses are mere substitutes for the primitive seas and similarly provide amino acids, nucleotides, and enzymes for use by naked viral molecules. An appreciation of this general correspondence between primitive seas and modern cells is an essential prerequisite to the comprehension of how bare replicating molecules gradually became replicating cells.
12. However, the existence of a messenger molecule would add very little of value to the evolving organisms if it could not itself be replicated and passed to the next generation. The development of a suitable mechanism for this purpose marks the beginnings of the level of evolutionary progress that follows.

10.3.2. A Second Major Developmental Level of Early Life

The establishment of replicable messengers introduces what appears to be a second series of major events in the evolution of life, a level represented by the RNA viruses. Fortunately, unlike the first stages which are exemplified by a paucity of known forms, a great number of viruses belonging to the present level has been described. Yet their number is at the same time unfortunate in a sense, for with their frequency comes diversity, a wide variety, which, like that of other living things, brings confusion to attempts to systematize them. For with diversification there can be no avoiding the production of side branches which go nowhere in the overall pattern of life's phylogenetic history but which can mislead those who would trace the pattern in detail. Thus many other interpretations of the data may prove to be at least as valid as that sketched below (Figure 10.2).

1. At first the RNA molecule serves directly for translation into proteins. Thus the relatively short genome is a plus strand, and, in this earliest period of its evolution, it is a single continuous molecule; no double-stranded replicative form is known to be produced during replication. On the basis of genome length and degree of overall complexity, several lesser levels of phylogenetic advancement appear to exist, as follows:

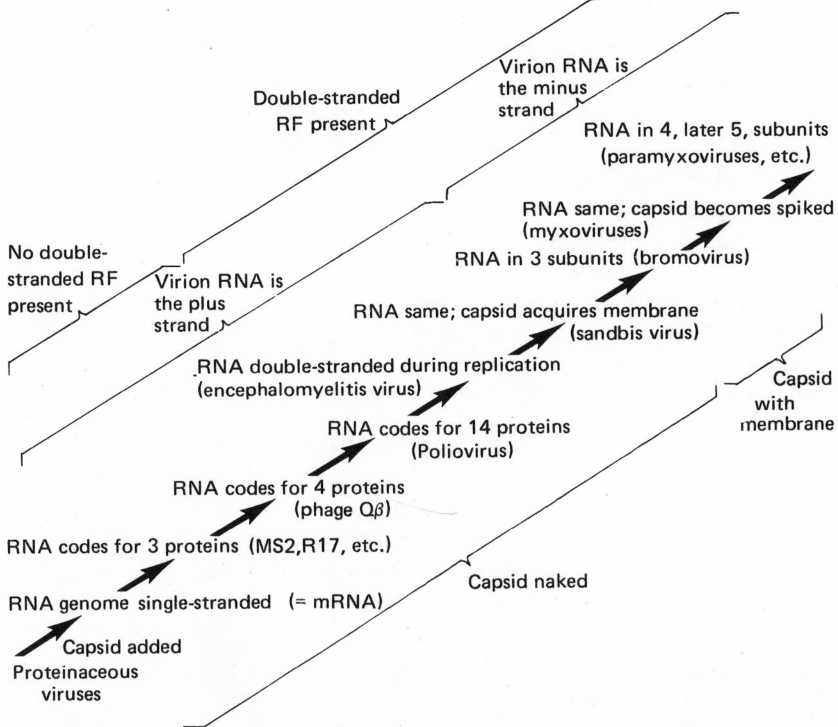

Figure 10.2. Stages in the evolution of simpler RNA viruses.

a. The RNA molecule has a molecular weight of 1.1×10^6 and contains codes for three genes, as in bacteriophages MS2, R17, and f2. These genes code for coat and maturation proteins and an RNA polymerase.
b. The RNA molecule, having a molecular weight of 1.3×10^6, codes for four genes; the products of these are as above with the addition of protein IIb (or A_1), of unknown function. This sublevel is represented by phage Qβ.
c. The RNA molecule has increased to $1.9–2.1 \times 10^6$ in weight. In the two representative forms (the two rigidoviruses), the RNA molecule accepts amino acids, that is, it contains tRNA equivalents. Thus it would appear probable that the RNA of the viruses at the above earlier levels also contained tRNA equivalents primitively but have lost them secondarily.
d. The RNA, of moelcular weight of 2.6×10^6, codes for four coat proteins and about ten nonstructural proteins (and tRNAs?). Polio- and mengoviruses, and possibly some others of the picornaviruses, represent this level.

2. As in #1 above, but a double-stranded replicative form appears during RNA replication, as exemplified by the encephalomyocarditis virus. The molecular weight of this single-chained molecule is approximately 3×10^6.
3. The RNA is still the plus single-stranded molecule, with a weight of around 3×10^6. Here, as in all RNA viruses that follow (with exception of the double-stranded types), a lipoprotein membrane surrounds the capsid. Represented by such arboviruses as Sandbis virus.
4. In all forms beyond this point, the RNA of the virions is the minus strand and consists of several segments, not a single continuous molecule. Precise information regarding the number of segments is not uniformly available, nor are total molecular weights of the RNA. Hence, the sequence proposed for these is even more tentative than that for the others.

   a. The bromoviruses have the RNA in three subunits totaling approximately 4×10^6 in molecular weight. Some tRNAs are known to be coded by the viral genome. The capsid is spherical, consists of 180 subunits, and apparently has the surface unarmed.
   b. The rhabdoviruses have slightly larger RNA molecules (molecular weight 4.2×10^6), but are three segmented as above. The capsid bears spikes on the envelope.
   c. The RNA of myxoviruses ranges in size from 3 to 5×10^6 in molecular weight and is broken into four or five segments. The envelope of the capsid is spiked.

d. The paramyxoviruses have still larger RNA molecules (on the order of 6.5×10^6 in Newcastle disease virus); however, the number of segments is unknown but probably will prove to be four or more. The capsid bears spikes on the envelope.

10.3.3. A Third Phylogenetic Level

The third major level of phylogenetic advancement found in early life is represented by a rather small number of viruses in which the RNA molecule of the virions is double-stranded and multisegmented. The segmented condition suggests that this line continued the preceding sequence; however, it is not clear in the few known types whether a lipoprotein membrane covers the capsid. When its presence or absence has been ascertained, and other macromolecular information becomes available, it may be possible to state the point of origin more precisely. It is not too unlikely that the double-stranded genome developed from the double-stranded replicatory form, but its insertion into the capsid would more than likely involve the development of a suitable enzyme, or at least a mutation in an already existing protein. However it may have evolved, the level represents a transitional stage that leads to a later advancement of unusual importance. Present state of knowledge indicates two substages (Figure 10.3A,B):

1. The capsid is simple, consisting of the usual single structure, as indicated by cytoplasmic polyhedrosis and bluetongue viruses; the genome is ten-segmented.

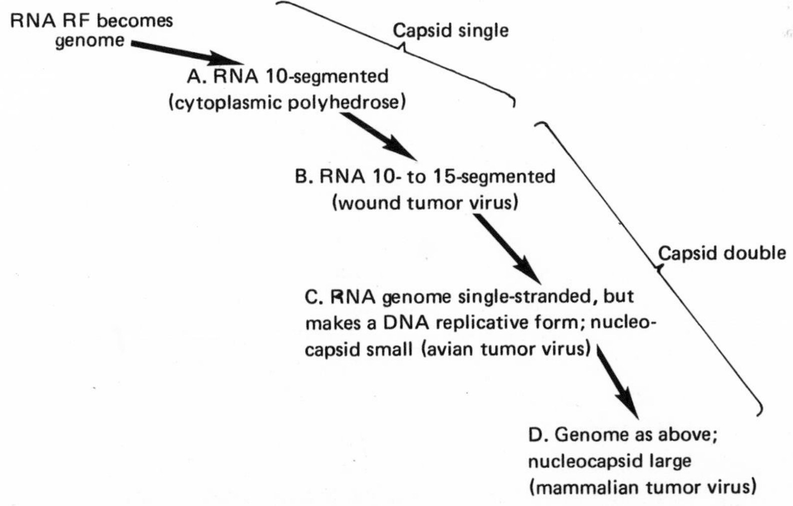

Figure 10.3. Preliminary phylogeny of advanced RNA viruses.

2. The ancestral capsid becomes enclosed by a second one of 92 capsomeres. The RNA has a total molecular weight of 16×10^6 and is ten-segmented in reoviruses, but 12–15 segments have been reported in wound tumor viruses.

10.3.4. A Fourth Level of Phylogenetic Advancement

In contrast to the third level of evolutionary development just described, the fourth has a rather abundant representation, for its membership includes many important RNA tumor viruses of vertebrates. The presence of a double capsid in these forms is strongly suggestive of relationships with the double-stranded RNA types. In the present case, the outer covering is sometimes described as a unit membrane, rather than as consisting of capsomeres as in the reoviruses. Consequently, the two may or may not be homologous conditions.

If actually related on the basis of capsid structure, the two lines soon diverged, for in the present lineage the virions contain only a single-stranded molecule. The major advancement made along this branch is the development of RNA-dependent DNA polymerase, so that DNA–RNA hybrid molecules serve as the replicatory intermediate form. The RNA is single-stranded, complexed with proteins, and consists of a number of separate size classes. Among the five classes that have been segregated are included ribosomal and transfer RNA species. The largest size class (60–70 S) probably represents the genome and also consists of three or more subunits. As the viruses of invertebrates and lower vertebrates remain rather poorly explored, only two sublevels or branches of evolution are known for this line, one in birds, the second in mammals (Figure 10.3C,D).

1. The avian line is marked by the greater length of the spikes on the capsid surface, but the major distinction is in the nucleocapsid's not completely filling the lumen of the capsid.
2. The capsid of the mammalian line is covered by low, rounded knobs, and the nucleocapsid fills the lumen completely.

10.3.5. A Single-Stranded-DNA Stage of Phylogeny

The DNA level of viral phylogeny seems to have had dual origins, if present viral structure truly reflects its past evolutionary history. In other words, two separate and distinct lines of ascent developed the same molecular processes independently. In one case, the expected single-stranded DNA forerunner is missing; in the second, the seemingly necessary transitional stage transcribing DNA from an RNA template as yet has not been found. Thus in neither case is the account fully documented by representative types; consequently, the avian and mammalian tremor viruses just described must serve as a model of the possible forebears of the second line, as well as of the present branch.

The first line is represented by single-stranded DNA viruses, all known representatives of which are small bacteriophages; these have the DNA molecule circular and are of a molecular weight between 1.7 and 2×10^6. In all the described forms, the virions are naked, lipo- and glycoproteins alike being absent. If capsid size and structure are valid considerations, this line then must be viewed as having arisen from a level represented by such early single-stranded RNA forms as phage Qβ (Chapter 9, Section 9.3.1) and the rigidoviruses (Chapter 9, Section 9.3.3). Two lesser levels are known, as follows (Figure 10.4A–C):

1. The capsid is "spherical" (icosohedral), ornamented with spikes, and characterizes a group of small bacteriophages, such as ϕX174, S12, and 1ϕ1. The DNA molecule has a weight of 1.7×10^6 and contains a single unit of 5-methylcytosine. Replication involves a double-stranded RF molecule and a rolling circle pattern.
2. The capsid is an elongate rod, naked, and unspiked, as in bacteriophage M13. On the basis of capsid structure, it is tempting to suggest a direct relationship between these phages and such rigidoviruses as tobacco mosaic virus, for these two types are the only ones known to have an elongate cylindrical plan of structure for the capsid. Unfortunately the proteins of the two do not seem to have been compared serologically. In the present group, the DNA molecule has a weight of 2×10^6.

In these organisms no evidence indicates that the DNA molecule arose from an RNA ancestral type, but since RNA messengers are used in the synthesis of its proteins, it seems reasonable to suppose that such a predecessor had been in the line of ascent.

10.3.6. Double-Stranded-DNA Levels of Evolution

As suggested above, two separate lines appear to have developed a double-stranded DNA genome independently. Although as knowledge of the viruses and their molecular biology increases, the duality of origin may be perceived to be more apparent than real, the data now available do not warrant other interpretations. Hence, the evidence is followed here, and independent origins of a like fundamental molecule are proposed, despite that practice's being against the author's principles. But a second principle is held more inviolable: in science, facts should be followed regardless of personal preferences and preexisting prejudices or dogmas.

1. The first of the two major lines apparently continues that of the single-stranded DNA viruses, for the capsid is naked and the DNA molecule small at the outset. In turn two separate branches seem to have developed from it early in the course of evolution, for the simplest represen-

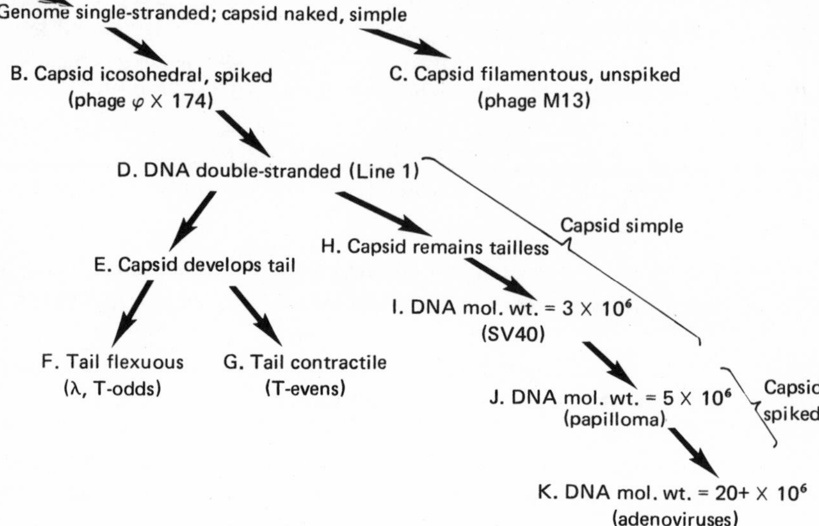

Figure 10.4. Some steps in the evolution of DNA viruses.

tative of each has DNA molecules close in molecular weight to those of the single-stranded type.

a. One branch developed tails on the capsid (Figure 10.4E–G), but this structure may have arisen only after cellular organisms had come into existence; possibly only the anlage of a tail had been present previously. Two types of tails are known. (Actually these structures are attachment and penetrating organelles, not tails in the usual biological sense.)

  (1) The group with the tail flexuous begins with such smaller species as λ, which has a DNA molecule with a weight of 3.8×10^6 (Figure 9.15). In T3 and T7 this is much larger (25×10^6) and in T5 it attains a weight of 80×10^6 (Fig. 10.5).
  (2) A group having the tail contractile developed later, perhaps from a short, flexuous-tailed forebear. A suggested sequence of events in the elongation of the tail is shown in Figure 9.15. Some of the T-even members of the branch have a DNA molecule of 130×10^6 molecular weight.

b. The second branch contains untailed spherical types (Fig. 10.4H–K), in the smallest of which (SV40 and polyoma) the DNA has a molecular weight of 3.6×10^6. Papilloma viruses have somewhat larger DNA molecules as reflected in a molecular weight of 5×10^6. At the end of the line stand such relatively large forms as the adenoviruses, whose DNA displays a molecular weight of $20–25 \times 10^6$. In the earlier members the capsid is unarmed, but the adenoviruses have long spikes at the vertices; it is naked in all.

2. The second major lineage of double-stranded DNA viruses shows resemblances to the type C oncogenic RNA viruses in several ways. First, the capsid is duplex (Roizman and Spear, 1971); in reality it consists of three capsids. Second, it is enclosed by a lipoprotein envelope, often bearing spikes. Third, the DNA is a toroid similar to that formed during development of type C oncogenic species. Finally, the DNA molecule is large, complexed with proteins, and breaks into multiple size classes when released from the capsid. Evolutionary progress along this line is currently poorly marked, as a consequence of the scanty macromolecular data that are available (Figure 10.6A–C).

   1. The herpesviruses, including the Epstein-Barr virus, have DNA molecules weighing 101×10^6, which fracture into seven size classes.
   6. The poxviruses have DNA molecules weighing between 150 and 200×10^6; that of vaccinia weighs 167×10^6, and of fowl pox, 192×10^6. Transcription of the DNA begins while still within the capsid, a point of considerable importance.

10.3.7. Earliest Cellular Level of Evolution

Above the last viral stage of phylogeny is still another, represented by the chlamydiae and rickettsias. Unfortunately, these latter two groups have not been popular as subjects for investigation by molecular biologists and biochemists. Because their processes of synthesizing macromolecules are consequently much more poorly documented than are those of viruses, only a crude suggestion of phylogenetic steps can be presented. Furthermore, this level does not appear to follow closely the last preceding viral stage, many transitional types evidently being missing at present. Despite the unbridged chasms that thus exist between the last and the present levels of evolution, relationships are nevertheless plainly demonstrated. According to current information, only two substages are now extant (Figure 10.6D–F):

1. At the lower stage of development are the chlamydiae, including the trachoma, inclusion conjunctivitis, psittacosis, ornithosis, and pneumonitis causative agents, often considered to be viruses (Rhodes and van Rooyen, 1968). These are small spherical bodies measuring be-

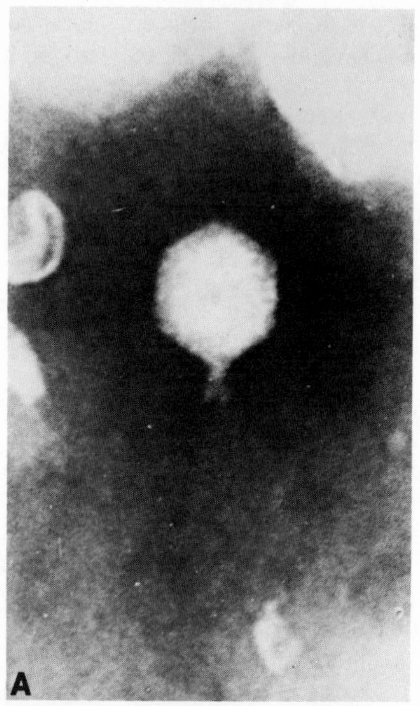

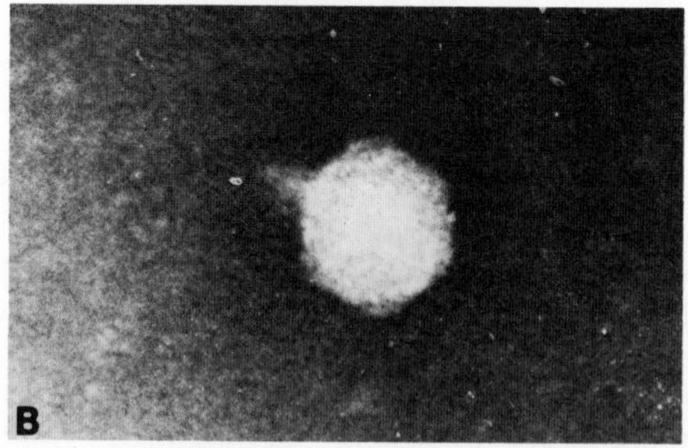

Figure 10.5. A miscellany of contractile-tailed viruses. (A) Phage 114. × 280,000. (B) Phage 17. × 260,000. (C) Phage 1. × 200,000. (D) Phage SP50, from *Bacillus subtilis*. (A–C, courtesy of Dr. Anna S. Tikhonenko; D, courtesy of Drs. B. Ten Eggeler and Michel Wurtz.)

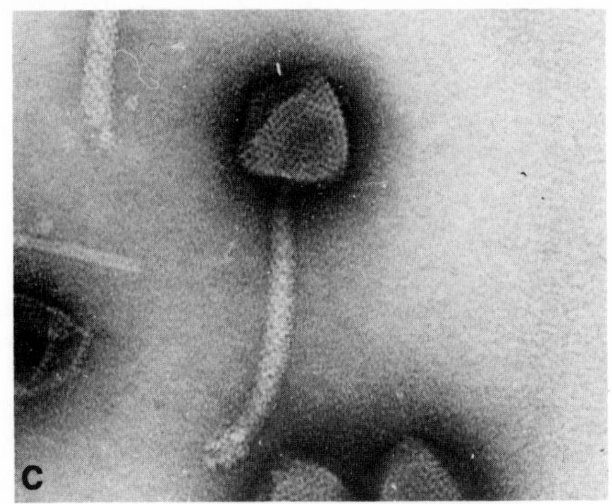

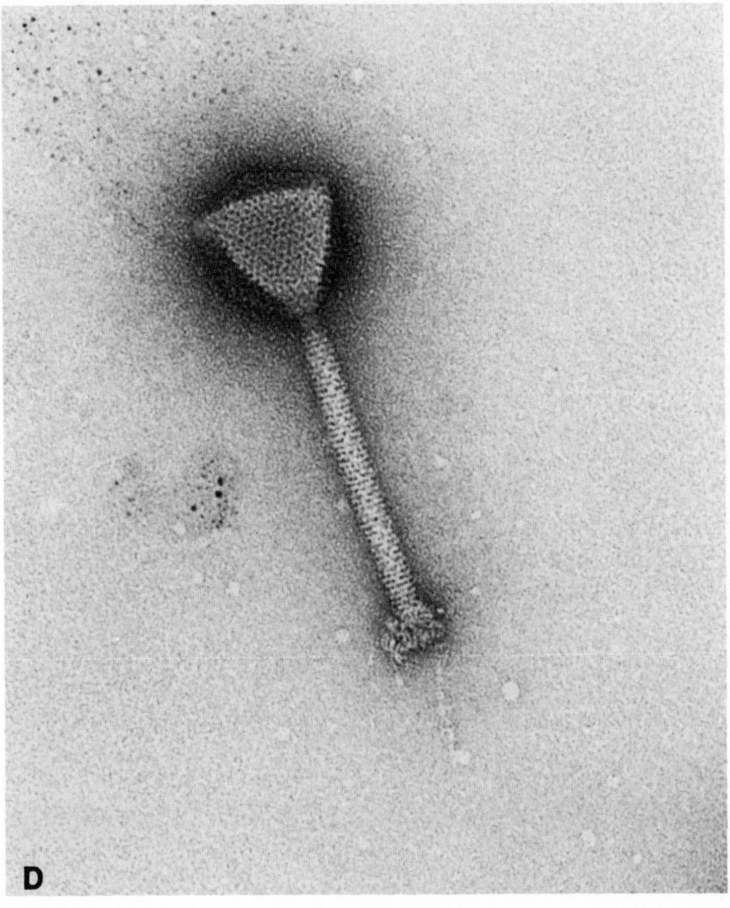

D

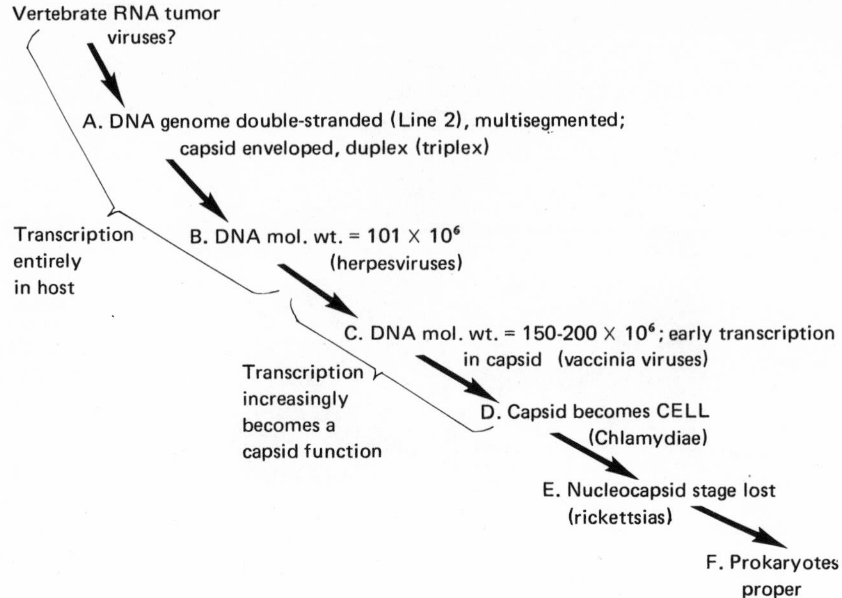

Figure 10.6. Stages leading to the advent of prokaryotes.

tween 0.3 and 1.0 μm in diameter, depending on the stage of development (Moulder, 1966). The smallest (0.3 μm) particles represent the elementary body form, which is the free and infectious type. Apparently these enter the host cell during phagocytosis; here they remain in the phagocytotic vesicles and become reorganized into the larger (0.5–1.0 μm) bodies. The latter, still in vesicles, undergo binary fission, the products of which are mostly the small infective type. Within the small particles is an electron-opaque nucleocapsid quite similar to that of the larger viruses, but this body is not visible in the mature form. The cell envelope consists of at least two distinct membranes, perhaps with an intermediate layer as in the rickettsias (Manire, 1966); the inner layer is composed of highly ordered arrays of units arranged in hexagons. Possibly these reflect the hexagons in the capsomeres of viruses. As in the colorless sulfur bacteria, cytochromes and flavoproteins are absent, and reduced NAD is not reoxidized (Weiss and Kiesow, 1966). Aside from an abundance of ribosomes and tRNAs, very little else appears to be known of the genetic mechanism in these organisms.

It is pertinent to note that the genetic mechanism always remains encapsulated and never becomes free in the host, as is consistently the case in viruses. Thus these organisms are completely cellular and even undergo binary cell division in a fashion comparable to that of many higher forms. Yet it is clear that two cellular bodies exist, one of which resembles the viral capsid in being the dispersal and infective body and in containing an electron-dense nucleocapsid. The secondary body derived from this elementary particle, in undergoing fission and in lacking an electron-dense nucleocapsid, therefore appears to be the earliest known cell; however, it should be noted to represent merely a modification of the elementary particle, not a new product. Thus it is reasonable to suggest that the cell is really a modified capsid in which all the life processes are carried out.

If this proposal is tentatively accepted as valid, it becomes possible to provide evidence of one of the earlier stages in the conversion of the capsid into a cell. First, it should be recalled that originally the capsid of virions is merely a temporary protective coat, within which none of the life processes occur. However, in such poxviruses as vaccinia virus (Chapter 9, Section 9.1.5), it was pointed out that early genomal portions of the DNA are transcribed into early mRNAs while still contained within the core of the complex capsid. Because some of the life processes are thus conducted within that container, it consequently corresponds to a cell to that extent. Continual increases in the number of activities carried out within the capsid ultimately eliminated the necessity of emerging from that protective coat, thereby bringing the archetypal cell into being.

2. The rickettsias are so obviously cellular that they are often considered to be degenerate bacteria (Rhodes and van Rooyen, 1968; Ormsbee, 1969). However, their fine structure resembles none of the known bacteria in detail, although it most closely approaches that of the Beggiotoales and blue-green algae. Electron micrographs show the cell envelope to be complex in consisting of two thin membranes and an intervening clear lamella (Schaechter *et al.*, 1957; Entwistle *et al.*, 1968). The presence of a pentose cycle in these and the Chlamydiae has been established (Weiss *et al.*, 1964), but the occurrence of a Krebs cycle is currently in doubt (Moulder, 1966). If one should be proven present, or if a series of cytochromes is found, the rickettsias must indeed be viewed as degenerate eubacteria. If cytochromes prove to be absent, this condition alone would be sufficient to indicate these organisms to be primitive, as viewed here.
3. From the level of the rickettsias, the cell underwent a series of evolutionary developments, the description of which is beyond the purpose of the present study, but which have been summarized preliminarily elsewhere (Dillon, 1962).

Aside from proposing a pathway that possibly was followed during the early stages of life's evolution, the foregoing synopsis discloses two generalizations of a broader nature. First it becomes amply evident that living things are chemical entities, whose fundamental properties are describable in ordinary physicochemical terms. Hence, to this degree the prevailing mechanistic view of the organic world is firmly supported.

But the preceding accounts substantiate even more strongly a second broad conclusion, that of a concept of life long advocated by Woese. Although living matter is physicochemical in large measure, especially at atomic and micromolecular levels, it is not simply and totally describable in deterministic terms. This is particularly manifest at the macromolecular levels of activity, in which organisms display properties found nowhere else in the physical world. For example, what other chemical system creates copies of its own molecules? Where is there another physical compound or mixture that, given time, will gain complexity by evolving new building blocks for its substance, as amino acids, sugars, and nucleotide bases have been evolved by living things? What besides live creatures can disobey the second law of thermodynamics by selecting specific individual molecules from the environment, thereby creating order out of disorder? And where outside of the organic world does an open system exist, one that never runs to equilibrium, as discussed in Chapter 2?

All these properties, and innumerable others, are unique to life, and all have arisen by means of evolutionary processes. Certainly the first simple interacting proteins that comprised the earliest protobionts were strictly a physicochemical phenomenon. But later, when those polyamino acid chains underwent mutational change, the structure of the original strands remained largely intact in the descendant species and thus predetermined their molecular primary structure to a major extent. In turn, the new species that ultimately were derived from these second forms carried the imprint of both of its ancestral stocks. Similarly the prokaryotes and eukaryotes of today carry remnants of their lineages, in some instances back to remote predecessors. These relicts remain, not necessarily because the original ancestral enzymes were the most efficient molecules possible—they merely were sufficiently successful to enable the primordial protobionts to survive and reproduce in ample numbers. The molecular activities in modern organisms thus appear to represent series of improvements in efficiencies still based on the primitive processes, made by accumulating minor changes that arose over the eons. Consequently, the genetic mechanism is not an outstandingly effective set of interacting molecules; it is a collection of fortuitous events that, like the genetic code, are frozen accidents. The old ways are too vital to be broken down and replaced by the slow evolutionary processes; hence, they are conserved and improved upon by the addition of this modifying enzyme or that new polymerase, until the genetic apparatus eventually attained its present amazing complexity.

The evolutionary processes that have brought into being the unique properties of the organic world are not to be confused with those active in the strictly physical universe which have given rise to the chemical elements and heavenly bodies. Those of living things result solely from interactions between the environment and a genetic mechanism. Since the latter has been shown to be basically the equivalent of life itself (Chapter 5, Section 5.3.1), the uniqueness of organisms stems from their being genetic systems, which can be said of nothing else known to exist. Because living things are thus physicochemical and simultaneously more than *just* physicochemical, a narrow mechanistic approach to their study no longer is adequate, and a new scientific philosophy needs to be developed to embrace the recent knowledge of life and thereby guide future investigations and interpretations in the field of biology and associated sciences. Since only an extension of the former point of view is essential, perhaps the name *biomechanism* might appropriately be applied to the concept. By whatever name it may ultimately become known, the evolutionary component irrevocably inculcated into all living systems must receive ample recognition, in order that a still greater comprehension can gradually be acquired of the nature and functionings of living matter and its bewildering genetic mechanism.

References

CHAPTER 1

Abelson, P. H. 1956. Amino acids formed in "primitive atmospheres." *Science* **124**:935.

Abelson, P. H. 1957. Some aspects of paleobiochemistry. *Ann. N.Y. Acad. Sci.* **69**:274–285.

Abelson, P. H. 1966. Chemical events on the primitive earth. *Proc. Nat. Acad. Sci. USA* **55**:1365–1372.

Agarwal, K. L., and Dahr, M. M. 1963. Nature of products formed by the action of polyphosphate ester on nucleotides. *Indian J. Chem.* **1**:451–452.

Akabori, S. 1959. Origin of the fore-protein. *In:* Oparin, A. I., ed., *The Origin of Life on the Earth,* London, Pergamon Press, p. 189–196.

Akabori, S., and Yamamoto, M. 1972. Model experiments on the prebiological formation of protein. *In:* Rohlfing, D. L., and A. I. Oparin, eds., *Molecular Evolution: Prebiological and Biological,* New York, Plenum Press, p. 189–197.

Armstrong, R. L., and Hein, S. M. 1973. Computer simulation of lead and strontium isotope evolution of the earth's crust and upper mantle. *Geochim. Cosmochim. Acta* **37**:1–18.

Bada, J. L., and Miller, S. L. 1968. Ammonia ion concentration in the primitive ocean. *Science* **159**:423–425.

Bahadur, K. 1954. Photosynthesis of amino-acids from paraformaldehyde and potassium nitrate. *Nature* **173**:1141.

Bahadur, K., and Raganayaki, S. 1958. *Proc. Nat. Inst. Sci. India* **27A**:292.

Bahadur, K., Ranganayaki, S., and Santamaria, L. 1958. Photosynthesis of amino-acids from paraformaldehyde involving the fixation of nitrogen in the presence of colloidal MbO as catalyst. *Nature* **182**:1668.

Barak, I., and Bar-Nun, A. 1975. The mechanisms of amino acids synthesis by high temperature shock-waves. *Origins Life* **6**:483–506.

Barker, S. A., Lloyd, I. R. L., and Stacy, M. 1962a. Polymerization of glucose induced by gamma radiation. *Radiation Res.* **16**:224–231.

Barker, S. A., Lloyd, I. R. L., and Stacy, M. 1962b. Structure of a radiation-induced polymer from glucose. *Radiation Res.* **17**:619–624.

Bar-Nun, A. 1975. Shock synthesis of amino acids. II. *Origins Life* **6**:109–115.

Bar-Nun, A., Bar-Nun, N., Bauer, S. H., and Sagan, C. 1970a. Shock synthesis of amino acids in simulated primitive environments. *Science* **168**:470–473.

Bar-Nun, A., Bar-Nun, N., Bauer, S. H., and Sagan, C. 1970b. [Reply to H. R. Hulett]. *Science* **170**:1001–1002.

Bar-Nun, A., Bar-Nun, N., Bauer, S. H., and Sagan, C. 1971. Shock synthesis of amino acids in simulated primitive environments. *In:* Buvet, R., and C. Ponnamperuma, eds., *Chemical Evolution and the Origin of Life,* Amsterdam, North-Holland Publishing Co., p. 114–122.

Beck, A., Lohrmann, R., and Orgel, L. E. 1967. Phosphorylation with inorganic phosphates at moderate temperatures. *Science* **157**:952.

Berkner, L. V., and Marshall, L. C. 1965. On the origin and rise of oxygen concentration in the earth's atmosphere. *J. Atmos. Sci.* **22**:225–231.

Berkner, L. V., and Marshall, L. C. 1966. Limitation on oxygen concentration in a primitive planetary atmosphere. *J. Atmos. Sci.* **23**:133–139.

Bernal, J. D. 1967. *The Origin of Life,* Cleveland, Ohio, World Publishing Company.

Beukers, R., and Berends, W. 1960. Isolation and identification of the irradiation product of thymine. *Biochim. Biophys. Acta* **41**:550–551.

Borsook, H., and Huffman, H. M. 1944. *In:*Schmidt, C. L. A., ed., *Chemistry of the Amino Acids and Proteins.* Charles C. Thomas, Springfield, Ill., p. 282.

Boutlerow, A. 1861. Formation synthétique d'une substance sucrée. *C. R. Acad. Sciences France* **53**:145–147.

Burton, F. G., Lohrmann, R. and Orgel, L. E. 1974. On the possible role of crystals in the origins of life. VII. The adsorption and polymerization of phosphoramidates by montmorillonite clay. *J. Mol. Evol.* **3**:141–150.

Butler, J. A. V. 1959. Changes induced in nucleic acids by ionizing radiations and chemicals. *Radiation Res., Suppl.* **1**:403–416.

Calvin, M. 1956. Chemical evolution and the origin of life. *Am. Sci.* **44**:248–263.

Calvin, M. 1961. The origin of life on earth and elsewhere. *Ann. Int. Med.* **54**:954–976.

Calvin, M. 1962. Evolution of photosynthetic mechanisms. *Perspect. Biol. Med.* **5**:147–172.

Calvin, M. 1965. Chemical evolution. *Proc. Roy. Soc. Lond.* **A 288**:441–466.

Calvin, M. 1967. Chemical evolution. *Evol. Biol.* **1**:1–25.

Calvin, M. 1969. *Chemical Evolution: Molecular Evolution towards the Origin of Living Systems on the Earth and Elsewhere*, Oxford, Oxford University Press.

Calvin, M. and Calvin, G. J. 1964. Atom to Adam. *Am. Sci.* **52**:163–186.

Chadha, M. S., Molton, P. M., and Ponnamperuma, C. 1975. Aminonitriles: Possible role in chemical evolution. *Origins Life* **6**:127–136.

Chadha, M. S., Replogle, L., Flores, J., and Ponnamperuma, C. 1971. Possible role of aminoacetonitrile in chemical evolution. *Bioorganic Chem.* **1**(3):269–274.

Chang, S., Flores, J., and Ponnamperuma, C. 1969. Peptide formation mediated by hydrogen cyanide tetramer: A possible prebiotic process. *Proc. Nat. Acad. Sci. USA* **64**:1011–1015.

Contreras, G., Esperjo, R., Mery, E., Ohlbaum, D., and Tohá, J. 1962. Polymerization of ribomononucleotides by γ-radiation. *Biochim. Biophys. Acta* **61**:718–727.

Cruikshank, D. P., Pilcher, C. B., and Morrison, D. 1976. Pluto: Evidence for methane frost. *Science* **194**:835–837.

Degens, E. T., and Matheja, J. 1971. Formation of organic polymers on inorganic templates. *In:* Kimball, A. P., and J. Oró, eds., *Prebiotic and Biochemical Evolution,* Amsterdam, North-Holland Publishing Company, p. 39–69.

Degens, E. T., Matheja, J., and Jackson, T. A. 1970. Template catalysis: Asymmetric polymerization of amino acids on clay minerals. *Nature* **227**:492–493.

Dhar, N. R., and Ram, A. 1933. Formaldehyde in the upper atmosphere. *Nature* **132**:819–820.

Dillon, L. S. 1974. Neovulcanism: A proposed replacement for the concepts of plate tectonics and continental drift. *Mem. Am. Assoc. Petrol. Geol.* **23**:167–239.

Dixon, M. A., and Webb, E. C. 1958. *Enzymes,* New York, Academic Press.

Dose, K. 1974. Chemical and catalytical properties of thermal polymers of amino acids (proteinoids). *Origins Life* **5**:239–252.

Dose, K., and Ettre, K. 1958. Die radiationschemische Syntheses von Aminosäuren und verwandten Verbindungen. *Zeit. Naturforsch.* **13b**:784–788.

Dose, K., and Ponnamperuma, C. 1967. The effect of ionizing radiation on *N*-acetyl glycine in the presence of ammonia. *Radiation Res.* **31**:650.

Dose, K., and Rauchfuss, H. 1972. On the electrophoretic behavior of thermal polymers of amino acids. *In*: Rohlfing, D. L., and A. I. Oparin, eds., *Molecular Evolution: Prebiological and Biological,* New York, Plenum Press, p. 199–231.

Dose, K., and Risi, S. 1968. Zur radiationschemischen Bildung von Aminosäuren durch Carboxylierung und Aminierung. *Zeit. Naturforsch.* **23b**:581–587.

Dose, K., and Zaki, L. 1971. Recent progress in the study and abiotic production of catalytically active polymers of α-amino acids. *In*: Buvet, R., and C. Ponnamperuma, eds., *Chemical Evolution and the Origin of Life*, Amsterdam, North-Holland Publishing Co., p. 263–278.

Ellenbogen, E. 1958. Photochemical synthesis of amino acids. *Abstr. 134th Nat. Mtg. Amer. Chem. Soc.*, Chicago., p. 47–48C.

Ellenbogen, E. 1959. Photochemical synthesis of amino acids. *Abstr. Am. Chem. Soc. Meeting, Chicago*, p. 47C.

Elmore, D. T., and Todd, A. R., 1952. Nucleotides. XVIII. The synthesis and properties of adenosine-5′-uridine-5′-phosphate. *J. Chem. Soc.* **1952**:3681–3686.

Euler, H., and Euler, A. 1906. Über die Bildung von i-aribinoketose aus Formaldehyd. *Chem. Berichte* **39**:45–51.

Farmer, C. B., Davies, D. W., and LaPorte, D. D. 1976. Mars: Northern summer icecap water vapor observations from Viking 2. *Science* **194**:1339–1341.

Fanale, F. P. 1971. A case for catastrophic early degassing of the Earth. *Chem. Geol.* **8**:79–105.

Ferris, J. P., and Orgel, L. E. 1965. Aminomalononitrile and 4-amino-5-cyanoimidazole in HCN polymerization and adenine synthesis. *J. Am. Chem. Soc.* **87**:4976–4977.

Ferris, J. P., and Ryan, T. J. 1973. Chemical evolution. XIV. Oxidation of diaminomaleonitrile and its possible role in hydrogen cyanide oligomerization. *J. Org. Chem.* **38**:3302–3307.

Ferris, J. P., Kuder, J. E., and Catalano, A. W. 1971. Photochemical reactions and the chemical evolution of purines and nicotinamide derivatives. *Science* **166**:765–766.

Ferris, J. P., Donner, D. B., and Lobo, A. P. 1973a. Possible role of HCN in chemical evolution: Investigation of the proposed direct synthesis of peptides from HCN. *J. Mol. Biol.* **74**:499–510.

Ferris, J. P., Donner, D. B., and Lobo, A. P. 1973b. Possible role of HCN in chemical evolution: The oligomerization and condensation of HCN. *J. Mol. Biol.* **74**:511–518.

Ferris, J. P., Zamek, O. S., Altbuch, A. M., and Freiman. H. 1974a. Chemical evolution. XVIII. Synthesis of pyrimidines from guanidine and cyanoacetaldehyde. *J. Mol. Evol.* **3**:301–309.

Ferris, J. P., Wos, J. D., Ryan, T. J., Lobo, A. P., and Donner, D. B. 1974b. Biomolecules from HCN. *Origins Life,* **5**:153–158.

Flores, J. J., and Leckie, J. O. 1973. Peptide formation mediated by cyanate. *Nature* **244**:435–437.

Flores, J. J., and Ponnamperuma, C. 1972. Polymerization of amino acids under primitive earth conditions. *J. Mol. Evol.* **2**:1–9.

Fox, S. W. 1960. How did life begin? *Science* **132**:200–208.

Fox, S. W. 1965. *The Origins of Prebiological Systems and of their Molecular Matrices*, New York, Academic Press.

Fox, S. W. 1974. Thermodynamic perspectives and the origin of life. *In*: Kursumoglu, B., S. L. Mintz, and S. M. Widmayer, eds., *Quantum Statistical Mechanics in the Natural Sciences,* New York, Plenum Press, p. 119–142.

Fox, S. W., and Dose, K. 1972. *Molecular Evolution and the Origin of Life,* San Francisco, W. H. Freeman and Company.

Fox, S. W., and Harada, K. 1958. Thermal copolymerization of amino acids to a product resembling protein. *Science* **128**:1214.

Fox, S. W., and Harada, K. 1960. The thermal copolymerization of amino acids common to proteins. *J. Am. Chem. Soc.* **82**:3745–3752.

Fox, S. W., and Harada, K. 1961. Synthesis of uracil under conditions of a thermal model of prebiological chemistry. *Science* **133**:1923–1924.

Fox, S. W., and Nakashima, T. 1967. Fractionation and characterization of an amidated thermal 1 : 1 : 1-proteinoid. *Biochem. Biophys. Acta* **140**:155–167.

Fox, S. W., and Windsor, C. R. 1970. Synthesis of amino acids by the heating of formaldehyde and ammonia. *Science* **170**:984–986.

Fox, S. W., and Yuyama, S. 1963. Effects of the Gram stain on microspheres from thermal polyamino acids. *J. Bacteriol.* **85**:279–283.

Fox, S. W., Johnson, J. E., and Middlebrook, M. 1955. Pyrosynthesis of aspartic acid and alanine from citric acid cycle intermediates. *J. Am. Chem. Soc.* **77**:1048–1049.

Fox, S. W., Johnson, J. E., and Vegotsky, A. 1956. On biochemical origins and optical activity. *Science* **124**:923–925.

Fox, S. W., Vegotsky, A., Harada, K., and Hoagland, P. D. 1957a. Spontaneous generation of anabolic pathways, protein, and nucleic acid. *Ann. N.Y. Acad. Sci.* **69**:328–337.

Fox, S. W., Vegotsky, A., Harada, K., and Hoagland, P. D. 1957b. Spontaneous generation of anabolic pathways, protein, and nucleic acid. *Ann. N.Y. Acad. Sci.* **69**:328–337.

Friedmann, N., Haverland, W. J., and Miller, S. L. 1971. Prebiotic synthesis of the aromatic and other amino acids. *In*: Buvet, R., and C. Ponnamperuma, eds., *Chemical Evolution and the Origin of Life*, Amsterdam, North-Holland Publishing Co., p. 123–135.

Friedmann, N., and Miller, S. L. 1969. Phenylalanine and tyrosine synthesis under primitive earth conditions. *Science* **166**:766–767.

Fuller, W. D., Sanchez, R. A., and Orgel, L. E. 1972. Studies in prebiotic synthesis. VII. Solid state synthesis of purine nucleosides. *J. Mol. Evol.* **1**:249–257.

Gabel, N. W., and Ponnamperuma, C. 1967. Model for origin of monosaccharides. *Nature* **216**:453–455.

Garrison, W. M., Morrison, D. C., Hamilton, J. G., Benson, A. A. and Calvin M. 1951. Reduction of CO_2 in aqueous solutions by ionising radiation. *Science* **114**:416–418.

Goldberg, E. D. 1954. Marine geochemistry. *J. Geol.* **62**:249–265.

Gottikh, B. P., and Slutsky, I. 1964. On polycondensation of ribonucleotides under the action of polyphosphoric ester. *Biochim. Biophys. Acta* **87**:163–165.

Groth, W. 1957. Photochemische Bildung von amino-säuren und anderen organischen Verbindungen aus Mischungen von $H_2O \cdot NH_3$ und der einfachsten Kohlenwasserstoffen. *Angew. Chem.* **69**:681.

Groth, W., and Suess, H. 1938. Bemerkungen zur Photochemie der Erdatmosphäre. *Naturwissenschaften 26*:77.

Groth, W., and von Weyssenhoff, H. 1957. Photochemische Bildung von Aminosäuren aus Mischungen einfacher Gase. *Naturwissenschaften* **44**:510–511.

Groth, W., and von Weyssenhoff, H. 1960. Photochemical formation of organic compounds from mixtures of simple gases. *Planet. Space Sci.* **2**:79–85.

Halmann, M. 1975. Models of prebiological phosphorylation. *Origins Life* **6**:169–174.

Handschuh, G. J., and Orgel, L. E. 1973a. Struvite and prebiotic phosphorylation. *Science* **179**:483–484.

Handschuh, G. J., and Orgel, L. E. 1973b. Precipitation of phosphates in a primeval sea. *Science* **181**:582.

Harada, K., and Fox, S. W. 1964. Thermal synthesis of natural amino acids from a postulated primitive terrestrial atmosphere. *Nature* **201**:335–336.

Hargreaves, W. R., Mulvihill, S. J., and Deamer, D. W. 1977. Synthesis of phospholipids and membranes in prebiotic conditions. *Nature* **266**:78–80.

Hart, M. H. 1974. A possible atmosphere for Pluto. *Icarus* **21**:242–247.

Hasselstrom, T., Henry, M. C., and Murr, B. 1957. Synthesis of amino acids by beta radiation. *Science* **125**:350–351.

Hayes, F. N., and Hansbury, E. 1964. The reaction of 5′-thymidylic acid with the condensation product of phosphorus pentoxide and ethyl ether. *J. Am. Chem. Soc.* **86**:4172–4175.

Hems, G. 1960. Chemical effects of ionizing radiation on DNA in dilute aqueous solution. *Nature* **186**:710–712.

Hess, S. L., Henry, R. M., Leovy, C. B., Mitchell, J. L., Ryan, J. A., and Tillman, J. E. 1976a. Early meteorological results from the Viking 2 lander. *Science* **194**:1352–1353.

Hess, S. L., Henry, R. M., Leovy, C. B., Mitchell, J. L., Ryan, J. A., and Tillman, J. E., 1976b. Preliminary meteorological results on Mars from the Viking 1 lander. *Science* **193**:788–791.

Heyns, K., Walter, W., and Meyer, E. 1957. Modelluntersuchungen zur Bildung organischer Verbindungen in Atmosphären einfacher Gase durch elektrische Entladungen. *Naturwissenschaften* **44**:385–389.

Hong, K. Y., Hong, J. H., and Becker, R. S. 1974. Hot hydrogen atoms: Initiators of reactions of interest in interstellar chemistry and evolution. *Science* **184**:984–987.

Hubbard, J. S., Hardy, J. P., and Horowitz, N. H. 1971. Photocatalytic production of organic compounds from CO and H_2O in a simulated martian atmosphere. *Proc. Nat. Acad. Sci. USA* **68**:574–578.

Hubbard, J. S., Hardy, J. P., Voecks, G. E., and Golub, E. E. 1973. Photocatalytic synthesis of organic compounds from CO and water: Involvement of surfaces in the formation and stabilization of products. *J. Mol. Evol.* **2**:149–166.

Hulett, H. R. 1969. Limitations on prebiological synthesis. *J. Theor. Biol.* **24**:56–72.

Hulett, H. R. 1970. Amino acid synthesis in simulated primitive environments. *Science* **170**:1000–1001.

Ibañez, J. D., Kimball, A. P., and Oró, J. 1971a. The effect of imidazole, cyanamide, and polyornithine on the condensation of nucleotides in aqueous systems. *In*: Buvet, R., and C. Ponnamperuma, eds., *Chemical evolution and the Origin of Life,* Amsterdam, North-Holland Publishing Co., p. 171–179.

Ibañez, J. D., Kimball, A. P., and Oró, J. 1971b. Possible prebiotic condensation of mononucleotides by cyanamide. *Science* **173**:444–446.

Ivanov, C. P., and Slavcheva, N. N. 1977. Formation of amino acids on heating glycine with alumina. *Origins Life* **8**:13–19.

Jacob, T. M., and Khorana, H. G. 1964. Studies on nucleotides. XXX. A comparative study of reagents for the synthesis of the $C_{3'}$–$C_{5'}$ internucleotidic linkage. *J. Am. Chem. Soc.* **86**:1630–1635.

Keldysh, M. V. 1977. Venus exploration with Venera 9 and Venera 10 spacecraft. *Icarus* **30**:605–625.

Keosian, J. 1968. *The Origin of Life. 2nd Ed.*, New York, Reinhold Book Corp.

Khorana, H. G. 1964. Carbodiimides. III. (A) A new method for the preparation of mixed esters of phosphoric acid. (B) Some observations on the base-catalyzed addition of alcohols to carbodiimides. *Can. J. Chem.* **32**:227–234.

Kieffer, H. H., Chase, S. C., Martin, T. Z., Miner, E. D., and Palluconi, F. D. 1976. Martian north pole summer temperatures: Dirty ice water. *Science* **194**:1341–1344.

Klein, H. P. 1972. Potential targets in the search for extraterrestrial life. *In:* Ponnamperuma, C., ed., *Exobiology,* Amsterdam, North-Holland Publishing Co., p. 449–464.

Kochetkov, N. K., Budowsky, E. I., Domkin, V. D., and Kuromov-Borissov, N. N. 1964. On the structure of polynucleotides obtained by condensation of nucleoside 2′(3′)phosphates with polyphosphoric ester. *Biochim. Biophys. Acta* **80**:145–148.

Lawless, J. G., and Boynton, C. D. 1973. Thermal synthesis of amino acids from a simulated primitive atmosphere. *Nature* **243**:405–407.

Levin, G. V., and Straat, P. A. 1976. Viking labeled release biology experiment: Interim results. *Science* **194**:1322–1329.

Loew, G. H., Chandra, M. S., and Chang, S. 1972. A molecular orbital and chemical study of aminoacetonitrile. *J. Theor. Biol.* **35**:359–371.

Loew, G. H., and Chang, S. 1975. Quantum chemical study of the thermodynamics, kinetics of formation and bonding of H_2CN: Relevance to prebiotic chemistry. *Origins Life* **6**:117–125.

Loew, O. 1886. *J. Prakt. Chem.* **33**(2):321.

Lohrmann, R., and Orgel, L. E. 1968. Prebiotic synthesis: Phosphorylation in aqueous solution. *Science* **161**:64–66.

Lohrmann, R., and Orgel, L. E. 1971. Urea-inorganic phosphate mixtures as prebiotic phosphorylating agents. *Science* **171**:490–494.

Lohrmann, R., and Orgel, L. E. 1973. Prebiotic activation processes. *Nature* **244**:418–420.

Lowe, C. U., Rees, M., and Markham, R. 1963a. Synthesis of complex organic compounds from simple precursors: formation of amino acids, amino acid polymers, fatty acids, and purines from NH_4CN. *Nature* **199**:219–222.

Lowe, C. U., Rees, M., and Markham, R. 1963b. NH_4CN and primitive earth synthesis. *Fed. Proc.* **22**:479.

Mahler, H. R., and Cordes, E. H. 1967. *Biological Chemistry*. New York, Harper and Row.

Margulis, L., Halvorson, H. O., Lewis, J., and Cameron, A. G. W. 1977. Limitations to growth of microorganisms on Uranus, Neptune, and Titan. *Icarus* **30**:793–808.

Mariani, E., and Torraca, G. 1953. The composition of formose. A chromatographic study. *Int. Sugar J.* **55**:309–311.

Mason, B. 1966. *Principles of Geochemistry,* New York, John Wiley & Sons.

Matthews, C.N. 1971. The origin of proteins: Heteropoly-peptides from hydrogen cyanide and water. *In*: Buvet, R., and C. Ponnamperuma, eds., *Chemical Evolution and the Origin of Life,* Amsterdam, North-Holland Publishing Co., p. 231–235.

Matthews, C. N. 1975. The origin of proteins: Heteropolypeptides from HCN and water. *Origins Life* **6**:155–162.

Matthews, C. N., and Moser, R. E. 1967. Peptide synthesis from HCN and water. *Nature* **215**:1230–1234.

McConnell, D. 1973. Precipitation of phosphates in a primeval sea. *Science* **181**:582.

McKelvey, V. E. 1967. Phosphate deposits. *Bull. U.S. Geol. Surv.* 1252-D:1–21.

McReynolds, J. H., Furlong, N. B., Birrell, P. J., Kimball, A. P., and Oró, J. 1971. Polymerization of deoxyribonucleotides by ultraviolet light. *In*: Kimball, A. P., and J. Oró, eds., *Prebiotic and Biochemical Evolution,* Amsterdam, North-Holland Publishing Co., p. 111–121.

Michelson, A. M. 1959. Polynucleotides. Pt. II. Homopolymers of cytidylic and pseudouridylic acid, copolymers with repeating subunits and the step-wise synthesis of polyribonucleotides. *J. Chem. Soc.* **1959**:3655.

Miller, S. L. 1953. A production of amino acids under possible primitive earth conditions. *Science* **117**:528–529.

Miller, S. L. 1955. Production of some organic compounds under possible primitive earth conditions. *J. Am. Chem. Soc.* **77**:2351–2352.

Miller, S. L. 1959. Formation of organic compounds on the primitive earth. *In*: Oparin, A. I., ed., *The Origin of Life on Earth*, London, Pergamon Press, p. 123–135.

Miller, S. L. 1974. The atmosphere of the primitive earth and the prebiotic synthesis of amino acids. *Origins Life* **5**:139–151.

Miller, S. L., and Orgel, L. E. 1974. *The Origins of Life on the Earth,* Englewood Cliffs, New Jersey, Prentice-Hall, Inc.

Miller, S. L., and Urey, H. C. 1959. Organic compound synthesis on the primitive earth. *Science* **130**:245–251.

Morávek, J., Kopecký, J., and Škoda, J. 1969. Thermic phosphorylations. VI. Formation of oligonucleotides from uridine 2′(3′)-phosphate. *Coll. Czech. Chem. Commun.* **33**:4120–4124.

Murthy, V. R., and Patterson, C. C. 1962. Primary isochron of zero age for meteorites and the earth. *J. Geophys. Res.* **67**:1161–1167.

Nagyvary, J., and Provenzale, R. 1971. Polymerization of nucleotides via displacement on carbon; its preparative and prebiotic significance. *In:* Kimball, A. P., and J. Oró, eds., *Prebiotic and Biochemical Evolution,* Amsterdam, North-Holland Publishing Co., p. 102–110.

Nier, A. O., and McElroy, M. B. 1976. Structure of the neutral upper atmosphere of Mars: Results from Viking 1 and Viking 2. *Science* **194**:1298–1300.

Noda, H., and Ponnamperuma, C. 1971. Polymer formation in a simulated Jovian atmosphere. *In:* Buvet, R., and C. Ponnamperuma, eds., *Chemical Evolution and the Origin of Life,* Amsterdam, North-Holland Publishing Co., p. 236–244.

Nooner, D. W., and Oró, J. 1974. Direct synthesis of polypeptides; polycondensation of α-amino acids by polymetaphosphate esters. *J. Mol. Evol* **3**:79–88.

Okawa, K. 1954. (Title in Japanese). *J. Chem. Soc. Japan* **75:**1199–1202.

Oparin, A. I. 1957. *The Origin of Life on the Earth. 3rd Ed.*, New York, Academic Press.

Oparin, A. I. 1968. *Genesis and Evolutionary Development of Life,* New York, Academic Press.

Oparin, A. I. 1971. Problem of the origin of life: Present state and prospects. *In:* Buvet, R., and C. Ponnamperuma, eds., *Chemical Evolution and the Origin of Life,* Amsterdam, North-Holland Publishing Co., p. 3–9.

Orgel, L. E., and Sulston, J. E. 1971. Polynucleotide replication and the origin of life. *In*: Kimball, A. J., and J. Oró, eds., *Prebiotic and biochemical evolution,* Amsterdam, North-Holland Publishing Co., p. 89–94.

Oró, J. 1963a. Synthesis of organic compounds by electric discharges. *Nature* **197**:862–867.

Oró, J. 1963b. Synthesis of organic compounds by high-energy electrons. Nature **197**:971–974.

Oró, J. 1965. Prebiological organic systems. *In*: Fox, S. W., ed., *The Origins of Prebiological Systems,* New York, Academic Press, p. 137–162.

Oró, J. and Cox, A. C. 1962. Non-enzymic synthesis of 2-deoxyribose. *Fed. Proc.* **21:**80.

Oró, J., and Guidry, C. L. 1960. A novel synthesis of polypeptides. *Nature* **186**:156–157.

Oró, J., and Guidry, C. L. 1961. Direct synthesis of polypeptides. I. Polycondensation of glycine in aqueous ammonia. *Arch. Biochem.* **93:**166–171.

Oró, J., and Kamat, S. S. 1961. Amino-acid synthesis from HCN under possible primitive earth conditions. *Nature* **190:**442–443.

Oró, J., and Kimball, A. P. 1962. Synthesis of purines under possible primitive earth conditions. II. Purine intermediates from HCN. *Arch. Biochem. Biophys.* **96**:293–313.

Oró, J., and Stephen-Sherwood, E. 1974. The prebiotic synthesis of oligonucleotides. *Origins Life* **5**:159–172.

Oró, J., Kimball, A. P., and McReynolds, J. 1969. *Sixth Fed. Eur. Biochem. Soc. Meeting, Madrid,* Abstr., p. 37.

Orton, G. S. 1975. The thermal structure of Jupiter. *Icarus,* **26**:125–141, 142–158.

Otroshchenko, V. A., and Vasilyeva, N. V. 1977. The role of mineral surfaces in the origin of life. *Origins Life* **8**:25–31.

Owen, T., and Biemann, K. 1976. Composition of the atmosphere at the surface of Mars: Detection of argon-36 and preliminary analysis. *Science* **193**:801–803.

Owen, T., Biemann, K. Ruchneck, D. R. Biller, J. E., Howarth, D. W. and LaFleur, A. L. 1976. The atmosphere of Mars: Detection of krypton and xenon. *Science* **194**:1293–1295.

Paecht-Horowitz, M. 1971. Polymerization of amino-acid phosphate anhydrides in the presence of clay minerals. *In:* Buvet, R., and C. Ponnamperuma, eds., *Chemical Evolution and the Origin of Life,* Amsterdam, North-Holland Publishing Co., p. 245–251.

Paecht-Horowitz, M. 1974. The possible role of clays in prebiotic peptide synthesis. *Origins Life* **5**:173–187.

Palm, C., and Calvin, M. 1962. Primordial organic chemistry. I. Compounds resulting from electron irradiation of $C^{14}H_4$. *J. Am. Chem. Soc.* **84**:2115–2121.

Paschke, R., Chang, R., and Young, D. 1957. Probable role of gamma irradiation in origin of life. *Science* **125**:881.

Patterson, C. C. 1956. Age of meteorites and the earth. *Geochim. Cosmochim. Acta* **10**:230–237.

Pavlovskaya, T. E., and Pasynskii, A. G. 1959. The original formation of amino acids under the action of ultraviolet rays and electrical discharges. *In:* Oparin, A. I., Pasynskii, A. G., Braunstein, T. E., and Pavlovskaya, T. E., eds., *The Origin of Life on the Earth.* London, Pergamon, p. 151.

Ponnamperuma, C. 1965. A biological synthesis of some nucleic acid constituents. *In*: Fox, S. W., ed., *The Origins of Prebiological Systems,* New York, Academic Press, p. 221–236.

Ponnamperuma C., and Flores, J. 1965. The gamma radiation of methane, ammonia and water. *Radiation Res.* **25**:229.

Ponnamperuma, C., and Mack, R. 1965. Nucleotide synthesis under possible primitive earth conditions. *Science* **148**:1221–1223.

Ponnamperuma, C. and Mariner, R. 1963. The formation of ribose and deoxyribose by U. V. irradiation of formaledehyde in water. *Radiation Res.* **19**:183.

Ponnamperuma, C., and Peterson, E. 1965. Peptide synthesis from amino acids in aqueous solutions. *Science* **147**:1572–1574.

Ponnamperuma, C., and Woeller, F. 1967. α-Aminonitriles formed by an electric discharge through a mixture of anhydrous methane and ammonia. *Curr. Mod. Biol.* **1**:156–158.

Ponnamperuma, C., and Sagan, C., and Mariner, R. 1963. Synthesis of ATP under possible primitive earth conditions. *Nature* **199**:222–226.

Rasool, S. I. 1972. Planetary atmospheres. *In:* Ponnamperuma, C., ed., *Exobiology*, Amsterdam, North-Holland Publishing Co., p. 369–399.

Raulin, F., and Toupance, G. 1975. Formating prebiochemical compounds in models of the primitive earth's atmosphere. II. CH_4-H_2S atmospheres. *Origins Life* **6**:91–97.

Ring, D., Wolman, Y., Friedmann, N., and Miller, S. L. 1972. Prebiotic synthesis of hydrophobic and protein amino acids. *Proc. Nat. Acad. Sci. USA* **69**:765–768.

Rohlfing, D. L., and Fouche, C. E. 1972. Stereo-enriched poly-α-amino acid synthesis under postulated prebiotic conditions. *In*: Rohlfing, D. L., and A. I. Oparin, eds., *Molecular Evolution: Prebiological and Biological,* New York, Plenum Press, p. 219–231.

Rubey, W. W. 1951. Geologic history of sea water: An attempt to state the problem. *Bull. Geol. Soc. Am.* **62**:1111–1148.

Rubey, W. W. 1955. Development of the hydrosphere and atmosphere, with special reference to probable composition of the early atmosphere. *Spec. Paper Geol. Soc. Am.* **62**:631–650.

Rutten, M. G. 1969. Sedimentary ores of the early and middle Precambrian and the history of atmospheric oxygen. *In: Sedimentary Ores: Ancient and Modern* (Revised), Leicester, Engl., Univ. of Leicester Press, p. 187–195.

Rutten, M. G. 1970. The history of atmospheric oxygen. *Space Life Sciences* **2**:5–17.

Saffhill, R. 1970. Selective phosphorylation of the *cis*-2′, 3′-diol of unprotected ribonucleosides with trimetaphosphate in aqueous solution. *J. Org. Chem.* **35**:2881–2883.

Sagan, C., and Khare, B. N. 1971. Long-wavelength ultraviolet photoproduction of amino acids on the primitive earth. *Science* **173**:417–420.

Sanchez, R. A., Ferris, J. P., and Orgel, L. E. 1966. Cyanoacetylene in prebiotic synthesis. *Science* **154**:784–785.

Saunders, M. A., and Rohlfing, D. L. 1972. Polyamino acids: Preparation from reported proportions of "prebiotic" and extraterrestrial amino acids. *Science* **176**:172–173.

Schmitz, E., 1913. Über den Mechanismus der Acrose-bildung. *Chem. Berich.* **46**:2327–2335.

Scholes, G., Stein, G., and Weiss, J. 1949. Action of X-rays on nucleic acids. *Nature* **164**:709–710.

Scholes, G., and Weiss, J. 1953a. Formation of labile phosphate esters by irradiation of nucleic acids with X-rays in aqueous systems. *Nature* **171**:920–921.

Scholes, G., and Weiss, J. 1953b. Chemical action of X-rays on nucleic acids and related substances in aqueous systems. *Biochem. J.* **53**:567–578.

Schramm, G. 1965. Synthesis of nucleosides and polynucleotides with metaphosphate esters. *In*: Fox, S. W., ed., *The Origins of Prebiological Systems,* New York, Academic Press, p. 299–315.

Schramm, G. 1971. Synthesis and properties of polyarabinonucleotides. *In*: Kimball, A. P., and J. Oró, *Prebiotic and Biochemical Evolution,* Amsterdam, North-Holland Publishing Co., p. 95–101.

Schramm, G., Grotsch, H., and Pollman, W. 1961. Nicht-enzymatische synthese von polysacchariden Nucleosiden und Nucleinsäuren, *Angew. Chem.* **73**:619.

Schramm, G., Grotsch, H., and Pollman, W. 1962. Nonenzymatic synthesis of polysaccharides, nucleosides, and nucleic acids and the origin of self-reproducing systems. *Angew. Chem.* **74**:53–60.

Schramm, G., and Ulmer-Schürnbrand, I. 1967. Synthesis of polyspongouridylic acids from uridylic acids and phenylpolyphosphate esters. *Biochim. Biophys. Acta* **145**:7–20.

Schramm, G., and Wissman, H. 1958. Peptidsynthesen mit Hilfe von Polyphosphorsäureestern. *Chem. Berich.* **91**:1073–1082.

Schwartz, A. W. 1969. Specific phosphorylation of the 2′ and 3′ positions in ribonucleosides. *Chem. Commun.* **23**:1393.

Schwartz, A. W. 1971. Phosphate: Solubilization and activation on the primitive earth. *In*: Buvet, R., and C. Ponnamperuma, eds. *Chemical Evolution and the Origin of Life,* Amsterdam, North-Holland Publishing Co., p. 207–215.

Schwartz, A. W. 1972a. The sources of phosphorus on the primitive earth—an inquiry. *In*: Rohlfing, D. L., and A. I. Oparin, eds., *Molecular Evolution: Prebiological and Biological,* New York, Plenum Publishing Co., p. 129–140.

Schwartz, A. W. 1972b. Prebiotic phosphorylation-nucleotide synthesis with apatite. *Biochim. Biophys. Acta* **281**:477–480.

Schwartz, A. W., Bradley, E., and Fox, S. W. 1965. Thermal condensation of cytidylic acid in the presence of polyphosphoric acid. *In*: Fox, S. W., ed., *The Origins of Prebiological Systems,* New York, Academic Press, p. 317–326.

Schwartz, A. W., and Deuss, H. 1971. Concentrative processes and the origin of biological phosphates. *In*: Schwartz, A. W., ed., *Theory and Experiment in Exobiology,* Groningen, Netherlands, Wolters-Noordhoff Publishing Co., p. 75–81.

Schwartz, A. W., and Fox, S. W. 1967. Condensation of cytidylic acid in the presence of polyphosphoric acid. *Biochim. Biophys. Acta* **134**:9–16.

Schwartz, A. W., and Ponnamperuma, CX. 1971. Phosphorylation of nucleosides by condensed phosphates in aqueous systems. *In*: Kimball, A. P., and J. Oró, eds., *Prebiotic and Biochemical Evolution,* Amsterdam, North-Holland Publishing Co., p. 78–82.

Schwartz, A. W., van der Veen. M., Bisseling, T., and Chittenden, G. J. F. 1973, Prebiotic phosphorylation II: Nucleotide synthesis in the reaction system apatiteacyanogen-water. *Bio-Systems* **5**:119–122.

Schwartz, A. W., van der Veen, M., Bisseling T., and Chittenden, G. J. F. 1975. Prebiotic nucleotide synthesis demonstration of a geologically plausible pathway. *Origins Life* **6**:162–168.

Shimizu, M. 1975. Molten earth and the origin of prebiological molecules. *Origins Life* **6**:15–21.

Sillén, L. G. 1967. The ocean as a chemical system. *Science* **156**:1189–1196.

Simionescu, C., Dénes, F., and Macoveanu, M. 1973. Synthesis of some amino acids, sugars, and peptides in cold plasma. Electron-microscopic studies on some proteid forms (III) *Biopolymers* **12**:237–241.

Škoda, J., and Morávek, J. 1966. Formation uridylyl (3′→5′) uridine, uridylyl (2′→5′) uridine, 6-azauridylyl (3′→5′) 6-azauridine and 6-azauridylyl (2′→5′)-6-azauridine by thermic phosphorylation of the corresponding nucleosides with inorganic phosphate. *Tetrahedron Lett.* **1966**:4167–4172.

Smoluchowski, R. 1975. Jupiter 1975. *Am. Sci.* **63**:638–648.

Soffen, G. A. 1976. Scientific results of the Viking missions. *Science* **194**:1274–1276.

Stabaugh, M. R., Harrey, A. J. and Nagyvary, J. 1974. The possible role of inorganic thiophosphate as a prebiotic phosphorylating agent. *J. Mol Evol.* **3**:317–321.

Steinman, G. 1971. Nonenzymatic synthesis of biologically pertinent peptides. *In*: Kimball, A. P., and J. Oró, eds., *Prebiotic and Biochemical Evolution,* Amsterdam, North-Holland Publishing Co., p. 31–38.

Steinman, G., and Cole, M. N. 1967. Synthesis of biologically pertinent peptides under possible primordial conditions. *Proc. Nat. Acad. Sci. USA* **58**:735–742.

Steinman, G., and Cole, M. N. 1968. Residue interactions within peptide systems. *Fed. Proc.* **27**:765.

Steinman, G., Kenyon, D. H., and Calvin, M. 1965. Dehydration concentration in aqueous solution. *Nature* **206**:707–708.

Steinman, G., Lemmon, R. M., and Calvin, M. 1964. Cyanamide: A possible key compound in chemical evolution. *Proc. Nat. Acad. Sci. USA* **52**:27–30.

Steinman, G., Smith, A. E., and Silver J. J. 1968. Synthesis of a sulfur-containing amino acid under simulated prebiotic conditions. *Science* **159**:1108–1109.

Stephen-Sherwood, E., Oró, J., and Kimball, A. P. 1971. Thymine: A possible prebiotic synthesis. *Science* **173**:446–447.

Stephen-Sherwood, E., Odom, D. G., and Oró, J. 1974. The prebiotic synthesis of deoxythymidine oligonucleotides. *J. Mol. Evol.* **3**:323–330.

Sulston, J., Lohrmann, R., Orgel, L. E., and Miles, H. T. 1968a. Nonenzymatic synthesis of oligoadenylates on a polyuridylic acid template. *Proc. Nat. Acad. Sci. USA* **59**:726–733.

Sulston, J., Lohrmann, R., Orgel, L. E., and Miles, H. T. 1968b. Specificity of oligonucleotide synthesis directed by polyuridylic acid. *Proc. Nat. Acad. Sci. USA* **60**:409–415.

Sulston, J., Lohrmann, R., Orgel, L. E., Schneider-Bernloehr, H., and Weimann, B. J. 1969. Non-enzymatic oligonucleotide synthesis on a polycytidylate template. *J. Mol. Biol.* **40**:227–234.

Sverdrup, H. U., Johnson, M. W., and Fleming, R. H. 1946. *The Oceans: Their Physics, Chemistry, and Biology.* New York, Prentice-Hall Publishing Co.

Toupance, G., Raulin, F., and Buvet, R. 1975. Formation of prebiochemical compounds in models of the primitive earth's atmosphere. I. $CH_4 - NH_3$ and $CH_4 - N_2$ atmospheres. *Origins Life* **6**:83–90.

Tseng. S. S., and Chang, S. 1975. Photochemical synthesis of simple organic free radicals on simulated planetary surfaces—an ESR study. *Origins Life* **6**:61–73.

Ts'o, P. O. P. 1970. Monomeric units of nucleic acids—bases, nucleosides, and nucleotides. *In*: Fasman, G. D., and S. N. Timasheff, eds., *Fine Structure of Proteins and Nucleic acids.* New York, Marcel Dekker, Inc., p. 49–190.

Urey, H. C. 1952. *The Planets.* New Haven, Conn., Yale University Press.

Urey, H. C. 1960. Primitive planetary atmospheres and the origin of life. *In*: Florkin, M., ed., *Aspects of the Origin of Life,* New York, Pergamon Press, p. 8–14.

Usher, D. A., and McHale, A. H. 1976. Nonenzymatic joining of oligoadenylates on a poly A template. *Science* **192**:53–54.

Van Trump, J. E., and S. L. Miller, 1972. Prebiotic synthesis of methionine. *Science* **178**:859–860.

Van Wazer, J. R. 1958. *Phosphorus and its Compounds. Vol. 1*, New York, Interscience.
Waehneldt, T. J., and Fox, S. W. 1967. Phosphorylation of nucleosides with polyphosphoric acid. *Biochim. Biophys. Acta* **134**:1–8.
Wakamatsu, H., Yamada, Y., Saito, T., Kumashiro, I., and Takenishi, T. 1966. Synthesis of adenine by oligomerization of HCN. *J. Org. Chem.* **31**:2035–2037.
Wald, G. 1964. The origins of life. *Proc. Nat. Acad. Sci. USA* **52**:595–611.
Wald, G. 1974. Fitness in the universe: Choices and necessities. *In*: Oró, J., S. L. Miller, C. Ponnamperuma, and R. S. Young, eds., *Cosmochemical Evolution and the Origins of Life. Vol. 1*, Dordrecht, Holland, D. Reidel Publishing Co., p. 7–27.
Yamamoto, O. 1972a. Radiation-induced binding of cysteine and cystine with aromatic amino acids and serum albumin in aqueous solution. *Int. J. Radiat. Phys. Chem.* **4**:227–236.
Yamamoto, O. 1972b. Radiation-induced binding of methionine with serum albumin, tryptophan or phenylalanine in aqueous solution. *Int. J. Radiat. Phys. Chem.* **4**:335–345.
Yamamoto, O. 1973a. Radiation-induced binding of nucleic acid constituents with protein constituents and with each other. *Int. J. Radiat. Phys. Chem.* **5**:213–229.
Yamamoto, O. 1973b. Radiation-induced binding of phenylalanine, trypotophan, and histidine mutually and with albumin. *Radiat. Res.* **54**:398–410.
Yang, C. C., and Orò, J. 1971. Synthesis of adenine, guanine, cytosine, and other nitrogen organic compounds by a Fischer-Tropsch-like process. *In*: Buvet, R., and C. Ponnamperuma, eds., *Chemical Evolution and the Origin of Life,* Amsterdam, North-Holland Publishing Co., p. 155–170.
Yuasa, S., and Ishigami, M. 1975. High frequency discharge experiment. I. Formation of organic compounds from methane and ammonia. *Origins Life* **6**:75–81.

CHAPTER 2

Allen, D. E., and Gillard, R. D. 1967. Stereoselective effects in peptide complexes. *Chem. Commun.* **1967**:1091–1092.
Allen, W. V., and Ponnamperuma, C. 1967. Possible prebiotic synthesis of monocarboxylic acids. *Curr. Mod. Biol.* **1**:24–28.
Aronoff, S. 1975. The number of biologically possible porphyrin isomers. *Ann. N.Y. Acad. Sci.* **244**:327–333.
Bada, J. L. 1972. The dating of fossil bones using the racemization of isoleucine. *Earth Planet. Sci. Lett.* **15**:223–231.
Bada, J. L., and Schroeder, R. A. 1972. Racemization of isoleucine in calcareous marine sediments—kinetics and mechanism. *Earth Planet. Sci. Lett.* **15**:1–7.
Bada, J. L., Schroeder, R. A., and Carter, G. F. 1974. New evidence for the antiquity of man in North America deduced from aspartic acid racemization. *Science* **184**:791–793.
Bernal, J. D. 1967. *The Origin of Life,* New York, World Publishing Co.
Bonner, W. A., and Flores, J. J., 1973. On the asymmetric adsorption of phenylalanine enantiomers by kaolin. *Curr. Mod. Biol.* **5**:103–113.
Bonner, W. A., and Flores, J. J. 1975. Experiments on the origins of optical activity. *Origins Life* **6**:187–194.
Bonner, W. A., Kavasmaneck, P. R., Martin, F. S., and Flores, J. J. 1974. Asymmetric adsorption of alanine by quartz. *Science* **186**:143–144.
Bullock, E., and Elton, R. A. 1972. Dipeptide frequencies in proteins and the CpG deficiency in vertebrate DNA. *J. Mol. Evol.* **1**:315–325.
Calvin, M. 1956. Chemical evolution and the origin of life. *Am. Sci.* **44**:248–263.

Calvin, M. 1969. *Chemical Evolution: Molecular Evolution Towards the Origin of Living Systems on the Earth and Elsewhere,* Oxford, Oxford University Press.

Calvin, M. 1975. Chemical evolution. *Am. Sci.* **63**:169–177.

Chibnall, A. C., and Westall, R. W. 1932. The estimation of glutamine in the presence of asparagine. *Biochem. J.* **26**:122–132.

Darge, W., Sass, R., and Thiemann, W. 1973. Enzymatic hydrolysis of poly-DL-lysine. *Z. Naturforsch.* **28**:116–119.

Degens. E. T., Matheja, J., and Jackson, T. A. 1970. Template catalysis: asymmetric polymerization of amino-acids on clay minerals. *Nature* **227**:492–493.

de Jong, H. G. B. 1932. Die Koazervation und ihre Bedeutung für die Biologie. *Protoplasma* **15**:110–176.

de Jong, H. G. B. 1947. Distribution of the complex component, which is present in excess, between complex coacervate and equilibrium liquid. *Proc. K. Nederland. Akad. Wetenschap.* **50**:707–711.

Dillon, L. S. 1974. Neovulcanism: A proposed replacement for continental drift. *Mem. Amer. Assoc. Petrol. Geol.* **23**:167–239.

Dillon, L. S. 1978. *Evolution: Concepts and Consequences. 2nd Ed.,* St. Louis, C. V. Mosby Co.

Dose, K. 1971. Catalysis. *In*: Schwartz, A. W., ed., *Theory and Experiment in Exobiology,* Gronigen, Wolters-Noordhoff Publishing Co., Vol. 1, p. 43–71.

Dose, K. 1974. Chemical and catalytical properties of thermal polymers of amino acids (proteinoids). *Origins Life* **5**:239–252.

Dose, K., and Zaki, L. 1971. Hämoproteinoide mit perodatischer and katalatischer Aktivitat. *Z. Naturforsch.* **26b**:144–148.

Evreinova, T. N. 1964. Distribution of nucleic acids in coacervate droplets. *Dokl. Akad. Nauk. SSSR* **141**:246–249.

Evreinova, T. N., and Kuznetsova, A. 1959. Determination of the weight of separate coacervate drops by interference microscopy. *Dokl. Akad. Nauk. SSSR* **124**:688–691.

Evreinova, T. N., and Kuznetsova, A. 1961. Application of interference microscopy to coacervates. *Biofizika* **6**:320–328.

Evreinova, T. N., and Kuznetsova, A. 1963. Histone-protamine nucleic acid coacervate drops. *Biofizika* **8**:459–463.

Evreinova, T. N., Pogosova, A., Chukanova, T., and Larinovoa, T. 1962. Introduction of amino acids into coacervates. *Naunchn. Dokl. Vysshei Shkoly* **1**:159–164.

Evreinova, T. N., Mamontova, T. W., and Karnaukhov, V. N. 1972. Coacervate systems and evolution of matter. *In*: Rohlfing, D. L., and A. I. Oparin, eds., *Molecular Evolution: Prebiological and Biological,* New York, Plenum Press, p. 361–370.

Evreinova, T. N., Mamontova, T. W., Karnaukhov, V. N., Stephanov, S. B., and Hrust, U. R. 1974. Coacervate systems and origin of life. *Origins Life* **5**:201–205.

Fisher, H., and Orth, H. 1972. *Die Chemie des Pyrrols.* Vol. 2. Leipzig, Akademische Verlag.

Flatmark, T. 1964. Studies on the peroxidase effect of cytochrome *c*. II. Purification of beef heart cytochrome *c* by gel filtrations. *Acta. Chem. Scand.* **18**:1517–1527.

Flatmark, T. 1967. Multiple molecular forms of bovine heart cytochrome *c*. V. A comparative study of their physiochemical properties and their reactions in biological systems. *J. Biol. Chem.* **242**:2454–2459.

Flatmark, T., and Sletten, K. 1968. Multiple forms of cytochrome *c* in the rat. *J. Bio. Chem.* **243**:1623–1629.

Flores, J. J., and Bonner, W. A. 1974. On the asymmetric polymerization of aspartic acid enantiomers by kaolin. *J. Mol. Evol.* **3**:49–50.

Fox, R. F. 1972. A non-equilibrium thermodynamical analysis of the origin of life. *In*: Rohlfing, D. L., and A. I. Oparin, eds., *Molecular Evolution: Prebiological and Biological,* New York, Plenum Press, p. 79–99.

Fox, S. W. 1964. Thermal polymerization of amino acids and production of formed microparticles on lava. *Nature* **201**:336–337.

Fox, S. W. 1965. A theory of macromolecular and cellular origins. *Nature* **205**:328–339.

Fox, S. W. 1968. Abiotic polymerization and self-organization. *In*: Mark, H. F., N. G. Gaylord, and N. M. Bikales, eds., *Encyclopedia of Polymer Science and Technology*, *Vol. 9*, New York, Interscience, p. 284–314.

Fox, S. W. 1975. Stereomolecular interactions and microsystems in experimental protobiogenesis. *BioSystems* **7**:22–36.

Fox, S. W. 1976. The evolutionary significance of phase-separated microsystems. *Origins Life* **7**:49–68.

Fox, S. W., and Dose, K. 1972. *Molecular Evolution and the Origin of Life,* San Francisco, W. H. Freeman and Company.

Fox, S. W., and Harada, K. 1958. Thermal copolymerization of amino acids to a product resembling protein. *Science* **128**:1214.

Fox, S. W., and Harada, K. 1960. The thermal polymerization of amino acids common to protein. *J. Am. Chem. Soc.* **82**:3745–3752.

Fox, S. W., and Harada, K. 1961. Synthesis of uracil under conditions of a thermal model of prebiological chemistry. *Science* **133**:1923–1924.

Fox, S. W., Harada, K., and Vegotsky, A. 1959. Thermal polymerization of amino acids and a theory of biochemical origins. *Experientia* **15**:81–84.

Fox, S. W., Harada, K., Woods, K. R., and Windsor, C. R. 1963. Amino acid compositions of proteinoids. *Arch. Biochem. Biophys.* **102**:439–445.

Fox, S. W., McCauley, R. J., and Wood, A. 1967. A model of primitive heterotrophic proliferation. *Comp. Biochem. Physiol.* **20**:773–778.

Fox, S. W., Jungck, J. R., and Nakashima, T. 1974. From proteinoid microsphere to contemporary cell: Formation of internucleotide and peptide bonds by proteinoid particles. *Origins Life* **5**:227–237.

Fox, S. W., and Krampitz, G. 1964. Catalytic decomposition of glucose in aqueous solution by thermal proteinoids. *Nature* **203**:1362–1364.

Fox, S. W., and Suzuki, F. 1976. Linkages in thermal copolymers of lysine. *BioSystems* **8**:40–44.

Fox, S. W., and Yuyama, S. 1963. Abiotic production of primitive protein and formed microparticles. *Ann. N.Y. Acad. Sci.* **108**:487–494.

Frydman, B., Frydman, R. B., Valasinas, A., Levy, S., and Feinstein, G. 1975. The mechanism of uroporphyrinogen biosynthesis. *Ann. N.Y. Acad. Sci.* **244**:371–395.

Gilbert, J. B., Price, V. E., and Greenstein, J. P. 1949. Effect of anions on the coenzymatic deamidation of glutamine. *J. Biol. Chem.* **180**:209–218.

Goldacre, R. J. 1958. Surface films, their collapse on compression, the shapes and sizes of cells and the origin of life. *In:* Danielli, J. R., K. G. A. Pankhurst, and A. C. Riddiford, eds., *Surface Phenomena in Chemistry and Biology,* London, Pergamon Press, p. 276–278.

Harada, K., and Fox, S. W. 1975. Characterization of functional groups of acidic thermal polymers of α-amino acids. *BioSystems* **7**:213–221.

Hardebeck, H. G., Krampitz, G., and Wulf, L. 1968. Decarboxylation of pyruvic acid in aqueous solution by thermal proteinoids. *Arch. Biochem. Biophys.* **123**:72–81.

Harfenist, E. J. 1953. The amino acid compositions of insulins isolated from beef, pork, and sheep glands. *J. Am. Chem. Soc.* **75**:5528–5533.

Herrera, A. L. 1924. *Biologia y plasmogenia.* Mexico City, H. Hermanos Sucesores.

Herrera, A. L. 1942. A new theory of the origin and nature of life. *Science* **96**:14.

Hodgson, G. W., and Ponnamperuma, C. A. 1968. Prebiological porphyrin synthesis: Porphyrins from electric discharge in methane, ammonia, and water vapor. *Proc. Nat. Acad. Sci. USA* **59**:22–28.

Hsu, L. L. 1972. Conjugation of proteinoid microspheres: A model of primordial recombination.

In: Rohlfing, D. L., and A. I. Oparin, eds., *Molecular Evolution*: *Prebiological and Biological*, New York, Plenum Press, p. 371–378.

Hsu, L. L., and Fox, S. W. 1976. Interactions between diverse proteinoids and microspheres in simulation of primordial evolution. *BioSystems* **8**:89–101.

Jackson, A. H., and Games, D. E. 1975. The later stages of porphyrin biosynthesis. *Ann. N.Y. Acad. Sci.* **244**:591–601.

Jackson, T. A. 1971. Evidence for the selective adsorption and polymerization of the L-optical isomers of amino acids relative to the D-optical isomers on the edge faces of kaolinite. *Experientia* **27**:242–243.

Josse, J., Kaiser, A. D., and Kornberg, A. 1961. Enzymatic synthesis of DNA. VIII. *J. Biol. Chem.* **236**:864–875.

Jukes, T. H. 1966. *Molecules and Evolution*, New York, Columbia University Press.

Kambe, M., Sakamoto, Y., and Kurahashi, K. 1971. Biosynthesis of tyrocidine by a cell-free enzyme system of *Bacillus brevis* ATCC8185. IV. Further separation of component II into two fractions. *J. Biochem.* **69**:1131–1133.

Keim, P., Vigna, R. A., Morrow, J. S., Marshall, R. C., and Gurd, F. R. N. 1973. Carbon 13 nuclear magnetic resonance of pentapeptides of glycine containing central residues serine, threonine, aspartic and glutamic acids, asparagine, and glutamine. *J. Bio. Chem.* **248**:7811–7818.

Kenyon, D. H., and Steinman, G. 1969. *Biochemical Predestination*, New York, McGraw-Hill Book Company.

King, G. A. M. 1977. Symbiosis and the origin of life. *Origins Life* **8**:39–53.

Klabunoswkii, E. I. 1959. Absolute asymmetric synthesis and asymmetric catalysis. *In*: Oparin, A. I., ed., *Origin of Life on Earth*, London, Pergamon Press, p. 158–168.

Kleinkauf, H., and Gevers, W. 1969. Nonribosomal polypeptide synthesis: The biosynthesis of a cyclic peptide antibiotic, gramidicin S. *Cold Spring Harbor Symp. Quant. Biol.* **34**:805–813.

Krampitz, G. 1959. Untersuchungen und Aminosäure-Kopolymerisaten. *Naturwissenschaften* **46**:558.

Krampitz, G., Diehl, S., and Nakashima, T. 1967. Aminotransferase-Aktivität von Polyanhydro-α-Aminosäuren (Proteinoiden). *Naturwissenschaften* **54**:516–517.

Krampitz, G., Haas, W., and Baars-Diehl, S. 1968. Glutaminsäure-Oxydoreduktase-Aktivität von Polyanhydro-α-Aminosäuren (Proteinoiden). *Naturwissenschaften* **55**:345–346.

Lederberg, J. 1960a. A view of genetics. *Science* **131**:269–276.

Lederberg, J. 1960b. Exobiology: Approaches to life beyond the earth. *Science* **132**:393–400.

Lipmann, F. 1971. Attempts to map a process evolution of peptide biosynthesis. *Science* **173**:875–884.

Lipmann, F. 1972. A mechanism for polypeptide synthesis on a protein template. *In*: Rohlfing, D. L., and A. I. Oparin, eds., *Molecular Biology: Prebiological and Biological*, New York, Plenum Press, p. 261–269.

Lipmann, F., Gevers, W., Kleinkauf, H., and Roskoski, R. 1971. Polypeptide synthesis on protein templates: The enzymatic synthesis of gramicidin S and tyrocidine. *Adv. Enzymol.* **35**:1–34.

McCullough, J. J., and Lemmon, R. M. 1974. The question of the possible asymmetric polymerization of aspartic acid on kaolinite. *J. Mol. Evol.* **3**:57–61.

Noda, H., Mizutani, H., and Okihana, H. 1975. Marcromolecules and the origin of life. *Origins Life* **6**:139–146.

Noguchi, J., and Saito, T. 1962. *In:* Stahmann, M., ed. *Polyamino Acids, Polypeptides, and Proteins*. University Wisconsin Press, Madison, p. 313.

Oparin, A. I. 1957. *The Origin of Life on the Earth. 3rd Ed.*, New York, Academic Press.

Oparin, A. I. 1968. *Genesis and Evolutionary Development of Life*, New York, Academic Press.

Oparin, A. I. 1971. Coacervate drops as models of prebiological systems. *In*: Kimball, A. P., and J. Oró, eds., *Prebiotic and Biochemical Evolution*, Amsterdam, North-Holland Publishing Co., p. 1–7.

Oparin, A. I. 1974. A hypothetical scheme for evolution of protobionts. *Origins Life* **5**:223–226.

Oparin, A. I., and Serebrovskaya, K. 1958. Activity of ribonuclease included into coacervate droplets. *Dokl. Akad. Nauk SSSR* **122**:197–200.

Oparin, A. I., Evreinova, T. N., Larionova, T. I., and Davydova, I. M. 1962. Synthesis and degradation of starch in coacervate droplets. *Dokl. Akad. Nauk SSSR* **143**: 980–983.

Oparin, A. I., Serebrovskaya, K. B., Pantskava, S., and Vasil'eva, N. 1963. Enzymic synthesis of polyadenylic acid in coacervate drops. *Biokhimiya* **28**:671–676.

Oparin, A. I., Serebrovskaya, K. B., Vasil'eva, N. V., and Balaevskaya, T. O. 1964. [The formation of coacervates from polypeptides and polynucleotides]. *Dokl. Akad. Nauk SSSR* **154**:471–472.

Pattee, H. H. 1965. The recognition of hereditary order in primitive chemical systems. *In*: Fox, S. W., ed., *The Origins of Prebiological Systems*, New York, Academic Press, p. 385–405.

Robertson, J. D. 1959. Molecular theory of cell membrane structure. *Verh. Internat. Kongr. E. M.* **4**:159–171.

Robertson, J. D. 1964. Unit membranes: A review with recent new studies of experimental alterations and a new subunit structure in synaptic membranes. *In*: Locke, M., ed., *Cellular Membranes in Development,* New York, Academic Press, p. 1–81.

Robinson, A. B. 1974. Evolution and the distribution of glutaminyl and asparaginyl residues in proteins. *Proc. Nat. Acad. Sci. USA* **71**:885–888.

Robinson, A. B., Irving, K., and McCrea, M. 1973a. Acceleration of the rate of deamidation of Gly Arg Asn Arg Gly and of human transferrin by addition of L-ascorbic acid. *Proc. Nat. Acad. Sci. USA* **70**:2122–2123.

Robinson, A. B., Scotchler, J. W., and McKerrow, J. H. 1973b. Rates of nonenzymatic deamidation of glutaminyl and asparaginyl residues in pentapeptides. *J. Am. Chem. Soc.* **95**:8156–8159.

Rohlfing, D. L. 1967. The catalytic decarboxylation of oxaloacetic acid by thermally prepared poly-α-amino acids. *Arch. Biochem. Biophys.* **118**:468–474.

Rohlfing, D. L. 1975. Coacervate-like microspheres from lysine-rich proteinoid. *Origins Life* **6**:203–209.

Rohlfing, D. L., and Fox, S. W. 1967. The catalytic activity of thermal polyanhydro-α-amino acids for the hydrolysis of *p*-nitrophenyl acetate. *Arch. Biochem. Biophys.* **118**:122–126.

Roskoski, R., Gevers, W., Kleinkauf, H., and Lipmann, F. 1970. Tyrocidine biosynthesis by three complementary fractions from *Bacillus brevis* (ATCC8185). *Biochemistry* **9**:4839–4845.

Rubey, W. W. 1951. Geologic history of sea water: An attempt to state the problem. *Bull. Geol. Soc. Am.* **62**:1111–1148.

Rubey, W. W. 1955. Development of the hydrosphere. *Spec. Geol. Soc. Am. Pap.* **62**:631–650.

Russell, C. S. 1974. Biosynthesis of porphyrins. II. *J. Theor. Biol.* **47**:145–151.

Rutten, M. G. 1971. *The Origin of Life by Natural Causes.* Amsterdam, Elsevier Publishing Co.

Schneider-Bernloehr, H., Lohrmann, R., Orgel, L. E., Sulston, J., and Weimann, B. J. 1968. Partial resolution of DL-adenosine by template synthesis. *Science* **162**:809–810.

Schoffeniels, E. 1967. *Cellular Aspects of Membrane Permeability*, New York, Pergamon Press.

Serebrovskaya, K. B., and Lozovaya, G. I. 1972. Modelling of structure and functional unity on coacervate systems. *In*: Rohlfing, D. L., and A. I. Oparin, eds., *Molecular Evolution: Prebiological and Biological*, New York, Plenum Press, p. 353–360.

Shemin, D. 1975. Porphyrin synthesis: Some particular approaches. *Ann. N.Y. Acad. Sci.* **244**:348–355.

Steinman, G. 1967a. Sequence generation in prebiological peptide synthesis. *Arch. Biochem. Biophys.* **119**:76–82.

Steinman, G. 1967b. Sequence generation in prebiological peptide synthesis. *Arch. Biochem. Biophys.* **121**:533–539.

Steinman, G., and Cole, M. N. 1967. Synthesis of biologically pertinent peptides under possible primordial conditions. *Proc. Nat. Acad. Sci. USA* **58**:735–742.

Subak-Sharpe, H., Bürk, R. R., Crawford, L. V., Morrison, J. M., Hay, J., and Keir, H. M. 1966. An approach to evolutionary relationships of mammalian DNA viruses through analysis of the pattern of nearest neighbor base sequences. *Cold Spring Harbor Symp. Quant. Biol.* **31**:737–748.

Swartz, M. N., Trautner, T. A., and Kornberg, A. 1962. Enzymatic synthesis of DNA. XI. Further studies on nearest neighbor base sequences in DNA. *J. Bio. Chem.* **237**:1961–1967.

Thiemann, W. 1974. The origin of optical activity. *Naturwissenschaften* **61**:1476–1483.

Thiemann, W., and Darge, W. 1974. Experimental attempts for the study of the origin of optical activity on earth. *Origins Life* **5**:263–283.

Usdin, V. R., Mitz, M. A., and Killos, J. 1967. Inhibition and reactivation of the catalytic activity of a thermal α-amino acid copolymer. *Arch. Biochem. Biophys.* **122**:258–261.

Vegotsky, A., and Fox, S. W. 1959. Pyropolymerization of amino acids to proteinoids with phosphoric acid or polyphosphoric acid. *Fed. Proc.* **18**:343.

West, E. S., and Todd, W. R. 1961. *Textbook of Biochemistry.* 3rd Ed., New York, Macmillan Company.

Wood, A., and Hardebeck, H. G. 1972. Light-enhanced decarboxylations by proteinoids. *Mol. Evol.* **1972**:233–245.

CHAPTER 3

Abelson, P. H. 1957. Some aspects of paleobiochemistry. *Ann. N.Y. Acad. Sci.* **69**:274–285.

Adams, R. L. P. 1974. Newly synthesised DNA is not methylated. *Biochim. Biophys. Acta* **335**:365–373.

Adams, R. L. P., and Hogarth, C. 1973. DNA methylation in isolated nuclei: Old and new DNAs are methylated. *Biochem. Biophys. Acta* **331**:214–220.

Adler, K., Beyreuther, K., Fanning, E., Geisler, N., Gronenborn, B., Klemm, A., Müller-Hill, B., Pfahl, M., and Schmitz, A. 1972. How *lac* repressor binds to DNA. *Nature* **237**:322–327.

Alberts, B., Amodio, F., Jenkins, M., Gutmann, E., and Ferris, F. 1968. DNA-binding proteins from *E. coli. Cold Spring Harbor Symp. Quant. Biol.* **33**:289–305.

Alberts, B., and Frey, L. 1970. T4 bacteriophage gene 32. *Nature* **227**:1313–1318.

Andersen, H. A. 1974. Replication of macronuclear DNA in the cytoplasm of *Tetrahymena pyriformis. J. Cell Sci.* **14**:289–300.

Antonov, A. C., Favorova, O. O., and Belozerskii, A. N. 1962. *Dokl. Akad. Nauk SSSR* **147**:1480.

Aposhian, H. V., and Kornberg, A. 1962. The polymerase formed after T2 bacteriophage infection of *E. coli*: A new enzyme. *J. Biol. Chem.* **237**:519–525.

Arber, W., and Linn, S. 1969. Messenger RNA. *Ann. Rev. Biochem.* **38**:647–676.

Arnberg, A. C., and Arwert, F. 1976. DNA-protein complex in circular DNA from *Bacillus* bacteriophage GA-1. *J. Virol.* **18**:783–784.

Arnott, S., and Hukins, D. W. L. 1972. Optimized parameters for A-DNA and B-DNA. *Biochem. Biophys. Res. Com.* **47**:1504–1509.

Arnott, S., and Hukins, D. W. L. 1973. Refinement of the structure of B-DNA and implications for the analysis of X-ray diffraction data from fibers of biopolymers. *J. Mol. Biol.* **81**:93–105.

Arrighi, F. E., and Hsu, T. C. 1971. Localization of heterochromatin in human chromosomes. *Cytogenetics* **10**:81–86.

Auld, D. S., Kawaguchi, H., Livingston, D. M., and Vallee, B. L. 1974. RNA-dependent DNA polymerase from avian myeloblastosis virus: A zinc metalloenzyme. *Proc. Nat. Acad. Sci. USA* **71**:2091–2095.

Baase, W. A., and Wang, J. C. 1974. An ω protein from *Drosophila melanogaster. Biochemistry* **13**:4299–4303.

Bada, J. L. 1972. The dating of fossil bones using the racemization of isoleucine. *Earth Planet. Sci. Lett.* **15**:223–231.

Bada, J. L., Luyendyk, B. P., and Maynard, J. B. 1970. Marine sediments: Dating by the racemization of amino acids. *Science* **170**:730–732.

Bada, J. L., and Protsch, R. 1973. Racemization reaction of aspartic acid and its use in dating fossil bones. *Proc. Nat. Acad. Sci. USA* **70**:1331–1334.

Balhorn, R., Chalkley, R., and Granner, D. 1972. Lysine-rich histone phosphorylation. A positive correlation with cell replication. *Biochemistry* **11**:1094–1098.

Balhorn, R., Jackson, V., Granner, D., and Chalkley, R. 1975. Phosphorylation of the lysine-rich histones throughout the cell cycle. *Biochemistry* **14**:2504–2511.

Balhorn, R., Rieke, W. O., and Chalkley, R. 1971. A reinvestigation of phosphorylation of lysine-rich histone during rat liver regeneration. *Biochemistry* **10**:3952–3958.

Balls, M., and F. S. Billett, eds. 1973. *The Cell Cycle in Development and Differentiation.* Cambridge, Cambridge University Press.

Baltimore, D. 1970. Viral RNA-dependent DNA polymerase. *Nature* **226**:1209–1211.

Banks, G. R., and Spanos, A. 1975. The isolation and properties of a DNA unwinding protein from *Ustilago maydis*. *J. Mol. Biol.* **93**:63–77.

Baril, E. F., Brown, O. E., Jenkins, M. D., and Laszlo, J. 1971. DNA polymerase with rat liver ribosomes and smooth membranes. *Biochemistry* **10**:1981–1992.

Baril, E. F., Jenkins, M. D., Brown, O. E., Laszlo, J., and Morris, H. P. 1973. DNA polymerase I and II in regenerating rat liver and Morris hepatomas. *Cancer Res.* **33**:1187–1193.

Baril, E. F., and Laszlo, J. 1971. Sub-cellular localization and characterization of DNA polymerases from rat liver and hepatomas. *Adv. Enzyme Regulation* **9**:183–204.

Barker, S. T., Kurtz, H., Taylor, B. A., and Ackerman, W. W. 1973. A covalently linked DNA-RNA molecule from human leukemia cells. *Biochem. Biophys. Res. Comm.* **50**:1068–1074.

Barry, J., Hama-Inaba, H., Moran, L., Alberts, B., and Wilberg, J. 1973. Proteins of the T4 bacteriophage replication apparatus. *In*: Wells, R. D., and R. B. Inman, eds., *DNA Synthesis in Vitro,* Baltimore, Md., Univ. Park Press, p. 195–214.

Barzilai, R., and Thomas, C. A. 1970. Spontaneous renaturation of newly-synthesized bacteriophage T7 DNA. *J. Mol. Biol.* **51**:145–156.

Baudy, P., Bram, S., Vastel, D., Lepault, J., and Kitzis, A. 1976. Chromatin subunit small angle neutron scattering. *Biochem. Biophys. Res. Comm.* **72**:176–183.

Bekkaring-Kuylaars, S. A. M., and Campagnari, F. 1974. Characterization and properties of a DNA polymerase partially purified from the nuclei of calf thymus cells. *Biochim. Biophys. Acta* **349**:277–295.

Berendes, H. D., and Keyl, H. G. 1967. Distribution of DNA in heterochromatin and euchromatin of polytene nuclei of *Drosophila hydei*. *Genetics* **57**:1–13.

Berezney, R., and Coffey, D. S. 1975. Nuclear protein matrix: Association with newly synthesized DNA. *Science* **189**:291–293.

Beridze, T. 1975. DNA nuclear satellites of the genus *Brassica*: Variation between species. *Biochim. Biophys. Acta* **395**:274–279.

Bernal, J. D. 1967. *The Origin of Life,* Cleveland, Ohio, World Publishing Co.

Bernard, O., Momparler, R. L., and Brent, T. P. 1974. Effect of DNA polymerase on nuclei from different phases of cell cycle. *Eur. J. Biochem.* **49**:565–571.

Bersier, D., and Braun, R. 1974. Pools of deoxyribonucleoside triphosphates in the mitotic cycle of *Physarum*. *Biochim. Biophys. Acta* **340**:463–471.

Bianchi, N. O., and Ayres, J. 1971. Polymorphic patterns of heterochromatin distribution in guinea pig chromosomes. *Chromosoma* **34**:254–260.

Billen, D. 1968. Methylation of the bacterial chromosome: An event at the "replication point". *J. Mol. Biol.* **31**:477–486.

Billett, M. A., and Barry, J. M. 1974. Role of histones in chromatin condensation. *Eur. J. Biochem.* **49**:477–484.

Birnboim, H. C., and Straus, N. A. 1975. DNA from eukaryotic cells contains unusually long pyrimidine sequences. *Can. J. Biochem.* **53**:640–643.

Birnstiel, M. L., and Hyde, B. B. 1963. Protein synthesis by isolated pea nucleoli. *J. Cell Biol.* **18**:41–50.

Bishop, J. M., Faras, A. J., Garapin, A. C., Goodman, H. M., Levinson, W. E., Stavneger, J., Taylor, J. M., and Varmus, H. E. 1973. Characteristics of the transcription of RNA by the DNA polymerase of Rous sarcoma virus. *In*: Wells, R. D., and R. B. Inman, eds., *DNA Synthesis in Vitro,* Baltimore, Md., Univ. Park Press, p. 341–359.

Blakley, R. L., and Vitols, E. 1968. The control of nucleotide biosynthesis. *Ann. Rev. Biochem.* **37**:201–224.

Boffa, L., Saccomani, G., Tamburro, A. M., Scatturin, A., and Vidali, G. 1971. Chromosomal nucleoproteins: C.D. studies on reconstituted nucleohistones from avian erythrocytes. *Int. J. Protein Res.* **3**:357–363.

Bolotin, M., Coen, D., Deutsch, J., Dujon, B., Netter, P., Petrochilo, E., and Slonimski, P. P. 1971. La recombinaison des mitochondries chez *S. cerevisiae*. *Bull. Inst. Pasteur* **69**:215–239.

Bonner, J., Dahmus, M. E., Fambrough, D., Huang, R. C., Marushige, K., and Tuan, D. Y. H. 1968. The biology of isolated chromatin. *Science* **159**:47–56.

Bonner, W. M. 1975. Protein migration into nuclei. *J. Cell Biol.* **64**:431–437.

Botchan, M. R., 1974. Bovine satellite I DNA consists of repetitive units 1,400 base pairs in length. *Nature* **251**:288–292.

Botchan, M. R., Kram, R., Schmid, C., and Hearst, J. E. 1971. Isolation and chromosomal localization of highly repeated DNA sequences in *D. melanogaster*. *Proc. Nat. Acad. Sci. USA* **68**:1125–1129.

Boyer, H. W., 1971. DNA restriction and modification mechanisms in bacteria. *Ann. Rev. Microbiol.* **25**:153–176.

Bayer, H. W., 1974. Restriction and modification of DNA: Enzymes and substrates. *Fed. Proc.* **33**:1125–1127.

Bradbury, E. M., Inglis, R. J., Matthews, H. R., and Langan, T. A. 1974. Molecular basis of control of mitotic cell division in eukaryotes. *Nature* **249**:553–555.

Brahms, J., and Mommaerts, W. H. F. M. 1964. A study of conformation of nucleic acids in solution by means of circular dichroism. *J. Mol. Biol.* **10**:73–88.

Bram, S. 1975. A double coil chromatin subunit model. *Biochimie* **57**:1301–1306.

Bram, S., Hellio, R., and Kouprach, S. 1975. Études préliminaires des sous-unités de la chromatine par cryodécapage. *C. R. Acad. Sci., Paris* **D281**:847–849.

Brandt, W. F., and von Holt, C. 1976. The occurrence of histone H3 and H4 in yeast. *FEBS Lett.* **65**:386–390.

Brasch, K., Adams, G. H. M., and Neelin, J. M. 1974. Evidence for erythrocyte-specific histone modification and structural changes in chromatin during goose erythrocyte maturation. *J. Cell Sci.* **15**:659–677.

Brasch, K., and Setterfield, G. 1974. Structural organization of chromosomes in interphase nuclei. *Expt. Cell Res.* **83**:175–185.

Brasch, K., Setterfield, G., and Neelin, J. M. 1972. Effects of sequential extraction of histone proteins on structural organization of avian erythrocyte and liver nuclei. *Exp. Cell Res.* **74**:27–41.

Brewer, E. N. 1972. DNA replication in *Physarum polycephalum*. *J. Mol Biol.* **68**:401–412.

Britten, R. J. 1968. Reassociation of nonrepeated DNA. *Carnegie Inst. Year Book* **67**:330–332.

Britten, R. J., and Kohne, D. E. 1968. Repeated sequences in DNA. *Science* **161**:529–540.

Brown, D. D., and Sugimoto, K. 1973. 5 S DNAs of *Xenopus laevis* and *Xenopus mulleri*: Evolution of a gene family. *J. Mol. Biol.* **78**:397–415.

Brown, I. R., and Church, R. B. 1971. RNA transcription from nonrepetitive DNA in mouse. *Biochem. Biophys. Res. Comm.* **42**:850–856.

Brown, R. L., Pathak, S., and Hsu, T. C. 1975. The possible role of histones in the mechanism of chromosomal G banding. *Science* **189**:1090–1091.

Brown, W. V., and Bertke, E. M. 1974. *Textbook of Cytology*, 2nd Ed., St. Louis, Missouri, The C. V. Mosby Company.

Brunk, C. F., and Hanawalt, P. C. 1967. Repair of damaged DNA in a eucaryotic cell: *Tetrahymena pyriformis*. *Science* **158**:663–664.

Brutlag, D., and Kornberg, A. 1972. A proofreading function for the 3′-5′ exonuclease activity in DNA polymerases. *J. Biol. Chem.* **247**:241–248.

Buchowicz, J. 1974. Is the cytoplasm a site of nuclear DNA synthesis? *Nature* **249**:350.

Burgoyne, L. A. 1972. DNA synthesis in mammalian systems. *Biochem. J.* **130**:959–964.

Byrd, E. W., and Kasinsky, H. E. 1973. Histone synthesis during early embryogenesis in *Xenopus laevis*. *Biochemistry* **12**:246–253.

Byrnes, J. J., Downey, K. M., Jurmark, B. S., and So, A. G. 1974a. Reticulocyte DNA polymerase. *Nature* **248**:687–689.

Byrnes, J. J., Downey, K. M., and So, A. G. 1974b. Metabolic regulation of cytoplasmic DNA synthesis. *Proc. Nat. Acad. Sci. USA* **71**:205–208.

Cabradilla, C. D., and Toliver, A. P. 1975. S-phase dependent forms of DNA–nuclear membrane complexes in HeLa cells. *Biochim. Biophys. Acta* **402**:188–198.

Cairns, J. 1963a. The bacterial chromosome and its manner of replication as seen by autoradiography. *J. Mol. Biol.* **6**:208–213.

Cairns, J. 1963b. The chromosome of *E. coli*. *Cold Spring Harbor Symp. Quant. Biol.* **28**:43–46.

Cairns-Smith, A. G. 1966. The origin of life and the nature of the primitive gene. *J. Theoret. Biol.* **10**:53–88.

Cairns-Smith, A. G. 1971. *The Life Puzzle: On Crystals and Organisms and on the Possibility of a Crystal as an Ancestor*, Toronto, University of Toronto Press.

Callan, H. G. 1973. Replication of DNA in eukaryotic organism. *Brit. Med. Bull.* **29**:192–195.

Cameron, I. L., and Nachtwey, D. S. 1967. DNA synthesis in relation to cell division in *Tetrahymena pyriformis*. *Exp. Cell Res.* **46**:385–395.

Capesius, I., Bierweiler, B., Bachmann, K., Rucker, W., and Nagl, W. 1975. An A + T-rich satellite DNA in a monocotyledonous plant, *Cymbidium*. *Biochim. Biophys. Acta* **395**:67–73.

Carlton, B. C., and Smith, M. P. W. 1974. Size distribution of the closed circular DNA molecules of *Bacillus megaterium*. *J. Bact.* **117**:1201–1209.

Case, S. T., and Baker, R. F. 1975. Position of regularly spaced single-stranded regions relative to 5-bromodeoxyuridine sensitive areas in sea urchin morula DNA. *Nature* **253**:64–66.

Case, S. T., Mongeon, R. L., and Baker, R. F. 1974. Single-stranded regions in DNA isolated from different developmental stages of the sea urchin. *Biochim. Biophys. Acta* **349**:1–12.

Cavalieri, L. F. 1963. Nucleic acids and information transfer. *J. Cell. Comp. Physiol.* **62**; *suppl.* **1**:111–122.

Cavalieri, L. F., and Carroll, E. 1970. RNA as a template with *E. coli* DNA polymerase. *Biochem. Biophys. Res. Comm* **41**:1055–1060.

Cavalieri, L. F., and Rosenberg, B. H. 1963. Nucleic acids and information transfer. *Progr. Neucleic Acid Res.* **2**:1–18.

Cech, T. R., and Hearst, J. E. 1976. Organization of highly repeated sequences in mouse main-band DNA. *J. Mol. Bio.* **100**:227–256.

Champoux, J. J., and Dulbecco, R. 1972. An activity from mammalian cells that untwists superhelical DNA—a possible swivel for DNA replication. *Proc. Nat. Acad. Sci. USA* **69**:143–146.

Champoux, J. J., and McConaughy, B. L. 1975. Priming of superhelical SV40 DNA by *E. coli* RNA polymerase for *in vitro* DNA synthesis. *Biochemistry* **14**:307–316.

Chandra, H. S., and Brown, S. W. 1975. Chromosome imprinting and the mammalian X chromosome. *Nature* **253**:165–168.

Chang, L. M. S., and Bollum, F. J. 1972a. Antigenic relationships in mammalian DNA polymerase. *Science* **175**:1116–1117.

Chang, L. M. S., and Bollum, F. J. 1972b. Low molecular weight DNA polymerase from rabbit bone marrow. *Biochemistry* **11**:1264–1272.

Chang, L. M. S., and Bollum, F. J. 1972c. Variation of DNA polymerase activities during rat liver regeneration. *J. Biol. Chem.* **247**:7948–7950.

Chang, L. M. S., Brown, M., and Bollum, F. J. 1973. Induction of DNA polymerase in mouse L cells. *J. Mol. Biol.* **74**:1–8.

Chapman, G. E., Hartman, P. G., and Bradbury, E. M. 1976. Studies on the role and mode of operation of the very-lysine-rich histone H1 in eukaryote chromatin. *Eur. J. Biochem.* **61**:69–75.

Chargoff, E. 1950. Chemical specificity of nucleic acids and mechanism of their enzymatic degradation. *Experientia* **6**:201–209.

Chargoff, E. 1951. Structure and function of nucleic acids as cell constituents. *Fed. Proc.* **10**:654–659.

Chiu, J. F., and Hnilica, L. S. 1977. Nuclear nonhistone proteins: Chemistry and function. *In*: Li, H. J., and R. A. Eckhardt, eds., *Chromatin and Chromosome Structure*. New York, Academic Press, p. 193–254.

Chin, B., and Bernstein, I. A. 1968. ATP and synchronous mitosis in *Physarum polycephalum*. *J. Bacteriol.* **96**:330–337.

Clark, R. J., and Felsenfeld, G. 1974. Chemical probes of chromatin structure. *Biochemistry* **13**:3622–3628.

Clark-Walker, G. D., and Miklos, G. L. G. 1974. Localization and quantification of circular DNA in yeast. *Eur. J. Biochem.* **41**:359–365.

Clowes, R. C. 1972. Molecular structure of bacterial plasmids. *Bacteriol. Rev.* **36**:361–405.

Comings, D. E., Avelino, E., Okada, T. A., and Wyandt, H. E. 1973. The mechanism of C- and G-banding of chromosomes. *Exp. Cell Res.* **77**:469–493.

Comings, D. E., and Kakefuda, T. 1968. Initiation of DNA replication at the nuclear membrane in human cells. *J. Mol. Biol.* **33**:225–229.

Comings, D. E., and Mattoccia, E. 1972. DNA of mammalian and avian heterochromatin. *Exp. Cell Res.* **71**:113–131.

Comings, D. E., and Okada, T. A. 1976. Fine structure of the heterochromatin of the kangaroo rat *Dipodomys ordii* and examination of the possible role of actin and myosin in heterochromatin condensation. *J. Cell Sci.* **21**:465–477.

Cook, J. S., and McGrath, J. R. 1967. Photoreactivating-enzyme activity in Metazoa. *Proc. Nat. Acad. Sci. USA* **58**:1359–1365.

Cooper, J. E. K., and Hsu, T. C. 1972. The C-band and G-band patterns of *Microtus agrestis* chromosomes. *Cytogenetics* **11**:295–304.

Coulter, M., Flintoff, W., Paetkan, V., Pulleyblank, D., and Morgan, A. R. 1974. *In vitro* synthesis and detection of DNAs with co-valently linked complementary sequences. *Biochemistry* **13**:1603–1609.

Courtois, Y., Dastugue, B., Kamiyama, M., and Kruk, J. 1975. Binding of chromosomal non-histone proteins to DNA and to nucleohistones. *FEBS Lett.* **50**:253–256.

Crick, F. H. C., 1971. General model for the chromosomes of higher organisms. *Nature* **234**:25–27.

Croes, A. F. 1966. Duplication of DNA during meiosis in baker's yeast. *Exp. Cell Res.* **41**:452–468.

Crothers, D. M. 1969. On the mechanism of DNA unwinding. *Acc. Chem. Res.* **2**:225–232.

Cummins, J. E., and Rusch, H. P. 1966. Limited DNA synthesis in the absence of protein synthesis in *Physarum polycephalum*. *J. Cell Biol.* **31**:577–583.

D'Anna, J. A., and Isenberg, I. 1974. A histone cross-complexing pattern. *Biochemistry* **13**:4992–4997.

Darlington, C. D. 1942. Chromosome chemistry and gene action. *Nature* **149**:66–69.

Davidson, J. N. 1963. Biochemical aspects of normal and malignant growth. *Scottish Med. J.* **8**:87–96.

Davidson, J. N. 1972. *The Biochemistry of the Nucleic Acids. 7th Ed.,* New York, Academic Press.

Davidson, R. H., Hough, B. R., Chamberlin, M. E., and Butten, R. J. 1974. Sequence repetition in the DNA of *Nassaria (Ilyanassa) obsoleta. Dev. Biol.* **25**:445–463.

Davies, D. R. 1967. X-ray diffraction studies of macromolecules. *Ann. Rev. Biochem.* **36**:321–364.

Davis, F. C. 1975. Unique sequence DNA transcripts present in mature oocytes of *Urechis canpo. Biochim. Biophys. Acta* **390**:33–45.

Dawid, I. B. 1974. 5-methylcytidylic acid: Absence from mitochondrial DNA of frogs and HeLa cells. *Science* **184**:80–81.

Day, R. O., Seeman, N. C., Rosenberg, J. M., and Rich, A. 1973. A crystalline fragment of the double helix. *Proc. Nat. Acad. Sci. USA* **70**:849–853.

Delius, H., Mantell, N. J., and Alberts, B. 1972. Characterization by electronmicroscopy of the complex formed between T4 bacteriophage gene 32-protein and DNA. *J. Mol. Biol.* **67**:341–350.

DeLucia, P., and Cairns, J. 1969. Isolation of an *E. coli* strain with a mutation affecting DNA polymerase. *Nature* **224**:1164–1166.

Deutscher, M. P., and Kornberg, A. 1969a. The pyrophosphate exchange and pyrophosphorolysis reactions of DNA polymerase. *J. Biol. Chem.* **244**:3019–3028.

Deutscher, M. P., and Kornberg, A. 1969b. Hydrolysis of DNA from the 5′-terminus by an exonuclease function of DNA polymerase. *J. Biol. Chem.* **244**:3029–3027.

De Waard, A., Paul, A. V., and Lehman, I. R. 1965. The structural gene for DNA polymerase in bacteriophages T4 and T5. *Proc. Nat. Acad. Sci. USA* **54**:1241–1248.

Dickson, R. C., Abelson, J., Barnes, W. M., and Reznikoff, W. S. 1974. Genetic regulation: The lac control region. *Science* **187**:27–35.

Dingman, C. W. 1974. Bidirectional chromosome replication. *J. Theor. Biol.* **43**:187–195.

Dingman, C. W., Fisher, M. P., and Ishizawa, M. 1974. DNA replication in *Escherichia coli:* Physical and kinetic studies of the replication point. *J. Mol. Biol.* **84**:275–295.

Dobner, P., and Flickinger, R. A. 1976. Repetitive DNA transcripts in frog embryo cytoplasm. *Biochim. Biophys. Acta* **432**:401–403.

Donnelly, G., and Sisken, J. E. 1967. RNA and protein synthesis required for entry of cells into mitosis and during the mitotic cycle. *Exp. Cell Res.* **46**:93–105.

Drets, M. E., and Shaw, M. W. 1971. Specific banding patterns of human chromosomes. *Proc. Nat. Acad. Sci. USA* **68**:2073–2077.

Dujon, B., Slonimski, P. P., and Weill, L. 1974. A model for recombination and segregation of mitochondrial genomes in *S. cerevisiae. Genetics* **78**:415–437.

Dunn, D. B., and Smith, J. D. 1958. The occurrence of 6-methylaminopurine in DNA. *Biochem. J.* **68**:627–636.

DuPraw, E. J. 1965a. The organization of nuclei and chromosomes in honeybee embryonic cells. *Proc. Nat. Acad. Sci. USA* **53**:161–168.

DuPraw, E. J. 1965b. Macromolecular organization of nuclei and chromosomes. *Nature* **206**:338–343.

DuPraw, E. J. 1968. *Cell and Molecular Biology,* New York, Academic Press.

DuPraw, E. J., and Bahr, G. F. 1968. Abstr. *Publ. 3rd Intern. Congr. Histochem. Cytochem., New York* (XII, XVII, XVIII).

Ehrlich, M., Ehrlich, K., and Mayo, J. A. 1975. Unusual properties of the DNA from *Xanthomonas* phage XP-12 in which 5-methylcytosine completely replaces cytosine. *Biochim. Biophys. Acta* **395**:109–119.

Eker, A. P. M., and Fichtinger-Schepman, A. M. J. 1975. Studies on a DNA photoreactivating enzyme from *Streptomyces griseus. Biochim. Biophys. Acta* **378**:54–63.

Elton, R. A. 1973a. The relationship of DNA base composition and individual protein composition in micro-organisms. *J. Mol. Evol.* **2**:263–276.

Elton, R. A. 1973b. Doublet frequencies in the DNA of genetic code limit organisms. *J. Mol. Evol.* **2**:293–302.

Engberg, J., Nilsson, J. R., Pearlman, R. E., and Leick, V. 1974. Induction of nucleolar and mitochondrial DNA replication in *Tetrahymena pyriformis. Proc. Nat. Acad. Sci. USA* **71**:894–898.

Englund, P. T. 1971. The initial step of *in vitro* synthesis of DNA by T4 DNA polymerase. *J. Biol. Chem.* **246**:5684–5687.

Ergle, D. R., and Katterman, F. R. H. 1961. DNA of cotton. *Plant Physiol.* **36**:811–815.

Ergle, D. R., Katterman, F. R. H., and Richmond, T. R. 1961. Aspects of nucleic acid composition in *Gossypium. Plant Physiol.* **39**:145–150.

Etkin, W. 1973. A representation of the structure of DNA. *BioScience* **23**:652–653.

Fambrough, D. M., Fujimura, F., and Bonner, J. 1968. Quantitative distribution of histone components in the pea plant. *Biochemistry* **7**:575–585.

Fasman, G. D., Schaffrausen, B., Goldsmith, L., and Adler, A. 1970. Conformational changes associated with f-1 histone-DNA complexes. *Biochemistry* **9**:2814–2822.

Fausler, B. S., and Loeb, L. A. 1974. Sea urchin nuclear DNA polymerase. *Methods Enzymol.* **29**:53–70.

Fazal, M., and Cole, R. D. 1977. Anomalies encountered in the classification of histones. *J. Biol. Chem.* **252**:4068–4072.

Felden, R. A., Sanders, M. M., and Morris, N. R. 1976. Presence of histones in *Aspergillus nidulans. J. Cell Biol.* **68**:430–439.

Flickinger, C. J. 1965. The fine structure of the nuclei of *Tetrahymena pyriformis* throughout the cell cycle. *J. Cell Biol.* **47**:619–630.

Fournier, M. J., Miller, W. L., and Doctor, B. P. 1974. Clustering of tRNA cistrons in *E. coli* DNA. *Biochem. Biophys. Res. Comm.* **60**:1148–1154.

Fox, S. W., and Dose, K. 1972. *Molecular Evolution and the Origin of Life,* San Francisco, W. H. Freeman and Company.

Franco, L., Johns, E. W., and Navlet, J. M. 1974. Histones from baker's yeast. *Eur. J. Biochem.* **45**:83–89.

Freese, E. B., and Freese, E. 1963. The rate of DNA strand separation. *Biochemistry* **2**:707–715.

Fridland, A. 1973. DNA precursors in eukaryotic cells. *Nature New Biol.* **243**:105–107.

Friedberg, E. C., and King, J. J. 1969. Endonucleolytic cleavage of UV-irradiated DNA controlled by the v^+ gene in phage T4. *Biochem. Biophys. Res. Comm.* **37**:646–651.

Friedman, D. L. 1974. On the mechanism of DNA replication in isolated nuclei from HeLa cells. *Biochim. Biophys. Acta* **353**:447–462.

Fry, K., Poon, R., Whitcome, P., Idriss, J., Salser, W., Mazrimas, J., and Hatch, F. 1973. Nucleotide sequence of HS-β satellite DNA from kangaroo rat, *Dipodomys ordii. Proc. Nat. Acad. Sci. USA* **70**:2642–2646.

Fry, M., and Weissbach, 1973. The utilization of synthetic RNA templates by a new DNA polymerase from cultured murine cells. *J. Biol. Chem.* **248**:2678–2683.

Fujita, H., Imamura, A., and Nogata, C. 1974. A molecular orbital study of stability and the conformation of double-stranded DNA-like polymers. *J. Theor. Biol.* **45**:411–433.

Fujiwara, Y. 1972. Effect of cycloheximide on regulatory protein for initiating mammalian DNA replication at the nuclear membrane. *Cancer Res.* **32**:2089–2095.

Fuller, W., Wilkins, M. H. F., Wilson, H. R., and Hamilton, L. D. 1965. The molecular configuration of DNA. *J. Mol. Biol.* **12**:60–80.

Gall, J. G. 1974. Repetitive DNA in *Drosophila. In:* Hamkalo, B. A., and J. Papaconstantinou, eds., *Molecular Cytogenetics,* New York, Plenum Publishing Corp., p. 59–74.

Gall, J. G., and Atherton, D. D. 1974. Satellite DNA sequences in *Drosophila virilis. J. Mol. Biol.* **85**:633–664.

Gallo, R. C., Sarin, P. S., Smith, R. G., Bobrow, S. N., Sarngadharan, M. G., Reitz, M. S., and Abrell, J. W. 1973. RNA-directed and -primed DNA polymerase activities in tumor viruses and human lymphocytes. *In:* Wells, R. D., and R. B. Inman, eds., *DNA synthesis in vitro,* Baltimore, Md., University Press, p. 250–286.

Gambarini, A. G., and Lara, F. J. S. 1974. Under-replication of ribosomal cistrons in polytene chromosomes of *Rhynchosciara. J. Cell Biol.* **62**:215–222.

Ganesan, A. T., Laipis, P. J., and Yehle, C. O. 1973. *In vitro* DNA synthesis and function of DNA polymerases in *Bacillus subtilis. In:* Wells, R. D., and R. B. Inman, eds., *DNA Synthesis in Vitro,* Baltimore, Md., University Park Press, p. 405–436.

Garcia-Herdugo, G., Fernandez-Gomez, M. E., Hidalgo, J., and Lopez-Saez, J. F. 1974. Effects of protein synthesis inhibition during plant mitosis. *Exp. Cell Res.* **89**:336–342.

Garrard, W. T. 1976. Two forms of rat liver histone H3. *FEBS Lett.* **64**:323–325.

Gautschi, J. R., and Clarkson, J. M. 1975. Discontinuous DNA replication in mouse P-815 cells. *Eur. J. Biochem.* **50**:403–412.

Gautschi, J. R., and Kern, R. M. 1973. DNA replication in mammalian cells in the presence of cycloheximide. *Exp. Cell Res.* **80**:15–26.

Gefter, M. L., Hirota, Y., Kornberg, T., Wechsler, J. A., and Barnoux, C. 1971. Analysis of DNA polymerase II and III in mutants of *E. coli* thermosensitive for DNA synthesis. *Proc. Nat. Acad. Sci. USA* **68**:3150–3153.

Gelderman, A. H., Rake, A. V., and Britten, R. J. 1971. Transcription of nonrepeated DNA in neonatal and fetal mice. *Proc. Nat. Acad. Sci. USA* **68**:172–176.

Georgiev, G. P., Ilyin, Y. V., Ryskov, A. P., Tehurikov, N. A., Yenikolopov, G. N., Gvozdev, V. A., and Ananiev, E. V. 1977. Isolation of eukaryotic DNA fragments containing structural genes and the adjacent sequences. *Science* **195**:394–397.

Geraci, D., Eremenko, T., Cocchiara, R., Granieri, A., Scarano, F., and Volpe, P. 1974. Correlation between synthesis and methylation of DNA in HeLa cells. *Biochem. Biophys. Res. Comm.* **57**:353–358.

Getz, M. J., Birnie, G. D., and Paul, J. 1974. Transcription of *in vitro* polyadenylated RNA with reverse transcriptase. *Biochemistry* **13**:2235–2240.

Gierer, A. 1966. Model for DNA and protein interactions and the function of the operator. *Nature* **212**:1480–1481.

Gilbert, W., and Maxam, A. 1973. The nucleotide sequence of the *lac* operator. *Proc. Nat. Acad. Sci. USA* **70**:3581–3584.

Gilmour, R. S., and Paul, J. 1970. Role of non-histone components in determining organ specificity of rabbit chromatins. *FEBS Lett.* **9**:242–244.

Glickman, B. W. 1974. The role of DNA polymerase I in pyrimidine dimer excision and repair replication in *E. coli* K12 following ultraviolet irradiation. *Biochim. Biophys. Acta* **335**:115–122.

Goebel, W. 1974. Studies on the initiation of plasmid DNA replication. *Eur. J. Biochem.* **41**:51–62.

Goebel, W., Royer-Pokora, B., Lindenmaier, W., and Bujard, H. 1974. Plasmids controlling synthesis of hemolysin in *E. coli:* Molecular properties. *J. Bact.* **118**:964–973.

Gold, M., and Hurwitz, J. 1964a. Purification and properties of the DNA-methylating activity of *E. coli*. *J. Biol. Chem.* **239**:3858–3865.

Gold, M., and Hurwitz, J. 1964b. Further studies on the properties of the DNA methylation reaction. *J. Biol. Chem.* **239**:3866–3874.

Gorovsky, M. A., Keevert, J. B., and Pleger, G. L. 1974. Histone F1 of *Tetrahymena* macronuclei. *J. Cell. Biol.* **61**:134–145.

Gottesfeld, J. M., Garrard, W. T., Bagi, G., Wilson, R. F., and Bonner, J. 1974. Partial purification of the template-active fraction of chromatin. *Proc. Nat. Acad. Sci. USA* **71**:2193–2197.

Gough, H. M., and Lederberg, S. 1966. Methylated bases in the host-modified DNA of *E. coli* and bacteriophage λ *J. Bact.* **91**:1460–1468.

Goulian, M., Lucas, Z. J., and Kornberg, A. 1968. Purification and properties of DNA polymerase induced by infection with phage T4. *J. Biol. Chem.* **243**:627–638.

Green, G., and Mahler, H. R. 1971. Conformational changes of DNA and polydeoxynucleotides in water and ethylene glycol. *Biochemistry* **10**:2200–2216.

Grippo, P., Locorotondo, G., and Caruso, A. 1975. Characterization of the two major DNA polymerase activities in oocytes and eggs of *Xenopus laevis*. *FEBS Lett.* **51**:137–142.

Gronow, M., and Thackrah, T. 1973. The nonhistone nuclear proteins of some rat tissues. *Arch. Biochem. Biophys.* **158**:377–386.

Grossman, L., Garvick, B. M., Ono, H. K., Braun, A. G., Hamilton, L. D. G., and Mahler, I. 1973. Mechanisms of excision-repair. *In*: Wells, R. D., and R. B. Inman, eds., *DNA Synthesis in Vitro,* Baltimore, Md., University Park Press, p. 27–34.

Gulati, S. C., Kacian, D. L., and Spiegelman, S. 1974. Conditions of using DNA polymerase I as an RNA-dependent DNA polymerase. *Proc. Nat. Acad. Sci. USA* **71**:1035–1039.

Hachmann, H. J., and Lezius, A. G. 1975. High-molecular-weight DNA polymerases from mouse myeloma. Purification and properties of three enzymes. *Eur. J. Biochem.* **50**:357–366.

Haggis, G. H., Michie, D., Muir, A. R., Roberts, K. B., and Walker, P. B. M. 1964. *Introduction to Molecular Biology*. London, Longmans.

Hall, R. H. 1971. *The Modified Nucleosides in Nucleic Acids*. New York, Columbia University Press.

Hamilton, L., Mahler, I., and Grossman, L. 1974. The biochemical and biological repair properties of a DNA polymerase from *Micrococcus luteus*. *Biochemistry* **13**:1886–1896.

Hamilton, R. T., and Wu, R. 1974. Conditions for the incorporation of ribonucleotides and deoxynucleotides into single-stranded areas of long double-stranded DNAs. *J. Biol. Chem.* **249**:2466–2472.

Hand, R., and Tamm, I. 1973. DNA replication. *J. Cell Biol.* **58**:410–418.

Hand, R., and Tamm, I. 1974. Initiation of DNA replication in mammalian cells and its inhibition by reovirus infection. *J. Mol. Biol.* **82**:175–183.

Harbers, E., and Vogt, M. 1966. Studies on the properties of nucleohistones. *In*: Busch, H., ed., *The Cell Nucleus,* New York, Academic Press, p. 165–177.

Hardin, J. A., Einem, G. E., and Lindsay, D. T. 1967. Simultaneous synthesis of histone and DNA in synchronously dividing *Tetrahymena pyriformis*. *J. Cell Biol.* **32**:709–717.

Harris, H. 1974. *Nucleus and Cytoplasm. 3rd Ed.* Oxford, Clarendon Press.

Hayes, W. 1966. Introduction: What *are* episomes and plasmids? *In*: Wolstenholme, G. E. W., and M. O'Connor, *Bacterial Episomes and Plasmids,* London, J. & A. Churchill, Ltd., p. 4–8.

Hayman, D. L., and Martin, P. G. 1965. Sex chromosome mosaicism in the marsupial genera *Isoodon and Perameles*. *Genetics* **52**:1201–1206.

Heitz, E. 1934. Über α- und β-heterochromatin sowie Konstanz und Bau der Chromomeren bei *Drosophila*. *Biol. Zentralbl.* **54**:588–609.

Heitz, E. 1942. Über mutative Intersexualität und Sexualität und Geschlecht-sunwandlung bei den Lebermoosen *Pellia neesiana* und *Sphaerocarpus donnellii*. *Naturwissenschaften* **30**:751.

Helinski, D. R., and Clewell, D. B. 1971. Circular DNA. *Ann. Rev. Biochem.* **40**:899–942.

Helinstetter, C. E. 1974. Initiation of chromosome replication in *E. coli*. *J. Mol. Biol.* **84**:1–19.

Hendler, R. W., Pereira, M., and Scharff, R. 1975. DNA synthesis involving a complexed form of DNA polymerase I in extracts of *E. coli*. *Proc. Nat. Acad. Sci. USA* **72**:2099–2103.

Henneberry, R. C., and Carlton, B. C. 1973. Characterization of the polydisperse closed circular DNA molecules of *Bacillus megaterium*. *J. Bact.* **114**:625–631.

Hennig, W. 1972. Highly repetitive DNA sequences in the genome of *Drosophila hydei*. *J. Mol. Biol.* **71**:419–432.

Hennig, W., Hennig, I., and Stern, H. 1970. Repeated sequences in DNA of *Drosophila* and their localization in giant chromosomes. *Chromosoma* **32**:31–63.

Hereford, L. M., and Hartwell, L. H. 1973. Role of protein synthesis in the replication of yeast DNA. *Nature New Biol.* **244**:129–131.

Hereford, L. M., and Hartwell, L. H. 1974. Sequential gene function in the initiation of *S. cerevisiae* DNA synthesis. *J. Mol. Biol.* **84**:445–461.

Hershfield, M. S., and Nossal, N. G. 1972. Hydrolysis of template and newly synthesized DNA by the 3′ to 5′ exonuclease activity of T4 DNA polymerase. *J. Biol. Chem.* **247**:3393–3403.

Hewish, D. R., and Burgoyne, L. A. 1973. Chromatin substructure. *Biochem. Biophys. Res. Comm.* **52**:504–510.

Heyden, H. W., and Zachau, H. 1971. Characterization of RNA in fractions of calf thymus chromatin. *Biochim. Biophys. Acta* **232**:651–660.

Hohmann, P., Tobey, R. A., and Gurley, L. R. 1976. Phosphorylation of distinct regions of f1 histone. *J. Biol. Chem.* **251**:3685–3692.

Holliday, R., and Pugh, J. E. 1975. DNA modification mechanisms and gene activity during development. *Science* **187**:226–232.

Holmes, D. S., Mayfield, J. E., Sander, G., and Bonner, J. 1972. Chromosomal RNA: Its properties. *Science* **177**:72–74.

Honda, B. M., Baillie, D. L., and Candido, E. P. M. 1974. The subunit structure of chromatin. *FEBS Lett.* **48**:156–159.

Hori, T., and Lark, K. G. 1973. Effect of puromycin on DNA replication in Chinese hamster cells. *J. Mol. Biol.* **77**:391–404.

Howard, A., and Pelc, S. R. 1953. Synthesis of DNA in normal and irradiated cells and its relation to chromosome breakage. *Heredity*, Suppl. **6**:261–273.

Huang, R. C. 1969. Effect of protein-bound RNA associated with chick embryo chromatin on template specificity of the chromatin. *J. Mol. Biol.* **39**:365–378.

Huberman, J. A. 1973. Structure of chromosome fibers and chromosomes. *Ann. Rev. Biochem.* **42**:335–378.

Huberman, J. A., Kornberg, A., and Albert, B. 1971. Stimulation of T4 bacteriophage DNA polymerase by the protein product of T4 gene 32. *J. Mol. Biol.* **62**:39–52.

Huberman, J. A., and Riggs, A. D. 1968. On the mechanism of DNA replication in mammalian chromosomes. *J. Mol. Biol.* **32**:327–341.

Hurwitz, J., and Wickner, S. 1974. Involvement of two protein factors and ATP in *in vitro* DNA synthesis catalyzed by DNA polymerase III of *E. coli*. *Proc. Nat. Acad. Sci. USA* **71**:6–10.

Huzyk, L., and Clark, D. J. 1971. Nucleoside triphosphate pools in synchronous cultures of *E. coli*. *J. Bact.* **108**:74–81.

Hwang, K. M., Murphree, S. A., Shansky, C. W., and Sartorelli, A. C. 1974. Sequential biochemical events related to DNA replication in the regenerating rat liver. *Biochim. Biophys. Acta* **366**:143–148.

Ihler, G., and Kawaii, Y. 1971. Alternate fates of the complementary strands of λ DNA after infection of *E. coli, J. Mol. Biol.* **61**:311–328.

Ilyin, Y. V., Varshavsky, A. Y., Mickelsaar, U. N., and Georgiev, G. P. 1971. Studies on deoxyribonucleoprotein structure. *Eur. J. Biochem.* **22**:235–245.

Imai, H. T. 1975. Evidence for nonrandom localization of the centromere on mammalian chromosomes, *J. Theor. Biol.* **49**:111–123.

Jackson, V., Shires, A., Tanphaichitr, N., and Chalkley, R. 1976. Modifications to histones immediately after synthesis. *J. Mol. Biol* **104**:471–483.

Jacob, F., and Monod, J. 1961. Genetic regulatory mechanisms in the synthesis of proteins. *J. Mol. Biol.* **3**:318–356.

Japha, G. 1939. Die Meiosis bei *Oenothera*. *Zeit. Bot.* **34**:321–369.

Jerzmanowski, A., Staron, K., Tyniec, B., Bernhardt-Smigielska, J., and Toczko, K. 1976. Subunit structure of *Physarum polycephalum* chromatin. *FEBS Lett.* **62**:251–254.

Jockusch, B. M., and Walker, I. O. 1974. The preparations and preliminary characterisation of chromatin from the slime mould *Physarum polycephalum*. *Eur. J. Biochem.* **48**:417–425.

Jones, D. D. 1975. Amino acid properties and side chain orientation in proteins. *J. Theor. Biol.* **50**:167–183.

Jones, K. W. 1970. Chromosomal and nuclear location of mouse satellite DNA in individual cells. *Nature* **225**:912–915.

Kalf, G. F., and Ch'ih, F. F. 1968. Purification and properties of DNA polymerase from rat liver mitochondria. *J. Biol. Chem.* **243**:4904–4916.

Kalt, M. R., and Gall, J. G. 1974. Observations on early germ cell development and premeiotic ribosomal DNA amplification in *Xenopus laevis*. *J. Cell Biol.* **62**:460–472.

Kalousek, F., and Morris, N. R. 1968. DNA methylase activity in rat spleen. *J. Biol. Chem.* **143**:2440–2442.

Kamiyama, M., Dastugue, B., Defer, N., and Kruh, J. 1972. Liver chromatin non-histone proteins. *Biochim. Biophys. Acta* **277**:576–583.

Kaplan, J. C., Kushner, S. R., and Grossman, L. 1969. Purification of two enzymes involved in the excision of thymine dimers from UV-irradiated DNA. *Proc. Nat. Acad. Sci. USA* **63**:144–151.

Kappler, J. W. 1970. The kinetics of DNA methylation in cultures of a mouse adrenal cell line. *J. Cell Physiol.* **75**:21–32.

Kaye, A. M., Salomon, R., and Fridlender, B. 1967. Base composition and presence of methylated bases in DNA from a blue-green alga *Plectronema boryanum*. *J. Mol. Biol.* **24**:479–484.

Keir, H. M. 1965. DNA polymerases from mammalian cells. *Prog. Nucleic Acid Res.* **4**:82–128.

Keir, H. M., Smellie, R. M. S., and Siebert, G. 1962. Intracellular location of DNA nucleotidyltransferase. *Nature* **196**:752–754.

Keller, W. 1972. RNA-primed DNA synthesis *in vitro*. *Proc. Nat. Acad. Sci. USA* **69**:1560–1564.

Kelly, R. B., Atkinson, M. R., Huberman, J. A., and Kornberg, A. 1969. Excision of thymine dimers and other mismatched sequences by DNA polymerase of *E. coli*. *Nature* **224**:495–501.

Kelner, A. 1949a. Effect of visible light on the recovery of *Streptomyces griseus* conidia from ultraviolet irradiation injury. *Proc. Nat. Acad. Sci. USA* **35**:73–79.

Kelner, A. 1949b. Photoreactivation of ultraviolet-irradiated *E. coli,* with special reference to the dose-reduction principle and to ultraviolet-induced mutation. *J. Bact.* **58**:511–522.

Kierzenbaum, A. L., and Tres, L. L. 1975. Structural and transcriptional features of the mouse spermatid genome. *J. Cell Biol.* **65**:258–270.

Kim, J. H., Gelbard, A. S., and Perez, A. G. 1968. Inhibition of DNA synthesis by actinomycin D and cycloheximide in synchronized HeLa cells. *Exp. Cell Res.* **53**:478–487.

Kirtikar, D. M., Slaughter, J., and Goldthwait, D. A. 1975. Endonuclease II of *E. coli:* Degradation of γ-irradiated DNA. *Biochemistry* **14**:1235–1244.

Kit, S. 1961. Equilibrium sedimentation in density gradients of DNA preparations from animal tissues. *J. Mol. Biol.* **3**:711–716.

Klein, A., and Sauerbier, W. 1965. Role of methylation in host controlled modification of phage T1. *Biochem. Biophys. Res. Comm.* **18**:440–445.

Klenow, H., and Overgaard-Hansen, K. 1973. Concerted effect of pancreatic DNase and the large fragment of DNA polymerase I. *In*: Wells, R. D., and R. B. Inman, eds., *DNA Synthesis in Vitro,* Baltimore, Md., University Park Press, p. 13–25.

Klett, R. P., Cerami, A., and Reich, E. 1968. Exonuclease VI, a new nuclease activity associated with *E. coli* DNA polymerase. *Proc. Nat. Acad. Sci. USA* **60**:943–950.

Knippers, R. 1970. DNA polymerase II. *Nature* **228**:1050–1053.

Koch. J. 1973. Cytoplasmic DNAs consisting of unique nuclear sequences in hamster cells. *FEBS Lett.* **32**:22–26.

Koch, J., and Götz, D. 1972. Transport of nuclear DNA into the cytoplasm in animal cells. *FEBS Lett.* **27**:9–12.

Koch, J., and von Pfeil, H. 1972. Transport of nuclear DNA into the cytoplasm in cultured animal cells. *FEBS Lett.* **24**:53–56.

Koike, K., and Wolstenholme, D. R. 1974. Evidence for discontinuous replication of circular mitochondrial DNA molecules from Novikoff rat ascites hepatoma cells. *J. Cell Biol.* **61**:14–25.

Kolata, G. B. 1974. Lac system: New research on how a protein binds to DNA. *Science* **184**:52–53.

Kornberg, A. 1957. Pyrophosphorylases and phosphorylases in biosynthetic reactions. *Adv. Enzymol.* **18**:191–240.

Kornberg, A. 1961. *The Enzymatic Synthesis of DNA,* London, John Wiley & Sons.

Kornberg, A. 1966. The biosynthesis of DNA. *In*: Koningsberger, V. V., and L. Bosch, eds., *Regulation of Nucleic Acid and Protein Synthesis,* Amsterdam, Elsevier Publishing Co., p. 22–38.

Kornberg, A. 1969. Active center of DNA polymerase. *Science,* **163**:1410–1418.

Kornberg, R. D. 1974. Chromatin structure: A repeating unit of histones and DNA. *Science* **184**:868–871.

Kornberg, R. D., and Thomas, J. O. 1974. Chromatin structure: Oligomers of the histones. *Science* **184**:865–868.

Kornberg, T., and Gefter, M. L. 1971. Properties of DNA polymerase III. *Proc. Nat. Acad. Sci. USA* **68**:761–764.

Kornberg, T., and Gefter, M. L. 1974. DNA polymerase II (*E. coli* K12). *Methods Enzymol.* **29**:22–26.

Kornberg, T., Lockwood, A., and Worcel, A. 1974. Replication of the *E. coli* chromosome with a soluble enzyme system. *Proc. Nat. Acad. Sci. USA* **71**:3189–3193.

Koslov, A. V., and Svetlikova, S. B. 1974. Properties and composition of DNP of *Escherichia coli* B. *J. Mol. Biol.* **7**:519–526.

Kössel, H., and Roychoudhury, R. 1974. Proofreading function of DNA polymerase I from *E. coli, J. Biol. Chem.* **249**:4049–4099.

Kostraba, N. C., Montagna, R. A., and Wang, T. Y. 1975. Study of the loosely bound non-histone chromatin proteins. *J. Biol. Chem.* **250**:1548–1555.

Kowalski, C. J., and Cheevers, W. P. 1976. Synthesis of high molecular weight DNA strands during S phase. *J. Mol. Biol.* **104**:603–615.

Kowalski, C. J., Nasjleti, C. E., and Harris, J. E. 1976. Human chromosomes. Evidence for autosomal sexual dimorphism. *Exp. Cell Res.* **100**:56–62.

Kriegstein, H. J., and Hogness, D. S. 1974. Mechanism of DNA replication in *Drosophila* chromosomes: Structure of replication forks and evidence for bidirectionality. *Proc. Nat. Acad. Sci. USA* **71**:135–139.

Kuprijanova, N. S., and Timofieva, M. J. 1974. Repeated nucleotide sequences in the loach genome. *Eur. J. Biochem.* **44**:59–65.

Kurashige, S., and Mitsuhashi, S. 1973. The possible presence of RNA-dependent DNA polymerase in the immune response. *Jpn. J. Microbiol.* **17**:105–109.

LaCour, L. F., and Wells, B. 1974. Fine structure and staining behaviour of heterochromatic segments in two plants. *J. Cell Sci.* **14**:505–521.

Laland, S. G., Overend, W. G., and Webb, M. 1952. The properties and composition of the DNAs from certain animal, plant, and bacterial sources. *J. Chem. Soc.* **1952**:3224–3231.

Lange, C. S. 1974. The organization and repair of mammalian DNA. *FEBS Lett.* **44**:153–156.

Langridge, R., Wilson, H. R., Hooper, C. W., and Wilkins, M. H. F. 1960a. X-ray diffraction study of a crystalline form of the lithium salt. *J. Mol. Biol.* **2**:19–37.

Langridge, R., Marvin, D. A., Seeds, W. E., Wilson, H. R., Hooper, C. W., Wilkins, M. H. F., and Hamilton, L. D. 1960b. Molecular models and their Fourier transforms. *J. Mol. Biol.* **2**:38–62.

Lark, C. 1968a. Studies on the *in vivo* methylation of DNA in *E. coli* $15T^-$ *J. Mol. Biol.* **31**:389–400.

Lark, C. 1968b. Effects of methionine analogs, ethionine and norleucine, on DNA synthesis in *E. coli* $15T^-$. *J. Mol. Biol.* **31**:401–414.

Leavitt, R. W., Wohlhieter, J. A., Johnson, E. M., Olson, G. E., and Baron, L. S. 1971. Isolation of circular DNA from *Salmonella typhosa* hybrids obtained from matings with *E. coli* Hfr donors. *J. Bact.* **108**:1357–1365.

Ledoux, L., Huart, R., and Jacobs, M. 1974. DNA-mediated genetic correction of thiamineless *Arabidopsis thaliana. Nature* **249**:17–21.

Lee, C. S., and Collins, L. 1977. Q- and C-bands in the metaphase chromosomes of *Drosophila nasutoides. Chromosoma* **61**:57–60.

Lehman, I. R. 1974a. T4 DNA polymerase. *Methods Enzymol.* **29**:46–53.

Lehman, I. R. 1974b. DNA ligase: Structure mechanism, and function. *Science* **186**:790–797.

Lehman, I. R., and Stevens, S. 1975. Postreplication repair of DNA in chick cells. *Biochim. Biophys. Acta* **402**:179–187.

Leis, J., Berkower, I., and Hurwitz, J. 1973. Characterization of AMV stimulatory protein and RNase H-associated activity. *In*: Wells, R. D., and R. B. Inman, eds., *DNA Synthesis in Vitro,* Baltimore, Md., University Park Press, p. 287–308.

Lesk, A. M. 1974. A combinatorial study of the effects of admitting non-Watson-Crick base pairings and of base composition on the helix-forming potential–polynucleotides of random sequence, *J. Theor. Biol.* **44**:7–17.

Lett, J. T., Klucis, E. S., and Sun, C. 1970. On the size of the DNA in the mammalian chromosome structural subunits. *Biophys. J.* **10**:277–292.

Levinthal, C., and Crane, H. R. 1956. On the unwinding of DNA. *Proc. Nat. Acad. Sci. USA* **42**:436–438.

Lezius, A. G., and Scheit, K. H. 1967. Enzymatic synthesis of DNA with 4-thio-thymidine triphosphate as substitute for dTTP. *Eur. J. Biochem.* **3**:85–94.

Li, H. J. 1977. Chromatin subunits. *In:* Li, H. J., and R. A. Eckhardt, eds., *Chromatin and Chromosome Structure,* New York, Academic Press, p. 143–166.

Lieberman, M. W., and Poirier, M. C. 1974a. Distribution of deoxyribonucleic acid repair synthesis among repetitive and unique sequences in the human diploid genome. *Biochemistry* **13**:3018–3023.

Lieberman, M. W., and Poirier, M. C. 1974b. Intragenomal distribution of DNA repair synthesis: Repair in satellite and mainband DNA in cultured mouse cells. *Proc. Nat. Acad. Sci. USA* **71**:2461–2465.

Lin, H. J. 1974. Isolation of a short, cytosine-rich repeating unit from the DNA of *E. coli, Biochim, Biophys. Acta* **349**:13–22.

Lin, M. S., and Davidson, R. L. 1974. Centric fusion, satellite DNA, and DNA polarity in mouse chromosomes. *Science* **185**:1179–1181.

Lindsay, J. G., Berryman, S., and Adams, R. L. P. 1970. Characteristics of DNA polymerase activity in nuclear and supernatant fractions of cultured mouse cells. *Biochem. J.* **119**:839–848.

Littlefield, J. W., and Jacobs, P. S. 1965. The relation between DNA and protein synthesis in mouse fibroblasts. *Biochim. Biophys. Acta* **108**:652–658.

Livingston, D. M., Hinkle, D. C., and Richardson, C. C. 1975. DNA polymerase III of *E. coli*. *J. Biol. Chem.* **250**:461–469.

Livingston, D. M., and Richardson, C. C. 1975. DNA polymerase III of *E. coli*. Characterization of associated exonuclease activities. *J. Biol. Chem.* **250**:470–478.

Lobeer, G. 1961. Struktur und Inhalt der Geschlectschromosomen. *Ber. Deuts. Bot. Gesell.* **59**:369–375.

Loeb, L. A. Tartof, K. D., and Travaglini, E. C. 1973. Copying natural RNAs with *E. coli* DNA polymerase I. *Nature New Biol.* **242**:66–69.

Lohr, D., and Van Holde, K. E. 1975. Yeast chromatin subunit structure. *Science* **188**:165–166.

Longuet-Higgins, H. C., and Zimm, B. H. 1960. Calculation of the rate of uncoiling of the DNA molecule. *J. Mol. Biol.* **2**:1–4.

Louarn, J.-M., and Bird, R. E. 1974. Size distribution and molecular polarity of newly replicated DNA in *E. coli. Proc. Nat. Acad. Sci. USA* **71**:329–333.

Low, R. L., Rashbaum, S. A., and Cozzarelli, N. R. 1976. Purification and characterization of DNA polymerase III from *B. subtilis*. *J. Biol. Chem.* **251**:1311–1325.

Luria, S. E. 1953. Host-induced modifications of viruses. *Cold Spring Harbor Symp. Quant. Biol.* **18**:237–244.

Luria, S. E., and Human, M. L. 1952. A nonhereditary, host-induced variation of bacterial viruses. *J. Bact.* **64**:557–569.

MacGillivray, A. J., and Rickwood, D. 1974. The heterogeneity of mouse-chromatin nonhistone proteins as evidenced by two-dimensional polyacrylamide-gel electrophoresis and ion-exchange chromatography. *Eur. J. Biochem.* **41**:181–190.

Macgregor, H. C., and Kezer, J. 1971. The chromosomal localization of a heavy satellite DNA in the testis of *Plethodon c. cinereus*. *Chromosoma* **33**:167–82.

Magnusson, G., Pigret, V., Winnacker, E. L., Abrams, R., and Reichard, P. 1973. RNA-linked short DNA fragments during polyoma replication. *Proc. Nat. Acad. Sci USA* **70**:412–415.

Maniloff, J. 1969. Collision lifetimes and recognition times for macromolecular synthesis. *J. Theor. Biol.* **25**:339–342.

Marinus, M. G., and Morris, N. R. 1974. Biological function for 6-methyladenine residues in the DNA of *E. coli* K12. *J. Mol. Biol.* **85**:309–322.

Marushige, Y., and Marushige, K. 1974. Properties of chromatin isolated from bull spermatozoa. *Biochim. Biophys. Acta* **340**:498–508.

Marvin, D. A., Spencer, M., Wilkins, M. H. F., and Hamilton, L. D. 1961. The molecular configuration of DNA: X-ray diffraction study of the C form of the lithium salt. *J. Mol. Biol.* **3**:547–565.

Marzluff, W. F., White, E. L., Benjamin, R., and Huang, R. C. C. 1975. Low molecular weight RNA species from chromatin. *Biochemistry* **14**:3715–3724.

Massie, H. R., Thompson, D. S., and Colarusso, L. J. 1975. Discontinuous DNA replication and molecular events preceding DNA replication in *B. subtilis*. *Arch. Biochem. Biophys.* **167**:213–229.

Masters, M., and Broda, P., 1971. Evidence for the bidirectional replication of the *E. coli* chromosome. *Nature New Biol.* **232**:137–140.

Mayer, R. J., Smith, R. G., and Gallo, R. C. 1974. Reverse transcriptase in normal rhesus monkey placenta. *Science* **185**:864–867.

Mayfield, J., and Bonner, J. 1972. A partial sequence of nuclear events in regenerating rat liver. *Proc. Nat. Acad. Sci. USA* **69**:7–10.

Mazia, D. 1974. The cell cycle. *Sci. Am.* **230(I)**:54–64.

McGavin, S. 1971a. Models of specifically paired like (homologous) nucleic acid structures. *J. Mol. Biol.* **55**:293–298.

McGavin, S. 1971b. A four strand nucleic acid model with a specific pairing of like Watson–Crick double helices and its properties. *First Eur. Biophys. Congr., Baden, Austria,* p. 259–262.

McGavin, S. 1973. An attitude to nucleic acid models. *Bull. Math. Biol.* **35**:407–409.

McGavin, S., Wilson, H. R., and Barr, G. C. 1966. Intercalated nucleic acid double helices: A sterochemical possibility. *J. Mol. Biol.* **22**:187–191.

McGhee, J. D., and Engel, J. D. 1975. Subunit structure of chromatin is the same in plants and animals. *Nature* **254**:449–450.

Meltz, M. L., and Painter, R. B. 1973. Distribution of repair replication in the HeLa cell genome. *Int. J. Radiat. Biol.* **23**:637–640.

Meselson, M., Yuan, R., and Heywood, J. 1972. Restriction and modification of DNA. *Ann. Rev. Biochem.* **41**:447–466.

Messing, J., Staudenbauer, W. L., and Hofschneider, P. H. 1974. Replication of the minicircular DNA of *E. coli* 15. *Eur. J. Biochem.* **44**:293–297.

Metz, C. W. 1938. Chromosome behavior, inheritance, and sex determination in *Sciara, Am. Nat.* **72**:485–520.

Meynell, E., Meynell, G. G., and Datta, N. 1968. Phylogenetic relationships of drug-resistant factors and other transmissible bacterial plasmids. *Bacteriol. Rev.* **32**:55–83.

Meynell, G. G. 1973. *Bacterial Plasmids*. Cambridge, Mass., The M. I. T. Press.

Michalke, H., and Bremer, H. 1969. RNA synthesis in *E. coli* after irradiation with ultraviolet light. *J. Mol. Biol.* **41**:1–23.

Miller, O. J., Schnedl, W., Allen, J., and Erlanger, B. F. 1974. 5-methylcytosine localised in mammalian constitutive heterochromatin. *Nature* **251**:636–637.

Miller, S. L., and Orgel, L. E. 1974. *The Origins of Life on the Earth,* Englewood Cliffs, New Jersey, Prentice-Hall, Inc.

Minato, S., and Werbin, H. 1971. Spectral properties of the chromophoric material associated with the DNA photoreactivating enzyme isolated from baker's yeast. *Biochemistry* **10**:4503–4508.

Mitchison, J. M. 1971. *The Biology of the Cell Cycle,* Cambridge, Cambridge University Press.

Mizuno, N. S., Stoop, C. E., and Sinha, A. A. 1971. DNA synthesis associated with the nuclear envelope. *Nature New Biol.* **229**:22–24.

Moll, R., and Wintersberger, E. 1976. Synthesis of yeast histones in the cell cycle. *Proc. Nat. Acad. Sci. USA* **73**:1863–1867.

Momparler, R. L., Rossi, M., and Labitan, A. 1973. Partial purification and properties of two forms of DNA polymerase from calf thymus. *J. Biol. Chem.* **248**:285–293.

Monahan, J. J., and Hall, R. H. 1973. Chromatin low molecular weight RNA components. *Can. J. Biochem.* **51**:903–912.

Monahan, J. J., and Hall, R. H. 1974a. Chromatin, and gene regulation in eukaryotic cells at the transcriptional level. *CRC Crit. Rev. Biochem.* **2**:67–112.

Monahan, J. J., and Hall, R. H. 1974b. An RNA fraction in chromatin of L-929 cells associated with DNA. *Nucleic Acid Res.* **1**:1421–1437.

Monahan, J. J., and Hall, R. H. 1975. A high molecular weight RNA fraction in chromatin. *Biochim. Biophys. Acta* **383**:40–55.

Monesi, V. 1969. DNA, RNA, and protein synthesis during the mitotic cell cycle. *In:* Lima-de-Faria, A., ed., *Handbook of Molecular Cytology,* Amsterdam, North-Holland Publishing Co., p. 472–499.

Monty, K. J., Litt, M., Kay, E. R. M., and Dounce, A. L. 1956. Isolation and properties of liver cell nucleoli. *J. Biophys. Biochem. Cyt.* **2**:127–145.

Moore, D. 1974. Dynamic unwinding of DNA helices: A mechanism of genetic recombination. *J. Theor. Biol.* **43**:167–186.

Moritz, K. B., and Roth, G. E. 1976. Complexity of germline and somatic DNA in *Ascaris. Nature* **259**:55–57.

Moses, R. E. 1974. The isolation and properties of DNA polymerase II from *E. coli. Methods Enzymol.* **29**:13–22.

Mowbray, S. L., Berbi, S. A., and Landy, A. 1975. Interdigitated repeated sequences in bovine satellite DNA. *Nature* **253**:367–370.

Mueller, G. C., Kajiwara, K., Stubblefield, E., and Rueckert, R. R. 1962. The effect of puromycin on the duplication of DNA. *Cancer Res.* **22**:1084–1090.

Nagl, W. 1974. DNA synthesis in tissue and cell cultures. *In:* Street, H. E., ed., *Tissue Culture and Plant Science,* New York, Academic Press, p. 19–42.

Nagl, W. 1975. Organization and replication of the eukaryotic chromosome. *Prog. Bot.* **37**:186–210.

Nath, K., and Hurwitz, J. 1974. Covalent attachment of polyribonucleotides to polydeoxyribonucleotides catalyzed by DNA ligase. *J. Biol. Chem.* **249**:3680–3688.

Nes, I. F., and Kleppe, K. 1974. Multiple DNA polymerases in *Actinobacter calcoaceticus. FEBS Lett.* **43**:180–184.

Nešković, B. A. 1968. Developmental phases in intermitosis and the preparation for mitosis of mammalian cells *in vitro. Int. Rev. Cytol.* **24**:71–97.

Neubort, S., and Bases, R. 1974. RNA-DNA covalent complexes in HeLa cells. *Biochim. Biophys. Acta* **340**:31–39.

Neuhard, J. 1966. On the regulation of the deoxyadenosine triphosphate and deoxycytidine triphosphate pools in *E. coli. Biochim. Biophys. Acta* **129**:104–115.

Newlon, C. S., Petes, T. D., Hereford, L. M., and Fangman, W. L. 1974. Replication of yeast chromosomal DNA. *Nature* **247**:32–35.

Nexø, B. A. 1975. Ribo- and deoxyribonucleotide triphosphate pools in synchronized populations of *Tetrahymena pyriformis. Biochim. Biophys. Acta* **378**:12–17.

Nilsson, J. R. 1970. Suggestive structural evidence for macronuclear ''subnuclei'' in *Tetrahymena pyriformis. J. Protozool.* **17**:539–548.

Noguti, T., and Kada, T. 1975. A cellular factor acting on γ-irradiated DNA and promoting its priming activity for DNA polymerase I. *Biochim. Biophys. Acta* **395**:284–293.

Noll, M. 1974. Subunit structure of chromatin. *Nature* **251**:249–251.

Novick, R. P. 1969. Extrachromosomal inheritance in bacteria. *Bact. Rev.* **33**:210–263.

Nüsslein, V., Otto, B., Bonhoeffer, F., and Schaller, H. 1971. Function of DNA polymerase III in DNA replication. *Nature New Biol.* **234**:285–286.

O'Brien, R. L., Sanyal, A. B., and Stanton, R. H. 1972. Association of DNA with the nuclear membrane of HeLa cells. *Exp. Cell Res.* **70**:106–112.

Okazaki, R., Okazaki, T., Sakabe, K., Sugimoto, K., Kainuma, R., Sugino, A., and Iwatsuki, N. 1968a. In vivo mechanism of DNA chain growth. *Cold Spring Harbor Symp. Quant. Biol.* **33**:129–143.

Okazaki, R., Okazaki, T., Sakabe, K., Sugimoto, K., and Sugino, A. 1968b. Possible discontinuity and unusual secondary structure of newly synthesized DNA chains. *Proc. Nat. Acad. Sci. USA* **59**:598–605.

Okazaki, T., and Kornberg, A. 1964. Purification and properties of a polymerase from *B. subtilis. J. Biol. Chem.* **239**:259–268.

Olins, A. L., and Olins, D. E. 1973. Spheroid chromatin units (ν bodies). *J. Cell Biol.* **59**:252a.

Olins, A. L., Senior, M. B., and Olins, D. E. 1976. Ultrastructural features of chromatin ν bodies. *J. Cell Biol.* **68**:787–792.

Olivera, B. M., and Bonhoeffer, F. 1974. Replication of *E. coli* requires DNA polymerase I. *Nature* **250:**513–514.

Pardon J. F., and Wilkins, M. H. F. 1972. A supercoil model for nucleohistone. *J. Mol. Biol.* **68**:115–124.

Pardue, M. L., and Gall, J. G. 1970. Chromosomal localization of mouse satellite DNA. *Science* **168**:1356–1358.

Pearson, C. K., Davis, P. B., Taylor, A., and Amos, N. A. 1976. The involvement of RNA in the initiation of DNA synthesis in mammalian cells. *Eur. J. Biochem.* **62**:451–459.

Pearson, G. D., and Hanawalt, P. C. 1971. Isolation of DNA replication complex from uninfected and adenovirus-infected HeLa cells. *J. Mol. Biol.* **62**:65–80.

Perlgut, L. E., Byers, D. L., Jope, R. S., and Khamvinwathna, V. 1975. Formation of triple-stranded bovine DNA *in vitro*. *Nature* **254**:86–87.

Pettersson, U. 1973. Some unusual properties of replicating adenovirus type 2 DNA. *J. Mol. Biol.* **81**,521–527.

Pettijohn, D. E., Hecht, R. M., Stonington, O. G., and Stamato, T. D. 1973. Factors stabilizing DNA folding in bacterial chromosomes. *In:* Wells, R. D., and R. B. Inman, eds., *DNA Synthesis in Vitro,* Baltimore, Md., University Park Press, p. 145–162.

Pflug, H. D., Meinel, W., Neumann, K. H., and Meinel, M. 1969. Entwicklungstendenzen des Frühen Lebens auf der Erde. *Naturwissenschaften* **56**:10–14.

Philippsen, P., Streeck, R. E., and Zachau, H. G. 1974. Defined fragments of calf, human, and rat DNA produced by restriction nucleases. *Eur. J. Biochem.* **45**:479–488.

Piessens, J. P., and Eker, A. P. M. 1975. Photoreactivation of template activity of UV-irradiated DNA in an RNA-polymerase system. *FEBS Lett.* **50**:125–129.

Pimpinelli, S., Gatti, M., De Marco, A. 1975. Evidence for heterogeneity in heterochromatin of *D. melanogaster*. *Nature* **256**:335–337.

Pivec, L., Horská, K., Vítek, A., and Doskočil, J. 1974. Plurimodal distribution of base composition in DNA of some higher plants. *Biochem. Biophys. Acta* **340**:199–206.

Platz, R., Stein, G. S., and Kleinsmith, L. J. 1973. Changes in the phosphorylation of nonhistone chromatin proteins during the cell cycle of HeLa S_3 cells. *Biochem. Biophys. Res. Comm.* **51**:735–740.

Poiesz, B. J., Seal, G., and Loeb, L. A. 1974. Reverse transcriptase: Correlation of zinc content with activity. *Proc. Nat. Acad. Sci. USA* **71**:4892–4896.

Pospelor, V. A., Svetlikova, S. B., and Vorob'ev, V. I. 1977. Heterogeneity of chromatin subunits. *FEBS Lett.* **74**:229–233.

Poulson, R., Krasny, J., and Zbarsky, S. H. 1974. Characterisation of nuclear and cytoplasmic DNA polymerases from rat intestinal mucosa. *Can. J. Biochem.* **52**:162–169.

Prescott, D. M. 1966. The syntheses of total macronuclear protein, histone, and DNA during the cell cycle in *Euplotes eurystomus*. *J. Cell Biol.* **3**:1–9.

Prescott, D. M. 1973. The cell cycle in amoebae. *In:* Jeon, K. W., ed., *The Biology of Amoeba,* New York, Academic Press, p. 467–478.

Price, T. D., Darmstadt, R. A., Hinds, H. A., and Zamenhof, S. 1967. Mechanism of synthesis of deoxyribonucleic acid *in vivo*. *J. Biol. Chem.* **242**:140–151.

Pulleyblank, D. E., and Morgan, A. R. 1975. The sense of naturally occurring super-helices and the unwinding angle of intercalated ethidium. *J. Mol. Biol.* **91**:1–13.

Rae, P. M. M. 1966. Whole mount electronmicroscopy of *Drosophila* salivary chromosomes. *Nature* **212**:139–142.

Rae, R. M. M. 1973. 5-hydroxymethyluracil in the DNA of a dinoflagellate. *Proc. Nat. Acad. Sci. USA* **70**:1141–1145.

Rastogi, A. K., and Koch, J. 1974. Generation in vivo of "covalently closed" circular mitochondrial DNA free of superhelical turns. *Eur. J. Biochem.* **46**:583–588.

Reeves, R., and Jones, A. 1976. Genomic transcriptional activitiy and the structure of chromatin. *Nature* **260**:495–500.

Richardson, C. C., Campbell, J. L., Chase, J. W., Hinkle, D. C., Livingston, D. M., Muleahy, H. L., and Shizuya, H. 1973. DNA polymerases of *E. coli*. *In:* Wells, R. D., and R. B. Inman, eds., *DNA Synthesis in Vitro,* Baltimore, Md., University Park Press, p. 65–69.

Ris, H. 1955. The submicroscopic structure of chromosomes. *In:* Hogeboom, G. H. *et al.*, eds., *Symposium on the Fine Structure of Cells,* New York, Interscience, p. 160–167.

Ris, H. 1959. Microstructure of the nucleus during spermiogenesis. *Colloq. Ges. Physiol. Chem.* **9**:1–30.

Ris, H. 1966. Fine structure of chromosomes. *Proc. Roy, Soc. Lond.* **B164**:246–257.

Ris, H., and Chandler, B. L. 1963. The ultrastructure of genetic systems in prokaryotes and eukaryotes. *Cold Spring Harbor Symp. Quant. Biol.* **28**:1–8.

Robbins, E., and Scharff, M. D. 1966. Some macromolecular characteristics of synchronized HeLa cells. *In:* Cameron, I., and G. Padilla, eds., *Cell Synchrony,* New York, Academic Press, Inc. p. 353–374.

Robson, B. 1974. Analysis of the code relating sequence to conformation in globular proteins. *Biochem. J.* **141**:853–867.

Robson, B., and Pain, R. H. 1974a. Analysis of the code relating sequence to conformation in the globular proteins. Development of a stereochemical alphabet on the basis of intra-residue information. *Biochem. J.* **141**:869–882.

Robson, B., and Pain, R. H. 1974b. An informational analysis of the residue in determining the conformation of its neighbors in the primary sequence. *Biochem. J.* **141**:883–897.

Robson, B., and Pain, R. H. 1974c. The distribution of residue pairs in turns and kinks in the backbone chain. *Biochem. J.* **141**:899–904.

Rodriguez, L. V., and Becker, F. F. 1976. Rat liver chromatin. Distribution of histone and nonhistone proteins in eu- and heterochromatin. *Arch. Biochem. Biophys.* **173**:438–447.

Rosenberg, E. 1965. D-mannose as a constituent of the DNA of a mutant strain of bacteriophage SP8. *Proc. Nat. Acad. Sci. USA* **53**:836–841.

Rosenberg, J. M., Seeman, N. C., Kim, J. J. P., Suddath, F. L., Nicholas, H. B., and Rich, A. 1973. Double helix at atomic resolution. *Nature* **243**:150–154.

Roth, R. 1973. Chromosome replication during meiosis. *Proc. Nat. Acad. Sci. USA* 70:3087–3091.

Roychoudhury, R. 1973. Transcriptional role in DNA replication. *J. Biol. Chem.* **248**:8465–8473.

Rubinow, S. I., and Yen, A. 1972. Quantitation of some DNA precursor data. *Nature New Biol.* **239**:73–74.

Rudner, R., Karkas, J. D., and Chargaff, E. 1968. Separation of *B. subtilis* DNA into complementary strands. *Proc. Nat. Acad. Sci. USA* **60**:630–635.

Rudner, R., Karkas, I. D., and Chargaff, E. 1969. Separation of microbial DNAs into complementary strands. *Proc. Nat. Acad. Sci. USA* **63**:152–159.

Russell, G. J., McGeoch, D. J., Elton, R. A., and Subak-Sharpe, J. H. 1973. Doublet frequency analysis of bacterial DNAs. *J. Mol. Evol.* **2**:277–292.

Sadowski, P., McGeer, A., and Becker, A. 1974. Terminal cross-linking of DNA catalyzed by an enzyme system containing DNA ligase, DNA polymerase, and exonuclease of bacteriophage T7. *Can J. Biochem.* **52**:525–535.

Sager, R. 1972. *Cytoplasmic Genes and Organelles,* New York, Academic Press.

Sager, R., and Kitchin, R. 1975. Selective silencing of eukaryotic DNA. *Science* **189**:426–433.

Sager, R., and Lane, D. 1972. Molecular basis of maternal inheritance. *Proc. Nat Acad. Sci. USA* **69**:2410–2413.

Sager, R., and Ramanis, Z. 1974. Mutations that alter the transmission of chloroplast genes in *Chlamydomonas. Proc. Nat. Acad. Sci. USA* **71**:4698–4702.

Saito, N., and Werbin, H. 1970. Purification of a blue-green algal DNA photo-reactivating enzyme. *Biochemistry* **9**:2610–2620.

Sakamaki, T., Fukuei, K., Takahashi, N., and Tanifuji, S. 1975. Rapidly labeled intermediates in DNA replication in higher plants. *Biochim. Biophys. Acta* **395**:314–321.

Salser, W., Fry, K., Brunk, C., and Poon, R. 1972. Nucleotide sequencing of DNA: Preliminary characterization of the products of specific cleavages at guanine, cytosine, or adenine residues. *Proc. Nat. Acad. Sci. USA* **69**:238–242.

Salser, W., Fry, K., Wesley, R. D., and Simpson, L. 1973. Use of nucleic acid fingerprints to estimate the complexity of minicircle DNA. *Biochim. Biophys. Acta* **319**:267–280.

Salzman, N., Moore, D., and Mendelsohn, J. 1966. Isolation and characterization of human metaphase chromosomes. *Proc. Nat. Acad. Sci. USA* **56**:1449–1456.

Sato, S., Ariake, S., Saito, M., and Sugimura, T. 1972. RNA bound to nascent DNA in Ehrlich ascites tumor cells. *Biochem. Biophys. Res. Comm.* **49**:827–834.

Saunders, G. F., Shirakawa, S., Saunders, P. P., Arrighi, F. E., and Hsu, T. C. 1972. Populations of repeated DNA sequences in the human genome. *J. Mol. Biol.* **63**:323–334.

Scarano, E. 1969. *Ann. Embryol. Morphogén. Suppl.* **1**:51–61.

Scarano, E., Iaccarino, M., Grippo, P., and Winckelmans, D. 1965. On methylation of DNA during development of the sea urchin embryo. *J. Mol. Biol.* **14**:603–607.

Schachman, H. K., Adler, J., Radding, C. M., Lehman, I. R., and Kornberg, A. 1960. Enzymatic synthesis of dexoyribonucleic acid *J. Biol. Chem.* **235**:3242–3249.

Scharff, M. D., and Robbins, E. 1965. Synthesis of ribosomal RNA in synchronized HeLa cells. *Nature,* **208**:464–466.

Scharff, M. D., and Robbins, E. 1966. Polyribosome disaggregation during metaphase. *Science* **151**:992–995.

Schekman, R., Wickner, W., Westergaard, O., Brutlag, D., Geider, K., Bertsch, L. L., and Kornberg, A. 1972. Synthesis of ϕX174 replicative form requires RNA synthesis resistant to rifampicin. *Proc. Nat. Acad. Sci. USA* **69**:2691–2695.

Schekman, R., Wickner, W., Westergaard, O., Brutlag, D., Geider K., Bertsch, L. L., and Kornberg, A. 1973. Initiation of DNA synthesis. *In:* Wells, R. D., and R. B. Inman, eds., *DNA Synthesis in Vitro,* Baltimore, Md., University Park Press, p. 175–183.

Schekman, R., Weiner, A., and Kornberg, A. 1974. Multienzyme systems of DNA replication. *Science* **186**:987–993.

Schweizer, D. 1973. Differential staining of plant chromosomes with Giemsa. *Chromosoma* **40**:307–320.

Scolnick, E. M., Aaronson, S. A., Todaro, G. J., and Parks, W. P. 1971. RNA dependent DNA polymerase activity in mammalian cells. *Nature* **229**:318–321.

Sedgwick, W. D., Wang, T. S., and Korn, D. 1972. Purification and properties of nuclear and cytoplasmic DNA polymerases from human KB cells. *J. Biol. Chem.* **247**:5026–5033.

Seki, S. and Mueller, G. C. 1975. A requirement for RNA protein and DNA synthesis in the establishment of DNA replicase activity in synchronized HeLa cells. *Biochim. Biophys. Acta* **378**:354–362.

Sen, A., and Levine, A. J. 1974. SV40 nucleoprotein complex activity unwinds superhelical turns in SV40 DNA. *Nature* **249**:343–344.

Senior, M. B., Olins, A. L., and Olins, D. E. 1975. Chromatin fragments resembling ν bodies. *Science* **181**:173–175.

Setlow, P. 1974. DNA polymerase I from *E. coli. Methods Enzymol.* **29**:3–12.

Setlow, R. B. 1970. The photochemistry, photobiology, and repair of polynucleotides. *Progr. Nucl. Acid. Res. Mol. Biol.* **8**:257–295.

Shea, M., and Kleinsmith, L. J. 1973. Template specific stimulation of RNA synthesis by phosphorylated nonhistone chromatin proteins. *Biochem. Biophys. Res. Comm.* **50**:473–477.

Sheehans, D. M., and Olins, D. E. 1974. The binding of nuclear nonhistone protein to DNA. *Biochim. Biophys. Acta* **353**:438–446.

Sheid, B., Srinivasan, P. R., and Borek, E. 1968. DNA methylase of mammalian tissues. *Biochemistry* **7**:280–285.

Shenkin, A., and Burdon, R. H. 1974. Deoxyadenylate-rich and deoxyguanylate-rich regions in mammalian DNA. *J. Mol. Biol.* **85**:19–39.

Shih, T. Y., and Fasman, G. D. 1971. C.D. studies of DNA complexes with arginine-rich histone IV. *Biochemistry* **10**:1675–1683.

Shiosaka, T., Aoki, M., and Tujii, S. 1974. Decrease in molecular weight of cytoplasmic DNA polymerase on treatment with bromelain plus RNase and DNase. *J. Biochem.* **75**:1399–1401.

Shires, A., Carpenter, M. P., and Chalkley, R. 1975. New histones found in mature mammalian testes. *Proc. Nat. Acad. Sci. USA* **72**:2714–2718.

Sinclair, J., Wells, R., Deumling, B., and Ingle, J. 1975. The complexity of satellite DNA in a higher plant. *Biochem. J.* **149**:31–38.

Sinonağlu, O. 1968. Solvent effects on molecular associations. *In:* Pullman, B., ed., *Molecular Associations in Biology,* New York, Academic Press, p. 427–445.

Sinsheimer, R. L. 1959. A single-stranded DNA from bacteriophage ϕX174. *J. Mol. Biol.* **1**:43–53.

Skinner, D. M., and Beattie, W. G. 1973. Cs_2SO_4 gradients containing both Hg^{2+} and Ag^+ effect the complete separation of satellite DNAs having identical densities in neutral CsCl gradients. *Proc. Nat. Acad. Sci. USA* **70**:3108–3110.

Skinner, D. M., Beattie, W. G., Blattner, F. R., Stark, B. P., and Dahlberg, J. E. 1974. The repeat sequence of a hermit crab satellite DNA is $(\text{–T–A–G–G–})_n \cdot (\text{–A–T–C–C})_n$. *Biochemistry* **13**:3930–3937.

Skinner, D. M., Beattie, W. G., Kerr, M. S., and Graham, D. E. 1970. Satellite DNAs in Crustacea: Two different satellites with the same density in neutral CsCl gradients. *Nature* **227**:837–839.

Smith, D. W., Schaller, H. E., and Bonhoeffer, F. J. 1970. DNA synthesis *in vitro*. *Nature* **226**:711–713.

Smith, H. W., and Halls, S. 1967. The transmissible nature of the genetic factor in *E. coli* that controls haemolysin production. *J. Gen. Microbiol.* **47**:153–161.

Smith-Sonneborn, J., and Klass, M. 1974. Changes in the DNA synthesis pattern of *Paramecium* with increased clonal age and interfission time. *J. Cell Biol.* **61**:591–598.

Smythies, J. R., Benington, F., Bradley, R. J., Morin, R. D., and Romine, W. O. 1974a. On the mechanism of interaction between histone I and DNA and histone III and DNA. *J. Theor. Biol.* **47**:309–315.

Smythies, J. R., Benington, F., Bradley, R. J., Morin, R. D., and Romine, W. O. 1974b. On the mechanisms of interaction between histone IIB_1 and DNA and histone IIB_2 and DNA. *J. Theor. Biol.* **47**:383–395.

Söderhäll, S., and Lindahl, T. 1976. DNA ligases of eukaryotes. *FEBS Lett.* **67**:1–8.

Sotelo, J. R., 1969. Ultrastructure of the chromosomes at meiosis. *In:* Lima-de-Faria, A., ed., *Handbook of Molecular Cytology,* Amsterdam, North-Holland Publishing Co., p. 412–434.

Southern, E. M. 1970. Base sequence and evolution of guinea-pig α-satellite DNA. *Nature* **227**:794–798.

Spadafora, C., and Geraci, G. 1975. The subunit structure of sea urchin sperm chromatin: A kinetic approach. *FEBS Lett.* **57**:79–82.

Spear, B. B., and Gall, J. G. 1973. Independent control of ribosomal gene replication in polytene chromosomes of *D. melanogaster*. *Proc. Nat. Acad. Sci. USA* **70**:1359–1363.

Sperling, R., and Bustin, M. 1974. Self assembly of histone F_2a_1. *Proc. Nat. Acad. Sci. USA,* **71**:4625–4629.

Spiegelman, S. 1971. DNA and the RNA viruses. *Proc. Roy. Soc. London,* **B177**:87–108.

Spiegelman, S., Burny, A., Das, M. R., Kaydar, J., Schlom, J., Travnicek, M., and Watson, K. 1970a. Characterization of the products of RNA-directed DNA polymerases in oncogenic RNA viruses. *Nature* **227**:563–567.

Spiegelman, S., Burny, A., Das, M. R., Kaydar, J., Schlom, J., Travnicek, M. and Watson, K. 1970b. Synthetic DNA-RNA hybrids and RNA-RNA duplexes as templates for the polymerases of the oncogenic RNA viruses. *Nature* **228**:430–432.

Spiegelman, S., Watson, K. B., and Kacian, D. L. 1971. Synthesis of DNA complements of natural RNAs: A general approach. *Proc. Nat. Acad. Sci. USA* **68**:2843–2845.

Spiker, S. 1975. An evolutionary comparison of plant histones. *Biochim. Biophys. Acta* **400**:461–467.

Srinivasan, P. R., and Borek, E. 1966. Enzymatic alteration of macromolecular structure. *Progr. Nucl. Acid Res.* **5**:157–189.

Srivastava, B. I. S. 1974. Deoxynucleotide-polymerizing enzymes in normal and malignant human cells. *Cancer Res.* **34**:1015–1026.

Srivastava, B. I. S., and Bardos, T. J. 1973. Inhibition of some DNA polymerase activities from cultured Burkitt cells by thiolated RNAs. *Life Sci.* **13**:47–53.

Srivastava, B. I. S., and Minowada, J. 1972. Ribonuclease-sensitive endogenous DNA polymerase activity and DNA-directed DNA polymerase in human tissue culture cell lines. *Cancer Res.* **32**:2481–2486.

Stack, S. M. 1975. Differential Giemsa staining of kinetochores in meiotic chromosomes of two higher plants. *Chromosoma* **51**:357–363.

Stack, S. M., and Clarke, C. R. 1973a. Pericentromeric chromosome banding in higher plants. *Can. J. Genet. Cyt.* **15**:367–369.

Stack, S. M., and Clarke, C. R. 1973b. Differential Giemsa staining of the telomeres of *Allium cepa* chromosomes. *Can. J. Genet. Cyt.* **15**:619–624.

Stack, S. M., Clarke, C. R., Cary, W. E., and Muffly, J. T. 1974. Different kinds of heterochromatin in higher plant chromosomes. *J. Cell Sci.* **14**:499–504.

Stein, G. S., and Baserga, R. B. 1970. The synthesis of acidic nuclear proteins in the prereplicative phase of the isoproterenol-stimulated salivary gland. *J. Biol. Chem.* **245**:6097–6105.

Stein, G. S., and Baserga, R. B. 1972. Nuclear proteins and the cell cycle. *Adv. Cancer Res.* **15**:287–319.

Stein, G. S., and Borun, T. W. 1972. The synthesis of acidic chromosomal proteins during the cell cycle of HeLa S-3 cells. *J. Cell Biol.* **52**:292–307.

Stein, G. S., and Farber, J. L. 1972. Role of nonhistone chromosomal proteins in the restriction of mitotic chromatin template activity. *Proc. Nat. Acad. Sci. USA* **69**:2918–2921.

Stein, G. S., and Matthews, D. E. 1973. Nonhistone chromosomal protein synthesis: Utilization of preexisting and newly transcribed mRNAs. *Science* **181**:71–73.

Stein, G. S., and Stein, J. 1976. Chromosomal proteins and their role in the regulation of gene expression. *BioScience* **26**:488–498.

Stern, H., and Hotta, Y. 1973. Biochemical controls of meiosis. *Ann. Rev. Gen.* **7**:37–66.

Sternglanz, H., and Bugg, C. E. 1973. Conformation of N^6-methyladenine, a base involved in DNA modification : restriction processes. *Science* **182**:833–834.

Sternglanz, R., Wang, H. F., and Donegan, J. J. 1976. Evidence that both growing DNA chains at a replication fork are synthesized discontinuously. *Biochemistry* **15**:1838–1843.

Stevens, L. 1970. The biochemical role of naturally occurring polyamines in nucleic acid synthesis. *Biol. Rev.* **45**:1–27.

Stone, L. B., Scolnick, E., Takemoto, K. K., and Aaronson, S. A. 1971. Visna virus: A slow virus with an RNA dependent DNA polymerase. *Nature* **229**:257–258.

Stonington, O. G., and Pettijohn. D. E., 1971. The folded genome of *E. coli* isolated in a protein-DNA-RNA complex. *Proc. Nat. Acad. Sci. USA* **68**:6–9.

Strauss, B. S. 1968. DNA repair mechanisms and their relation to mutation and recombination. *Curr. Top. Microbiol. Immunol.* **44**:1–85.

Sueoka, N. 1961. Variation and heterogeneity of base composition of DNAs: A compilation of old and new data. *J. Mol. Biol.* **3**:31–40.

Sundaralingam, M. 1969. Stereochemistry of nucleic acids and their constituents. IV. Allowed and preferred conformations of nucleosides, nucleoside mono-, di-, tri-, tetraphosphates, nucleic acids, and polynucleotides. *Biopolymers* **7**:821–860.

Sutherland, B. M., Chamberlin, M. J., and Sutherland, J. C. 1973. DNA photoreactivating enzyme from *E. coli*. *J. Biol. Chem.* **248**:4200–4205.

Sutherland, B. M., Runge, P., and Sutherland, J. C. 1974. DNA photoactivating enzyme from placental mammals. *Biochemistry* **13**:4710–4715.

Tait, R. C., Harris, A. L., and Smith, D. W. 1974. DNA repair in *E. coli* mutants deficient in DNA polymerases I, II, and/or III. *Proc. Nat. Acad. Sci. USA* **71**:675–679.

Tait, R. C., and Smith, D. W. 1974. Roles for *E. coli* polymerases I, II, and III in DNA replication. *Nature* **249**:116–120.

Tavitian, A., Hamelin, R., Tchen, P., Olofsson, B., and Boiron, M. 1974. Extent of transcription of mouse sarcoma-leukemia virus by RNA-directed DNA polymerase. *Proc. Nat. Acad. Sci. USA* **71**:755–759.

Taylor, J. H., 1974. Units of DNA replication in chromosomes of eukaryotes. *Int. Rev. Cytol.* **37**:1–20.

Temin, H. M., and Mizutani, S. 1970. RNA-dependent DNA polymerase in virions of Rous sarcoma virus. *Nature* **226**:1211–1213.

Temussi, P. A. 1975. Automatic comparison of the sequences of calf thymus histones. *J. Theor. Biol.* **50**:25–33.

Thomas, A. J., and Sherratt, H. S. A. 1956. The isolation of nucleic acid fractions from plant leaves and their purine and pyrimidine composition. *Biochem. J.* **62**:1–4.

Thrall, C. L., Park, W. D., Rashba, H. W., Stein, J. L., Mans, R. J., and Stein, G. S. 1974. *In vitro* synthesis of DNA complementary to polyadenylated histone messenger RNA. *Biochem. Biophys. Res. Comm.* **61**:1443–1449.

Tobin, R. S., and Seligy, V. L. 1975. Characterization of chromatin-bound erythrocyte histone V (f2c). *J. Biol. Chem.* **250**:358–364.

Tolstoshev, P., and Wells, J. R. E. 1974. Nature and origin of chromatin-associated RNA of avian reticulocytes. *Biochemistry* **13**:103–111.

Tomilin, N. V., and Svetlova, M. P. 1974. On the mechanism of postreplication repair in *E. coli* cells. *FEBS Lett.* **43**:185–188.

Travaglini, E. C., and Loeb, L. A. 1974. RNA dependent DNA synthesis by *E. coli* DNA polymerase I. *Biochemistry* **13**:3010–3017.

Tsuruo, T., and Ukita, T. 1974. Purification and further characterization of three DNA polymerases of rat ascites hepatoma cells. *Biochim. Biophys. Acta* **353**:146–159.

Tunis, M. J. B., and Hearst, J. E. 1968. On the hydration of DNA. *Biopolymers* **6**:1325–1344.

Tunis-Schneider, M. J. B., and Maestre, M. F. 1970. C.D. spectra of oriented and unoriented DNA films. *J. Mol. Biol.* **52**:521–541.

Vallentyne, J. R. 1956. Thermal degradation of amino acids. *Carn. Inst. Year Book* **56**:185–186.

Vandergrift, V., Serra, M., Moore, D. S., and Wagner, T. E. 1974. The role of the arginine-rich histones in the maintenance of DNA conformation in chromatin. *Biochemistry* **13**:5087–5092.

van de Sande, J. H., Lin, C. C., and Jorgenson, K. F. 1977. Reverse banding on chromosomes produced by a guanosine-cytosine specific DNA binding antibiotic. *Science* **195**:400–402.

van der Vleiet, P. C., and Sussenbach, G. S. 1972. The mechanism of adenovirus DNA synthesis in isolated nuclei. *Eur. J. Biochem.* **33**:584–592.

van Holde, K. E., Sahasrabuddhe, C. G., Shaw, B. R., Van Bruggen, E. F. J., and Arnberg, A. C. 1974. Electron microscopy of chromatin subunit particles. *Biochem. Biophys. Res. Comm.* **60**:1365–1370.

Vanyushin, B. F., and Belozerskii, A. N. 1959. A comparative study of the composition of RNA in higher plants. *Dokl. Akad. Nauk SSSR* **127**:196–199.

Vincent, W. S. 1952. The isolation and chemical properties of the nucleoli of starfish oocytes. *Proc. Nat. Acad. Sci. USA* **38**:139–145.

Viola, M. V. 1973. Reverse transcriptase and 70 S RNA in supernatant from a human cell line. *J. Nat. Cancer Inst.* **50**:1175–1178.

Voet, D., and Rich, A. 1970. The crystal structures of purines, pyrimidines and their intermolecular complexes. *Prog. Nucl. Acid. Res. Mol. Biol.* **10**:183–265.

Volpe, P., and Eremenko, T. 1974. Preferential methylation of regulatory genes in HeLa cells. *FEBS Lett.* **44**:121–126.

Vorob'ev, V. I., and Kosjuk, G. N. 1974. Distribution of repetitive and non-repetitive nucleotide sequences in the DNA of sea urchin. *FEBS Lett.* **47**:43–46.

Vosa, C. G., and Marchi, P. 1972. Quinacrine fluorescence and Giemsa staining in plants. *Nature New Biol.* **237**:191–192.

Votavová, H., and Šponar, J. 1974. Satellite components of calf thymus DNA. Chromatographic isolation of the native components and characterization of their sequence heterogeneity. *Col. Czechoslov. Chem. Comm.* **39**:2312–2324.

Votavová, H., and Šponar, J. 1975a. Reassociation kinetics of three satellite components of calf thymus DNA. *Nucl. Acid Res.* **2**:185–196.

Votavová, H., and Šponar, J. 1975b. Identification and separation of components of calf thymus DNA using a CsCl-netrospsin density gradient. *Nucl. Acid. Res.* **2**:431–446.

Wainfan, E., Srinivasan, P. R., and Borek, E. 1965. Alterations in tRNA methylases after bacteriophage infection or induction. *Biochemistry* **4**:2845–2848.

Walker, P. M. B. 1969. The specificity of molecular hybridization in relation to studies on higher organisms. *Progr. Nucl. Acid. Res. Mol. Biol.* **9**:301–328.

Walker, P. M. B. 1971. "Repetitive" DNA in higher organisms. *Progr. Biophys. Mol. Biol.* **23**:145–190.

Wallace, P. G., Hewish, D. R., Venning, M. M., and Burgoyne, L. A. 1971. Multiple forms of mammalian DNA polymerase. *Biochem. J.* **125**:47–54.

Wallace, R. D., and Kates, J. 1972. State of adenovirus 2 DNA in the nucleus and its mode of transcription. *J. Virol.* **9**:627–635.

Walton, S. W., 1971. Sex-chromosome mosaicism in pouch young of marsupials *Perameles* and *Isoodon*. *Cytogenetics* **10**:115–120.

Wang, H. F., and Popenoe, E. A. 1977. Variation of DNA polymerase activities during avian erythropoiesis. *Biochim. Biophys. Acta* **474**:98–108.

Wang, H. F., and Sternglanz, R. 1974. Thymine-labelled deoxyoligonucleotide involved in DNA chain growth in *B. subtilis*. *Nature* **248**:147–150.

Wang, J. C. 1971. Interaction between DNA and an *E. coli* protein ω. *J. Mol. Biol.* **55**:523–534.

Wang, J. C. 1973. Protein ω: A DNA. *In:* Wells, R. D., and R. B. Inman, eds., *DNA Synthesis in Vitro,* Baltimore, Md., University Park Press, p. 163–174.

Waqar, M. A., and Huberman, J. A. 1973. Evidence for attachment of RNA to pulse-labeled DNA in slime mold, *Physarum polycephalum*. *Biochem. Biophys. Res. Comm.* **51**:174–181.

Ward, D. C., Humphries, D. C., and Weinstein, I. B. 1972. Synthetic RNA-dependent DNA polymerase activity in normal rat liver and hepatomas. *Nature* **237**:499–503.

Warnecke, P., Kruse, K., and Harbers, E. 1973. Isolation and characterization of nonhistone proteins from euchromatic and heterochromatic deoxyribonucleo-protein of rat liver. *Biochim. Biophys. Acta* **331**:295–304.

Warner, H. P., and Barnes, J. E. 1966a. DNA synthesis in *E. coli* infected with some DNA polymerase-less mutants of bacteriophage T4. *Virology* **28**:100–107.

Warner, H. R., and Barnes, J. E. 1966b. Evidence for a dual role for the bacteriophage T4-induced deoxycytidine triphosphate nucleotidohydrolase. *Proc. Nat. Acad. Sci. USA* **56**:1233–1240.

Watson, J. D., and Crick, F. H. C. 1953a. A structure for DNA. *Nature* **171**:736–738.

Watson, J. D., and Crick, F. H. C. 1953b. Genetic implications of the structure of DNA. *Nature* **171**:964–967.

Weintraub, H., and Holtzer, H. 1972. Fine control of DNA synthesis in developing chick red blood cells. *J. Mol. Biol.* **66**:13–35.

Weisbach, A., Schlabach, A., Friedlender, R. and Bolden, A. 1971. DNA polymerases from human cells. *Nature New Biol.* **231:**167–170.

Werner, R. 1971a. Mechanism of DNA replication. *Nature* **230**:570–572.

Werner, R. 1971b. Nature of DNA precursors. *Nature New Biol.* **233**:99–103.

West, E. S., and Todd, W. R. 1961. *Textbook of Biochemistry. 3rd Ed.,* New York, The Macmillan Company.

Westergaard, O., Brutlag, D., and Kornberg, A. 1973. Incorporation of RNA primer into the phage replicative form. *J. Biol. Chem.* **248**:1361–1364.

Whitlock, J. P., and Simpson, R. T. 1976. Removal of histone Hl exposes a fifty base pair DNA segment between nucleosomes. *Biochemistry* **15**:3307–3314.

Williams, C. A., and Ockey, C. H. 1970. Distribution of DNA replicator sites in mammalian nuclei after different methods of cell synchronization. *Exp. Cell Res.* **63**:365–372.

Wintersberger, E. 1974. DNA polymerases from yeast. Further purification and characterization of DNA-dependent DNA polymerases A and B. *Eur. J. Biochem.* **50**:41–47.

Wintersberger, U. 1974. Absence of a low-molecular-weight DNA polymerase from nuclei of the yeast *S. cerevisiae*. *Eur. J. Biochem.* **50**:197–202.

Woese, C. R. 1967. *The Genetic Code,* New York, Harper and Row.

Woese, C. R. 1970a. Concerning the accuracy of codon recognition. The allosteric ribosome model. *J. Theor. Biol.* **26**:83–88.

Woese, C. R. 1970b. The problem of evolving a genetic code. *BioScience* **20**:471–485.

Woese, C. R., and Bleyman, M. A. 1972. Genetic code limit organisms—do they exist? *J. Mol. Evol.* **1**:223–229.

Wolstenholme, D. R., Koike, K., and Cochran-Fouts, P. 1973. Replication of mitochondrial DNA. *Cold Spring Harbor Symp. Quant. Biol.* **38**:267–280.

Wouters-Tyrou, D., Sautiere, P., and Biserte, G. 1976. Covalent structure of the sea urchin histone H_4. *FEBS Lett.* **65**:225–228.

Wu, A. M., and Gallo, R. C. 1975. Reverse transcriptase. *Crit. Rev. Biochem.* **3**:289–347.

Wyatt, G. R. 1951. The purine and pyrimidine composition of deoxypentose nucleic acids. *Biochem. J.* **48**:584–590.

Yamada, M., and Hanaoka, F. 1973. Periodic changes in the association of mamalian DNA with the membrane during the cell cycle. *Nature New Biol.* **243**:227–230.

Yamagishi, H. 1970. Nucleotide distribution in the DNA of *E. coli*. *J. Mol. Biol.* **49**:603–608.

Yamagishi, H. 1974. Nucleotide distribution in bacterial DNAs differing in G + C content. *J. Mol. Evol.* **3**:239–242.

Yamagishi, H., and Takahashi, I. 1971. Heterogeneity in nucleotide composition of *B. subtilis* DNA. *J. Mol. Biol.* **57**:369–372.

Yamashita, T., and Shimojo, H. 1973. Replication of adenovirus 12 DNA in association with the nuclear membrane. *Jpn. J. Microbiol.* **17**:419–423.

Yang, J. T., and Samejima, T. 1969. Optical rotatory dispersion and circular dichroism of nucleic acids. *Progr. Nucl. Acid. Res. Mol. Biol.* **9**:223–300.

Yoshida, S., Kondo, T., and Ando, T. 1974. Multiple molecular species of cytoplasmic DNA polymerase from calf thymus. *Biochim. Biophys. Acta* **353**:463–474.

Young, C. W., Hendler, F. J., and Karnofsky, D. A. 1969. Synthesis of protein for DNA replication and cleavage events in the sand dollar embryo. *Exp. Cell. Res.* **58**:15–26.

Yudelevich, A., Ginsberg, B., and Hurwitz, J. 1968. Discontinuous synthesis of DNA during replication. *Proc. Nat. Acad. Sci. USA* **61**:1129–1136.

Zimmerman, S. B., Cohen, G. H., and Davies, D. R. 1975. X-ray fiber diffraction and model-building study of polyguanylic acid and polyinosinic acid. *J. Mol. Biol.* **92**:181–192.

Zusman, D. R., Carbonell, A., and Haga, J. Y. 1973. Nucleoid condensation and cell division in *E. coli* MX74T2ts52 after inhibition of protein synthesis. *J. Bact.* **115**:1167–1178.

CHAPTER 4

Abraham, D. J. 1971. Proposed detailed structural model for tRNA and its geometric relationship to a messenger. *J. Theor. Biol.* **30**:83–91.

Adams, J. M., and Cory, S. 1970. Untranslated nucleotide sequence at the 5′-end of R17 bacteriophage RNA. *Nature* **227**:570–574.

Adams, J. M., and Cory, S. 1975. Modified nucleosides and bizarre 5′-termini in mouse myeloma mRNA. *Nature* **255**:28–33.

Adesnik, M., and Darnell, J. E. 1972. Biogenesis and characterization of histone mRNA in HeLa cells. *J. Mol. Biol.* **67**:397–406.

Ajtkhozhin, M. A., and Akhanov, A. U. 1974. Release of mRNP-particles of the informosome type from polyribosomes of higher plant embryos. *FEBS Lett.* **41**:275–279.

Allende, J. E., Monro, R., and Lipmann, F. 1964. Resolution of the *E. coli* amino acyl sRNA transfer factor into two complementary fractions. *Proc. Nat. Acad. Sci. USA* **51**:1211–1216.

Ames, B. N., and Hartman, P. E. 1963. The histidine operon. *Cold Spring Harb. Symp. Quant. Biol.* **28**:349–356.

Anderson, J. S., Bretscher, M. S., Clark, B. F. C., and Marcher, K. A. 1967. A GTP requirement for binding initiator tRNA to ribosomes. *Nature* **215**:490–492.

Arias, I. M., Doyle, D., and Schimke, R. T. 1969. Studies on the synthesis and degradation of proteins of the endoplasmic reticulum of rat liver. *J. Biol. Chem.* **244**:3303–3315.

Attardi, G., Huang, P. C., and Kabat, S. 1965. Recognition of rRNA sites in DNA. *Proc. Nat. Acad. Sci. USA* **53**:1490–1498.

Averner, M. J., and Pace, N. R. 1972. The nucleotide sequence of marsupial 5 S ribosomal RNA. *J. Biol. Chem.* **247**:4491–4493.

Ayuso-Parilla, M., Henshaw, E. C., and Hirsch, C. A. 1973a. The ribosome cycle in mammalian protein synthesis. *J. Biol. Chem.* **248**:4386–4393.

Ayuso-Parilla, M., Hirsch, C. A., and Henshaw, E. C. 1973b. Release of the nonribosomal proteins from the mammalian native 40S ribosomal subunit by aurintricarboxylic acid. *J. Biol. Chem.* **248**:4394–4399.

Bag, J., and Sarkar, S. 1975. Cytoplasmic nonpolysomal mRNP containing actin mRNA in chicken embryonic muscles. *Biochemistry* **14**:3800–3807.

Ball, L. A. 1973. Mutual influence of the secondary structure and informational content of a mRNA. *J. Theor. Biol.* **4**:243–247.

Barrieux, A., Ingraham, H. A., David, D. N., and Rosenfeld, M. G. 1975. Isolation of messenger-like RNPs. *Biochemistry* **14**:1815–1821.

Beaudet, A. L., and Caskey, C. T. 1971. Mammalian peptide chain termination. *Proc. Nat. Acad. Sci. USA* **68**:619–624.

Bell, E., and Reeder, R. 1965. Short- and long-lived mRNA in embryonic chick lens. *Science* **150**:71–72.

Bellmare, G., Jordan, B. R., Rocca-Serra, J., and Monier, R. 1972. Accessibility of *E. coli* 5 S RNA base residues to chemical reagents. *Biochimie* **54**:1453–1466.

Bellmare, G., Vigne, R., and Jordan, B. R. 1973. Interaction between *E. coli* ribosomal proteins and 5 S RNA molecules. *Biochimie* **55**:29–35.

Benhamou, J., and Jordan, B. R. 1976. Nucleotide sequence of *D. melanogaster* 5 S RNA. *FEBS Lett.* **62**:146–149.

Benhamou, J., Jourdan, R., and Jordan, B. R. 1977. Sequence of *Drosophila* 5 S RNA synthesized by cultured cells and by the insect at different developmental stages. *J. Mol. Evol.* **9**:279–298.

Berns, A. J. M., Strous, G. J. A. M., and Bloemendal, H. 1972. Heterologous *in vitro* synthesis of lens α-crystallin polypeptide. *Nature New Biol.* **236**:7–9.

Billeter, M. A., Dahlberg, J. E., Goodman, H. E., Hindley, J., and Weissmann, C. 1969. Sequence of first 175 nucleotides from the 5′-terminus of Qβ RNA synthesized *in vitro*. *Nature* **224**:1083–1086.

Bishop, J. O., Morton, J. G., Rosbach, M., and Richardson, M. 1974. Three abundance classes in HeLa cell mRNA. *Nature* **250**:199–204.

Bitar, K. G. 1975. The primary structure of the ribosomal protein L29 from *E. coli*. *Biochim. Biophys. Acta* **386**:99–106.

Bitar, K. G., and Wittmann-Liebold, B. 1975. The primary structure of the 5 S rRNA binding protein L25 of *E. coli*. *Hoppe-Seyler's Zeit. Phys. Chem.* **356**:1343–1352.

Blobel, G. 1972. Protein tightly bound to globin mRNA. *Biochem. Biophys. Res. Comm.* **47**:88–95.

Blobel, G. 1973. A protein of molecular weight 78,000 bound to the poly(A) region of eukaryotic mRNAs. *Proc. Nat. Acad. Sci. USA* **70**:924–928.

Bloemendal, H. 1977. The vertebrate eye lens. *Science* **197**:127–138.

Blundell, M., Craig, E., and Kennell, D. 1972. Decay rates of different mRNA in *E. coli* and models of decay. *Nature New Biol.* **238**:46–49.

Bollini, R., Soffientini, A. N., Bertani, A., and Lanzani, G. A. 1974. Some molecular properties of the elongation factor EF-1 from wheat embryos. *Biochemistry* **13**:5421–5425.

Bonnet, J., and Ebel, J. P. 1972. Interpretation of incomplete reactions in tRNA aminoacylation. Aminoacylation of yeast $tRNA^{Val}$ with yeast valyl-tRNA synthetase. *Eur. J. Biochem.* **31**:335–344.

Bonnet, J., and Ebel, J. P. 1974. Correction of aminoacylation errors: evidence for a nonsignificant role of the aminoacyl-tRNA synthetase catalyzed deacylation of aminoacyl-tRNAs. *FEBS Lett.* **39**:259–262.

Bonnet, J., Giegé, R., and Ebel, J. P. 1972. Lack of specificity in the aminoacyl-tRNA synthetase-catalysed deacylation of aminoacyl-tRNA. *FEBS Lett.* **27**:139–144.

Bostock, C. J., Prescott, D. M., and Lauth, M. 1971. Lability of 26 S rRNA in *Tetrahymena pyriformis*. *Exp. Cell Res.* **66**:260–262.

Branlant, C., and Ebel, J.-P. 1977. Studies on the primary structure of *E. coli* 23 S RNA. Nucleotide sequence of the ribonuclease T_1 digestion products containing more than one uridine residue. *J. Mol. Biol.* **111**:215–256.

Brenner, S., Jacob, F., and Meselson, M. 1961. An unstable intermediate carrying information from genes to ribosomes for protein synthesis. *Nature* **190**:576–581.

Bretscher, M. S. 1965. Fractionation of oligolysyl-adenosine complexes derived from polylysine attached to tRNA. *J. Mol. Biol.* **12**:913–919.

Bretscher, M. S. 1966. Polypeptide chain initiation and the characterization of ribosomal binding in *E. coli*. *Cold Spring Harbor Symp. Quant. Biol.* **31**:289–296.

Bretscher, M. S. 1968. Direct translation of a circular mRNA. *Nature* **220**:1088–1091.

Bretscher, M. S. 1969. Direct translation of bacteriophage fd DNA in the absence of neomycin B. *J. Mol. Biol.* **42**:595–598.

Bretscher, M. S., Goodman, H. M., Menninger, J. R., and Smith, J. D. 1965. Polypeptide chain termination using synthetic polynucleotides. *J. Mol. Biol.* **14**:634–639.

Brinacombe, R., Nierhaus, K. H., Garrett, R. A., and Wittmann, H. G. 1976. The ribosome of *E. coli*. *Progr. Nucl. Acid Res. Mol. Biol.* **18**:1–44, 323–325.

Brosius, J., and Chen, R. 1976. The primary structure of protein L16 located at the peptidyl-transferase center of *E. coli* ribosomes. *FEBS Lett.* **68**:105–109.

Brosius, J., Schiltz, E., and Chen, R. 1975. The primary structure of the 5 S RNA binding protein L18 from *E. coli* ribosomes. *FEBS Lett.* **56**:359–361.

Brot, N., Tate, W. P., Caskey, C.-T., and Weissbach, H. 1974. The requirement for ribosomal proteins L7 and L12 in peptide-chain termination. *Proc. Nat. Acad. Sci. USA* **71**:89–92.

Brouwer, J., and Planta, R. J. 1975. The origin of high molecular weight proteins in ribosomal preparations of *Bacillus licheniformis*. *FEBS Lett.* **53**:73–75.

Brown, D. D., and Weber, C. S. 1968. Unique DNA sequences homologous to 4 S RNA, 5 S RNA, and rRNA. *J. Mol. Biol.* **34**:661–680.

Brown, D. D., Wensink, P. C., and Jordan, E. 1971. Purification and some characteristics of 5 S DNA from *Xenopus laevis*. *Proc. Nat. Acad. Sci. USA* **68**:3175–3179.

Brown, G. L. 1963. Preparation, fractionation, and properties of sRNA. *Progr. Nucl. Acid Res.* **2**:259–310.

Brown, J. C., and Doty, P. 1968. Protein factor requirement for binding of mRNA to ribosomes. *Biochem. Biophys. Res. Comm.* **30**:284–291.

Brownlee, G. G., Cartwright, E., McShane, T., and Williamson, R. 1972. The nucleotide sequence of somatic 5 S RNA from *Xenopus laevis*. *FEBS Lett.* **25**:8–12.

Brownlee, G. G., Sanger, F., and Barrell, B. G. 1967. Nucleotide sequence of 5 S rRNA from *E. coli*. *Nature* **215**:735–736.

Brownlee, G. G., Sanger, F., and Barrell, B. G. 1968. The sequence of 5 S rRNA. *J. Mol. Biol.* **34**:379–412.

Bryan, R. N., and Hayashi, M. 1973. Two proteins are bound to most species of polysomal mRNA. *Nature New Biol.* **244**:271–274.

Buckingham, M. E., Caput, D., Cohen, A., Whalen, R. G., and Gros, F. 1974. The synthesis and stability of cytoplasmic mRNA during myoblast differentiation in culture. *Proc. Nat. Acad. Sci. USA* **71**:1466–1470.

Burr, H., and Lingrel, J. B. 1971. Poly A sequences at the 3′ termini of rabbit globin mRNAs. *Nature New Biol.* **233**:41–43.

Burstein, Y., Kantor, F., and Schechter, I. 1976. Partial amino-acid sequence of the precursor of an immuno-globulin light chain containing NH_2-terminal pyroglutamic acid. *Proc. Nat. Acad. Sci. USA* **73**:2604–2608.

Busby, W. F., Hele, P., and Chang, M. C. 1974. Apparent amino acid incorporation by ejaculated rabbit spermatozoa. *Biochim. Biophys. Acta* **330**:246–259.

Busiello, E., and DiGirolamo, M. 1973. Aminoacyl-tRNA binding sites in *E. coli* and reticulocyte ribosomes. *FEBS Lett.* **35**:341–343.

Campo, M. S., and Bishop, J. O. 1974. Two classes of mRNA in cultured rat cells. *J. Mol. Biol.* **90**:649–663.

Cann, A., Gambino, R., Banks, J., and Bank, A. 1974. Poly(A) sequences and biological activity from human globin mRNA. *J. Biol. Chem.* **249**:7536–7540.

Cantor, C. R. 1967. Possible conformations of 5 S rRNA. *Nature* **216**:513–514.

Cantor, C. R. 1968. The extent of base pairing in 5 S rRNA. *Proc. Nat. Acad. Sci. USA* **59**:478–483.

Cashion, L. M., and Stanley, W. M. 1974. Two eukaryotic initiation factors (IF-I and IF-II) of protein synthesis that are required to form an initiation complex with rabbit reticulocyte ribosomes. *Proc. Nat. Acad. Sci. USA* **71**:436–440.

Caskey, C. T. 1973. Peptide chain termination. *Adv. Prot. Chem.* **27**:243–276.

Caskey, C. T., Beaudet, A. L., and Tate, W. P. 1974. Mammalian release factor. *Methods Enzymol.* **30**:293–303.

Caskey, T., Leder, P., Moldave, K., and Schlessinger, D. 1972. Translation: Its mechanism and control. *Science* **176**:195–197.

Ceccarini, C., and Maggio, R. 1968. Studies on the ribosomes from the cellular slime molds, *Dictyostelium discoideum* and *D. purpureum*. *Biochim. Biophys. Acta* **166**:134–141.

Cedergren, R. J., and Sankoff, D. 1976. Evolutionary origin of 5.8 S rRNA. *Nature* **260**:74–75.

Chambers, R. W. 1971. On the recognition of tRNA by its aminoacyl-tRNA ligase. *Progr. Nucl. Acid Res.* **11**:489–525.

Chatterjee, S. K., Kazemie, M., and Matthaei, H. 1973. Separation of the ribosomal proteins by two-dimensional electrophoresis. *Hoppe-Seyler's Z. Phys. Chem.* **354**:481–486.

Chen, R., and Ehrke, G. 1976. The primary structure of protein L34 from the large ribosomal subunit of *E. coli*. *FEBS Lett.* **63**:215–217.

Chen, R., Mende, L., and Arfsten, U. 1975. The primary structure of protein L27 from the peptidyl-tRNA binding site of *E. coli* ribosomes. *FEBS Lett.* **59**:96–99.

Chen, R., and Wittmann-Liebold, B. 1975. The primary structure of protein S9 from the 30 S subunit of *E. coli* ribosomes. *FEBS Lett.* **52**:139–140.

Cimadevilla, J. M., and Hardesty, B. 1975. Isolation and partial characterization of a 40 S ribosomal subunit-tRNA binding factor from rabbit reticulocytes. *J. Biol. Chem.* **250**:4389–4397.

Clark, B. F. C., and Marcker, K. A. 1966. The role of *N*-formyl-methionyl-sRNA in protein biosynthesis. *J. Mol. Biol.* **17**:394–406.

Cole, P. E., and Crothers, D. H. 1972. Conformational changes of tRNA. *Biochemistry* **11**:4368–4374.

Cole, P. E., Young, S. K., and Crothers, D. M. 1972. Conformational changes of tRNA. Equilibrium phase diagrams. *Biochemistry* **11**:4358–4368.

Comb, D. G., and Sarkar, N. 1967. The binding of 5 S rRNA to ribosomal subunits. *J. Mol. Biol.* **25**:317–330.

Contreras, R., Vandenberghe, A., Min Jou, W., de Wachter, R., and Fiers, W. 1971. Studies on the bacteriophage MS2 nucleotide sequence of a 3′-terminal fragment ($n = 104$). *FEBS Lett.* **18**:141–144.

Contreras, R., Ysebaert, M., Min Jou, W., and Fiers, W. 1973. Bacteriophage MS_2 RNA: nucleotide sequence of the end of the A protein gene on the intercistronic region. *Nature New Biol.* **241**:99–101.

Conway, T. W., and Lipmann, F. 1964. Characterization of a ribosome-linked guanosine triphosphatase in *E. coli* extracts. *Proc. Nat. Acad. Sci. USA* **52**:1462–1469.

Cornick, G. G., and Kretsinger, R. H. 1977. The 30 S subunit of the *E. coli* ribosome. *Biochim. Biophys. Acta* **474**:398–410.

Corry, M. J., Payne, P. I., and Dyer, T. A. 1974a. The nucleotide sequence of 5 S rRNA from the blue-green alga *Anacystis nidulans*. *FEBS Lett.* **46**:63–66.

Corry, M. J., Payne, P. I., and Dyer, T. A. 1974b. A sequence analysis of 5 S rRNA from the blue-green alga *Oscillatoria tenuis* and a comparison of blue-green alga 5 S rRNA with those of bacterial and eukaryotic origin. *FEBS Lett.* **46**:67–70.

Cox, R. A. 1970. A spectrophotometric study of the secondary structure of RNA isolated from the smaller and larger ribosomal subparticles of rabbit reticulocytes. *Biochem. J.* **117**:101–118.

Cox, R. A., Gould, H., and Kanagalingam, K. 1968. A study of the alkaline hydrolysis of fractionated reticulocyte rRNA and its relevance to secondary structure. *Biochem. J.* **106**:733–741.

Cramer, F. 1969. Three-dimensional structure of tRNA. *Progr. Nucl. Acid. Res.* **11**:391–421.

Cramer, F., and Erdman, V. A. 1968. Amount of adenine and uracil base pairs in *E. coli* 23 S, 16 S, and 5 S rRNA. *Nature* **218**:92–93.

Cremer, K., and Schlessinger, D. 1974. Ca^{2+} ions inhibit mRNA degradation but permit mRNA transcription and translation in DNA-coupled systems from *E. coli*. *J. Biol. Chem.* **249**:4730–4736.

Crick, F. H. C. 1966. The genetic code: Yesterday, today, and tomorrow. *Cold Spring Harbor Symp. Quant. Biol.* **31**:3–10.

Crick, F. H. C. 1967. The genetic code. *Proc. Roy. Soc. London* **B167**:331–347.

Crick, F. H. C., Barnett, L., Brenner, S., and Watts-Tobin, R. J. 1961. General nature of the genetic code for proteins. *Nature* **192**:1227–1232.

Crystal, R. A., and Anderson, W. F. 1972. Initiation of hemoglobin synthesis. *Proc. Nat. Acad. Sci. USA* **69**:706–711.

Crystal, R. A., Elson, N. A., and Anderson, W. F. 1974. Initiation of globin synthesis. *Methods Enzymol.* **30**:101–127.

Czernilofsky, A. P., Collatz, E. E., Stöffler, G., and Kuechler, E. 1974. Proteins at the tRNA binding sties of *E. coli* ribosomes. *Proc. Nat. Acad. Sci. USA* **71**:230–234.

Dahlberg, A. E. 1974. Two forms of the 30 S ribosomal subunit of *E. coli*. *J. Biol. Chem.* **249**:7673–7678.

Darnell, J. E., Wall, R., and Tushinski, R. J. 1971. An adenylic acid-rich sequence in mRNA of HeLa cells and its possible relationship to reiterated sites in DNA. *Proc. Nat. Acad. Sci. USA* **68**:1321–1325.

Davidson, J. N. 1972. *The Biochemistry of the Nucleic Acids. 7th Ed.*, New York, Academic Press.

Daya-Grosjean, L., Reinbolt, J., Pongs, O., and Garrett, R. A. 1974. A study of the regions of ribosomal proteins S4, S8, S15, and S20 that interact with 16 S RNA of *E. coli*. *FEBS Lett.* **44**:253–256.

Dayhoff, M. O. 1972. *Atlas of Protein Sequence and Structure 1972*. Washington, National Biomedical Research Foundation.

Diez, J., and Brawerman, G. 1974. Elongation of the poly(A) segment of mRNA in the cytoplasm of mammalian cells. *Proc. Nat. Acad. Sci. USA* **71**:4091–4095.

Dillon, L. S. 1962. Comparative cytology and the evolution of life. *Evolution* **16**:102–117.

Dillon, L. S. 1963. A reclassification of the major groups of organisms based upon comparative cytology. *Syst. Zool.* **12**:71–82.

Dillon, L. S. 1973. The origins of the genetic code. *Bot. Rev.* **39**:301–345.

Dillon, L. S. 1978. *Evolution: Concepts and Consequences. 2nd Ed.*, St. Louis, Mo., C. V. Mosby Co.

Dina, D., Crippa, M., and Beccari, E. 1973. Hybridization properties and sequence arrangement in a population of mRNAs. *Nature New Biol.* **242**:101–105.

Dina, D., Meza, I., and Crippa, M. 1974. Relative positions of the 'repetitive,' 'unique' and poly(A) fragments of mRNA. *Nature* **248**:486–490.

Dohme, F., and Nierhaus, K. H. 1976. Role of 5 S RNA in assembly and function of the 50 S subunit from *E. coli*. *Proc. Nat. Acad. Sci. USA* **73**:2221–2225.

Dovgas, N. V., Markova, L. F., Mednikova, T. A., Vinokurov, L. M., Alakhov, Y. B., and Ovchinnihov, Y. A. 1975. The primary structure of the 5 S RNA binding protein L25 from *E. coli* ribosomes. *FEBS Lett.* **53**:351–354.

Dovgas, N. V., Vinokurov, L. M., Velmoga, I. S., Alakhov, Y. B., and Ovchinnikov, Y. A. 1976. The primary structure of protein L10 from *E. coli* ribosomes. *FEBS Lett.* **67**:58–61.

Drews, J., Bednarik, K., and Grasmuck, H. 1974. Elongation factor 1 from Krebs II mouse ascites cells. *Eur. J. Biochem.* **41**:217–227.

Dube, S. K. 1973. Recognition of tRNA by the ribosome. A possible role of 5 S RNA. *FEBS Lett.* **36**:39–42.

Dubuy, B., and Weissman, S. M. 1971. Nucleotide sequence of *Pseudomonas fluorescens* 5 S RNA. *J. Biol. Chem.* **246**:747–761.

Dworkin, M. B., Rudensey, L. M., and Infante, A. A. 1977. Cytoplasmic nonpolysomal RNP particles in sea urchin embryos and their relationship to protein synthesis. *Proc. Nat. Acad. Sci. USA* **74**:2231–2235.

Ebel, J. P., Geigé, R., Bonnet, J., Kern, D., Befort, N., Bollack, C., Fasiolo, F., Gangloff, J., and Dirheimer, G. 1973. Factors determining the specificity of the tRNA aminoacylation reaction. *Biochimie* **55**:547–557.

Edmunds, M., Vaughan, M. H., and Nakazato, H. 1971. Poly(A) sequences in the hnRNA and rapidly-labelled polyribosomal RNA of HeLa cells. *Proc. Nat. Acad. Sci. USA* **68**:1336–1340.

Edström, J.-E., and Tanguay, R. 1974. Cytoplasmic RNAs with messenger characteristics in salivary gland cells of *Chironomus tentans*. *J. Mol. Biol.* **84**:569–583.

Ehrenfeld, E., and Summers, D. 1972. Adenylate-rich sequences in vesicular stomatitis virus mRNA. *J. Virol.* **10**:683–688.

Ehresmann, C., Fellner, P., and Ebel, J. P. 1970. Nucleotide sequences of sections of 16 S rRNA. *Nature* **227**:1321–1323.

Ehresmann, C., Steigler, P., Fellner, P., and Ebel, J. P. 1975. The determination of the primary structure of the 16 S ribosomal RNA of *E. coli*. III. Further studies. *Biochimie* **57**:711–748.

Eisenstadt, J. M., and Brawerman, G. 1967. The role of the native subribosomal particles of *E. coli* in polypeptide chain initiation. *Proc. Nat. Acad. Sci. USA* **58**:1560–1565.

Erdmann, V. A. 1976. Structure and function of 5 S and 5.8 S RNA. *Progr. Nucl. Acid Res. Mol. Biol.* **18**:45–90.

Erdmann, V. A., Sprinzl, M., and Pongs, O. 1973. The involvement of 5 S RNA in the binding of tRNAs to ribosomes. *Biochem. Biophys. Res. Comm.* **54**:942–948.

Eremenko, T., and Volpe, P. 1975. Polysome translational state during the cell cycle. *Eur. J. Biochem.* **52**:203–210.

Fakunding, J. L., Traut, R. R., and Hershey, J. W. B. 1973. Dependence of initiation factor IF-2 activity on proteins L7 and L12 from *E. coli* 50 S ribosomes. *J. Biol. Chem.* **248**:8555–8559.

Fakunding, J. L., Trough, J. A., Traut, R. R., and Hershey, J. W. B. 1974. Purification and phosphorylation of initiation factor IF2. *Meth. Enzym.* **30**:24–31.

Favre, A., Morel, C., and Scherrer, K. 1975. The secondary structure and poly(A) content of globin mRNA as a pure RNA and in polyribosome-derived RNP complexes. *Eur. J. Biochem.* **57**:147–157.

Fellner, P., and Ebel, J. P. 1970. Observations on the primary structure of the 23 S rRNA from *E. coli. Nature* **225**:1131–1132.

Fellner, P., Ehresmann, C., and Ebel, J. P. 1970. Nucleotide sequences present within the 16 S rRNA of *E. coli. Nature* **225**:26–29.

Fellner, P., Ehresmann, C., and Ebel, J. P. 1972a. The determination of the primary structure of the 16 S rRNA of *E. coli*. I. Nucleotide sequence analysis of T_1 and pancreatic RNase digestion products. *Biochimie* **54**:853–900.

Fellner, P., Ehresmann, C., Stiegler, P., and Ebel, J. P. 1972b. Partial nucleotide sequence of 16 S rRNA from *E. coli. Nature New Biol.* **239**:1–5.

Fellner, P., and Sanger, F. 1968. Sequence analysis of specific areas of the 16 S and 23 S rRNAs. *Nature* **219**:236–238.

Fiers, W., Contreras, R., de Wachter, R., Haegeman, G., Merregaert, J., Min Jou, W., and Vandanberghe, A. 1971. Recent progress in the sequence determination of bacteriophage MS2 RNA. *Biochimie* **53**:495–506.

Fiers, W., Contreras, R., Duerinck, F., Haegeman, G., Iserentant, D., Merregaert, J., Min Jou, W., Molemans, F., Racymackers, A., Van der Bergh, A., Volckaert, G., and Ysebaert, M. 1976. Complete nucleotide sequence of bacteriophage MS2 RNA: Primary and secondary structure of the replicase gene. *Nature* **260**:500–507.

Firtel, R. A., Jacobson, A., and Lodish, H. F. 1972. Isolation and hybridization kinetics of mRNA from *Dictyostelium discoideum. Nature New Biol.* **239**:225–228.

Fischel, J. L., and Ebel, J. P. 1975. Sequence studies on the 5 S RNA of *Proteus vulgaris:* Comparison with the 5 S of *E. coli. Biochimie* **57**:899–904.

Fiser, I., Scheit, K. H., Stöffler, G., and Kuechler, E. 1974. Identification of protein S1 at the mRNA binding site of the *E. coli* ribosome. *Biochem. Biophys. Res. Comm.* **60**:1112–1118.

Florendo, N. T. 1969. Ribosome substructure in intact mouse liver cells. *J. Cell Biol.* **41**:335–339.

Ford, P. J., and Southern, E. M. 1973. Different sequences for 5 S RNA in kidney cells and ovaries of *Xenopus laevis. Nature New Biol.* **241**:7–12.

Forget, B. G., and Jordan, B. 1969. 5 S RNA synthesized by *E. coli* in presence of chloramphenicol. *Science* **167**:382–384.

Forget, B. G., and Weissman, S. M. 1967. Nucleotide sequence of KB cell 5 S RNA. *Science* **158**:1695–1699.

Forget, B. G., and Weissman, S. M. 1969. The nucleotide sequence of ribosomal 5 S RNA from KB cells. *J. Biol. Chem.* **244**:3148–3165.

Forget, B. G., Housman, D., Benz, E. J., and McCaffrey, R. P. 1975. Synthesis of DNA complementary to separated human α and β globin mRNAs. *Proc. Nat. Acad. Sci. USA* **72**:984–988.

Fouquet, H., and Sauer, H. W. 1975. Variable redundancy in RNA transcripts isolated in S and G_2 phase of the cell cycle of *Physarum. Nature* **255**:253–255.

Fox, G. E., and Woese, C. R. 1975. 5 S RNA secondary structure. *Nature* **256**:505–507.

Fromson, D., and Duchastel, A. 1975. Poly(A)-containing polyribosomal RNA in sea urchin embryos. *Biochim. Biophys. Acta* **378**:394–404.

Fromson, D., and Verma, D. P. S. 1976. Translation of nonpolyadenylated mRNA of sea urchin embryos. *Proc. Nat. Acad. Sci. USA* **73**:148–151.

Fujisawa, T., and Eliceiri, G. L. 1975. Ribosomal proteins of hamster, mouse, and hybrid cells. *Biochim. Biophys. Acta* **402**:238–243.

Funatsu, G., Yaguchi, M., and Wittmann-Liebold, B. 1977. Primary structure of protein S 12 from the small *E. coli* ribosomal subunit. *FEBS Lett.* **73**:12–17.

Ganoza, M. C., and Fox, J. L. 1974. Isolation of a soluble factor needed for protein synthesis with various messenger ribonucleic acids other than poly(U). *J. Biol. Chem.* **249**:1037–1043.

Garen, A. 1968. Sense and nonsense in the genetic code. *Science* **160**:149–159.

Garrett, R. A., Schulte, C., Stöffler, G., Gray, P., and Monier, R. 1974. The release of proteins and 5 S RNA during the unfolding of *E. coli* ribosomes. *FEBS Lett.* **49**:1–4.

Garrett, R. A., and Wittmann, H. G. 1973. Structure of bacterial ribosomes. *Adv. Prot. Chem.* **27**:277–347.

Gasior, E., and Moldave, K. 1972. Evidence for a soluble protein factor specific for the interaction between aminoacylated tRNAs and the 40 S subunit of mammalian ribosomes. *J. Mol. Biol.* **66**:391–402.

Georgiev, G. P., and Samarina, O. P. 1971. D-RNA containing RNP particles. *Adv. Cell Biol.* **2**:47–110.

Gesteland, R. F. 1966. Unfolding of *E. coli* ribosomes by removal of magnesium. *J. Mol. Biol.* **18**:356–371.

Ghosh, H. P., and Khorana, H. G. 1967. On the role of ribosomal subunits in protein synthesis. *Proc. Nat. Acad. Sci. USA* **58**:2455–2461.

Ghosh, H. P., Söll, D., and Khorana, H. G. 1967. Initiation of protein synthesis *in vitro* as studied by using ribopolynucleotide with repeating nucleotide sequences as messengers. *J. Mol. Biol.* **25**:275–298.

Gillespie, D., Takemoto, K., Robert, M., and Gallo, R. C. 1973. Poly(A) in visna virus RNA. *Science* **179**:1328–1330.

Ginzburg, I., and Zanier, A. 1975. Characterization of different conformational forms of 30 S ribosomal subunits in isolated and associated states. *J. Mol. Biol.* **93**:465–476.

Glazier, K., and Schlessinger, D. 1974. Magic spot metabolism in an *E. coli* mutant temperature sensitive in elongation factor Ts. *J. Bact.* **117**:1195–1200.

Glick, B. R. 1977. The role of *E. coli* ribosomal proteins L7 and L12 in peptide chain elongation. *FEBS Lett.* **73**:1–5.

Glick, B. R., and Ganoza, M. C. 1976. Characterization and site of action of a soluble protein that stimulates peptide-bond synthesis. *Eur. J. Biochem.* **71**:483–491.

Goldberg, A. L., and Wittes, R. E. 1966. Genetic code: Aspects of organization. *Science* **153**:420–424.

Goodman, H. M., Billeter, M. A., Hindley, J., and Weissmann, C. 1970. The nucleotide sequence at the 5′-terminus of the Qβ RNA minus strand. *Proc. Nat. Acad. Sci. USA* **67**:921–928.

Gordon, J., Schweiger, M., Krisko, I., and Williams, C. A. 1969. Specificity and evolutionary divergence of the antigenic structure of the polypeptide chain elongation factors. *J. Bact.* **100**:1–4.

Gormly, J. R., Yang, C. H., and Horowitz, J. 1971. Further studies on ribosome unfolding. *Biochim. Biophys. Acta* **247**:80–90.

Gorski, J., Morrison, M. R., Merkel, C. G., and Lingrel, J. B. 1975. Poly(A) size class distribution in globin mRNAs as a function of time. *Nature* **253**:749–751.

Gottlieb, M., Lubsen, N. H., and Davis, B. D. 1974. Ribosome dissociation factors. *Methods Enzymol.* **30**:87–94.

Gould, R. M., Thornton, M. P., Liepkalns, V., and Lennarz, W. J. 1968. Participation of aminoacyl transfer ribonucleic acid in aminoacyl phosphatidylglycerol synthesis. *J. Biol. Chem.* **243**:3096–3104.

Gralla, J., and De Lisi, C. 1974. mRNA is expected to form stable secondary structures. *Nature* **248**:330–332.

Gralla, J., Steitz, J. A., and Crothers, D. M. 1974. Direct physical evidence for secondary structure in an isolated fragment of R17 bacteriophage mRNA. *Nature* **248**:204–208.

Grasmuck, H., Nolan, R. D., and Drews, J. 1974. Elongation factor 1 from ascites tumor cells. *Eur. J. Biochem.* **48**:485–493.

Grasmuk, H., Nolan, R. D., and Drews, J. 1976. A new concept of the function of EF-1 in peptide chain elongation. *Eur. J. Biochem.* **71**:271–279.

Gray, M. W. 1974. The presence of O^2-methylpseudouridine in the 18 S + 26 S ribosomal ribonucleates of wheat embryo. *Biochemistry* **13**:5453–5463.

Grayson, S., and Berry, S. J. 1973. Estimation of the half-life of a secretory protein message. *Science* **180**:1071–1072.

Greenberg, J. R. 1975. Messenger RNA metabolism of animal cells. *J. Cell Biol.* **64**:269–288.

Greenberg, J. R. 1976. Isolation of L-cell mRNA which lacks poly(A). *Biochemistry* **15**:3516–3552.

Greenberg, J. R., and Perry, R. P. 1972. Relative occurrence of poly(A) sequences in messenger and hnRNA of L cells as determined by poly(U)-hydroxylapatite chromatography. *J. Mol. Biol.* **72**:91–98.

Grierson, D. 1974. Characterization of RNA components from leaves of *Phaseolus aureus*. *Eur. J. Biochem.* **44**:509–515.

Gross, P. R., and Cousineau, G. H. 1963. Effects of actinomyin D on macromolecule synthesis and early development in sea urchin eggs. *Biochem. Biophys. Res. Comm.* **10**:321–326.

Groves, W. E., and Kempner, E. S. 1967. Amino acid coding in *Sarcina lutea* and *Saccharomyces cerevisiae*. *Science* **156**:387–390.

Grozdanovič, J., and Hradec, J. 1975. Different binding sites of poly(A)-containing and poly(A)-free fractions of nuclear ribonucleic acid to ribosomes from rat liver. *Biochem. Biophys. Acta* **402**:69–82.

Gualerzi, C., Janda, H. G., Passow, H., and Stöffler, G. 1974. Studies on the protein moiety of plant ribosomes. *J. Biol. Chem.* **249**:3347–3355.

Gualerzi, C., Pon, C. L., and Kaji, A. 1971. Initiation factor dependent release of aminoacyl-tRNAs from complexes of 30 S ribosomal subunits, synthetic polynucleotide and aminoacyl tRNA. *Biochem. Biophys. Res. Comm.* **45**:1312–1319.

Gurdon, J. B., Lane, C. D., Woodland, H. R., and Marbaix, G. 1971. Use of frog eggs and oocytes for the study of mRNA and its translation in living cells. *Nature* **233**:177–182.

Haenni, A. L., and Lucas-Lenard, J. 1970. Function of the elongation factors T and G. *In*: Ochoa, S., C. F. Heredin, C. Asensio, and D. Nachmansohn, eds., *Macromolecules: Biosynthesis and Function,* New York, Academic Press, p. 97–108.

Haguenau, F. 1958. The ergastoplasm: Its history, ultrastructure and biochemistry. *Internat. Rev. Cytol.* **7**:425–483.

Haines, M. E., Carey, N. H., and Palmiter, R. D. 1974. Purification and properties of ovalbumin mRNA. *Eur. J. Biochem.* **43**:549–560.

Hall, N. D., and Arnstein, H. R. V. 1973. Specificity of reticulocyte initiation factors for the translation of globin mRNA. *Biochem. Biophys. Res. Comm.* **54**:1489–1497.

Hamel, E., and Cashel, M. 1973. Role of guanine nucleotides in protein synthesis. *Proc. Nat. Acad. Sci. USA* **70**:3250–3254.

Hampel, A., and Enger, M. D. 1973. Subcellular distribution of aminoacyl-transfer RNA synthetases in Chinese hamster ovary cell culture. *J. Mol. Biol.* **79**:285–293.

Hardy, S. J. S., Kurland, C. G., Voynow, P., and Mora, G. 1969. The ribosomal proteins of *E. coli*. *Biochemistry* **8**:2897–2905.

Hasselkorn, R., and Rothman-Denes, L. B. 1973. Protein synthesis. *Ann. Rev. Biochem.* **42**:397–438.

Hatfield, D. 1972. Recognition of nonsense codons in mammalian cells. *Proc. Nat. Acad. Sci. USA* **69**:3014–3018.

Hatlen, L., and Attardi, G. 1971. Proportion of the HeLa cell genome complementary to tRNA and 5 S RNA. *J. Mol. Biol.* **56**:535–554.

Hattman, S., and Hofschneider, P. H. 1968. Influence of T4 on the formation of RNA phage-specific polyribosomes and polymerase. *J. Mol. Biol.* **35**:513–522.

Hawley, D. A., Miller, M. J., Slobin, L. I., and Wahba, A. J. 1974. The mechanism of action of initiation factor 3 in protein synthesis. *Biochem. Biophys. Res. Comm.* **61**:329–337.

Held, W. A., Getle, W. R., and Nomura, M. 1974. Role of 16 S ribosomal RNA and the 30 S ribosomal protein S12 in the initiation of natural mRNA translation. *Biochemistry* **13**:2115–2122.

Hellerman, J. G., and Shafritz, D. A. 1975. Initiation of poly(A) and mRNA with eukaryotic initiator Met $tRNA_f$ binding factor. *Proc. Nat. Acad. Sci. USA* **72:**1021–1025.

Hemminki, K. 1974. Poly(A) in RNA extracted by thermal phenol fractionation from chick embryo brain and liver. *Biochim. Biophys. Acta* **340**:262–268.

Henriksen, O., Robinson, E. A., and Maxwell, E. S. 1975a. Interaction of guanosine nucleotides with EF-2. I. Equilibrium dialysis studies. *J. Biol. Chem.* **250**:720–724.

Henriksen, O., Robinson, E. A., and Maxwell, E. S. 1975b. Interaction of guanosine nucleotides with EF-2. II. Effects of ribosomes and magnesium ions on guanosine diphosphate and guanosine triphosphate binding to the enzyme. *J. Biol. Chem.* **250**:725–730.

Henshaw, E. C., Guiney, D. G., and Hirsch, C. A. 1973. The ribosome cycle in mammalian protein synthesis. *J. Biol. Chem.* **248**:4367–4376.

Herr, W., and Noller, H. F. 1975. A fragment of 23 S RNA containing a nucleotide sequence complementary to a region of 5 S RNA. *FEBS Lett.* **53**:248–252.

Herrington, M. D., and Hawtrey, A. O. 1971. Differences in the ribosomes prepared from lactating and non-lactating bovine mammary gland. *Biochem. J.* **121**:279–285.

Hershey, A. D., Dixon, J., and Chase, M. 1953. Nucleic acid economy in bacteria infected with bacteriophage T2. *J. Gen. Physiol.* **36**:777–789.

Hershey, J. W. B., Dewey, K. F., and Thach, R. E. 1969. Purification and properties of IF-1. *Nature* **222**:944–947.

Highland, J. H., Bodley, J. W., Gordon, J., Hasenbank, R., and Stöffler, G. 1973. Identity of the ribosomal proteins involved in the interaction with EF-G. *Proc. Nat. Acad. Sci. USA* **70**:147–150.

Highland, J. H., Ochsner, E., Gordon, J., Hasenbank, R., and Stöffler, G. 1974. Inhibition of phenylalanyl-tRNA binding and EF-Tu-dependent GTP hydrolysis by antibodies specific for several ribosomal proteins. *J. Mol. Biol.* **86**:175–178.

Hindley, J., and Page, S. M. 1972. Nucleotide sequence of yeast 5 S rRNA. *FEBS Lett.* **26**:157–160.

Hindley, J., and Staples, D. H. 1969. Sequence of a ribosome binding site in bacteriophage Qβ-RNA. *Nature* **224**:964–967.

Hirashima, A., Wang, S., and Inouye, M. 1974. Cell-free synthesis of a specific lipoprotein of the *E. coli* outer membrane directed by purified mRNA. *Proc. Nat. Acad. Sci. USA* **71**:4149–4153.

Hirsch, C. A., Cox, M. A., von Venrooij, W. J., and Henshaw, E. C. 1973. The ribosome cycle in mammalian protein synthesis. *J. Biol. Chem.* **248**:4377–4385.

Hitz, H., Schäfer, D., and Wittmann-Liebold, B. 1975. Primary structure of ribosomal protein S6 from the wild type and a mutant of *E. coli*. *FEBS Lett.* **56**:259–262.

Hodge, L. D., Robbins, E., and Scharff, M. D. 1969. Persistence of mRNA through mitosis in HeLa cells. *J. Cell Biol.* **40**:497–507.

Holland, M. J., Hager, G. L., and Rutter, W. J. 1977. Characterization of purified poly(A)-containing mRNA from *S. cerevisiae*. *Biochemistry* **16**:8–16.

Houdebine, L. M. 1976. Absence of poly(A) in a large part of newly synthesized casein mRNAs. *FEBS Lett.* **66**:110–113.

Howard, G. A., and Herbert, E. 1975. Ribosomal subunit localization of hemoglobin mRNA. *Eur. J. Biochem.* **54**:75–80.

Howard, G. A., Traugh, J. A., Croser, E. A., and Traut, R. A. 1975. Ribosomal proteins from rabbit reticulocytes. *J. Mol. Biol.* **93**:391–404.

Hsu, W. T., and Weiss, S. B. 1969. Selective translation of T4 template RNA by ribosomes from T4-infected *E. coli. Proc. Nat. Acad. Sci. USA* **64**:345–351.

Huynh-Van-Tan, Delaunay, J., and Schapira, G. 1971. Eukaryotic ribosomal proteins. *FEBS Lett.* **17**:163–167.

Iatrou, K., and Dixon, G. H. 1977. The distribution of poly(A)$^+$ and poly(A)$^-$ protamine mRNA sequences in the developing trout testis. *Cell* **10**:433–441.

Igarashi, K., Sugawara, K., Izumi, I., Nagayama, C., and Hirose, S. 1974. Effect of polyamines on polyphenylalanine synthesis by *E. coli* and rat liver ribosomes. *Eur. J. Biochem.* **48**:495–502.

Ilan, Jo. 1969. The role of tRNA in translational control of specific mRNA during insect metamorphosis. *Cold Spring Harb. Symp. Quant. Biol.* **34**:787–791.

Ilan, Jo. and Ilan, Ju. 1973a. Sequence homology at the 5′-termini of insect mRNA. *Proc. Nat. Acad. Sci. USA* **70**:1355–1358.

Ilan, Ju., and Ilan, Jo. 1973b. An mRNA bound initiation factor and its role in translation of natural message. *Nature New Biol.* **241**:176–180.

Ilan, Jo., Ilan, Ju., and Quastel, J. H. 1966. Effects of actinomysin D on nucleic acid metabolism and protein biosynthesis during metamorphosis of *Tenebrio molitor* L. *Biochem. J.* **100**:441–447.

Inoue-Yokosawa, N., Ishikawa, C., and Kagiro, Y. 1974. The role of guanosine triphosphate in translocation reaction catalyzed by EF-G. *J. Biol. Chem.* **249**:4321–4323.

Ishikura, H., Yamada, Y., and Nishimura, S. 1971. Structure of serine tRNA from *E. coli* and purification of serine tRNAs with different codon responses. *Biochim. Biophys. Acta* **228**:471–481.

Ishizuka, S., Kawakami, M., Ejiri, S., and Shimura, K. 1974. The initiator amino acid in silk fibroin biosynthesis. *FEBS Lett.* **47**:318–322.

Isono, K., and Isono, S. 1976. Lack of ribosomal protein S1 in *Bacillus stearothermophilus. Proc. Nat. Acad. Sci. USA* **73**:767–770.

Isono, S., and Isono, K. 1975. Purification and characterization of 30 S ribosomal proteins from *Bacillus stearothermophilus. Eur. J. Biochem.* **50**:483–488.

Iwasaki, K., Motoyoshi, K., Nagata, S. and Kaziro, Y. 1976. Purification and properties of a new polypeptide chain elongation factor, EF-1β, from pig liver. *J. Biol. Chem.* **251**:1843–1845.

Jacob, F., and Monod, J. 1961. Genetic regulatory mechanisms in the synthesis of proteins. *J. Mol. Biol.* **3**:318–356.

Jacobson, A., Firtel, R. A., and Lodish, H. F. 1974. Synthesis of messenger and ribosomal RNA precursors in isolated nuclei of the cellular slime mold *Dictyostelium discoideum. J. Mol. Biol.* **82**:213–230.

Jantzen, H. 1974. Polyadenylsäure-enthaltende RNA und Genaklivitätsmuster während der Entwicklung von *Acanthamoeba castellanii. Biochim. Biophys. Acta* **374**:38–51.

Jay, G., and Kaempfer, R. 1975. Initiation of protein synthesis. Binding of mRNA. *J. Biol. Chem.* **250**:5742–5748.

Jeffery, W. R., and Brawerman, G. 1974. Characterization of the steady-state population of messenger RNA and its poly(A) segment in mammalian cells. *Biochemistry* **13**:4633–4637.

Jerez, C., Sandoval, A., Allende, J., Henes, C., and Ofengand, J. 1969. Specificity of the interaction of aminoacyl RNA with a protein-GTP complex from wheat embryo. *Biochemistry* **8**:3006–3014.

Johnson, T. C., and Luttges, M. W. 1966. The effects of maturation on *in vitro* protein synthesis by mouse brain cells. *J. Neurochem.* **13**:545–552.

Johnston, R. E., and Bose, H. R. 1972. An adenylate-rich segment in the virion RNA of Sindbis virus. *Biochem. Biophys. Res. Comm.* **46**:712–718.

Jordan, B. R., and Galling, G. 1973. Nucleotide sequence of *Chlorella* cytoplasmic 5 S RNA. *FEBS Lett.* **37**:333–334.

Jordan, B. R., Galling, G., and Jourdan, R. 1974. Sequence and conformation of 5 S RNA from *Chlorella* cytoplasmic ribosomes: Comparison with other 5 S RNA molecules. *J. Mol. Biol.* **87**:205–225.

Jukes, T. H. 1969. Recent advances in studies of evolutionary relationships between proteins and nucleic acids. *Space Life Sci.* **1**:469–494.

Kabat, D. 1975. Potentiation of hemoglobin mRNA. *J. Biol. Chem.* **250**:6085–6092.

Kaempfer, R. 1968. Ribosome subunit exchange during protein synthesis. *Proc. Nat. Acad. Sci. USA* **61**:106–113.

Kaempfer, R. 1969. Ribosome subunit exchange in the cytoplasm of a eucaryote. *Nature* **222**:950–953.

Kaempfer, R., and Meselson, M. 1968. Permanent association of 5 S RNA molecules with 50 S ribosomal subunits in growing bacteria. *J. Mol. Biol.* **34**:703–708.

Kaempfer, R., Meselson, M., and Raskas, H. 1968. Cyclic dissociation into stable subunits and reformation of ribosomes during bacterial growth. *J. Mol. Biol.* **31**:277–289.

Kaji, A., Kaji, H., and Novelli, G. D. 1965. Soluble amino acid-incorporating system. *J. Biol. Chem.* **240**:1185–1191.

Kaltschmidt, E., Kahan, L., and Nomura, M. 1974. *In vitro* synthesis of ribosomal proteins directed by *E. coli* DNA. *Proc. Nat. Acad. Sci. USA 71*:446–450.

Kates, J. 1970. Transcription of the vaccinia virus genome and the occurrence of poly(A) sequences in mRNA. *Cold Spring Harbor Symp. Quant. Biol.* **35**:743–752.

Kawakita, M., Arai, K.-I., and Kaziro, Y. 1974. Interactions between EF-Tu-guanosine triphosphate and ribosomes and the role of ribosome-bound tRNA in guanosine triphosphate reaction. *J. Biochem.* **76**:801–809.

Kay, A., Sander, G., and Grunberg-Manago, M. 1973. Effect of ribosomal protein L12 upon initiation factor IF-2 activities. *Biochem. Biophys. Res. Comm.* **51**:979–986.

Kaziro, Y., Inoue-Yokosawa, N., and Kawakita, M. 1972. Studies on polypeptide EF-G from *E. coli*. *J. Biochem.* **72**:853–863.

Kemp, D. J. 1975. Unique and repetitive sequences in multiple genes for feather keratin. *Nature* **254**:573–575.

Kim, S.-H., and Rich, A. 1969. Crystalline transfer RNA: The three-dimensional Patterson function of 12-Ångstrom resolution. *Science* **166**:1621–1624.

Kim, W. S. 1969. N-formylseryl-tRNA. *Science* **163**:947–949.

Kimura, M., and Ohta, T. 1973. Eukaryotes-prokaryotes divergence estimated by 5 S ribosomal RNA sequences. *Nature New Biol.* **243**:199–200.

Kischa, K., Möller, W., and Stöffler, G. 1971. Reconstitution of a GTPase activity by a 50 S ribosomal protein from *E. coli*. *Nature New Biol.* **233**:62–63.

Kiselev, N. A., Stel'mashchuk, V. Y., Lerman, M. I., and Abakumova, O. Y. 1974. On the structure of liver ribosomes. *J. Mol. Biol.* **86**:577–586.

Kiss, A., Sain, B., and Venetianer, P. 1977. The number of rRNA genes in *E. coli*. *FEBS Lett.* **79**:77–79.

Klagsbrun, M. 1973. An evolutionary study of the methylation of tRNA and rRNA in prokaryote and eukaryote organisms. *J. Biol. Chem.* **248**:2612–2620.

Klein, W. H., Murphy, W., Attardi, G., Britten, R. J., and Davidson, E. H. 1974. Distribution of repetitive and nonrepetitive sequence transcripts in HeLa mRNA. *Proc. Nat. Acad. Sci. USA* **71**:1785–1789.

Knight, E. J. R., and Darnell, J. E. 1967. Distribution of 5 S RNA in HeLa cells. *J. Mol. Biol.* **28**:491–502.

Kohler, R. E., Ron, E. Z., and Davis, B. D. 1968. Significance of the free 70S ribosomes in *E. coli* extracts. *J. Mol. Biol.* **36**:71–82.

Kolakofsky, D., Dewey, K., and Thack, R. E. 1969. Purification and properties of initiation factor f_2. *Nature* **223**:694–697.

Koser, R. B., and Collier, J. R. 1971. The molecular weight and thermolability of *Ilyanassa* rRNA. *Biochim. Biophys. Acta* **254**:272–277.

Koteliansky, V. E., Domogatsky, S. P., Gudkov, A. T., and Spirin, A. S. 1977. Elongation factor-dependent reactions on ribosomes deprived of proteins L7 and L12. *FEBS Lett.* **73**:6–11.

Krauss, S. W., and Leder, P. 1975. Turnover of protein synthetic elongation and initiation factors in *E. coli*. *J. Biol. Chem.* **250**:4714–4717.

Krystosek, A., Cawthon, M. L., and Kabat, D. 1975. Improved methods for purification and assay of eukaryotic mRNAs and ribosomes. *J. Biol. Chem.* **250**:6077–6084.

Küntzel, H. 1969. Proteins of mitochondrial and cytoplasmic ribosomes. *Nature* **222**:142–146.

Kurland, C. G. 1970. Ribosome structure and function emergent. *Science* **169**:1171–1177.

Kwan, S. W., and Brawerman, G. 1972. A particle associated with the poly(A) segment in mammalian mRNA. *Proc. Nat. Acad. Sci. USA* **69**:3247–3250.

Lai, M. M. C., and Duesberg, P. H. 1972. Adenylic acid-rich sequence in RNA of Rous sarcoma virus and Rausche mouse leukaemia virus. *Nature* **235**:383–386.

Lake, J. A. 1976. Ribosome structure determined by electron microscopy of *E. coli* small subunits, large subunits and monomeric ribosomes. *J. Mol. Biol.* **105**:131–159.

Lanzani, G. A., Bollini, R., and Soffientini, A. N. 1974. Heterogeneity of EF-1 from wheat embryos. *Biochim. Biophys. Acta* **335**:275–283.

Lawrence, F. 1973. Effect of adenosine on methionyl-tRNA synthetase. *Eur. J. Biochem.* **40**:493–500.

Lawrence, F., Blanquet, S., Poiret, M., Robert-Gero, M., and Waller, J.-P. 1973. The mechanism of action of methionyl-tRNA synthetase. *Eur. J. Biochem.* **36**:234–243.

Lawrence, F., Shire, D. J., and Waller, J.-P. 1974. The effect of adenosine analogues on ATP-pyrophosphate exchange reaction catalysed by methionyl-tRNA synthetase. *Eur. J. Biochem.* **41**:73–81.

Laycock, A. G., and Hunt, J. A. 1969. Synthesis of rabbit globin by a bacterial cell-free system. *Nature* **221**:1118–1122.

Leaver, C. J., and Ingle, J. 1971. The molecular integrity of chloroplast rRNA. *Biochem. J.* **123**:235–243.

LeBleu, B., Nudel, U., Falcoff, E., Prives, C., and Revel, M. 1972. A comparison of the translation of Mengo virus RNA and globin mRNA in Krebs ascites cell-free extracts. *FEBS Lett.* **25**:97–103.

Lebleu, G., Marbaix, G., Huez, G., Timmerman, J., Burny, A., and Chantrenne, N. 1971. Characterization of the mRNP released from reticulocyte polyribosomes by EDTA treatment. *Eur. J. Biochem.* **19**:264–269.

Leder, P. 1973. The elongation factors in protein synthesis. *Adv. Prot. Chem.* **27**:213–240.

Lee-Huang, S., and Ochoa, S. 1973. Purification of two messenger-discriminating species of IF-3 from *E. coli*. *Methods Enzymol.* **30**:45–53.

Lee-Huang, S., and Ochoa, S. 1974a. Preparation and properties of crystalline initiation factor 1 (IFI) from *Escherichia coli*. *Methods Enzymol.* **30:**31–39.

Lee-Huang, S., and Ochoa, S. 1974b. Pruification of two messenger-discriminating species of initiation factor 3 (IF3) from *E. coli*. *Methods Enzymol.* **30:**45–53.

Leffler, S. and Szer, W. 1974a. Purification and properties of initiation factor IF-3 from *Caulobacter crescentus*. *J. Biol. Chem.* **249**:1458–1464.

Leffler, S., and Szer, W. 1974b. Polypeptide chain initiation in *Caulobacter crescentus* without initiation factor IF-1. *J. Biol. Chem.* **249**:1465–1468.

Leibowitz, M. J., and Soffer, R. L. 1969. A soluble enzyme from *E. coli* which catalyzes the transfer of leucine and phenylalanine from tRNA to acceptor proteins. *Biochem. Biophys. Res. Comm.* **36**:47–53.

Leighton, T. 1974. Further studies on the stability of sporulation mRNA in *B. subtilis. J. Biol. Chem.* **249**:7808–7812.

Levin, D. H., Kyner, D., and Acs, G. 1973. Protein initiation in eukaryotes. *Proc. Nat. Acad. Sci. USA* **70**:41–45.

Levinthal, C., Hosoda, J., and Shub, D. 1967. The control of protein synthesis after phage infection. *In:* Colter, S. J., and W. Paranchych, eds., *The Molecular Biology of Viruses,* New York, Academic Press, p. 71–87.

Levinthal, C., Keywan, A., and Higa, A. 1962. Messenger RNA turnover and protein synthesis in *B. subtilis* inhibited by actinomycin D. *Proc. Nat. Acad. Sci. USA* **48**:1631–1638.

Levitt, M. 1969. Detailed molecular model for tRNA. *Nature* **224**:759–763.

Lewin, B. M. 1970. *The Molecular Basis of Gene Expression.* New York, Wiley-Interscience.

Liautard, J. P., Setyono, B., Spindler, E., and Köhler, K. 1976. Comparison of proteins bound to the different functional classes of mRNA. *Biochim. Biophys. Acta* **425**:373–383.

Lim, L., and Canellakis, E. S. 1970. Adenine-rich polymer associated with rabbit reticulocyte mRNA. *Nature* **227**:710–712.

Lim, L., Canellakis, Z. N. and Canellakis, E. S. 1970. Metabolism of naturally occurring homopolymers. *Biochim. Biophys. Acta* **209**:128–138.

Lin, J. Y., Tsung, C. M., and Fraenkel-Conrat, J. 1967. The coat protein of the RNA bacteriophage MS2. *J. Mol. Biol.* **24**:1–14.

Lipmann, F. 1969. Polypeptide chain elongation in protein biosynthesis. *Science* **164**:1024–1031.

Loeb, J. N., Howell, R. R., and Tomkins, G. M. 1965. Turnover of rRNA in rat liver. *Science* **149**:1093–1095.

Loening, U. E. 1968. Molecular weights of rRNA in relation to evolution. *J. Mol. Biol.* **38:**355–365.

Loening, U. E., Grierson, D., Rogers, M. E., and Sartirana, M. L. 1972. Properties of rRNA precursor. *In:* Cox, R. A., and A. A. Hadjiolov, eds., *Functional Units in Protein Biosynthesis,* New York, Academic Press, p. 395–405.

Loftfield, R. B. 1972. The mechanism of aminoacylation of tRNA. *Progr. Nucl. Acid. Res. Mol. Biol.* **12**:87–128.

Lucas-Lenard, J., and Lipmann, F. 1966. Separation of three microbial amino acid polymerization factors. *Proc. Nat. Acad. Sci. USA* **55**:1562–1566.

Lucas-Lenard, J., and Lipmann, F. 1971. Protein biosynthesis. *Ann. Rev. Biochem.* **40**:409–448.

Lukanidin, E. M., Zalmanzon, E. S., Komaromi, L., Samarina, O. P., and Georgiev, G. P. 1972. Structure and function of informofers. *Nature New Biol.* **238**:193–197.

MacInnes, J. W. 1972. Differences between ribosomal subunits from brain and those from other tissues. *J. Mol. Biol.* **65**:157–162.

MacInnes, J. W. 1973. Mammalian brain ribosomes are behaviourly and structurally heterogeneous. *Nature New Biol.* **241**:244–246.

MacLeod, M. C. 1975. Comparisons of the properties of cytoplasmic poly(A)-containing RNA from polysomal and nonpolysomal fractions of murine myeloma cells. *Biochemistry* **14**:4011–4018.

Maden, B. E. H., Forbes, J., de Jong, P., and Klootwijk, J. 1975. Presence of a hypermodified nucleotide in HeLa cell 18 S and *Saccharomyces carlsbergensis* 17 S ribosomal RNAs. *FEBS Lett.* **59**:60–63.

Madison, J. T. 1968. Primary structure of RNA. *Ann. Rev. Biochem.* **37**:131–148.

Maelicke, A., Engel, G., Cramer, F., and Staehelin, M. 1974. ATP-induced specificity of the binding of serine tRNAs from rat liver to seryl-tRNA sythetase from yeast. *Eur. J. Biochem.* **42**:311–314.

Mainwaring, W. I. P., Wilce, P. A., and Smith, A. E. 1974. Studies on the form and synthesis of mRNA in the rat ventral prostate gland, including its tissue-specific stimulation by androgens. *Biochem. J.* **137**:513–524.

Maizels, N. 1974. *E. coli* lactose operon ribosome binding site. *Nature* **249**:647–649.

Majumdar, A., Bose, K. K., and Gupta, N. K. 1976. Specific binding of *E. coli* chain IF-2 to fMet-tRNA$^{met}_{f}$. *J. Biol. Chem.* **251**:137–140.

Mangiarotti, G., and Schlessinger, D. 1966. Extraction of polyribosomes and ribosomal subunits from fragile growing *E. coli. J. Mol. Biol.* **20**:123–143.

Mangiarotti, G., and Schlessinger, D. 1967. Formation and lifetime of mRNA molecules, ribosome subunit couples and polyribosomes. *J. Mol. Biol.* **29**:355–418.

Mansbridge, J. N., Crossley, J. A., Lanyon, W. G., and Williamson, R. 1974. The poly(A) sequence of mouse globin mRNA. *Eur. J. Biochem.* **44**:261–269.

Marcus, A., Seal, S. N., and Weeks, D. P. 1974. Protein chain initiation in wheat embryo. *Methods Enzymol.* **30**:94–101.

Marshall, R. E. 1967. Fine structure of RNA codewords recognized by bacterial, amphibian, and mammalian transfer RNA. *Science* **155**:820–826.

Mathews, M. B., Osburn, M., Berns, A. J. M., and Bloemandal, H. 1972a. Translation of two mRNAs from lens in a cell-free system from Krebs II ascites cells. *Nature New Biol.* **236**:5–7.

Mathews, M. B., Pragnell, I. B., Osburn, M., and Arnstein, H. R. V. 1972b. Stimulation by reticulocyte initiation factors of protein synthesis in a cell-free system from Krebs II ascites cells. *Biochim. Biophys. Acta* **287**:113–123.

Maugh, T. H. 1975. Ribosomes (II): A complicated structure begins to emerge. *Science* **190**:258–260.

Mazumder, R. 1971. Studies on polypeptide chain initiation factors F_1 and F_2. *FEBS Lett.* **18**:64–66.

Mazumder, R. 1972. IF-2-dependent ribosomal binding of *N*-formylmethionyl-tRNA without added GTP. *Proc. Nat. Acad. Sci. USA* **69**:2770–2773.

McConkey, E. H., and Hauber, E. J. 1975. Evidence for heterogeneity of ribosomes within the HeLa cell. *J. Biol. Chem.* **250**:1311–1318.

McCorquodale, D. J., Oleson, A. E., and Buchanan, J. M. 1967. Control of virus-induced enzyme synthesis in bacteria. *In:* Colter, S. J., and W. Paranchych, eds., *The Molecular Biology of Viruses,* New York, Academic Press, p. 31–54.

McCroskey, R. P., Zasloff, M., and Ochoa, S. 1972. Polypeptide chain initiation and stepwise elongation with *Artemia* ribosomes and factors. *Proc. Nat. Acad. Sci. USA* **69**:2451–2455.

McKnight, G. S., and Schimke, R. T. 1974. Ovalbumin mRNA. *Proc. Nat. Acad. Sci. USA* **71**:4327–4331.

McLaughlin, C. S., Warner, J. R., Edmunds, M., Nakagato, H., and Vaughan, M. H. 1973. Poly(A) sequences in yeast mRNA. *J. Biol. Chem.* **248**:1466–1471.

McNeil, R. G., and McLaughlin, C. S. 1974. Differential biological activity of three species of methionyl-tRNA in yeast. *Biochim. Biophys. Acta* **374**:176–186.

Meier, D., Lee-Huang, S., and Ochoa, S. 1973. Factor requirements for initiation complex formation with natural and synthetic messengers in *E.coli* systems. *J. Biol. Chem.* **248**:8613–8615.

Merkel, C. G., Wood, T. G., and Lingal, J. B. 1976. Shortening of the poly(A) region of mouse globin mRNA. *J. Biol. Chem.* **251**:5512–5515.

Mertes, M., Peters, M. A., Mahoney, W., and Yarus, M. 1972. Isoleucylation of tRNA$^{met}_{f}$ (*E. coli*) by isoleucyl-tRNA synthetase from *E. coli. J. Mol. Biol.* **71**:671–685.

Mescher, A., and Humphreys, T. 1974. Activation of maternal mRNA in the absence of poly(A) formation in fertilised sea urchin eggs. *Nature* **249**:138–139.

Metafora, S., Terada, M., Dow, L. W., Marks, P. A., and Bank, A. 1972. Increased efficiency of exogenous mRNA translation in a Krebs ascites cell lystate. *Proc. Nat. Acad. Sci. USA* **69**:1299–1303.

Meyer, M., Bout, W. S., de Vries, M., and Nanninga, N. 1974. Electron microscopic and sedimentation studies on rat-liver ribosomal subunits. *Eur. J. Biochem.* **42**:259–268.

Miller, D. L., and Weissbach, H. 1974. Elongation factor Tu and the aminoacyl-tRNA · EF-Tu · GTP complex. *Methods Enzymol.* **30:**219–232.

Miller, R. V., and Sypherd, P. S. 1973. Topography of the *E. coli* 30 S ribosome revealed by the modification of ribosomal proteins. *J. Mol. Biol.* **78**:539–550.

Milman, G., Goldstein, J., Scolnick, E., and Caskey, T. 1969. Peptide chain termination. *Proc. Nat. Acad. Sci. USA* **63**:183–190.

Min Jou, W., Haegeman, G., Ysebaert, M., and Fiers, W. 1972. Nucleotide sequence of the gene coding for the bacteriophage MS2 coat protein. *Nature* **237**:82–88.

Miyazaki, M. 1974. Studies on the nucleotide sequence of pseudoruidine-containing 5 S RNA from *S. cerevisiae. J. Biochem.* **75**:1407–1410.

Mizumoto, K., Iwasaki, K. Kazior, Y., Nojiri, C., and Yamada, Y. 1974. Studies on peptide EF-2 from pig liver. *J. Biochem.* **75**:1057–1062.

Modolell, J. 1974. The initial steps in protein synthesis. *Methods Enzymol.* **30**:79–86.

Monroy, A., Maggio, R., and Rinaldi, A. M. 1965. Experimentally induced activation of the ribosomes of the unfertilized sea urchin egg. *Proc. Nat. Acad. Sci. USA* **54**:107–111.

Montagnier, L., Collandre, H., De Maeyer-Guiguard, J., and De Maeyer, E. 1974. Two forms of mouse interferon mRNA. *Biochem. Biophys. Res. Comm.* **59**:1031–1038.

Moore, P. B., Engelman, D. M., and Schoenborn, B. P. 1974. Asymmetry in the 50 S ribosomal subunit of *E. coli. Proc. Nat. Acad. Sci. USA* **71**:172–176.

Moore, V. G., Atchison, R. E., Thomas, G., Moran, M., and Noller, H. F., 1975. Identification of a ribosomal protein essential for peptidyl transferase activity. *Proc. Nat. Acad. Sci. USA* **72**:844–848.

Morel, C., Kayibanda, B. and Scherrer, K. 1971. Proteins associated with globin mRNA in avian erythroblasts. *FEBS Lett.* **18**:84–88.

Morell, P., and Marmur, J. 1968. Association of 5 S RNA to 50 S subunits of *E. coli* and *B. subtilis. Biochemistry* **7**:1141–1152.

Morikawa, N., and Imamoto, F. 1969. On the degradation of mRNA for the tryptophan operon in *E. coli. Nature* **223**:37–40.

Morinaga, T., Funatsu, G., Funatsu, M., and Wittmann, H. G. 1976. Primary structure of the 16 S rRNA binding protein S15 from *E. coli* ribosomes. *FEBS Lett.* **64**:307–309.

Munsche, D., and Wollgiehn, R. 1974. Altersabhängige labilität der ribosomalen RNA aus chloroplasten von *Nicotiana rustica. Biochim. Biophys. Acta* **340**:437–445.

Murthy, M. R. V. 1972. Free and membrane bound ribosomes of rat cerebral cortex. *J. Biol. Chem.* **247:**1936–1943.

Musso, R. E., de Crombrugghe, B., Pastan, I., Sklar, J., Yot, P., and Weissman, S. 1974. The 5′-terminal nucleotide sequence of galactose mRNA of *E. coli. Proc. Nat. Acad. Sci. USA* **71**:4940–4944.

Muthukrishnan, S., Both, G. W., Furuichi, Y., and Shatkin, A. J. 1975. 5′-terminal 7-methylguanosine in eukaryotic mRNA is required for translation. *Nature* **255**:33–37.

Muto, A. 1970. Nucleotide distribution of *E. coli* 16 S rRNA. *Biochemistry* **9**:3683–3693.

Naaktgeboren, N., Roobol, K., and Voorma, H. O. 1977. The effect of the initiation factor IF-1 on the dissociation of 70-S ribosomes of *E. coli. Eur. J. Biochem.* **72**:49–56.

Nakamoto, T., Conway, T. W., Allende, J. E., Spyrides, G. J., and Lipmann, F. 1963. Formation of peptide bonds. *Cold Spring Harbor Symp. Quant. Biol.* **28**:227–232.

Nakazato, H., Venkatesan, S., and Edmonds, M. 1975. Poly(A) sequences in *E. coli* mRNA. *Nature* **256**:144–146.

Nanninga, A. 1973. Structural aspects of ribosomes. *Int. Rev. Cytol.* **35**:135–188.

Natale, P. J., and Buchanan, J. M. 1974. Initiation characteristics for the synthesis of five T4 phage-specific mRNAs *in vitro*. *Proc. Nat. Acad. Sci. USA* **71**:422–426.

Nathans, D., and Lipmann, F. 1961. Amino acid transfer from aminoacyl-RNAs to protein on ribosomes of *E. coli*. *Proc. Nat. Acad. Sci. USA* **47**:497–504.

Nazar, R. N., Sitz, T. O., and Busch, H. 1975. Tissue specific differences in the 2′-*O*-methylation of eukaryotic 5.8 S ribosomal RNA. *FEBS Lett.* **59**:83–87.

Nazar, R. N., Sitz, T. O., and Busch, H. 1976. Sequence homologies in mammalian 5.8 S rRNA. *Biochemistry* **15**:505–508.

Nesbitt, J. A., and Lennarz, W. J. 1968. Participation of aminoacyl tRNA in aminoacyl phosphatidylglycerol synthesis. *J. Biol. Chem.* **243**:3088–3095.

Nishikawa, K., and Takemura, S. 1974. Nucleotide sequence of 5 S RNA from *Torulopsis utilis*. *FEBS Lett.* **40**:106–109.

Nishizuka, Y., and Lipmann, F. 1966. Comparison of guanosine triphosphate split and polypeptide synthesis with a purified *E. coli* system. *Proc. Nat. Acad. Sci. USA* **55**:212–219.

Nolan, R. D., Grasmuk, H., Högenauer, G., and Drews, J. 1974. EF-1 from Krebs II mouse ascites cells. *Eur. J. Biochem.* **45**:601–609.

Nolan, R. D., Grasmuk, H., and Drews, J. 1975. The binding of tritiated EF-1 and EF-2 to ribosomes from Krebs II mouse ascites tumor cells. *Eur. J. Biochem.* **50**:391–402.

Noller, H. F. 1974. Topography of 16 S RNA in 30 S ribosomal subunits. *Biochemistry* **13**:4694–4703.

Noller, H. F., and Herr, W. 1974. Nucleotide sequence of the 3′-terminus of *E. coli* 16 S rRNA. *Mol. Biol. Repts.* **1**:437–439.

Nomura, M. 1973. Assembly of bacterial ribosomes. *Science* **179**:864–873.

Nudel, U., LeBleu, B., Zehavi-Willner, T., and Revel, M. 1973. Messenger RNP and initiation factors in rabbit-reticulocyte polyribosomes. *Eur. J. Biochem.* **33**:314–322.

Ohta, N., Sanders, M., and Newton, A. 1975. Poly(A) sequences in the RNA of *Caulobacter crescentus*. *Proc. Nat. Acad. Sci. USA* **72**:2343–2346.

Olsnes, S. 1970. Characterization of protein bound to rapidly-labeled RNA in polyribosomes from rat liver. *Eur. J. Biochem.* **15**:464–471.

Ono, Y., Skoultchi, A., Klein, A., and Lengyel, P. 1968. Discrimination against the initiator tRNA by microbial amino-acid polymerization factors. *Nature* **220**:1304–1307.

Otaka, T., and Kaji, A. 1974. Inhibitory effect of EF-G and GMPPCP on peptidyl transferase. *FEBS Lett.* **44**:324–329.

Ouellette, A. J., and Malt, R. A. 1976. Accumulation and decay of mRNA in mouse kidney. *Biochemistry* **15**:3358–3361.

Pace, N. R. 1973. Structure and synthesis of the rRNA of prokaryotes. *Bact. Rev.* **37**:562–603.

Pace, N. R., Walker, T. A., and Pace, B. 1974. The nucleotide sequence of chicken 5 S rRNA. *J. Mol. Evol.* **3**:151–159.

Pain, V. N., and Clemens, M. J. 1973. The role of soluble protein factors in the translational control of protein synthesis in eukaryotic cells. *FEBS Lett.* **32**:205–212.

Paradies, H. H., Franz, A., Pon, C. L., and Gualerzi, C. 1974. Conformational transition of the 30 S ribosomal subunit induced by IF-3. *Biochem. Biophys. Res. Comm.* **59**:600–607.

Payne, P. I., Woledge, J., and Corry, M. J., 1973. No evidence for tissue-specific sequences of cytoplasmic 5 S and 5.8 S ribosomal RNAs in the broad bean. *FEBS Lett.* **35**:327–330.

Peeters, B., Vanduffel, L., Depuydt, A., and Rombauts, W. 1973. The number and size of the proteins in the subunits of human placental ribosomes. *FEBS Lett.* **36**:217–221.

Pemberton, R. E., Housman, D., Lodish, H., and Baglioni, C. 1972. Isolation of duck haemoglobin mRNA and its translation by rabbit reticulocyte cell-free system. *Nature New Biol.* **235**:99–102.

Pene, J. J., Knight, E., and Darnell, J. E. 1968. Characterization of a new low molecular weight RNA in HeLa cell ribosomes. *J. Mol. Biol.* **33**:609–624.

Penman, S., Scherrer, K., Becker, Y., and Darnell, J. E. 1963. Polyribosomes in normal and poliovirus infected HeLa cells and their relationship to mRNA. *Proc. Nat. Acad. Sci. USA* **49**:654–662.

Perlman, S., Abelson, H., and Penman, S. 1973. Mitochondrial protein synthesis: RNA with the properties of eukaryotic mRNA. *Proc. Nat. Acad. Sci. USA* **70**:350–353.

Perlman, S., Hirsch, M., and Penman, S. 1972. Utilization of messenger in adenovirus-2-infected cells at normal and elevated temperatures. *Nature New Biol.* **238**:143–144.

Perry, R. P., and Kelley, D. E. 1974. Existence of methylated mRNA in mouse L cells. *Cell* **1**:37–42.

Perry, R. P., Kelley, D. E., and La Torre, J. 1972. Lack of poly(A) sequences in the mRNA of *E. coli. Biochem. Biophys. Res. Comm.* **48**:1593–1600.

Perry, R. P., and Scherrer, K. 1975. The methylated constituents of globin mRNA. *FEBS Lett.* **57**:73–78.

Person, S., and Osburn, M. 1968. The conversion of *amber* suppressors to *ochre* suppressors. *Proc. Nat. Acad. Sci. USA* **60**:1030–1038.

Petersen, N. S., and McLaughlin, C. S. 1973. Monocistronic mRNA in yeast. *J. Mol. Biol.* **81**:33–45.

Petrissant, G. 1973. Evidence for the absence of the G-T-ψ-C sequence from two mammalian initiator transfer RNAs. *Proc. Nat. Acad. Sci. USA* **70**:1046–1049.

Philipson, L., Wall, R., Glickman, G., and Darnell, J. E. 1971. Addition of poly(A) sequences to virus-specific RNA during adenovirus replication. *Proc. Nat. Acad. Sci. USA* **68**:2806–2809.

Pieczenik, G., Horiuchi, K., Model, P., McGill, C., Mazur, B. J., Vorts, G. F., and Zinder, N. D. 1975. Is mRNA transcribed from the strand complementary to it in a DNA duplex? *Nature* **253**:131–132.

Polya, G. M., and Phillips, D. R. 1976. The occurrence in amino acid sequences of extensive informational symmetries based on possible codon–codon complementarity in the encoding polynucleotides. *Biochem. J.* **153**:681–690.

Pribula, C. D., Fox, G. E., and Woese, C. R. 1974. Nucleotide sequence of *Bacillus megaterium* 5 S RNA. *FEBS Lett.* **44**:322–323.

Pribula, C. D., Fox, G. E., and Woese, C. R. 1976. Nucleotide sequence of *Clostridium pasteurianum* 5 S rRNA. *FEBS Lett.* **64**:350–352.

Procunier, J. D., and Tartof, K. D. 1976. Restriction map of 5 S RNA genes of *D. melanogaster. Nature* **263**:255–257.

Puckett, L., Chambers, S., and Darnell, J. E. 1975. Short-lived mRNA in HeLa cells and its impact on the kinetics of accumulation of cytoplasmic polyadenylate. *Proc. Nat. Acad. Sci. USA* **72**:389–393.

Rawson, J. R., and Stutz, E. 1968. Characterization of *Euglena* cytoplasmic ribosomes and rRNA by zone velocity sedimentation in sucrose gradients. *J. Mol. Biol.* **33**:309–314.

Reid, B. R., Einarson, B., and Schmidt, J. 1972. Loop accessibility in transfer RNA. *Biochimie* **54**:325–332.

Reinbolt, J., and Schiltz, E. 1973. The primary structure of ribosomal protein S4 from *E. coli. FEBS Lett.* **36**:250–252.

Retel, J., and Planta, R. J. 1968. Investigation of the rRNA sites in yeast DNA by the hybridization technique. *Biochim. Biophys. Acta* **169**:416–429.

Revel, M., Lelong, J. C., Brawerman, G., and Gros, F. 1968a. Function of three protein factors and ribosomal subunits in the initiation of protein synthesis in *E. coli. Nature* **219**:1016–1020.

Revel, M., Herzberg, H., Becarevic, A., and Gros, F. 1968b. Role of a protein factor in the functional binding of ribosomes to natural mRNA. *J. Mol. Biol.* **33**:231–249.

Rho, H. M., and Green, M. 1974. The homopolyadenylate and adjacent nucleotides at the 3′-terminus of 30–40 S RNA subunits in the genome of murine sarcoma-leukemia virus. *Proc. Nat. Acad. Sci. USA* **71**:2386–2390.

Ricard, B., and Salser, W. 1974. Size and folding of the messenger for phage T4 lysozyme. *Nature* **248**:314–317.

Ricard, B., and Salser, W. 1975. Secondary structures formed by random RNA sequences. *Biochem. Biophys. Res. Comm.* **63**:548–554.

Richter, D., Erdmann, V. A., and Sprinzl, M. 1973. Specific recognition of GTψC loop (Loop IV) of tRNA by 50 S ribosomal subunits. *Nature New Biol.* **246**:132–135.

Riley, W. T. 1973. Amino acid sequences and double-stranded messages—a means of directing the site of mutation? *J. Theor. Biol.* **40**:285–300.

Ringer, D., and Chládek, S. 1974. Ribosomal peptidyl transferase: recognition points on the 3′-terminus of AA-tRNA. *FEBS Lett.* **39**:75–78.

Ritter, E., and Wittmann-Liebold, B. 1975. The primary structure of protein L30 from *E. coli* ribosomes. *FEBS Lett.* **60**:153–155.

Roberts, R. J. 1972. Structure of two glycyl-tRNAs from *Staphylococcus epidermidis*. *Nature New Biol.* **237**:44–45.

Robinson, E. A., Henriksen, O., and Maxwell, E. S. 1974. Elongation factor 2. *J. Biol. Chem.* **249**:5088–5093.

Rohrbach, M. S., Dempsey, M. E., and Bodley, J. W. 1974. Preparation of homogeneous EF-G and examination of the mechanism of guanosine triphosphate hydrolysis. *J. Biol. Chem.* **249**:5094–5101.

Ron, E. Z., Kohler, R. E., and Davis, B. D. 1968. Magnesium ion dependence of free and polysomal ribosomes from *E. coli*. *J. Mol. Biol.* **36**:83–90.

Rosbash, M., and Ford, P. J. 1974. Poly(A)-containing RNA in *Xenopus laevis*. *J. Mol. Biol.* **85**:87–101.

Rosen, J. M., Woo, S. L. C., Holder, J. W., Means, A. R., and O'Malley, B. W. 1975. Preparation and preliminary characterization of purified ovalbumin mRNA from the hen oviduct. *Biochemistry* **14**:69–78.

Rubin, G. M. 1973. The nucleotide sequence of *S. cerevisiae* 5.8 S rRNA. *J. Biol. Chem.* **248**:3860–3875.

Rubin, G. M. 1974. Three forms of the 5.8 S ribosomal RNA species in *S. cerevisiae*. *Eur. J. Biochem.* **41**:197–202.

Sabol, S., and Ochoa, S. 1974. Preparation of radioactive IF-3. *Methods Enzymol.* **30**:39–53.

Sabol, S., Sillero, M. A. G., Iwasaki, K. and Ochoa, S. 1970. Purification and properties of IF-3. *Nature* **228**:1269–1273.

Sadowski, P. D., and Howden, J. A. 1968. Isolation of two distinct classes of polysomes from a nuclear fraction of rat liver. *J. Cell. Biol.* **37**:163–181.

Sager, R., and M. G. Hutchinson. 1967. Cytoplasmic and chloroplast ribosomes of *Chlamydomonas*. *Science* **157**:709–711.

Sagher, D., Edelman, M., and Jakob, K. M. 1974. Poly(A)-associated RNA in plants. *Biochim. Biophys. Acta* **349**:32–38.

Sampson, J., Mathews, M. B., Osburn, M., and Borghetti, A. F. 1972. Hemoglobin mRNA translation in cell-free systems from rat and mouse liver and Landschutz ascites cells. *Biochemistry* **11**:3636–3640.

Sanger, F. 1971. Nucleotide sequences in bacteriophage RNA. *Biochem. J.* **124**:833–843.

Sankoff, D., Morel, C., and Cedergren, R. J. 1973. Evolution of 5 S RNA and the nonrandomness of base replacement. *Nature New Biol.* **245**:232–234.

Santi, D. V., and Danenberg, P. V. 1971. Phenylalanyl tRNA synthetase from *E. coli*. Analysis of the phenylalanine binding site. *Biochemistry* **10**:4813–4820.

Santi, D. V., Danenberg, P. V., and Satterly, P. 1971. Phenylalanyl tRNA synthetase from *E. coli*. Reaction parameters and order of substrate addition. *Biochemistry* **10**:4804–4812.

Scharff, M. D., and Robbins, E. 1966. Polyribosome disaggregation during metaphase. *Science* **151**:922–995.

Schedl, P. D., Singer, R. E., and Conway, T. W. 1970. A factor required for the translation of bacteriophage f2 RNA in extracts of T4-infected cells. *Biochem. Biophys. Res. Comm.* **38**:631–637.

Schiff, N., Miller, M. J. and Wahba, A. J. 1974. Purification and properties of chain IF-3 from T4-infected and uninfected *E. coli* MRE600. *J. Biol. Chem.* **249**:3797–3802.

Schiltz, E., and Reinbolt, J. 1975. Determination of the complete amino-acid sequence of protein S4 from *E. coli* ribosomes. *Eur. J. Biochem.* **56**:467–481.

Shimotohno, K., Kodama, Y., Hashimoto, J., and Miura, K. 1977. Importance of 5′-terminal blocking structure to stabilize mRNA in eukaryotic protein synthesis. *Proc. Nat. Acad. Sci. USA* **74**:2734–2738.

Schlessinger, D., Marchesi, V. T., and Kwan, B. C. K. 1965. Binding of ribosomes to cytoplasmic reticulum of *Bacillus megaterium*. *J. Bact.* **90**:456–466.

Schrier, P. I., Maassen, J. A., and Möller, W. 1973. Involvement of 50 S ribosomal proteins L6 and L10 in the ribosome dependent GTPase activity of EF-G. *Biochem. Biophys. Res. Comm.* **53**:90–98.

Schrier, P. I., and Möller, W. 1975. The involvement of 50 S ribosomal protein L11 in the EF-G dependent GTP hydrolysis of *E. coli* ribosomes. *FEBS Lett.* **54**:130–134.

Scolnick, E., Tompkins, R., Caskey, T., and Nirenberg, M. 1968. Release factors differing in specificity for terminator codons. *Proc. Nat. Acad. Sci. USA* **61**:768–774.

Seal, S. N., and Marcus, A. 1973. Translation of the initial codons of satellite tobacco necrosis virus RNA in a cell-free system from wheat embryo. *J. Biol. Chem.* **248**:6577–6582.

Shine, J., and Dalgarno, L. 1973. Occurrence of heat-dissociable rRNA in insects. *J. Mol. Biol.* **75**:57–72.

Shine, J., and Dalgarno, L. 1974a. The 3′-terminal sequence of *E. coli* 16 S ribosomal RNA. *Proc. Nat. Acad. Sci. USA* **71**:1342–1346.

Shine, J., and Dalgarno, L. 1974b. Identical 3′-terminal octanucleotide sequence in 18 S ribosomal ribonucleic acid from different eukaryotes. *Biochem. J.* **141**:609–615.

Shine, J., and Dalgarno, L. 1975. Determinant of cistron specificity in bacterial ribosomes. *Nature* **254**:34–38.

Shine, J., Hunt, J. A., and Dalgarno, L. 1974. Studies on the 3′-terminal sequences of the large rRNA of different eukaryotes and those associated with 'hidden' breaks in heat-dissociable insect 26 S RNA. *Biochem. J.* **141**:617–625.

Siddiqui, M. A. Q., and Hosokawa, N. 1968. Role of 5 S rRNA in polypeptide synthesis. *Biochem. Biophys. Res. Comm.* **32**:1–8.

Siegert, W., Bauer, G., and Hofschneider, P. H. 1973. Direct evidence for messenger activity of influenza virion RNA. *Proc. Nat. Acad. Sci. USA* **70**:2960–2963.

Simsek, M., Petrissant, G., and Rajbhandary, U. L. 1973a. Replacement of the sequence G-T-ψ-C-G(A) by G-A-U-C-G in initiator transfer RNA of rabbit liver cytoplasm. *Proc. Nat. Acad. Sci. USA* **70**:2600–2604.

Simsek, M., Ziegenmeyer, J., Heckman, J., and Rajbhandary, U. L. 1973b. Absence of the sequence G-T-ψ-C-G(A) in several eukaryotic cytoplasmic initiator transfer RNAs. *Proc. Nat. Acad. Sci. USA* **70**:1041–1045.

Singer, R. E., and Conway, T. W. 1973. Defective initiation of f2 RNA translation by ribosomes from bacteriophage T4-infected cells. *Biochim. Biophys. Acta* **331**:102–116.

Skogerson, L., and Wakatama, E. 1976. A ribosome-dependent GTPase from yeast distinct from EF-2. *Proc. Nat. Acad. Sci. USA* **73**:73–76.

Slack, J. M. W., and Loening, U. E. 1974. 28 S RNA from *Xenopus laevis* contains a sequence of three adjacent 2′-*O*-methylations. *Eur. J. Biochem.* **43**:69–72.

Slater, I., Gillespie, D., and Slater, D. W. 1973. Cytoplasmic adenylation and processing of maternal RNA. *Proc. Nat. Acad. Sci. USA* **70**:406–411.

Slayter, H., Kiho, Y., Hall, C. E., and Rich, A. 1968. An electron microscopic study of large bacterial polyribosomes. *J. Cell Biol.* **37**:583–590.

Smith, I., Dubnau, D., Morrell, P., and Marmur, J. 1968. Chromosomal location of DNA base sequences complementary to tRNA and to 5 S, 16 S, and 23 S ribosomal RNA in *B. subtilis*. *J. Mol. Biol.* **33**:123–140.

Smith, K. E., and Henshaw, E. C. 1975. Binding of met-tRNA$_f$ to native and derived 40 S ribosomal subunits. *Biochemistry* **14**:1060–1067.

Spencer, M., Pigram, W. J., and Littlechild, J. 1969. Studies on rRNA structure. *Biochim. Biophys. Acta* **179**:348–359.

Spierer, P., Zimmermann, R. A., and Mackie, G. A. 1975. RNA-protein interactions in the ribosome. *Eur. J. Biochem.* **52**:459–468.

Spirin, A. S. 1969. Informosomes. *Eur. J. Biochem.* **10**:20–35.

Stadler, H. 1974. The primary structure of the 16 S rRNA binding protein S8 from *E. coli* ribosomes. *FEBS Lett.* **48**:114–116.

Stadler, H., and Wittmann-Liebold, B. 1976. Determination of the amino-acid sequence of the ribosomal protein S8 of *E. coli*. *Eur. J. Biochem.* **66**:49–56.

Stavnezer, J., and Juang, R. C. C. 1971. Synthesis of a mouse immunoglobin light chain in a rabbit reticulocyte cell-free system. *Nature New Biol.* **230**:172–176.

Steitz, J. A. 1969. Polypeptide chain initiation. *Nature* **224**:957–964.

Steitz, J. A. 1973. Discriminatory ribosome rebinding of isolated regions of protein synthesis initiation from the RNA of bacteriophage R17. *Proc. Nat. Acad. Sci. USA* **70**:2605–2609.

Stevens, A. R., and Pachler, P. F. 1972. Discontinuity of 26 S rRNA in *Acanthamoeba castellani*. *J. Mol. Biol.* **66**:225–237.

Steward, D. L., Shaeffer, J. R., and Humphrey, R. M. 1968. Breakdown and assembly of polyribosomes in synchronized Chinese hamster cells. *Science* **161**:791–793.

Stewart, J. W., Sherman, F., Shipman, N. A., and Jackson, M. 1971. Identification and mutational relocation of the AUG codon initiating translation of iso-1-cytochrome *c* in yeast. *J. Biol. Chem.* **246**:7429–7445.

Stiles, C. D., Lee, K.-L., and Kenney, F. T. 1976. Differential degradation of mRNAs in mammalian cells. *Proc. Nat. Acad. Sci. USA* **73**:2634–2638.

Stöffler, G., Hasenbank, R., Bodley, J. W., and Highland, J. H. 1974. Inhibition of protein L7/L12 binding to 50 S ribosomal cores by antibodies specific for L6, L10, and L18. *J. Mol. Biol.* **86**:171–174.

Stöffler, G., Wool, I. G., Lin, A., and Rak, K.-H. 1974. The identification of the eukaryotic ribosomal proteins homologous with *E. coli* proteins L7 and L12. *Proc. Nat. Acad. Sci. USA* **71**:4723–4726.

Stoltzfus, C. M., Shatkin, A. J., and Banerjee, A. K. 1973. Absence of poly(A) from reovirus mRNA. *J. Biol. Chem.* **248**:7993–7998.

Strycharz, W. A., Ranki, M., and Dahl, H. H. M. 1974. A high-molecular-weight protein component required for natural messenger translation in ascites tumor cells. *Eur. J. Biochem.* **48**:303–310.

Subramanian, A. R. 1974. Sensitive separation procedure for *E. coli* ribosomal proteins and the resolution of high-molecular-weight components. *Eur. J. Biochem.* **45**:541–546.

Subramanian, A. R., and Davis, B. D. 1970. Activity of IF$_3$ in dissociating *E. coli* ribosomes. *Nature* **228**:1273–1275.

Subramanian, A. R., Ron, E. Z., and Davis, B. D. 1968. A factor required for ribosome dissociation in *E. coli*. *Proc. Nat. Acad. Sci. USA* **61**:761–767.

Sundaric, R. M., Stringer, E. A. Schulman, L. D. H., and Maitra, U. 1976. Interaction of bacterial IF-2 with initiator tRNA. *J. Biol. Chem.* **251**:3338–3345.

Sundquist, B., Persson, T., and Lindberg, U. 1977. Characterization of mRNA–protein complexes from mammalian cells. *Nucl. Acids Res.* **4**:899–915.

Sussman, M. 1966. Protein synthesis and the temporal control of genetic transcription during slime mold development. *Proc. Nat. Acad. Sci. USA* **55**:813–818.

Szer, W., Hermoso, J. M., and Leffler, S. 1975. Ribosomal protein S1 and polypeptide chain initiation in bacteria. *Proc. Nat. Acad. Sci. USA* **72**:2325–2329.

Szer, W., and Leffler, S. 1974. Interaction of *E. coli* 30 S ribosomal subunits with MS2 phage RNA in the absence of initiation factors. *Proc. Nat. Acad. Sci. USA* **71**:3611–3615.

Takagi, M., Tanaka, T., and Ogatu, K. 1970. Chromosome activity and cell function in polytenic cells. *Biochim. Biophys. Acta* **217**:108–119.

Tate, W. P., Beaudet, A. L., and Caskey, C. T. 1973. Influence of guanine nucleotides and elongation factors on interaction of release factors with the ribosome. *Proc. Nat. Acad. Sci. USA* **70**:2350–2352.

Terao, K., and Ogata, K. 1975. Studies on structural proteins of the rat liver ribosomes. *Biochim. Biophys. Acta* **402**:214–229.

Teraoka, H., and Tanaka, K. 1973. Effect of polyamines on the binding of dihydrostreptomycin and *N*-acetylphenylalanyl-tRNA to ribosomes from *E. coli. Eur. J. Biochem.* **40**:423–429.

Terhorst, C., Möller, W., Laursen, R., and Wittmann-Liebold, B. 1973. The primary structure of an acidic protein which is involved in GTP hydrolysis dependent on elongation factors G and T. *Eur. J. Biochem.* **34**:138–152.

Terhorst, C., Wittmann-Liebold, B., and Möller, W. 1972. 50 S ribosomal proteins. *Eur. J. Biochem.* **25**:13–19.

Toivonen, J. E., and Nierlich, D. P. 1974. Biological decay of the 5′-triphosphate termini of the RNA of *E. coli. Nature* **252**:74–76.

Träeger, L. 1970. Termination der Proteinsynthese. *Naturwissenschaften* **57**:560–564.

Tsiapalis, C. M., Dorson, J. W., De Sante, D. M., and Bollum, F. J. 1973. Terminal riboadenylate transferase. *Biochem. Biophys. Res. Comm.* **50**:737–743.

Tsurugi, K., Morita, T., and Ogata, K. 1974. Mode of degradation of ribosomes in regenerating rat liver *in vivo. Eur. J. Biochem.* **45**:119–126.

Ulbrich, B., and Nierhaus. K. H., 1975. Pools of ribosomal proteins in *E. coli. Eur. J. Biochem.* **57**:49–54.

Van, N. T., Holder, J. W., Woo, S. L. C., Means, A. R., and O'Malley, B. W. 1976. Secondary structure of ovalbumin mRNA. *Biochemistry* **15**:2054–2062.

Vandekerckhove, J., Francq, H., and Van Montagu, M. 1969. The amino acid sequence of the coat protein of the bacteriophage MS-2 and localization of the amber mutation in the coat mutants growing on a su_3^+ suppressor. *Arch. Intern. Physiol. Biochim.* **77**:175–180.

Vandekerckhove, J., Rombauts, W., Peeters, B., and Wittmann-Liebold, B. 1975. Determination of the complete amino-acid sequence of protein S21 from *E. coli. Hoppe-Seyler's Z. Phys. Chem.* **356**:1955–1976.

Vandekerckhove, J., Rombauts, B., and Wittmann-Liebold, B. 1977. The primary structure of protein S16 from *E. coli* ribosomes. *FEBS Lett.* **73**:18–21.

Van de Walle, C. 1973. Poly(A) sequences in plant RNA. *FEBS Lett.* **34**:31–34.

Van Dieijen, G., Van der Laken, C. J., van Knippenberg, P. H., and Van Duin, J. 1975. Function of *E. coli* ribosomal protein S1 in translation of natural and synthetic mRNA. *J. Mol. Biol.* **93**:351–366.

Van Duin, J., and van Knippenberg, P. H. 1974. Requirement of protein S1 for translation. *J. Mol. Biol.* **84**:185–195.

Van Duin, J., van Knippenberg, P. H., Dieben, M., and Kurland, C. G. 1972. Functional heterogeneity of the 30 S ribosomal subunit of *E. coli. Mol. Gen. Genetics* **116**:181–191.

van Knippenberg, P. H. 1975. A possible role of the 5′-terminal sequence of 16 S rRNA in the recognition of initiation sequences for protein synthesis. *Nucl. Acids Res.* **2**:79–85.

van Knippenberg, P. H., Hooykass, P. J. J., and Van Duin, J. 1974. The stoichiometry of *E. coli* 30S ribosomal protein S1 on *in vivo* and *in vitro* polyribosomes. *FEBS Lett.* **41**:323–326.

Vaquero, C., Reibel, L., Delaunay, J., and Schapiro, G. 1973. Translation of globin mRNA among eukaryotes. *Biochem. Biophys. Res. Comm.* **54**:1171–1177.

Vaughn, M. H., and Hansen, D. S., 1973. Control of initiation of protein synthesis in human cells. *J. Biol. Chem.* **248**:7087–7096.

Vigne, R., Jordan, B. R., and Monier, R. 1973. A common conformational feature in several prokaryotic and eukaryotic 5 S RNAs. *J. Mol. Biol.* **76**:303–311.

Visentin, L. P., Matheson, A. T., and Yaguchi, M. 1974. Homologies in procaryotic ribosomal proteins. *FEBS Lett.* **41**:310–314.

Volckaert, G., and Fiers, W. 1973. Studies on the bacteriophage MS2. G-U-G as the initiation codon of the A-protein cistron. *FEBS Lett.* **35**:91–96.

Volkin, E., and Astrachan, L. 1956. Phosphorus incorporation in *E. coli* RNA after infection with bacteriophage T2. *Virology* **2**:149–161.

Vournakis, J., and Rich, A. 1972. Ribosomal transformations during protein synthesis. *In*: Cox, R. A., and A. A. Hadjiolov, eds., *Functional Units in Protein Biosynthesis,* New York, Academic Press, p. 287–299.

Wade, M., Laursen, R. A., and Miller, D. L. 1975. Amino acid sequence of EF-Tu. *FEBS Lett.* **53**:37–39.

Wahba, A. J., Mazumder, R., Iwasaki, K., Choe, Y. B., Miller, M. J., Sillero, M. A. G., and Ochoa, S. 1968. Role of ribosome factors in polypeptide chain initiation. *Abstr. Fed. Eur. Biochem. Soc., Madrid,* p. 9.

Wahba, A. J., and Miller, M. J. 1974. Chain initiation factors from *E. coli. Meth. Enzym.* **30**:3–18.

Walker, T. A., Betz, J. L., Olah, J., and Pace, N. R. 1975. The nucleotide sequence of dolphin and bovine 5 S rRNA. *FEBS Lett.* **54**:241–244.

Weber, K., and Koenigsberg, W. 1967. Amino acid sequence of the f_2 coat protein. *J. Biol. Chem.* **242**:3563–3578.

Wegnez, M., Monier, R., and Denis, H. 1972. Sequence heterogeneity of 5 S RNA in *Xenopus laevis. FEBS Lett.* **25**:13–20.

Wei, C. M., Gershowitz, A., and Moss, B. 1976. 5′-terminal and internal methylated nucleotide sequences in HeLa cell mRNA. *Biochemistry* **15**:397–401.

Weiner, A. M., and Weber, K. 1973. A single UGA codon functions as a natural termination signal in the coliphage Qβ coat protein cistron. *J. Mol. Biol.* **80**:837–855.

Weissbach, H., Redfield, B., and Moon, H. M. 1973. Further studies on the interactions of EF-1 from animal tissues. *Arch. Biochem. Biophys.* **156**:267–275.

Weissmann, C., Billeter, M. A., Goodman, H. M., Hindley, J., and Weber, H. 1973. Structure and function of phage RNA. *Ann. Rev. Biochem.* **42**:303–328.

Welfle, H., Stahl, J., and Bielka, H. 1972. Studies on proteins of animal ribosomes. *FEBS Lett.* **26**:228–232.

Wells, G. N., and Beevers, L. 1974. Protein synthesis in the cotyledons of *Pisum sativum* L. *Biochem. J.* **139**:61–69.

Wen, W. N., León, P. E., and Hague, D. R. 1974. Multiple gene sites for 5 S and 18 + 28 S RNA on chromosomes of *Glyptotendipes barvipes. J. Cell Biol.* **62**:132–144.

Westover, K. C., and Jacobson, L. A. 1974. Control of protein synthesis in *E. coli. J. Biol. Chem.* **249**:6272–6279.

White, H. B., Laux, B. E., and Dennis, D. 1972. Messenger RNA structure: Compatibility of hairpin loops with protein sequence. *Science* **175**:1264–1266.

Wice, M., and Kennell, D. 1974. Decay of mRNA from the tryptophan operon of *E. coli* as a function of growth temperature. *J. Mol. Biol.* **84**:649–652.

Wigle, D. T., and Smith, A. E. 1973. Specificity in initiation of protein synthesis in a fractionated mammalian cell-free system. *Nature New Biol.* **242**:136–140.

Williamson, A. R., and Schweet, R. 1964. Role of the genetic message in initiation and release of the polypeptide chain. *Nature* **202**:435–437.

Williamson, R. 1973. The protein moieties of animal messenger ribonucleoproteins. *FEBS Lett.* **37**:1–6.

Williamson, R., and Brownlee, G. G. 1969. The sequence of 5 S ribosomal RNA from two mouse cell lines. *FEBS Lett.* **3**:306–308.

Williamson, R., Morrison, M., Lanyon, G., Eason, R., and Paul, J. 1971. Properties of mouse globin mRNA and its preparation in milligram quantities. *Biochemistry* **70**:3014–3020.

Wittmann, H. G. 1976. Structure, function and evolution of ribosomes. *Eur. J. Biochem.* **61**:1–13.

Wittmann-Liebold, B. 1973. Studies on the primary structure of 20 proteins from *E. coli* ribosomes by means of an improved protein sequenator. *FEBS Lett.* **36**:247–249.

Wittmann-Liebold, B., and Dzionara, M. 1976a. Comparison of amino acid sequences among ribosomal proteins of *E.coli*. *FEBS Lett.* **61**:14–19.

Wittman-Liebold, B., and Dzionara, M. 1976b. Studies on the significance of sequence homologies among proteins from *E. coli* ribosomes. *FEBS Lett.* **65:**281–283.

Wittmann-Liebold, B., Greuer, B., and Pannenbecker, R. 1975. The primary structure of protein L32 from the 50 S subunit of *E. coli* ribosomes. *Hoppe-Seyler's Z. Phys. Chem.* **356**:1977–1979.

Wittman-Liebold, B., Marzinzig, E., and Lehmann, A. 1976. Primary structure of protein S20 from the small ribosomal subunit of *E. coli*. *FEBS Lett.* **68**:110–114.

Wittmann-Liebold, B., and Pannenbecker, R. 1976. Primary structure of protein L33 from the large subunit of the *E. coli* ribosome. *FEBS Lett.* **68**:115–118.

Woese, C. R. 1972. Evolution of macromolecular complexity. *J. Theor. Biol.* **33**:29–34.

Woledge, J., Corry, M. J., and Payne, P. I. 1974. Ribosomal RNA homologies in flowering plants. *Biochim. Biophys. Acta* **349**:339–350.

Wong, J. T. F. 1975. A co-evolution theory of the genetic code. *Proc. Nat. Acad. Sci. USA* **72**:1909–1912.

Wong, K. L., Bolton, P. H., and Kearns, D. R. 1975. Tertiary structure in *E. coli* $tRNA^{Arg}$ and $tRNA^{Val}$. *Biochim. Biophys. Acta* **383**:446–451.

Wong, Y. P., Reid, B. R., and Kearns, D. R. 1973. Conformation of charged and uncharged $tRNA^{Phe}$. *Proc. Nat. Acad. Sci. USA* **70**:2193–2195.

Woodley, C. L., Chen, Y. C., and Gupta, N. K. 1974. Purification and properties of the peptide chain initiation factors from rabbit reticulocytes. *Methods Enzymol.* **30**:141–153.

Yaguchi, M. 1975. Primary structure of protein S18 from the small *E. coli* ribosomal subunit. *FEBS Lett.* **59**:217–220.

Yaguchi, M., Matheson, A. T., and Visentin, L. P. 1974. Procaryotic ribosomal proteins: N-terminal sequence homologies and structural correspondence of 30 S ribosomal proteins from *E. coli* and *Bacillus stearothermophilus*. *FEBS Lett.* **46**:296–300.

Yamada, Y., Whitaker, P. A., and Nakada, D. 1974. Functional instability of T7 early mRNA. *Nature* **248**:335–338.

Yanofsky, C., and Ito, J. 1966. Nonsense codons and polarity in the tryptophan operon. *J. Mol. Biol. 21*:313–334.

Yarus, M. 1972. Phenylalanyl-tRNA synthetase and isoleucyl-$tRNA^{Phe}$: A possible verification mechanism for aminoacyl-tRNA. *Proc. Nat. Acad. Sci. USA* **69**:1915–1919.

Yarus, M., and Barrell, B. G. 1971. The sequence of nucleotides in $tRNA^{Ile}$ from *E. coli* B. *Biochem. Biophys. Res. Comm.* **43**:729–733.

Yogo, Y., and Wimmer, E. 1972. Poly(A) at the 3′-terminus of poliovirus RNA. *Proc. Nat. Acad. Sci. USA* **69**:1877–1882.

Yokosawa, H., Inoue-Yokosawa, N., Arai, K.-i., Kawakita, M., and Kaziro, Y. 1973. The role of guanosine triphosphate hydrolysis in EF-Tu-promoted binding of aminoacyl tRNA to ribosomes. *J. Biol. Chem.* **248**:375–377.

Yu, R. S. T., and Wittmann, H. G. 1973. The sequence of steps in the attachment of 5 S RNA to cores of *E. coli* ribosomes. *Biochim. Biophys. Acta* **324**:375–385.

Zalik, S., and Jones, B. L. 1973. Protein biosynthesis. *Ann. Rev. Plant Physiol.* **24**:47–68.

Zasloff, M., and Ochoa, S. 1971. A supernatant factor involved in initiation complex formation with eukaryotic ribosomes. *Proc. Nat. Acad. Sci. USA* **68**:3059–3063.

Zasloff, M., and Ochoa, S. 1973. Polypeptide chain initiation in eukaryotes. *J. Mol. Biol.* **73**:65–76.

Zinder, N. 1963. Properties of a bacteriophage containing RNA. *Perspectives Virol.* **3**:58–67.

Zylber, E. A., and Penman, S. 1971. Synthesis of 5 S and 4 S RNA in metaphase-arrested HeLa cells. *Science* **172**:947–949.

CHAPTER 5

Aarstad, K., and Øyen, T. B. 1975. On the distribution of 5 S RNA cistrons on the genome of *S. cerevisiae*. *FEBS Lett.* **51**:227–231.

Abraham, K. A., and Bhagava, P. M. 1963. The uptake of radioactive amino acids by spermatozoa. *Biochem. J.* **86**:308–313.

Adesnik, M., and Levinthal, C. 1969. Synthesis and maturation of rRNA in *E. coli*. *J. Mol. Biol.* **46**:281–304.

Akabori, S., and Yamamoto, M. 1972. Model experiments on the probiological formation of protein. *In*: Rohlfing, D. L., and A. I. Oparin, eds., *Molecular Evolution: Prebiological and Biological*, New York, Plenum Press, p. 189–197.

Altman, S. 1971. Isolation of tyrosine tRNA precursor molecules. *Nature New Biol.* **229**:19–21.

Altman, S., and Smith, J. D. 1971. Tyrosine tRNA precursor molecule polynucleotide sequence. *Nature New Biol.* **233**:35–39.

Arce, C. A., Barra, H. S., Rodriguez, J. A., and Caputto, R. 1975. Tentative identification of the amino acid that binds tyrosine as a single unit into soluble brain proteins. *FEBS Lett.* **50**:5–7.

Arion, V. Y., and Georgiev, G. P. 1967. On functional heterogeneity of chromosomal information RNA. *Proc. Acad. Sci. USSR* **172**:716–719.

Aronson, A., and Wilt, F. H. 1969. Properties of nRNA in sea urchin embryos. *Proc. Nat. Acad. Sci. USA* **62**:186–193.

Attardi, G., Parnas, H., Hwang, M. I. H., and Attardi, B. 1966. Giant-size rapidly labeled nRNA and cytoplasmic mRNA in immature duck erythrocytes. *J. Mol. Biol.* **20**:145–182.

Baltimore, D., and Franklin, R. M. 1963. Properties of the mengovirus and poliovirus RNA polymerases. *Cold Spring Harbor Symp. Quant. Biol.* **28**:105–108.

Barra, H. S., Arce, C. A., Rodriguez, J. A., and Caputto, R. 1973a. Incorporation of phenylalanine as a single unit into rat brain protein. *J. Neurochem.* **21**:1241–1251.

Barra, H. S., Rodriguez, J. A., Arce, C. A., and Caputto, R. 1973b. A soluble preparation from rat brain that incorporates into its own proteins (^{14}C) arginine by a ribonuclease-sensitive system and (^{14}C) tyrosine by a ribonuclease-insensitive system. *J. Neurochem.* **20**:97–108.

Barra, H. S., Arce, C. A., Rodriguez, J. A., and Caputto, R. 1974. Some common properties of the enzyme that incorporates tyrosine as a single unit and the microtubule proteins. *Biochem. Biophys. Res. Comm.* **60**:1384–1390.

Barrell, B. G., Seidman, J. G., Guthrie, C., and McClain, W. H. 1974. The nucleotide sequence of a precursor to serine and proline tRNAs. *Proc. Nat. Acad. Sci. USA* **71**:413–416.

Bautz, E. K. F. 1972. Regulation of RNA synthesis. *Progr. Nucl. Acid. Res.* **12**:129–160.

Berg, D., Barrett, K., and Chamberlain, M. 1971. Purification of two forms of *E. coli* RNA polymerase and of sigma component. *Methods Enzymol.* **21**:506–519.

Bhargava, P. M., Bishop, M. W. H., and Work, T. S. 1956. Incorporation of (^{14}C) amino acids into the proteins of bull spermatozoa. *Biochem. J.* **73**:247–256.

Blatt, B., and Feldmann, H. 1973. Characterization of precursors to tRNA in yeast. *FEBS Lett.* **37**:129–133.

Blatti, S. P., and Ingles, C. J., Lindell, T. J., Morris, P. W., Weaver, R. F., Weinberg, F., and Rutter, W. J. 1970. Structure and regulatory properties of eukaryotic RNA polymerase. *Cold Spring Harbor Symp. Quant. Biol.* **35**:649–657.

Body, B. A., and Brownstein, B. H. 1976. Ribosomal precursor particles of *Bacillus megaterium. J. Bact.* **126**:1149–1155.

Bramwell, M. E. 1974. The behavior of hnRNA in partially and completely denaturing conditions. *Biochem. J.* **141**:477–484.

Bramwell, M. E., and Harris, H. 1967. The origin of polydispersity in sedimentation patterns of rapidly labelled nRNA. *Biochem. J.* **103**:816–830.

Brentani, R., and Brentani, M. 1969. Messenger RNA in the nucleolus. *Genetics, Suppl.* **61**:391–399.

Brentani, R., Brentani, M., and Raw, I. 1964. Role of nucleolar RNA on the incorporation of ribosomal amino acid. *Nature* **201**:1130.

Brentani, R., Brentani, M., and Raw, I. 1967. Messenger activity of purified RNA from rat liver nucleoli. *Nature* **214**:1122–1123.

Brishammar, S., and Juntti, N. 1975. A poly(U) polymerase in tobacco leaves. *Biochim. Biophys. Acta.* **383**:351–358.

Brown, D. D., and Gurdon, J. B. 1966. Size distribution and stability of DNA-like RNA synthesized during development of anucleolate embryos of *Xenopus laevis. J. Mol. Biol.* **19**:399–422.

Brown, D. D., and Weber, C. S. 1968. Unique DNA sequences homologous to 4 S, 5 S, and rRNA. *J. Mol. Biol.* **34**:681–697.

Burdon, R. H. 1971. RNA maturation in animal cells. *Progr. Nucleic Acid. Res. Mol. Biol.* **11**:33–76.

Burgess, R. R., and Travers, A. A. 1971. Purification of the RNA polymerase sigma factor. *Methods Enzymol.* **21**:500–506.

Burgess, R. R., Travers, A. A., Dunn, J. J., and Bautz, E. K. F. 1969. Factor stimulating transcription by RNA polymerase. *Nature* **221**:43–46.

Busby, W. F., Hele, P., and Chang, M. C. 1974. Apparent amino acid incorporation by ejaculated rabbit spermatozoa. *Biochim. Biophys. Acta* **330**:246–259.

Calvin, M. 1969. *Chemical Evolution: Molecular Evolution towards the Origin of Living Systems on the Earth and Elsewhere.* Oxford, Oxford University Press.

Carre, D. S., and Chapeville, F. 1974. Study of the *E. coli* tRNA nucleotidyl transferase; effect of inorganic ions and thiol blocking reagents on enzyme activity. *Biochim. Biophys. Acta* **361**:176–184.

Carre, D. S., Litvak, S., and Chapeville, F. 1974. Study of *E. coli* tRNA nucleotidyl transferase. Interactions of the enzyme with tRNA. *Biochim. Biophys. Acta* **361**:185–197.

Chamberlin, M., and Berg, P. 1962. DNA-directed synthesis of RNA by an enzyme from *E. coli. Proc. Nat. Acad. Sci. USA* **48**:81–94.

Chamberlin, M., and Berg, P. 1964. Mechanism of RNA polymerase action: Formation of DNA-RNA hybrids with single-stranded templates. *J. Mol. Biol.* **8**:297–313.

Chambon, P. 1974. Eukaryotic RNA polymerases. *In:* Boyer, D. D., ed., *The Enzymes,* New York, Academic Press, **10**:261–331.

Chambon, P., Gissinger, F., Kedinger, C., Mandel, J. L., and Meilhac, M. 1973. Structural and functional properties of three mammalian nuclear DNA-dependent RNA polymerases. *Stud. Biophys.* **31**:29–32.

Chambon, P., Mandel, J. L., Gissinger, F., Kedinger, C., Gross-Bellard, M., and Hossenlopp, L.

1974. Transcription of double-stranded viral and cellular DNAs by purified mammalian DNA-dependent RNA polymerases. *In:* Bisivas, B. B., R. K. Mandel, A. Stevins, and W. E. Cohn, eds., *Control of Transcription,* New York, Plenum Publishing Co., p. 257–268.

Chan, L., Harris, S. E., Rosen, J. M., Means, A. R., and O'Malley, B. W. 1977. Processing of nuclear heterogeneous RNA: Recent developments. *Life Sci.* **20**:1–16.

Chang, C. N., and Chang, F. N. 1974. Methylation of ribosomal proteins *in vitro. Nature* **251**:731–733.

Chang, C. N., and Chang, F. N. 1975. Nature and stoichiometry of the methylated amino acids in 50 S ribosomal proteins. *Biochemistry* **14**:468–477.

Clarkson, S. G., Birnstiel, M. L., and Serra, V. 1973a. Reiterated tRNA genes of *Xenopus laevis. J. Mol. Biol.* **79**:391–410.

Clarkson, S. G., Birnstiel, M. L., and Purdom, I. F. 1973b. Clustering of transfer RNA genes of *Xenopus laevis. J. Mol. Biol.* **79**:411–429.

Cline, M. J., Eason, R., and Smellie, R. M. S. 1963. The biosynthesis of RNA following infection with a RNA virus. *J. Biol. Chem.* **238**:1788–1792.

Cooper, H. L., and Gibson, E. M. 1969. Control of synthesis and wastage of rRNA in lymphocytes. *J. Biol. Chem.* **246**:5059–5066.

Cramer, J. H., Bhargava, M. M., and Halvorson, H. O. 1972. Isolation and characterization of γ DNA of *S. cerevisiae. J. Mol. Biol.* **71**:11–20.

Cramer, J. H., Sebastian, J., Rownd, R. H., and Halvorson, H. O. 1974. Transcription of *S. crevisiae* rDNA *in vivo* and *in vitro. Proc. Nat. Acad. Sci. USA* **71**:2188–2192.

Cranston, J. W., Silber, R., Malathi, V. G., and Hurwitz, J. 1974. Characterization of adenosine triphosphate-inorganic pyrophosphate exchange reaction and demonstration of an enzyme-adenylate complex with T4-induced enzyme. *J. Biol. Chem.* **249**:7447–7456.

Dahlberg, A., and Peacock, A. C. 1971. Studies of 16 and 23 S rRNA of *E. coli,* using composite gel electrophoresis. *J. Mol. Biol.* **55**:61–74.

Daneholt, B. 1973. The giant RNA transcript in balbiani ring of *Chironomus tentans. In*: Hamkalo, B. A., and J. Pajaconstantinou, eds., *Molecular Cytogenetics,* New York, Plenum Publishing Corp., p. 155–165.

Daneholt, B., and Hosick, H. 1973. Evidence for transport of 75 S RNA from a discrete chromosome region via nuclear sap to cytoplasm in *Chironomus tentans. Proc. Nat. Acad. Sci. USA* **70**:442–446.

Darlix, J. L. 1974. Rho, a factor causing the modulation of early T7 genes transcription. *Biochimie* **56**:693–701.

Darlix, J. L. 1975. Simultaneous purification of *E. coli* termination factor rho, RNase III, and RNase H. *Eur. J. Biochem.* **51**:369–376.

Darlix, J. L., and Fromageot, P. 1974. Restriction of gene transcription by nucleotide analogs. *Biochimie* **56**:703–710.

Darnell, J. E., Jelinek, W. R., and Molloy, G. R. 1973. Biogenesis of mRNA: Genetic regulation in mammalian cells. *Science* **181**:1215–1221.

Davidson, J. N. 1972. *The Biochemistry of the Nucleic Acids. 7th Ed.*, New York, Academic Press.

Dennis, P. P., and Nomura, M. 1975. Stringent control of the transcriptional activities of ribosomal protein genes in *E. coli. Nature* **255**:460–465.

Desrosiers, R., Friderici, K., and Rottman, F. 1974. Identification of methylated nucleosides in mRNA from Novikoff hepatoma cells. *Proc. Nat. Acad. Sci. USA* **71**:3971–3975.

Deutscher, M. P. 1973. Synthesis and functions of the -C-C-A terminus of tRNA. *Progr. Nucleic Acid Res. Mol. Biol.* **13**:51–92.

Deutscher, M. P., Foulds, J., and McClain, W. H. 1974. Transfer RNA nucleotidyltransferase plays an essential role in the normal growth of *E. coli* and in the biosynthesis of some bacteriophage T4 tRNAs. *J. Biol. Chem.* **249**:6696–6699.

Dhar, R., Weissman, S. M., Zain, B. S., Pan, J., and Lewis, A. M. 1974. Nucleotide sequence preceding an RNA polymerase initiation site on SV_{40} DNA. *Nucleic Acid Res.* **1**:595–613.

Dijk, J., and Singhal, R. P. 1974. Precursor molecules of tRNAs in *E. coli. J. Biol. Chem.* **249**:645–648.

Downey, K. M., Byrnes, J. J., Jurmark, B. S., and So, A. G. 1973. Reticulocyte RNA-dependent RNA polymerase. *Proc. Nat. Acad. Sci. USA* **70**:3400–3404.

Dunn, J. J., and Studier, F. W. 1973a. T7 early RNAs are generated by site-specific cleavages. *Proc. Nat. Acad. Sci. USA* **70**:1559–1563.

Dunn, J. J., and Studier, F. W. 1973b. T7 early RNAs and *E. coli* rRNAs are cut from large precursor RNAs *in vivo* by ribonuclease III. *Proc. Nat. Acad. Sci. USA* **70**:3296–3300.

Earp, H. S. 1974. Glucocorticoid regulation of transcription. *Biochim. Biophys. Acta* **340**:95–107.

Eason, R., and Smellie, R. M. S. 1964. Observation on the biosynthesis of polyribonucleotides *in vitro. Biochem. J.* **94**:7P.

Edlin, G., and Broda, P. 1968. Physiology and genetics of the "RNA control" locus in *E. coli. Bacteriol. Rev.* **32**:206–226.

Edmonds, M., and Caramela, M. G. 1969. The isolation and characterization of adenosine monophosphate-rich polynucleotides synthesized by Ehrlich ascites cells. *J. Biol. Chem.* **244**:1314–1324.

Edmonds, M., Vaughan, M. H., and Nakazato, H. 1971. Poly(A) sequences in hnRNA and rapidly-labeled polyribosomal RNA of HeLa cells. *Proc. Nat. Acad. Sci. USA* **68**:1336–1340.

Eström, J. E., and Daneholt, B. 1967. Sedimentation properties of the newly synthesized RNA from isolated nuclear components of *Chironomus tentans* salivary gland cells. *J. Mol. Biol.* **28**:331–343.

Eikhom, T. S., and Spiegelman, S. 1967. The dissociation of Qβ-replicase and the relation of one of the components to a poly-C-dependent poly-G-polymerase. *Proc. Nat. Acad. Sci. USA* **57**:1833–1840.

Faiferman, I., and Pogo, A. O. 1975. Isolation of a ribonucleoprotein network that contains hnRNA and is bound to the nuclear membrane. *Biochemistry* **14**:3808–3816.

Farashyan, V. R., Ryskov, A. P., and Georgiev, G. P. 1973. Short poly(A) sequences at the 3′-end of nuclear D-RNAs. *Molecular Biol.* **7**:362–371.

Firtel, R. A., and Lodish, H. F. 1973. A small nuclear precursor of mRNA in the cellular slime mold *Dictyostelium discoideum. J. Mol. Biol.* **79**:295–314.

Firtel, R. A., and Pederson, T. 1975. Ribonucleoprotein particles containing hnRNA in the cellular slime mold *Dictyostelium discoideum. Proc. Nat. Acad. Sci. USA* **72**:301–305.

Flint, S. J., de Pomerai, D. I., Chesterton, C. J., and Butterworth, P. H. W. 1974. Template specificity of eucaryotic DNA dependent RNA polymerases. *Eur. J. Biochem.* **42**:567–579.

Ford, P. J., and Southern, E. M. 1973. Different sequences for 5 S RNA in kidney cells and ovaries of *Xenopus laevis. Nature New Biol.* **241**:7–12.

Fouquet, H., and Sauer, H. W. 1975. Variable redundancy in RNA transcripts isolated in S and G_2 phase of the cell cycle of *Physarum. Nature* **255**:253–255.

Fox, S. W. 1960. How did life begin? *Science* **132**:200–208.

Fox, S. W. 1971. Self-assembly of the protocell from a self-ordered polymer. *In*: Kimball, A. P., and J. Oró, eds., *Prebiotic and Biochemical Evolution,* Amsterdam, North-Holland Publishing Co., p. 8–30.

Fox, S. W., and Dose, K. 1972. *Molecular Evolution and the Origin of Life,* San Francisco, W. H. Freeman and Company.

Fox, S. W., Harada, K., and Kendrick, J. 1959. Production of spherules from synthetic proteinoid and hot water. *Science* **129**:1221–1223.

Fox, S. W., and Yuyama, S. 1963. Abiotic production of primitive protein and formed microparticles. *Ann. N.Y. Acad. Sci.* **108**:487–494.

Fromson, D., and Duchastel, A. 1975. Poly(A)-containing polyribosomal RNA in sea urchin embryos: Changes in proportion during development. *Biochim. Biophys. Acta* **378**:394–404.

Fujisawa, T., Abe, S., Kawada, T., Satake, M., and Ogata, K. 1973a. Studies on the processing of 45-S RNA in rat liver nucleolus, with specific reference to 29.5-S RNA. *Biochim. Biophys. Acta* **324**:226–240.

Fujisawa, T., Abe, S., Satake, M., and Ogata, K. 1973b. Conversion of rat liver nucleolar 29.5-S RNA to 28-S RNA *in vitro*. *Biochim. Biophys. Acta* **324**:241–253.

Furth, J. J., and Austin, G. E. 1970. RNA polymerase of lymphoid tissue. *Cold Spring Harbor Symp. Quant. Biol.* **35**:641–648.

Geiduschek, E. P., Brody, E. N., and Wilson, D. L. 1968. Some aspects of RNA transcription. *In*: Pullman, B., ed., *Molecular Associations in Biology,* New York, Academic Press, p. 163–182.

Ghysen, A., and Celis, J. E. 1974. Joint transcription of two $tRNA_1^{Tyr}$ genes from *E. coli*. *Nature* **249**:418–421.

Gill, D. M. 1967. Incorporation of (^{14}C) arginine into rat liver proteins catalyzed by soluble enzymes only. *Biochim. Biophys. Acta* **145**:792–805.

Gissinger, F., Kedinger, C., and Chambon, P. 1974. General enzymatic properties of purified calf thymus RNA polymerases AI & B. *Biochimie* **56**:319–333.

Giudice, G., Pirrone, A. M., Roccheri, M., and Trapani, M. 1973. Maturational cleavage of nucleolar rRNA precursor can be catalyzed by nonspecific endonuclease. *Biochim. Biophys. Acta* **319**:72–80.

Goldberger, R. F. 1974. Autogenous regulation of gene expression. *Science* **183**:810–816.

Gorini, L. 1970. Informational suppression. *Ann. Rev. Genetics* **4**:107–134.

Grankowski, N., Kudlicki, W., and Gascor, E. 1974. Ribosome-associated protein kinase from *S. cerevisiae*. *FEBS Lett.* **47**:103–106.

Greene, M., and Cartas, M. 1972. The genome of RNA tumor viruses contains poly(A) sequences. *Proc. Nat. Acad. Sci. USA* **69**:791–794.

Greenberg, J. R., and Perry, R. P. 1972. Relative occurrence of poly(A) sequences in messenger and hnRNA of L cells as determined by poly(U)-hydroxylapatite chromatography. *J. Mol. Biol.* **72**:91–98.

Grienenberger, J. M., and Simon, D. 1975. Structure and biosynthesis of rRNAs from the oncogenic bacterium *Agrobacterium tumefasciens*. *Biochem. J.* **149**:23–30.

Grierson, D. 1974. Characterisation of RNA components from leaves of *Phaseolus aureus*. *Eur. J. Biochem.* **44**:509–515.

Grierson, D., and Loening, U. 1974. Ribosomal RNA precursors and the synthesis of chloroplast and cytoplasmic rRNA in leaves of *Phaseolus aureus*. *Eur. J. Biochem.* **44**:501–507.

Gross, R. H., and Beer, M. 1975. The RNA polymerases from *Drosophila melanogaster*. *Biochemistry* **14**:4024–4031.

Grummt, I. 1975. Synthesis of RNA molecules larger than 45 S by isolated rat liver nucleoli. *Eur. J. Biochem.,* **57**:159–167.

Gurley, W. B., Lin, C. -Y., Guilfoyle, T. J., and Nagao, R. T. 1976. Analysis of plant RNA polymerase I transcript in chromatin and nuclei. *Biochim. Biophys. Acta.* **425**:168–174.

Guthrie, C. 1975. The nucleotide sequences of the dimeric precursor to glutamine and leucine tRNAs coded by bacteriophage T4. *J. Mol. Biol.* **95**:529–547.

Guthrie, C., Seidman, J. G., Altman, S., Barrell, B. G., Smith, J. D., and McClain, W. H. 1973. Identification of tRNA precursor molecules made by phage T4. *Nature New Biol.* **246**:6–11.

Hackett, P. B., and Sauerbier, W. 1974. Radiological mapping of the rRNA transcription unit in *E. coli*. *Nature* **251**:639–641.

Hackett, P. B., and Sauerbier, W. 1975. The transcriptional organization of the rRNA genes in mouse L cells. *J. Mol. Biol.* **91**:235–256.

Hager, G. L., Holland, M. J., and Rutter, W. J. 1977. Isolation of RNA polymerases I, II, and III from *S. cerevisiae*. *Biochemistry* **16**:1–8.

Hames, B. D., and Perry, R. P. 1977. Homology relationship between the mRNA and hnRNA of mouse L cells. *J. Mol. Biol.* **109**:437–453.

Hanson, E. D. 1966. Evolution of the cell from primordial living systems. *Quart. Rev. Biol.* **41**:1–12.

Harris, H. 1974. *Nucleus and Cytoplasm. 3rd Ed.*, Oxford, Clarendon Press.

Hayashi, M., Hayashi, M. N., and Spiegelman, S. 1963a. Restriction of an *in vivo* genetic transcription to one of the complementary strands of DNA. *Proc. Nat. Acad. Sci. USA* **50**:664–672.

Hayashi, M., Hayashi, M. N., and Spiegelman, S. 1963b. Replicating form of a single-stranded DNA virus: Isolation and properties. *Science* **140**:1313–1316.

Hayes, F., and Vasseur, M. 1974. *In vitro* maturation of a 16 S RNA precursor. *FEBS Lett.* **46**:364–367.

Hayes, F., and Vasseur, M. 1976. Processing of the 17-S *E. coli* precursor RNA in the 27-S preribosomal particle. *Eur. J. Biochem.* **61**:433–442.

Hayes, F., Vasseur, M., Nikolaev, N., Schlessinger, D., Widada, J. S., Krol, A., and Branlant, C. 1975. Structure of a 30 S pre-ribosomal RNA of *E. coli*. *FEBS Lett.* **56**:85–91.

Hecht, N., and Woese, C. 1968. Separation of bacterial rRNA from its macromolecular precursors by polyacrylamide gel electrophoresis. *J. Bact.* **95**:986–990.

Helser, T. L., Davies, J. E., and Dahlberg, J. E. 1972. Mechanism of kasugamycin resistance in *E. coli*. *Nature New Biol.* **235**:6–9.

Hemminki, K. 1974. Poly(A) in RNA extracted by thermal phenol fractionation from chick embryo brain and liver. *Biochim. Biophys. Acta* **340**:262–268.

Hercules, K., Schweiger, M., and Sauerbier, W. 1974. Cleavage by RNase III converts T3 and T7 early precursor RNA into translatable message. *Proc. Nat. Acad. Sci. USA* **71**:840–844.

Higashinakagawa, T., and Muramatsu, M. 1974. Ribosome precursor particles in the nucleolus of rat liver. *Eur. J. Biochem.* **42**:245–258.

Hirsch, M., and Penman, S. 1974. Post-transcriptional addition of poly(A) to mitochondrial RNA by a cordycepin-insensitive process. *J. Mol. Biol.* **83**:131–142.

Hoffman, D. J., and Niyogi, S. K. 1973. RNA initiation with dinucleoside monophosphates during transcription of bacteriophage T4 DNA with RNA polymerase of *E. coli*. *Proc. Nat. Acad. Sci. USA* **70**:574–578.

Holland, M. J., Hager, G. L., and Rutter, W. J. 1977. Selective transcription of ribosomal genes by RNA polymerase I. *Biochemistry* **16**:16–24.

Honjo, T., and Reeder, R. H. 1973. Preferential transcription of *Xenopus laevis* rRNA in interspecies hybrids between *Xenopus laevis* and *X. mulleri*. *J. Mol. Biol.* **80**:217–228.

Honjo, T., and Reeder, R. H. 1974. Transcription of *Xenopus* chromatin by homologous RNA polymerase. *Biochemistry* **13**:1896–1899.

Horinishi, H., Hashizume, S., Seguchi, M., and Takahashi, K. 1975. Incorporation of methionine by a soluble enzyme system from *E. coli*. *Biochem. Biophys. Res. Comm.* **67**:1136–1143.

Huet, J., Degélée, S., Iboua, F., Buhler, J.-M., Sentenac, A., and Fromageot, P. 1976. Further characterization of yeast RNA polymerases. Effect of subunits removal. *Biochimie* **58**:71–80.

Hurwitz, J., Furth, J. J., Anders, M., and Evans, A. 1962. The role of DNA in RNA synthesis. *J. Biol. Chem.* **237**:3752–3759.

Ingwall, J. S., Weiner, C. D., Morales, M. F., Davis, E., and Stockdale, F. E. 1974. Specificity of creatine in the control of muscle protein synthesis. *J. Cell Biol.* **63**:145–151.

Ishikawa, K., Sato, T., Sato, S., and Ogata, K. 1974. RNP complexes containing nascent DNA-like RNA in the crude chromatin fraction of rat liver. *Biochim. Biophys. Acta* **357**:420–437.

Issinger, O. G., and Traut, R. R. 1974. Selective phosphorylation from GTP of proteins L7 and

L12 of *E. coli* 50 S ribosomes by a protein kinase from rabbit reticulocytes. *Biochem. Biophys. Res. Comm.* **59**:829–836.

Jacob, S. T. 1973. Mammalian RNA polymerases. *Nucleic Acid Res. Mol. Biol.* **13**:93–126.

Jacobson, A., Firtel, R. A., and Lodish, H. F., 1974. Synthesis of messenger and ribosomal RNA precursors in isolated nuclei of the cellular slime mold, *Dictyostelium discoideum. J. Mol. Biol.* **82**:213–230.

Jaehning, J. A., Stewart, C. C., and Roeder, R. G. 1975. DNA-dependent RNA polymerase levels during the response of human peripheral lymphocytes to phytohaemogglutinin. *Cell* **4**:51–58.

Jänne, O., Bardin, C. W., and Jacob, S. T. 1975. Effect of polyamines on the *in vitro* transcription of DNA and chromatin. *Biochemistry* **14**:3589–3597.

Jantzen, H. 1974. Polyadenylasäure-enthaltende RNA und Genaklivitätsmuster während der Entwicklung von *Acanthamoeba castellanii. Biochim. Biophys. Acta* **374**:38–51.

Jay, G., and Kaempfer, R. 1974. Sequence of events in initiation of translation: A role for initiator transfer RNA in the recognition of messenger RNA. *Proc. Nat. Acad. Sci. USA* **71**:3199–3203.

Jeffrey, W. R., and Brawerman, G. 1974. Characterization of the steady-state population of messenger RNA and its poly(A) segment in mammalian cells. *Biochemistry* **13**:4633–4637.

Jones, K. W. 1965. The role of the nucleolus in the formation of ribosomes. *J. Ultrastruct. Res.* **13**:257–262.

Jones, K. W., and Truman, D. E. S., 1964. A hypothesis for DNA transcription and mRNA synthesis *in vivo. Nature* **202**:1264–1267.

Jordan, B. R., Forget, B. G., and Monier, R. 1971. A low molecular weight RNA synthesized by *E. coli* in the presence of chloramphenicol. *J. Mol. Biol.* **55**:407–421.

Jordan, B. R., Jourdan, R. and Jacq, B. 1976. Late steps in the maturation of 26 S ribosomal RNA: Generation of 5.8 S and 2 S RNAs by cleavages occurring in the cytoplasm. *J. Mol. Biol.* **101**:85–105.

Kaback, D. B., Bhargava, M. M., and Halvorson, H. O. 1973. Location and arrangement of genes coding for ribosomal RNA in *S. cerevisiae. J. Mol. Biol.* **79**:735–739.

Kabat, D. 1970. Phosphorylation of ribosomal proteins in rabbit reticulocytes. *Biochemistry* **9**:4160–4175.

Kabat, D. 1971. Phosphorylation of ribosomal proteins in rabbit reticulocytes. A cell-free system with ribosomal protein kinase activity. *Biochemistry* **10**:197–203.

Kaji, A., Kaji, H., and Novelli, G. D. 1963. A soluble amino-acid incorporating system. *Biochem. Biophys. Res. Comm.* **10**:406–409.

Kaji, A., Kaji, H., and Novelli, G. D. 1965. Soluble amino acid-incorporating system. I. Preparation of the system and nature of the reaction. *J. Biol. Chem.* **240**:1185–1191.

Kaji, H., Novelli, G. D., and Kaji, A. 1963. A soluble amino acid-incorporating system from rat liver. *Biochim. Biophys. Acta* **76**:474–477.

Kapitza, E. L., Stukacheva, E. A., and Shemyakin, M. F. 1976. The effect of the termination Rho factor and ribonuclease III on the transcription of bacteriophage ϕX174 DNA *in vitro. FEBS Lett.* **64**:81–84.

Kedinger, C., Gissinger, F., and Chambon, P. 1974. Molecular structures and immunological properties of calf-thymus enzyme AI and of calf-thymus and rat-liver enzymes B. *Eur. J. Biochem.* **44**:421–436.

Kedinger, C., Nuret, P., and Chambon, P. 1971. Structural evidence for two α-amantin sensitive RNA polymerases in calf thymus. *FEBS Lett.* **15**:169–174.

Keller, W., and Goor, R. 1970. Mammalian RNA polymerases: Structural and functional properties. *Cold Spring Harbor Symp. Quant. Biol.* **35**:671–680.

Kemper, B., and Haberner, J. F. 1974. Non-ribosomal incorporation into a specific protein by a cell-free extract of parathyroid tissue. *Biochim. Biophys. Acta* **349**:235–239.

Kerjan, P., and Szulmajster, J. 1974. Intracellular ribonuclease activity in stationary phase cells of *B. subtilis*. *Biochem. Biophys. Res. Comm.* **59**:1079–1087.

Kierszenbaum, A. L., and Tres, L. T. 1974. Nucleolar and perichromosomal RNA synthesis during meiotic prophase in the mouse testis. *J. Cell Biol.* **60**:39–53.

Kiss, A., Sain, B., and Venetianer, P. 1977. The number of rRNA genes in *E. coli*. *FEBS Lett.* **79**:77–79.

Klagsbrun, M. 1973. An evolutionary study of the methylation of tRNA and rRNA into nucleic acid in prokaryote and eukaryote organisms. *J. Biol. Chem.* **248**:2612–2620.

Klee, C. B. 1967. Structural alterations of polynucleotide phosphorylase leading to primer dependence. *J. Biol. Chem.* **242**:3579–3580.

Klee, C. B. 1969. The proteolytic conversion of polynucleotide phosphorylase to a primer-dependent form. *J. Biol. Chem.* **244**:2558–2566.

Klootwijk, J., and Planta, R. J. 1973a. Analysis of the methylation sites in yeast ribosomal RNA. *Eur. J. Biochem.* **39**:325–333.

Klootwijk, J., and Planta, R. J. 1973b. Modified sequence in yeast ribosomal RNA. *Mol. Biol. Reports* **1**:187–191.

Knöchel, W., and Tiedemann, H. 1975. Size distribution and cell-free translation of globin-coding hnRNA from avaian erythroblasts. *Biochim. Biophys. Acta* **378**:383–393.

Kolata, G. B. 1974a. Control of protein synthesis (I): poly(A) in the cytoplasm. *Science* **185**:517–518.

Kolata, G. B. 1974b. Control of protein synthesis (II): RNA in the nucleus. *Science* **185**:603–604.

Kolata, G. B. 1977. Overlapping genes: More than anomalies? *Science* **196**:1187–1188.

Kossman, C. R., Stamato, T. D., and Pettijohn, D. E. 1971. Tandem snythesis of the 16 S and 23 S rRNA sequences of *E. coli*. *Nature New Biol.* **234**:102–104.

Krakow, J. S., and Ochoa, S. 1963. RNA polymerase of *Azotobacter vinelandii*. *Proc. Nat. Acad. Sci. USA* **49**:88–94.

Kumar, A., and Wu, R. S. 1973. Role of rRNA transcription in ribosome processing in HeLa cells. *J. Mol. Biol.* **80**:265–276.

Kurek, E., Grankonski, N., and Gasior, E. 1972a. On the phosphorylation of *E. coli* ribosomes I. An *in vivo* labeling of ribosomes. *Acta Microbiol. Polonica* **A4**:171–176.

Kurek, E., Grankonski, N., and Gasior, E. 1972b. On the phosphorylation of *E. coli*· ribosomes. II. Reaction in cell-free system. *Acta Microbiol. Polonica* **A4**:177–183.

Lacey, J. C., and Mullins, D. W. 1972. Proteins and nucleic acids in prebiotic evolution. *In:* Rohlfing, D. L., and A. I. Oparin, eds., *Molecular Evolution: Prebiological and Biological*, New York, Plenum Press, p. 171–188.

Lacey, J. C., and Pruitt, K. M. 1969. Origin of the genetic code. *Nature* **223**:799–804.

Landy, A., Foeller, C., and Ross, W. 1974. DNA fragments carrying genes for $tRNA_1^{Tyr}$. *Nature* **249**:738–742.

Lavi, S., and Shatkin, A. J. 1975. Methylated simian virus 40-specific RNA from nuclei and cytoplasm of infected BSC-1 cells. *Proc. Nat. Acad. Sci. USA* **72**:2012–2016.

Leaver, C. J., and Key, J. L. 1970. Ribosomal RNA synthesis in plants. *J. Mol. Biol.* **49**:671–680.

Lee, S. Y., Mendecki, J., and Brawerman, G. 1971. A polynucleotide segment rich in adenylic acid in the rapidly labeled polyribosomal RNA component of mouse sarcoma 180 ascites cells. *Proc. Nat. Acad. Sci. USA* **68**:1331–1335.

Leibowitz, M. J., and Soffer, R. L. 1969. A soluble enzyme from *E. coli* which catalyzes the transfer of leucine and phenylalanine from tRNA to acceptor proteins. *Biochem. Biophys. Res. Comm.* **36**:47–53.

Letendre, C. H., and Singer, M. F. 1974. Studies on primer-independent polynucleotide phosphorylase of *Micrococcus luteus*. *J. Biol. Chem.* **249**:7383–7389.

Levitan, I. B., and Webb, T. E. 1970. Posttranscriptional control in the steroid-mediated induction of hepatic tyrosine transaminase. *Science* **167**:283–285.

Lewin, B. M. 1970. *The Molecular Basis of Gene Expression,* New York, Wiley–Interscience.

Liau, M. C., Craig, N. C., and Perry, R. P. 1968. Factors which influence the ability of isolated nucleoli to process 45-S RNA. *Biochim. Biophys. Acta* **169**:196–205.

Liebl, V., and Leiblova, J. 1968. Coacervate systems and life. *J. Brit. Interplan. Soc.* **21**:295–312.

Lim, L., and Canellakis, E. S. 1970. Adenine-rich polymer associated with rabbit reticulocyte mRNA. *Nature* **227**:710–712.

Lindahl, L. 1975. Intermediates and time kinetics of the *in vivo* assembly of *E. coli* ribosomes. *J. Mol. Biol.* **92**:15–37.

Lindell, T. J., Weinberg, F., Morris, P. W., Roeder, R. G., and Rutter, W. J. 1970. Specific inhibition of nuclear RNA polymerase II by α-amantin. *Science* **170**:447–448.

Link, G., and Richter, G. 1975. Properties and subunit composition of RNA polymerase II from plant cell cultures. *Biochim. Biophys. Acta* **395**:337–346.

Littauer, U. Z., and Inouye, H. 1973. Regulation of tRNA. *Ann. Rev. Biochem.* **42**:439–471.

Loeb, J. F., and Blat, C. 1970. Phosphorylation of some rat liver ribosomal proteins and its activation by cyclic AMP. *FEBS Lett.* **10**:105–108.

Loening, U. E. 1972. Ribosomal RNA in evolution. *Biochem. J.* **129**:35*p*.

Lowery-Goldhammer, C., and Richardson, J. P. 1974. An RNA-dependent nucleoside triphosphate phosphohydrolase (ATPase) associated with rho termination factor. *Proc. Nat. Acad. Sci. USA* **71**:2003–2007.

Lowry, C. V., and Dahlberg, J. E. 1971. Structural differences between the 16 S rRNA of *E. coli* and its precursor. *Nature New Biol.* **232**:52–54.

Lukanidin, E. M., Zalmanzon, E. S., Komaromi, L., Samarina, O. P., and Georgiev, G. P. 1972. Structure and function of informofers. *Nature New Biol.* **238**:193–197.

Luria, S. E. 1965. Asymmetrical transcription of T4 phage DNA by purified RNA polymerase. *Biochem. Biophys. Res. Comm.* **18**:735–742.

Maden, B. E. H. 1968. Ribosome formation in animal cells. *Nature* **219**:685–688.

Maden, B. E. H., and Salim, M. 1974. The methylated nucleotide sequences in HeLa cell rRNA and its precursors. *J. Mol. Biol.* **88**:133–164.

Maden, B. E. H., and Tartof, K. 1974. Nature of the rRNA transcribed from the X and Y chromosomes of *D. melanogaster. J. Mol. Biol.* **90**:51–64.

Maio, J. J., and Kurnit, D. M. 1974. Transcription of mammalian satellite DNAs by homologous DNA-dependent RNA polymerases. *Biochim. Biophys. Acta* **349**:305–319.

Mangiarotti, G., Apirion, D., Schlessinger, D., and Silengo, L. 1968. Biosynthetic precursors of 30 S and 50 S ribosomal particles in *E. coli. Biochemistry* **7**:456–472.

Mangiarotti, G., Turco, E., Pongetto, A., and Altruda, F. 1974. Precursor 16 S RNA in active 30 S ribosomes. *Nature* **247**:147–148.

Marmur, J., and Greenspan, C. M. 1963. Transcription *in vivo* of DNA from bacteriophage SP8. *Science* **142**:387–389.

Martin, F., and Brachet, J. 1959. Autoradiographic studies on the incorporation of amino acids into spermatozoa. *Exp. Cell Res.* **17**:399–404.

Marzluff, W. F., Murphy, E. C., and Huang, R. C. C. 1974. Transcription of the genes for 5 S rRNA and tRNA in isolated mouse myeloma cell nuclei. *Biochemistry* **13**:3689–3696.

Matzura, H. 1973. Joint propagation of the β and β' subunits of RNA polymerase in *E. coli. Nature New Biol.* **244**:262–264.

Mayo, V. S., and DeKloet, S. R. 1971. Disaggregation of "giant" heterogeneous nuclear RNA of mouse Ehrlich ascites cells by thermal denaturation in the presence of formaldehyde. *Biochim. Biophys. Acta* **247**:74–79.

McClain, W. H., Barrell, B. G., and Seidman, J. G. 1975. Nucleotide alterations in bacteriophage T4 serine tRNA that affect the conversion of precursor RNA into tRNA. *J. Mol. Biol.* **99**:717–732.

McClain, W. H., Guthrie, C., and Barrell, B. G. 1972. Eight tRNAs induced by infection of *E. coli* with phage T4. *Proc. Nat. Acad. Sci. USA* **69**:3703–3707.

Melli, M., Ginelli, E., Corneo, G., and di Lernia, R. 1975. Clustering of the DNA sequences complementary to repetitive nuclear RNA of HeLa cells. *J. Mol. Biol.* **93**:23–38.

Meyhack, B., Meyhack, I., and Apirion, D. 1974. Processing of precursor particles containing 17 S rRNA in a cell free system. *FEBS Lett.* **49**:215–219.

Milanino, R., and Chargoff, E. 1973. A purine polyribonucleotide synthetase from *E. coli. Proc. Nat. Acad. Sci. USA* **70**:2558–2562.

Miller, D. L., and Hamkalo, B. A. 1972. Visualization of RNA synthesis on chromosomes. *Int. Rev. Cyt.* **33**:1–25.

Mills, D. R., Peterson, R. L. and Spiegelman, S. 1967. An extracellular Darwinian experiment with a self-duplicating nucleic acid molecule. *Proc. Nat. Acad. Sci. USA* **58**:217–224.

Monjardine, J. P., and Crawford L. V. 1970. RNA polymerase from mouse embryo cells. *Cold Spring Harbor Symp. Quant. Biol.* **35**:659–662.

Morgan, T. H. 1914. No crossing over in the male of *Drosophila* of genes in the second and third pairs of chromosomes. *Biol. Bull.* **26**:195–204.

Moses, R. E., and Singer, M. F. 1970. Studies on the polymerization reaction catalyzed by primer-dependent and primer-independent enzymes. *J. Biol. Chem.* **245**:2414–2422.

Mosmose, K., and Kaji, A. 1966. Soluble amino acid-incorporating system. III. Further studies on the product and its relation to the ribosomal system for incorporation. *J. Biol. Chem.* **241**:3294–3307.

Mueller, H. J. 1955. Life. *Science* **121**:1–9.

Müller, W. E. G., Totsuka, A., Kroll, M., Nusser, I., and Zahn, R. K. 1975. Poly(A) polymerase in quail oviduct: Changes during estrogen induction. *Biochim. Biophys. Acta* **383**:147–159.

Müller, W. E. G., Totsuka, A., and Zahn, R. K. 1974. Association of an estradiol receptor with the DNA-dependent RNA polymerase I from immature quail. *Biochim. Biophys. Acta* **366**:224–233.

Muramatsu, M., and Fujisawa, T. 1968. Methylation of rRNA precursor and tRNA in rat liver. *Biochim. Biophys. Acta* **157**:476–492.

Naito, S., and Ishihama, A. 1975. Isolation and properties of the transcription complex of *E. coli* RNA polysomes. *Biochim. Biophys. Acta* **402**:88–104.

Nakanishi, S., Adhya, S., Gottesman, M., and Pastan, I. 1974. Activation of transcription at specific promoters by glycerol. *J. Biol. Chem.* **247**:4050–4056.

Needham, A. E. 1959. The origination of life. *Quart. Rev. Biol.* **34**:189–209.

Nierlich, D. P., Lamfrom, H., Sarabhai, A., and Abelson, J. 1973. Transfer RNA synthesis *in vitro. Proc. Nat. Acad. Sci. USA* **70**:179–182.

Niessing, J., and Sekeris, C. 1972. A homoribopolynucleotide synthetase in rat liver nuclei associated with RNP particles containing DNA-like RNA. *FEBS Lett.* **22**:83–88.

Nikolaev, N., and Schlessinger, D. 1974. Binding of ribosomal proteins to 30 S preribosomal RNA of *E. coli. Biochemistry* **13**:4272–4278.

Nikolaev, N., Schlessinger, D., and Wellauer, P. K. 1974. 30 S pre-rRNA of *E. coli* and products of cleavage by ribonuclease III. *J. Mol. Biol.* **86**:741–747.

Nikolaev, N., Silengo, L., and Schlessinger, D. 1973. A role for ribonuclease III in processing of rRNA and mRNA precursors in *E. coli. J. Biol. Chem.* **248**:7967–7969.

Olson, M. O. J., Prestayko, A. W., Jones, C. E., and Busch, H. 1974. Phosphorylation of proteins of ribosomes and nucleolar preribosomal particles from Novikoff hepatoma ascites cells *J. Mol. Biol.* **90**:161–168.

Oparin, A. I. 1957. *The Origin of Life on the Earth. 3rd Ed.*, New York, Academic Press.

Oparin, A. I. 1968. *Genesis and Evolutionary Development of Life.* New York, Academic Press.

Oparin, A. I. 1971. Coacervate drops as models of prebiological systems. *In*: Kimball, A. P., and J. Oró, eds., *Prebiotic and Biochemical Evolution,* Amsterdam, North-Holland Publishing Co., p. 1–7.

Orgel, L. E. 1968. Evolution of the genetic apparatus. *J. Mol. Biol.* **38**:381–394.

Pace, B., Peterson, R., and Pace, N. 1970. Formation of all stable RNA species in *E. coli* by post-transcriptional modification. *Proc. Nat. Acad. Sci. USA* **65**:1097–1104.

Pace, N. R. 1973. Structure and synthesis of the rRNA of prokaryotes. *Bact. Rev.* **37**:562–603.

Pace, N. R., Pato, M. L., McKibbin, J., and Radcliffe, C. W. 1973. Precursors of 5 S rRNA in *B. subtilis*. *J. Mol. Biol.* **75**:619–631.

Pardue, N. L., Brown, D. D., and Birnstiel, M. L. 1973. Location of the genes for 5 S rRNA in *Xenopus laevis*. *Chromosoma* **42**:191–203.

Parker, C. S., and Roeder, R. C. 1977. Selective and accurate transcription of *Xenopus laevis* 5 S RNA genes in isolated chromatin by purified RNA polymerase III. *Proc. Nat. Acad. Sci. USA* **74**:44–48.

Pastan, I., and Perlman, R. L. 1972. Regulation of gene transcription in *E. coli* by cyclic AMP. *In*: Greengard, P., G. A. Robison, and R. Paoletti, eds., *Advances in Cyclic Nucleotide Research,* New York, Raven Press, p. 11–16.

Paul, J., Gilmur, R. S., and Thomou, T. 1970. Organ-specificity of transcription from mammalian chromatin. *In*: Ochoa, S., C. F. Heredia, C. Asensio, and D. Nachmansohn, eds., *Macromolecules: Biosynthesis and Function,* New York, Academic Press, p. 237–242.

Pederson, T. 1974. Proteins associated with hnRNA in eukaryotic cells. *J. Mol. Biol.* **83**:163–183.

Perry, R. P., Cheng, T. Y., Freed, J. J., Greenberg, J. R., Kelley, D. E., and Tartof, K. D. 1970a. Evolution of the transcription unit of rRNA. *Proc. Nat. Acad. Sci. USA* **65**:609–616.

Perry, R. P., Greenberg, J. R., and Tartof, K. D. 1970b. Transcription of rRNA, hnRNA, and mRNA in eucaryotes. *Cold Spring Harbor Symp. Quant. Biol.* **35**:577–587.

Perry, R. P. and Kelley, D. E. 1974. Existence of methylated mRNA in mouse L cells. *Cell* **1**:37–42.

Perry, R. P., and Scherrer, K. 1975. The methylated constituents of globin mRNA. *FEBS Lett.* **57**:73–78.

Petranyi, P., Jendrisak, J. J., and Burgess, R. R. 1977. RNA polymerase II from wheat germ contains tightly bound zinc. *Biochem. Biophys. Res. Comm.* **74**:1031–1038.

Pettijohn, D. E. 1972. Ordered and preferential initiation of rRNA synthesis *in vivo*. *Nature New Biol.* **235**:204–206.

Philipson, L., Wall, R., Glukman, G., and Darnell, J. E. 1971. Addition of poly(A) to virus-specific RNA during adenovirus replication. *Proc. Nat. Acad. Sci. USA* **68**:2806–2809.

Prescott, D. M., Bostock, C., Gamow, E., and Lauth, M. 1971. Characterization of rapidly labeled RNA in *Tetrahymena pyriformis*. *Exp. Cell Res.* **67**:124–128.

Prestayko, A. W., Klomp, G. R., Schmoll, D. J., and Busch, H. 1974a. Comparison of proteins of ribosomal subunits and nucleolar preribosomal particles from Novikoff hepatoma ascites cells by two-dimensional polyacrylamide gel electrophoresis. *Biochemistry* **13**:1945–1951.

Prestayko, A. W., Olson, M. O. J., and Busch, H. 1974b. Phosphorylation of proteins of ribosomes and nucleolar preribosomal particles *in vivo* in Novikoff hepatoma ascites cells. *FEBS Lett.* **44**:131–135.

Proudfoot, N. J., and Brownlee, G. G. 1974. Nucleotide sequence adjacent to poly(A) in globin mRNA. *FEBS Lett.* **38**:179–183.

Quincey, R. V., and Wilson, S. H. 1969. The utilization of genes for rRNA, 5 S RNA, and tRNA in liver cells of adult rats. *Proc. Nat. Acad. Sci. USA* **64**:981–988.

Quinlan, T. J., Billings, P. B., and Martin, T. E. 1974. Nuclear RNP complexes containing poly(A) from mouse ascites cells. *Proc. Nat. Acad. Sci. USA* **71**:2632–2636.

Racevskis, J., and Webb, T. E. 1974. Processing and release of rRNA from isolated nuclei. *Eur. J. Biochem.* **49**:93–100.

Reeder, R. H., and Roeder, R. G. 1972. Ribosomal RNA synthesis in isolated nuclei. *J. Mol. Biol.* **67**:433–441.

Retèl, J., and Planta, R. J. 1970. On the mechanism of the biosynthesis of rRNA in yeast. *Biochim. Biophys. Acta* **224**:458–469.

Rether, B., Gangloff, J., and Ebel, J. P. 1974. Replacement of the terminal CCA sequence in yeast-tRNA^{Phe} by several unusual sequences. *Eur. J. Biochem.* **50**:289–295.

Rho, J. H., and Bonner, J. 1961. The site of RNA synthesis in the isolated nucleus. *Proc. Nat. Acad. Sci. USA* **47**:1611–1619.

Richardson, J. P. 1970. Rho factor function in T4 transcription. *Cold Spring Harbor Symp. Quant. Biol.* **35**:127–134.

Richardson, J. P. 1974. Effects of supercoiling on transcription from bacteriophage PM2 DNA. *Biochemistry* **3**:3164–3169.

Richardson, J. P. 1975. Initiation of transcription by *E. coli* RNA polymerase from supercoiled and non-supercoiled bacteriophage PM2 DNA. *J. Mol. Biol.* **91**:477–487.

Richardson, J. P., Grimely, C., and Lowery, C. 1975. Transcription termination factor rho activity is altered in *E. coli* with *suA* gene mutations. *Proc. Nat. Acad. Sci. USA* **72**:1725–1728.

Ritossa, F. M., Atwood, K. C., and Spiegelman, S. 1966. A molecular explanation of the bobbed mutants of *Drosophila* as partial deficiencies of "ribosomal" DNA. *Genetics* **54**:819–834.

Rizzo, A. J., and Webb, T. E. 1972. Regulation of ribosome formation in regenerating rat liver. *Eur. J. Biochem.* **27**:136–144.

Roberts, J. W. 1970. The ρ factor: Termination and anti-termination in λ. *Cold Spring Harbor Symp. Quant. Biol.* **35**:121–126.

Robertson, H. D., Altman, S., and Smith, J. D. 1972. Purification and properties of a specific *E. coli* ribonuclease which cleaves a tyrosine tRNA precursor. *J. Biol. Chem.* **247**:5243–5251.

Roeder, R. G. 1974. Multiple forms of DNA-dependent RNA polymerase in *Xenopus laevis*. *J. Biol. Chem.* **249**:241–248.

Roeder, R. G., and Rutter, W. J. 1969. Multiple forms of DNA-dependent RNA polymerase in eukaryotic organisms. *Nature* **224**:234–237.

Roeder, R. G., and Rutter, W. J. 1970. Specific nucleolar and nucleoplasmic RNA polymerases. *Proc. Nat. Acad. Sci. USA* **65**:675–682.

Rogers, M. E., Loening, U. E., and Fraser, R. S. S. 1970. Ribosomal RNA precursors in plants. *J. Mol. Biol.* **49**:681–692.

Rogerson, A. C., and Ezekiel, D. H. 1974. Decay of RNA synthesis in amino acid-starved *E. coli* after rifampin treatment. *J. Bact.* **117**:987–993.

Roth, J. S. 1964. Biological information in a single strand of DNA. *Nature* **202**:182–183.

Roufa, D. J., and Axelrod, D. 1971. The repeated tRNA genes of animal cells in culture. *Biochim. Biophys. Acta* **254**:429–439.

Roy-Burman, P. 1970. *Analogs of Nucleic Acid Components*, Berlin, Springer Verlag.

Russell, P. J., Hammett, J. R., and Selker, E. U. 1976. *Neurospora crassa* cytoplasmic ribosomes: rRNA synthesis in the wild type. *J. Bact.* **127**:785–793.

Ryan, A. M., and Borek, E. 1971. The relaxed control phenomenon. *Progr. Nucl. Acid Res. Mol. Biol.* **11**:193–228.

Ryskov, A. P., Saunders, G. F., Farashyan, V. R., and Georgiev, G. P. 1973. Double-helical regions in nuclear precursor of mRNA (pre-mRNA). *Biochim. Biophys. Acta* **312**:152–164.

Ryskov, A. P., Yenikolopov, G. N., and Limborska, S. A. 1974. Complementary regions of the nuclear precursor of mRNA. *FEBS Lett.* **47**:98–102.

Sagan, C. 1957. Radiation and the origin of the gene. *Evolution* **11**:40–55.

Saito, K., and Mitsuhashi, S. 1973. RNA-dependent RNA replicase in the immune response. *Jpn. J. Microbiol.* **17**:117–121.

Salim, M., and Maden, B. E. H. 1973. Early and late methylations in HeLa cell ribosome maturation. *Nature* **244**:334–336.

Samarina, O. P., Krichevskaya, A. A., and Georgiev, G. P. 1966. Nuclear RNP particles containing mRNA. *Nature* **210**:1319–1322.

Samarina, O. P., Lukanidin, E. M., Molnar, J., and Georgiev, G. P. 1968. Structural organization of nuclear complexes containing DNA-like RNA. *J. Mol. Biol.* **33**:251–263.

Sanderson, K. E. 1967. Revised linkage map of *Salmonella typhimurium*. *Bacter. Rev.* **31**:354–372.

Scarpulla, R. C., C. E. Deutsch, and R. L. Soffer. 1976. Transfer of methionyl residues by leucyl, phenylalanyl-tRNA-protein transferase. *Biochem. Biophys. Res. Comm.*, **71**:584–589.

Scherrer, K., Latham, H., and Darnell, J. E. 1963. Demonstration of an unstable RNA and of a precursor to rRNA in HeLa cells. *Proc. Nat. Acad. Sci. USA* **49**:240–248.

Scherrer, K., Marcaud, L., Zajdela, F., London, I. M., and Gros, F. 1966. A rapidly labelled, unstable 60 S RNA with messenger properties in duck erythroblasts. *Proc. Nat. Acad. Sci. USA* **56**:1571–1578.

Schmidt, F. J. 1975. A novel function of *E. coli* tRNA nucleotidyl-transferase. *J. Biol. Chem.* **250**:8399–8403.

Schwartz, L. B., Sklar, V. E. F., Jaehning, J. A., Weinmann, R., and Roeder, R. G. 1974. Isolation and partial characterization of the multiple forms of DNA-dependent RNA polymerase in the mouse myeloma, MOPC315. *J. Biol. Chem.* **249**:5889–5897.

Schweiger, A., and Schmidt, D. 1974. Isolation of RNA-binding proteins from rat liver nuclear 30 S-particles. *FEBS Lett.* **41**:17–20.

Scott, N. S. 1973. Ribosomal RNA cistrons in *Euglena gracilis*. *J. Mol. Biol.* **81**:327–336.

Seidman, J. G., and McClain, W. H. 1975. Three steps in conversion of large precursor RNA into serine and proline tRNAs. *Proc. Nat. Acad. Sci. USA* **72**:1491–1495.

Seidman, J. G., Barrell, B. G., and McClain, W. H. 1975. Five steps in the conversion of a large precursor RNA into bacteriophage proline and serine tRNAs. *J. Mol. Biol.* **99**:733–760.

Seifart, K. H., and Benecke, B. J. 1975. DNA-dependent RNA polymerase C. *Eur. J. Biochem.* **53**:293–300.

Seitz, Ur., and Seitz, Ul. 1973. Biosynthese der ribosomalen RNS bei der blaugrünen Algae *Anacystis nidulans*. *Arch. Mikrobiol.* **90**:213–222.

Serfling, E., Maximovsky, L. F., and Wobus, U. 1974. Synthesis and processing of rRNA in salivary general cells of *Chironomus thummi*. *Eur. J. Biochem.* **45**:277–289.

Siegel, R. B., and Summers, W. C. 1970. Control of phage-specific RNA synthesis *in vivo* by early phage genes. *J. Mol. Biol.* **49**:115–123.

Sirlin, J. L., Jacob, J., and Kato, K. 1962. The role of messenger to nucleolar RNA. *Exp. Cell Res.* **27**:355–359.

Sklar, V. E. F., Schwartz, L. B., and Roeder, R. G. 1975. Distinct molecular structures of nuclear class I, II, and III DNA-dependent RNA polymerases. *Proc. Nat. Acad. Sci. USA* **72**:348–352.

Slack, J. M. W., and Loening, U. E. 1974. 5′-ends of ribosomal and ribosomal precursor RNAs from *Xenopus laevis*. *Eur. J. Biochem.* **43**:59–67.

Slater, D. W., Slater, I., and Gillespie, D. 1972. Post-fertilization synthesis of poly(A) in sea urchin embryos. *Nature* **240**:333–337.

Slater, D. W., Slater, I., Gillespie, D., and Gillespie, S. 1974. Post-fertilization polyadenylylation during transcriptive and translational inhibition. *Biochem. Biophys. Res. Comm.* **60**:1222–1228.

Slater, I., and Slater, D. W. 1974. Polyadenylylation and transcription following fertilization. *Proc. Nat. Acad. Sci. USA* **71**:1103–1107.

Smith, A. E., Bellware, B. T., and Silver, J. J. 1967. Formation of nucleic acid coacervates by dehydration and rehydration. *Nature* **214**:1038–1040.

Smith, I., Dubnau, D., Morrell, P., and Marmur, J. 1968. Chromosomal location of DNA base sequences complementary to tRNA and to 5 S, 16 S, and 23 S rRNA in *B. subtilis*. *J. Mol. Biol.* **33**:123–140.

Smith, M. J., Hough, B. R., Chamberlin, M. E. and Davidson, E. H. 1974. Repetitive and non-repetitive sequences in sea urchin hnRNA. *J. Mol. Biol.* **85**:103–126.

Soeiro, R., Vaughan, M. H., Warner, J. R., and Darnell, J. E. 1968. The turnover of nuclear DNA-like RNA in HeLa cells. *J. Cell Biol.* **39**:112–118.

Soffer, R. L. 1968. The arginine transfer reaction. *Biochim. Biophys. Acta* **155**:228–240.

Soffer, R. L. 1973. Peptide acceptors in the leucine-phenylalanine transfer reaction. *J. Biol. Chem.* **248**:8424–8428.

Soffer, R. L., and Horinishi, H. 1969. General characteristics of the arginine-transfer reaction in rabbit liver cytoplasm. *J. Mol. Biol.* **43**:163–175.

Soffer, R. L., and Mendelsohn, N. 1966. Incorporation of arginine by a soluble system from sheep thyroid. *Biochem. Biophys. Res. Comm.* **23**:252–258.

Sogin, M., Pace, B., Pace, N. R., and Woese, C. R. 1971. Primary structural relationship of p16 to m16 rRNA. *Nature New Biol.* **232**:48–49.

Somers, D. G., Pearson, M. L., and Ingles, C. J. 1975. Regulation of RNA polymerase II activity in a mutant rat myeloblast cell line resistant to α-amantin. *Nature* **253**:372–374.

Spiegelman, S., Mills, D. R., and Kramer, F. R. 1975. The extracellular evolution of structure in replicating RNA molecules. *In:* Miller, I. R., ed., *Stability and Origin of Biological Information,* New York, John Wiley & Sons, p. 123–172.

Spohr, G., Imaizumi, T., and Scherrer, K. 1974. Synthesis and processing of nuclear precursor-mRNA in avian erythroblasts and HeLa cells. *Proc. Nat. Acad. Sci. USA* **71**:5009–5013.

Squires, C., Konrad, B., Kirschbaum, J., and Carbon, J. 1973. Three adjacent tRNA genes in *E. coli. Proc. Nat. Acad. Sci. USA* **70**:438–441.

Steggles, A. W., Wilson, G. N., Kantor, J. A., Picciano, D. J., Falvey, A. K., and Anderson, W. F. 1974. Cell-free transcription of mammalian chromatin. *Proc. Nat. Acad. Sci. USA* **71**:1219–1223.

Stein, G. S., Chandhuri, S. C., and Baserga, R. B. 1972. Gene activation in WI-38 fibroblasts stimulated to proliferate. *J. Biol. Chem.* **247**:3918–3922.

Stetter, K. O., and Zillig, W. 1974. DNA-dependent RNA polymerase from *Lactobacillus curvatus. Eur. J. Biochem.* **48**:527–540.

Stoof, T. J., de Regt, V. C. H. F., Raué, H. A., and Planta, R. J. 1974. Two precursor 5 S RNA species in *Bacillus licheniformis. FEBS Lett.* **49**:237–241.

Studier, F. W. 1973. Analysis of bacteriophage T7 early RNAs and proteins on slab gels. *J. Mol. Biol.* **79**:237–248.

Sulston, J., Lohrman, R., Orgel, L. E., Schneider-Bernloehr, H., Weimann, B. J., and Miles, H. T. 1969. Non-enzymatic oligonucleotide synthesis on a poly(C) template. *J. Mol. Biol.* **40**:227–234.

Sypherd, P. S. 1968. Ribosome development and the methylation of rRNA. *J. Bact.* **95**:1844–1850.

Szybalski, W., Kubinski, H., and Sheldrick, P. 1966. Pyrimidine clusters on the transcribing-strand of DNA and their possible role in the initiation of RNA synthesis. *Cold Spring Harbor Symp. Quant. Biol.* **31**:123–127.

Tartof, K. D. 1971. Increasing the multiplicity of ribosomal RNA genes in *Drosophila melanogaster. Science* **171**:294–297.

Tartof, K. D., and Perry, R. P. 1970. The 5 S RNA genes of *Drosophila melanogaster. J. Mol. Biol.* **51**:171–184.

Telles, N. C., and Coble, D. W. 1968. The nucleolus—A morphological and possible functional relationship to the nuclear membrane. *Fed. Proc.* **27**:836.

Terhune, M. W., and Sandstead, H. H. 1972. Decreased RNA polymerase activity in mammalian zinc deficiency. *Science* **177**:68–69.

Thammana, P., and Held, W. A. 1974. Methylation of 16 S RNA during ribosome assembly *in vitro. Nature* **251**:682–686.

Timberlake, W. E., and Turian, G. 1974. Multiple DNA-dependent RNA polymerases of *Neurospora. Experientia* **30**:1236–1238.

Tissières, A., Mitchell, H. K., and Tracy, U. M. 1974. Protein synthesis in salivary glands of *Drosophila melanogaster. J. Mol. Biol.* **84**:389–398.

Torelli, U., and Torelli, G. 1973. Poly(A)-containing molecules in hnRNA of normal PhA-stimulated lymphocyte and acute leukemiablast cells. *Nature New Biol.* **244**:134–136.

Trapman, J., de Jonge, P., and Planta, R. J. 1975. On the biosynthesis of 5.8 S rRNA in yeast. *FEBS Lett.* **57**:26–30.

Travers, A. A., and Burgess, R. R. 1967. Cyclic re-use of the RNA polymerase sigma factor. *Nature* **222**:536–540.

Tsai, M. J., and Saunders, G. F. 1974. Isolation and characterization of human DNA-dependent RNA polymerase. *Biochim. Biophys. Acta* **366**:61–69.

Van de Walle, C., and Deltour, R. 1974. Presence of hnRNA in a plant: *Zea mays. FEBS Lett.* **49**:87–91.

Van Keulen, H., Planta, R. J., and Retèl, J. 1975. Structure and transcription specificity of yeast RNA polymerase A. *Biochim. Biophys. Acta* **395:**179–190.

Vickers, T. G., and Midgley, J. E. M. 1971. Evidence for tRNA precursors in bacteria. *Nature New Biol.* **233**:210–212.

Walton, G. M., Gill, G. N., Abrams, I. B., and Garren, L. D. 1971. Phosphorylation of ribosome-associated protein by an adenosine 3′, 5′-cyclic monophosphate dependent protein kinase. *Proc. Nat. Acad. Sci. USA* **68**:880–884.

Warner, J. R., Soeiro, R., Birnboim, H. C., Girard, M., and Darnell, J. E. 1966. Rapidly labelled HeLa cell nuclear RNA. *J. Mol. Biol.* **19**:349–361.

Weinberg, R. A. 1973. Nuclear RNA metabolism. *Ann. Rev. Biochem.* **42**:330–354.

Weinberg, R. A., and Penman, S. 1970. Processing of 45 S nucleolar RNA. *J. Mol. Biol.* **47**:169–178.

Weinmann, R., and Roeder, R. G. 1974. Role of DNA-dependent RNA polymerase III in the transcription of the tRNA and 5 S RNA genes. *Proc. Nat. Acad. Sci. USA* **71**:1790–1794.

Weisbeek, P. J., Borrias, W. E., Langeveld, S. A., Baas, P. D., and van Arkel, G. A. 1977. Bacteriophage ϕX174: Gene A overlaps gene B. *Proc. Nat. Acad. Sci. USA* **74**:2504–2508.

Wellauer, P. K., and Dawid, I. B. 1973a. Secondary structure maps of RNA. *Proc. Nat. Acad. Sci. USA* **70**:2827–2831.

Wellauer, P. K., and Dawid, I. B. 1973b. Secondary structure maps of rRNA and its precursors as determined by electron microscopy. *Cold Spring Harbor Symp. Quant. Biol.* **38**:525–535.

Wilkie, N., and Smellie, R. M. S. 1968. Polyribonucleoside synthesis by subfractions of microsomes from rat liver. *Biochem. J.* **109**:229–238.

Willems, M., Penman, M., and Penman, S. 1969. The regulation of RNA synthesis and processing in the nucleolus during inhibition of protein synthesis. *J. Cell Biol.* **41**:177–187.

Wilson, J. H., Kim, J. S., and Abelson, J. N. 1972. Bacteriophage T4 tRNA. *J. Mol. Biol.* **71**:547–556.

Wimber, D. E., and Steffensen, D. M. 1970. Localization of 5 S RNA genes on *Drosophila* chromosomes by RNA–DNA hybridization. *Science* **170**:639–641.

Winicov, I. 1976. Alternate temporal order in ribosomal RNA maturation. *J. Mol. Biol.* **100**:141–155.

Winicov, I., and Perry, R. P. 1974. Characterization of a nucleolar endonuclease possibly involved in rRNA maturation. *Biochemistry* **13**:2908–2914.

Winters, M. A., and Edmonds, M. 1973. A poly(A) polymerase from calf thymus. *J. Biol. Chem.* **248**:4763–4768.

Wood, W. B., and Berg, P. 1963. Studies on the "messenger" activity of RNA synthesized with RNA polymerase. *Cold Spring Harbor Symp. Quant. Biol.* **28**:237–246.

Wykes, J., and Smellie, R. M. S. 1966. The synthesis of polyribonucleotides by cytoplasmic enzymes. *Biochem. J.* **99**:347–355.

Yankofsky, S. A., and Spiegelman, S. 1962. Identification of the ribosomal RNA cistron by sequence complementarity. *Proc. Nat. Acad. Sci. USA* **48**:1069–1078.

Yankofsky, S. A., and Spiegelman, S. 1963. Distinct cistrons for the two rRNA components. *Proc. Nat. Acad. Sci. USA* **49**:538–544.

Yoshida, N., Inoue, H., Sasaki, A., and Otsuka, H. 1971. Ribonuclease from *Streptomyces erythreus. Biochim. Biophys. Acta* **228**:636–647.

Young, H. A., and Whiteley, H. R. 1975. DNA-dependent RNA-polymerase in the dimorphic fungus *Mucor rouxii. J. Biol. Chem.* **250**:479–487.

Zain, B. S., Weissman, S. M., Dhar, R., and Pan, J. 1974. The nucleotide sequence preceding an RNA polymerase initiation site on SV_{40} DNA. Part I. The sequence of the late strand transcript. *Nucleic Acid Res.* **1**:577–594.

Zimmer, S. G., and Millette, R. L. 1975. DNA-dependent RNA polymerase from *Pseudomonas* BAL-31. *Biochemistry* **14**:290–299.

Zingales, B., and Colli, W. 1977. Ribosomal RNA genes in *B. subtilis*. Evidence for a cotranscriptional mechanism. *Biochim. Biophys. Acta* **474**:562–577.

Zylber, E. A., and Penman, S. 1971. Products of RNA polymerases in HeLa cell nuclei. *Proc. Nat. Acad. Sci. USA* **68**:2861–2865.

CHAPTER 6

Abraham, D. J. 1971. Proposed detail structural model for tRNA and its geometric relationship to a messenger. *J. Theor. Biol.* **30**:83–91.

Arfmann, H. A., Labitzke, R., Lawaczeck, R., and Wagner, K. G. 1974. Aromatic amino acid-lysine copolymers. *Biochimie* **56**:53–60.

Barker, H. A., and Beck, J. V. 1941. The fermentative decomposition of purines by *Clostridium acidi-urici* and *C. cylindrosporum*. *J. Biol. Chem.* **141**:3–27.

Barker, H. A., Ruben, S., and Beck, J. V. 1940. Synthesis of acetic acid from CO_2 by *Clostridium acidi-urici*. *Proc. Nat. Acad. Sci. USA* **26**:477–482.

Besson, J., and Gavaudan, P. 1967. Sur l'organisation logarithmique du code génétique. *C.R. Acad. Sci. Paris* **D264**:1311–1314.

Bishop, M. J., Lohrmann, R., and Orgel, L. E. 1972. Prebiotic phosphorylation of thymidine at 65° C in simulated desert conditions. *Nature* **237**:162–164.

Brack, A., and Orgel, L. E. 1975. β structures of alternating peptides and their possible prebiotic significance. *Nature* **256**:383–387.

Breed, R. S., Murray, E. G. D., and Smith, J. N. R. 1957. *Bergey's Manual for Determinative Bacteriology*, Baltimore, Williams and Wilkins, p. 837–853.

Bruenn, J., and Jacobson, K. B. 1972. New species of tyrosine tRNA in nonsense suppressor strains of yeast. *Biochim. Biophys. Acta* **287**:68–76.

Burton, S. D., Morita, R. Y., and Miller, W. 1966. Utilization of acetate by *Beggiatoa*. *J. Bact.* **91**:1192–1200.

Calvin, M. 1975. Chemical evolution. *Am. Sci.* **63**:169–177.

Carter, C. W., and Kraut, J. 1974. A proposed model for interaction of polypeptides with RNA. *Proc. Nat. Acad. Sci. USA* **71**:283–287.

Chan, T., and Garen, A. 1969. Leucine insertion by the $Su6^+$ suppressor gene. *J. Mol. Biol.* **45**:545–548.

Chan, T., and Garen, A. 1970. Tryptophan insertion by the $Su9^+$ gene, a suppressor of UGA nonsense triplet. *J. Mol. Biol.* **49**:231–234.

Chan, T., Webster, R. E., and Zinder, N. D. 1971. Suppression of UGA codon by a tryptophan tRNA. *J. Mol. Biol.* **56**:101–116.

Conrad, M. 1970. A mechanism for the evolution of the genetic code. *Curr. Mod. Biol.* **3**:260–264.

Contreras, R., Ysebaert, M., Min Jou, W., and Fiers, W. 1973. Bacteriophage MS2 RNA: Nucleotide sequence of the end of the A protein gene and the intercistronic region. *Nature New Biology* **241**:99–101.

Crick, F. H. C. 1968. The origin of the genetic code. *J. Mol. Biol.* **38**:367–379.

Dayhoff, M. O. 1971. Evolution of proteins. *In:* Buvet, R., and C. Ponnamperuma, eds., *Chemical Evolution and the Origin of Life,* Amsterdam, North-Holland Publishing Co., p. 392–419.

Dillon, L. S. 1962. Comparative cytology and the evolution of life. *Evolution* **16**:102–117.
Dillon, L. S. 1963. A reclassification of the major groups of organisms based upon comparative cytology. *Syst. Zool.* **12**:71–82.
Dillon, L. S. 1973. Origins of the genetic code. *Bot. Rev.* **39**:301–345.
Fox, S. W. 1974. Origins of biological information and the genetic code. *Mol. Cell. Biochem.* **3**:129–142.
Fox, S. W., Harada, K., and Vegotsky, A. 1959. Thermal polymerization of amino acids and a theory of biochemical origins. *Experientia* **15**:81–84.
Fox, S. W., and Nakashima, T. 1967. Fractionation and characterization of an amidated thermal 1:1:1-proteinoid. *Biochim. Biophys. Acta* **140**:155–167.
Fox, S. W., Yuki, A., Waehneldt, T. V., and Lacey, J. C. 1971. The primordial sequence—ribosomes, and the genetic code. *In:* Buvet, R., and C. Ponnamperuma, eds., *Chemical Evolution and the Origin of Life*, Amsterdam, North-Holland Publishing Co., p. 252–262.
Gatlin, L. L. 1972. *Information Theory and the Living System*, New York, Columbia University Press.
Gavaudan, P. 1971a. [The internal logic of the genetic coding table]. *C.R. Hebd. Séances Acad. Sci.* **272**:1672–1675.
Gavaudan, P. 1971b. The genetic code and the origin of life. *In:* Buvet, R., and C. Ponnamperuma, eds., *Chemical Evolution and the Origin of Life*, Amsterdam, North-Holland Publishing Co., p. 432–445.
Goodman, H. M., Abelson, J. N., Landy, A., Brenner, S., and Smith, J. D. 1968. Amber suppression: A nucleotide change in the anticodon of a tyrosine tRNA. *Nature* **217**:1019–1024.
Goodman, H. M., Abelson, J. N., Landy, A., Zadrazil, S., and Smith, J. D. 1970. The nucleotide sequences of tyrosine tRNAs of *E. coli*. *Eur. J. Biochem.* **13**:461–483.
Harpold, M. A., and Calvin, M. 1973. Amino acid-nucleotide interactions on an insoluble solid support. *Biochim. Biophys. Acta* **308**:117–128.
Hartman, H. 1975. Speculations on the evolution of the genetic code. *Origins Life* **6**:423–427.
Hasegawa, M., and Yano, T. A. 1975. Entropy of the genetic information and evolution. *Origins Life* **6**:219–227.
Hashimoto, S., Miyazaki, M., and Takemura, S. 1969. Nucleotide sequence of tyrosine tRNA from *Torulopsis utilis*. *J. Biochem.* **65**:659–661.
Hélène, C. 1971. Role of aromatic amino-acid residues in the binding of enzymes and proteins to nucleic acids. *Nature New Biol.* **234**:120–121.
Higa, A. I., de Forchetti, S. R. M., and Cazzulo, J. J. 1976. CO_2-fixing enzymes in *Pseudomonas fluorescens*. *J. Gen. Microb.* **93**:69–74.
Hirsch, D. 1971. Tryptophan tRNA as the UGA suppressor. *J. Mol. Biol.* **58**:439–458.
Huxley, J. 1963. *Evolution: The Modern Synthesis*, London, Allen and Unwin.
Ishigami, M., and Nagano, K. 1975. The origin of the genetic code. *Origins Life* **6**:551–560.
Jeppesen, P. G. N., Nichols, J. L., Sanger, F., and Barrell, B. G. 1970. Nucleotide sequences from bacteriophage R17 RNA. *Cold Spring Harbor Symp. Quant. Biol.* **35**:13–20.
Jett, M., and Jamieson, G. A. 1971. A homology between codon sequence and the linkage in glycoproteins. *Carbohydr. Res.* **18**:446–468.
Jorré, R. P., and Curnow, R. N. 1975. The evolution of the genetic code. *Biochimie* **57**:1147–1154.
Jukes, T. H. 1966. *Molecules and Evolution*, New York, Columbia University Press.
Jukes, T. H. 1966. Recent advances in studies of evolutionary relationships between proteins and nucleic acids. *Space Life Sci.* **1**:469–494.
Jukes, T. H., and Gatlin, L. 1971. Recent studies concerning the coding mechanism. *Progr. Nucleic Acid Res. Mol. Biol.* **11**:303–350.
Kaplan, R. W. 1971. The problem of chance in formation of protobionts by random aggregation of

macromolecules. *In:* Buvet, R., and C. Ponnamperuma, eds., *Chemical Evolution and the Origin of Life*, Amsterdam, North-Holland Publishing Co., p. 319–329.

Keil, F. 1912. Beiträge zur Physiologie der farblosen Schwefelbakterien. *Beiträg. Biol. Pflanzen* **11**:335–372.

Krzanowska, H. 1970. [Genetic code and evolution]. *Wszechswiat* **7/8**:169–174.

Lacey, J. C., and Pruitt, K. M. 1969. Origin of the genetic code. *Nature* **223**:799–804.

Lacey, J. C., Weber, A. L., and White, W. E. 1975. A model for the coevolution of the genetic code and the process of protein synthesis: Review and assessment. *Origin Life* **6**:273–283.

Lesk, A. M. 1970. On the origin of the genetic code: Photochemical interaction between amino acids and nucleic acids not requiring adaptors. *J. Theor. Biol.* **27**:171–173.

Ljungdahl, L., and Wood, H. G. 1969. Total synthesis of acetate from CO_2 by heterotropic bacteria. *Ann. Rev. Microbiol.* **23**:515–538.

Maier, S., and Murray, R. G. E. 1965. The fine structure of *Thioploca ingrica* and a comparison with *Beggiatoa*. *Can. J. Microb.* **11**:645–655.

Mednikov. B. M. 1971. The origin of ribosomes and the evolution of rRNA. *In:* Buvet, R., and C. Ponnamperuma, eds., *Chemical Evolution and the Origin of Life*, Amsterdam, North-Holland Publishing Co., p. 231–235.

Melcher, G. 1970. A new hypothesis on the evolution of the genetic code. *Biophysics* **7**:25–28.

Miklos, J. 1971. Notes on genetic code: I. Analyzing Claviere's data: Anticodon-amino acid assignments and miscoding through amino acid substitution. *Stud. Biophys.* **28**:223–230.

Min Jou, W., Haegeman, G., Ysebaert, M., and Fiers, W. 1972. Nucleotide sequence of the gene coding for bacteriophage MS2 coat protein. *Nature* **237**:82–88.

Model, P., Webster, R. E., and Zinder, N. D. 1969. The UGA codon *in vitro*: Chain termination and suppression. *J. Mol. Biol.* **43**:177–190.

Moore, G. W., Barnabas, J., and Goodman, M. 1973. A method for constructing maximum parsimony ancestral amino acid sequences on a given network. *J. Theor. Biol.* **56**:63–82.

Nagyvary, J., and Fendler, J. H. 1974. Origin of the genetic code: A physical–chemical model of primitive codon assignments. *Origins Life* **5**:357–362.

Nichols, J. L. 1970. Nucleotide sequence from the polypeptide chain termination region of the coat protein cistron in phage R17 RNA. *Nature* **225**:147–151.

Orgel, L. E. 1968. Evolution of the genetic apparatus. *J. Mol. Biol.* **38**:381–393.

Orgel, L. E. 1972. A possible step in the origin of the genetic code. *ISR J. Chem.* **10**:287–292.

Papentin, F. 1973. Experiments on protein evolution and evolutionary aspects of the genetic code. *J. Theor. Biol.* **39**:417–430.

Parker, D. J., Wu, T.-F., and Wood, H. G. 1971. Total synthesis of acetate from CO_2:Methyltetrahydrofolate, an intermediate, and a procedure of separation of the folates. *J. Bact.* **108**:770–776.

Parker, D. J., Wood, H. G., Ghambeer, R. K., and Ljungdahl, L. G. 1972. Total synthesis of acetate from CO_2 during carboxylation of trideuteriomethyl-cobalamin. *Biochemistry* **11**:3074–3080.

Pringsheim, E. G. 1964. Heterotrophism and species concepts in *Beggiatoa*. *Am. J. Bot.* **51**:898–913.

Raszka, M., and Mandel, M. 1971. Interaction of aromatic amino acids with neutral poly(A). *Proc. Nat. Acad. Sci. USA* **68**:1190–1191.

Raszka, M., and Mandel, M. 1972a. Interaction of amino acids and related compounds with neutral poly A. *First Eur. Biophys. Congr., Baden* **1**:263–268.

Raszka, M., and Mandel, M. 1972b. Is there a physical chemical basis for the present genetic code? *J. Mol. Evol.* **2**:38–43.

Ratner, V. A., and Bachinskii, A. G. 1972a. [Population model of occurrence of codon stable ambiguity in a genetic code]. *Genetika* **8**:153–160.

Ratner, V. A., and Bachinskii, A. G. 1972b. [Population models of degeneracy arising in genetic code. II. Competition of 2 series for free nonsense]. *Genetika* **8**:179–184.

Rich, A. 1974. Transfer RNA and the translation apparatus in the origin of life. *Origins Life* **5**:207–219.

Sagers, R. D., Benziman, M., and Gunsalus, I. C. 1961. Acetate formation in *Clostridium acidiurici:* Acetokinase. *J. Bact.* **82**:233–238.

Salthe, S. N. 1972. *Evolutionary Biology,* New York, Holt, Rinehart and Winston, Inc.

Sambrook, J. F., Fan, D. P., and Brenner, S. 1967. A strong suppressor specific for UGA. *Nature* **214**:452–453.

Saxinger, C., and Ponnamperuma, C. 1971. Experimental investigation on the origin of the genetic code. *J. Mol. Evol.* **1**:63–73.

Saxinger, C., and Ponnamperuma, C. 1974. Interactions between amino acids and nucleotides in the prebiotic milieu. *Origins Life* **5**:189–200.

Saxinger, C., Ponnamperuma, C., and Woese, C. 1971. Evidence for the interaction of nucleotides with immobilized amino acids and its significance for the origin of the genetic code. *Nature New Biol.* **234**:172–174.

Schapp, T. 1971. Dual information in DNA and evolution of genetic code. *J. Theor. Biol.* **32**:293–298.

Schulman, M., Chamber, R. K., Ljungdahl; L. G., and Wood, H. G. 1973. Total synthesis of acetate from CO_2. VII. Evidence with *Clostridium thermoaceticum* that the carboxyl of acetate is derived from the carboxyl of pyruvate by transcarboxylation and not by fixation of CO_2. *J. Biol. Chem.* **248**:6255–6261.

Schulman, M., Parker, D., Ljungdahl, L. G., and Wood, H. G. 1972. Total synthesis of acetate from CO_2. V. Determination by mass analysis of the different types of acetate formed from $^{13}CO_2$ by heterotrophic bacteria. *J. Bact.* **109**:633–644.

Schutzenberger, M. P., Gavaudan, P., and Besson, J. 1969. Sur l'existence d'une certaine corrélation entre lepoids moléculaire d'acides aminés et le nombre de triplets intervenane dans leur codage. *C.R. Acad. Sci. Paris* **D268**:1342–1344.

Simpson, G. G. 1949. *The Meaning of Evolution: A Study of the History of Life and Its Significance for Man*, New Haven, Conn., Yale University Press.

Smith, J. D., Abelson, J. N., Goodman, H. M., Landy, A., and Brenner, S. 1968. Amber suppressor tRNA. *In:* Fröholm, L. O., and S. G. Laland, eds. *Structure and Function of tRNA and 5 S-RNA*, New York, Academic Press, p. 37–51.

Smith, J. M. 1966. *The Theory of Evolution. 2nd Ed.*, Harmondsworth, England, Penguin Books.

Smith, K. C. 1968. The biological importance of U.V.-induced DNA–protein cross-linking *in vivo* and its probable chemical mechanism. *Photochem. Photobiol.*, **7**:651–660.

Smith, K. C. 1969. Photochemical addition of amino acids to ^{14}C-uracil. *Biochem. Biophys. Res. Comm.* **34**:354–357.

Smith, K. C., and Meun, D. H. C. 1968. Kinetics of the photochemical addition of [^{35}S] cysteine to polynucleotides and nucleic acids. *Biochemistry* **7**:1033–1037.

Steitz, J. A. 1969. Polypeptide chain initiation: Nucleotide sequences of the three ribosomal binding sites in phage R17 RNA. *Nature* **224**:957–964.

Sun, A. Y., Ljungdahl, L., and Wood, G. H. 1969. Total synthesis of acetate from CO_2. II. Purification and properties of formyltetrahydrofolate synthetase from *Clostridium thermoaceticum. J. Bact.* **98**:842–844.

West, E. S., and Todd, W. R. 1961. *Textbook of Biochemistry. 3rd Ed.*, New York, The Macmillan Company.

Woese, C. R. 1968. The fundamental nature of the genetic code: Prebiotic interactions between polynucleotides and polyamino acids or their derivatives. *Proc. Nat. Acad. Sci. USA* **59**:110–117.

Woese, C. R. 1973. Evolution of nucleic acid replication: The possible role of simple repeating sequence polypeptides therein. *J. Mol. Evol.* **2**:205–208.

Woese, C. R., and Bleyman, M. A. 1972. Genetic code limit organisms—do they exist? *J. Mol. Evol.* **1**:223–229.

Wong, J. T. F. 1975. A co-evolution theory of the genetic code. *Proc. Nat. Acad. Sci. USA* **72**:1909–1912.

Wong, J. T. F. 1976. The evolution of a universal genetic code. *Proc. Nat. Acad. Sci. USA* **73**:2336–2340.

Yockey, H. P. 1973. Information theory into applications to biogenesis and evolution. *In:* Locker, A., ed., *Biogenesis, Evolution, Homeostasis*, Berlin, Springer-Verlag, p. 9–23.

Zipser, D. 1967. UGA: A third class of suppressible polar mutants. *J. Mol. Biol.* **29**:441–445.

CHAPTER 7

Alden, C. J., and Arnott, S. 1973. Nucleotide conformations in codon–anticodon interactions. *Biochem. Biophys. Res. Comm.* **53**:806–811.

Armstrong, D. J., Burrows, W. J., Skoog, F., Roy, K. L., and Söll, D. 1969a. Cytokinins: Distribution in tRNA species of *E. coli*. *Proc. Nat. Acad. Sci. USA* **63**:834–841.

Armstrong, D. J., Skoog, F., Kirkegard, L. H., Hampel, A. E., Bock, R. M., Gillam, I., and Tener, G. M. 1969b. Cytokinins: Distribution in species of yeast tRNA. *Proc. Nat. Acad. Sci. USA* **63**:504–511.

Arnold, H., and Kersten, H. 1973. The occurrence of ribothymidine, 1-methyladenosine, methylated guanosines and the corresponding methyltransferases in *E. coli* and *B. subtilis*. *FEBS Lett.* **36**:34–38.

Arnold, H., Schmidt, W., and Kersten, H. 1975. Occurrence and biosynthesis of ribothymidine in tRNAs of *B. subtilis*. *FEBS Lett.* **52**:62–65.

Atkins, J. F., and Ryce, S. 1974. UGA and non-triplet suppressor reading of the genetic code. *Nature* **249**:527–530.

Baczynskyj, L., Biemann, K., and Hall, R. H. 1968. Sulfur-containing nucleoside from yeast tRNA: 2-thio-5 (or 6)-uridine acetic acid methyl ester. *Science* **159**:1481–1483.

Bartz, J., Söll, D., Burrows, W. J., and Skoog, F. 1970. Identification of the cytokinin-active ribonucleosides in pure *E. coli* tRNA species. *Proc. Nat. Acad. Sci. USA* **67**:1448–1453.

Bhargava, P. M., Pallaiah, T., and Premkumar, E. 1970. Aminoacyl-tRNA synthetase recognition code-words in yeast tRNAs: A proposal. *J. Theor. Biol.* **29**:447–469.

Blobstein, S. H., Gebert, R., Grunberger, D., Nakanishi, K., and Weinstein, I. B. 1975. Structure of the fluorescent nucleoside of yeast $tRNA^{Phe}$. *Arch. Biochem. Biophys.* **167**:668–673.

Carbon, J., Squires, C., and Hill, C. W. 1970. Glycine tRNA of *E. coli*. II. Improved GGA-recognition in strains containing a genetically altered tRNA; Reversal by a secondary suppressor mutation. *J. Mol. Biol.* **52**:571–584.

Caskey, C. T., Beaudet, A., and Nirenberg, M. 1968. Codons and protein synthesis. 15. Dissimilar responses of mammalian and bacterial tRNA fractions to mRNA codons. *J. Mol. Biol.* **37**:99–118.

Cedergren, R. J., Cordeau, J. R., and Robillard, P. 1972. On the phylogeny of tRNAs. *J. Theor. Biol.* **37**:209–220.

Celis, J. E., Hooper, M. L., and Smith, J. D. 1973. Amino acid acceptor stem of *E. coli* suppressor $tRNA^{Tyr}$ is a site of synthetase recognition. *Nature New Biol.* **244**:261–264.

Chambers, R. W. 1971. On the recognition of tRNA by its aminoacyl-tRNA ligase. *Progr. Nucl. Acid Res. Mole. Biol.* **11**:489–525.

Chinali, G., Sprinzl, M., Parmeggiani, A., and Cramer, F. 1974. Participation in protein biosynthesis of tRNAs bearing altered 3′-terminal ribosyl residues. *Biochemistry* **13**:3001–3010.

Crick, F. H. C. 1966. Codon–anticodon pairing: The wobble hypothesis. *J. Mol. Biol.* **19**:548–555.

Dayhoff, M. O., ed. 1969. *Atlas of Protein Sequence and Structure 1969*, Vol. 4. Silver Spring, Md., Nat. Biomed. Res. Found.

Dudock, B. S., Katz, G., Taylor, E. K., and Holley, W. 1969. Primary structure of wheat germ phenylalanine transfer RNA. *Proc. Nat. Acad. Sci. USA* **62**:941–945.

Dudock, B. S., DiPeri, C., and Michael, M. S. 1970. On the nature of the yeast phenylalanine transfer ribonucleic acid synthetase recognition site. *J. Biol. Chem.* **245**:2465–2468.

Dudock, B. S., DiPeri, C., Scileppi, K., and Reszelback, R. 1971. The yeast phenylalanyl–tRNA synthetase recognition site: The region adjacent to the dihydrouridine loop. *Proc. Nat. Acad. Sci. USA* **68**:681–684.

Dugré, M., and Cedergren, R. J. 1974. Origine de l'inosine dans les tRNA de levure. *Can. J. Biochem.* **52**:417–422.

Eisinger, J., and Gross, N. 1974. The anticodon–anticodon complex. *J. Mol. Biol.* **88**:165–174.

Fiers, W., Contreras, R., Duerinck, F., Haegeman, G., Iserentant, D., Merregaert, J., Min Jou, W., Molemans, F., Raegmaekers, A., Van der Bergh, A., Volckaert, G., and Ysebaert, M. 1976. Complete nucleotide sequence of bacteriophage MS2 RNA: Primary and secondary structure of the replicase gene. *Nature* **260**:500–507.

Geller, M., Pohorille, A. and Jaworski, A., 1973. Electronic structure of thiouracils and their interaction with adenine. *Biochim. Biophys. Acta* **331**:1–8.

Genter, N., and Berg, P. 1971. Occurrence of a glycyl-lipopolysaccharide structure in *E. coli* and its enzymatic formation from glycyl-tRNA. *Fed. Proc.* **30**:1218.

Ghysen, A., and Celis, J. E. 1974. Mischarging single and double mutants of *E. coli sup* 3 tyrosine tRNA. *J. Mol. Biol.* **83**:333–351.

Gillam, I., Millward, S., Blew, D., von Tigerstrom, M., Wimmer, E., and Tener, G. M. 1967. The separation of soluble RNAs on benzoylated dimethylaminoethylcellulose. *Biochemistry* **6**:3043–3056.

Gould, R. M., Thornton, M. P., Liepkalns, V., and Lennarz, W. J. 1968. Participation of aminoacyl transfer ribonucleic acid in aminoacyl phosphatidylglycerol synthesis. II. Specificity of alanyl phosphatidylglycerol synthesis. *J. Biol. Chem.* **243**:3096–3104.

Griffin, G. D., Yang, W. K., and Novelli, G. D. 1976. Transfer RNA species in human lymphocytes stimulated by mitogens and in leukemic cells. *Arch. Biochem. Biophys.* **176**:187–196.

Hall, R. H. 1971. *The Modified Nucleosides in Nucleic Acids*, Columbia University Press, N.Y., p. 257–280.

Hirsch, D. 1971, Tryptophan tRNA as the UGA suppressor. *J. Mol. Biol.* **58**:439–458.

Holmes, W. M., Hurd, R. E., Reid, B. R., Rimerman, R. A., and Hatfield, G. W. 1975. Separation of tRNA by sepharose chromatography using reverse salt gradients. *Proc. Nat. Acad. Sci. USA* **72**:1068–1071.

Inokuchi, H., Celis, J. E., and Smith, J. D. 1974. Mutant tyrosine tRNAs of *E. coli:* Construction by recombination of a double mutant A1G82 chargeable with glutamine. *J. Mol. Biol.* **85**:187–192.

Isham, K. R., and Stulberg, M. P. 1974. Modified nucleosides in undermethylated phenylalanine tRNA from *E. coli. Biochim. Biophys. Acta* **340**:177–182.

Jacobson, B. K. 1971. Role of an isoacceptor tRNA as an enzyme inhibitor: Effect on tryptophan pyrrolase of *Drosophila. Nature New Biol.* **231**:17–19.

Johnson, H., Hayashi, H., and Söll, D. 1970. Isolation and properties of a tRNA deficient in ribothymidine. *Biochemistry* **9**:2823–2831.

Kaji, A., Kaji, H., and Novelli, G. D. 1965a. Soluble amino acid-incorporating system. I. Preparation of the system and nature of the reaction. *J. Biol. Chem.* **240**:1185–1191.

Kaji, A., Kaji, H., and Novelli, G. D. 1965b. Soluble amino acid-incorporating system. II. Soluble nature of the system and the characterization of the radioactive product. *J. Biol. Chem.* **240**:1192–1199.

Kaminek, M. 1974. Evolution of tRNA and the origin of two positional isomers of zeatin. *J. Theor. Biol.* **48**:489–492.

Körner, A., and Söll, D. 1974. *N*-(purin-6-ylcarbamoyl)threonine: Biosynthesis *in vitro* in tRNA by an enzyme purified from *E. coli. FEBS Lett.* **39**:301–306.

Kruppa, J., and Zachau, H. G. 1972. Multiplicity of serine-specific transfer RNAs of brewers' and bakers' yeast. *Biochim. Biophys. Acta* **277**:499–512.

Leibowitz, M. J., and Soffer, R. L. 1969. A soluble enzyme from *Escherichia coli* which catalyzes the transfer of leucine and phenylalanine from tRNA to acceptor proteins. *Biochem. Biophys. Res. Comm.* **36**:47–53.

Li, H. J., Nakanishi, K., Grunberger, D., and Weinstein, I. B., 1973. Biosynthetic studies of the Y base in yeast phenylalanine tRNA. Incorporation of guanosine. *Biochem. Biophys. Res. Comm.,* **55**:818–823.

Lindahl, T., Adams. A., Geroch, M., and Fresco, J. R. 1967. Selective recognition of the native conformation of tRNAs by enzymes. *Proc. Nat. Acad. Sci. USA* **57**:178–185.

Liu, L. P., and Ortwerth, B. J. 1972. Specificity of rat liver lysine tRNA for codon recognition. *Biochemistry* **11**:12–17.

Matsuhashi, M., Dietrich, C. P., and Strominger, J. L. 1967. Biosynthesis of the peptidoglycan of bacterial cell walls. III. The role of soluble ribonucleic acid and of lipid intermediates in glycine incorporation in *Staphylococcus aureus*. *J. Biol. Chem.* **242**:3191–3206.

Maugh, T. H. 1974. Rous sarcoma virus: A new role for transfer RNA. *Science* **186**:41.

Melera, P. W., Momeni, C., and Rusch, H. L. 1974. Analysis of isoaccepting tRNAs during the growth phase mitotic cycle of *Physarum polycephalum*. *Biochemistry* **13**:4139–4142.

Morikawa, K., Torii, K., Iitaka, Y., and Tsuboi, M. 1974. Crystal and molecular structure of the methyl ester of uridin-5-oxyacetic acid: A minor constituent of *E. coli* tRNAs. *FEBS Lett.* **48**:279–282.

Münch, H. J., and Thiebe, R. 1975. Biosynthesis of the nucleoside Y in yeast $tRNA^{Phe}$: Incorporating of the 3-amino-3-carboxypropyl-group from methionine. *FEBS Lett.* **51**:257–258.

Murao, K., Saneyoshi, M., Harada, F., and Nishimura, S. 1970a. Uridin-5-oxyacetic acid: A new minor constituent from *E. coli* valine tRNA. *Biochem. Biophys. Res. Comm.* **38**:657–662.

Neidhardt, F. C. 1966. Roles of amino acid activating enzymes in cellular physiology. *Bacteriol. Rev.* **30**:701–719.

Nesbitt, J. A., and Lennarz, W. J. 1968. Participation of aminoacyl tRNA in aminoacyl phosphatidylglycerol synthesis. *J. Biol. Chem.* **243**:3088–3095.

Nishimura, S. 1972. Minor components in tRNA: Their characterization, location, and function. *Progr. Nucleic Acid Res. Mol. Biol.* **12**:49–85.

Nishimura, S., Taya, Y., Kuchino, Y., and Ohashi, Z. 1974. Enzymatic synthesis of 3-(3-amino-3-carboxypropyl)uridine in *E. coli* phenylalanine tRNA: Transfer of the 3-amino-3-carboxypropyl group from 5-adenosylmethionine. *Biochem. Biophys. Res. Comm.* **57**:702–708.

Nishimura, S., and Weinstein, I. B. 1969. Fractionation of rat liver tRNA. Isolating tyrosine, valine, serine, and phenylalanine tRNAs and their coding properties. *Biochemistry* **8**:832–842.

Nishimura, S., Yamada, Y., and Ishikura. H. 1969. The presence of 2-methylthio-N(Δ^2-isopentenyl)adenosine in serine and phenylalanine tRNAs from *E. coli*. *Biochim. Biophys. Acta* **179**:517–520.

Odom, O. W., Hardesty, B., Wintermeyer, W., and Zachau, H. G. 1974. The effect of removal or replacement with proflavin of the Y base in the anticodon loop of yeast $tRNA^{Phe}$ on binding into the acceptor or donor sites of reticulocyte ribosomes. *Arch. Biochem. Biophys.* **162**:536–551.

Ofengand, J., and Henes, C. 1969. The function of pseudouridylic acid in tRNA. *J. Biol. Chem.* **244**:6241–6253.

Ohashi, Z., Maeda, M., McCloskey, J. A., and Nishimura, S. 1974. 3-(3-amino-3-carboxypropyl)uridine: A novel modified nucleoside isolated from *E. coli* phyenylalanine tRNA. *Biochemistry* **13**:2620–2625.

Parthasarathy, R., Ohrt, J. M., and Chkeda, G. B. 1974a. Conformation of N-(purin-6-ylcarbamoyl)glycine, a hypermodified base in tRNA. *Biochem. Biophys. Res. Comm.* **57**:649–653.

Parthasarathy, R., Ohrt, J. M., and Chkeda, G. B. 1974b. Conformation and possible role of hypermodified nucleosides adjacent to 3′-end of anticodon in tRNA: N-(purin-6-ylcarbamoyl)-L-threonine riboside. *Biochem Biophys. Res. Comm.* **60**:211–218.

Pearson, R. L., Hancher, C. W., Weiss, J. F., Holladay, D. W., and Klemers, A. D. 1973. Preparation of crude tRNA and chromatographic purification of five tRNAs from calf liver. *Biochim. Biophys. Acta* **294**:236–249.

Pearson, R. L., Weiss, J. F., and Kelmers, A. D. 1971. Improved separation of tRNAs on polychlorotrifluoroethylene-supported reversed-phase chromatography columns. *Biochim. Biophys. Acta* **228**:770–774.

Peterkofsky, A., and Jesensky, C. 1969. The localization of $N^6(\Delta^2$-isopentenyl)-adenosine among the acceptor species of tRNA of *Lactobacillus acidophilus*. *Biochemistry* **8**:3798–3807.

Petit, J. F., Strominger, J. L., and Söll, D. 1968. Biosynthesis of the peptidoglycan of bacterial cell walls. II. Incorporation of serine and glycine into interpeptide bridges in *Staphylococcus epidermidis*. *J. Biol. Chem.* **243**:757–767.

Petrissant, G. 1973. Evidence for the absence of G-T-Ψ-C sequence from two mammalian initiator tRNAs. *Proc. Nat. Acad. Sci. USA* **70**:1046–1049.

Powers, D. M., and Peterkofsky, A. 1972. Biosynthesis and specific labeling of N(purin-6-ylcarbamoyl)threonine of *E. coli* tRNA. *Biochem. Biophys. Res. Comm.* **46**:831–838.

Randerath, E., Chia, L.-L. S. Y., Morris, H. P., and Randerath, K. 1974. Base analysis of RNA by ^{3}H postlabelling—a study of ribothymidine content and degree of base methylation of 4 S RNA. *Biochim. Biophys. Acta* **366**:159–167.

Rao, P. M.,and Kaji, H. 1974. Utilization of isoaccepting leucyl-tRNA in the soluble incorporation system and protein synthesizing systems from *E. coli*. *FEBS Lett.* **43**:199–202.

Richter, D., Erdmann, V. A., and Sprinzl, M. 1973. Specific recognition of GTΨC loop (loop IV) of tRNA by 50 S ribosomal subunits from *E. coli*. *Nature New Biol.* **246**:132–135.

Roberts, W. S. L., Strominger, J. L., and Söll, D. 1968a. Biosynthesis of the peptidoglycan of bacterial cell walls. VI. Incorporation of L-threonine into interpeptide bridges in *Micrococcus roseus*. *J. Biol. Chem.* **243**:749–756.

Roberts, W. S. L., Petit, J. F., and Strominger, J. L. 1968b. Biosynthesis of the peptidoglycan of bacteria. VIII. Specificity in the utilization of L-alanyl tRNA for interpeptide bridge synthesis in *Arthrobacter crystallopoietes*. *J. Biol. Chem.* **243**:768–772.

Roe, B., and Dudock, B. 1972. The role of the fourth nucleotide from the 3′-end in the yeast phenylalanyl tRNA synthetase recognition site: Requirement for adenosine. *Biochem. Biophys. Res. Comm.* **49**:399–406.

Roth, J. R., and Ames, B. N. 1966. Histidine regulatory mutants in *Salmonella typhimurium*. II. Histidine regulatory mutants having layered histidyl-tRNA sythetase. *J. Mol. Biol.* **22**:325–334.

Saneyoshi, M., Ohashi, Z., Harada, F., and Nishimura, S. 1972. Isolation and characterization of 2-methyl-adenosine from *E. coli* $tRNA_2^{Glu}$, $tRNA_1^{Asp}$, $tRNA_1^{His}$, and $tRNA^{Arg}$. *Biochim. Biophys. Acta* **262**:1–10.

Schlesinger, S., and Magasanik, B. 1964. Effect of α-methylhistidine on the control of histidine synthesis. *J. Mol. Biol.* **9**:670–682.

Schulman, LaD. H. 1972. Structure and function of *E. coli* formylmethionine tRNA: Loss of methionine acceptor activity by modification of a specific guanosine residue in the acceptor stem of formylmethionine tRNA from *E. coli*. *Proc. Nat. Acad. Sci. USA* **69**:3594–3597.

Sekiya, T., Takeishi, K., and Ukita, T. 1969. Specificity of yeast glutamic acid tRNA for codon recognition. *Biochim. Biophys. Acta* **182**:411–426.

Sen, G. C., and Ghosh. H. P., 1973. Coding properties of isoaccepting lysine tRNA species from baker's yeast. *Biochim. Biophys. Acta* **308**:106–116.

Silbert, D. F., Fink, G. R., and Ames, B. N. 1966. Histidine regulatory mutants in *Salmonella typhimurium*. III. A class of regulatory mutants deficient in tRNA for histidine. *J. Mol. Biol.* **22**:335–347.

Simsek, M., Ziegenmeyer, J., Heckman, J., and RajBhandary, U. L. 1973. Absence of the sequence GTΨCG(A)− in several eukaryotic cytoplasmic initiator tRNAs. *Proc. Nat. Acad. Sci. USA* **70**:1041–1045.

Smith, J. D., and Celis, J. E. 1973. Mutant tyrosine tRNA that can be charged with glutamine. *Nature New Biol.* **243**:66–71.

Söll, D. 1971. Enzymatic modification of tRNA. *Science* **173**:293–299.

Söll, D., Cherayil, J., Jones, D. S., Faulkner, R. D., Hampel, A., Bock, R. M., and Khorana, H. C. 1966. sRNA specificity for codon recognition as studied by the ribosomal binding technique. *Cold Spring Harbor Symp. Quant. Biol.* **31**:51–62.

Sprinzl, M., and Cramer, F. 1973. Accepting site for aminoacylation of $tRNA^{Phe}$ from yeast. *Nature New Biol.* **245**:3–5.

Taglang, R., Waller, J. P., Befort, N., and Fasciolo, F. 1970. Amino-acylation du $tRNA_1^{Val}$ de *E. coli* par la phénylalanyl-tRNA synthétase de levure. *Eur. J. Biochem.* **12**:550–557.

Tal, J., Deutscher, M. P., and Littauer, U. Z. 1972. Biological activity of *E. coli* $tRNA^{Phe}$ modified in its -CCA terminus. *Eur. J. Biochem.* **28**:478–491.

Thiebe, R., and Poralla, K. 1973. Origin of the nucleoside Y in yeast $tRNA^{Phe}$. *FEBS Lett.* **38**:27–28.

Wallace, R. B., and Freeman, K. B. 1974. Multiple species of methionyl-tRNA from mouse liver mitochondria. *Biochem. Biophys. Res. Comm.* **60**:1440–1445.

White, B. N., and Tener, G. M. 1973. Properties of $tRNA^{Phe}$ from *Drosophila*. *Biochim. Biophys. Acta* **312**:267–275.

White, B. N., Dunn, R., Gillam, I., Tener, G. M., Armstrong, D. J., Skoog, F., Frihart, C. R., and Leonard, N. J. 1975. An analysis of five serine tRNAs from *Drosophila*. *J. Biol. Chem.* **250**:515–521.

Wittig, B., Reuter, S., and Gottschling, H. 1973. Purification of the four lysine-specific tRNAs from chick embryos. *Biochim. Biophys. Acta* **331**:221–230.

Woodward, W. R., and Herbert, E. 1972. Coding properties of reticulocyte lysine tRNAs in hemoglobin synthesis. *Science* **177**:1197–1199.

Wu, M., Davidson, N., Attardi, G., and Aloni, Y. 1972. Expression of the mitochondrial genome in HeLa Cells. XIV. The relative positions of the 4 S RNA genes and of the rRNA genes in mitochondrial DNA. *J. Mol. Biol.* **71**:81–93.

Yamada, Y., Nishimura, S., and Ishikura, H. 1971. The presence of 2-methylthio-N^6-(Δ^2-isopentenyl)adenosine in leucine, tryptophan, and cysteine tRNAs from *E. coli*. *Biochim. Biophys. Acta* **247**:170–174.

Yang. W. K., Hellman, A., Martin, D. H., Hellman, K. B., and Novelli, G. D. 1969. Iso-accepting tRNAs of L–M cells in culture and after tumor induction in C_3H mice. *Proc. Nat. Acad. Sci. USA* **64**:1411–1418.

Yang, W. K., and Novelli, G. D. 1971. Analysis of isoaccepting tRNAs in mammalian tissues and cells. *Methods Enzymol.* **20**:44–55.

CHAPTERS 7 AND 8: PRIMARY STRUCTURE OF tRNAs

Glycine tRNAs

Barrell, B. G., Coulson, A. R., and McClain, W. H. 1973. Nucleotide sequence of a glycine tRNA coded by bacteriophage T4. *FEBS Lett.* **37**:64–69.

Hill, C. W., Combriato, G., Sternhart, W., Riddle, D. L., and Carbon, J. 1973. The nucleotide sequence of the GGG-specific glycine tRNA of *E. coli* and *Salmonella typhimurium*. *J. Biol. Chem.* **248**:4252–4262.

Hill, C. W., Squires, C., and Carbon, J. 1970. Structural genes for two glycine tRNA species. *J. Mol. Biol.* **52**:557–569.

Marcu, K., Mignery, R., Reszelbach, R., Roe, B., Sirover, M., and Dudock, B. 1973. The absense of ribothymidine in specific tRNAs. I. Glycine and threonine tRNAs of wheat embryo. *Biochem. Biophys. Res. Comm.* **55**:477–483.

Riddle, D. L., and Carbon, J. 1973. Frameshift suppression: A nucleotide addition in the anticodon of a glycine tRNA. *Nature New Biol.* **242**:230–234.

Roberts, R. J. 1972. Structures of two glycyl-tRNAs from *Staphylococcus epidermidis. Nature New Biol.* **237**:44–46.

Roberts, R. J. 1974. Staphylococcal tRNAs. II. Sequence analysis of isoaccepting glycine tRNA IA and IB from *Staphylococcus epidermidis* Texas 26. *J. Biol. Chem.* **249**:4787–4796.

Roberts, R. J., Lovinger, G. G., Tamura, T., and Strominger, J. L. 1974. Staphylococcal tRNAs. I. Isolation and purification of the isoaccepting glycine tRNAs from *Staphylococcus epidermidis* Texas 26. *J. Biol. Chem.* **249**:4781–4786.

Squires, C., and Carbon, J. 1971. Normal and mutant glycine tRNAs. *Nature New Biol.* **233**:274–277.

Stahl, S., Paddock, G., and Abelson, J. 1973. T4 bacteriophage $tRNA^{Gly}$. *Biochem. Biophys. Res. Comm.* **54**:567–569.

Stewart, T. S., Roberts, R. J., and Strominger, J. L. 1971. Novel species of RNA. *Nature* **230**:36–38.

Yoshida, M. 1973. The nucleotide sequence of $tRNA^{Gly}$ from yeast. *Biochem. Biophys. Res. Comm.* **50**:779–784.

Zachau, H. G. 1969. Transfer ribonucleic acids. *Angew. Chem.* (*Int. Ed.*) **8**:711–727.

Alanine tRNAs

Holley, R. W., Apgar, J., Everett, G. A., Madison, J. T., Marquisee, M., Merrill, S. H., Penswrick, J. R., and Zamir, A. 1965a. Structure of a ribonucleic acid. *Science,* **147**:1462–1465.

Holley, R. W., Everett, G. A., Madison, J. T., and Zamir, A. 1965b. Nucleotide sequences in the yeast alanine tRNA. *J. Biol. Chem.* **240**:2122–2127.

Merrill, C. R. 1968. Reinvestigation of the primary structure of yeast alanine tRNA. *Biopolymers* **6**:1727–1735.

Murao, K., Hasegawa, T., and Ishikura, H. 1976. 5-methoxyuridine: A new minor constituent located in the first position of the anticodon of $tRNA^{Ala}$, $tRNA^{Thr}$, and $tRNA^{Val}$ from *B. subtilis. Nucl. Acids Res.* **3**:2851–2857.

Penswick, J. R., Martin, R., and Dirheimer, G. 1975. Evidence supporting a revised sequence for yeast alanine tRNA. *FEBS Lett.* **50**:28–31.

Takemura, S., and Ogawa, K. 1973. The primary structure of alanine $tRNA_1$ from *Torulopsis utilis. J. Biochem.* **74**:323–333.

Takemura, S., Ogawa, K., and Nakazawa, K. 1972. Nucleotide sequence of alanine $tRNA_1$ from *Torulopsis utilis. FEBS Lett.* **25**:29–32.

Takemura, S., Ogawa, K., and Nakazawa, K. 1973. The primary structure of alanine $tRNA_1$ from *Torulopsis utilis. J. Biochem.* **74**:313–322.

Williams, R. J., Nagel, W., Roe, B., and Dudock, B. 1974. Primary structure of *E. coli* alanine tRNA: Relation to the yeast phenylalanyl tRNA synthetase recognition site. *Biochem. Biophys. Res. Comm.* **60**:1215–1221.

Yoshida, M., Kaziro, Y., and Ukita, T. 1968. Evidence for the important role of inosine residue in codon recognition of yeast alanine tRNA. *Biochim. Biophys. Acta* **166**:646–655.

Aspartic Acid tRNAs

Gangloff, J., Keith, G., Ebel, J. P., and Dirheimer, G. 1971. Structure of asparate-tRNA from brewer's yeast. *Nature New Biol.* **230**:125–126.

Gangloff, J., Keith, G., Ebel, J. P., and Dirheimer, G. 1972a. The primary structure of aspartate tRNA from brewer's yeast. Complete digestion with pancreatic ribonuclease and T_1 ribonuclease. *Biochim. Biophys. Acta* **259**:198–209.

Gangloff, J., Keith, G., Ebel, J. P., and Dirheimer, G. 1972b. The primary structure of aspartate tRNA from brewer's yeast. II. Partial digestions with pancreatic ribonuclease and T_1 ribonuclease and derivation of complete sequence. *Biochim. Biophys. Acta* **259**:210–222.

Harada, F., and Nishimura, S. 1972. Possible anticodon sequences of $tRNA^{His}$, $tRNA^{Asn}$, and $tRNA^{Asp}$ from *E. coli* B. Universal presence of nucleoside Q in the first position of the anticodon of these transfer ribonucleic acids. *Biochemistry* **11**:301–308.

Harada, F., Yamaizumi, K., and Nishimura, S. 1972. Oligonucleotide sequences of RNase T_1 and pancreatic RNase digests of *E. coli* aspartic acid tRNA. *Biochem. Biophys. Res. Comm.* **49**:1605–1609.

Keith, G., Gangloff, J., Ebel, J. P., and Dirheimer, G. 1970. Etablissement de la séquence de nucléotides de l'aspartate-t-RNA de levure de bière. *C. R. Acad. Sci. Paris,* **271**:613–616.

Glutamic Acid tRNAs

Kobayashi, T., Irie, T., Yoshida, M., Takeishi, K., and Ukita, T. 1974. The primary structure of yeast glutamic acid tRNA specific to the GAA codon. *Biochim. Biophys. Acta* **366**:168–181.

Munninger, K. O., and Chang, S. H. 1972. A fluorescent nucleoside from glutamic acid tRNA of *E. coli* K12. *Biochem. Biophys. Res. Comm.* **46**:1837–1842.

Ohashi, Z., Murao, K., Yahagi, T., von Minden, D. L., McCloskey, J. A., and Nishimura, S. 1972. Characterization of C^+ located in the first position of the anticodon of *E. coli* $tRNA^{Met}$ as N^4-acetylcytidine. *Biochim. Biophys. Acta* **262**:209–213.

Ohashi, Z., Saneyoshi, M., Harada, F., Hara, H., and Nishimura, S. 1970. Presumed anticodon structure of glutamic acid tRNA from *E. coli:* A possible location of a 2-thiouridine derivative in the first position of the anticodon. *Biochem. Biophys. Res. Comm.* **40**:866–872.

Singhal, R. P. 1971. Modification of *E. coli* glutamate tRNA with bisulfite. *J. Biol. Chem.* **246**:5848–5851.

Yoshida, M., Takeishi K., and Ukita, T. 1970. Anticodon structure of GAA-specific glutamic acid tRNA from yeast. *Biochem. Biophys. Res. Comm.* **39**:852–857.

Yoshida, M., Takeishi, K., and Ukita, T. 1971. Structural studies on a yeast glutamic acid tRNA specific to GAA codon. *Biochim. Biophys. Acta* **228**:153–166.

Valine tRNAs

Bayev, A. A., Venkstern, T. V., Mirzabekov, A. D., Krutilina, A. I., Li, L., and Axelrod, V. D. 1967. Primary structure of the valine tRNA. *Mol. Biol.* **I**:754–758.

Bonnet, J., Ebel, J. P., and Dirheimer, G. 1971. Primary structure of $tRNA_2^{Val.}$ from brewer's yeast. *FEBS Lett.* **15**:286–290.

Bonnet, J., Ebel, J. P., Dirheimer, G., Shershneva, L. P., Krutilina, A. I., Venkstern, T. V., and Bayer, A. A. 1974. The corrected nucleotide sequence of valine tRNA from baker's yeast. *Biochimie* **56**:1211–1213.

Harada, F., Kimura, F., and Nishimura, S. 1969. Nucleotide sequence of valine tRNA from *E. coli* B. *Biochim. Biophys. Acta* **195**:590–592.

Harada, F., Kimura, F., and Nishimura, S. 1971. Primary sequence of $tRNA^{Val}$, from *E. coli* B. *Biochemistry* **10**:3269–3283.

Kimura-Harada, F., Saneyoshi, M., and Nishimura, S. 1971. 5-methyl-2-thiouridine: A new sulfur-containing minor constituent from rat liver glutamic acid and lysine tRNAs. *FEBS Lett.* **13**:335–338.

Mirzavekov, A. D., Lastit, D., Leoina, E. S., Undritsov, I. M., and Baev, A. A. 1972. The acceptor activity of dissected baker's yeast $tRNA_1^{Val}$: Localization of two possible recognition sites of valyl-tRNA ligase. *Mol. Biol.* **6**:69–84.

Mizutani, T., Miyazaki, M., and Takemura, S. 1968. The primary structure of valine-I tRNA from *Torulopsis utilis*. *J. Biochem.* **64**:839–848.

Murao, K., Saneyoshi, M., Harada, F., and Nishimura, S. 1970. Uridin-5-oxyacetic acid: A new minor constituent from *E. coli* tRNA I. *Biochem. Biophys. Res. Comm.* **38**:657–662.

Piper, P. W. 1975b. The primary structure of the major cytoplasmic valine tRNA of mouse myeloma cells. *Eur. J. Biochem.* **51**:295–304.

Piper, P. W., and Clark, B. F. 1974a. The nucleotide sequences of cytoplasmic methionine and valine tRNAs from mouse myeloma cells. *FEBS Lett.* **47**:56–59.

Takada-Gurrier, C., Grosjean, H. G., Dirheimer, G., and Keith, G. 1976. The primary structure of $tRNA_2^{Val}$ from *B. stearothermophilus*. *FEBS Lett.* **62**:1–3.

Takemura, S., Mizutani, T., and Miyazaki, M. 1968a. The primary structure of valine-I tRNA from *Torulopsis utilis*. *J. Biochem.* **63**:277–278.

Takemura, S., Mizutani, T., and Miyazaki, M. 1968b. The primary structure of valine-I tRNA from *Torulopsis utilis*. I. Complete digestion with pancreatic ribonuclease and ribonuclease T_1. *J. Biochem.* **64**:827–837.

Yaniv, M., and Barrell, B. G. 1969. Nucleotide sequence of *E. coli* B $tRNA^{Val}$. *Nature* **222**:278–279.

Yaniv, M., and Barrell, B. G. 1971. Sequence relationship of three valine acceptor tRNAs from *E. coli*. *Nature New Biol.* **233**:113–114.

Zachau, H. G. 1972. Transfer ribonucleic acids. *In:* Bosch, L., ed., *The Mechanism of Protein Synthesis and its Regulation,* Amsterdam, North-Holland Publishing Co., p. 173–217.

II Leucine tRNAs

Blank, H. U., and Söll, D. 1971. The nucleotide sequence of two leucine tRNA species from *E. coli* K12. *Biochem. Biophys. Res. Comm.* **43**:1192–1197.

Dube, S. K., Marcker, K. A., and Yudelevich, A. 1970. The nucleotide sequence of a leucine tRNA from *E. coli*. *FEBS Lett.* **9**:168–170.

Histidine tRNAs

Harada, F., Sato, S., and Nishimura, S. 1972. Unusual CCA-stem structure of *E. coli* B $tRNA_1^{His}$. *FEBS Lett.* **19**:352–355.

Singer, C. E., and Smith, G. R. 1972. Histidine regulation in *Salmonella typhimurium* XIII. Nucleotide sequence of histidine tRNA. *J. Biol. Chem.* **247**:2983–3000.

Singer, C. E., Smith, G. R., Cortese, R., and Ames, B. N. 1972. Mutant $tRNA^{His}$ ineffective in repression and lacking two pseudouridine modifications. *Nature New Biol.* **238**:73–74.

Glutamine tRNAs

Comer, M. M., Foss, K., and McClain, W. H. 1975. A mutation of the wobble nucleotide of a bacteriophage T4 tRNA. *J. Mol. Biol.* **99**:283–293.

Folk, W. R., and Yaniv, M. 1972. Coding properties and nucleotide sequences of *E. coli* glutamine tRNAs. *Nature New Biol.* **237**:165–166.

Seidman, J. G., Comer, M. M., and McClain, W. H. 1974. Nucleotide alterations in the bacteriophage T4 glutamine tRNA that affect ochre suppressor activity. *J. Mol. Biol.* **90**:677–689.

II Arginine tRNAs

Murao, K., Tanabe, T., Ishii, F., Namiki, M., and Nishimura, S. 1972. Primary sequence of arginine tRNA from *E. coli*. *Biochem. Biophys. Res. Comm.*, **47**:1332–1337.

Weissenbach, J., Martin, R., and Dirheimer, G. 1972. Nucleotide sequence of $tRNA_{II}^{Arg}$ from brewer's yeast. *FEBS Lett.* **28**:353–355.

Weissenbach, J., Martin, R., and Dirheimer, C. 1975a. The primary structure of $tRNA_{II}^{Arg}$ from brewer's yeast. *Eur. J. Biochem.* **56**:521–526.

Weissenbach, J., Martin, R., and Dirheimer, G., 1975. Partial digestion with T_1 RNase and primary sequence of yeast $tRNA_{II}^{Arg}$. *Eur. J. Biochem.* **56**:527–532.

Proline tRNAs

Barrell, B. G., Seidman, J. G., Guthrie, C., and McClain, W. H. 1974. Transfer RNA biosynthesis: The nucleotide sequence of a precursor to serine and proline tRNAs. *Proc. Nat. Acad. Sci. USA* **71**:413–416.

Seidman, J. G., Barrell, B. G., and McClain, W. H. 1975. Five steps in the conversion of a large precusor RNA into bacteriophage proline and serine tRNAs. *J. Mol. Biol.* **99**:733–760.

Lysine tRNAs

Madison, J. T., and Boguslawski, S. J. 1974. Partial digestion of a yeast lysine tRNA and reconstruction of the nucleotide sequence. *Biochemistry* **13**:524–527.

Madison, J. T., Boguslawski, S. J., and Teetor, G. H. 1972. Nucleotide sequence of a lysine tRNA from baker's yeast. *Science* **176**:687–689.

Madison, J. T., Boguslawski, S. J., and Teetor, G. H. 1974. Oligonucleotide composition of a yeast lysine tRNA. *Biochemistry* **13**:518–523.

Smith, C. J., Ley, A. N., D'Obrenan, P., and Mitra, S. K. 1971. The structure and coding specificity of a lysine tRNA from the haploid yeast *S. cerevisiae* αS288C. *J. Biol. Chem.*, **246**:7817–7829.

Smith, C. J., Teh, H. S., Ley, A. N., and D'Obrenan, P. 1973. The nucleotide sequences of the major and minor lysine tRNAs from the haploid yeast *S. cerevisiae* αS288C. *J. Biol. Chem.* **248**:4475–4485.

Isoleucine tRNAs

Harada, F., and Nishimura, S. 1974. Purification and characterization of AUA specific isoleucine tRNA from *E. coli* B. *Biochemistry* **13**:300–307.

Takemura, S., Murakami, M., and Miyazaki, M. 1969a. Nucleotide sequence of isoleucine tRNA from *Torulopsis utilis*. *J. Biochem.* **65**:489–491.

Takemura, S., Murakami, M., and Miyazaki, M., 1969b. The primary structure of isoleucine tRNA from *Torulopsis utilis*. *J. Biochem.* **65**:553–566.

Yarus, M., and Barrell, B. G. 1971. The sequence of nucleotides in $tRNA^{Ile}$ from *E. coli* B. *Biochem. Biophys. Res. Comm.* **43**:729–733.

Methionine tRNAs

Cory, S., and Marcker, K. A. 1970. The nucleotide sequence of methionine $tRNA_M$. *J. Biochem.* **12**:177–194.

Cory, S., Marcker, K. A., Dube, S. K., and Clark, B. F. C. 1968. Primary structure of a methionine tRNA from *E. coli*. *Nature* **220**:1039–1040.

Dube, S. K., and Marcker, K. A. 1969. The nucleotide sequence of N-formyl-methionyl-transfer RNA. *Eur. J. Biochem.* **8**:256–262.

Dube, S. K., Marcker, K. A., Clark, B. F. C., and Cory, S. 1968. Nucleotide sequence of N-formyl-methionyl-transfer RNA. *Nature* **218**:233–234.

Dube, S. K. Marcker, K. A., Clark, B. F. C., and Cory, S. 1969. The nucleotide sequence of N-formyl-methionyl-transfer RNA. *Eur. J. Biochem.* **8**:244–255.

Ecarot-Charrier, B., and Cedergren, R. J. 1976. The preliminary sequence of $tRNA_F^{Met}$ from *Anacystis nidulans* compared with other initiator tRNAs. *FEBS Lett.* **63**:287–290.

Egan, B. Z., Weiss, J. F., and Kelmers, A. D. 1973. Separation and comparison of primary structures of three formylmethionine tRNAs from *E.coli* K-12MO. *Biochem. Biophys. Res. Comm.* **55**:320–327.

Gruhl, H., and Feldmann, H. 1975. The primary structure of a noninitiating methionine specific tRNA from brewer's yeast. *FEBS Lett.* **57**:145–148.

Högenauer, G., Turnowsky, F., and Unger, F. M. 1972. Codon–anticodon interaction of methionine specific tRNAs. *Biochem. Biophys. Res. Comm.* **46**:2100–2106.

Ishikura, H., Yamada, Y., Murao, K., Saneyoshi, M., and Nishimura, S. 1969. The presence of N-[9-(β-D-ribofuranosyl)purine-6yl-carbomoyl]threonine in serine, methionine, and lysine tRNAs from *E. coli*. *Biochem. Biophys. Res. Comm.* **37**:990–995.

Koiwai, O., and Miyazaki, M. 1976. The primary structure of non-initiator $tRNA_M^{Met}$ from baker's yeast. *J. Biochem.* **80**:951–959.

Petrissant, G., and Boisnard, M. 1974. Particularités structurales du méthionine $tRNA_m^{Met}$ de foie de lapin. *Biochimie* **56**:787–789.

Piper, P. W. 1975a. The nucleotide sequence of a methionine tRNA which functions in protein elongation in mouse myeloma cells. *Eur. J. Biochem.* **51**:283–293.

Piper, P. W., and Clark, B. F. C. 1974b. Primary structure of a mouse myeloma cell initiator tRNA. *Nature* **247**:516–518.

Simsek, M., and RajBhandary, U. L. 1972. The primary structure of yeast initiator tRNA. *Biochem. Biophys. Res. Comm.* **49**:508–515.

Simsek, M., RajBhandary, U. L., Boisnard, M. and Petrissant G. 1974. Nucleotide sequence of rabbit liver and sheep mammary gland cytoplasmic initiator tRNAs. *Nature* **247**:518–520.

Yamada, Y., and Ishikura, H. 1975. Nucleotide sequence of initiator tRNA from *B. subtilis*. *FEBS Lett.* **54**:155–158.

III Serine tRNAs

Ish-Horowicz, D., and Clark, B. F. C. 1973. The nucleotide sequence of a serine tRNA from *E. coli*. *J. Biol. Chem.* **248**:6663–6673.

Rogg, H., Müller, P., and Staehelin, M. 1975. Nucleotide sequences of rat liver serine tRNA. *Eur. J. Biochem.* **53**:115–127.

Yamada, Y., and Ishikura, H. 1973. Nucleotide sequence of $tRNA_3^{Ser}$ from *E. coli*. *FEBS Lett.* **29**:231–234.

Threonine tRNAs

Clarke, L., and Carbon, J. 1974. The nucleotide sequence of a threonine tRNA from *E. coli*. *J. Biol. Chem.* **249**:6874–6885.

III Arginine tRNAs

Kuntzel, B., Weissenbach, J. and Dirheimer, G. 1972. The sequence of nucleotides in $tRNA^{Arg}_{III}$ from brewer's yeast. *FEBS Lett.* **25**:189–191.

Kuntzel, B., Weissenbach, J., and Dirheimer, G. 1974. Structure primaire des $tRNA^{Arg}_{III}$ de levure de bière. *Biochimie* **56**:1069–1087.

Phenylalanine tRNAs

Blobstein, S. H., Grunberger D., Weinstein, I. B., and Nakanishi, K. 1973. Isolation and structure determination of the fluorescent base from bovine liver phenylalanine tRNA. *Biochemistry* **12**:188–193.

Dudock, B. S., and G. Katz. 1969. Large oligonucleotide sequences in wheat germ phenylalanine tRNA. *J. Biol. Chem.,* **244**:3069–3074.

Dudock, B. S., Katz, G., Taylor, E. K., and Holley, R. W. 1969. Primary structure of wheat germ phenylalanine tRNA. *Proc. Nat. Acad. Sci. USA* **62**:941–945.

Everett, G. A., and Madison, J. T. 1976. Nucleotide sequence of $tRNA^{Phe}$ from pea (*Pisum sativum,* Alaska). *Biochemistry* **15**:1016–1021.

Guerrier-Takada, C., Dirheimer, G., Grosjean, H., and Keith, G. 1975. The primary structure of $tRNA^{Phe}$ from *Bacillus stearothermophilus*. *FEBS Lett.* **60**:286–289.

Harbers, K., Thiebe, R., and Zachau, H. G. 1972. Preparation and characterization of fragments from yeast $tRNA^{Phe}$. *Eur. J. Biochem.* **26**:132–143.

Keith, G., Ebel, J. P., and Dirheimer, G. 1974. The primary structure of two mammalian $tRNAs^{Phe}$: Identity of calf liver and rabbit liver $tRNAs^{Phe}$. *FEBS Lett.* **48**:50–52.

Keith, G., Picaud, F., Weissenbach, J., Ebel, J. P., Petrissant, G., and Dirheimer, G. 1973. The primary structure of rabbit liver $tRNA^{Phe}$ and its comparison with known $tRNA^{Phe}$ sequences. *FEBS Lett.* **31**:345–347.

Kim, S. H., Quigley, G. J., Suddath, F. L., McPherson, A., Sneden, D., Kim, J. J., Weinzierl, J., and Rich, A. 1973. Three-dimensional structure of yeast phenylalanine tRNA: Folding of the polynucleotide chain. *Science* **179**:285–288.

Kim, S. H., Suddath, F. L., Quigley, G. J., McPherson, A., Sussman, J. L., Wang, A. H. J., Seeman, N. C., and Rich, A. 1974. Three-dimensional tertiary structure of yeast phenylalanine tRNA. *Science* **185**:435–440.

Nakanishi, K., Furutachi, N., Funamizu, M., Grunberger, D., and Weinstein, I. B. 1970. Structure of the fluorescent Y base from yeast phenylalanine tRNA. *J. Am. Chem. Soc.* **92**:7617–7619.

Philippsen, P., and Zachau, H. G. 1971. Fragments of yeast $tRNA^{Phe}$ and $tRNA^{Ser}$ prepared by partial digestion with spleen phosphodiesterase. *FEBS Lett.* **15**:69–74.

Pongo, O., Bald, R., and Reinwald, E. 1973. On the structure of yeast $tRNA^{Phe}$: Complementary-oligonucleotide binding studies. *Eur. J. Biochem.* **32**:117–125.

RajBhandary, U. L., Chang, S. H., Sneider, J., and Davis, D. 1968. Yeast phenylalanine tRNA: Partial digestion with ribonuclease T_1, and derivation of the total primary structure. *J. Biol. Chem.* **243**: 598–608.

RajBhandary, U. L., Chang, S. H., Stuart, A., Faulkner, R. D., Hoskinson, R. M., and Khorana,

H. G. 1967. The primary structure of yeast phenylalanine tRNA. *Proc. Nat. Acad. Sci. USA* **57**:751–758.

Rosenfeld, A., Stevens, C. L., and Printz, M. P. 1970. Studies on the secondary structure of phenylalanyl tRNA. *Biochemistry* **9**:4971–4980.

Takemura, S., Kasai, H., and Goto, M. 1974. Nucleotide sequence of the anticodon region of *Torulopsis* phenylalanine tRNA. *J. Biochem.* **75**:1169–1172.

Uziel, M., and Gassen, H. G. 1969. Structure of $tRNA^{Phe}$. *Fed. Proc.* **28**:409.

Wong, Y. P., Kearns, D. R., Shulman, R. G., Yamane, T., Chang, S., Chirskyian, J. G., and Fresco, J. R. 1973. High resolution NMR study of base pairing in the native and denatured conformers of tRNA. *J. Mol. Biol.* **74**:403–406.

Tyrosine tRNAs

Altman, S., and Smith, J. D. 1971. Tyrosine tRNA precursor molecule polynucleotide sequence. *Nature New Biol.* **233**:35–39.

Doctor, B. P., Loebel, J. E., and Kellog, D. A. 1966. Studies on the species specificity of yeast and *E. coli* tyrosine $tRNA_2$. *Cold Spring Harbor Symp. Quant. Biol.* **31**:543–548.

Doctor, B. P., Loebel, J. E., Sodd, M. A. and Winter, D. B. 1969. Nucleotide sequence of *E. coli* tyrosine tRNA. *Science* **163**:693–695.

Goodman, H. M., Abelson, J. N., Landy, A., Brenner, S., and Smith, J. D. 1968. Amber suppression: A nucleotide change in the anticodon of a tyrosine tRNA. *Nature* **217**:1019–1024.

Goodman, H. M., Abelson, J. N., Landy, A., Zadrazil, S., and Smith, J. D. 1970. The nucleotide sequences of tyrosine tRNAs of *E. coli*. *Eur. J. Biochem.* **13**:461–483.

Harada, F., Gross, H. J., Kimura, F., Chang, S. H., Nishimura, S., and RajBhandary, U. L. 1968. 2-methylthio N^6-(Δ^2-isopentenyl) adenosine: A component of *E. coli* tyrosine tRNA. *Biochem. Biophys. Res. Comm.* **33**:299–306.

Hashimoto, S., Miyazaki, M., and Takemura, S. 1969. Nucleotide sequence of tyrosine tRNA from *Torulopsis utilis*. *J. Biochem.* **65**:659–661.

Hachimoto, S., Takemura, S., and Miyazaki, M. 1972. Partial digestion with ribonuclease T_1 and derivation of the complete sequence of $tRNA^{Tyr}$ from *Torulopsis utilis*. *J. Biochem.* **72**:123–134.

Madison, J. T., Everett, G. A., and Kung, H. 1966a. Nucleotide sequence of a yeast tyrosine tRNA. *Science* **153**:531–534.

Madison, J. T. Everett, G. A., Kung, H. 1966b. On the nucleotide sequence of yeast tyrosine tRNA. *Cold Spring Harbor Symp. Quant. Biol.* **31**:409–416.

Madison, J. T., Everett, G. A., Kung, H. 1967. Oligonucleotides from yeast tyrosine tRNA. *J. Biol. Chem.* **242**:1318–1323.

Madison, J. T., and Kung, H.-K. 1967. Large oligonucleotides isolated from yeast tyrosine tRNA after partial digestion with ribonuclease T_1. *Science* **242**:1324–1330.

Seno, T., and Nishimura, S. 1971. Cleavage of *E. coli* tyrosine $tRNA_2$ in S-region and its effects on the structure and function of the reconstituted molecules. *Biochim. Biophys. Acta* **228**:141–152.

Takemura, S., Hashimoto, S., and Miyazaki, M. 1972. Complete digestion of tyrosine tRNA from *Torulopsis utilis* with pancreatic and T_1 ribonucleases. *J. Biochem.* **72**:111–121.

IV Serine tRNAs

Ginsberg, T., Rogg, H., and Staehelin, M. 1971. Nucleotide sequences of rat liver serine-tRNA. *Eur. J. Biochem.* **21**:249–257.

Hentzen, D., and Garel, J. P. 1976. Anticodon loop sequences of $tRNA^{Ser}_{CGA}$ and $tRNA^{Ser}_{IGA}$ from the posterior silkgland of *Bombyx mori* L. *Biochem. Biophys. Res. Comm.* **71**:241–248.

Ishikura, H., Yamada, Y., and Nishimura, S. 1971a. The nucleotide sequence of a serine tRNA from *E. coli*. *FEBS Lett.* **16**:68–70.

Ishikura, H., Yamada, Y., and Nishimura, S. 1971b. Structure of serine tRNA from *E. coli*. *Biochim. Biophys. Acta* **228**:471–481.

McClain, W. H., Barrell, B. G., and Seidman, J. G. 1975. Nucleotide alterations in bacteriophage T4 serine tRNA that affect the conversion of precursor RNA into tRNA. *J. Mol. Biol.* **99**:717–732.

Rogg, H., and Staehelin, M. 1971a. Nucleotide sequences of rat liver serine-tRNA. 1. Products of digestion with pancreatic ribonuclease. *Eur. J. Biochem.* **21**:235–242.

Rogg, H., and Staehelin, M. 1971b. Nucleotide sequence of rat liver serine-tRNA. 2. The products of digestion with ribonuclease T_1. *Eur. J. Biochem.* **21**:243–248.

Staehelin, M. 1971. The primary structure of tRNA. *Experientia,* **27**:1–11.

Staehelin, M., Rogg, H., Baguley, B. C., Ginsberg, T., and Wehrli, W. 1968. Structure of a mammalian serine tRNA. *Nature* **219**:1363–1365.

Zachau, H. G., Dütting, D., and Feldmann, H. 1966a. Nucleotidsequenzen zweier serinspezifischer Transfer-Ribonucleinsäuren. *Angew. Chem.*, **78**:392–393.

Zachau, H. G., Dütting, D., and Feldmann, H. 1966b. Serine specific tRNAs. *Hoppe-Seyler's Z. Phys. Chem.* **347**:229–235.

Tryptophan tRNAs

Hirsch, D. 1970. Tryptophan tRNA of *E. coli*. *Nature* **228**:57.

Hirsch, D. 1971. Tryptophan tRNA as the UGA suppressor. *J. Mol. Biol.* **58**:439–458.

Keith, G., Roy, A., Ebel, J. P., and Dirheimer, G. 1971. The nucleotide sequences of two tryptophan-tRNAs from brewer's yeast. *FEBS Lett.* **17**:306–308.

Maugh, T. H. 1974. Rous sarcoma virus: A new role for tRNA. *Science* **186**:41.

IV Leucine tRNAs

Chang, S. H., Kuo, S., Hawkins, E., and Miller, N. R., 1973. The corrected nucleotide sequence of yeast leucine tRNA. *Biochem. Biophys. Res. Comm.* **51**:951–955.

Chang, S. H., and Miller, N. 1971. The nucleotide sequence of yeast leucine tRNA. *Fed. Proc.* **30**:1101.

Kowalski, S., Yamane, T., and Fresco, J. R. 1971. Nucleotide sequence of the "denaturable" leucine tRNA from yeast. *Science* **172**:385–387.

Pinkerton, T. C., Paddock, G., and Abelson, J. 1972. Bacteriophage T4 $tRNA^{Leu}$. *Nature New Biol.* **240**:88–90.

Pinkerton, T. C., Paddock, G., and Abelson, J. 1973. Nucleotide sequence determination of bacteriophage T4 leucine tRNA. *J. Biol. Chem.* **248**:6348–6365.

Cysteine tRNAs

Holness, N.J., and Atfield, G. 1974. Nucleotide sequence of $tRNA^{Cys}$ from baker's yeast. *FEBS Lett.* **46**:268–270.

Holness, N. J., and Atfield, G. 1976a. The extraction and purification of a cysteine tRNA from baker's yeast. *Biochem. J.* **153**:429–435.

Holness, N. J., and Atfield, G. 1976b. The nucleotide sequence of cysteine tRNA from baker's yeast. *Biochem. J.* **153**:447–454.

CHAPTER 8

Abraham, D. J. 1971. Proposed detailed structural model for tRNA and its geometric relationship to a messenger. *J. Theor. Biol.* **30**:83–91.

Allende, J. E., and Allende, C. C. 1971. Detection and isolation of complexes between aminoacyl-tRNA synthetases and their substrates. *Meth. Enzym.* **20**:210–220.

Bhargava, P. M. 1971. Aminoacyl-tRNA synthetase recognition code-words in yeast tRNAs—a proposal. *J. Theor. Biol.* **29**:447–469.

Bina-Stein, M., and Crothers, D. M. 1974. Conformational changes of tRNA. *Biochemistry* **13**:2771–2775.

Blanquet, S., Fayat, G., Poiret, M., and Waller, J. P. 1975. The mechanism of action of methionyl-RNA synthetase from *E. coli*. *Eur. J. Biochem.* **51**:567–571.

Bolton, P. H., and Kearns, D. R. 1975. NMR evidence for common tertiary structure base pairs in yeast and *E. coli* tRNA. *Nature* **255**:347–349.

Briand, J. P., Jonard, G., Guilley, H., Richards, K., and Hirth, L. 1977. Nucleotide sequence ($n = 159$) of the amino-acid-accepting 3′-OH extremity of TYMV RNA. *Eur. J. Biochem.*, **72**:453–463.

Briand, J. P., Richards, K. E., Bouley, J. P., Witz, J., and Hirth, L. 1976. Structure of the amino-acid accepting 3′-end of high-molecular-weight eggplant mosaic virus RNA. *Proc. Nat. Acad. Sci. USA* **73**:737–741.

Budzik, G. P., Lam, S. S. M., Schoemaker, H. J. P., and Schimmel, P. R. 1975. Two photo cross-linked complexes of isoleucine specific tRNA with aminoacyl tRNA synthetases. *J. Biol. Chem.* **250**:4433–4439.

Carbon, J., and Curry, J. B. 1968. Genetically and chemically derived missense suppressor tRNAs with altered enzymatic aminoacylation rates. *J. Mol. Biol.* **38**:201–216.

Carbon, J., and Fleck, E. W. 1974. Genetic alteration of structure and function in glycine tRNA of *E. coli*: Mechanism of suppression of the tryptophan synthetase A78 mutation. *J. Mol. Biol.* **85**:371–391.

Caron, M., Brisson, N., and Dugas. H. 1976. Evidence for a conformational change in $tRNA^{Phe}$ upon aminoacylation. *J. Biol. Chem.* **251**:1529–1530.

Cedergren, R. J., Cordeau, J. R., and Robillard, P. 1972. On the phylogeny of tRNAs. *J. Theor. Biol.* **37**:209–220.

Celis, J. E., Hooper, M. L., and Smith, J. D. 1973. Amino acid acceptor stem of *E. coli* suppressor $tRNA^{Tyr}$ is a site of synthetase recognition. *Nature New Biol.* **244**:261–264.

Chambers, R. W. 1971. On the recognition of tRNA by its aminoacyl-tRNA ligase. *Progr. Nucl. Acid Res. Mol. Biol.* **11**:489–525.

Chambers, R. W., Aoyagi, S., Furukawa, Y., Zawadzka, H., and Bhanot, O. S. 1973. Inactivation of valine acceptor activity by a C→U missense change in the anticodon of yeast valine tRNA. *J. Biol. Chem.* **248**:5549–5551.

Chapeville, F., Lipmann, F., von Ehrenstein, G., Weisblum, B., Ray, W. J., and Benzes, S. 1962. On the role of soluble RNA in coding for amino acids. *Proc. Nat. Acad. Sci. USA* **48**:1086–1092.

Chapeville, F., and Rouget, P. 1972. Aminoacyl-tRNA synthetases. *Frontiers Biol.* **27**:5–32.

Chatterjee, S. K. and Kaji, H. 1970. Conformational changes of tRNA on aminoacylation. *Biochim. Biophys. Acta* **224**:88–98.

Chinali, G., Sprinzl, M., Parmeggioni, A., and Cramer, F. 1974. Participation in protein biosynthesis of tRNA bearing altered 3′-terminal ribosyl residues. *Biochemistry* **13**:3001–3010.

Cole, P. E., and Crothers, D. M. 1972. Conformational changes of tRNA. *Biochemistry* **11**:4368–4374.

Cole, P. E., Yang, S. K., and Crothers, D. M. 1972. Conformational changes of tRNA. Equilibrium phase diagrams. *Biochemistry* **11**:4358–4368.

Cramer, F. 1971. Three-dimensional structure of tRNA. *Progr. Nucl. Acid Res. Mol. Biol.* **11**:391–421.

Dayhoff, M. O., and McLaughlin, P. J. 1972. Early evolution: Transfer RNA. *In*: Dayhoff, M. O., ed., *Atlas of Protein Sequence and Structure. Vol. 5*. National Biomedical Research Foundation, Washington, D.C., p. 111–118.

Delaney, P., Bierbaum, J., and Ofengand, J. 1974. Conformational changes in thiouridine region of *E. coli* tRNA as assessed by photochemically induced cross-linking. *Arch. Biochem. Biophys.* **161**:260–267.

Dube, S. K. 1973. Evidence for "three-point" attachment of tRNA to methionyl tRNA synthetase. *Nature New Biol.* **243**:103–105.

Eisinger, J., and Gross, N. 1975. Conformers, dimers, and anticodon complexes of $tRNA_2^{Glu}$ (*E. coli*). *Biochemistry* **14**:4031–4040.

Elder, K. T., and Smith, A. E. 1973. Methionine tRNA of avian myeloblastosis virus. *Proc. Nat. Acad. Sci. USA* **70**:2823–2826.

Evans, J. A., and Nazario, M. 1974. *Neurospora* arginyl tRNA ligase binding and dissoaciation of tRNA. *Biochemistry* **13**:3092–3098.

Fasiolo, F., Befort, N., Boulanger, Y., and Ebel, J. P. 1970. Purification et quelques propriétés de la phénylalanyl-tRNA synthétase de levure de boulangerie. *Biochim. Biophys. Acta* **217**:305–318.

Fasiolo, F., and Ebel, J. P. 1974. Yeast phenylalanyl tRNA synthetase. *Eur. J. Biochem.* **49**:257–263.

Fasiolo, F., Renny, P., Pouyet, J., and Ebel, J. P. 1974. Yeast phenylanlanyl-tRNA synthetase. *Eur. J. Biochem.* **50**:227–236.

Gamble, R. C., and Schimmel, P. R. 1974. Transfer RNA conformation in solution investigated by isotope labeling. *Proc. Nat. Acad. Sci. USA* **71**:1356–1360.

Gangloff, J., Dirheimer, G., and Gangloff, M. L. 1973. Studies on aspartyl-tRNA synthetase from baker's yeast. *Biochim. Biophys. Acta* **294**:263–272.

Glick, J. M., and Leboy, P. S. 1977. Purification and properties of tRNA (adenine-1)-methyltransferase from rat liver. *J. Biol. Chem.* **252**:4790–4795.

Gros, C., Lemaire, G., Rapenbusch, R. V., and Labouesse, B. 1972. The subunit structure of tryptophanyl tRNA synthetase from beef pancreas. *J. Biol. Chem.* **247**:2931–2943.

Gross, H. J. 1973. Transfer RNA: Evidence for decreasing size variation during evolution. *J. Mol. Evol.* **2**:339–342.

Guilley, H., Jonard, G., and Hirth, L. 1975. Sequence of 71 nucleotides at the 3′-end of tobacco mosaic virus RNA. *Proc. Nat. Acad. Sci. USA* **72**:864–868.

Haines, J. A., and Zamecnik, P. C. 1967. Chemical modification of aminoacyl ligases and the effect on formation of aminoacyl-tRNAs. *Biochim. Biophys. Acta* **146**:227–238.

Hanke, T., Bartmann, P., Hennecke, H., Kosakowski, H. M., Jaenicke, R., Holler, E., and Boeck, A. 1974. L-phenylalanyl-tRNA synthetase of *E. coli* K-10; A reinvestigation of molecular weight and subunit structure. *Eur. J. Biochem.* **43**:601–607.

Harpold, M. A., and Calvin, M. 1973. A simple model of the amino acid acceptor terminus of a tRNA. *Biochim. Biophys. Acta* **308**:117–128.

Hashimoto, S., Kawata, M., and Takemura, S. 1972a. Reconstitution of an active acceptor complex which lacks the anticodon of *Torulopsis* tyrosine tRNA. *J. Biochem.* **72**:1339–1349.

Hashimoto, S., Takemura, S., and Miyazaki, M. 1972b. Partial digestion with ribonuclease T_1 and derivation of the complete sequence of tyrosine tRNA from *Torulopsis utilis*. *J. Biochem.* **72**:123–134.

Hashimoto, S., Takemura, S., Yabuki, S., Konishi, K., and Samejima, T. 1972c. Physicochemical studies on conformation of a complex reconstituted from half molecules of *Torulopsis utilis* tyrosine tRNA. *J. Biochem.* **72**:1185–1195.

Hecht, S. M., Kozarich, J. W., and Schmidt, F. J. 1974. Isomeric phenylalanyl-tRNAs. Position of the aminoacyl moiety during protein biosynthesis. *Proc. Nat. Acad. Sci. USA* **71**:4317–4321.

Heider, H., Gottschalk, E., and Cramer, F. 1971. Isolation and characterization of seryl-tRNA synthetase from yeast. *Eur. J. Biochem.* **20**:144–152.

Hennecke, H., and Böck, A. 1975. Altered α subunits in phenylalanyl-tRNA synthetases from *p*-fluorophenylalanine-resistant strains of *E. coli*. *Eur. J. Biochem.* **55**:431–437.

Hirshfield, I. N., and Bloemers, H. P. J. 1969. The biochemical characterization of two mutant arginyl tRNA synthetases from *E. coli* K-12. *J. Biol. Chem.* **244**:2911–2916.

Holler, E., Hammer-Rober, B., Hanke, T., and Bartmann, P. 1975. The catalytic mechanism of amino acid: tRNA ligases. *Biochemistry* **14**:2496–2503.

Holmquist, R., and Jukes, T. H. 1973. No evidence for a common evolutionary origin of 5 S rRNA and tRNA. *Nature New Biol.* **245**:127.

Isham, K. R., and Stulberg, M. P. 1974. Modified nucleosides in under-methylated phenylalanine tRNA from *E. coli*. *Biochim. Biophys. Acta* **340**:177–182.

Jones, C. R., and Kearns, D. R. 1974. Investigations of the structure of yeast $tRNA^{Phe}$ by nuclear magnetic resonance: Paramagnetic rare earth ion probes of structure. *Proc. Nat. Acad. Sci. USA* **71**:4237–4240.

Jukes, T. H., and Holmquist, R. 1972. Evolution of tRNA molecules as a repetitive process. *Biochem. Biophys. Res. Comm.* **49**:212–216.

Kim, S. H. 1975. Symmetry recogniton hypothesis model for tRNA binding to aminoacyl tRNA synthetase. *Nature* **256**:679–681.

Kim, S. H., Quigley, G. J., Suddath, F. L., McPherson, A., Sneden, D., Kim, J. J., Weinzierl, J., and Rich, A. 1973. Three dimensional structure of yeast phenylalanine tRNA; folding of the polynucleotide chain. *Science* **179**:285–288.

Kim, S. H., Suddath, F. L., Quigley, G. J., McPherson, A., Sussman, J. L., Wang, A. H. J., Seeman, N. C., and Rich, A. 1974a. Three dimensional tertiary structure of yeast phenylalanine tRNA. *Science* **185**:435–440.

Kim, S. H., Sussman, J. L., Suddath, F. L., Quigley, G. J., McPherson, A., Wang, A. H., Seeman, N. C., and Rich, A. 1974b. The general structure of tRNA molecules. *Proc. Nat. Acad. Sci. USA* **71**:4970–4974.

Kiselev, L. L., and Favorova, O. O. 1974. Aminoacyl-tRNA synthetases: Some recent results and achievements. *Adv. Enzymol.* **40**:141–238.

Koch, G. L. E., Boulanger, Y., and Hartley, B. S. 1974. Repeating sequences in aminoacyl-tRNA synthetases. *Nature* **249**:316–320.

Lamy, D., Jonard, G., Guilley, H., and Hirth, L. 1975. Comparison between the 3′OH end RNA sequence of two strains of TMV which may be aminoacylated. *FEBS Lett.* **60**:202–204.

Lapointe, J., and Söll, D. 1972a. Glutamyl tRNA synthetase of *E. coli*. I. Purification and properties. *J. Biol. Chem.* **247**:4966–4974.

Lapointe, J., and Söll, D. 1972b. Glutamyl tRNA synthetase of *E. coli*. II. Interaction with intact glutamyl tRNA. *J. Biol. Chem.* **247**:4975–4981.

Lawrence, F. 1973. Effect of adenosine on methionyl-tRNA synthetase. *Eur. J. Biochem.* **40**:493–500.

Lawrence, F., Blanquet, S., Poiret, M., Robert-Gero, M., and Waller, J. P. 1973. Ion requirements and kinetic parameter of the ATP-PP_i exchange and methionine-transfer reactions catalyzed by the native and trypsin-modified enzymes. *Eur. J. Biochem.* **36**:234–243.

Lawrence, F., Shire, D. J., and Waller, J. P., 1974. The effect of adenosine analogues on the ATP-pyrophosphate exchange reaction catalysed by methionyl-tRNA synthetase. *Eur. J. Biochem.* **41**:73–81.

Lemaire, G., Gros, C., Epely, S., Kaminski, M., and Labouesse, B. 1975. Multiple forms of tryptophanyl-tRNA synthetase from beef pancreas. *Eur. J. Biochem.* **51**:237–252.

Lengyel, P., and Söll, D. 1969. Mechanism of protein biosynthesis. *Bacteriol. Rev.* **33**:264–301.

Levitt, M. 1969. Detailed molecular model for tRNA. *Nature* **224**:759–763.

Maelicke, A., Sprinzl, M., van der Haar, F., Khwaja, T. A., and Cramer, F. 1974. Structural studies on phenylalanine tRNA from yeast with the spectroscopic label formycin. *Eur. J. Biochem.* **43**:617–625.

Marcu, K., Mignery, R., Reszelbach, R., Roe, B., Sirover, M., and Dudock, B. 1973. The absence of ribothymidine in specific eukaryotic tRNAs I. Glycine and threonine tRNAs of wheat embryo. *Biochem. Biophys. Res. Comm.* **55**:477–483.

Marcu, K., *et al.* 1974. Personal communication.

Mehler, A. H., and Mitra, S. K. 1967. The activation of arginyl tRNA synthetase by tRNA. *J. Biol. Chem.* **242**:5495–5499.

Muench, K. H., Lipscomb, M. S., Lee, M., and Kuehl, G. V. 1975. Homologous cysteine-containing sequences in tryptophanyl-tRNA synthetases from *E. coli* and human placentas. *Science* **187**:1089–1091.

Mullins, D. W., Lacey, J. C., and Hearn, R. A. 1973a. 5 S rRNA and tRNA—evidence for a common evolutionary origin. *Nature New Biol.* **242**:80–81.

Mullins, D. W., Lacey, J. C., and Hearn, R. A. 1973b. Reply. *Nature New Biol.* **245**:127–128.

Murayama, A., Raffin, J. P., Remy, P., and Ebel, J. P. 1975a. Yeast phenylalanyl-tRNA synthetase: Properties of the sulfhydryl groups; evidence for -SH requirements in tRNA acylation. *FEBS Lett.* **53**:15–22.

Murayama, A., Raffin, J. P., Remy, P., and Ebel, J. P. 1975b. Yeast phenylalanyl-tRNA synthetase; isolation of subunits on organomercurial-sepharose columns. *FEBS Lett.* **53**:23–25.

Nazario, M., and Evans, J. A. 1974. Physical and kinetic studies of arginyl tRNA ligase of *Neurospora. J. Biol. Chem.* **249**:4934–4942.

Novelli, G. D. 1967. Amino acid activation for protein synthesis. *Ann. Rev. Biochem.* **36**:449–484.

Öberg, B., and Philipson, L. 1972. Binding of histidine to TMV RNA. *Biochem. Biophys. Res. Comm.* **48**:927–932.

Odom, O. W., Hardesty, B., Wintermeyer, W., and Zachau, H. G. 1974. The effect of removal or replacement with proflavine of the Y base in the anticodon loop of yeast $tRNA^{Phe}$ on binding into the acceptor or donor sites of reticulocyte ribosomes. *Arch. Biochem. Biophys.* **162**:536–551.

Ofengand, J., Chládek, S., Robilard, G., and Bierbaum, J. 1974. Enzymatic acylation of oxydized reduced tRNA by *E. coli,* yeast, and rat liver synthetases occurs almost exclusively at the 2′ hydroxyl. *Biochemistry* **13**:5425–5432.

Ofengand, J., and Henes, C. 1969. The function of pseudouridylic acid in tRNA. *J. Biol. Chem.* **244**:6241–6253.

Papas, T. S., and Peterkofsky, A. 1972. A random sequential mechanism for arginyl tRNA synthetase of *E. coli. Biochemistry* **11**:4602–4608.

Parfait, R., and Grosjean, H. 1972. Arginyl-tRNA synthetase from *Bacillus stearothermophilus. Eur. J. Biochem.* **30**:242–249.

Penneys, N. S., and Muench, K. H. 1974. Human placental tryptophanyl tRNA synthetase. *Biochemistry 13*:560–565.

Petrissant, G. 1973. Evidence for the absence of the G-T-ψ-C sequence from two mammalian initiator tRNAs. *Proc. Nat. Acad. Sci. USA* **70**:1046–1049.

Ravel, J. M., Wang, S. F., Heinemeyer, C., and Shive, W. 1965. Glutamyl and glutaminyl RNA synthetases of *E. coli. J. Biol. Chem.* **240**:432–438.

Reid, B. R., Einarson, B., and Schmidt, J. 1972. Loop accessibility in tRNA. *Biochimie* **54**:325–332.

Riesner, D., Maass, G., Thiebe, R., Philippsen, P., and Zachau, H. G. 1973. The conformational transitions in yeast $tRNA^{Phe}$ as studied with $tRNA^{Phe}$ fragments. *Eur. J. Biochem.* **36**:76–88.

Roberts, J. W., and Carbon, J. 1974. Molecular mechanism for missense suppression in *E. coli*. *Nature* **250**:412–414.

Roberts, R. J., Lovinger, G. G., Tamura, T., and Strominger, J. L. 1974. Staphylococcal tRNAs. I. Isolation and purification of the isoaccepting tRNA from *Staphylococcus epidermidis* Texas 26. *J. Biol. Chem.* **249**:4781–4786.

Roe, B., Michael, M., and Dudock, B. 1973. Function of N^2-methylguanine in phenylalanine tRNA. *Nature New Biol.* **246**:135–138.

Rymo, L., Lundvik, L., and Lagerkvist, U. 1972. Subunit structure and binding properties of three amino acid tRNA ligases. *J. Biol. Chem.* **247**:3888–3899.

Santi, D. V., Danenberg, P. V., and Satterly, P. 1971. Phenylalanyl tRNA synthetase from *E. coli*. Reaction parameters and order of substrate addition. *Biochemistry* **10**:4804–4812.

Schmidt, J., Wang, R., Stanfield, S., and Reid, B. R. 1971. Yeast phenylalanyl tRNA synthetase. *Biochemistry* **10**:3264–3268.

Schoemaker, H. J. P., Budzik, G. P., Giegé, R., and Schimmel, P. R. 1975. Three photocross-linked complexes of yeast phenylalanine specific tRNA with aminoacyl tRNA synthetases. *J. Biol. Chem.* **250**:4440–4444.

Schoemaker, H. J. P., and Schimmel, P. R. 1974. Photo-induced joining of a tRNA with its cognate amino-acid-tRNA synthetase. *J. Mol. Biol.* **84**:503–513.

Seno, T., Agris, P. F., and Söll, D. 1974. Involvement of the anticodon region of *E. coli* $tRNA^{Gln}$ and $tRNA^{Glu}$ in the specific interaction with cognate aminoacyl-tRNA synthetase. *Biochim. Biophys. Acta* **349**:328–338.

Shulman, R. G., Hilbers, C. W., Kearns, D. R., Reid, B. R., and Wong, Y. P. 1973. Ring-current shifts in the 300 MHz NMR spectra of six purified tRNA molecules. *J. Mol. Biol.* **78**:57–69.

Simsek, M., Petrissant, G., and RajBhandary, U. L. 1973a. Replacement of the sequence G-T-ψ-C-G(A)- by G-A-U-C-G in initiator tRNA of rabbit liver cytoplasm. *Proc. Nat. Acad. Sci. USA* **70**:2600–2604.

Simsek, M., Ziegenmeyer, J., Heckman, J., and RajBhandary, U. L. 1973b. Absence of the sequence G-T-ψ-C-G(A)- in several eukaryotic cytoplasmic initiator rRNAs. *Proc. Nat. Acad. Sci. USA* **70**:1041–1045.

Singer, C. E., Smith, G. R., Cortese, R., and Ames, B. N. 1972. Mutant $tRNA^{His}$ ineffective in repression and lacking two pseudouridine modifications. *Nature New Biol.* **238**:72–74.

Singhal, R. P. 1971. Modification of *E. coli* glutamate tRNA with bisulfite. *J. Biol. Chem.* **246**:5848–5851.

Singhal, R. P. 1974. Chemical probe of structure and function of tRNAs. *Biochemistry* **13**:2924–2932.

Sprinzl, M., and Cramer, F. 1973. Accepting site for aminoacylation of $tRNA^{Phe}$ from yeast. *Nature New Biol.* **245**:3–5.

Squires, C., and Carbon, J. 1971. Normal and mutant glycine tRNAs. *Nature New Biol.* **233**:274–277.

Steinberg, W. 1974. Temperature-induced depression of tryptophan biosynthesis in a tryptophanyl-tRNA synthetase mutant of *B. subtilis*. *J. Bact.* **117**:1023–1034.

Suddath, F. L., Quigley, G. J., McPherson, A., Sneden, D., Kim, J. J., Kim, S. H., and Rich, A. 1974. Three dimensional structure of yeast phenylalanine tRNA at 3.0 Å resolution. *Nature* **248**:20–24.

Taglang, R., Waller, J. P., Befort, N., and Fasiolo, F. 1970. Amino-acylation du $tRNA^{Val}$ de *E. coli* par la phénylalanyl–tRNA synthétase de levure. *Eur. J. Biochem.* **12**:550–557.

Tal, J., Deutscher, M. P., and Littauer, U. Z. 1972. Biological activity of *E. coli* $tRNA^{Phe}$ modified in its C-C-A terminus. *Eur. J. Biochem.* **28**:478–491.

Thiebe, R. 1975. Aminoacylation of tRNA. Magnesium requirement and spermidine effect. *FEBS Lett.* **51**:259–261.

Thiebe, R., and Zachau, H. G. 1968. A special modification next to the anticodon of phenylalanine tRNA. *Eur. J. Biochem.* **5**:546–555.

Thomas, G. J., Chen, M. C., and Hartman, K. A. 1973. Raman studies of nucleic acids. X. Conformational structure of *E. coli* tRNAs in aqueous solution. *Biochim. Biophys. Acta* **324**:37–49.

von Ehrenstein, G., Weisblum, B., and Benzer, S. 1963. The function of sRNA as amino acid adaptor in the synthesis of hemoglobin. *Proc. Nat. Acad. Sci. USA* **49**:669–675.

White, B. N., and Tener, G. M. 1973. Properties of $tRNA^{Phe}$ from *Drosophila. Biochim. Biophys. Acta* **312**:267–275.

Williams, R. J., Nagel, W., Roe, B., and Dudock, B. 1974. Primary structure of *E. coli* alanine tRNA: Relation to the yeast phenylalanyl tRNA synthetase recognition site. *Biochem. Biophys. Res. Comm.* **60**:1215–1221.

Wolfenden, R., Rammler, D. H., and Lipmann, F. 1964. On the site of esterification of amino acids to soluble RNA. *Biochemistry* **3**:329–338.

Wong, Y. P., Reid, B. R., and Kearns, D. R. 1973. Conformation of charged and uncharged tRNAs. *Proc. Nat. Acad. Sci. USA* **70**:2193–2195.

Yang, C. H., and Söll, D. 1974. Studies of tRNA tertiary structure by singlet–singlet energy transfer. *Proc. Nat. Acad. Sci. USA* **71**:2838–2842.

Yang, S. K., and Crothers, D. M. 1972. Conformational changes of tRNA. Comparison of the early melting transition of two tyrosine-specific tRNAs. *Biochemistry* **11**:4375–4381.

Yem, D. W., and Williams, L. S. 1973. Evidence for the existence of two arginyl-tRNA synthetase activities in *E. coli. J. Bact.* **113**:891–894.

CHAPTER 9

Aaslestad, H. G., Clark, H. F., Bishop, D. H. L., and Koprowski, H. 1971. Comparison of the RNA polymerases of two rhabdoviruses, Kern Canyon virus and vesicular stomatitis virus. *J. Virol.* **7**:726–735.

Abelson, J., and Thomas, C. A. 1966. The anatomy of the T5 bacteriophage DNA molecule. *J. Mol. Biol.* **18**:262–287.

Abram, D., and Koffler, H. 1964. The *in vitro* formation of flagella-like filaments and other structures from flagellin. *J. Mol. Biol.* **9**:168–185.

Acheson, N. H., Buetti, E., Scherrer, K., and Weil, R. 1971. Transcription of the polyoma virus genome: Synthesis and cleavage of late polyoma-specific RNA. *Proc. Nat. Acad. Sci. USA* **68**:2231–2235.

Adams, J. M. 1972. Nucleotide sequence from the 5′-end to the first cistron of R17 bacteriophage ribonucleic acid. *Biochemistry* **11**:976–988.

Adams, M. H., and Wade, E. 1954. Classification of bacterial viruses: The relationship of two *Serratia* viruses to coli-dysentery phages T3, T7, and D44. *J. Bact.* **68**:320–325.

Air, G. M., Sanger, F., and Coulson, A. R. 1976. Nucleotide and amino acid sequences of gene G of ϕX174. *J. Mol. Biol.* **108**:519–533.

Almeida, J. D., and Waterson, A. P. 1970. Two morphological aspects of influenza virus. *In*: Barry, R. D., and B. W. J. Mahy, eds., *The Biology of Large RNA Viruses,* New York, Academic Press, p. 27–51.

Aloni, Y. 1972. Extensive symmetrical transcription of SV40 DNA in virus-yielding cells. *Proc. Nat. Acad. Sci. USA* **69**:2404–2409.

Aloni, Y. 1975. Methylated SV40 mRNAs. *FEBS Lett.* **54**:363–367.

Alper, T., Cramp, W. A., Haig, D. A., and Clarke, M. C. 1967. Does the agent of scrapie replicate without nucleic acid? *Nature* **214**:764–766.

Alper, T., Haig, D. A., and Clarke, M. C. 1966. The exceptionally small size of the scrapie agent. *Biochem. Biophys. Res. Comm.* **22**:278–284.

Anderson, T. 1960. On the fine structure of the temperate bacteriophages P1, P2, and P22. *Proc. Eur. Reg. Conf. Elect. Micro.* **2**:10008–10012.

Anraku, N., and Tomizawa, J. 1965. Joining of parental polynucleotides of phage T4 in the presence of 5-fluorideoxyuridine. *J. Mol. Biol.* **11**:501–508.

August, J. T., Banerjee, A. K., Eoyang, L., de Fernandez, M. T. F., Hori, K., Kuo, C. H., Rensing, U., and Shapiro, L. 1968. Synthesis of bacteriophage Qβ RNA. *Cold Spring Harbor Symp. Quant. Biol.* **33**:73–81.

Auld, D. S., Kawaguchi, H., Livingston, D. M., and Vallee, B. L. 1974. Reverse transcriptase from avian myeloblastosis virus: A zinc metalloenzyme. *Biochem. Biophys. Res. Comm.* **57**:967–972.

Baas, P. D., and Jansz, H. S. 1972a. Asymmetric information transfer during φX174 DNA replication. *J. Mol. Biol.* **63**:557–568.

Baas, P. D., and Jansz, H. S. 1972b. φX174 replicative form DNA replication, origin and direction. *J. Mol. Biol.* **63**:569–576.

Bachenheimer, S. L., and Roizman, B. 1972. RNA synthesis in cells infected with herpes simplex virus. *J. Virol.* **10**:875–879.

Baldwin, R. L., Barrand, P., Fritsch, A., Goldthwait, D. A., and Jacob, F. 1966. Cohesive sites on the DNAs from several temperate coliphages. *J. Mol. Biol.* **17**:343–357.

Baliga, B. S., Borek, E., Weinstein, I. B., and Srinivasan, P. R. 1969. Differences in the tRNAs of normal liver and Novikoff hepatoma. *Proc. Nat. Acad. Sci. USA* **62**:899–905.

Ball, L. A. 1973. Mutual influence of the secondary structure and information content of a mRNA. *J. Theor. Biol.* **41**:243–247.

Baltimore, D. 1976. Viruses, polymerases, and cancer. *Science* **192**:632–636.

Baltimore, D., Huang, A., Manly, K. F., Rekosh, D., and Stampfer, M. 1971. The synthesis of protein by mammalian RNA viruses. *In*: Wolstenholme, G. E. W., and M. O'Connor, eds., *Strategy of the Viral Genome,* Edinburgh, Churchill Livingston, p. 101–110.

Baltimore, D., Huang A., and Stampfer, M. 1970. RNA synthesis of vesicular stomatitis virus. *Proc. Nat. Acad. Sci. USA* **66**:572–576.

Baltimore, D., and Smoler, D. 1971a. Primer requirement and template specificity of the RNA tumor virus DNA polymerase. *Proc. Nat. Acad. Sci. USA* **68**:1507–1511.

Baltimore, D., and Smoler, D. 1971b. Template and primer requirements for the avian myeloblastosis DNA polymerase. *In*: Ribbons, D. W., J. F. Woessner, and J. Schultz, eds., *Nucleic Acid–Protein Interaction and Nucleic Acid Synthesis in Virus Infection.* Vol. 2, Amsterdam, North-Holland Publishing Co., p. 328–332.

Bancroft, J. B., Hiebert, E., Rees, M. W., and Markham, R. 1968. Properties of cowpea chlorotic mottle virus, its protein and nucleic acid. *Virology* **34**:224–239.

Barrell, B. G., Coulson, A. R., and McClain, W. H. 1973. Nucleotide sequence of a glycine tRNA coded by bacteriophage T4. *FEBS Lett.* **37**:64–69.

Barrell, B. G., Seidman, J. G., Guthrie, C., and McClain, W. H. 1974. Transfer RNA biosynthesis: The nucleotide sequence of a precursor to serine and proline tRNAs. *Proc. Nat. Acad. Sci. USA* **71**:413–416.

Barrell, B. G., Weith, H. L., Donelson, J. E., and Robertson, H. D. 1975. Sequence analysis of the ribosome-protected bacteriophage φX174 DNA fragment containing the gene *G* initiation site. *J. Mol. Biol.* **92**:377–393.

Battula, N., and Loeb, L. A. 1975. Characterization of polynucleotides with errors in base-pairing synthesized by AMV DNA polymerase. *J. Biol. Chem.* **250**:4405–4409.

Battula, N., and Loeb, L. A. 1976. On the fiedlity of DNA replication. *J. Biol. Chem.* **251**:982–986.

Bautz, E. K. F., Kasai, T., Reilly, E., and Bautz, F. A. 1966. Regulation of mRNA synthesis in *E. coli* after infection with bacteriophage T4. *Proc. Nat. Acad. Sci. USA* **55**:1081–1088.

Bautz, E. K. F., and Reilly, E. 1966. Gene-specific mRNA: Isolation by the deletion method. *Science* **151**:328–330.

Beijerinck, M. W. 1899. Über ein Contagium vivum fluidum als Ursache der Flecken-krankheit der Tabaksblätter. *Zentr. Bakter. Parasit.* **5**:27–35.

Benjamin, T. L. 1966. Virus-specific RNA in cells productively infected or transformed by polyoma virus. *J. Mol. Biol.* **16**:359–373.

Bergoin, M., and Dales, S. 1971. Comparative observations on poxviruses of invertebrates and vertebrates. *In*: Maramorosch, K., and E. Kurstak, eds., *Comparative Virology,* New York, Academic Press, p. 169–205.

Bijlenga, R. K. L., Broek, R. v.d., and Kellenberger, E. 1974. The transformation of γ-particles into T4 heads. 1. Evidence for the conservative mode of this transformation. *J. Supramol. Struct.* **2**:45–59.

Billeter, M. A., Dahlberg, J. E., Goodman, H. M., Hindley, J., and Weissmann, C. 1969. Sequence of first 175 nocleotides from the 5′-terminus of Qβ RNA synthesized *in vitro. Nature* **224**:1083–1086.

Bils, R. F., and Hall, C. E. 1962. Electron microscopy of wound-tumor virus. *Virology* **17**:123–130.

Bishop, D. H. L., and Roy, P. 1971. Kinetics of RNA synthesis by vesicular stomatitis virus particles. *J. Mol. Biol.* **58**:799–814.

Boedtker, H. 1959. Some physical properties of infective RNA isolated from TMV. *Biochim. Biophys. Acta* **32**:519–531.

Bolle, A., Epstein, R. H., Salser, W., and Geiduschek, E. P. 1968. Transcription during bacteriophage T4 development. *J. Mol. Biol.* **31**:325–348.

Bolognesi, D. P. 1974. Structural components of RNA tumor viruses. *Adv. Virus. Res.* **19**:315–359.

Bolognesi, D. P., and Obara, T. 1970. Minor RNA and other components of host origin intrinsic to ALV particles. *Bibl. Haematol.* **36**:126–139.

Bonar, R. A., Sverak, A., Bolognesi, D. P., Langlois, A. J., Beard, D., and Beard, J. W. 1967. RNA components of BAI strain A (myeloblastosis) avian tumor virus. *Cancer Res.* **27**:1138–1157.

Bonhoeffer, F., and Gierer, A. 1963. On the growth mechanism of the bacterial chromosome. *J. Mol. Biol.* **7**:534–540.

Bordier, C., and Dubochet, J. 1974. Electron microscopic localization of the binding sites of *E. coli* RNA polymerase in the early promoter region of T7 DNA. *Eur. J. Biochem.* **44**:617–624.

Borland, R., and Mahy, B. W. J. 1968. DNA-dependent RNA polymerase activity in cells infected with influenza virus. *J. Virol.* **2**:33–39.

Botchan, P., Wang, J. C., and Echols, H. 1973. Effect of circularity and superhelicity on transcription from bacteriophage λ DNA. *Proc. Nat. Acad. Sci. USA* **70**:3077–3081.

Both, G. W., Banerjee, A. K., and Shatkin, A. J. 1975. Methylation-dependent translation of viral mRNAs *in vitro. Proc. Nat. Acad. Sci. USA* **72**:1189–1193.

Boyce, R. P., Kraiselburd, E., Ryan, S., and Chessin, H. 1969. Ring opening of covalent λ phage DNA circles after thermal induction of superinfected lysogens. *Virology* **37**:679–681.

Bradley, D. E. 1967. Ultrastructure of bacteriophages and bacteriocins. *Bact. Rev.* **31**:230–314.

Bradley, D. E. 1971. A comparative study of the structure and biological properties of bacteriophages. *In:* Maramorosch, K., and E. Kurstak, eds., *Comparative Virology,* New York, Academic Press, p. 207–253.

Brandner, G., and Mueller, N. 1974. Persistence of late SV40 genome transcription after inhibition of DNA replication by cytosine arabinoside. *FEBS Lett.* **42**:124–126.

Breindl, M., and Holland, J. J. 1975. Coupled *in vitro* transcription and translation of vesicular stomatitis virus messenger RNA. *Proc. Nat. Acad. Sci. USA* **72**: 2545–2549.

Breitenfield, P. M., and Schäfer W. 1957. The formation of fowl plague virus antigens in infected cells, as studied with fluorescent antibodies. *Virology* **4**:328–345.

Brenner, S., Jacob, F., and Meselson, M. 1961. An unstable intermediate carrying information from genes to ribosomes for protein synthesis. *Nature* **190**:576–581.

Briand, J. P., Richards, K. E., Bouley, J. P., Witz, J., and Hirth, L. 1976. Structure of the amino-acid accepting 3′-end of high molecular weight eggplant mosaic virus RNA. *Proc. Nat. Acad. Sci. USA* **73**:737–741.

Broers, A. M., Panessa, B. J., Gennaro, J. F. 1975. High-resolution scanning electron microscopy of bacteriophages 3C and T4. *Science* **189**:637–639.

Brown, L. R., and Dowell, C. E. 1968a. Replication of coliphage M-13. I. Effects on host cells after synchronized infection. *J. Virol.* **2**:1290–1295.

Brown, L. R., and Dowell, C. E. 1968b. Replication of coliphage M-13. II. Intracellular DNA forms associated with M-13 infection of mitomycin C-treated cells. *J. Virol.* **2**:1296–1307.

Brown, N. L., and Smith, M. 1977. DNA sequence of a region of the ϕX174 genome coding for a ribosome binding site. *Nature* **265**:695–698.

Brown, R. M. 1972. Algal viruses. *Adv. Virus Res.,* **17**:243–277.

Brown, W. V., and Berthke, E. M. 1974. *Textbook of Cytology. 2nd Ed.,* St. Louis, Missouri, C. V. Mosby Co.

Buchan, A., and Watson, D. H. 1969. The immunological specificity of thymidine kinases in cells infected by viruses of the herpes group. *J. Gen. Virol.* **4**:461–463.

Buetti, E. 1974. Characterization of late polyoma mRNA. *J. Virol.* **14**:249–260.

Bujard, H. 1969. Location of single-strand interruptions in the DNA of bacteriophage T5. *Proc. Nat. Acad. Sci. USA* **62**:1167–1174.

Burnet, F. M. 1945. *Virus as Organism,* Cambridge, Harvard University Press.

Butel, J. S., and Rapp, F. 1965. The effect of arabinofuranosylcytosine on the growth cycle of SV40. *Virology* **27**:490–495.

Carnegie, J. W., Deeney, A. O. C., Olson, K. C., and Beaudreau, G. S. 1969. An RNA fraction from myeloblastosis virus having properties similar to tRNA. *Biochim. Biophys. Acta* **190**:274–284.

Caro, L. G. 1965. The molecular weight of λ DNA. *Virology* **25**:226–236.

Carriquiry, E., and Litvak, S. 1974. Further studies on the enzymatic aminoacylation of TMV-RNA by histidine. *FEBS Lett.* **38**:287–291.

Carroll, R. B., Neet, K., and Goldthwait, D. A. 1975. Studies on the self-association of bacteriophage T4 gene 32 protein by equilibrium sedimentation. *J. Mol. Biol.* **91**:275–291.

Cartwright, B., Smale, C. J., and Brown, F. 1970. Structural and biological relations in vesicular stomatitis virus. *In*: Barry, R. D., and B. W. J. Mahy, eds., *The Biology of Large RNA Viruses*, New York, Academic Press, p. 115–132.

Casak, J. 1971. Arboviruses: Incorporation in a general system of virus classification. *In*: Maramorosch, K., and E. Kurstak, eds., *Comparative Virology*, New York, Academic Press, p. 307–333.

Casjens, S., Hohn, T., and Kaiser, A. D. 1972. Head assembly steps controlled by genes *F* and *W* in bacteriophage λ. *J. Mol. Biol.* **64**:551–563.

Casals, J. 1971. *In*: Maramorosch, K., and H. Kurstak, eds., *Comparative Virology*, New York, Academic Press, p. 307–333.

Cascino, A., Geiduschek, E. P., Gafferata, R. L., and Haskelkorn, R. 1971. T4 DNA replication and viral gene expression. *J. Mol. Biol.* **61**:357–367.

Chakraborty, P. P., Brandyopadhyaz, P., Huang, H. H., and Maitra, U. 1974. Fidelity of *in vitro* transcription of T3 DNA by bacteriophage T3-induced RNA polymerase and by *E. coli* RNA polymerase. *J. Biol. Chem.* **249**:6901–6909.

Chamberlin, M., McGrath, J., and Waskell, L. 1970. New RNA polymerase from *E. coli* infected with bacteriophage T7. *Nature* **228**:227–231.

Champeil, P., and Brahams, J. 1974. Conformational properties of some viral DNAs. *Eur. J. Biochem.* **45**:253–259.

Champness, J. N., Bloomer, A. C., Bricogne, G., Butler, P. J. G., and Klug, A. 1976. The structure of the protein disk of tobacco mosaic virus to 5 Å resolution. *Nature* **259**:20–24.

Chandler, R. L. 1963. Experimental scrapie in the mouse. *Res. Vet. Sci.* **4**:276–285.

Chang, S. H., Hefti, E., Obijeski, J. F., and Bishop, D. H. L. 1974. RNA transcription by the virion polymerases of five rhabdoviruses. *J. Virol.* **13**:652–661.

Chen, M. J., Locker, J., and Weiss, S. B. 1976. Physical mapping of bacteriophage T5 tRNAs. *J. Biol. Chem.* **251**:536–547.

Chen, J. M., Shiau, R. P., Hwang, L.-T., Vaughan, J., and Weiss, S. B. 1975. Methionine and formylmethionine specific tRNAs coded by bacteriophage T5. *Proc. Nat. Acad. Sci. USA* **72**:558–562.

Cho, H. J. 1976. Is the scrapie agent a virus? *Nature* **262**:411–412.

Clark, S., Losick, R., and Pero, J. 1974. New RNA polymerase from *B. subtilis* infected with phage PBS2. *Nature* **252**:21–24.

Clarke, M. C., and Millson, G. C. 1976. The membrane location of scrapie infectivity. *J. Gen. Virol.* **31**:441–445.

Clements, J. B., and Sinsheimer, R. L. 1975. RNA metabolism in ϕX174-infected cells. *J. Virol.* **15**:151–160.

Cohen, J. A., 1967. Chemistry and structure of nucleic acids of bacteriophages. *Science* **158**:343–351.

Cohen, P. S., and Ennis, H. L. 1965. The requirement for potassium for bacteriophage T4 protein and DNA synthesis. *Virology* **27**:282–289.

Cohen, S. S. 1948. The synthesis of bacterial viruses. *J. Biol. Chem.* **174**:281–293.

Colonno, R. J., and Stone, H. O. 1975. Methylation of mRNA of Newcastle disease virus *in vitro* by a virion-associated enzyme. *Proc. Nat. Acad. Sci. USA* **72**:2611–2615.

Comer, M. M., Guthrie, C., and McClain, W. H. 1974. An ochre suppressor of bacteriophage T4 that is associated with a tRNA. *J. Mol. Biol.* **90**:665–676.

Compans, R. W., and Choppin, P. W. 1968. The nucleic acid of the parainfluenza virus SV5. *Virology* **35**:289–296.

Compans, R. W., and Choppin, P. W. 1971. The structure and assembly of influenza and parainfluenza viruses. *In:* Maramorosch, K., and E. Kurstak, eds., *Comparative Virology,* New York, Academic Press, p. 407–432.

Compans, R. W., Dimmock, N. J., and Meier-Ewert, H. 1970. An electron microscopic study of the influenza virus-infected cell. *In*: Barry, R. D., and B. W. J. Mahy, eds., *The Biology of Large RNA Viruses,* New York, Academic Press, p. 87–108.

Crawford, L. V. 1965. A study of human papilloma virus DNA. *J. Mol. Biol.* **13**:362–372.

Crick, J., and Brown, F. 1970. Small immunizing subunits in rabies virus. *In*: Barry, R. D., and B. W. J. Mahy, eds., *The Biology of Large RNA Viruses*, New York, Adademic Press, p. 133–140.

Daniel, V., Sarid, S. and Littauer, U. Z. 1970. Bacteriophage induced tRNA in *E. Coli*. *Science* **167**:1682–1688.

Darlix, J. L., and Dausse, J. P. 1975. Localization of *E. coli* RNA polymerase initiation sites on T7 DNA early promoter region. *FEBS Lett.,* **50**:214–218.

Darlix, J. L., and Horaist, M. 1975. Existence and possible roles of transcriptional barriers in T7 DNA early region as shown by electron microscopy. *Nature* **256**:288–292.

Dausse, J. P., Sentenac, A., and Fromageot, P. 1975. Interaction of RNA polymerase from *E. coli* with DNA. *Eur. J. Biochem.* **57**:569–578.

Davidson, N., and Szybalski, W. 1971. Physical and chemical characteristics of λ DNA. *In*: Hershey, A. D., ed., *The Bacteriophage Lambda*, Cold Spring Harbor, L. I., Cold Spring Harbor Laboratory, p. 45–82.

Davis, R. W., and Hyman, R. W. 1971. A study in evolution: The DNA base sequence homology between coliphages T7 and T3. *J. Mol. Biol.* **62**:287–301.

Dawson, P., Skalka, A., and Simon, L. D. 1975. Bacteriophage λ head morphogenesis: Studies on the role of DNA. *J. Mol. Biol.* **93**:167–180.

de Zoeten, G. A., and Schlegel, D. E. 1967a. Broadbean mottle virus in leaf tissue. *Virology* **31**:173–176.

de Zoeten, G. A., and Schlegel, D. E. 1967b. Nucleolar and cytoplasmic uridine-^{3}H incorporation in virus-infected plants. *Virology* **32**:416–427.

Dottin, R. P., Cutler, L. S., and Pearson, M. L. 1975. Repression and autogenous stimulation in vitro by bacteriophage λ repressor. *Proc. Nat. Acad. Sci. USA* **72**:804–808.

Dottin, R. P., and Pearson, M. L. 1973. Regulation by *N* gene protein of phage λ of anthranilate synthetase synthesis *in vitro. Proc. Nat. Acad. Sci. USA* **70**:1078–1082.

Dressler, D., and Wolfson, J. 1970. Rolling circle for φX174 DNA replication. 3. Synthesis of supercoiled duplex rings. *Proc. Nat. Acad. Sci. USA* **67**:456–461.

Drohan, W. N., Shoyab, M., Wall, R., and Baluda, M. A. 1975. Interspersion of sequences in AMV RNA that rapidly hybridize with leukemic chicken cell DNA. *J. Virol.* **15**:550–555.

Duesberg, P. H. 1968a. The RNAs of influenza virus. *Proc. Nat. Acad. Sci. USA* **59**:930–937.

Duesberg, P. H. 1968b. Physical properties of Rous sarcoma virus. *Proc. Nat. Acad. Sci. USA* **60**:1511–1518.

Duesberg, P. H., Martin, G. S., and Vogt, P. K. 1970. Glycoprotein components of avian and murine RNA tumor viruses. *Virology* **41**:631–646.

Duesberg, P. H., and Robinson, W. S. 1967. On the structure and replication of influenza virus. *J. Mol. Biol.* **25**:383–406.

Dunn, J., and Studier, F. W. 1973a. T7/early RNAs are generated by site-specific cleavages. *Proc. Nat. Acad. Sci. USA* **70**:1559–1563.

Dunn, J., and Studier, F. W. 1973b. T7/early RNAs and *E. coli* rRNAs are cut from large precursor RNAs *in vivo* by ribonuclease III. *Proc. Nat. Acad. Sci. USA* **70**:3296–3300.

Dunnebacke, T. H., and Kleinschmidt, A. K. 1967. RNA from reovirus as seen in protein monolayers by electron microscopy. *Z. Naturforsch.* **22B**:159–164.

Dyson, R. D., and Van Holde, K. E. 1967. An investigation of bacteriophage λ, its protein ghosts and subunits. *Virology* **33**:559–566.

Earhart, C. F. 1970. The association of host and phage DNA with the membrane of *E. coli. Virology* **42**:429–436.

Earhart, C. F., Sauri, D. J., Fletcher, G., and Wulff, J. L. 1973. Effect of inhibition of macromolecule synthesis on the association of bacteriophage T4 DNA with membrane. *J. Virol.* **11**:527–534.

Edgell, M. H., Hutchinson, C. A., and Sinsheimer, R. L. 1969. The process of infection with bacteriophage φX174. XXVIII. Removal of the spike proteins from the phage capsid. *J. Mol. Biol.* **42**:547–558.

Eiden, J. J., Quade, K., and Nichols, J. L. 1976. Interaction of tRNA^{Trp} with Rous sarcoma virus 35 S RNA. *Nature* **259**:245–247.

Eisen, H. A., Fuerst, C. R., Siminovitch, L., Thomas, R., Lambert, L., Pereira da Silva, L., and Jacob, F. 1966. Genetics and physiology of defective lysogeny in K12 (λ). *Virology* **30**:224–241.

Eisenberg, S., and Denhardt, D. T. 1974. The mechanism of replication of ϕX174 single-stranded DNA. *Biochem. Biophys. Res. Comm.* **61**:532–537.

Elder, K. T., and Smith, A. E. 1973. Methionine tRNAs of avian myeloblastosis virus. *Proc. Nat. Acad. Sci. USA* **70**:2823–2826.

Elder, K. T., and Smith, A. E., 1974. Methionine tRNAs associated with avian oncornavirus 70 S RNA. *Nature* **247**:435–437.

Epstein, R. H., Bolle, A., Steinberg, C. M., Kellenberger, E., Boy de la Tour, E., Chevally, R., Edgar, R. S., Sussman, M., Denhardt, G. H., Leilauçis, A. 1963. Physiological studies of conditional lethal mutants of bacteriophage T4D. *Cold Spring Harbor Symp. Quant. Biol.* **28**:375–394.

Erickson, R. J. 1975. The binding of polynucleotides to the DNA polymerase of AMV. *Arch. Biochem. Biophys.* **167**:238–246.

Erickson, R. J., and Grosch, J. C. 1974. The inhibition of AMV DNA polymerase by synthetic polynucleotides. *Biochemistry* **13**:1987–1993.

Erickson, R. J., Janik, B., and Sonuner, R. G. 1973. The inhibition of the AMV DNA polymerase by poly(U) fraction of varying chain lengths. *Biochem. Biophys. Res. Comm.* **52**:1475–1482.

Erikson, E., and Erikson, R. L. 1972. Transfer RNA-synthetase activity associated with AMV. *J. Virol.* **9**:231–233.

Everitt, E., Sundquist, B., Pettersson, U., and Philipson, L. 1973. Isolation and topography of low molecular weight antigens from the virion of adenovirus type 2. *Virology* **52**:130–147.

Faras, A. J., Taylor, J. M., Levinson, W. E., Goodman, H. M., and Bishop, J. M. 1973. RNA-directed DNA polymerase of Rous sarcoma virus. *J. Mol. Biol.* **79**:163–183.

Farber, F. E., and Rawls, W. E. 1975. Isolation of ribosome-like structures from Pichinde virus. *J. Gen. Virol.* **26**:21–31.

Fareed, G. C., Wilt, E. M., and Richardson, C. C. 1971. Enzymatic breakage and joining of DNA. *J. Biol. Chem.* **246**:925–932.

Feix, G., Pollet, R., and Weissmann, C. 1968. Enzymatic synthesis of infectious viral RNA with noninfectious Qβ minus strands as template. *Proc. Nat. Acad. Sci. USA* **59**:145–152.

Fenner, F. 1976. The classification and nomenclature of viruses. *Intervirology* **6**:1–12.

Fidanian, H. M., Drohán, W. N., and Baluda, M. A. 1975. RNA of simian sarcoma-associated virus type I produced in human tumor cells. *J. Virol.* **15**:449–457.

Fiers, W., Contreras, R., Duerinck, F., Haegeman, G., Merregaert, J., Min Jou, W., Raeymaekers, A., Volckaert, G., Ysebaert, M., Van de Kerckhove, J., Nolf, F., and Van Montagu, M. 1975. A-protein gene of bacteriophage MS2. *Nature* **256**:273–278.

Fiers, W., Contreras, R., Duerinck, F., Haegeman, C., Iserentant, D., Merregaert, J., Min Jou, W., Molemans, F., Raeymaekers, A., Van der Berghe, A., Volckaert, G., and Ysebaert, M. 1976. Complete nucleotide sequence of bacteriophage MS2RNA: primary and secondary structure of the replicase gene. *Nature* **260**:500–507.

Fiers, W., Contreras, R., de Watcher, R., Haegemàn, G., Merregaert, J., Min Jou, W., Vandenberghe, A., Volckaert, G., and Ysebaert, M. 1973. Structure and function of the RNA of bacteriophage MS2. *Proc. 2nd. Duran-Reynals Symp.*, 1973, Barcelona, Spain, p. 35–50.

Finch, J. T. 1965. Preliminary X-ray diffraction studies on tobacco rattle and barley striped mosaic viruses. *J. Mol. Biol.* **12**:612–619.

Finch, J. T., and Gibbs, A. J. 1970. The structure of the nucleocapsid filaments of the paramyxoviruses. *In*: Barry, R. D., and B. W. J. Mahy, eds., *The Biology of Large RNA Viruses*, New York, Academic Press, p. 109–114.

Finch, J. T., and Klug, A. 1965. Structure of rabbit papilloma virus. *J. Mol. Biol.* **13**:1–12.

Flaks, J. G., and Cohen, S. S. 1959. Virus-induced acquisition of metabolic function. *J. Biol. Chem.*, **234**:1501–1506.

Flamand, A., and Bishop, D. H. L. 1974. *In vivo* synthesis of RNA by vesicular stomatitis virus and its mutants. *J. Mol. Biol.* **87**:31–53.

Fleckenstein, B., Bornkamm, G. W., and Ludwig, H. 1975. Repetitive sequences in complete and defective genomes of *Herpesvirus saimiri*. *J. Virol.* **15**:398–406.

Fraenkel-Conrat, H., and Singer, B. 1957. Virus reconstitution: Combination of protein and nucleic acid from different strains. *Biochim. Biophys. Acta* **24**:541–548.

Francke, B., and Roy, D. S. 1972. Cis-limited action of the gene A product of bacteriophage φX174 and the essential bacterial site. *Proc. Nat. Acad. Sci. USA* **69**:475–479.

Frankel, F. R. 1968. Evidence for long DNA strands in the replicating pool after T4 infection. *Proc. Nat. Acad. Sci. USA* **59**:131–138.

Franklin, R. E., Klug, A., and Holmes, K. C. 1957. X-ray diffraction studies of the structure and morphology of TMV virus. *Ciba Found. Symp. Nature Viruses*, p. 39–51.

Fraser, D. 1967. *Viruses and Molecular Biology,* New York, The Macmillan Co.

Frearson, P. M., and Crawford, L. V. 1972. Polyoma virus basic proteins *J. Gen. Virol.* **14**:141–155.

Frenkel, N., and Roizman, B. 1972. Separation of the herpesvirus DNA on sedimentation in alkaline gradients. *J. Virol.* **10**:565–572.

Frisby, D., Smith, J., Jeffers, V., and Porter, A. 1976. Size and location of poly(A) in encephalomyocarditis virus RNA. *Nucl. Acids Res.* **3**:2789–2810.

Fujimura, R. K., and Roop, B. C. 1976. Characterization of DNA polymerase induced by bacteriophage T5 with DNA containing single strand breaks. *J. Biol. Chem.* **251**:2168–2175.

Furlong, D., Swift, H., and Roizman, B. 1972. Arrangement of herpesvirus DNA in the core. *J. Virol* **10**:1071–1074.

Furuichi, Y. 1974. Methylation-coupled transcription by virus associated transcriptase of cytoplasmic polyhedrosis-virus containing double-stranded RNA. *Nuc. Acids Res.* **1**:809–822.

Furuichi, Y., and Miura, K.-I. 1975. A blocked structure at the 5′-terminus of mRNA from cytoplasmic polyhedrosis virus. *Nature* **253**:374–375.

Furuichi, Y., Morgan, M., Muthukrishnan, S., and Shatkin, A. J. 1975. Reovirus mRNA contains a methylated blocked 5′-terminal structure: $m^7G(5')ppp(5')G^m$ pCp-. *Proc. Nat. Acad. Sci. USA* **72**:362–366.

Furuichi, Y., Muthukrishnan, S., Tomasz, J., and Shatkin, A. J. 1976. Mechanism of formation of reovirus mRNA 5′-terminal blocked and methylated sequence, m^7GpppG^mpC. *J. Biol. Chem.* **251**:5043–5053.

Gajdusek, D. C. 1967. Slow-virus infections of the nervous system. *New England J. Med.* **276**:392–400.

Gamow, R. I. 1969. Thermodynamic treatment of bacteriophage T4B adsorption kinetics. *J. Virol.* **4**:113–115.

Gantt, R., Stromberg, K., and Julian, B. 1972. Absence of RNA methylase in the AMV core. *J. Virol.* **9**:1057–1058.

Gefter, M. L., Hausmann, R., Gold, M., and Hurwitz, J. 1966. The enzymatic methylation of RNA and DNA. *J. Biol. Chem.* **241**:1995–2006.

Geider, K., and Kornberg, A. 1974. Conversion of the M13 viral single strand to the double-stranded replicative forms by purified proteins. *J. Biol. Chem.* **249**:3999–4005.

Gelfand, D. H., and Hayashi, M. 1969. Electrophoretic characterization of φX174-specified proteins. *J. Mol. Biol.* **44**:501–516.

Gibbons, R. A., and Hunter, G. D. 1967. Nature of the scrapie agent. *Nature* **215**:1041–1043.

Gibbs, A. J., and McIntyre, G. A. 1970. A method for assessing the size of a protein from its composition. *J. Gen. Virol.* **9**:51–67.

Gibson, W., and Roizman, B. 1971. Compartmentalization of spermine in herpes simplex virion. *Proc. Nat. Acad. Sci. USA* **68**:2818–2821.

Gilbert, W., and Dressler, D. 1968. DNA replication: The rolling circle model. *Cold Spring Harbor Symp. Quant. Biol.* **33**:473–483.

Gilden, R. V., and Oroszlan, S. 1972. Group-specific antigens of RNA tumor viruses as markers for subinfectious expression of the RNA virus genome. *Proc. Nat. Acad. Sci. USA* **69**:1021–1025.

Gillespie, D., and Gallo, R. C. 1975. RNA processing and RNA tumor virus origin and evolution. *Science* **188**:802–811.

Glover, D. M. 1974. Coupling of polyoma DNA and RNA synthesis. *Biochem. Biophys. Res. Comm.* **57**:1137–1143.

Gold, M., Hausmann, R. L., Maitra, U., and Hurwitz, J. 1964. Effects of bacteriophage infection on the activity of the methylating enzymes. *Proc. Nat. Acad. Sci. USA* **52**:292–297.

Gold, P., and Dales, S. 1968. Localization of nucleotide phosphohydrolase activity within vaccinia. *Proc. Nat. Acad. Sci. USA* **60**:845–852.

Goldman, E., and Lodish, H. 1972. Specificity of protein synthesis by bacterial ribosomes and initiation factors: Absence of change after phage T4 infection. *J. Mol. Biol.* **67**:35–48.

Goldstein, N. O., Pardoe, I. U., and Burness, A. H. T. 1976. Requirement of an adenylic-acid-rich segment for the infectivity of encephalomyocarditis virus RNA. *J. Gen. Virol.* **31**:1–6.

Goodman, H. M., Billeter, M. A., Hindley, J., and Weissmann, C. 1970. The nucleotide sequence at the 5′-terminus of the Qβ RNA minus strand. *Proc. Nat. Acad. Sci. USA* **67**:921–928.

Gottesman, M. E., and Weisberg, R. A. 1971. Prophage insertion and excision. *In*: Hershey, A.D., ed., *The Bacteriophage* λ, Cold Spring Harbor, L.I., Cold Spring Harbor Laboratory, p. 113–138.

Graham, F. L., van der Eb, A. J., and Heijneker, H. L. 1974. Size and location of the transforming region in human adenovirus type 5 DNA. *Nature*, **251**:687–691.

Granboulan, N., and Girard, M. 1969. Molecular weight of poliovirus RNA. *J. Virol.* **4**:475–479.

Green, M. 1970. Oncogenic viruses. *Ann. Rev. Biochem.* **39**:701–756.

Green, M., and Cartas, M. 1972. The genome of RNA tumor viruses contains poly(A) sequences. *Proc. Nat. Acad. Sci. USA* **69**:791–794.

Green, M., Pina, M., Kimes, R., Wensink, P. C., MacHattie, L. A., and Thomas, C. A. 1967. Adenovirus DNA, I. Molecular weight and conformation. *Proc. Nat. Acad. Sci. USA* **57**:1302–1309.

Green, R. G. 1935. On the nature of the filterable viruses. *Science* **82**:443–445.

Green, R. W., Bolognesi, D. P., Schäfer, W., Pister, L., Hunsmann, G., and de Noronha, F. 1973. Polypeptides of mammalian oncornaviruses. *Virology* **56**:565–579.

Griffith, J. S. 1967. Self-replication and scrapie. *Nature* **215**:1043–1044.

Grippo, P., and Richardson, C. C. 1971. DNA polymerase of bacteriophage T7. *J. Biol. Chem.* **246**:6867–6873.

Groner, Y., Pollack, Y., Berissi, H., and Revel, M. 1972a. Cistron specific translation control protein in *E. coli. Nature New Biol.* **239**:16–19.

Groner, Y., Scheps, R., Kamen, R., Kolakofsky, D., and Revel, M. 1972b. Host subunit of Qβ replicase is translation control factor i. *Nature New Biol.* **239**:19–20.

Gross, L. 1970. *Oncogenic Viruses. 2nd Ed.*, Oxford, Pergamon Press.

Grunberger, D., Weinstein, I. B., and Mushinski, J. F. 1975. Deficiency of the Y base in a hepatoma phenylalanine tRNA. *Nature* **253**:66–68.

Guha, A., and Szybalski, W. 1968. Fractionation of the complementary strands of coliphage T4 DNA based on the asymmetric distribution of the poly U and poly UG binding sites. *Virology* **34**:608–618.

Guilley, H., Jonard, G., and Hirth, L. 1974. A TMV RNA nucleotide sequence specifically recognized by TMV protein. *Biochimie* **56**:181–185.

Guilley, H., Janard, G., and Hirth, L. 1975. Sequence of 71 nucleotides at the 3′-end of TMV RNA. *Proc. Nat. Acad. Sci. USA* **72**:864–868.

Gulati, S. C., Kacian, D. L., and Spiegelman, S. 1974. Conditions for using DNA polymerase I as an RNA-dependent DNA polymerase. *Proc. Nat. Acad. Sci. USA* **71**:1035–1039.

Gupta, R. S., and Schlessinger, D. 1975. Differential modes of chemical decay for early and late λ mRNA. *J. Mol. Biol.* **92**:311–318.

Gussin, G. N. 1966. Three complementation groups in bacteriophage R17. *J. Mol. Biol.* **21**:435–453.

Guthrie, C., and McClain, W. H. 1973. Conditionally lethal mutants of bacteriophage T4 defective in production of a tRNA. *J. Mol. Biol.* **81**:137–155.

Hall, T. C., Shih, D. S., and Kaesberg, P. 1972. Enzyme-mediated binding of tyrosine to brome-mosaic-virus RNA. *Biochem. J.* **129**:969–976.

Haruna, I., Itoh, Y. H., Yamane, K., Miyake, T., Shiba, T., and Watanabe, I. 1971. Isolation and properties of RNA replicases induced by SP and FI phages. *Proc. Nat. Acad. Sci. USA* **68**:1778–1779.

Haruna, I., and Spiegelman, S. 1965. Specific template requirements of RNA replicase. *Proc. Nat. Acad. Sci. USA* **54**:579–587.

Harvey, C. L., Wright, R., and Mussbaum, A. L. 1973. Lambda phage DNA: Joining of a chemically synthesized cohesive end. *Science* **179**:291–293.

Hassenlopp, P., Oudet, P., and Chambon, P. 1974. Animal DNA-dependent RNA polymerases. Studies on the binding of mammalian RNA polymerases AI and B to simian virus 40 DNA. *Eur. J. Biochem.* **41**:397–411.

Hatanaka, M., Twiddy, E., and Gilden, R. V. 1972. Protein kinase associated with RNA tumor viruses and other budding RNA viruses. *Virology* **47**:536–538.

Hausmann, R. 1967. Synthesis of an S-adenosyl-methionine-cleaving enzyme in T3-infected *E. coli* and its disturbance by co-infection with enzymatically incompetent bacteriophages. *J. Virol.* **1**:57–63.

Hausmann, R., and Gold, M. 1966. The enzymatic methylation of RNA and DNA. *J. Biol. Chem.* **241**:1985–1994.

Hausmann, R., and Gomez, B. 1967. Amber mutants of bacteriophages T3 and T7 defective in phage-directed DNA synthesis. *J. Virol.* **1**:779–792.

Hay, J., Perera, P. A. J., Morrison, J. M., Gentry, G. A., and Subak-Sharpe, J. H. 1971. Herpes virus-specified proteins. *In*: Wolstenholme, G. E. W., and M. O'Connor, eds., *Strategy of the Viral Genome*, Edinburgh, Churchill Livingstone, p. 355–372.

Hayashi, Y., and Hayashi, M. 1970. Fractionation of ϕX174 specific messenger RNA. *Cold Spring Harb. Symp. Quant. Biol.* **35**:171–177.

Hayashi, Y., Hayashi, M., and Spiegelman, S. 1963. Restriction of *in vivo* genetic transcription to one of the complementary strands of DNA. *Proc. Nat. Acad. Sci. USA* **50**:664–672.

Hayes, W. 1968. Trends and methods in virus research. *In*: Society for General Microbiology, *The Molecular Biology of Viruses,* Cambridge, Cambridge University Press, p. 1–14.

Heine, J. W., Spear, P. G., and Roizman, B. 1972. The proteins specified by herpes simplex virus. *J. Virol.* **9**:431–439.

Hendrickson, H. E., and McCorquodale, D. J. 1971. The relationship between phage DNA synthesis and protein synthesis in T5-infected cells. *Biochem. Biophys. Res. Comm.* **43**:735–740.

Hendrix, R. W., and Casjens, S. R. 1974. Protein fusion: A novel reaction in bacteriophage λ head assembly. *Proc. Nat. Acad. Sci. USA* **71**:1451–1455.

Henry, T. J., and Pratt, D. 1969. The proteins of bacteriophage M13. *Proc. Nat. Acad. Sci. USA* **62**:800–807.

Hercules, K., Schweiger, M., and Sauerbier, W. 1974. Cleavage by RNase III converts T3 and T7 early precursor RNA into translatable message. *Proc. Nat. Acad. Sci. USA* **71**:840–844.

Hershey, A. D. 1955. An upper limit to the protein content of the germinal substance of bacteriophage T2. *Virology* **1**:108–127.

Hershey, A. D., Dixon, J., and Chase, M. 1953. Nucleic acid economy in bacteria infected with bacteriophage T2. *J. Gen. Physiol.* **36**:777–789.

Hershey, A. D., Garen, A., Fraser, D. K., and Hudis, J. D. 1954. Growth and inheritance in bacteriophage. *Yearbook Carn. Inst. Wash.* **53**:210–241.

Hirth, L. 1971. Comparative properties of rod-shaped viruses. *In*: Maramorosch, K., and E. Kurstak, eds., *Comparative Virology*, New York, Academic Press, p. 335–360.

Hofstettler, H., Monstein, H.-J., and Weissmann, C. 1974. The readthrough protein A_1 is essential for the formation of viable Qβ particles. *Biochim. Biophys. Acta* **374**:238–251.

Hogness, D. S., Doerfler, W., Egan, J. B., and Black, L. W. 1967. The position and orientation of genes in λ and λdg DNA. *In*: Colter, J. S., and W. Paranchych, eds., *The Molecular Biology of Viruses*, New York, Academic Press, p. 91–110.

Hogness, D. S., and Simmons, J. R. 1964. Breakage of λdg DNA: Chemical and genetic characterization of each isolated half-molecule. *J. Mol. Biol.* **9**:411–438.

Hohn, B., Wurtz, M., Klein, B., Lustig, A., and Hohn, T. 1974. Phage λ DNA packaging *in vitro*. *J. Inframol. Struct.* **2**:302–317.

Hori, K., Eoyang, L., Banerjee, A. K., and August, J. T. 1967. Template activity of synthetic ribopolymers in the Qβ RNA polymerase reaction. *Proc. Nat. Acad. Sci. USA* **57**:1790–1797.

Horiuchi, K., and Matsuhashi, S. 1970. Three cistrons in bacteriophage Qβ. *Virology* **42**:49–60.

Horiuchi, K., Webster, R. E., and Matsuhashi, S. 1971. Gene products of bacteriophage Qβ. *Virology* **45**:429–439.

Horne, R. W., Brenner, S., Waterson, A. P., and Wildy, P. 1959. The icosahedral form of an adenovirus. *J. Mol. Biol.* **1**:84–86.

Howatson, A. F. 1971. Oncogenic viruses: A survey of their properties. *In*: Maramorosch, K., and E. Kurstak, eds., *Comparative Virology*, New York, Academic Press, p. 509–537.

Howley, P. M., Mullarkey, M. F., Takemoto, K. K., and Martin, M. A. 1975. Characterization of human papovavirus BK DNA. *J. Virol.* **15**:173–181.

Huang, A. S., and Wagner, R. R. 1966. Comparative sedimentation coefficients of RNA extracted from plaque-forming and defective particles of vesicular stomatitis virus. *J. Mol. Biol.* **22**:381–384.

Huang, A. S., Baltimore, D., and Stampfer, M. 1970. RNA synthesis of vesicular stomatitis virus. *Virology* **42**:946–957.

Huang, A. S., Baltimore, D., and Bratt, M. 1971. RNA polymerase in virions of Newcastle disease virus. *J. Virol.* **7**:389–394.

Huang, W. M., and Buchanan, J. M. 1974. Synergistic interactions of T4 early proteins concerned with their binding to DNA. *Proc. Nat. Acad. Sci. USA* **71**:2226–2230.

Huberman, J. 1968. Visualization of replicating mammalian and T4 bacteriophage DNA. *Cold Spring Harbor Symp. Quant. Biol.* **33**:509–523.

Hull, R. 1970. Large RNA plant-infecting viruses. *In*: Barry, R. D., and B. W. J. Mahy, eds., *The Biology of Large RNA Viruses*, New York, Academic Press, p. 153–164.

Hummeler, K. 1971. Bullet-shaped viruses. *In*: Maramorosch, K., and E. Kurstak, eds., Comparative Virology, New York, Academic Press, p. 361–386.

Humphries, E. H., and Temin, H. 1972. Cell cycle-dependent activation of Rous sarcoma virus-infected stationary chicken cells: Avian leukosis virus group-specific antigens and RNA. *J. Virol.* **10**:82–87.

Huppert, J., Gresland, L., and Hillova, J. 1970. Newcastle disease virus RNA synthesis in cells infected with unirradiated or UV-irradiated virus. *In*: Barry, R. D., and B. W. J. Mahy, eds., *The Biology of Large RNA Viruses*, New York, Academic Press, p. 482–492.

Hurst, R. E., and Incardona, N. L. 1969. Molecular weights of viruses from isopycnic centrifugation with Schlieren optics. *Virology* **37**:62–73.

Hyde, J. M., Gafford, L. G., and Randall, C. C. 1967. Molecular weight determination of fowlpox virus DNA by electron microscopy. *Virology* **33**:112–120.

Hyman, L. 1940. *The Invertebrates: Protozoa through Ctenophora*, New York, McGraw-Hill Book Company Inc.

Hyman, R. W. 1971. Physical mapping of T7 mRNA. *J. Mol. Biol.* **61**:369–376.

Isacson, P., and Koch, A. E. 1965. Association of host antigens with a parainfluenza virus. *Virology* **27**:129–138.

Iwanowsky, D. 1892. Über die Mosaikkrankheit der Tabakspflanze. *Bull. Acad. Imp. Sci.* St. Petersburg, **3**:67–72.

Iype, P. T., Abraham, K. A., and Bhargava. P. M. 1963. Further evidence for a positive role of acrosome in the uptake of labelled amino acids by bovine and avian spermatozoa. *J. Reprod. Fertil.* **5**:151–158.

Jacquemin-Sablon, A., and Richardson, C. C. 1970. Analysis of the interruptions in bacteriophage T5 DNA. *J. Mol. Biol.* **47**:477–494.

Jarrett, O., Pitts, J. D., Whalley, J. M., Clason, A. E., and Hay, J. 1971. Isolation of the nucleic acid of feline leukemia virus. *Virology* **43**:317–320.

Johnson, P. H., and Sinsheimer, R. L. 1974. Structure of an intermediate in the replication of bacteriophage ϕX174 DNA. *J. Mol. Biol.* **83**:47–61.

Juarez, H., Juarez, D., and Hedgcoth, C. 1974. Seven lysine isoaccepting tRNAs from polyoma virus-transformed cells. *Biochem. Biophys. Res. Comm.* **61**:110–116.

Jungwirth, C., and Joklik, W. K. 1965. Studies on "early" enzymes in HeLa cells infected with vaccinia virus. *Virology* **27**:80–93.

Kacian, D. L., Watson, K. F., Burney, A., and Spiegelman, S. 1971. Purification of the DNA polymerase of AMV. *Biochim. Biophys. Acta* **246**:365–383.

Kaiser, A. D., and Wu, R. 1968. Structure and function of DNA cohesive ends. *Cold Spring Harbor Symp. Quant. Biol.* **33**:729–734.

Kaiser, D. 1971. Lambda DNA replication. *In*: Hershey, A. D., ed., *The Bacteriophage Lambda,* Cold Spring Harbor, L.I., Cold Spring Harbor Laboratory, p. 195–210.

Kaiser, D., and Dworkin, M. 1975. Gene transfer to a myxobacterium by *E. coli* phage Pl. *Science* **187**:653–654.

Kamen, R., Kondo, M., Römer, W., and Weissmann, C. 1972. Reconstitution of Qβ replicase lacking subunit α with protein-synthesis interference factor i. *Eur. J. Biochem.* **31**:44–51.

Kang, C. Y., and Prevec, L. 1969. Proteins of vesicular stomatitis virus. *J. Virol.* **3**:404–413.

Kasai, T., and Bautz, E. K. F. 1969. Regulation of gene-specific RNA synthesis in bacteriophage T4. *J. Mol. Biol.* **41**:401–418.

Kasai, T., Bautz, E. K. F., Guha, A., and Szybalski, W. 1968. Identification of the transcribing DNA strand for the *rII* and endolysin genes of coliphage T4. *J. Mol. Biol.* **34**:709–712.

Kates, J. R., and McAuslan, B. R. 1967a. Messenger RNA synthesis by a "coated" viral genome. *Proc. Nat. Acad. Sci. USA* **57**:314–320.

Kates, J. R., and McAuslan, B. R. 1967b. Poxvirus DNA-dependent RNA polymerase. *Proc. Nat. Acad. Sci. USA* **58**:134–140.

Kates, J. R., and McAuslan, B. R. 1967c. Interrelation of protein synthesis and viral DNA synthesis. *J. Virol.* **1**:110–114.

Katsura, I., and Kühl, P. W. 1974. A regulator protein for the length determination of bacteriophage λ tail. *J. Supramol. Struct.* **2**:239–253.

Katsura, I., and Kühl, P. W. 1975a. Morphogenesis of the tail of bacteriophage λ. II. *In vitro* formation and properties of phage particles with extra long tails. *Virology* **63**:238–251.

Katsura, I., and Kühl, P. W. 1975b. Morphogenesis of the tail of bacteriophage λ. III. Morphogenetic pathway. *J. Mol. Biol.* **91**:257–273.

Katz, E., and Moss, B. 1970. Formation of a vaccinia virus structural polypeptide from a higher molecular weight precursor: Inhibition by rifampicin. *Proc. Nat. Acad. Sci. USA* **66**:677–684.

Katze, J. R. 1975. Alterations in SVT2 cell tRNAs in response to cell density and serum type. *Biochim. Biophys. Acta* **383**:131–139.

Keir, H. M. 1968. Virus-induced enzymes in mammalian cells infected with DNA viruses. *In*: Society for General Microbiology, *Molecular Biology of Viruses*, Cambridge, Cambridge University Press, p. 67–99.

Keir, H. M., Subak-Sharpe, H., Shedden, W. I. H., Watson, D. H., and Wildy, P. 1966. Immunological evidence for a specific DNA polymerase produced after infection by herpes simplex virus. *Virology* **30**:154–157.

Keith, J., and Fraenkel-Conrat, H. 1975. TMV RNA carries 5′-terminal triphosphorylated guanosine blocked by 5′-linked 7-methylguanosine. *FEBS Lett.* **57**:31–33.

Keith, J., Gleason, M., and Fraenkel-Conrat, H. 1974. Characterization of the end groups of RNA of Rous sarcoma virus. *Proc. Nat. Acad. Sci. USA* **71**:4371–4375.

Kellenberger, E. 1961. Vegetative bacteriophage and the maturation of the virus particles. *Adv. Virus Res.* **8**:1–62.

Kellenberger, E. 1966. The genetic control of the shape of a virus. *Sci. Am.,* **215** (Dec.):32–39.

Kellenberger, E., and Edgar, R. S. 1971. Structure and assembly of phage particles. *In*: Hershey, A. D., ed., *The Bacteriophage Lambda*, Cold Spring Harbor, N.Y., Cold Spring Harbor Laboratory, p. 271–295.

Kelly, T. J., and Thomas, C. A. 1969. Intermediate in the replication of bacteriophage T7 DNA molecules. *J. Mol. Biol.* **44**:459–475.

Kieff, E. D., Bachenheimer, S. L., and Roizman, B. 1971. Size, composition, and structure of the DNA of herpes simplex subtypes 1 and 2. *J. Virol.* **8**:125–132.

Kiger, J. A., and Sinsheimer, R. L. 1969. Fractionation of replicating λ DNA on benzoylated-naphthoylated DEAE cellulose. *J. Mol. Biol.* **40**:467–490.

King, A. M. Q., and Wells, R. D. 1976. All intact subunit RNAs from Rous sarcoma virus contain poly(A). *J. Biol. Chem.* **251**:150–152.

King, J. 1970. Steps in T4 tail core assembly. *FEBS Symp.* **21**:171–180.

King, J., and Wood, W. B. 1969. Assembly of bacteriophage T4 tail fibers. *J. Mol. Biol.* **39**:583–602.

Kingsbury, D. W. 1966. Newcastle disease virus RNA. *J. Mol. Biol.* **18**:204–214.

Klagsbrun, M. 1971. Changes in the methylation of tRNA in vaccinia infected HeLa cells. *Virology* **44**:153–167.

Klenk, H. D., and Choppin, P. W. 1969. Chemical composition of the parainfluenza virus SV5. *Virology* **37**:155–157.

Kleppe, K., van de Sande, J. H., and Khorana, H. G. 1970. Polynucleotide ligase-catalyzed joining of deoxyribo-oligonucleotides on ribopolynucleotide templates and of ribo-oligonucleotides on deoxyribopolynucleotide templates. *Proc. Nat. Acad. Sci. USA* **67**:68–73.

Klug, A. 1965. Structure of viruses of the papilloma-polyoma type. *J. Mol. Biol.* **13**:749–756.

Knippers, R., Whalley, J. M., and Sinsheimer, R. L. 1969. The process of infection with bacteriophage φX174. *Proc. Nat. Acad. Sci. USA* **64**:275–282.

Knolle, P., and Hohn, T. 1975. Morphogenesis of RNA phages. *In*: Cold Spring Harbor Laboratory, *RNA Phages*, Cold Spring Harbor, N.Y., p. 147–201.

Knolle, P., and Kaudewitz, P. 1963. Effect of RNase pre-treatment of cells of *E. coli* K12 on plaque yields from subsequent infection with RNA phage fr. *Biochem. Biophys. Res. Comm.* **11**:383–387.

Knopf, K.-W., and Bujard, H. 1975. Structure and function of the genome of coliphage T5. *Eur. J. Biochem.* **53**:371–385.

Korant, B. D., and Pootjies, C. F. 1970. Physiochemical properties of *Agrobacterium tumefaciens* phage LV-1 and its DNA. *Virology* **40**:48–54.

Korn, D., and Weissbach, A. 1963. The effect of lysogenic induction on the deoxyribonucleases of *E. coli* K12 λ. *J. Biol. Chem.* **238**:3390–3394.

Kühl, P. W., and Katsura, I. 1975. Morphogenesis of the tail of bacteriophage λ. *Virology* **63**:221–237.

Kuno, S., and Lehman, I. R. 1962. Gentiobiose, a constituent of DNA from coliphage T6. *J. Biol. Chem.* **237**:1266–1270.

Küppers, B., and Sumper, M. 1975. Minimal requirements for template recognition by bacteriophage Qβ replicase: Approach to general RNA-dependent RNA synthesis. *Proc. Nat. Acad. Sci. USA* **72**:2640–2643.

Labaw, L. W. 1951. The origin of phosphorus in *E. coli* bacteriophages. *J. Bact.* **62**:169–173.

Labaw, L. W. 1953. The origin of phosphorus in the T1, T5, T6, and T7 bacteriophages of *E. coli*. *J. Bact.* **66**:429–436.

LaColla, P., and Weissbach, A. 1975. Vaccinia virus infection of HeLa cells. *J. Virol.* **15**:305–315.

Lai, M. M. C., and Duesberg, P. H. 1972. Adenylic acid-rich sequence in RNAs of Rous sarcoma virus and Rauscher mouse leukaemia virus. *Nature* **235**:383–386.

Lamy, D., Jonard, G., Guilley, H., and Hirth, L. 1975. Comparison between the 3′OH end RNA sequence of two strains of TMV which may be aminoacylated. *FEBS Lett.* **60**:202–204.

Lane, L. C. 1974. The bromoviruses. *Adv. Virus Res.* **19**:152–220.

Lane, L. C., and Kaesberg, P. 1971. Multiple genetic components in bromegrass mosaic virus. *Nature New Biol.* **232**:40–43.

Lanni, F., and Lanni, Y. T. 1966. Genetic suppressors of phage T5 amber mutants. *J. Bact.* **92**:521–523.

Lanni, Y. T. 1968. First-step-transfer DNA of bacteriophage T5. *Bact. Rev.* **32**:227–242.

Lanni, Y. T. 1969. Functions of two genes in the first-step-transfer DNA of bacteriophage T5. *J. Mol. Biol.* **44**:173–184.

Lanni, Y. T., and Szybalski, W. 1969. Transcription patterns for coliphage T5. *Bact. Proc.* **1969**:192.

Laver, W. G. 1970. Isolation of an arginine-rich protein from particles of adenovirus type 2. *Virology* **41**:488–500.

Lebeurier, G., Nicolaieff, A., and Richards, K. E. 1977. Inside-out model for self-assembly of tobacco mosaic virus. *Proc. Nat. Acad. Sci. USA* **75**:149–153.

Lee Huang, S., and Ochoa, S. 1971. Messenger discriminating species of initiation factor F_3. *Nature New Biol.* **234**:236–239.

Lee Huang, S., and Ochoa, S. 1973. Purification and properties of two messenger-discriminating species of *E. coli* initiation factor 3. *Arch. Biochem. Biophys.* **156**:84–96.

Lehman, I. R., and Pratt, E. A. 1960. On the structure of the glucosylated hydroxymethyl cytosine nucleotides of coliphages T2, T4, and T6. *J. Biol. Chem.* **235**:3254–3259.

Leis, J. P., Berkower, I., and Hurwitz, J. 1973. Mechanism of action of ribonuclease H isolated from AMV and *E. coli*. *Proc. Nat. Acad. Sci. USA* **70**:466–470.

Leis, J. P., and Hurwitz, J. 1972. Isolation and characterization of a protein that stimulates DNA synthesis from AMV. *Proc. Nat. Acad. Sci. USA* **69**:2331–2335.

Leis, J. P., Schincariol, A., Ishizaki, R., and Hurwitz, J. 1975. RNA-dependent DNA polymerase activity of RNA tumor viruses. *J. Virol.* **15**:484–487.

Levinson, W. E., Varmus, H. E., Garapin, A. C., and Bishop, J. M. 1972. DNA of Rous sarcoma virus: Its nature and significance. *Science* **175**:76–78.

Levitt, J., and Becker, Y. 1967. The effect of cytosine arabinoside on the replication of herpes simplex virus. *Virology* **31**:129–134.

Lindquist, B. H., and Sinsheimer, R. L. 1967a. Process of infection with bacteriophage ϕX174. XIV. Studies on macromolecular synthesis during infection with a lysis-defective mutant. *J. Mol. Biol.* **28**:87–94.

Lindquist, B. H., and Sinsheimer, R. L. 1967b. The process of infection with bacteriophage ϕX174. XV. Bacteriophage DNA synthesis in abortive infections with a set of conditional lethal mutants. *J. Mol. Biol.* **30**:69–80.

Lindstrom, D. M., and Dulbecco, R. 1972. Strand orientation of SV40 transcription in productively infected cells. *Proc. Nat. Acad. Sci. USA* **69**:1517–1520.

Litvak, S., Tarragó, A., Tarragó-Litvak, L., and Allende, J. E. 1973. Elongation factor-viral genome interaction dependent on the aminoacylation of TYMV and TMV RNAs. *Nature New Biol.* **241**:88–90.

Loh, P. C., and Shatkin, A. J. 1968. Preparation and properties of the internal capsid components of reovirus. *J. Gen. Virol.* **3**:233–257.

Loh, P. C., and Soergel, M. 1967. Macromolecular synthesis in cells infected with reovirus type 2 and the effect of Ara-C. *Nature* **214**:622–623.

Luftig, R. B., and Haselkorn, R. 1968. Comparison of blue-green algae virus LPP-1 and the morphologically related viruses G_{111} and coliphage T7. *Virology* **34**:675–678.

Luftig, R. B., and Kilham, S. S. 1971. An electron microscope study of Rauscher leukemia virus. *Virology* **46**:277–297.

Luria, S. E., and Darnell, J. E. 1967. *General Virology, 2nd Ed.*, New York, John Wiley & Sons.

MacFarlane, E. S. 1969. Properties of the *in vitro* soluble RNA methylase activity of hamster tumors induced by adenovirus 12. *Can. J. Microbiol.* **15**:189–192.

MacHattie, L. A., Ritchie, D. A., Thomas, C. A., and Richardson, C. C. 1967. Terminal repetition in permuted T2 bacteriophage DNA molecules. *J. Mol. Biol.* **23**:355–364.

Mahadik, S. P., Dharmgrongartama, B., and Srinivasan, P. R. 1972. An inhibitory protein of *E. coli* RNA polymerase in bacteriophage T3-infected cells. *Proc. Nat. Acad. Sci. USA* **69**:162–166.

Mahadik, S. P., Dharmgrongartama, B., and Srinivasan, P. R. 1974. Regulation of host RNA synthesis in bacteriophage T3-infected cells. *J. Biol. Chem.* **249**:1787–1791.

Mahy, B. W. J. 1970. The replication of fowl plaque virus RNA. *In*: Barry, R. D., and B. W. J. Mahy, eds., *The Biology of Large RNA Viruses*, New York, Academic Press, p. 392–415.

Makover, S. 1968. A preferred origin for the replication of λ DNA. *Cold Spring Harbor Symp. Quant. Biol.* **33**:621–622.

Mandel, J.-L., and Chambon, P. 1974a. Studies on the reaction parameters of transcription *in vitro* of simian virus 40 DNA by mammalian RNA polymerases AI and B. *Eur. J. Biochem.* **41**:367–378.

Mandel, J.-L., and Chambon, P. 1974b. Analysis of the RNAs synthesized on simian virus 40 superhelical DNA by mammalian RNA polymerases AI and B. *Eur. J. Biochem.* **41**:379–395.

Mandel, M., and Berg, A. 1968. Cohesive sites and helper phage function of P2, λ, and 186's DNAs. *Proc. Nat. Acad. Sci. USA* **60**:265–268.

Marchin, G. L., Müller, U. R., and Al-Khateeb, G. H. 1974. The effect of tRNA on virally modified valyl tRNA synthetase of *E. coli*. *J. Biol. Chem.* **249**:4705–4711.

Marvin, D. A., and Hohn, B. 1969. Filamentous bacterial viruses. *Bact. Rev.* **33**:172–209.

Marvin, D. A., and Wachtel, E. J. 1975. Structure and assembly of filamentous bacterial viruses. *Nature* **253**:19–23.

Marvin, D. A., Wiseman, R. L., and Wachtel, E. J. 1974. Molecular architecture of the class II (Pfl, Xf) virion. *J. Mol. Biol.*, **82**:121–138.

Mathews, C. K. 1971. *Bacteriophage Biochemistry*, New York, Van Nostrand Reinhold Co.

McAuslan, B. R. 1971. Enzymes specified by DNA-containing animal viruses. *In:* Wolstenholme, G. E. W., and M. O'Connor, eds., *Strategy of the Viral Genome*, Edinburgh, Churchill Livingstone, p. 25–38.

McAuslan, B. R., Herde, P., Pett, D., and Rose, J. 1965. Nucleases of virus-infected animal cells. *Biochem. Biophys. Res. Comm.* **20**:586–591.

McClain, W. H., Guthrie, C., and Barrell, B. G. 1972. Eight tRNAs induced by infection of *E. coli* with phage T4. *Proc. Nat. Acad. Sci. USA* **69**:3703–3707.

McClain, W. H., Guthrie, C., and Barrell, B. G. 1973. The *psu*$_1^+$ amber suppressor gene of bacteriophage T4: Identification of its amino acid and tRNA. *J. Mol. Biol.* **81**:157–171.

McCorquodale, D. J., and Buchanan, J. M. 1968. Patterns of protein synthesis in T5-infected *E. coli*. *J. Biol. Chem.* **243**:2550–2559.

McCorquodale, D. J., and Lanni, Y. T. 1970. Patterns of protein synthesis in *E. coli* infected by amber mutants in the first-step-transfer DNA of T5. *J. Mol. Biol.* **48**:133–144.

McGregor, S., and Mayor, H. D. 1971. Chemical and hydrodynamic analysis of the rhinovirion. *J. Virol.* **7**:41–46.

Mechler, B., and Arber, W. 1969. Parental fd DNA is efficiently transferred into progeny bacteriophage particles even at low multiplicity of infection. *J. Mol. Biol.* **45**:443–450.

Miller, R. C., and Kozinsky, A. W. 1970. Newly synthesized proteins in the T4 protein-DNA complex. *J. Virol.* **5**:502–506.

Mills, D. R., Peterson, R. L., and Spiegelman, S. 1967. An extracellular Darwinian experiment with a self-duplicating nucleic acid molecule. *Proc. Nat. Acad. Sci. USA* **58**:217–224.

Millward, S., and Graham, A. F. 1970. Structural studies on reovirus: Discontinuities in the genome. *Proc. Nat. Acad. Sci. USA* **65**:422–429.

Millward, S., and Graham, A. F. 1971. Structure and transcription of the genomes of double-stranded RNA viruses. *In*: Maramorosch, K., and E. Kurstak, eds., *Comparative Virology,* New York, Academic Press, p. 387–406.

Min Jou, W. and Fiers, W. 1976. Studies on bacteriophage MS2 XXXIII. Comparison of the nucleotide sequences in related bacteriophage RNAs. *J. Mol. Biol.* **106**:1047–1060.

Min Jou, W., Haegeman, G., Ysebaert, M., and Fiers, W. 1972. Nucleotide sequence of the gene coding for the bacteriophage MS2 coat protein. *Nature* **237**:82–88.

Minagawa, T. 1961. Some characteristics of the internal protein phage T2. *Virology* **13**:515–527.

Miura, K., Watanabe, K., Sugiura, M., and Shatkin, A. J. 1974. The 5′-terminal nucleotide sequences of double-stranded RNA of human reovirus. *Proc. Nat. Acad. Sci. USA* **71**:3979–3983.

Miyake, T., Haruna, I., Shiba, T., Itoh, Y. H., Yamane, K., and Watanabe, I. 1971. Grouping of RNA phages based on the template specificity of their RNA replicases. *Proc. Nat. Acad. Sci. USA* **68**:2022–2024.

Mölling, K., Bolognesi, D. P., Bauer, H., Busen, W., Plassman, H. W., and Hausen, P. 1971. Association of viral reverse transcriptase with an enzyme degrading the RNA moiety of RNA-DNA hybrids. *Nature New Biol.* **234**:240–243.

Montagnier, L. 1968. The replication of viral RNA. *In*: Crawford, L. V., and M. G. P. Stocker, eds., *The Molecular Biology of Viruses,* Cambridge, Cambridge University Press, p. 125–147.

Montagnier, L., and Sanders, F. K. 1963a. Sedimentation properties of infective RNA extracted from encephalomyocarditis virus. *Nature* **197**:1178–1181.

Montagnier, L., and Sanders, F. K. 1963b. Replicative form of encephalomyocarditis virus RNA. *Nature* **199**:664–667.

Moore, C. H., Barron, F. F., Bohnert, D., and Weissmann, C. 1971. Possible origin of a minor virus specific protein (A_1) in Qβ particles. *Nature New Biol.* **234**:204–206.

Moroni, C. 1972. Structural proteins of Rauscher leukemia virus and Harvey sarcoma virus. *Virology* **47**:1–7.

Morris, C. F., Sinha, N. K., and Alberts, B. M. 1975. Reconstruction of bacteriophage T4 DNA replication apparatus from purified components. *Proc. Nat. Acad. Sci. USA* **72**:4800–4804.

Moss, B., Rosenblum, E. N., and Gershowitz, A. 1975. Characterization of a poly(A) polymerase from vaccinia virions. *J. Biol. Chem.* **250**:4722–4729.

Mudd, J. A., and Summers, D. F. 1970a. Protein synthesis in vesicular stomatitis virus-infected HeLa cells. *Virology* **42**:328–340.

Mudd, J. A., and Summers, D. F., 1970b. Polysomal RNA of vesicular stomatitis virus-infected HeLa cells. *Virology* **42**:958–968.

Müller, U. R., and Marchin, G. L. 1975. Temporal appearance of bacteriophage T4-modified valyl tRNA synthetase in *E. coli*. *J. Virol.* **15**:238–243.

Munyon, W., Paoletti, E., Ospina, J., and Grace, J. T. 1968. Nucleotide phosphohydrolase in purified vaccinia virus. *J. Virol.* **2**:167–172.

Murialdo, H., and Siminovitch, L. 1971. The morphogenesis of bacteriophage λ. *In:* Hershey, A. D., ed., *The Bacteriophage Lambda,* Cold Spring Harbor, New York, Cold Spring Harbor Laboratory, p. 711–723.

Murialdo, H. and Siminovitch, L., 1972. Identification of gene products and control of the expression of the morphogenetic information. *Virology* **48**:785–823.

Murray, K., and Murray, U. E. 1973. Terminal nucleotide sequences of DNA from temperate coliphages. *Nature New Biol.* **243**:134–139.

Nakajima, H., and Obara, J. 1967. Physicochemical studies of Newcastle disease virus. *Arch. Ges. Virusforsch.* **20**:287–295.

Nayak, D. P., and Baluda, M. A. 1967. Isolation and partial characterization of nucleic acid of influenza virus. *J. Virol.* **1**:1217–1223.

Nermut, M. V., Frank, H., and Schäfer, W. 1972. Properties of mouse leukemia viruses. *Virology* **49**:345–358.

Newman, J. F. E., and Brown, F. 1970. RNA of vesicular stomatitis virus. *In*: Barry, R. D., and B. W. J. Mahy, eds., *The Biology of Large RNA Viruses*, New York, Academic Press, p. 360–368.

Niles, E. G., Coulon, S. W., and Summers, W. C. 1974. Purification and physical characterization of T7 RNA polymerase from T7-infected *E. coli* B. *Biochemistry* **93**:3904–3912.

Norrby, E. 1966. The relationship between the soluble antigens and the virion of adenovirus type 3. *Virology* **28**:236–248.

Norrby, E. 1969. The structural and functional diversity of adenovirus capsid components. *J. Gen. Virol.* **5**:221–236.

Norrby, E. 1971. Adenoviruses. *In*: Maramorosch, K., and E. Kurstak, eds., *Comparative Virology,* New York, Academic Press, p. 105–134.

Nowinski, R. C., Sarker, N. H., Old, L. J., Moore, D. H., Scheer, D. I., and Kilgers, J. 1971. Characteristics of the structural components of the mouse mammary tumor virus. *Virology* **46**:21–38.

Obara, T., Bolognesi, D. P., and Bauer, H. 1971. Ribosomal RNA in avian leukosis virus particles. *Int. J. Cancer* **7**:535–541.

Öberg, B., and Philipson, L. 1972. Binding of histidine to TMV RNA. *Biochem. Biophys. Res. Comm.* **48**:927–932.

Öberg, B., Saborio, J., Persson, T., Everitt, E., and Philipson, L. 1975. Identification of the in vitro translation products of adenovirus mRNA by immunoprecipitation. *J. Virol.* **15**:199–207.

Oda, K., and Joklik, W. K. 1967. Hybridization and sedimentation studies on "early" and "late" vaccinia mRNA. *J. Mol. Biol.* **27**:395–419.

Offord, R. E. 1966. Electron microscopic observations on the substructure of tobacco rattle virus. *J. Mol. Biol.* **17**:370–375.

Ohe, K. 1972. Virus-coded origin of low–molecular weight RNA from KB cells infected with adenovirus type 2. *Virology* **47**:726–733.

Ohe, K., and Weissman, S. M. 1970. Nucleotide sequence of an RNA from cells infected with adenovirus 2. *Science* **167**:879–881.

Ohno, T., Nozu, Y., and Okada, Y. 1971. Polar reconstitution of TMV. *Virology* **44**:510–516.

Ortwerth, B. J., and Liu, L. P. 1973. Correlation between a specific isoaccepting lysyl tRNA and cell division in mammalian tissues. *Biochemistry* **12**:3978–3984.

Ortwerth, B. J., Yonuschot, G. R., and Carlson, J. V. 1973. Properties of $tRNA_4^{Lys}$ from various tissues. *Biochemistry* **12**:3985–3991.

Otto, B., and Reichard, P. 1975. Replication of polyoma DNA in isolated nuclei. *J. Virol.* **15**:259–267.

Paddock, G. V., and Abelson, J. 1973. Sequence of T4, T2, and T6 bacteriophage species I RNA and specific cleavage by an *E. coli* endonuclease. *Nature New Biol.* **246**:2–5.

Paddock, G. V., and Abelson, J. 1975a. Nucleotide sequence determination of bacteriophage T4 species I RNA. *J. Biol. Chem.* **250**:4185–4206.

Paddock, G. V., and Abelson, J. 1975b. Nucleotide sequence determination of bacteriophage T2 and T6 species I RNA. *J. Biol. Chem.* **250**:4207–4219.

Pal, B. K., and Roy-Burman, P. 1975. Phosphoproteins: Structural components of oncornaviruses. *J. Virol.* **15**:540–549.

Palmenberg, A., and Kaesberg, P. 1973. Amber mutant of bacteriophage Qβ capable of causing overproduction of Qβ replicase. *J. Virol.* **11**:603–605.

Palmenberg, A., and Kaesberg, P. 1974. Synthesis of complementary strands of heterologous RNAs with Qβ replicase. *Proc. Nat. Acad. Sci. USA* **71**:1371–1375.

Parkinson, J. S. 1968. Genetics of the left arm of the chromosomes of bacteriophage λ. *Genetics* **59**:311–325.

Pattison, I. H., and Jones, K. M. 1967. The possible nature of the transmissible agent of scrapie. *Vet. Rec.* **80**:2–9.

Paul, H. L. 1961. Physikalische und chemische untersuchungen am broad bean Mottle-Virus. *Z. Naturforsch.* **B16**:786–791.

Pedersen, I. R. 1971. Lymphocytic choriomeningitis virus RNAs. *Nature New Biol.* **234**:112–114.

Pedersen, I. R. 1973. Different classes of RNA isolated from lymphocytic choriomeningitis virus. *J. Virol.* **11**:416–423.

Pereira da Silva, L. H., and Jacob, F. 1967. Induction of C_{11} and O functions in early defective λ phages. *Virology* **33**:618–624.

Peters, G. G., and Hayward, R. S. 1974a. The 3′-terminal sequence of coliphage T7 "early" RNA. *Biochem. Biophys. Res. Comm.* **61**:759–766.

Peters, G. G., and Hayward, R. S. 1974b. Dinucleotide sequences in the regions of T7 DNA coding for termination of early transcription. *Eur. J. Biochem.* **48**:199–208.

Pett, D. M., Estes, M. K., and Pagano, J. S. 1975. Structural proteins of SV40. *J. Virol.* **15**:379–385.

Pettersson, U., Philipson, L., and Höglund, S. 1968. Purification and characterization of adenovirus type 2 fiber antigen. *Virology* **35**:204–215.

Pfefferkorn, E. R., and Hunter, H. S. 1963. The source of the RNA and phospholipid of Sindbis virus. *Virology* **20**:446–456.

Philipson, L., and Petterson, U. 1973. Structure and function of virion proteins of adenoviruses. *Prog. Exp. Tumor Res.* **18**:1–55.

Philipson, L., Wall, R., Glickman, G., and Darnell, J. E. 1971. Addition of poly(A) to virus-specific RNA during adenovirus replication. *Proc. Nat. Acad. Sci. USA* **68**:2806–2809.

Phillips, L. A., Park, J. J., and Hollis, V. W. 1974. Polyriboadenylate sequences at the 3′-termini of RNA obtained from mammalian leukemia and sarcoma viruses. *Proc. Nat. Acad. Sci. USA* **71**:4366–4370.

Pinkerton, T. C., Paddock, G., and Abelson, J. 1972. Bacteriophage T4 $tRNA^{Leu}$. *Nature New Biol.* **240**:88–90.

Pinkerton, T. C., Paddock, G., and Abelson, J. 1973. Nucleotide sequence determinations of bacteriophage T4 leucine tRNA. *J. Biol. Chem.* **248**:6348–6365.

Pirrotta, V. 1975. Sequence of the O_R operator of phage λ. *Nature* **254**:114–117.

Poglazov, B. F. 1970. Self-assembly of T-even bacteriophages in morphogenesis. *FEBS Symp.* **21**:181–193.

Pons, M. W. 1967. Some characteristics of double-stranded influenza virus RNA. *Arch. Ges. Virusforsch.* **22**:203–209.

Pons, M. W. 1972. Studies on the replication of influenza virus RNA. *Virology* **47**:823–832.

Pons, M. W., and Hirst, G. K. 1968. Polyacrylamide gel electrophoresis of influenza virus RNA. *Virology* **34**:385–388.

Porter, A., Carey, N., and Fellner, P. 1974. Presence of a large poly(rC) tract within the RNA of encephalomyocarditis virus. *Nature* **248**:675–678.

Prage, L. and Pettersson, U. 1971. Purification and properties of an arginine-rich core protein from adenovirus type 2 and type 3. *Virology* **45**:364–373.

Prage, L., Pettersson, U., Höglund, S., Lonberg-Holm, K., and Philipson, L. 1970. Sequential degradation of the adenovirus type 2 virion. *Virology* **42**:341–358.

Prevec, L., and Graham, A. F. 1966. Reovirus-specific polyribosomes in infected L-cells. *Science* **154**:522–524.

Pribnow, D. 1975. Nucleotide sequence of an RNA polymerase binding site at an early T7 promoter. *Proc. Nat. Acad. Sci. USA* **72**:784–788.

Price, R., and Penman, S. 1972. A distinct RNA polymerase activity synthesizing 5.5 S, 5 S, and 4 S RNA in nuclei from adenovirus 2-infected HeLa cells. *J. Virol.* **9**:621–626.

Pricer, W. E., and Weissbach, A. 1967. The synthesis of DNA exonucleases associated with the formation of temperate inducible bacteriophages. *J. Biol. Chem.* **242**:1701–1704.

Pritchett, R. F., Hayward, S. D., and Kieff, E. D. 1975. DNA of Epstein-Barr virus. *J. Virol.* **15**:556–569.

Putnam, F. W., Miller, D., Palm, L., and Evans, E. A. 1952. Biochemical studies of virus reproduction. *J. Biol. Chem.* **199**:177–191.

Radding, C. M., Szpirer, J., and Thomas, R. 1967. The structural gene for λ exonuclease. *Proc. Nat. Acad. Sci. USA* **57**:277–283.

Radloff, R. J., and Kaesberg, P. 1973. Electrophoretic and other properties of bacteriophage Qβ. *J. Virol.* **11**:116–125.

Randerath, E., Chia, L.-L.S.Y., Morris, H. P., and Randerath, K. 1974. Transfer RNA base composition studies in Morris hepatomas and rat liver. *Cancer Res.* **34**:643–653.

Ray, D. S., Bscheider, H. P., and Hofschneider, P. H. 1966. Replication of the single-stranded DNA of the male-specific bacteriophage M13. *J. Mol. Biol.* **21**:473–484.

Ray, D. S., and Schekman, R. W. 1969. Replication of bacteriophage M13. III. Identification of the intracellular single-stranded DNA. *J. Mol. Biol.* **43**:645–649.

Ray, P. N., and Pearson, M. L. 1974. Evidence for post-transcriptional control of the morphogenetic gene of bacteriophage λ. *J. Mol. Biol.* **85**:163–175.

Ray, P. N., and Pearson, M. L. 1975. Functional inactivation of bacteriophage λ morphogenetic gene mRNA. *Nature* **253**:647–650.

Razin, A., Sadat, J. W., and Sinsheimer, R. L. 1973. *In vivo* methylation of replicating bacteriophage φX174 DNA. *J. Mol. Biol.* **78**:417–425.

Rensing, U. F. E., and August, J. T. 1969. The 3′-terminus and the replication of phage RNA. *Nature* **224**:853–856.

Rensing, U. F. E., and Schoenmakers, J. G. G., 1973. A sequence of 50 nucleotides from coliphage R17 RNA. *Eur. J. Biochem.* **33**:8–18.

Rho, H. M., Grandgenett, D. P., and Green, M. 1975. Sequence relatedness between the subunits of AMV reverse transcriptase. *J. Biol. Chem.* **250**:5278–5280.

Říman, J., and Beaudreau, G. S. 1970. Viral DNA-dependent DNA polymerase and the properties of thymidine labelled material in virions of an oncogenic RNA virus. *Nature* **228**:427–430.

Říman, J., Korb, J., and Michlová, A. 1972. Specific ribosomes, components of an oncogenic RNA virus. *FEBS Symp.* **22**:99–114.

Ritchie, D. A., Thomas, C. A., MacHattie, L. A., and Wensink, P. C. 1967. Terminal repetition in non-permuted T3 and T7 bacteriophage DNA molecules. *J. Mol. Biol.* **23**:365–376.

Riva, S., and Geiduschek, E. P. 1969. Replication-coupled transcription in T4 development. *Fed. Proc.* **28**:660.

Robberson, D. L., Crawford, L. V., Syrett, C., and James, A. W. 1975. Unidirectional replication of a minority of polyoma virus and SV40 DNAs. *J. Gen. Virol.* **26**:59–69.

Roberts, J. W. 1969. Promoter mutation in vitro. *Nature* **223**:480–482.

Robertson, H. D. 1975. Isolation of specific ribosome binding sites from single-stranded DNA. *J. Mol. Biol.* **92**:363–375.

Roizman, B. and Roane, P. R. 1964. The multiplication of herpes simplex virus. *Virology,* **22**:262–269.

Roizman, B., and Spear, P. G. 1971. Herpesviruses: Current information on the composition and structure. *In*: Maramorosch, K., and E. Kurstak, eds., *Comparative Virology,* New York, Academic Press, p. 135–168.

Rosenkranz, H. S. 1973. RNA in coliphage T5. *Nature* **242**:327–329.

Ross, J., Tronick, S. R., and Scolnick, E. M. 1972. Poly(A) rich RNA in the 70 S RNA of murine leukemia-sarcoma virus. *Virology* **49**:230–235.

Rott, R., Drzeniek, R., and Frank, H. 1970. On the structure of influenza viruses. *In:* Barry, R. D., and B. W. J. Mahy, eds., *The Biology of Large RNA Viruses,* New York, Academic Press, p. 75–85.

Rouvière, J., Lederberg, S., Granboulan, P., and Gros, F. 1969. Structural sites of RNA synthesis in *E. coli. J. Mol. Biol.* **46**:413–430.

Roy, P., Clark, H. F., Madore, H. P., and Bishop, D. H. L. 1975. RNA polymerase associated with virions of pike fry rhabdovirus. *J. Virol.* **15**:338–347.

Roy-Burman, P. 1971. DNA polymerase associated with feline leukemia and sarcoma viruses: properties of the enzyme and its product. *Int. J. Cancer* **7**:409–415.

Roy-Burman, Pal, B. K., Gardner, M. B., and McAllister, R. M. 1974. Structural polypeptides of primate derived type C RNA tumor viruses. *Biochem. Biophys. Res. Comm.* **56**:543–551.

Rubenstein, I. 1968. Heat-stable mutants of T5 phage. *Virology* **36**:356–376.

Rubin, H., and Colby, C. 1968. Early release of growth inhibition in cells infected with Rous sarcoma cells. *Proc. Nat. Acad. Sci. USA* **60**:482–488.

Rueckert, R. R., 1971. Picornaviral architecture. *In*: Maramorosch, K., and E. Kurstak, eds., *Comparative Virology,* New York, Academic Press, p. 256–306.

Rymo, L., Parsons, J. T., Coffin, J. M., and Weissmann, C. 1974. *In vitro* synthesis of Rous sarcoma virus-specific RNA is catalyzed by a DNA-dependent RNA polymerase. *Proc. Nat. Acad. Sci. USA* **71**:2782–2786.

Saigo, K., and Uchida, H. 1974. Connection of the right-hand terminus of DNA to the proximal end of the tail in bacteriophage λ. *Virology* **61**:524–536.

Sakurai, T., Mikaye, T., Shiba, T., and Watanabe, I. 1968. Isolation of a possible fourth group of RNA phage. *Jpn. J. Microb.* **12**:544–546.

Sambrook, J., Sugden, B., Keller, W., and Sharp, P. A. 1973. Transcription of SV40. *Proc. Nat. Acad. Sci. USA* **70**:3711–3715.

Sanger, F., Air, G. M., Barrell, B. G., Brown, N. L., Coulson, A. R., Fiddes, J. C., Hutchison, C. A., Slocombe, P. M., and Smith, M. 1977. Nucleotide sequence of bacteriophage ϕX174 DNA. *Nature* **265**:687–695.

Sano, H., and Feix, G. 1974. RNA ligase activity of DNA ligase from phage T4 infected *E. coli. Biochemistry* **13**:5110–5115.

Sarkar, N. H., Nowinski, R. C., and Moore, D. H. 1971. Helical nucleocapsid structure of the oncogenic RNA viruses (Oncornaviruses). *J. Virol.* **8**:564–572.

Sarkar, N. H., Sarkar, S., and Kozloff, L. M. 1964. Tail components of T2 bacteriophage. *Biochemistry* **3**:511–516.

Sarov, I., and Becker, Y. 1967. Studies on vaccinia virus DNA. *Virology* **33**:369–375.

Sato, T., Friend, C., and de Harven, E. 1971. Ultrastructural changes in Friend erythroleukemia cells treated with dimethyl sulfoxide. *Cancer Res.* **31**:1402–1417.

Sauer, G. 1971. Apparent differences in transcriptional control in cells productively infected and transformed by SV40. *Nature New Biol.* **231**:135–138.

Savage, T., Roizman, B., and Heine, J. W. 1972. The proteins specified by herpes simplex virus. *J. Gen. Virol.* **17**:31–48.

Schäfer, W., Lange, J., Bolognesi, D. P., de Noronha, F., Post, J., and Richard, C. 1971. Isolation and characterization of two group-specific antigens from feline leukemia virus. *Virology* **44**:73–82.

Schäfer, W., Lange, J., Fischinger, P. J., Frank, H., Bolognesi, D. P., and Pister, L. 1972. Properties of mouse leukemia viruses. *Virology* **47**:210–228.

Scherberg, N. H., and Weiss, S. B. 1970. Detection of bacteriophage T4- and T5-coded tRNAs. *Proc. Nat. Acad. Sci. USA* **67**:1164–1171.

Scherberg, N. H., and Weiss, S. B. 1972. T4 tRNAs: Codon recognition and translational properties. *Proc. Nat. Acad. Sci. USA* **69**:1114–1118.

Schiff, N., Miller, M. J., and Wahba, A. J. 1974. Purification and properties of chain IF-3 from T4-infected and uninfected *E. coli* MRE600. *J. Biol. Chem.* **249**:3797–3802.

Schochetman, G., and Schlom, J. 1975. RNA subunit structure of Mason-Pfizer monkey virus. *J. Virol.* **15**:423–427.

Schröder, C. H., Erben, E., and Kaerner, H.-C. 1973. A rolling circle model of the *in vivo* replication of bacteriophage ϕX174 replicative form DNA. *J. Mol. Biol.* **79**:599–613.

Schröder, C. H., and Kaerner, H. C. 1972. Replication of bacteriophage ϕX174 replicative form DNA *in vivo*. *J. Mol. Biol.* **71**:351–362.

Schuster, R. C., and Weissbach, A. 1968. Evidence for a new endonuclease synthesized by λ bacteriophage. *J. Virol.* **2**:1096–1101.

Seal, G., and Loeb, L. A. 1976. Enzyme activities associated with DNA polymerases from RNA tumor viruses. *J. Biol. Chem.* **251**:975–981.

Sebring, E. D., and Salzman, N. P. 1967. Metabolic properties of early and late vaccinia virus mRNA. *J. Virol.* **1**:550–558.

Seidman, J. G., Comer, M. M., and McClain, W. H. 1974. Nucleotide alterations in the bacteriophage T4 glutamine tRNA that affect ochre suppressor activity. *J. Mol. Biol.* **90**:677–689.

Sela, I. 1972. Tobacco enzyme-cleaved fragments of TMV-RNA specifically accepting serine and methionine. *Virology* **49**:90–94.

Sela, I., and Antignus, Y. 1971. Spectrophotometric determination of nucleic acids and nucleoproteins at the far-UV region. *Anal. Biochem.* **43**:217–226.

Semal, J., and Kummert, J. 1971. Sequential synthesis of double-stranded and single-stranded RNA by cell-free extracts of barley leaves infected with brome mosaic virus. *J. Gen. Virol.* **10**:79–89.

Serwer, P. 1976. Internal proteins of bacteriophage T7. *J. Mol. Biol.* **107**:271–291.

Sgaramella, V., and H. G. Khorana. 1972. A further study of the T4 ligase-catalyzed joining of DNA at base-paired ends. *J. Mol. Biol.*, 72:493–502.

Sharma, O. K., Mays, L. L., and Borek, E. 1975. Functional differences in protein synthesis between rat liver tRNA and tRNA from Novikoff hepatoma. *Biochemistry* **14**:509–514.

Shatkin, A. J., Sipe, J. D., and Loh, P. 1968. Separation of ten reovirus genome segments by polyacrylamide gel electrophoresis. *J. Virol.* **2**:986–991.

Sheldon, R., and Kates, J. 1974. Mechanism of poly(A) synthesis by vaccinia virus. *J. Virol.* **14**:214–224.

Shiba, T., and Miyake, T. 1975. New type of infectious complex of *E. coli* RNA phage. *Nature* **254**:157–158.

Shih, D. S., and Kaesberg, P. 1973. Translation of brome mosaic viral RNA in a cell-free system derived from wheat embryo. *Proc. Nat. Acad. Sci. USA* **70**:1799–1803.

Shih, D. S., Kaesberg, P., and Hall, T. C. 1974. Messenger and aminoacylation functions of brome mosaic virus RNA after chemical modification of the 3′-terminus. *Nature* **249**:353–355.

Shildkraut, C. L., Wierzchowski, K. L., Marmur, J., Green, D. M., and Doty, P. 1962. A study of the base sequence homology among the T series of bacteriophages. *Virology* **18**:43–55.

Showe, M. K., and Kellenberger, E. 1975. Control mechanisms in virus assembly. *In*: Burke, D. C., and W. C. Russel, eds., *Control Processes in Virus Multiplication,* Cambridge, Cambridge University Press, p. 407–438.

Siegel, R. B., and Summers, W. C. 1970. The process of infection with coliphage T7. *J. Mol. Biol.* **49**:115–123.

Siegert, W., Bauer, G., and Hofschneider, P. H. 1973. Direct evidence for messenger activity of influenza virion RNA. *Proc. Nat. Acad. Sci. USA* **70**:2960–2963.

Siegert, W., and Hofschneider, P. H. 1973. A direct approach to study the messenger properties of influenza-virion RNA. *FEBS Lett.* **34**:145–146.

Simon, L. D., and Anderson, T. F. 1967. The infection of *E. coli* by T2 bacteriophages as seen in the electron microscope. *Virology* **32**:298–305.

Simon, M. N., and Studier, F. W. 1973. Physical mapping of the early region of bacteriophage T7 DNA. *J. Mol. Biol.* **79**:249–265.

Simpson, R. W., and Hirst, G. K. 1968. Temperature-sensitive mutants of influenza A virus. *Virology* **35**:41–49.

Sinsheimer, R. L. 1968. Bacteriophage ϕX174 and related viruses. *Progr. Nucl. Acid Res. Mol. Biol.* **8**:115–169.

Sinsheimer, R. L., Hutchison, C. A., and Linquist, B. H. 1967. Bacteriophage ϕX174: Viral functions. *In*: Colter, J. S., and W. Paranchych, eds., *The Molecular Biology of Viruses*, New York, Academic Press, p. 175–192.

Sinsheimer, R. L., Knippers, R., and Komano, T. 1968. Stages in the replication of bacteriophage ϕX174 DNA *in vivo*. *Cold Spring Harbor Symp. Quant. Biol.* **33**:443–447.

Sirover, M. A., and Loeb, L. A. 1974. Infidelity of DNA synthesis: A general property of RNA tumor viruses. *Biochem. Biophys. Res. Comm.* **61**:360–364.

Skalka, A. 1969. Nucleotide distribution and functional orientation in the DNA of phage ϕ80. *J. Virol.* **3**:150–156.

Skalka, A., Burgi, E., and Hershey, A. D. 1968. Segmental distribution of nucleotides in the DNA of bacteriophage λ. *J. Mol. Biol.* **34**:1–16.

Skare, J., Niles, E. G., and Summer, W. C. 1974. Localization of the leftmost initiation site for T7 late transcription *in vivo* and *in vitro*. *Biochemistry* **13**:3912–3916.

Skogerson, L., Roufa, D., and Leder, P. 1971. Characterization of the initial peptides of Qβ RNA polymerase and control of its synthesis. *Proc. Nat. Acad. Sci. USA* **68**:276–279.

Smith, G. R., and Hedgpeth, J. 1975. Oligo(A) not coded by DNA generating 3′-terminal heterogeneity in a λ phage RNA. *J. Biol. Chem.* **250**:4818–4821.

Smith, K. O., Gehle, W. D., and Trousdale, M. D. 1965. Architecture of the adenovirus capsid. *J. Bact.* **90**:254–261.

Smith, M., Brown, N. L., Air, G. M., Barrell, B. G., Coulson, A. R., Hutchison, C. A., and Sanger, F. 1977. DNA sequence at the C termini of the overlapping genes A and B in bacteriophage ϕX174. *Nature* **265**:702–705.

Smith, M., and Skalka, A. 1966. Some properties of DNA from phage-infected bacteria. *J. Gen. Physiol.* **49**:127–142.

Smith, R. E., Zweerink, H. J., and Joklik, W. K. 1969. Polypeptide components of virions, top component and cores of reovirus type 3. *Virology* **39**:791–810.

Snyder, L., and Geiduschek, E. P. 1968. In vitro synthesis of T4 late mRNA. *Proc. Nat. Acad. Sci. USA* **59**:459–466.

Sokol, F., Clark, H. F., Wiktor, T. J., McFalls, M. L., Bishop, D. H. L., and Obijeski, J. F. 1974. Structural phosphoproteins associated with ten rhabdoviruses. *J. Gen. Virol.* **24**:433–455.

Sokol, F., and Koprowski, H. 1975. Structure-function relationships and mode of replication of animal rhabdoviruses. *Proc. Nat. Acad. Sci. USA* **72**:933–936.

Sonnabend, J. A., Martin, E. M., and Mécs, E. 1967. Viral specific RNAs in infected cells. *Nature* **213**:365–367.

Spear, P. G., and Roizman, B. 1970. The proteins specified by herpes simplex virus. IV. The site of glycosylation and accumulation of viral membrane proteins. *Proc. Nat. Acad. Sci. USA* **66**:730–737.

Spear, P. G., and Roizman, B. 1972. The proteins specified by herpes simplex virus. V. Purification of structural proteins of the herpesvirion. *J. Virol.* **9**:143–159.

Spiegelman, G. B., and H. R. Whiteley. 1974a. Purification of RNA polymerase from SP82-infected *B. subtilis. J. Biol. Chem.,* **249**:1476–1482.

Spiegelman, G. B., and Whiteley, H. R. 1974b. *In vivo* and *in vitro* transcription by ribonucleic acid polymerase from SP82-infected *B. subtilis. J. Biol. Chem.* **249**:1483–1489.

Spiegelman, S., Burny, A., Das, M. R., Keyder, J., Schlom, J., Travnicek, M., and Watson, K. 1970. Synthetic DNA-RNA hybrids and RNA-RNA duplexes as templates for the polymerases of oncogenic RNA viruses. *Nature* **228**:430–432.

Spiegelman, S., Pace, N. R., Mills, D. R., Levisohn, R., Eikhorn, T. S., Taylor, M. M., Peterson, R. L., and Bishop, D. H. L. 1968. The mechanism of RNA replication. *Cold Spring Harbor Symp. Quant. Biol.* **33**:101–124.

Spiegelman, S., Watson, K. F., and Kacian, D. L. 1971. Synthesis of DNA complements of natural RNAs: A general approach. *Proc. Nat. Acad. Sci. USA* **68**:2843–2845.

Spremulli, L. L., Haralson, M. A., and Ravel, J. M. 1974. Effect of T4 infection on initiation of protein synthesis and messenger specificity of IF-3. *Arch. Biochem. Biophys.* **165**:581–587.

Springgate, C. F., Battula, N., and Loeb, L. A. 1973. Infidelity of DNA synthesis by reverse transcriptase. *Biochem. Biophys. Res. Comm.* **52**:400–406.

Stahl, F. W., Murray, N. E., Nakata, A., and Crasemann, J. M. 1966. Intergenic *cis–trans* position effects in bacteriophage T4. *Genetics* **54**:223–232.

Stahl, S., Paddock, G. and Abelson, J. 1973. T4 bacteriophage $tRNA^{Gly}$. *Biochem. Biophys. Res. Comm.* **54**:567–569.

Staudenbauer, W. L. 1974. Involvement of DNA polymerases I and III in the replication of bacteriophage M-13. *Eur. J. Biochem.* **49**:249–256.

Stavis, R. L., and August, J. T. 1970. The biochemistry of RNA bacteriophage replication. *Ann. Rev. Biochem.* **39**:527–560.

Steitz, J. A. 1968. Identification of the A protein from bacteriophage R17. *J. Mol. Biol.* **33**:923–936.

Steitz, J. A. 1970. The reconstitution of RNA bacteriophages. *FEBS Symp.* **21**:203–212.

Stone, A. B. 1970. General inhibition of *E. coli* macromolecular synthesis by high multiplicities of bacteriophage ϕX174. *J. Mol. Biol.* **47**:215–229.

Storer, G. B., Shepherd, M. G., and Kalmakoff, J. 1973. Enzyme activities associated with cytoplasmic polyhedrosis virus from *Bombyx mori. Intervirology* **2**:87–94.

Stott, E. J., and Killington, R. A. 1972. Rhinoviruses. *Ann. Rev. Microbiol.* **26**:503–524.

Strand, M., and August, J. T. 1976. Structural proteins of RNA tumor viruses. *J. Biol. Chem.* **251**:559–564.

Strauss, E. G., and Kaesberg, P. 1970. Acrylamide gel electrophoresis of bacteriophage Qβ. *Virology* **42**:437–452.

Studier, F. W. 1969. The genetics and physiology of bacteriophage T7. *Virology,* **39**:562–574.

Studier, F. W. 1972. Bacteriophage T7. *Science* **176**;367–376.

Studier, F. W. 1973. Analysis of bacteriophage T7 early RNAs and proteins on slab gels. *J. Mol. Biol.,* **79**:237–248.

Studier, F. W., and Hausmann, R. 1969. Integration of two sets of T7 mutants. *Virology,* **39**:587–588.

Studier, F. W., and Maizel, J. V. 1969. T7-directed protein synthesis. *Virology* **39**:575–586.

Stussi, C., Guilley, H., Lebeurier, G., and Hirth, L. 1972. Some recent advances in the comprehension of *in vitro* morphogenesis of TMV. *Biochimie* **54**:287–296.

Subak-Sharpe, H. 1968. Virus-induced changes in translation mechanisms. *In:* The Society for General Microbiology, *The Molecular Biology of Viruses,* Cambridge, Cambridge University Press, p. 47–66.

Summers, D. F., Roumiantzeff, M., and Maizel, J. V. 1971. The translation and processing of poliovirus proteins. *In:* Wolstenholme, G. E. W., and M. O'Connor, eds., *Strategy of the Viral Genome,* Edinburgh, Churchill Livingstone, p. 111–124.

Summers, W. C., and Siegel, R. B. 1970. Transcription of late phage RNA by T7 RNA polymerase. *Nature* **228**:1160–1162.

Swetley, P., and Watanabe, Y. 1974. Cell cycle dependent transcription of SV40 DNA in SV40-transformed cells. *Biochemistry* **13**:4122–4126.

Syvanen, M. 1975. Processing of bacteriophage λ DNA during its assembly into heads. *J. Mol. Biol.* **91**:165–174.

Szabo, C., Dharmgrongartama, B., and Moyer, R. W. 1975. The regulation of transcription in bacteriophage T5-infected *E. coli*. *Biochemistry* **14**:989–997.

Szabo, C., and Moyer, L. W. 1975. Purification and properties of a bacteriophage T5-modified form of *E. coli* RNA polymerase. *J. Virol.* **15**:1042–1046.

Szekely, M., and Loviny, T. 1975. 5′-terminal phosphorylation and secondary structure of double-stranded RNA from a fungal virus. *J. Mol. Biol.* **93**:79–87.

Tai, H. T., Smith, C. A., Sharp, P. A., and Vinograd, J. 1972. Sequence heterogeneity in closed SV40 DNA. *J. Virol.* **9**:317–325.

Takahashi, I., and Marmur, J. 1963. Replacement of thymodylic acid by deoxyuridylic acid in the DNA of a transducing phage for *B. subtilis*. *Nature* **197**:794–795.

Takahashi, S. 1974. The rolling-circle replicative structure of a bacteriophage λ DNA. *Biochem. Biophys. Res. Comm.* **61**:607–613.

Takemoto, K. K., Mattern, C. F. T., and Murakami, W. T. 1971. The papovavirus group. *In:* Maramorosch, K., and E. Kurstak, eds., *Comparative Virology,* New York, Academic Press, p. 81–104.

Taketo, A. 1973. Sensitivity of *E. coli* to viral nucleic acid. *Mol. Gen. Genet.* **122**:15–22.

Tal, J., Craig, E. A., and Raskas, H. J. 1975. Sequence relationships between adenovirus 2 early RNA and viral RNA size classes synthesized at 18 hours after infection. *J. Virol.* **15**:137–144.

Tannock, G. A., Gibbs, A. J., and Cooper, P. D. 1970. A re-examination of the molecular weight of poliovirus RNA. *Biochem. Biophys. Res. Comm.* **38**:298–304.

Tanyashin, V. I. 1968. Use of phenol fractionation for the isolation of the replicative form of DNA T5 phage. *Biokhimiya* **33**:713–720.

Taube, S. E., McGuire, P. M., and Hodge, L. D. 1974. RNA synthesis specific for an integrated adenovirus genome during the cell cycle. *Nature* **250**:416–418.

Temin, H. M. 1964. Nature of the provirus of Rous sarcoma. *Nat. Cancer Inst. Monogr.,* 17:557–570.

Temin, H. M. 1967. Studies on carcinogenesis by avian sarcoma viruses. *J. Cell Physiol.* **69**:53–64.

Temin, H. M., and Baltimore, D. 1972. RNA-directed DNA synthesis and RNA tumor viruses. *Adv. Virus Res.* **17**:129–186.

Teramoto, Y. A., Puentes, M. J., Young, L. J. T., and Cardiff, R. D. 1973. Structure of the mouse mammary tumor virus: Polypeptides and glycoproteins. *J. Virol.* **13**:411–418.

Thach, S. S., Dobbertin, D., Lawrence, C., Golini, F., and Thach, R. E. 1974. Structure of replication complexes of encephalomyocarditis virus. *Proc. Nat. Acad. Sci. USA* **71**:2549–2553.

Thermes, C., Daegelen, P., de Franciscis, V., and Bordy, E. 1976. *In vitro* system for induction of delayed early RNA of bacteriophage T4. *Proc. Nat. Acad. Sci. USA* **73**:2569–2573.

Thomas, C. A., and MacHattie, L. A. 1964. Circular T2 DNA molecules. *Proc. Nat. Acad. Sci. USA* **52**:1297–1301.

Thomas, C. A., and Rubenstein, I. 1964. The arrangements of nucleotide sequences in T2 and T5 bacteriophage DNA molecules. *Biophys. J.* **4**:93–106.

Thouvenel, J. C., Guilley, H., Stussi, C., and Hirth, L. 1971. Evidence for polar reconstitution of TMV. *FEBS Lett.* **16**:204–206.

Tikhonenko, A. S. 1961. Comparative study of the morphology of phage particles by means of shadowing and negatives staining in phosphotungstic acid. *Biofizika* **6**:410–413.

Tikhonenko, A. S. 1970. *Ultrastructure of Bacterial Viruses*. Trans. from Russian by B. Haigh. New York, Plenum Press.

Tikhonenko, A. S., and Zavarzina, N. B. 1966. Morphology of a lytic agent of *Chlorella pyrenoidosa*. *Mikrobiologiya* **35**:850–853.

Tobey, R. A. 1964. Mengo virus replication. II. Isolation of polyribosomes containing the infecting viral genome. *Virology* **23**:23–29.

Tomich, P. K., Chiu, C. S., Wovcha, M. G., and Greenberg, C. R. 1974. Evidence for a complex regulating the *in vivo* activities of early enzymes induced by bacteriophage T4. *J. Biol. Chem.* **249**:7613–7622.

Tomich, P. K., and Greenberg, G. R. 1973. On the effect of a dCMP hydroxymethylase mutant of bacteriophage T4 showing enzyme activity in extracts. *Biochem. Biophys. Res. Comm.* **50**:1032–1038.

Tomizawa, J., and Ogawa, T. 1968. Replication of phage λ DNA. *Cold Spring Harbor Symp. Quant. Biol.* **33**:525–532.

Tooze, J., ed. 1973. *The Molecular Biology of Tumour Viruses,* Cold Spring Harbor, New York, Cold Spring Harbor Laboratory.

Trávníček, M. 1969. Some properties of amino-acceptor RNA isolated from avian tumour virus bai strain A (avian myeloblastosis). *Biochim. Biophys. Acta* **182**:427–439.

Trávníček, M., and Říman, J. 1973. Occurrence of aminoacyl-tRNA synthetase in an RNA oncogenic virus. *Nature New Biol.* **241**:60–62.

Tronick, S. R., Scolnick, E. M., and Parks, W. P. 1972. Reversible inactivation of the DNA polymerase of Rauscher leukemia virus. *J. Virol.* **50**:885–888.

Tsuruo, T., Hirayama, K., and Ukita, T. 1975. Three DNA polymerases of rat ascites hepatoma cells: Properties of the enzymes and effect of RNA synthesis on the reactions. *Biochim. Biophys. Acta* **383**:274–281.

Tsuruo, T., Satoh, H., and Ukita, T. 1972a. DNA polymerases of ascites hepatoma cells. I. Purification and properties of a DNA polymerase from soluble fraction. *Biochem. Biophys. Res. Comm.* **48**:769–775.

Tsuruo, T., Tomita, Y., Satoh, H., and Ukita, T. 1972b. DNA polymerases of ascites hepatoma cells. II. Purification and properties of DNA polymerases from nuclear membrane-chromatin fraction. *Biochem. Biophys. Res. Comm.* **48**:776–782.

Tsuruo, T., and Ukita, T. 1974. Purification and further characterization of three DNA polymerases of rat ascites hepatoma cells. *Biochim. Biophys. Acta* **353**:146–159.

Urushibara, T., Furiuchi, Y., Nishimura, C., and Miura, K.-I. 1975. A modified structure at the 5′-terminus of mRNA of vaccinia virus. *FEBS Lett.* **49**:385–389.

Valentine, R. C., Engelhardt, D. L., and Zinder, N. D. 1964. Host-dependent mutants of the bacteriophage f2. *Virology* **23**:159–163.

Valentine, R. C., and Pereira, H. G. 1965. Antigens and structure of the adenovirus. *J. Mol. Biol.* **13**:13–20.

Valentine, R. C., and Wedel, H. 1965. The extracellular stages of RNA bacteriophage infection. *Biochem. Biophys. Res. Comm.* **21**:106–112.

Vandekerckhove, J., and Van Montagu, M. 1974. Sequence analysis of fluorescamine-stained peptides and proteins purified on a nanomole scale. Applications to proteins of bacteriophage MS2. *Eur. J. Biochem.* **44**:279–288.

Vandenberghe, A., Min Jou, W., and Fiers, W. 1975. 3′-terminal nucleotide sequence ($n = 361$) of bacteriophage MS2 RNA. *Proc. Nat. Acad. Sci. USA* **72**:2559–2561.

van der Eb, A. J., van Kesteren, L. W., and van Bruggen, E. F. J. 1969. Structural properties of adenovirus DNAs. *Biochim. Biophys. Acta* **182**:530–541.

Van der Marel, P., Tasseron-de Jong, J., and Bosch, L. 1975. The proteins associated with mRNA from uninfected and adenovirus type 5-infected KB cells. *FEBS Lett.* **51**:330–334.

van Mansfeld, A. D. M., Vereijken, J. M., and Jansz, H. S. 1976. The nucleotide sequence of a

DNA fragment, 71 base pairs in length, near the origin of DNA replication of bacteriophage φX174. *Nucl. Acids Res.* **3**:2827–2844.

Varshavsky, A. J., Bakayev, V. V., Chumackov, P. M., and Georgiev, G. P. 1976. Minichromosome of SV40-presence of histone H1. *Nucl. Acids Res.* **3**:2101–2113.

Vasquez, C., Granboulan, N., and Franklin, R. M. 1966. Structure of the RNA bacteriophage R17. *J. Bact.* **92**:1779–1786.

Verma, I. M., Meuth, N. L., Bromfield, E., Manly, K., and Baltimore, D. 1971. Covalently linked RNA-DNA molecule as initial product of RNA tumour virus DNA polymerase *Nature New Biol.* **233**:131–134.

Vigier, P. 1974. Replication and integration of the genome of oncornaviruses. In: Kurstak, E. and K. Maramorosch, eds., *Viruses, Evolution, and Cancer,* New York, Academic Press, p. 209–233.

Wagner, R. R., and Schnaitman, C. A. 1970. Proteins of vesicular stomatitis virus. *In*: Barry, R. D., and B. W. J. Mahy, eds., *The Biology of Large RNA Viruses,* New York, Academic Press, p. 655–672.

Wagner, R. R., Schnaitman, T., and Snyder, R. M. 1969a. Structural proteins of vesicular stomatitis viruses. *J. Virol.* **3**:395–403.

Wagner, R. R., Schnaitman, T., Snyder, R. M., and Schnaitman, C. A. 1969b. Protein composition of the structural components of vesicular stomatitis virus. *J. Virol.* **3**:611–618.

Wagner, R. R., Snyder, R. M., and Yamazaki, S. 1970. Proteins of vesicular stomatitis virus: Kinetics and cellular sites of synthesis. *J. Virol.* **5**:548–558.

Wahl, R., and Kozloff, L. M. 1962. The nucleoside triphosphate content of various bacteriophages. *J. Biol. Chem.* **237**:1953–1960.

Wainfan, E. 1968. Development of tRNA methylating enzymes with altered properties during heat induction of *E. coli* K12 (λ $C_1$857). *Virology* **35**:282–288.

Walls, P. A., and Pootjes, C. F. 1975. Host-phage interaction in *Agrobacterium tumefaciens*. *J. Virol.* **15**:372–378.

Walsh, M. L., and Cohen, P. S. 1974a. Polyribosome metabolism in bacteriophage T4 infected *E. coli*. General properties. *Arch. Biochem. Biophys.* **162**:369–373.

Walsh, M. L., and Cohen, P. S. 1974b. Polyribosome metabolism in bacteriophage T4 infected *E. coli*. Isolation and characterization of two classes of polyribosomes. *Arch. Biochem. Biophys.* **162**:374–384.

Walter, G., Seifert, W., and Zillig, W. 1968. Modified DNA-dependent RNA polymerase from *E. coli* infected with bacteriophage T4. *Biochem. Biophys. Res. Comm.* **30**:240–247.

Walz, A., and Pirrotta, V. 1975. Sequence of the P_r promoter of phage λ. *Nature* **254**:118–121.

Wang, J. C., and Kaiser, A. D. 1973. Evidence that the cohesive ends of mature λ DNA are generated by the gene *A* product. *Nature New Biol.* **241**:16–17.

Wang, J. C., and Schwartz, H. 1967. Noncomplementarity in base sequences between the cohesive ends of coliphages 186 and λ and the formation of interlocked rings between the two DNAs. *Biopolymers* **5**:953–966.

Wang, L., and Duesberg, P. H. 1973. DNA polymerase of murine sarcoma-leukemia virus. *J. Virol.* **12**:1512–1521.

Ward, R. L., and Stevens, J. G. 1975. Lifetimes of mRNA molecules directing the synthesis of viral proteins in herpes simplex virus-infected cells. *J. Virol.* **15**:81–89.

Warner, H. R., and Barnes, J. E. 1966. Evidence for a dual role for the bacteriophage T4-induced deoxycytidine triphosphate nucleotidohydrolase. *Proc. Nat. Acad. Sci. USA* **56**:1233–1240.

Watanabe, I., Miyake, T., Sakurai, T., Shiba, T., and Ohno, T. 1967. Isolation and grouping of RNA phages. *Proc. Jpn. Acad.* **43**:204–209.

Watanabe, Y., and Graham, A. F. 1967. Structural units of reovirus RNA and their possible functional significance. *J. Virol.* **1**:665–677.

Waters, L. C. 1975. Transfer RNAs associated with the 70 S RNA of AKR murine leukemia virus. *Biochem. Biophys. Res. Comm.* **65**:1130–1136.

Waters, L. C., Yang, W. K., Mullin, B. C., and Nichols, J. L. 1975. Purification of tryptophan tRNA from chick cells and its identity with "spot 1" RNA of Rous sarcoma virus. *J. Biol. Chem.* **250**:6627–6629.

Watson, J. D., 1971. The structure and assembly of murine leukemia virus. *Virology* **45**:586–597.

Watson, J. D., and Littlefield, J. W. 1960. Some properties of DNA from Shope papilloma virus. *J. Mol. Biol.* **2**:161–165.

Weber, H., and Weissmann, C. 1970. The 3′-termini of bacteriophage Qβ plus and minus strands. *J. Mol. Biol.* **51**:215–224.

Weber, K. 1967. Amino acid sequence studies on the tryptic peptides of the coat protein of the bacteriophage R17. *Biochemistry* **6**:3144–3154.

Weber, K., and Konigsberg, W. 1967. Amino acid sequence of the f_2 coat protein. *J. Biol. Chem.* **242**:3563–3578.

Wei, C. M., and Moss, B. 1975. Methylated nucleotides block 5′-terminus of vaccinia virus messenger RNA. *Proc. Nat. Acad. Sci. USA* **72**:318–322.

Weiner, A. M., and Weber, K. 1971. Natural and read-through at the UGA termination signal of Qβ coat protein cistron. *Nature New Biol.* **234**:206–209.

Weiss, E., and Kiesow, L. A. 1966. Incomplete citric acid cycle in agents of the psittacosis-trachoma group (*Chlamydia*). *Bact. Proc.* **1966**:85.

Weiss, E., Myers, W. F., Dressler, H. R., and Chun-Hoon, H. 1964. Glucose metabolism by agents of the psittacosis-trachoma group. *Virology* **22**:551–562.

Weiss, R. A. 1970. Studies on the loss of growth inhibition in cells infected with Rous sarcoma virus. *Int. J. Cancer* **6**:333–341.

Weissmann, C., Billeter, M. A., Goodman, H. M., Hindley, J., and Weber, H. 1973. Structure and function of phage RNA. *Ann. Rev. Biochem.* **42**:303–328.

Weissmann, C., and Ochoa, S. 1967. Replication of phage RNA. *Progr. Nucl. Acid. Res. Mol. Biol.* **6**:353–399.

Weith, H. L., and Gilham, P. T. 1969. Polynucleotide sequence analysis by sequential base elimination: 3′-terminus of phage Qβ RNA. *Science* **166**:1004–1005.

Werner, R. 1968a. Initiation and propagation of growing points in DNA of phage T4. *Cold Spring Harbor Symp. Quant. Biol.* **33**:501–508.

Werner, R. 1968b. Distribution of growing points in DNA of bacteriophage T4. *J. Mol. Biol.* **33**:679–692.

Westphal, H., and Kiehn, E. D. 1970. The *in vitro* product of SV40 DNA transcription and its specific hybridization with DNA of SV40-transformed cells. *Cold Spring Harbor Symp. Quant. Biol.* **35**:819–822.

Wickner, S., and Hurwitz, J. 1974. Conversion of ϕX174 viral DNA to double-stranded form by purified *E. coli* proteins. *Proc. Nat. Acad. Sci. USA* **71**:4120–4124.

Wildy, P., Russell, W. C., and Horne, R. W. 1960. The morphology of herpes viruses. *Virology* **12**:204–222.

Wilson, J. H., and Kells, S. 1972. Bacteriophage T4 tRNA. I. Isolation and characterization of two phage-coded nonsense suppressors. *J. Mol. Biol.* **69**:39–56.

Wilson, J. H., Kim, J. S., and Abelson, J. B., 1972. Clustering of the genes for the T4 tRNAs. *J. Mol. Biol.* **71**:547–556.

Winocour, E. 1965. Attempts to detect an integrated polyoma genome by nucleic acid hybridization. *Virology* **25**:276–288.

Witter, R., Frank, H., Moennig, V., Hunsmann, G., Lange, J., and Schäfer, W. 1973. Properties of mouse leukemia viruses. *Virology* **54**:330–345.

Wittmann-Liebold, B., and Wittmann, H. G. 1967. Coat proteins of strains of two RNA viruses: Comparison of their amino acid sequences. *Mol. Gen. Genetics* **100**:358–363.

Wong-Staal, F., Gillespie, D., and Gallo, R. C. 1976. Proviral sequences of baboon endogenous type C RNA virus in DNA of human leukaemic tissues. *Nature* **262**:190–195.

Wood, W. B., Edgar, R. S., King, J., Lielausis, I., and Henninger, M. 1968. Bacteriophage assembly. *Fed. Proc.* **27**:1160–1166.

Wu, R., and Taylor, E. 1971. Complete nucleotide sequence of the cohesive ends of bacteriophage λ DNA. *J. Mol. Biol.* **57**:491–512.

Yamamoto, M., and Uchida, H. 1973. Organization and function of bacteriophage T4 tail. 1. Isolation of heat-sensitive T4 tail mutants. *Virology* **52**:234–245.

Yamamoto, M., and Uchida, H. 1975. Organization and function of the tail of bacteriophage T4. II. Structural control of the tail contraction. *J. Mol. Biol.* **92**:207–223.

Yamamoto, N., and Anderson, T. F. 1961. Genomic masking and recombination between serologically unrelated phages P22 and P221. *Virology* **14**:430–439.

Yamazaki, H., Bancroft, J. B., and Kaesburg, P. 1961. Biophysical studies of broad bean mottle virus. *Proc. Nat. Acad. Sci. USA* **47**:979–983.

Yogo, Y., Teng, M. H., and Wimmer, E. 1974. Poly(U) in poliovirus minus RNA is 5′-terminal. *Biochem. Biophys. Res. Comm.* **61**:1101–1109.

Yogo, Y., and Wimmer, E. 1975. Poly(U) and poly(A) as components of the purified poliovirus replicative intermediate. *J. Mol. Bio.* **92**:467–477.

Young, E. T., and Sinsheimer, R. L. 1967. Vegetative bacteriophage λ DNA. I. Infectivity in a spheroplast assay. *J. Mol. Biol.* **30**:147–164.

Young, E. T., and Sinsheimer, R. L. 1968. Vegetative λ DNA. III. Pulse-labeled components. *J. Mol. Biol.* **33**:49–60.

Zetter, B. R., and Cohen, P. S. 1974. Post-transcriptional regulation of T4 enzyme synthesis. *Arch. Biochem. Biophys.* **162**:560–567.

Zlotnik, I., and Rennie, J. C. 1965. Experimental transmission of mouse passaged scrapie to goats, sheep, rats and hamsters. *J. Comp. Path.* **75**:147–157.

CHAPTER 10

Bonar, R. A., Sverak, A. L., Bolognesi, D. P., Langlois, A. J., Beard, D., and Beard, J. W. 1967. Ribonucleic components of BA1 strain A (myeloblastosis) avian tumor virus. *Cancer Res.* **27**:1138–1157.

Burton, S. D., Morita, R. Y., and Miller, W. 1966. Utilization of acetate by *Beggiatoa*. *J. Bacteriol.* **91:**1192–1200.

Dillon, L. S. 1962. Comparative cytology and the evolution of the living world. *Evolution* **16**:102–117.

Dillon, L. S. 1974. Neovulcanism: A proposed replacement for the concepts of plate tectonics and continental drift. *Mem. Amer. Assoc. Petr. Geol.* **23**:167–239.

Entwistle, P. F., Robertson, J. S., and Juniper, B. E. 1968. The ultrastructure of a rickettsia pathogenic to a saturnid moth. *J. Gen. Microbiol.* **54**:97–104.

Gay, F., Clarke, L. K., and Dermott, E. 1970. Morphogenesis of Bittner virus. *J. Virol.* **5**:801–806.

Lwoff, A., and Tournier, P. 1971. Remarks on the classification of viruses. *In:* Maramorosch, K., and E. Kurstak, eds., *Comparative Virology,* New York, Academic Press, p. 2–42.

Manire, G. P. 1966. Structure of purified cell walls of dense forms of meningopneumonitis organisms. *J. Bact.* **91**:409–413.

McClain, W. H., Guthrie, C., and Barrell, B. G. 1972. Eight tRNAs induced by infection of *E. coli* with phage T4. *Proc. Nat. Acad. Sci. USA* **69**:3703–3707.

Moulder, J. W. 1966. The relation of the psittacosis group (Chlamydiae) to bacteria and viruses. *Ann. Rev. Microbiol.* **20**:107–130.

Ormsbee, R. A. 1969. Rickettsiae (as organisms). *Ann. Rev. Microbiol.* **23**:275–292.

Rhodes, A. J., and van Rooyen, C. E. 1968. *Textbook of Virology. 5th Ed.*, Baltimore, Williams & Wilkins Co.

Roizman, B., and Spear, P. G. 1971. Herpesviruses: Current information on the composition and structure. *In:* Maramorosch, K., and E. Kurstak, eds., *Comparative Virology,* New York, Academic Press, p. 135–168.

Rosenkranz, H. S. 1973. RNA in coliphage T5. *Nature* **242**:327–329.

Rubey, W. W. 1955. Development of the hydrosphere and atmosphere, with special reference to probable composition of the early atmosphere. *Spec. Paper Geol. Soc. Amer.* **62**:631–650.

Schaechter, M., Tousimis, A. J., Cohn, Z. A., Campbell, J., and Hahn, F. E. 1957. Morphological, chemical and serological studies of the cell walls of *Rickettsia mooseri. J. Bact.* **74**:822–829.

Schoenheimer, R. 1942. *The dynamic state of body constituents.* Cambridge, Mass., Harvard University Press.

Sigurdsson, B. 1954. Observations on three slow infections of sheep. *Brit. Vet. J.* **110**:255, 307, 341.

Weiss, E. and Kiesow, L. A. 1966. Incomplete citric acid cycles in the agents of the psittacosis-trachoma group (Chlamydia). *Bact. Proc.* **1966**:85.

Weiss, E., Myers, W. F., Dressler, H. R., and Chun-Hoon, H. 1964. Glucose metabolism by agents of the psittacosis-trachoma group. *Virology* **22**:551–562.

Woese, C. R. 1970. The problem of evolving a genetic code. *Bioscience* **20**:471–485.

Yčas, M. 1974. On earlier states of the biochemical system. *J. Theor. Biol.* **44**:145–160.

Index

*Illustrations are indicated by boldface numbers